McGRAW-HILL
YEARBOOK OF
SCIENCE &
TECHNOLOGY

2013

McGRAW-HILL
YEARBOOK OF
SCIENCE &
TECHNOLOGY

2013

Comprehensive coverage of recent events and research as compiled by
the staff of the McGraw-Hill Encyclopedia of Science & Technology

New York   Chicago   San Francisco   Lisbon   London   Madrid   Mexico City

Milan   New Delhi   San Juan   Seoul   Singapore   Sydney   Toronto

On the front cover

Computer-aided picture from the CMS (Compact Muon Solenoid) detector at the Large Hadron Collider (LHC), showing a candidate event for the decay of a Higgs boson into two gamma-ray photons ($H \rightarrow \gamma\gamma$). *(Courtesy of J. Incandela, Status of the CMS SM Higgs Search, CERN Webcast seminar, July 4, 2012; https://indico.cern.ch/conferenceDisplay.py?confId=197461)*

ISBN   978-0-07-180140-9
MHID   0-07-180140-5
ISSN   0076-2016

This book was printed on acid-free paper.

It was set in Garamond Book and Neue Helvetica Black Condensed by Aptara, New Delhi, India. The art was prepared by Aptara. The book was printed and bound by RR Donnelley.

Contents

Editorial Staff

Editing, Design, and Production Staff

Consulting Editors

Dr. Roger M. Rowell. *Professor Emeritus, Department of Biological Systems Engineering, University of Wisconsin, Madison.* FORESTRY.

Dr. Thomas C. Royer. *Department of Ocean, Earth, and Atmospheric Sciences, Old Dominion University, Norfolk, Virginia.* OCEANOGRAPHY.

Prof. Ali M. Sadegh. *Director, Center for Advanced Engineering Design and Development, Department of Mechanical Engineering, The City College of the City University of New York.* MECHANICAL ENGINEERING.

Prof. Joseph A. Schetz. *Fred D. Durham Endowed Chair Professor of Aerospace & Ocean Engineering, Virginia Polytechnic Institute and State University, Blacksburg.* FLUID MECHANICS.

Dr. Alfred S. Schlachter. *Advanced Light Source, Lawrence Berkeley National Laboratory, Berkeley, California.* ATOMIC AND MOLECULAR PHYSICS.

Prof. Ivan K. Schuller. *Department of Physics, University of California, San Diego, La Jolla.* CONDENSED-MATTER PHYSICS.

Jonathan Slutsky. *Naval Surface Warfare Center, Carderock Division, West Bethesda, Maryland.* NAVAL ARCHITECTURE AND MARINE ENGINEERING.

Dr. Arthur A. Spector. *Department of Biochemistry, University of Iowa, Iowa City.* BIOCHEMISTRY.

Dr. Anthony P. Stanton. *Tepper School of Business, Carnegie Mellon University, Pittsburgh, Pennsylvania.* GRAPHIC ARTS AND PHOTOGRAPHY.

Dr. Michael R. Stark. *Department of Physiology, Brigham Young University, Provo, Utah.* DEVELOPMENTAL BIOLOGY.

Prof. John F. Timoney. *Maxwell H. Gluck Equine Research Center, Department of Veterinary Science, University of Kentucky, Lexington.* VETERINARY MEDICINE.

Dr. Daniel A. Vallero. *Adjunct Professor of Engineering Ethics, Pratt School of Engineering, Duke University, Durham, North Carolina.* ENVIRONMENTAL ENGINEERING.

Prof. Pao K. Wang. *Department of Atmospheric and Oceanic Sciences, University of Wisconsin, Madison.* METEOROLOGY AND CLIMATOLOGY.

Dr. Nicole Y. Weekes. *Department of Psychology, Pomona College, Claremont, California.* NEUROPSYCHOLOGY.

Prof. Mary Anne White. *Department of Chemistry, Dalhousie University, Halifax, Nova Scotia, Canada.* MATERIALS SCIENCE AND METALLURGICAL ENGINEERING.

Dr. Thomas A. Wikle. *Department of Geography, Oklahoma State University, Stillwater.* PHYSICAL GEOGRAPHY.

Article Titles and Authors

Preface

The ever-accelerating growth of scientific knowledge is both one of the most thrilling aspects of life in the twenty-first century and one of the most daunting. Scholarly estimates suggest that the number of scientific journals is currently well over 25,000 and that the number of new research papers they publish annually is in the millions. Keeping up with all the developments even in single disciplines long since stopped being feasible for any individual. That mismatch between the cascading rivers of new information and the modest capacities of the human mind is especially unfortunate because deeper understandings of so many of the world's problems—along with solutions to them—lie within that scientific bounty.

Nevertheless, certain discoveries and technical innovations stand out every year as being more remarkable, significant, or influential than others, and anyone determined to try to stay current with science and technology would do well simply to stay well-informed about those. This volume was conceived with that end in mind. The 2013 edition of the *McGraw-Hill Yearbook of Science & Technology* continues its 50-year mission of keeping professionals and nonspecialists alike abreast of key research and developments with a broad range of concise reviews invited by a distinguished panel of consulting editors and written by international leaders in science and technology.

In this edition, we report on the physicists' detection of what may be the long-sought Higgs boson particle; the surprisingly complex eyes of a huge predator of the Cambrian seas; developments in active traffic management; applications of microbiology to geotechnical engineering; recent discoveries in alternative RNA splicing; the dangers that wind turbines pose to bats; intelligent microgrids; cognitive technology; cancer stem cells; the use of magnetic nanoparticles in cancer therapies; deep seabed mining; the discovery of the ancient Denisovan people in Asia contemporary with Europe's Neandertals; the fabrication of flexible solar cells; factors relating to the risk of autism; neuroscientific insights into free will; the genomics of depression; the controversy over the threat from the H5N1 bird flu virus; discoveries of planets around other stars by the Kepler space telescope; mineral evolution; ocean acidification; progress toward the eradication of polio; psychological research on repressed memories; a discussion of the most recent transit of Venus; the 2012 Nobel Prizes; and an update on space flight, including the rise of commercial space activities, among many other fascinating topics.

Each contribution to the Yearbook is the result of a well-informed collaboration. Our consulting editors, whose expertise covers a full spectrum of disciplines, select the topics in consultation with our editorial staff based on the present significance and potential applications of recent findings. One or more authorities are then invited to write concise yet thorough articles that explore the new work in each field. Through careful editing and extensive use of specially prepared graphics, McGraw-Hill strives to make every article as readily understandable as possible to nonspecialists.

Librarians, students, teachers, the scientific community, journalists and other communicators, and general readers continue to find in the *McGraw-Hill Yearbook of Science & Technology* the information they need to follow the rapid pace of advances in science and technology and to understand the developments in these fields that will shape the world of the twenty-first century.

John Rennie
EDITORIAL DIRECTOR

Acheulean

The Acheulean (alternatively Acheulian) is an archeological culture defined by its characteristic stone tools, particularly the handaxe, found in Africa and Eurasia from approximately 1.76 million years ago (~1.76 MYA) to approximately 150,000 years ago (~150 KYA) [see **illustration**]. The term is derived from the French village of St. Acheul, where numerous handaxes were reported in the nineteenth century. Handaxes associated with fossilized bones of extinct mammals found in northern France and England in the eighteenth and nineteenth centuries were essential evidence in demonstrating the evolution of human and other animal communities. This marked a major break with the prior Western academic tradition of biblical interpretations of world origins.

Characteristic stone tools. Handaxes are the most recognizable of the Acheulean stone tools. As shown in the illustration, a handaxe is a large, teardrop-shaped implement, often more than 10 cm (4 in.) in length. Handaxes are crafted by carefully sculpting or shaping the piece, accomplished by the careful removal of numerous small flakes (stone splinters) from a cobble or other large piece of stone. Making a handaxe is a complex skill that requires precise hand–eye coordination, a detailed understanding of geometry, and the ability to foresee and anticipate the effects of sequential actions toward accomplishing a predetermined goal. The flakes that shape a handaxe are removed by repeatedly striking the edge of the in-progress handaxe using river cobbles and other "hammers" made of stone, hardwood, or antler. The flakes removed during handaxe shaping were often used as tools themselves because they have razor-sharp edges, can be held in the hand, and are suitable for a range of cutting or scraping tasks.

The working edge of a handaxe is bifacial, that is, an edge formed by the intersection of two curved, flaked surfaces or "faces" (see illustration). This edge is less sharp but much more durable than those of the flakes removed during handaxe manufacture. This durable edge provides one clue about some of the functions likely served by handaxes: large tools with tough edges are useful for tasks such as heavy-duty woodworking and the butchery of game obtained by hunting or scavenging. Microscopic examination of some handaxe edges confirms that they were used for woodworking. Some handaxes were made, used, and discarded in the same place, as observed, for example, at the horse butchery site at Boxgrove, England (dated to approximately 500 KYA). Other handaxes were transported considerable distances from their site of manufacture, often more than 10 km (6.2 mi) but rarely more than 60 km (37.2 mi). This transport distance is determined by comparing the location of tool discard (and recovery by archeologists) and the geological outcrops where the rock, from which the handaxe was made, occurs. Long-distance transport of handaxes demonstrates their importance to early humans, who would have carried these ready-made tools in anticipation of future needs. This portability and multifunctionality, namely, the presence of a durable edge and the

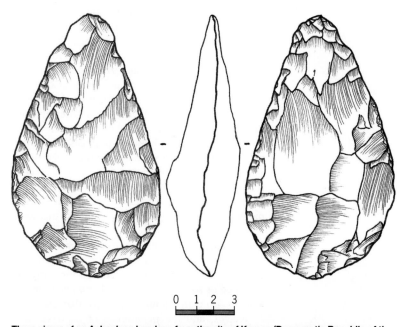

0   1   2   3

Three views of an Acheulean handaxe from the site of Kamoa (Democratic Republic of the Congo). The bifacial edge is visible in profile and is the intersection of the two worked faces, shaped by the scars resulting from multiple flake removals. Scale bar is given in centimeters. [*Redrawn from D. Cahen, Le Site Archéologique de la Kamoa (Région du Shaba, République du Zaïre) de l'Age de la Pierre Ancien à l'Age du Fer, Annales de la Musée Royal de l'Afrique Centrale, Series Humaines no. 84, Tervuren, Belgium, 1975*]

ability to produce razor-sharp flakes during shaping and resharpening, have resulted in Acheulean implements being likened to the "Swiss Army knife of the Paleolithic."

Associated hominins. Although classically linked with *Homo erectus*, handaxes and other elements of Acheulean technology have been found associated with a number of different types of early humans (termed hominins) in Africa and Eurasia. Although attributing early hominin fossil remains to particular species is a complex and contentious issue, handaxes have been found with hominin fossils of *H. erectus*, *H. ergaster*, *H. rhodesiensis*, *H. heidelbergensis*, and perhaps *H. sapiens*. Based on current evidence, it does not appear that Acheulean technology is particular to a single hominin species.

Age and distribution. Although not unique to a particular species, recent finds in northern Kenya near the shores of Lake Turkana have shown that handaxes and *H. erectus* appear more or less at the same time in eastern Africa, around 1.76 MYA. This would appear to suggest that changes in behavior (represented in the archeological record) may have occurred at the same time as changes in biology (represented in the fossil record). However, fossils of *H. erectus* also appear at approximately 1.6–1.8 MYA at sites across Asia (in Georgia and Indonesia) that lack handaxes. These Asian *H. erectus* fossils are the oldest direct evidence for the migration of hominins out of Africa. However, the importance of handaxes and related technologies to this dispersal out of Africa remains unclear.

In Europe, the oldest handaxes date to approximately 1 MYA, but remain extremely rare until about 500 KYA, and fossils attributed to *H. erectus* are absent. Handaxes are relatively abundant after approximately 500 KYA, perhaps reflecting major shifts in hominin population abundance, density, and geographic distribution. Curiously, although Acheulean archeological sites are widespread across Africa and much of Eurasia, there are important gaps in their distribution. Handaxes are rare in regions located east of about 90°E longitude and north of about 53°N latitude. It is unclear why Acheulean sites are rare in these eastern areas, but this may relate to the presence of different hominin populations with alternative toolmaking strategies or the replacement of stone handaxes by tools made of other materials (such as bamboo) that are not preserved in the archeological record. The northern limit likely tracks harsh glacial and periglacial environments that were uninhabitable by early hominins.

Handaxes were made from approximately 1.76 MYA to perhaps as recently as approximately 150 KYA. These early dates attest to the early appearance of sophisticated mental and manipulative abilities in early hominin groups that included *H. erectus*, and the persistence of handaxes and similar implements over such a long time span attests to the usefulness of this technology.

Subsistence. Investigators have a very limited understanding of the types of foods eaten by Acheulean tool-using populations. Because bones preserve relatively well in the archeological record, the evidence is skewed toward animal products, although a variety of plant foods (including tubers) likely made an important contribution to the diet. Evidence from some sites, including Elandsfontein, South Africa, suggests that Acheulean populations may have hunted only rarely and perhaps often scavenged kills made by carnivores. In contrast, sites such as Lehringen, Germany, and Boxgrove, England, suggest active group hunting of herds of horses (in the former case with large wooden javelins or spears).

Meat obtained by hunting or scavenging may have been cooked, although early evidence for the use and control of fire remains rare and controversial. Some but not all Acheulean groups may have used fire. Cooking would have increased the nutritional value of meat, made it easier to digest, and would have increased the accessibility of some plant foods that require processing before eating. Fire may also have increased the range of habitats inhabitable by early hominins and, as has been suggested, would have increased the effective daylight hours for face-to-face socialization.

Social and symbolic life. Little is directly known about the social and symbolic aspects of the lives of the early humans who made Acheulean artifacts. Similarities among handaxes probably reflect at least some degree of social learning, with the necessary skills and techniques transferred among generations and probably among many members of a single community. The butchery and likely hunting of large game (including, for example, horses and, rarely, elephants) imply access to substantial amounts of meat, necessitating sharing among different members of a social group, which is an act that among recent foragers is subject to complex rules of relatedness and status. More indirectly, primate models have been used to examine the relationships among brain size, body size, group size, and language among ancient hominins. They suggest that hominins associated with Acheulean technology may have lived in extended groups of 100 or more, and needed at least rudimentary language to navigate the resulting numerous and complex social relations among groups of that size.

Unfortunately, language does not fossilize, so other forms of indirect evidence must be used. Archeologists typically rely on symbols, that is, abstractions that stand for something else. There is little evidence from Acheulean archeological sites for the sorts of symbolism, such as artwork, found during later periods. There are a few ambiguous specimens from Africa and Eurasia, including some possible engraved pieces from Wonderwerk Cave, South Africa, and a piece of volcanic rock that may have been shaped into a crude representation of human form from Berekhat Ram, Israel. More definitive but difficult-to-interpret evidence comes from an approximately 600-KYA hominin skull from Bodo, Ethiopia, with a number of cut marks made by stone tools. Although the marks on the skull were clearly made by hominins, their location is inconsistent with cannibalism because they are on parts of the skull (such as the browridges) with little meat or nutritional value, and are instead more similar to ritual defleshing (excarnation, or removing of the flesh) of the dead found among some recent human populations. This may signal a rich belief system among some populations who used Acheulean technology, but it remains a singular example.

End of the Acheulean. By 150 KYA and often before, Acheulean sites are replaced by those attributed to the Middle Paleolithic in Eurasia and the Middle Stone Age in Africa, with hominins such as the Neandertals and *H. sapiens* replacing *H. erectus* and other early species. Acheulean technology is an evolutionary success: it persisted for more than 1.5 million years and was employed across three continents. However, in a scenario all too familiar to us in the twenty-first century, the Acheulean, like most technologies, was ultimately replaced by something different, which presumably was better suited to the lives of our ancestors in Africa and elsewhere.

For background information *see* ANTHROPOLOGY; ARCHEOLOGICAL CHRONOLOGY; ARCHEOLOGY; EARLY MODERN HUMANS; FOSSIL; FOSSIL HUMANS; PALEOLITHIC; PHYSICAL ANTHROPOLOGY; PREHISTORIC TECHNOLOGY; SOCIOBIOLOGY in the McGraw-Hill Encyclopedia of Science & Technology.

Christian A. Tryon

Bibliography. S. C. Antón and C. C. Swisher III, Early dispersals of *Homo* from Africa, *Annu. Rev. Anthropol.*, 33:271–296, 2004, DOI:10.1146/annurev.anthro.33.070203.144024; R. Klein, *The Human Career*, 3d ed., University of Chicago Press, Chicago, 2008; S. J. Lycett and J. A. J. Gowlett, On questions surrounding the Acheulean "tradition," *World Archaeol.*, 40:295–315, 2008, DOI:10.1080/00438240802260970; C. Tryon, B. Pobiner, and R. Kauffman, Archaeology and human evolution, *Evol. Educ. Outreach*, 3:377–386, 2010, DOI:10.1007/s12052-010-0246-9; B. Wood, *Human Evolution: A Very Short Introduction*, Oxford University Press, Oxford, U.K., 2006.

Acoustical properties of nontraditional woods

Historically, people have taken advantage of the natural qualities of particular species of wood in the creation of wooden musical instruments. For example, Sitka spruce is selected for soundboards in pianos because of its tight, uniform grain, which helps ensure consistent vibrational properties of the wood across the entire piece. Ebony is used for woodwinds such as clarinets, oboes, and bassoons because of its black color and its high content of waxes, oils, and resins, which can reduce the rate of moisture sorption. Rosewood and bubinga (known as African rosewood) are used to make professional recorders. However, high-quality woods of some species are becoming rarer, and they are expensive. Thus, the wooden musical instrument industry, although continuing to prefer traditional woods, has sought

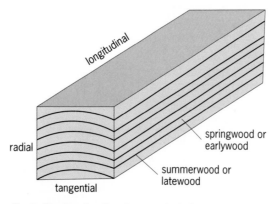

Fig. 1. Growing directions in a tree, including earlywood and latewood.

alternative species that are less expensive and more available. For example, maple is often used today for woodwinds, though mainly for student-grade instruments and not for professional-grade instruments. Still, as the availability of high-quality tropical hardwoods goes down and the cost of traditional wood species goes up, the wooden musical instrument industry has started to look at common, inexpensive, native woods as substitutes for the old traditional favorites.

Wood performance and acoustical properties. Wood can be defined as an anisotropic, viscoelastic, and hygroscopic material. All three of these properties have a profound effect on the acoustical properties of wood. As an anisotropic composite, the properties of wood differ in all three growing directions of the tree: lengthwise (longitudinal or transverse), from the center out (radial), and along the annual rings (tangential) [**Fig. 1**]. The density of wood also varies from lower values in the earlywood or springwood (cells added in the spring) to higher values in the latewood or summerwood (cells added in the summer and early fall).

In general, the higher the density of the wood, the more swelling will occur when moisture is added to the cell wall. In the presence of moisture, the tangential grain direction in wood swells about twice as much as the grain in the radial direction. In most wood species, there is very little, if any, swelling in the longitudinal or transverse direction. When moisture is lost in wood, shrinkage occurs—more in the tangential grain direction compared to the radial direction, and almost none in the longitudinal direction. Given this property, tangential grain direction is kept to a minimum to minimize any changes in the wooden member dimensions and the resulting sound properties. Because of the variability in swelling in different wood directions and in different species, it is typically important to avoid mixing grain directions or species when constructing a wooden musical instrument.

As a viscoelastic composite, the wood cell wall is made up of an elastic phase, which is the crystalline cellulose polymer, and a viscous phase, which is the noncrystalline cellulose phase, including the amorphous hemicellulose polymers and the lignin

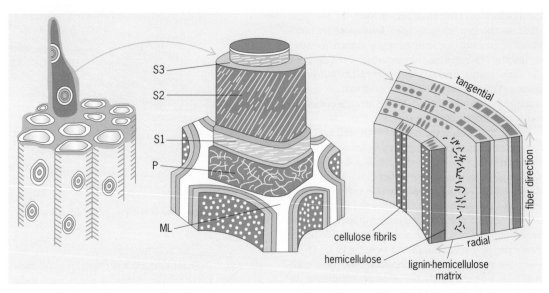

Fig. 2. Architecture of the wood cell wall showing the various layers of the cell wall and the distribution of the polymers in the cell wall. Terms: S1–S3, secondary cell wall layers; P, primary wall; ML, middle lamella.

polymer (**Fig. 2**). What this means in acoustical terms is that sound travels faster through the more elastic or crystalline portion of the wood and more slowly through the more viscous or noncrystalline portion. When a vibration is propagated in wood, the vibration quickly passes through the elastic crystalline cellulose; however, the vibration is held up or delayed in the viscous phase. This mixed pattern of variation in vibrations is the factor that confers a mellow or soft tone to wooden musical instruments. In contrast, instruments made of silver, for example, have only one elastic phase and vibrations travel faster through the metal, resulting in a bright sound. Instruments made of plastic, for example, have only one viscous phase, and the sound from these instruments is often referred to as muddy or dark.

Because the orientation (fibril angle) of the crystalline cellulose is somewhat parallel to the longitudinal or transverse direction in the wood, sound travels faster in this direction compared to the tangential and radial directions.

Viscoelasticity in wood is studied using dynamic mechanical analysis, in which an oscillatory force (stress) is applied to the wood and the resulting displacement (strain) is measured. In purely elastic materials, the stress and strain occur in phase, so the response of one occurs simultaneously with the other. In purely viscous materials, there is a phase difference between stress and strain, with the strain lagging behind the stress by a $90°$ phase lag. Wood exhibits behavior somewhere in between that of a purely viscous material and a purely elastic material, resulting in some phase lag in strain.

As a hygroscopic composite, the moisture content in the wood depends on the relative humidity of the environment to which the wood is exposed. This occurs because the wood cell wall contains hydroxyl groups and other oxygen-containing groups that attract moisture through hydrogen bonding. This moisture swells the cell wall, and the wood expands until the cell wall is saturated with water. Moisture beyond the fiber saturation point is free water in the void portion within and between cells and does not contribute to further expansion. This process is reversible, and the wood shrinks as it loses moisture below the fiber saturation point.

The sorption of water molecules in the wood cell-wall polymers acts as a plasticizer to loosen the cell-wall microstructure, which increases the viscous properties of the wood. This affects the tone quality of wooden musical instruments because the acoustical properties of wood are reduced or dulled as the moisture content increases; in particular, the higher the moisture content, the greater is the effect on the viscous phase. Therefore, a person playing a musical instrument made of wood tries to keep the instrument as dry as possible and may change instruments during a long passage of music. As mentioned previously, water molecules in the cell wall also cause the wood to swell, which increases the deformation of wooden parts. This is particularly bad from a durability and sound-quality perspective for parts that are under stress (for example, the strings of guitars and violins, the pinblock in a piano, and the joints of instruments).

Nontraditional approach. One approach to the selection of nontraditional woods for musical instruments is simply to substitute different species for the traditional standards. Certainly, there are other hardwoods that could be substituted for ebony; however, they might not be black, and black is a very traditional color for woodwind instruments. Still, though, the problems of supply, cost, and properties (anisotropic, viscoelastic, and hygroscopic) remain the same.

A different approach is to use a common domestic hardwood that can be acquired at a reasonable cost and improve its anisotropic, viscoelastic, and hygroscopic properties. Because the acoustical properties of wood are adversely affected by moisture, they can be stabilized and improved if the wood can be stabilized. It is possible to reduce the hygroscopicity of wood by modifying the hydroxyl groups in cell-wall polymers so that they do not bond with water. Dimensional stability can be improved by permanently bulking the cell wall in such a way that the elastic limit of the cell wall is reached, so additional moisture does not result in swelling.

There are several approaches to achieve this goal. For example, one of the most common, clear, uniform, and affordable hardwoods is hard maple. It grows in many parts of the United States. It is presently used in the musical instrument industry for student-grade instruments, so it is not an unknown species for this type of use.

Hard maple can be chemically altered at the molecular level to greatly improve the acoustical properties of the wood. The simplest chemistry, which is now commercially available, is the process of acetylation. The wood is reacted with acetic anhydride, and the accessible hydroxyl groups are replaced by an acetate group with a by-product of acetic acid (vinegar). Because two things cannot be in the same place at the same time, the acetate groups occupy space in the cell wall and swell it to almost the volume size of its green wood (unseasoned, newly harvested wood) after drying. The result is a modified wood with greatly reduced moisture content, which indicates that the hygroscopicity of the wood has been dramatically reduced and the dimensional stability has been greatly increased.

Vibrational properties of acetylated wood. One way to study the acoustical properties of wood is through vibrational analysis. A simple harmonic stress results in a phase difference between the stress and strain. The ratio of the dynamic Young's modulus E' to specific gravity γ, that is, E'/γ = specific modulus, and the internal friction (the tangent of the phase angle, $\tan \delta$) are important properties to measure in relation to the acoustical properties of wood; E'/γ is related to sound velocity, and $\tan \delta$ is related to sound absorption or damping within the wood.

In a dynamic vibrational analysis, using a simple free-free vibrational beam test using acetylated and nonacetylated Sitka spruce, sound velocity and sound absorption were determined at several different relative humidities. The conclusions of this research showed that acetylation of wood slightly reduced both sound velocity and sound absorption compared to unreacted wood. Acetylation greatly reduced variability in the moisture content of the cell-wall polymers, thereby stabilizing the physical dimensions of the wood and thus its acoustical properties. These tests did not determine whether acetylation enhanced sound quality.

Musical instruments made of acetylated wood. Several wooden instruments and parts for musical instruments have been made from acetylated wood and have been evaluated by both amateur and professional musicians. Two violins were made (one in Japan and one in Sweden) from acetylated wood. Both violins were played by professional violinists. Based on their experience, the musical quality of

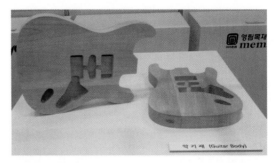

Fig. 3. Solid-body guitar made in Japan from acetylated wood.

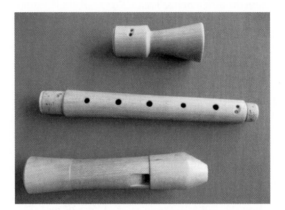

Fig. 4. Recorder made from acetylated maple.

both violins was recorded as excellent. A guitar was also made in Japan using acetylated wood and played by several amateur guitarists. The guitar was very responsive, with good note separation. A solid-body guitar has been made in Japan as well (**Fig. 3**). In addition, a piano soundboard was made and evaluated in Japan. The quality of the sound was deemed excellent. Also, a thin wooden diaphragm speaker system was made in Japan.

An acetylated wood recorder was played at the Boston Early Music Festival in June 1993 (**Fig. 4**). Its acoustical properties were sustained through a wide range of moisture changes, and it retained its tuning and tone quality without modification.

Trumpet and trombone mouthpieces have been made in Sweden and played by both amateur and professional players. The feedback was very positive. Clarinet reeds have also been made using acetylated wood, and it was recorded that the vibrational properties and the anticreep properties of the reeds were enhanced by the acetylation process.

Conclusions. As a result of several years of experimentation on the effects of acetylation on the acoustical properties of wood, it can be concluded that acetylation of wood slightly increases density and slightly reduces (by approximately 5%) both sound velocity and sound absorption in comparison to unreacted wood. Acetylation does not change the acoustic converting efficiency. In short, acetylation slightly increases the viscous properties of the wood. Acetylation reduces the amount of moisture in the cell wall, decreasing the effect of moisture on the viscous properties of wood. This allows a wooden musical instrument to be played longer without having to let it dry out. Thus, an instrument made from acetylated wood can be played under a greater range of moisture conditions without losing tone quality. Acetylation also greatly stabilizes the physical dimensions of the wood. Therefore, the major effect of acetylation of wood is to stabilize its acoustical properties.

For background information *see* ACOUSTICS; CELL WALLS (PLANT); MOISTURE-CONTENT MEASUREMENT; MUSICAL ACOUSTICS; MUSICAL INSTRUMENTS; SOUND; VIBRATION; WOOD ANATOMY; WOOD ENGINEERING DESIGN; WOOD PROPERTIES in the McGraw-Hill Encyclopedia of Science & Technology.   Roger M. Rowell

Bibliography. H. Akitsu et al., Effect of humidity on vibrational properties of chemically modified wood, *Wood Fiber Sci.*, 25:250-260, 1993; V. Bucur, *Acoustics of Wood*, 2d ed., Springer-Verlag, Berlin/Heidelberg, 2006; T. Ono, Transient response of wood for musical instruments and its mechanism in vibrational property, *J. Acoust. Soc. Jpn.*, 20:117-124, 1999; R. M. Rowell, Acetylation of wood: A journey from analytical technique to commercial reality, *For. Prod. J.*, 56:4-12, 2006; R. M. Rowell, Chemical modification of wood, chap. 14, pp. 381-420, in R. M. Rowell (ed.), *Handbook of Wood Chemistry and Wood Composites*, CRC Press, Boca Raton, FL, 2005; T. Sasaki et al., Effect of moisture on the acoustical properties of wood, *J. Jpn. Wood Res. Soc.*, 34:794-803, 1988; C. Tanaka, T. Najia, and A. Takahashi, Acoustical property of wood, *J. Jpn. Wood Res. Soc.*, 33:811-817, 1987; U. G. K. Wegst, Wood for sound, *Am. J. Bot.*, 93:1439-1448, 2006; H. Yano, M. Norimoto, and R. M. Rowell, Stabilization of acoustical properties of wooden musical instruments by acetylation, *Wood Fiber Sci.*, 25:395-403, 1993.

Active traffic management

Active Traffic Management (ATM) is the latest in a long line of U.S. federal government initiatives to improve the productivity and efficiency of the transportation system. Where ATM differs from previous initiatives is in the use of advanced technologies and improved intelligence in road, vehicle, and traffic management centers to actively adjust traffic controls on a real-time basis to respond to and anticipate special events, incidents, and weather. The latest evolution of ATM combines ATM and Transportation Demand Management (TDM) into a comprehensive program of Active Transportation and Demand Management (ATDM) that manages both the demand and supply sides of the transportation problem. The use of road pricing and sophisticated travel information systems in ATDM greatly expands the ability of public agencies to deliver a dependable, efficient, and reliable transportation system that cost-effectively moves people and goods.

Context. The U.S. federal government has been promoting better management of the transportation system and the road subsystem, in particular, since the 1960s and 1970s, starting with Transportation System Management (TSM) and following with Transportation Demand Management (TDM) and Integrated Congestion Management (ICM). However, it was a technical tour of Europe in mid-2005 that opened U.S. experts' eyes to the potential of a comprehensive and coordinated management approach using the latest technologies to actively manage traffic congestion. While the fundamental concepts of Active Traffic Management (ATM) were known, this tour demonstrated what can be accomplished with a comprehensive ATM approach and led to the evolution of Active Transportation and Demand Management (ATDM).

Purpose and strategies. ATM/ATDM is defined as the dynamic management, control, and influence of travel demand and flow on transportation facilities. The purpose of ATM/ATDM is to increase the productivity of the transportation system in terms of the number of people and goods moved, while improving the efficiency and the reliability of the system by reducing recurring and nonrecurring delay. ATM/ATDM strategies are implemented in the form of special event, work zone, incident, and weather management plans that specify the facility-monitoring data requirements and the desired control response to each situation.

ATM can be employed on both freeways and urban streets, but has the greatest benefits when applied to an entire corridor. ATM strategies employ a variety of capacity and demand management measures to increase facility productivity and efficiency.

Capacity management measures include locally or system dynamic ramp metering, traffic responsive and adaptive signal control, and dynamic lane management, such as variable speed limits, temporary shoulder lane use, and dynamic lane closures. High-occupancy vehicle (HOV) lanes and truck-only lanes may be dynamically opened to general traffic in response to demand surges, incidents, or weather, for example.

Demand management measures include dynamic congestion pricing, dynamic parking management, and traveler information systems. Employer-based TDM programs that dynamically respond to existing or predicted conditions on the highway system may be part of an ATM system. Dynamic parking management systems vary rates according to occupancy and transmit that information in real-time to travelers.

Technology. Technology is the key to effective ATM/ATDM. ATM/ATDM requires the continuous and comprehensive monitoring of facility performance and the ability to change tolls, lane assignments, speed limits, message signs, in-vehicle traffic information, metering rates, and traffic signal timings on a real-time basis.

System monitoring is accomplished through the use of roadside detectors such as inductive loops cut into the pavement, overhead or side-mounted radar detectors, overhead closed-circuit television cameras (CCTV), and transponders. A high-capacity communications system, such as fiber-optic cable, conveys the extensive information back to a central

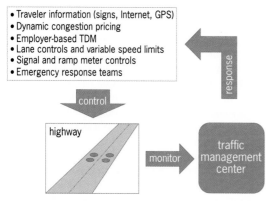

- Traveler information (signs, Internet, GPS)
- Dynamic congestion pricing
- Employer-based TDM
- Lane controls and variable speed limits
- Signal and ramp meter controls
- Emergency response teams

response

control

highway

monitor

traffic management center

Active traffic management system.

traffic management center (TMC). The trained personnel and computer algorithms used by the TMC identify surges in demand, incidents, and weather, and quickly determine the appropriate response. The response may consist of adjustments to ramp metering rates, speed limits, lane closures, message signs, and signal timings. Information on facility status is conveyed to the driving public via the Internet, cell phones, roadside message signs, or in-vehicle navigation devices (see **illustration**).

Institutional cooperation. ATM is most effective when all of the public and private agencies and individuals that use, affect, or control the operation of the highway facility are included in and part of the ATM strategy. This means that while technology is a key component of ATM, institutional coordination and cooperation determine its success. State and local agencies, which operate the freeways and urban streets in a metropolitan area, must be part of the ATM strategy, along with providers of traveler information (such as 511.org, INRIX®, TomTom®, and ETCETERA in Europe), major employers, transit operators, emergency responders (such as tow trucks, hazmat teams, fire, police, and highway patrol), and the driving public. All of these institutions and individuals must be part of the ATM solution.

Applications. There are many ATM/ATDM deployments in various stages of design and implementation throughout the United States. The two installations that currently represent the most comprehensive and advanced applications of ATM/ATDM in the United States are I-5 in Seattle, Washington, and I-35W in Minneapolis, Minnesota. The city of San Francisco, California has an advanced dynamic parking management application that is currently in operation.

The I-5 ATM system in Seattle consists of an integrated system of dynamic ramp metering, variable speed limits, variable message signs, and overhead lane closure warning signs on 7 mi of freeway in downtown Seattle, Washington. It is operated by the Washington State DOT and the Seattle Traffic Management Center. Mainline loop detectors and overhead gantries capable of displaying speed limits and advanced warning of lane closures by individual lane are spaced, on average, every half mile. The variable speed limits are regulatory. (They may be used as

prima facie evidence of unsafe driving speeds.) Plans are to expand the ATM system to the I-90 and State Route 520 freeways and include variable pricing.

The I-35W Price Dynamic Shoulder Lane (PDSL) project in Minneapolis, Minnesota operates in the northbound direction between 46th Street and downtown. The left-hand shoulder lane is open to high-occupancy vehicles and toll-paying single-occupant vehicles from 6–10 a.m. and 2–7 p.m. weekdays. The shoulder is closed to traffic at other times. The PDSL project opened in September 2009. Every half mile, overhead gantries with variable message signs give the status of the PDSL, toll rates, advisory speed limits, and advanced warnings of lane closures. Tolls are collected electronically using MnPASS toll tag readers.

SFPark in the city of San Francisco is an example of an active parking management system. Parking-space occupancy sensors on-street for curb parking and within city parking garages determine parking availability. Parking rates are adjusted at the wireless connected electronic meters in the field based on demand to encourage travelers to switch to off-peak hours or go to less used public garages. Drivers are able to pay remotely by phone and credit card, enabling them to pay for extended stays, as needed.

Prospects. ATM/ATDM installations in the United States have been hindered by the high initial technology investment costs and the lack of experience and analysis tools for determining the benefits of ATM/ATDM to agencies and the traveling public. There is limited design and planning guidance available.

The Federal Highway Administration is developing a program to increase awareness and knowledge of ATDM develop, test, and evaluate ATDM strategies; provide tools and methods for predicting the performance benefits of ATDM investments; train agencies on the effective deployment of ATDM systems; and provide guidance to Federal Highway Administration (FHWA) division offices, which in turn will provide technical support to local and state agencies considering the implementation of ATDM. The program focuses on foundational research, analysis tool development and guidance, and outreach and training. Through the publication of information briefs, preparation of public resource guides, development of informational websites, and sponsorship of peer exchange workshops, the FHWA will disseminate the information on ATDM benefits to the profession.

For background information *see* TRAFFIC-CONTROL SYSTEMS; TRANSPORTATION ENGINEERING; HIGHWAY ENGINEERING in the McGraw-Hill Encyclopedia of Science & Technology.   Richard G. Dowling

Bibliography. C. Fuhs, *Synthesis of Active Traffic Management Experiences in Europe and the United States*, FHWA-HOP-10-031, Federal Highway Administration, Washington, D.C., 2010, http://ops.fhwa.dot.gov/publications/fhwahop10031/listcont.htm; B. Kuhn, *Efficient Use of Highway Capacity, Summary Report to Congress*, FHWA-HOP-10-023, Federal Highway Administration, Washington,

D.C., 2010, http://ops.fhwa.dot.gov/publications/ fhwahop10023/index.htm#toc; M. Mirshahi et al., *Active Traffic Management: The Next Step in Congestion Management*, FHWA-PL-07-012, Federal Highway Administration, Washington, D.C., 2007, http://international.fhwa.dot.gov/pubs/ pl07012/; E. N. Schreffler, *Integrating Active Traffic and Travel Demand Management: A Holistic Approach to Congestion Management*, FHWA-PL-11-011, Federal Highway Administration, Washington, D.C., 2011, http://www.trb.org/Main/Blurbs/ 165100.aspx.

All Taxa Biodiversity Inventory

Composed of more than 500,000 acres (200,000 hectares) of wilderness and serving as a refuge for some of the richest and most diverse communities of life in the temperate world, the Great Smoky Mountains National Park (located in Tennessee and North Carolina) is considered to be the most diverse national park in America. Because of its expanse of elevations and habitats, environments similar to those found in areas ranging from Arkansas to Canada can be studied without ever leaving its boundaries. As a result, the Great Smoky Mountains National Park was designated an International Biosphere Reserve and a World Heritage Site in 1976 and 1983, respectively.

The All Taxa Biodiversity Inventory (ATBI) is a concentrated effort to identify, within a relatively short time frame, every species of plant and animal living within the Great Smoky Mountains National Park. It is the first comprehensive biological inventory ever undertaken in North America. The ATBI in the Smokies was conceived in late 1997, in part as a prototype for other reserves. Basic approaches for sampling were worked out by late 1998, and funding was sought for a pilot program that began in the autumn of 2000. The goal of the ATBI is not just to compile lists of what occurs in the park, but to (1) discover the parkwide distribution of each species, (2) determine the relative abundance of each species, and (3) gather data on the seasonality and ecological relationships of all species in the park. Information resulting from this project allows the attention of the park's management to be focused on organisms and habitats with special needs, as well as on more efficient maintenance of healthy populations of species and their ecological surroundings. It also provides a baseline record for the examination of global factors, including acid rain, climate change, and pollution—knowledge that is essential if this park's biodiversity is to be preserved for future generations to enjoy.

The ATBI is a world-renowned project and has generated a strong interest in biodiversity, natural history, taxonomy, and conservation. Hundreds of scientists from around the world are contributing their expertise. In addition, the involvement of nonscientist citizens and students as "citizen scientists" has been significant. Efforts are currently under way to organize an alliance of ATBI parks and reserves to promote biodiversity research, funding, education, communication, and exchange of data. Discover Life in America (DLIA) is a nonprofit organization founded in 1998 that works under a cooperative agreement with the U.S. National Park Service and the Great Smoky Mountains National Park to conduct the ATBI. DLIA administers grants to scientists; organizes volunteers; coordinates development, marketing, and public relations; and assists park partners with education programs.

Estimates of the number of nonmicrobial species in the park vary from 50,000 to 75,000 or more. About 16,000 are currently known. As a result of continuing ATBI investigations, many new species are being added to the records of the park's flora and fauna. For example, an extremely rare lichen, newly described and named *Leioderma cherokeense*, was discovered near the park boundary by Tor Tønsberg, a lichenologist from Norway. Repeated searches failed to find this species anywhere else in the park. Until this discovery in the park, only one other *Leioderma* species in North America was known (specifically, another rare species from the Puget Sound area in Washington State).

Researchers use a variety of techniques appropriate to their particular group of organisms. In addition to hand collecting, dip nets, seines, and electroshocking are used by fisheries biologists; large pieces of tin are placed on the ground to serve as cover for snakes and lizards; birds and bats are captured with mist (nylon) nets; mammals are sampled with mist nets, live traps, and pitfall traps; and insects are lured by malaise traps (tentlike structures made of fine mesh material). A technique utilized by some groups of researchers is a "bio-quest" or "bio-blitz." A bio-blitz is a coordinated sampling effort by groups of experts and volunteers that fan out over the park and collect as many of the appropriate organisms as possible in a specific time period. Bio-blitzes have been held for several groups, including moths and butterflies, beetles, snails, bats, slime molds, millipedes, protists, fungi, and bryophytes. At the second Lepidoptera Bio-Blitz in June 2002, 860 species of moths, butterflies, and skippers in the Smokies were identified in a 24-hour period. This broke the previous North American record for this type of time-constrained sampling, set in the park in 2000, in which 720 species were documented. The 2002 survey located 138 species that were previously unknown in the park, as well as 51 species that were believed to be new to science.

One interesting technique is a tree canopy biodiversity research project, which was first attempted during the summer of 2000. Researchers from seven universities, including bryologists (liverwort and moss experts), a lichenologist, an ecologist, a mycologist (fungi), a myxomycetologist (slime molds), and a flowering plant systematist, served as mentors and as the "ground crew" for a group of student climbers who accessed the tree canopy using a double-rope climbing technique to explore and collect myxomycetes, macrofungi, mosses, liverworts, and lichens. Their objectives were to (1) initiate the

Great Smoky Mountains National Park species tally

Taxon	Old records (prior to ATBI)	New to park (since ATBI began)	New to science	Total records	Estimated total
Microbes					
Bacteria	0	191	270	461	
Archaea	0	0	44	44	
Microsporidia	0	3	5	8	
protists	1	39	2	42	
viruses	0	15	7	22	
Slime molds	128	143	18	289	300
Plants					
vascular	1598	87	0	1685	1750
nonvascular (mosses, etc.)	472	10	0	482	520
Algae	358	566	78	1002	1300–1500
Fungi	2157	583	57	2797	
Lichens	370	408	26	804	
Cnidaria (jellyfish, hydra)	0	2	0	2	
Platyhelminthes (flatworms)	6	19	1	26	
Bryozoa (moss animals)	0	1	0	1	
Acanthocephala (spiny-headed worms)	0	1	0	1	
Nematomorpha (horsehair worms)	1	3	0	4	
Nematodes (roundworms)	13	8	2	23	500
Nemertea (ribbon worms)	0	1	0	1	
Mollusks (snails, mussels, etc.)	121	49	8	178	
Annelids					
aquatic oligochaetes	1	43	1	45	
earthworms	19	8	4	31	
branchiobdellids (crayfish worms)	1	9	0	10	
Tardigrades (waterbears)	3	58	21	82	96
Arachnids					
mites	22	226	31	279	
ticks	7	4	0	11	
harvestmen	1	14	3	18	30
spiders	226	266	41	533	600
scorpions, pseudoscorpions	2	1	0	3	
Crustaceans (crayfish, copepods, etc.)	17	68	26	111	
Chilopoda (centipedes)	20	9	0	29	
Symphyla (symphylans)	0	0	2	2	6
Pauropoda (pauropods)	7	25	17	49	65
Diplopoda (millipedes)	41	24	3	68	78
Protura (proturans)	11	5	10	26	40
Collembola (springtails)	44	116	60	220	300
Diplura (diplurans)	3	5	4	12	
Microcoryphia (jumping bristletails)	0	2	1	3	5
Thysanura (silverfish)	3	0	0	3	3
Ephemeroptera (mayflies)	75	51	8	134	
Odonata (dragonflies, damselflies)	61	34	0	95	130
Orthoptera (grasshoppers, crickets, etc.)	65	37	2	104	
Orthopteroids (cockroaches, mantids, walking sticks)	6	7	0	13	
Dermaptera (earwigs)	2	0	0	2	
Plecoptera (stoneflies)	70	46	3	119	131
Hemiptera (true bugs, hoppers)	272	351	3	626	
Thysanoptera (thrips)	0	47	0	47	100
Psocoptera (barklice)	16	52	7	75	85

Great Smoky Mountains National Park species tally (cont.)

Taxon	Old records (prior to ATBI)	New to park (since ATBI began)	New to science	Total records	Estimated total
Phthiraptera (lice)	8	41	0	49	
Coleoptera (beetles)	887	1575	56	2518	3000
Neuroptera (lacewings, antlions, etc.)	12	33	0	45	
Hymenoptera (bees, ants, etc.)	266	522	23	811	
Trichoptera (caddisflies)	155	78	5	238	275
Lepidoptera (butterflies, moths, skippers)	891	944	36	1871	2222
Siphonaptera (fleas)	17	9	1	27	
Mecoptera (scorpionflies)	17	2	1	20	23
Diptera (flies)	587	503	38	1128	4500
Vertebrates					
fish	70	6	0	76	86
amphibians	41	2	0	43	45
reptiles	38	2	0	40	45
birds	237	10	0	247	260
mammals	64	1	0	65	70
TOTALS	9510	7365	925	17,800	

first survey and inventory of tree canopy biodiversity for myxomycetes, macrofungi, mosses, liverworts, and lichens in the Great Smoky Mountains National Park, (2) compare the assemblages of tree canopy groups of cryptogams (plant or plantlike organisms that do not produce seeds, including algae, ferns, fungi, and mosses) with those on ground sites, (3) compare species diversity of the targeted organisms among tree species, and (4) search for undescribed taxa that are new to science in all of the targeted groups of organisms. Targeted samples were collected from more than 160 trees representing 25 different tree species. These samples were scanned for specimens directly on the bark surface using a dissecting microscope. Bark samples were then placed in moist chambers in order to culture the organisms. Hundreds of species were identified, with more than 100 species being new to the park.

Since the beginning of the ATBI in 1998 through January 2012, the following numbers of new species have been found: 925 species that are new to science, and 7365 species that are new to the park (see **table**). Species that are new to science are ones that have never been formally identified and described. New records for the park include species that were previously known to science, but that until now had not been found in the Smokies.

Only 11 of the new park records are of terrestrial vertebrates: two amphibians, two turtles, six birds, and one mammal. This limited number was fully anticipated because previous work in the park had concentrated on the larger and most highly visible forms. Although most of the new species are small in size, they are often of great importance with regard to the health of ecosystems. For example, they play crucial roles in the cycling of nutrients, they function as predators or prey, and they may even improve the vigor of trees and other plants by fixing nitrogen in the soil.

Diatoms (a group of microscopic algae) are especially important. As the base of the food chain, diatoms are cosmopolitan in their abundance within almost all types of aquatic habitats, and their global biological importance is dramatic. Both freshwater and marine diatoms are estimated to remove nearly half of all the carbon dioxide (a greenhouse gas) from the Earth's atmosphere by photosynthesis. This is a greater amount than that absorbed by all the tropical rainforests, temperate forests, and grasslands combined. In addition, diatoms produce globally at least 25% of the oxygen on Earth. Significantly, diatoms are used as environmental indicators in streams and rivers to assess a watershed's biological integrity. Many state and federal Environmental Protection Agency staff use diatoms, along with fish and aquatic insects, to biologically monitor streams. As observed with fish and aquatic insects, the presence of individual species of diatoms can signal pollution. Certain species are pollution-tolerant, whereas others are more pollution-sensitive. Temperature, light levels, nutrient resources, pH, and toxic materials can also dictate diatom community distribution, and these organisms respond quickly to environmental changes. Diatoms are an excellent biomonitor because of their short generation time, doubling their population approximately once per day. As a result of this short generation time, diatoms are one of the first to recolonize an area after an environmental disturbance.

Above all, the ATBI is a continuing, ongoing effort. For example, despite the history of botanical work in the park over the last century, there are still large or easily recognizable species of vascular plants in the park that remain unrecorded. Wherever comprehensive inventories are undertaken in any region with complex landscapes, it is usually the rare species that are last to be found because of their limited distributions. Therefore, the ATBI will continue to provide park managers with information on new exotic species that invade as well as on rare native species.

For background information *see* BACILLAR-IOPHYCEAE; BIODIVERSITY; CONSERVATION OF RESOURCES; ECOLOGICAL COMMUNITIES; ECOLOGY; ECOSYSTEM; ENVIRONMENTAL MANAGEMENT; POPULATION ECOLOGY; RESTORATION ECOLOGY; SPECIATION in the McGraw-Hill Encyclopedia of Science & Technology.   Donald W. Linzey

Bibliography. K. J. Gaston and J. I. Spicer, *Biodiversity: An Introduction*, 2d ed., Blackwell, Oxford, U.K., 2004; M. J. Jeffries, *Biodiversity and Conservation*, 2d ed., Routledge, Abingdon, Oxon, U.K., 2005; D. W. Linzey, *A Natural History Guide to Great Smoky Mountains National Park*, University of Tennessee Press, Knoxville, 2008; T. E. Lovejoy and L. Hannah (eds.), *Climate Change and Biodiversity*, Yale University Press, New Haven, CT, 2006; M. J. Novacek (ed.), *The Biodiversity Crisis: Losing What Counts*, New Press, New York, 2001.

Allergenicity of cyanobacteria

Prokaryotic planktonic organisms are the oldest photosynthetic, oxygenic life forms that gave sharp rise to atmospheric oxygen about 2.5 billion years ago. It is estimated that they still contribute up to 30% of the yearly oxygen production on Earth. Most photosynthetic prokaryotes are single cells without a nucleus that have been classified as cyanobacteria and comprise about 165 genera and 1500 species. Although they were originally referred to as blue-green algae, their shared characteristics of bacteria better fit the modern nomenclature of cyanobacteria. Cyanobacteria are found in all illuminated environments on Earth, including hot springs, ice fields, soil, and even the fur of animals. They play key roles in the carbon and nitrogen cycles of the biosphere. Two nontoxic species, *Arthrospira platensis (Spirulina)* and *Aphanizomenon flos-aquae*, are mass produced or naturally harvested and consumed throughout the world as protein-rich nutritional supplements. However, about 40 cyanobacterial genera produce toxins, called cyanotoxins, with dermatotoxic, neurotoxic, and/or hepatotoxic properties, which have been implicated in animal and, rarely, human deaths. Because they are gram negative, they also contain lipopolysaccharide or endotoxin as a major component of their cell walls. Slight structural differences in the lipid A acyl moieties of cyanobacterial endotoxins are thought to attenuate the severe biotoxic effects caused by the endotoxins of classical gram-negative bacteria, such as *Escherichia coli* and *Salmonella*.

Since cyanobacteria and green algae occur as blooms on freshwater surfaces, it is plausible to speculate that there could be alternative routes of human exposure to these organisms other than simple dermal contact (see **illustration**). Airborne dissemination of algae was not considered a risk of respiratory sensitization until a number of investigators demonstrated that they could be isolated at altitudes as high as 500 m and that the overall algal content of outdoor dust can be very high, frequently exceeding fungal counts in blowing dust. The presence of cyanobacterial activity in aerial biota of both outdoor and indoor environments is well documented by stationary and moving samplers. Airborne dispersal patterns of cyanobacteria are variable, based on proximity to soil algae populations and meteorological conditions, particularly wind direction and elevation. These factors were established during two consecutive summers on the isolated island of Oahu, Hawaii. Seasonal cyanobacterial atmospheric dispersal patterns from September to November were found over a two-year sampling period in Delhi, India. In a more recent survey of the subtropical city of Varnasi, India, it was observed that soil algae were dominant sources of aeroalgal flora and that airborne cyanobacteria were more common in warmer months of the year.

Many cyanobacterial species have also been isolated from bioaerosols sampled at the human breathing zone (2.5 m). It is thought that nonmarine microalgae are projected into the air from thermal updrafts of soil and bodies of freshwater at wind elevations of 30° up to 30° down. Their presence within indoor environments is most likely due to contamination from outdoor sources as well as aerogenic dissemination from indoor water sources (that is, aquaria or humidifiers). For example, it was previously demonstrated that viable algal organisms

Cyanobacteria blooms.

could be cultured from dust samples collected from 41 homes in southwest Ohio. In those studies, a total of 20 algal species were isolated and identified, but the most frequently observed organisms were four green and two cyanobacterial genera, *Anabaena* and *Schizothrix*. Subsequently, it was demonstrated that a variety of eukaryotic green algae caused skin sensitization as well as upper and lower airway symptoms after controlled nasal and bronchoprovocation testing. Further, an epidemiologic prospective cohort study several years ago revealed a significant increase in self-reported respiratory symptoms in humans exposed to higher levels of cyanobacteria sampled from public recreational pools.

The three possible exposure routes through which cyanobacterial sensitization can occur are (1) dermal, (2) respiratory, and (3) gastrointestinal. Anecdotal reports of cutaneous reactions include pruritic rashes, hives, and papulovesicles. Contact dermatitis was documented by a positive patch test to an extract of the genus *Anabaena* and its phycocyanin derivative. Positive patch tests to lichens, which contain cyanobacterial endosymbionts, were also reported in three patients exposed to these organisms. A single positive patch test to aqueous cyanobacterial suspensions was reported in a patient with atopic dermatitis. There were no other reactions in a group of 20 patients presenting for diagnostic patch testing. Taken together, these reports indicate that epicutaneous sensitization may cause cellular-mediated contact dermatitis. A prospective cohort study of freshwater exposure to cyanobacteria revealed that skin symptoms were more likely to occur 7 days after subjects were exposed to high levels of cyanobacterial species (≤5000 cells/ml) including *Microcystis* genera. The delayed nature of these responses suggested an allergic mechanism. It was also noted that allergenic manifestations did not correlate with toxicity of organisms assayed concurrently in the specific blooms collected during this study.

Oral exposure as a possible route of sensitization to cyanobacteria has not been explored. Two nontoxic species, *Arthrospira platensis* and *Aphanizomenon flos-aquae*, are marketed widely as health-food supplements in North America and other parts of the world. Because these and other supplements are exempt from the U.S. Food and Drug Administration supervisory mandate, there is very little published information about possible adverse side effects of these agents. There are anecdotal reports of acute rhabdomyolysis, bullous pemphigoid, and activation of underlying autoimmune disease in several patients. However, the vast majority of published reports extol their possible benefits, which include immunosuppressive effects on humoral and cell-mediated immune responses, inhibition of interleukin-4 (IL-4) production from mitogen-stimulated peripheral blood monocytes, and clinical efficacy in the treatment of allergic rhinitis.

In 1979, a skin-test survey in India using a panel of 10 aerogenically derived eukaryotic and prokaryotic organisms revealed evidence of immunoglobin

E (IgE) skin sensitization to several cyanobacterial species in patients with allergic symptoms. Whether these test reagents contained cyanotoxins was not stipulated. A major obstacle to interpreting the possible allergenic effects of cyanobacteria in previous reports is that they focused primarily on toxic effects by analyzing whole cells and cyanotoxin concentrations as surrogates for exposure. Nevertheless, some of these studies suggested that allergic rather than toxic factors could have been responsible for human reactions encountered after exposure to these organisms.

Perhaps the most cogent reason for studying possible allergenicity of cyanobacteria in humans is an exploration of a recent hypothesis proposed by C. Emanuelsson and M. D. Spangfort, who searched 391 known allergenic protein sequences and found few or no bacterial homologs to the known allergen sequences selected for comparison. They concluded that allergens are usually eukaryotic proteins that lack bacterial homology, implying that bacterial proteins are relatively nonallergenic, with the possible exception of *Staphylococcus aureus* in patients with atopic dermatitis.

J. A. Bernstein and coworkers (2011) recently prepared detoxified cyanobacteria skin-test reagents for nine species of cyanobacteria species that are ubiquitous in freshwater blooms, soil, and air samples. These species included *M. aeruginosa, A. patensis (Spirulina), Aphanizomenon flos-aquae, Oscillatoria, Synechocystis, Pseudanabaena, Scytonema, Lyngbya,* and *Synechococcus*. Skin-prick testing was administered at a concentration of 1/20 weight/volume in conjunction with a negative saline and positive histamine control. All patients recruited for this study had a physician diagnosis of allergic, nonallergic, or mixed rhinitis based on correlation of history with skin-prick test (SPT) responses or serum-specific IgE to common aeroallergens. A total of 259 patients between the ages of 7 and 78 years old were tested to the cyanobacteria species. Of note, 74 patients (29% of the population) were SPT positive to at least one of the cyanobacterial strains. The most common SPT responses were to *M. aeruginosa* and *Aphanizomenon flos-aquae*. There was a strong correlation between atopy and the presence of a positive SPT response to one or more species of cyanobacteria. Skin-test sensitization to cyanobacteria was more prevalent in physician-diagnosed allergic rhinitis than in patients with mixed or nonallergic rhinitis. This investigation was the first to demonstrate that cyanobacteria allergenicity resides in the non-toxin-containing component of the organism.

Several animal hypersensitivity models confirm that toxic microalgae may be allergenic. One of these evaluated contact sensitivity to mice; the other compared toxic algal blooms to laboratory, axenic cultures in guinea pig, maximal intracutaneous sensitivity, and rabbit eye irritant tests. The latter study demonstrated that induction of maximum intracutaneous sensitivity was inversely related to microcystin content. Further, it was shown that nontoxic *Aphanizomenon* genera exhibited greater allergenicity than toxic cyanobacterial strains. In addition, axenic, nontoxic, or toxic strains were nonallergic for unexplained reasons. The results from J. A. Bernstein's study discussed earlier, which found the greatest percent skin sensitization to *M. aeruginosa* and *A. flos-aquae* cyanobacterial species, supports this observation.

Most case reports and epidemiologic field studies have focused on the occurrence of toxic symptoms in humans. However, in two large population-based prospective cohort studies, respiratory symptoms associated with higher concurrent algal water content were among the most prominent patient-reported symptoms. In one of these investigations, respiratory symptoms were more significant 7 days after exposure, suggesting a possible allergic mechanism. A recent prospective investigation in the United States did not observe a similar effect, but far fewer subjects participated in this study and they also did not appear to be willing participants, because of a misperception that their favorite recreational lakes might be declared "off limits."

Oral contact with either pathogenic or nonpathogenic cyanobacterial species is an obvious route of exposure. Oral ingestion of small amounts of water contaminated with toxic cyanobacterial blooms is unavoidable during recreational activities (such as swimming and water skiing) in bodies of water. A recent risk analysis estimated the number of water-borne illnesses in the United States to be 19.5 million/year. In searching for possible causality, the potential risk of cyanobacterial toxins is often considered to be a health concern. However, their possible allergenic effects have not been seriously considered up to this point. More significant is the fact that several "nontoxic" cyanobacterial species are regularly consumed as nutritional supplements. One of these, *Spirulina* (*A. platensis*), is produced in large outdoor ponds, dried in solar evaporators, and consumed throughout the world. Although it is nontoxic, the extent of possible allergenicity of this agent is unknown. On the other hand, another edible species, *A. flos-aquae*, is harvested from Klamath Lake, Oregon, which experiences periodic extensive contaminating blooms of *M. aeruginosa* containing potent hepatotoxic microcystins. Because microcystins were as high 10.89 µg/g in some consumer products, the Oregon Health Department adopted a regulatory standard of 1 µg/g as a safe microcystin level for this product. Similar to *Spirulina*, possible allergenic effects of *Aphanizomenon* species have not been determined prior to this current investigation.

The fact that human allergenicity to several cyanobacterial species has been determined appears to contradict a recent hypothesis that most eukaryotic proteins lack bacterial homologs, thereby suggesting that prokaryotic organisms are relatively nonallergenic. There are a number of cogent exceptions to this hypothesis among both gram-negative and -positive heterotrophic bacteria. Some bacterial exotoxins and endotoxins also are potent adjuvants and/or may enhance IgE-mediated histamine

release. Assuming this biologic principle to be correct, it is possible that cyanobacterial allergenicity can be partially explained by concurrent adjuvant effects of lipopolysaccharide (LPS) constituents within plasma membranes of these gram-negative autotrophs. Although endotoxin potency is variable among cyanobacterial species and strains compared to classic gram-negative heterotrophs (for example, *E. coli*), LPS could exert adjuvant effects by stimulation of the innate immune system, ultimately leading to and perhaps augmenting adaptive IgE-specific immune responses. For example, transient flulike syndromes have occurred a few hours after taking a bath, showering, or exposure to humidifiers with water containing elevated levels of cyanobacterial-derived endotoxin. It is possible that similar reactions in susceptible atopic populations could stimulate the cascade of IgE-mediated allergenicity.

Based on the growing evidence supporting the allergenicity of cyanobacteria, long-term, controlled, cohort studies should be designed to investigate the prevalence of cyanobacteria sensitization in the general population. Such studies will likely assume greater importance as increased human exposure to cyanobacteria parallels widespread implementation of future unique applications of these organisms, such as biofertilization, biofuels, hydrogen production, and space habitation.

For background information *see* ALLERGY; CYANOBACTERIA; DERMATITIS; ENDOTOXIN; EPIDEMIOLOGY; IMMUNOLOGY; PROKARYOTAE in the McGraw-Hill Encyclopedia of Science & Technology.

Jonathan A. Bernstein; I. Leonard Bernstein

Bibliography. W. B. Anderson, D. G. Dixon, and C. I. Mayfield, Estimation of endotoxin inhalation from shower and humidifier exposure reveals potential risk to human health, *J. Water Health*, 5:553–572, 2007, DOI:10.2166/wh.2007.043; H. Annadotter et al., Endotoxins from cyanobacteria and gram-negative bacteria as the cause of an acute influenza-like reaction after inhalation of aerosols, *EcoHealth*, 2:209–221, 2005, DOI:10.1007/s10393-005-5874-0; L. C. Backer et al., Recreational exposure to microcystins during algal blooms in two California lakes, *Toxicon*, 55(5):909–921, 2010, DOI:10.1016/j.toxicon.2009.07.006; C. Benaim-Pinto, Airborne algae as a possible etiologic factor in respiratory allergy in Caracas, Venezuela, *J. Allergy Clin. Immunol.*, 49(6):356–358, 1972; I. L. Bernstein et al., Immune responses in farm workers after exposure to *Bacillus thuringiensis* pesticides, *Environ. Health Perspect.*, 107:575–582, 1999, DOI:10.1289/ehp.99107575; I. L. Bernstein and R. S. Safferman, Clinical sensitivity to green algae demonstrated by nasal challenge and in vitro tests of immediate hypersensitivity, *J. Allergy Clin. Immunol.*, 51:22–28, 1973; I. L. Bernstein and R. S. Safferman, Viable algae in house dust, *Nature*, 227:851–852, 1970, DOI:10.1038/227851a0; J. A. Bernstein et al., Cyanobacteria: An unrecognized ubiquitous sensitizing allergen?, *Allergy Asthma Proc.*, 32:106–110, 2011, DOI:10.2500/aap.2011.32.3434; D. C. Blanchard and L. Syzdek,

Mechanism for the water-to-air transfer and concentration of bacteria, *Science*, 170:626–628, 1970; R. M. Brown, Jr., Studies of Hawaiian fresh-water and soil algae. I. The atmospheric dispersal of algae and fern spores across the island of Oahu, Hawaii, *Contrib. Phycol.*, pp. 175–188, September 1971; R. M. Brown, Jr., D. D. Larson, and H. C. Bold, Airborne algae: Their abundance and heterogeneity, *Science*, 143:583–585, 1964; W. Carmichae, A world overview—one-hundred-twenty-seven years of research on toxic cyanobacteria—where do we go from here?, *Adv. Exp. Med. Biol.*, 619:105–125, 2008, DOI:10.1007/978-0-387-75865-7; T. Cavalier-Smith, Cell evolution and Earth history: Stasis and revolution, *Phil. Trans. Roy. Soc. Lond. B*, 361:969–1006, 2006, DOI:10.1098/rstb.2006.1842; R. H. Champion, Atopic sensitivity to algae and lichens, *Br. J. Dermatol.*, 85:551–557, 1971, DOI:10.1111/j.1365-2133.1971.tb14081.x; C. Cingi et al., The effects of spirulina on allergic rhinitis, *Eur. Arch. Otorhinolaryngol.*, 265:1219–1223, 2008, DOI:10.1007/s00405-008-0642-8; P. Clementsen et al., Endotoxin from *Haemophilus influenzae* enhances IgE-mediated and non-immunological histamine release, *Allergy*, 45:10–17, 1990, DOI:10.1111/j.1398-9995.1990.tb01078.x; S. G. Cohen and C. B. Reif, Cutaneous sensitization to blue-green algae, *J. Allergy*, 24:452–457, 1953, DOI:10.1016/0021-8707(53)90047-1; W. K. Czaja et al., The future prospects of microbial cellulose in biomedical applications, *Biomacromolecules*, 8:1–12, 2007, DOI:10.1021/bm060620d; G. Doekes et al., IgE sensitization to bacterial and fungal biopesticides in a cohort of Danish greenhouse workers: The BIOGART study, *Am. J. Ind. Med.*, 46:404–407, 2004, DOI:10.1002/ajim.20086; H. Durand-Chastel, Production and use of *Spirulina* in Mexico, in G. Shelef and C. J. Soeder (eds.), *Algae Biomass*, Elsevier/North-Holland Biomedical Press, pp. 51–64, 1980; D. Dutta et al., Hydrogen production by Cyanobacteria, *Microb. Cell Fact.*, 4:36, 2005, DOI:10.1186/1475-2859-4-36; O. E. el Saadi, Murray River water, raised cyanobacterial cell counts, and gastrointestinal and dermatological symptoms, *Med. J. Aust.*, 162:122–125, 1995; C. Emanuelsson and M. D. Spangfort, Allergens as eukaryotic proteins lacking bacterial homologues, *Mol. Immunol.*, 44:3256–3260, 2007, DOI:10.1016/j.molimm.2007.01.019; G. E. Fogg et al., *Blue-green Algae*, Academic Press, London, U.K., 1973; D. J. Gilroy et al., Assessing potential health risks from microcystin toxins in blue-green algae dietary supplements, *Environ Health Perspect.*, 108:435–439, 2000, DOI:10.1289/ehp.00108435; B. J. Hales et al., Anti-bacterial IgE in the antibody responses of house dust mite allergic children convalescent from asthma exacerbation, *Clin. Exp. Allergy*, 39:1170–1178, 2009, DOI:10.1111/j.1365-2222.2009.03252.x; Health Canada announces results of blue-green algal products testing—Only *Spirulina* found Microcystin-free, Health Canada news release, Sept. 17, 1997; H. A. Heise, Symptoms of hay fever caused by algae. II. Microcystis, another form of algae producing allergenic reactions, *Ann. Allergy*, 9:100–101, 1951; M. F. Hofer et al., Upregulation of IgE synthesis by staphylococcal toxic shock syndrome toxin-1 in peripheral blood mononuclear cells from patients with atopic dermatitis, *Clin. Exp. Allergy*, 25:1218–1227, 1995, DOI:10.1111/j.1365-2222.1995.tb03046.x; L. L. Kjaergard et al., Basophil-bound IgE and serum IgE directed against *Haemophilus influenzae* and *Streptococcus pneumoniae* in patients with chronic bronchitis during acute exacerbations, *APMIS*, 104:61–67, 1996, DOI:10.1111/j.1699-0463.1996.tb00687.x; A. H. Knoll, Cyanobacteria and earth history, in A. Herrero and E. Flores (eds.), *The Cyanobacteria: Molecular Biology, Genomics and Evolution*, Caister Academic Press, 2008; U. Kosecka et al., Pertussis toxin stimulates hypersensitivity and enhances nerve-mediated antigen uptake in rat intestine, *Am. J. Physiol.*, 267:G745–G753, 1994; O. Kraigher et al., A mixed immunoblistering disorder exhibiting features of bullous pemphigoid and pemphigus foliaceus associated with *Spirulina* algae intake, *Int. J. Dermatol.*, 47:61–63, 2008, DOI:10.1111/j.1365-4632.2007.03388.x; A. N. Lee and V. P. Werth, Activation of autoimmunity following use of immunostimulatory herbal supplements, *Arch Dermatol.*, 140(6):723–727, 2004, DOI:10.1001/archderm.140.6.723; P. S. C. Leung et al., Induction of shrimp tropomyosin-specific hypersensitivity in mice, *Int. Arch. Allergy Immunol.*, 147:305–314, 2008, DOI:10.1159/000144038; M. Lobner et al., Enhancement of human adaptive immune responses by administration of a high-molecular-weight polysaccharide extract from the cyanobacterium *Arthrospira platensis*, *J. Med. Food*, 11:313–322, 2008, DOI:10.1089/jmf.2007.564; T. K. Mao, J. Van de Water, and M. E. Gershwin, Effects of a *Spirulina*-based dietary supplement on cytokine production from allergic rhinitis patients, *J. Med. Food*, 8:27–30, 2005, DOI:10.1089/jmf.2005.8.27; E. E. Mazokopakis et al., Acute rhabdomyolysis caused by *Spirulina* (*Arthrospira platensis*), *Phytomedicine*, 15:525–527, 2008, DOI:10.1016/j.phymed.2008.03.003; A. Mittal, M. K. Agarwal, and D. N. Shivpuri, Respiratory allergy to algae: Clinical aspects, *Ann. Allergy*, 42:253–256, 1979; A. Mittal, M. K. Agarwal, and D. N. Shivpuri, Studies on allergenic algae of Delhi area: Botanical aspects, *Ann. Allergy*, 42:248–252, 1979; A. Oehling et al., Potentiation of histamine release against inhalant allergens (*Dermatophagoides pteronyssinus*) with bacterial antigens in bronchial asthma, *J. Investig. Allergol Clin. Immunol.*, 7:211–216, 1997; M. Olaizola, Commercial development of microalgal biotechnology: From the test tube to the marketplace, *Biomol. Eng.*, 20:459–466, 2003, DOI:10.1016/S1389-0344(03)00076-5; P. Y. Ong et al., Association of staphylococcal superantigen-specific immunoglobulin E with mild and moderate atopic dermatitis, *J. Pediatr.*, 153:803–806, 2008, DOI:10.1016/j.jpeds.2008.05.047; R. Pauwels, G. Verschraegen, and M. van der Straeten, IgE antibodies to bacteria in patients with bronchial asthma, *Allergy*, 35:665–669, 1980, DOI:10.1111/j.1398-9995.1980.

tb02019.x; L. S. Pilotto et al., Health effects of exposure to cyanobacteria (blue-green algae) during recreational water-related activities, *Aust. N. Z. J. Public Health*, 21:562–566, 1997, DOI:10.1111/j.1467-842X.1997.tb01755.x; M. Plaza et al., Screening for bioactive compounds from algae, *J. Pharm. Biomed. Anal.*, 51:450–455, 2010, DOI:10.1016/j.jpba.2009.03.016; R. Raj et al., A perspective on the biotechnological potential of microalgae, *Crit. Rev. Microbiol.*, 34:77–88, 2008; M. Rasool and E. P. Sabina, Appraisal of immunomodulatory potential of *Spirulina fusiformis:* An in vivo and in vitro study, *J. Nat. Med.*, 63:169–175, 2009, DOI:10.1007/s11418-008-0308-2; S. Raziuddin, H. W. Siegelman, and T. G. Tornabene, Lipopolysaccharides of the cyanobacterium *Microcystis aeruginosa, Eur. J. Biochem.*, 137:333–336, 1983, DOI:10.1111/j.1432-1033.1983.tb07833.x; S. Rechter et al., Antiviral activity of *Arthrospira*-derived spirulan-like substances, *Antiviral Res.*, 72:197–206, 2006, DOI:10.1016/j.antiviral.2006.06.004; S. Rellan et al., First detection of anatoxin-a in human and animal dietary supplements containing cyanobacteria, *Food Chem. Toxicol.*, 47(9):2189–2195, 2009, DOI:10.1016/j.fct.2009.06.004; K. A. Reynolds, K. D. Mena, and C. P. Gerba, Risk of waterborne illness via drinking water in the United States, *Rev. Environ. Contam. Toxicol.*, 192:117–158, 2008, DOI:10.1007/978-0-387-71724-1˙4; P. Rupa et al., A neonatal swine model of allergy induced by the major food allergen chicken ovomucoid (Gal d 1), *Int. Arch. Allergy Immunol.*, 146:11–18, 2008, DOI:10.1159/000112498; H. E. Schlichting, Jr., Ejection of microalgae into the air via bursting bubbles, *J. Allergy Clin. Immunol.*, 53:185–188, 1974, DOI:10.1016/0091-6749(74)90006-2; H. E. Schlichting, Jr., Meterological conditions affecting the dispersal of airborne algae and protozoa, *Lloydia*, 27:64–78, 1964; H. E. Schlichting, Jr., Viable species of algae and protozoa in the atmosphere, *Lloydia*, 24:81–88, 1961; N. W. Schroder et al., Innate immune responses during respiratory tract infection with a bacterial pathogen induce allergic airway sensitization, *J. Allergy Clin. Immunol.*, 122:595–602.e5, 2008, DOI:10.1016/j.jaci.2008.06.038; N. K. Sharma, S. Singh, and A. K. Rai, Diversity and seasonal variation of viable algal particles in the atmosphere of a subtropical city in India, *Environ Res.*, 102:252–259, 2006, DOI:10.1016/j.envres.2006.04.003; N. A. Sorkhoh et al., Establishment of oil-degrading bacteria associated with cyanobacteria in oil-polluted soil, *J. Appl. Bacteriol.*, 78:194–199, 1995, DOI:10.1111/j.1365-2672.1995.tb02842.x; I. Stewart et al., Cutaneous hypersensitivity reactions to freshwater cyanobacteria—Human volunteer studies, *BMC Dermatol.*, 6:6, 2006, DOI:10.1186/1471-5945-6-6; I. Stewart et al., Epidemiology of recreational exposure to freshwater cyanobacteria—An international prospective cohort study, *BMC Public Health*, 6:93, 2006, DOI:10.1186/1471-2458-6-93; I. Stewart et al., Primary irritant and delayed-contact hypersensitivity reactions to the freshwater cyanobacterium *Cylindrospermopsis raciborskii* and its associated toxin cylindrospermopsin, *BMC Dermatol.*, 6:5, 2006, DOI:10.1186/1471-5945-6-5; K. Takayama and N. Quereshi, Chemical structure of lipid A, in D. C. Morrison and J. L. Ryan (eds.), *Bacterial Endotoxin Lipopolysaccharides*, vol. 1, *Molecular Biochemistry and Cellular Biology*, CRC Press, pp. 43–65, 1992; R. D. Tee and J. Pepys, Specific serum IgE antibodies to bacterial antigens in allergic lung disease, *Clin. Allergy*, 12:439–450, 1982, DOI:10.1111/j.1365-2222.1982.tb01642.x; E. Tiberg et al., Detection of *Chlorella*-specific IgE in mould-sensitized children, *Allergy*, 45:481–486, 1990, DOI:10.1111/j.1398-9995.1990.tb00523.x; E. Tiberg, S. Dreborg, and B. Bjorksten, Allergy to green algae (*Chlorella*) among children, *J. Allergy Clin. Immunol.*, 96:257–259, 1995; A. Torokne, A. Palovics, and M. Bankine, Allergenic (sensitization, skin and eye irritation) effects of freshwater cyanobacteria—Experimental evidence, *Environ Toxicol.*, 16:512–516, 2001, DOI:10.1002/tox.10011; A. A. Tsygankov et al., Hydrogen production by cyanobacteria in an automated outdoor photobioreactor under aerobic conditions, *Biotechnol. Bioeng.*, 80:777–783, 2002, DOI:10.1002/bit.10431; M. A. van Overeem, A sampling apparatus for aeroplankton, *Proc. Roy. Acad. Amsterdam*, 34:981–990, 1936; A. H. Woodcock, Note concerning human respiratory irritation associated with high concentration of plankton and mass mortality of marine organisms, *J. Marine Res.*, 7:56, 1948.

Alternative RNA splicing

To convert the genetic information stored in deoxyribonucleic acid (DNA) to proteins, the DNA is transcribed to produce an intermediate called precursor messenger ribonucleic acid (pre-mRNA). Only parts of the pre-mRNA, the exons (which comprise less than 5% of the average human gene), are exported into the cytosol as mature mRNA. Prior to the formation of mRNA, the exons are spliced together, which removes intermediate (intronic) sequences. Notably, this process can be alternative, which allows an organism to determine which parts of the DNA are translated into proteins (**Fig. 1**). Alternative splicing affects more than 92% of human genes, making it one of the most versatile mechanisms to increase the use of genetic information stored in DNA.

Definition of alternative pre-mRNA splicing. Pre-mRNA splicing is an RNA maturation event in which parts of the pre-mRNA sequences are joined and subsequently exported into the cytosol as exons, and the intervening sequences are removed as introns. Alternative splicing is the alternate inclusion of exons or introns from the primary transcript (pre-mRNA) into mature mRNA. The majority of human genes are alternatively spliced; that is, they contain sequences that can be used either as an exon or as an intron. Alternative splicing creates different mRNA isoforms from a single gene. Mostly, but not always, these mRNA isoforms encode distinct protein products. Based on the arrangement of splice sites,

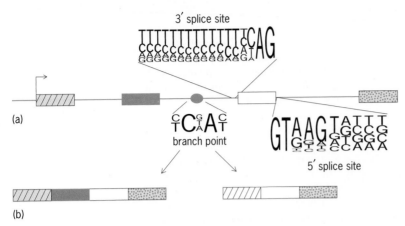

(a)

(b)

Fig. 1. RNA elements that control alternative splicing. (*a*) A pre-mRNA is shown schematically. The transcriptional start site is shown by the arrow. Exons are shown as boxes; introns are shown as lines. The consensus sequences for the branch point, the 3′ splice site, and the 5′ splice site are indicated. A consensus sequence describes the frequency of the nucleotides at each position. The consensus sequences are shown as sequence logos. The total height of all bases at a given position is 100%, and the height of each letter (A, C, G, T) indicates the percent usage. (*b*) The pre-mRNA generates two mRNAs by joining the exons and removing the introns. Note that the solid-blue alternative exon is used alternatively because it is present in one mRNA but absent in the other.

tive exons. Because their mutation to the consensus sequence leads to constitutive exons (that is, the inclusion of those exons in all derived mRNAs), they are referred to as weak splice sites. Furthermore, because the splice sites are so variable, auxiliary sequences on the pre-mRNAs help in their recognition. These sequences either enhance or silence an exon, and they can be intronic or exonic.

Alternative exon plasticity. Cells can regulate the exons that they use alternatively, which is a process that involves known signal transduction pathways. For example, a carbohydrate-containing meal will change alternative splicing of protein kinase C (an enzyme regulating important physiological functions) through a signaling cascade emanating from insulin.

Function of alternative splicing. Alternative splicing plays an important physiological role because almost all human protein-coding genes undergo alternative splicing. The overall function of alternative splicing is to increase the diversity of mRNAs expressed from the genome, which increases the number of proteins that the cell can make. By making different protein isoforms, alternative splicing regulates binding between proteins, binding between proteins and nucleic acids, binding between proteins and membranes, and protein location within the cell. Alternative splicing generates proteins with different enzymatic properties and ligand interactions. In most cases, the changes caused by individual splicing isoforms are small. However, cells typically coordinate numerous changes in splicing programs, where modifications in the activity of a regulatory factor can control hundreds of alternative exons. Such a coordinated change of alternative exons can have strong effects on cell proliferation, cell survival, and properties of the nervous system (**Fig. 3**).

One of the best-understood examples for the physiological role of alternative splicing is the sex determination of *Drosophila*. A cascade of splicing regulatory proteins (Sex-lethal, transformer, and

five basic types of alternative splicing can be discriminated: cassette exons, the alternative 5′ splice site, the alternative 3′ splice site, mutually exclusive exons, and retained introns (**Fig. 2**).

Splice site selection. Exons are determined by three crucial elements: the 3′ splice site, the 5′ splice site, and the branch point (Fig. 1). These sequences are short and follow only loose consensus sequences that can be described as a matrix (in general, a consensus sequence describes the frequency of the nucleotides at each position). The only invariant positions are the GT-AG dinucleotides (that is, guanine-thymine and adenine-guanine dinucleotides) flanking 98% of all introns. The other positions of the splice sites can differ significantly from consensus sequences. Splice sites, which deviate more strongly from the consensus sequences, characterize alterna-

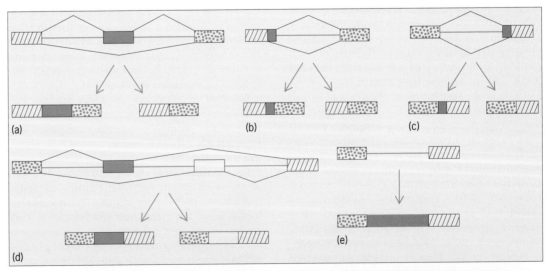

(a)

(b)

(c)

(d)

(e)

Fig. 2. Types of alternative splicing: (*a*) cassette exon; (*b*) alternative 5′ end; (*c*) alternative 3′ end; (*d*) mutually exclusive exons; (*e*) retained intron.

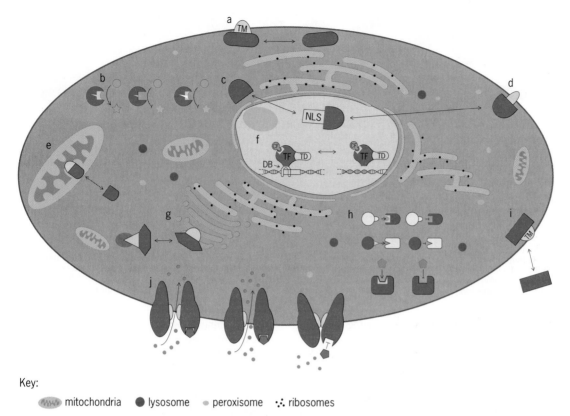

Key:

🌊 mitochondria   ● lysosome   ● peroxisome   ⋰ ribosomes

Fig. 3. Function of alternative exons. A cell is shown schematically. Proteins that undergo alternative splicing are shown in red; their alternative parts are shown in yellow. (*a*) Generation of soluble isoforms by eliminating a transmembrane domain (**TM**). (*b*) Generation of isoforms with different enzymatic activities. (*c, d*) Generation of proteins with different localizations in the nucleus, cytosol, and plasma membrane. (*e*) Change of localization between mitochondria and cytosol by incorporating or deleting a mitochondrial localization signal. (*f*) Generation of transcription factors with different DNA-binding domains. Terms: DB, DNA-binding domain; CF, cofactor; NLS, nuclear localization signal; TF, transcription factor; TD, transactivation domain. (*g*) Change in localization between different membranes. (*h*) Change of binding activities. (*i*) Generation of membrane-bound or secreted isoforms. (*j*) Generation of receptors with different properties.

transformer-2) determines the sex-specific alternative splicing of the transcription factor doublesex, which results in male- or female-specific doublesex isoforms. This cascade of splicing decisions ultimately results in sex-specific development. Another well-characterized system is programmed cell death (apoptosis). A number of programmed cell-death regulatory genes undergo alternative splicing. The resulting protein isoforms are functionally distinct and sometimes act antagonistically.

Nonsense-mediated decay (NMD) is an RNA surveillance mechanism that is responsible for the degradation of transcripts with premature translation termination codons (PTCs). Approximately one-quarter of alternative exons contain stop codons that either create shortened proteins or lead to degradation of the RNA through nonsense-mediated RNA decay. The coupling of alternative splicing and NMD allows alternative splicing to control mRNA abundance independently from transcriptional activity.

Alternative splicing is connected to other parts of gene expression. Alternative splicing is coordinated with other mechanisms of gene expression, including, most notably, transcription. Human RNA polymerase contains an extended carboxy-terminal domain (CTD) that binds to several proteins, including splicing factors (**Fig. 4**). Therefore, the polymerase

at a promoter region where transcription is initiated can be primed with splicing factors, which will predetermine alternative splicing decisions of the transcript. An additional regulatory mechanism involves RNA polymerase speed. Generally, alternative exons need additional proteins that bind to pre-mRNA to facilitate exon inclusion. This binding requires time, and a polymerase that is processing fast will prevent binding of auxiliary proteins, which leads to skipping of alternative exons. RNA polymerase speed is regulated in part by the structure of chromatin (the deoxyribonucleoprotein complex forming the major portion of the nuclear material and of the chromosomes). An open chromatin structure allows a fast-acting polymerase, which reduces the inclusion of alternative exons. In contrast, closed chromatin results in a slow-acting polymerase, which favors inclusion of alternative exons.

In support of the role of chromatin, it has been observed that exonic regions are preferentially marked by increased nucleosome occupancy and specific methylation of histones (certain proteins associated with DNA to form nucleoproteins) relative to introns. These markings are also splicing-related, with slightly lower levels of methylation observed for alternatively spliced exons. The relationship between alternative splicing and chromatin is reciprocal,

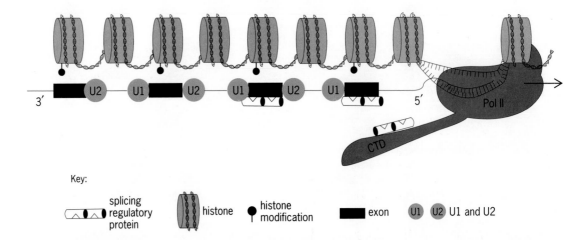

Key:

splicing regulatory protein   histone   histone modification   ■ exon   U1 U2 U1 and U2

Fig. 4. Alternative splicing is coupled to chromatin structure. Chromatin is shown as DNA wrapped around nucleosomes (tinted gray). RNA polymerase II (Pol II) moves along the RNA as indicated by the arrow. The boundaries of exons correlate with nucleosomes. During the splicing reaction, the spliceosome (the nucleoprotein particle that aids in the splicing of RNA), indicated schematically by its components U1 and U2, changes histone modifications. CTD = carboxy-terminal domain.

because splicing and alternative splicing can also change chromatin marks.

Splicing and disease. Mutations that alter the RNA sequence or changes in *trans*-acting factors that comprise or affect the splicing machinery can alter pre-mRNA splicing. These changes can range from a modulation of the degree to which an existing exon is included in the pre-mRNA to complete exon skipping or constitutive exon usage. In addition, new exons can be created by the generation of new splice sites when these splice sites are located next to an authentic splice site or a cryptic splice site (which is used only when use of an authentic splice site is disrupted by mutation). In most cases, mutations lead to exon skipping, and it has been estimated that approximately 60% of disease-causing point mutations change splice site selection. Examples of diseases caused by changes in pre-mRNA splicing include various forms of thalassemias, frontotemporal dementia with parkinsonism linked to chromosome 17 (FTDP-17), Hutchinson-Gilford progeria, hypercholesterolemia, medium-chain acyl-coenzyme A dehydrogenase (MCAD) deficiency, familial dysautonomia, and myotonic dystrophy.

Intronic mutations located hundreds of nucleotides away from known exons can cause aberrant alternative splicing, as shown by a case of neurofibromatosis type 1 (NF1). This indicates that mutations can cause diseases by regulating alternative exons over long distances.

The possibility to take advantage of alternative splicing in the treatment of human disease is illustrated in the case of Duchenne muscular dystrophy. In this condition, deletions within the large Duchenne gene cause a complete loss of the gene product dystrophin, because the reading frame is destroyed. However, if this reading frame is restored, a shortened dystrophin protein can be produced, which results in the milder Becker-type dystrophy. The selective removal of exons in the dystrophin gene is one way to restore the reading frame. Oligonucleotides have been designed that

cause skipping of defined exons when delivered to muscle; this type of therapy is now in advanced clinical trials.

Conclusions and perspectives. The full appreciation of the importance of alternative splicing in cellular functions and disease has just begun. The progress in sequencing of all exons of an individual will likely identify more diseases that are caused by alternative splicing and identify exons that are characteristic for certain individuals. Moreover, the repertoire of substances that change splice site selection is increasing and will likely allow future treatment of some diseases caused by missplicing.

For background information *see* CHROMOSOME; DEOXYRIBONUCLEIC ACID (DNA); EXON; GENE; GENETICS; HISTONE; INTRON; MUSCULAR DYSTROPHY; NUCLEOSOME; PROTEIN; RIBONUCLEIC ACID (RNA); TRANSCRIPTION in the McGraw-Hill Encyclopedia of Science & Technology.

Paolo Convertini; Stefan Stamm

Bibliography. B. J. Blencowe and B. R. Graveley (eds.), *Alternative Splicing in the Postgenomic Era*, Springer, New York, 2007; T. A. Cooper, L. Wan, and G. Dreyfuss, RNA and disease, *Cell*, 136:777–793, 2009, DOI:10.1016/j.cell.2009.02.011; R. F. Luco et al., Epigenetics in alternative pre-mRNA splicing, *Cell*, 144:16–26, 2011, DOI:10.1016/j.cell.2010.11.056; S. Stamm, C. Smith, and R. Lührmann (eds.), *Alternative pre-mRNA Splicing: Theory and Protocols*, Wiley, Weinheim, Germany, 2012; S. D. Wilton and S. Fletcher, RNA splicing manipulation: Strategies to modify gene expression for a variety of therapeutic outcomes, *Curr. Gene Ther.*, 11:259–275, 2011, DOI:10.2174/156652311796150381.

Aluminum shipbuilding

While most metal ships have been constructed from steel, in the last century aluminum has played an increasing role in the marine industry. Aluminum's comparatively low density and high strength have made it an attractive option for both full and partial ship structures where weight control is a critical parameter.

Aluminum's beginnings. Though aluminum is the most abundant metal in the Earth's crust (~8% by weight), it was not discovered until very early in the nineteenth century because of the difficulty in reducing it from its ore. Even after its discovery, aluminum remained rare until 1888 when the advent of the Hall-Héroult process made industrial scale production cost effective. In fact, the material was so precious during this time that, in 1884 when a 6.28-lb (2.85-kg) aluminum cap was placed on top of the Washington monument, it was the largest known casting in the world. By contrast it is estimated that world aluminum production in 1918, just 34 years later, was 180,000 metric tons. In 2010, Alcoa Inc., a single producer of aluminum, produced 4,500,000 metric tons of the material, an increase of 25 times.

Early applications. Unalloyed in its pure form, aluminum is not a particularly interesting engineering material. AA1060-O (99.60% pure aluminum) has a yield strength of approximately 4 ksi (4000 lb/in^2, or 28 MPa), but when alloyed with relatively low concentrations of copper, magnesium, silicon, manganese, lithium or other elements, the material's strength can exceed 90 ksi (621 MPa). Given that aluminum is approximately one-third the density of steel, the high specific strength possible with aluminum alloys makes it attractive to designers seeking reductions in weight.

Aluminum's initial use in the marine industry came in the 1890s. The French sloop *Vendenesse* (1892), and the American built *Defender* (1895), utilized aluminum to increase speed. In fact, *Defender* won the America's Cup in 1895. These early designs achieved remarkable performance, but because of the limited choice of alloys available at the time they suffered from corrosion issues. The aluminum alloys used on the *Defender* were alloyed with approximately 6% copper, equivalent to today's 2xxx series. Nevertheless, these early applications demonstrated the material's potential to improve ship performance.

Aluminum becomes practical. New alloys, tempers, and fabrication technologies, in conjunction with the 1922 Washington Disarmament Conference limitation on overall ship displacement, encouraged naval architects to reintegrate the material into ship designs. In 1928 this resulted in USS *Houston* (CA-30) utilizing aluminum alloy deckhouses made from 2017 "Duralumin." In 1935, aluminum alloy was introduced as a superstructure material for *Sims*-class destroyers. In 1936, it was introduced throughout the entire deckhouse of the *Gleaves*-class destroyers. Records show over 100 US Navy ships with aluminum alloy superstructure applications in 1940. In parallel, the reduced weight of the material was being leveraged by commercial industry. In 1939, fourteen tons of aluminum successfully replaced forty tons of steel in the Norwegian commercial vessel M.V. *Fernplant*. This was successful enough to motivate the conversion of three sister ships to the same configuration. Improvements in alloying, tempering, and the adoption of gas tungsten arc welding (GTAW) and gas metal arc welding (GMAW) in the 1940s led to the material being practical for all aspects of ship construction.

In 1952, S.S. *United States* utilized approximately 2000 tons of aluminum to save approximately 8000 tons of weight, thereby enabling the establishment of transatlantic speed records. By the 1960s, it was common for both naval and commercial ships to contain aluminum in both their superstructures and hulls. Additionally, most of the common shipbuilding alloys in use today had already been developed, including AA5052, AA5083, AA5086, AA5454, AA5456, and AA6061. AA6082, which is common in Europe, was registered with the Aluminum Association in 1972.

Modern alloys and tempers for shipbuilding. Modern aluminum products are defined by their chemistry or "alloy" and their material processing or "temper." Chemistries are grouped into series by the primary alloying element (see **table**). When a new chemistry is identified and commercialized it is given a designation from the Aluminum Association within this naming convention. Casting alloys follow a different standard, but are typically less common in shipbuilding. With some unique exceptions, there are two temper families: T for heat-treatables and H for strain-hardenable alloys. The T or H is followed by a numeric designator denoting the processing technique. Plate, extrusions, and forgings of the same chemistry and similar engineering performance will likely have different temper designations since they are usually process specific. This convention does have some notable exceptions, for example, where the temper designation can refer to an intended application or testing to which the material must conform. Specifically, H116 and H321 are commonly used for high Mg 5xxx-series marine plate products that have passed inter-granular and exfoliation testing.

5xxx-series material is typically considered best for the marine environment. These materials exhibit good weld strength and corrosion resistance even when directly exposed to seawater. The most common material, AA5083-H116, has a wrought yield strength of 29 ksi (200 MPa) and a welded yield strength of 18 ksi. Some 5xxx-series materials used in marine applications approach 40 ksi (276 MPa) yield strength. Unfortunately, high strength 5xxx-series

Aluminum Association series definitions (casting alloys excluded)	
Series	Primary alloying element
1xxx	Pure aluminum (99% or greater)
2xxx	Copper
3xxx	Manganese
4xxx	Silicon
5xxx	Magnesium
6xxx	Magnesium and silicon (Mg$_2$Si)
7xxx	Zinc
8xxx	Other elements
9xxx	Unused series

alloys with magnesium content of greater than ~3.5% are potentially susceptible to stress corrosion cracking (SSC). SSC is managed by reducing environmental temperatures, managing stress below threshold limits, and proper manufacturing techniques. This drives a design balance between strength and operating environment. For example, AA5456 has a nominal magnesium content of 5.1% giving it relatively good wrought and welded yield strengths of 33 ksi and 19 ksi, respectively, but the material is more susceptible to SSC than AA5083, which has a nominal magnesium content of 4.5%, but lower strength. With this in mind, alloys with magnesium levels as high as 6.0% are in use, but in general, industry has focused on AA5083 as a good compromise. Properly designed aluminum structures are commonly surviving more than 30 years of marine service.

6xxx-series material is also commonly used in ship and boat construction. It is seldom used for large hull construction designed for extremely long service lives. It is common in small boat building and is used above the waterline in large projects. Some sources suggest that low copper 6xxx-series alloys can survive almost as long in a marine environment as 5xxx-series materials. Because 6xxx-series materials are heat treatable, they are easily manufactured and available in a wide variety of extruded shapes, but lose a portion of their strength when welded. Welded properties of AA6061 can drop from 40 ksi (276 MPa) yield strength in the T6 condition to 15 ksi (103 MPa), or 38% of the original strength, when welded. Some classing agents require the use of completely annealed properties in and near welds, for example, 8 ksi (55 MPa) for AA6061. This is a significant disadvantage because many ship designers, especially of large projects, assume repairs will be necessary and design to the welded strength versus wrought strength. This can be overcome by utilizing other construction or repair techniques such as fasteners or bonding, but these have not been adopted by large shipyards.

Other alloy series are used in shipbuilding to a much lower degree. Of particular interest are very high strength 7xxx-series and 2xxx-series materials. These can have strengths exceeding 78 ksi (538 MPa). Some benefit could be gained in internal "dry" applications where corrosion is less of a concern. The ship building industry has not adopted this approach broadly due to the increased complexity of supply and lack of performance experience.

Cost. Aluminum is more expensive than steel on a per unit mass basis. The total cost increase is reduced by the fact that an aluminum ship of equivalent performance is less massive, but even with this efficiency a 5–10% increase in acquisition cost typically remains. The selection of aluminum is usually driven by other life-cycle considerations, including reduction of weight high in a ship, reduction of draft, reduction of maintenance, downsizing propulsion, or other weight-related performance parameters. Typically, this means that certain classes of ship are more likely to utilize aluminum than others.

Aluminum is used extensively in ferries, warships, cruise ships, work boats, and other applications where costs are driven by speed or fuel consumption over a wide range of operating speeds. Bulk carriers with predictable, low speeds and extremely sensitive acquisition costs are typically constructed of steel.

Shipbuilding considerations with aluminum. Design and fabrication with aluminum is no more challenging than with steel. Issues are encountered by those new to the material because some of the design practices and lessons learned from steel do not directly translate to aluminum.

Modulus of elasticity, coefficient of thermal expansion, and fatigue. Many issues are encountered by not properly accounting for differences in "second order" material properties. This is particularly evident when mating two materials, common in shipbuilding at the interface between steel hulls and aluminum superstructures. Aluminum's modulus is approximately one third that of steel while its coefficient of thermal expansion is higher. These mismatches necessitate a detailed fatigue assessment accounting for thermal growth, fatigue endurance limits, and stress concentrations natural at structural transitions. Often, an explosively bonded bi-metallic strip is utilized to connect the aluminum structure to the steel hull. These strips typically have good fatigue and corrosion resistance as long as they are not located in standing water.

Processing technique. Shipyards require a multitude of unique skill sets to construct ships and boats. Limiting the number of required fabrication skills leads to a more flexible and productive workforce. Though processing specifics can differ, aluminum utilizes many traditional metal fabrication techniques, such as bending, welding, brazing, and cutting. Often the changes required are procedural and design related. Changes in fabricating equipment are typically limited to welding. Nondestructive testing, bending, and cutting can all be completed with traditional tools. This is in stark contrast to composites that require completely different equipment, processes, and workforce skills for fabrication and quality control. For small-scale applications of composites this may be acceptable, but when working with materials for large portions of large ships and boats these factors can become limiting.

Product form. Aluminum is commercially available in sheet, plate, standard shapes, forgings, and castings. An often overlooked design advantage of aluminum is its ability to be procured relatively cheaply in custom profiles via extrusion. Given the volumes typically necessary for even minor shipbuilding applications, it is often economical to have custom shapes manufactured for specific applications. This frees designers from using standard shapes that often have to be up-sized, leaving unnecessary weight in designs.

The feasibility and freedom of extrusion design varies with the alloys typical in marine construction. The 6xxx-series materials are heat-treated to achieve their strength. In general, they can be extruded in a soft condition allowing complex shapes with hollow or even multiple-hollow sections.

The 5xxx-series materials are strain-hardened, thus limiting the amount the material can be deformed without over hardening and creating too much resistance for the extrusion press or die. Typically, 5xxx-series materials can produce a wide variety of I, T, and L shapes and tubing. Custom hollows, multi-hollows, and overall size will be limited due to strain-hardening and equipment limitations. Because of the multitude of tradeoffs, it is typically best practice to involve suppliers early in the profile design process.

Coating. 5xxx and 6xxx-series aluminums typical in marine construction can be left uncoated in service. Many builders utilize an anti-fouling coating where the hull is in continuous contact with sea water, leaving the rest of the hull and superstructure bare (**Fig. 1**). All of the structure can be coated when branding or other cosmetic concerns are important. Studies have shown significant weight and labor savings can be achieved by eliminating coatings. One specific study showed a weight savings of 8 tons, or 0.5% for a typical small frigate (lightweight ship displacing approximately 1600 tons).

Fire barriers. Aluminum does not burn under atmospheric conditions, but like any metal, it loses strength when heated. At temperatures above 300°F (~149°C), aluminum's strength falls below acceptable limits for many structural applications. Ship classing and design agencies have established fire barrier (insulation) codes that protect structural aluminum bulkheads during shipboard fire. The International Convention for the Safety of Life at Sea (SOLAS) as well as others, provide guidance as to which bulkheads must be protected and to what degree. These standards typically limit bulk temperatures to approximately 284°F (~140°C) for specified amounts of time, allowing for the control of shipboard fire. Military standards typically include a shock requirement before the fire event to better simulate weapons effects on installed insulation.

Lift capability. Aluminum is approximately one third the density of steel, but because of stiffness considerations, aluminum structures typically have to be designed at one half the weight of steel. In practical terms, this means that existing crane capacity will be able to lift either larger modules or more completely fit-out modules. This capability can improve flexibility of workflow within shipyards.

Friction stir welding. Friction stir welding (FSW) is a relatively new solid-state joining process that allows certain materials, including aluminum, to be joined without transitioning through a liquid phase. Because the process is solid-state, it produces less distortion and generally higher mechanical properties. Additionally, it can be used in materials that are considered unweldable by traditional means. The increased mechanical properties are rarely exploited during design because the friction stir welding process cannot currently be performed in the field.

This new technology allows more efficient use of integrally stiffened panels. Using traditional joining methods, a panel consisting of four stiffeners requires eight welds. A friction stir welded panel utilizing H or inverted T extrusions requires only

Fig. 1. Black antifouling coating with balance of hull and superstructure bare aluminum on USS *INDEPENDENCE* (LCS-2). (*U.S. Navy Photo by Lt. Jan Shultis, 2012*)

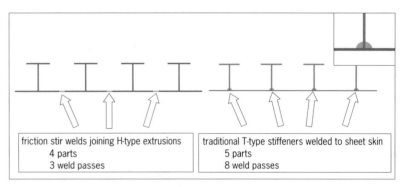

friction stir welds joining H-type extrusions	traditional T-type stiffeners welded to sheet skin
4 parts	5 parts
3 weld passes	8 weld passes

Fig. 2. Comparison of new technology and traditional joining methods, showing how aggressive use of extrusions and friction stir welding has reduced welding effort, part count, and distortion. (*Courtesy of Alcoa Inc.*)

three joints (**Fig. 2**). Studies have concluded that friction stir welding and extrusion-intensive construction techniques can reduce the weight of an aluminum structure by 16% and welding labor by 31–50%.

For background information *see* ALUMINUM; ALUMINUM (METALLURGY); ALLOY; HEAT TREATMENT (METALLURGY); METAL, MECHANICAL PROPERTIES OF; MERCHANT SHIP; NAVAL SURFACE SHIP; SHIPBUILDING in the McGraw-Hill Encyclopedia of Science & Technology.   Kyle A. Crum

Bibliography. Aluminum Association, *Aluminum Design Manual*, 9th ed., Aluminum Association, Washington, D.C., 2010; American Bureau of Shipping (ABS), *High Speed Craft*, ABS, Houston, TX, 2012; S. F. Pollard, *Boatbuilding with Aluminum*, International Marine/McGraw-Hill, Camden, ME, 2d ed., 2007; Ship Structure Committee, *Aluminum Structure Design and Fabrication Guide*, SSC-452, National Technical Information Service, Springfield, VA, 2008.

Animal camouflage

Visual camouflage is recognized as one of the commonest and most powerful forces in natural selection and is found in nearly every ecosystem on Earth, yet curiously it is one of the least-studied phenomena in biology. Camouflage patterns have evolved in animals of every size and shape, including both invertebrates and vertebrates, and occur in aquatic and terrestrial habitats. Camouflage is even employed at night, a seemingly odd notion until it is realized that many nocturnal predators have specialized visual systems for seeing well under lighting conditions that humans cannot perceive. In recent years, camouflage has been studied in considerable experimental detail, and some exciting new discoveries are reshaping our views of the visual mechanisms by which camouflage works.

Hide in plain sight. Camouflage may be achieved in three ways: background matching, disruptive coloration, and masquerade. Background matching is thought to operate by avoiding detection by the predator, whereas masquerade retards recognition of prey by the predator. Disruptive coloration seems to influence both detection and recognition processes, although few experimental data exist. There is general agreement that background matching, where the appearance generally resembles the color, lightness, and pattern of one or several background types, is one of the most common mechanisms of camouflage. For background matching to be effective, the light and dark patches of the body pattern (regardless of the type of animal) need to generally resemble the size scale and contrast of the light and dark background patches (of course, color and physical texture have to generally match as well). Except for the rare cases of a high-fidelity match, camouflage is not simply looking exactly like the background. Disruptive coloration uses a set of markings that creates the appearance of false edges and boundaries, and it hinders the detection or recognition of an object's, or part of an object's, true outline and shape, thus distracting the predator's attention. The third camouflage tactic is masquerade, in which the prey may be detected as distinct from the visual background but not recognized as edible or

interesting (for example, by resembling a leaf or a stick). In this deceptive resemblance, unlike background matching, animals do not generally resemble the background, but rather they actively choose to generally match visual features of objects beyond the immediate surroundings.

Camouflage body pattern types. The ubiquity of animal camouflage begs a basic question that has not been adequately addressed: How many body patterns for camouflage are there? One way to approach the problem is to study the animals that can best change their appearance for camouflage in a wide variety of visual habitats: that is, the cephalopods (octopus, squid, and cuttlefish in the phylum Mollusca). Extensive studies of cuttlefish in the ocean and laboratory indicate, surprisingly and counterintuitively, only three basic patterning templates: namely, uniform, mottle, and disruptive. A survey of animal body patterns among thousands of camouflage images reveals similar pattern categories (**Fig. 1**). Of course, there is variation within each broad pattern class. A chief characteristic of uniform body patterns is little or no contrast; that is, there are no light–dark demarcations that produce spots, lines, stripes, or other configurations within the body pattern. Mottle body patterns are defined as small- to moderate-scale light and dark patches (or mottles) distributed somewhat evenly and repeatedly across the body surface. Disruptive body patterns are characterized by large-scale light and dark components of multiple shapes, orientations, scales, and contrasts.

Such classification into three categories is relatively new and partly controversial, but recently developed quantitative methods support the morphological and photographic evidence of just a few basic camouflage pattern types, at least in the cuttlefish (see below). Moreover, the pattern types correlate to the visual mechanisms involved with background matching and disruptive coloration, and the basic tenets of deceiving predators by interfering with

uniform

mottle

disruptive

Fig. 1. Three body pattern types (uniform, mottle, and disruptive) are universal for camouflage in the animal kingdom. These patterns have evolved in animals of every size and shape, including both invertebrates and vertebrates, and occur in aquatic and terrestrial habitats.

Second:frame 0:00   0:08 (270 ms)   2:02 (2070 ms)

Fig. 2. *Octopus vulgaris* reacting to a diver (predator). The initial change from camouflaged to conspicuous takes only milliseconds (ms) as a result of direct neural control of the skin. Full expression of the threat display (*right*) is 2 s. Video frame rate is 30 frames per second. [*Adapted from R. Hanlon, Cephalopod dynamic camouflage, Curr. Biol., 17(11):R400–R404, 2007; http://hermes.mbl.edu/mrc/hanlon/video.html*]

their perceptual abilities for detection and recognition of prey. The overall hypothesis, based upon the concept of parsimony (the principle that the simplest scientific explanation is best), is that there is a relatively simple "visual sampling rule" for each of the few basic camouflage body pattern types.

Cephalopod dynamic camouflage. Cephalopod mollusks possess soft bodies, diverse behavior, elaborate skin patterning capabilities, and a sophisticated visual system that controls body patterning for communication and camouflage. Although most animals have a fixed or slowly changing camouflage pattern, cephalopods have evolved a different defense tactic: they use their keen vision and sophisticated skin, with direct neural control for rapid change and fine-tuned optical diversity, to rapidly adapt their body

pattern for appropriate camouflage against a staggering array of visual backgrounds, including colorful coral reefs, temperate rock reefs, kelp forests, sand or mud plains, seagrass beds, and others (**Fig. 2**).

The eye as a sensor of diverse visual backgrounds. Testing the visual cues that drive the adjustment of body patterning and posture is possible with cephalopods. European cuttlefish, *Sepia officinalis*, are particularly suited for this task because they are well adapted to laboratory environments and they are, like many shallow-water benthic (bottom-dwelling) cephalopods, behaviorally driven to camouflage themselves on almost any background; thus, both natural and artificial backgrounds can be presented to cuttlefish in order to observe their camouflaging response (**Fig. 3**).

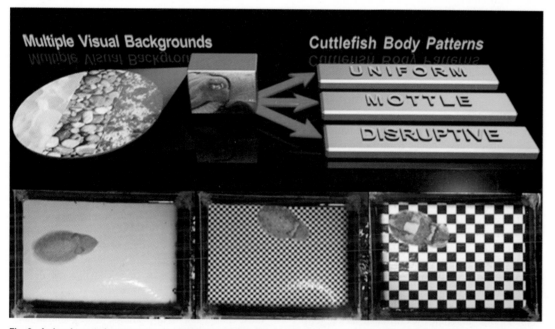

Fig. 3. A visual sensorimotor assay for probing cuttlefish (*Sepia officinalis*) perception and subsequent dynamic camouflage. *Top row*: Visual backgrounds with different size, contrast, edge characteristics, and arrangement are perceived by the cuttlefish, which quickly translates the information into a complex, highly coordinated body pattern type of uniform, mottle, or disruptive. *Bottom row*: Simple visual stimuli (such as uniformity or small to large high-contrast checkerboards) can elicit uniform, mottle, or disruptive camouflage patterns (left to right) in cuttlefish. The chief difference in the latter two backgrounds is the scale of the checker. Both the visual background and the body pattern can be quantified so that correlations can be made between visual input and motor output. [*Adapted from R. Hanlon, Cephalopod dynamic camouflage, Curr. Biol., 17(11):R400–R404, 2007*]

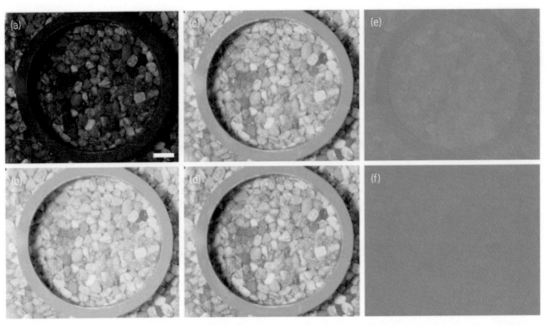

Fig. 4. Color matching of camouflaged cuttlefish when viewed by hypothetical dichromatic and trichromatic fish predators. (a) The pseudocolor image for simulating the human view of the cuttlefish at near-surface. This composite image was formed by using three frames (650, 550, and 450 nm) of the hyperspectral imaging data multiplied by the irradiance spectrum. (b) The monochromatic image for representing the luminance information of dichromatic and trichromatic predators. (c, d) The composite images for simulating the dichromatic and trichromatic predator views of the cuttlefish, respectively. (e, f) The isoluminant chromatic images for representing the color information of dichromatic and trichromatic predators, respectively. Scale bar: 2 cm (0.8 in.). [*Adapted from C. C. Chiao et al., Hyperspectral imaging of cuttlefish camouflage indicates good color match in the eyes of fish predators, Proc. Natl. Acad. Sci. USA, 108(22):9148–9153, 2011*]

Which properties of the background determine whether a cuttlefish will produce a uniform, mottle, or disruptive pattern? This issue has received much attention over the past decade. Three of the most important factors are (1) the spatial frequency content of the background, (2) the contrast of the background, and (3) whether or not the background contains any bright elements of roughly the same size as the cuttlefish's "white square" (a rectangular skin patch on the dorsal mantle; lower right panel in Fig. 3). In addition, increasing substrate luminance tends to attenuate the production of disruptive patterns. The edge of objects and visual depth also provide salient cues in evoking disruptive patterns.

Camouflage may benefit from both optical and physical texture, with the latter being chiefly a result of the changeable skin papillae. Furthermore, arm postures of cuttlefish are often associated with three-dimensional structures (corals, algae, or kelp), and these field observations suggest that this is a visually driven response for camouflage.

Camouflage in the eyes of the beholder. Many visual predators have keen color perception, and thus camouflage patterns should provide some degree of color matching in addition to other visual factors such as pattern, contrast, and texture. However, most cephalopods, including the cuttlefish, lack color perception; thus, the vexing question of how they achieve color-blind camouflage still remains. Nevertheless, their color resemblance to natural visual backgrounds appears to be excellent. This is not surprising, as many of their predators, including teleost fishes, diving birds, and marine mammals, typically have dichromatic, trichromatic, or even tetrachromatic vision.

Although quantifying camouflage effectiveness in the eyes of the predator is challenging, the use of hyperspectral imaging (HSI) has proved to be a powerful tool because it records full-spectrum light data to simultaneously assess color match and pattern match in the spectral and spatial domains, respectively. Application of HSI on camouflaged cuttlefish on natural backgrounds has recently shown that most reflectance spectra of individual cuttlefish and substrates are similar, rendering the color match possible. Modeling color vision of potential dichromatic and trichromatic fish predators of cuttlefish corroborated the spectral match analysis and demonstrated that camouflaged cuttlefish show good color match as well as pattern match in the eyes of fish predators (**Fig. 4**). These findings provide supporting evidence that cuttlefish can produce color-coordinated camouflage on natural substrates, despite lacking color vision.

Conclusions. Extrapolating what has been learned from cephalopods may uncover more universal concepts of camouflage as practiced by animals that have a fixed body pattern or limited capability for changeability. The finding that cephalopods, the most changeable in appearance among animal taxa, appear to have as few as three basic pattern classes (uniform, mottle, and disruptive) for camouflage is surprising, counterintuitive, and provocative. It may be oversimplified and there are some morphological and neural constraints on the cephalopod system. Yet, the idea (1) suggests a parsimonious solution to a complex problem, (2) is testable, and (3) may stimulate new ways to view the complex sensory world of visual predator–prey interactions in nature.

For background information *see* BEHAVIORAL ECOLOGY; CEPHALOPODA; COLEOIDEA; COLOR VISION; ECOLOGY; PHYSIOLOGICAL ECOLOGY (ANIMAL); PIGMENTATION; PREDATOR-PREY INTERACTIONS; PROTECTIVE COLORATION; SEPIOIDEA in the McGraw-Hill Encyclopedia of Science & Technology.

Chuan-Chin Chiao; Roger T. Hanlon

Bibliography. C. C. Chiao et al., Hyperspectral imaging of cuttlefish camouflage indicates good color match in the eyes of fish predators, *Proc. Natl. Acad. Sci. USA*, 108(22):9148–9153, 2011, DOI:10.1073/pnas.1019090108; R. Hanlon, Cephalopod dynamic camouflage, *Curr. Biol.*, 17(11):R400–R404, 2007; R. T. Hanlon et al., Cephalopod dynamic camouflage: Bridging the continuum between background matching and disruptive coloration, *Philos. Trans. R. Soc. Lond. B Biol. Sci.*, 364(1516):429–437, 2009, DOI:10.1098/rstb.2008.0217; M. Stevens and S. Merilaita, Animal camouflage: Current issues and new perspectives, *Philos. Trans. R. Soc. Lond. B Biol. Sci.*, 364(1516):423–427, 2009, DOI:10.1098/rstb.2008.0217.

Anoiapithecus and Pierolapithecus

Hominoids constitute a natural group of primates that includes gibbons and siamangs (hylobatids or lesser apes), great apes (orangutans and African apes), and humans. Apes currently display a low biodiversity and limited geographic distribution (restricted to southeastern Asia and equatorial Africa). However, in the Miocene [23–5 million years ago (MYA)], hominoids were much more diverse and widely distributed across Africa, Asia, and Europe. Unfortunately, ape fossil remains are usually scarce, given their low population densities and frequent association with (sub)tropical environments, which are unfavorable for fossil preservation. This frequently hinders an accurate assessment of hominoid paleobiodiversity or a conclusive decipherment of hominoid phylogenetic relationships. Two new genera of large-bodied apes from the Miocene of Spain, *Pierolapithecus* (**Fig. 1**) and *Anoiapithecus* (**Fig. 2**), were described during the 2000s, having significant implications for understanding the role of European apes in the radiation of the great ape–human clade.

Abocador de Can Mata. In 2002, excavations of Miocene sediments were undertaken with heavy machinery in the fossiliferous (fossil-rich) area of els Hostalets de Pierola (Vallès-Penedès Basin, Catalonia, Spain) in order to enlarge the landfill of Can Mata [Abocador de Can Mata (ACM) in Catalan]. Given current laws for the preservation of archeological and paleontological heritage, a team of paleontologists was assembled to control the diggers' activities and to recover any fossil remains that might be unearthed. Over the years, thousands of Middle Miocene vertebrate remains have been recovered from more than 200 localities, roughly spanning from 12.5 to 11.5 MYA. Among primates, a partial skeleton of *Pierolapithecus* (described in 2004) and a partial face of *Anoiapithecus* (described in 2009) stand out.

Pierolapithecus and Anoiapithecus. The holotype (type specimen) of *Pierolapithecus catalaunicus*, from locality Barranc de Can Vila 1 (ACM/BCV1; 11.9 MYA), comprises the face and more than 80 other bone fragments (including hand and foot bones, limb bone fragments, vertebrae, ribs, and clavicle) of a single adult male individual, with an estimated body mass of approximately 30–35 kg (66–77 lb). The face displays a long profile coupled with a modern, great ape-like configuration. The morphology of the vertebrae, ribs, and clavicle indicates that *Pierolapithecus* displayed a modern hominoid-like, orthograde (upright) body plan—characterized by the possession of a dorsoventrally shallow and mediolaterally wide thorax, with dorsally positioned scapulae (shoulder blades). This agrees well with the presence of a single articulation between the forearm and the wrist (that is, there is no ulnocarpal articulation), which also is observed in extant hominoids, but is at odds with the primitive morphology of the *Pierolapithecus* hand phalanges, which are short and stout, and most closely resemble those of pronograde (horizontally oriented), earlier Miocene apes.

The holotype of *Anoiapithecus brevirostris*, from the similarly aged locality of ACM/C3-Aj (11.9 MYA), consists of the partial face of a young adult male individual. As seen in *Pierolapithecus*, *Anoiapithecus* shows a modern great ape-like facial configuration, although it differs from the former by displaying a strikingly short (orthognathous) facial profile.

Dryopithecine paleobiodiversity. *Pierolapithecus* and *Anoiapithecus*, together with other European fossil apes, are currently grouped into the subfamily Dryopithecinae. In order to provide a natural classification of living and past organisms, taxonomists rely on groups descended from a single ancestor (clades); however, given phylogenetic uncertainties, taxonomic disagreements often emerge among scientists. Before the discovery of the ACM hominoids, a single genus *Dryopithecus* was recognized for both Middle and Late Miocene Spanish apes; now, however, it is evident that Late Miocene forms are distinct enough to warrant a distinction at the genus level (*Hispanopithecus*). Some investigators remain unconvinced that *Pierolapithecus* and *Anoiapithecus* also belong to genera different from *Dryopithecus*, which is also recorded at ACM. However, *Dryopithecus* differs from *Anoiapithecus* and *Pierolapithecus* in several craniodental features. Moreover, *Anoiapithecus* differs from *Pierolapithecus* by displaying a much shorter face (as mentioned previously) and possessing a frontal sinus, among other features. The distinction of as many as three different genera and species in a single basin close to 12 MYA is evidence of a previously unsuspected diversity of Middle Miocene apes in Europe, leading to the suspicion that their actual paleobiodiversity had long gone unnoticed as a result of insufficient sampling.

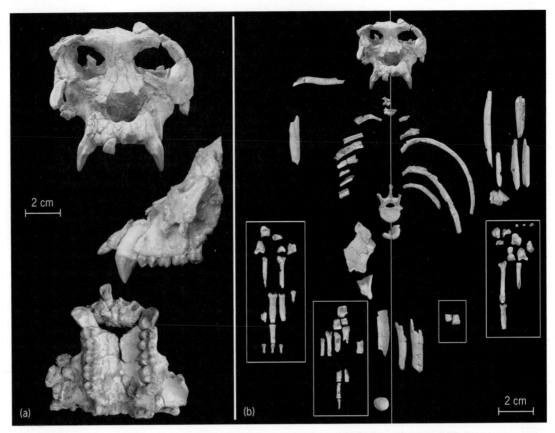

Fig. 1. Partial cranium (*a*) and skeleton (*b*) of *Pierolapithecus catalaunicus*. (*Modified from S. Moyà-Solà et al., Pierolapithecus catalaunicus, a new Middle Miocene great ape from Spain, Science, 306:1339–1344, 2004*)

Phylogeny and paleobiogeography. Dryopithecines are generally considered stem hominids (that is, basal members of the great ape–human clade), preceding the divergence between Asian pongines (orangutans and closely allied extinct forms) and African hominines (African apes, australopiths, and humans). It is currently uncertain whether dryopithecines constitute a natural group of their own or whether some of them are more closely related than others to extant hominids (**Fig. 3**). In any case, the great ape-like cranial morphology of dryopithecines (including a wide nasal aperture widest at the base, a high face, and a flat midfacial region) indicates that they postdate the hylobatid-hominid divergence. Nonetheless, *Anoiapithecus* and *Pierolapithecus* still share several primitive features (for example, thick tooth enamel and a restricted maxillary sinus) with earlier Middle Miocene kenyapithecines (such as *Griphopithecus* and *Kenyapithecus*), which by approximately 14 MYA were present in both Africa and Eurasia. This suggests that dryopithecines evolved in Europe, following an initial dispersal or range extension event of thick-enamelled hominoids out of Africa into Eurasia. It has been

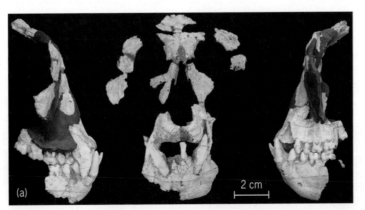

Fig. 2. Face (*a*) and reconstruction (*b*) of *Anoiapithecus brevirostris*. (*Part a: Modified from S. Moyà-Solà et al., A unique Middle Miocene European hominoid and the origins of the great ape and human clade, Proc. Natl. Acad. Sci. USA, 106:9601–9606, 2009, copyright © by the authors. Part b: Artwork by M. Palmero, copyright © Institut Català de Paleontologia Miquel Crusafont*)

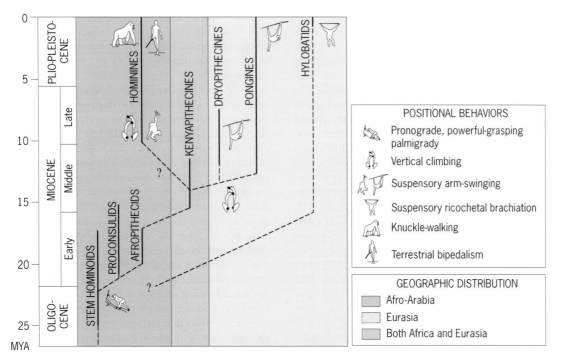

Fig. 3. Phylogenetic tree depicting the position of dryopithecines relative to living and extinct hominoids, together with their geographic distribution and main locomotor innovations.

alternatively proposed that dryopithecines are more closely related to hominines. However, available cranial evidence for *Pierolapithecus* (including the absence of a frontal sinus) suggests that dryopithecines might ultimately prove to be more closely related to Asian pongines. Given the meager Middle Miocene hominoid record from Africa, it is currently conceivable that Eurasian and African hominoids independently evolved from similar kenyapithecine-like ancestors, following a vicariance event (that is, a geographical isolation of populations) during the Middle Miocene. If so, dryopithecines might merely constitute an extinct group with no living descendants, which is a sister-taxon of pongines instead of hominines as a whole.

Paleobiology and evolutionary implications. The thick-enameled molars of *Pierolapithecus* and *Anoiapithecus*, together with results of dental microwear analysis, suggest that these fossil apes displayed adaptations for the consumption of hard food items (sclerocarpy). In contrast, *Dryopithecus* and *Hispanopithecus* share a thinner-enameled condition, indicating a shift towards a softer frugivorous (fruit-eating) diet by the Late Miocene. Sclerocarpy might have been the key adaptation that enabled hominoids to initially disperse out of Africa, being subsequently inherited by some dryopithecines and most pongines, but independently modified along several hominoid lineages from the Late Miocene onwards.

With regard to positional behaviors, the skeletal remains of *Pierolapithecus* show that it displayed a locomotor repertoire unknown among extant great apes. The absence of ulnocarpal articulation and the possession of an upright body plan reflect an adaptation to verticalized trunk postures during locomotion. In contrast, the phalangeal morphology of *Pierolapithecus* indicates the retention of adaptations for above-branch, powerful-grasping palmigrady (that is, four-limbed locomotion on top of the branches with the hands and feet strongly held onto them), with no specific adaptations to below-branch arm-swinging. Orthogrady has been interpreted as an adaptation to both vertical climbing and suspensory behaviors, which are displayed to some extent by all extant apes. Nevertheless, the retention of palmigrady-related features and the lack of suspensory adaptations in *Pierolapithecus* suggest that, originally, orthogrady was only an adaptation to vertical climbing and clambering, being later independently co-opted in several lineages for suspensory behaviors. Given its great ape status, the combination of locomotor adaptations displayed by *Pierolapithecus* strengthens the view that suspensory behaviors were independently acquired, at the very least, between hylobatids and hominids.

This is not surprising, especially because most investigators agree that homoplasy (false homology or independent evolution) played a major role in the radiation of hominoids, with their locomotor apparatus having evolved in a mosaic fashion—with new adaptations being progressively and repeatedly superimposed upon previous ones with the passing of time. Nevertheless, the unique combination of postcranial features displayed by *Pierolapithecus* should warn paleontologists against reconstructing the last common ancestors of extant hominoids merely on the basis of the biased evidence provided by the few remaining living apes, without taking the fossil evidence into account.

For background information *see* APES; BIODIVER-SITY; DENTAL ANTHROPOLOGY; FOSSIL; FOSSIL APES; FOSSIL HUMANS; FOSSIL PRIMATES; PALEONTOLOGY; PHYLOGENY; PHYSICAL ANTHROPOLOGY; PRIMATES in the McGraw-Hill Encyclopedia of Science & Technology.   David M. Alba

Bibliography. D. M. Alba et al., Enamel thickness in the Middle Miocene great apes *Anoiapithecus, Pierolapithecus* and *Dryopithecus*, *Proc. R. Soc. B*, 277:2237–2245, 2010, DOI:10.1098/rspb.2010.0218; S. Almécija et al., *Pierolapithecus* and the functional morphology of Miocene ape hand phalanges: Paleobiological and evolutionary implications, *J. Hum. Evol.*, 57:284–297, 2009, DOI:10.1016/j.jhevol.2009.02.008; I. Casanovas-Vilar et al., Updated chronology for the Miocene hominoid radiation in Western Eurasia, *Proc. Natl. Acad. Sci. USA*, 108:5554–5559, 2011, DOI:10.1073/pnas.1018562108; S. Moyà-Solà et al., A unique Middle Miocene European hominoid and the origins of the great ape and human clade, *Proc. Natl. Acad. Sci. USA*, 106:9601–9606, 2009, DOI:10.1073/pnas.0811730106; S. Moyà-Solà et al., *Pierolapithecus catalaunicus*, a new Middle Miocene great ape from Spain, *Science*, 306:1339–1344, 2004, DOI:10.1126/science.1103094.

Application of microbiology in geotechnical engineering

Have you ever wondered how a high-rise building can be constructed on soft ground? The answer is that the soft soil can be improved before the construction of the high-rise building. In fact, bacteria, can be used to improve the engineering properties of soil. In this article, the application of this microbial approach to geotechnical engineering, a discipline that deals with soils and foundations, is discussed.

There are a number of ways to strengthen soft or weak soil. One of the common ones is to use cement or chemicals to increase the load-bearing capacity or the so-called shear strength of the soil. The same process can be used to reduce the water conductivity of soil or the rate of water flow in soil. This is necessary when there is a need to prevent water from flowing in the ground, for example, to cut off the flow of contaminated groundwater. In this case, cement or chemicals are used as binders and mixed with soil to either increase the shear strength or reduce the water conductivity of the soil. However, the use of cement or chemicals for soil improvement is not sustainable in the long run, because cement and chemical production require a considerable amount of natural resource (for example, limestone) and energy. The production process also generates carbon dioxide, dust, and possibly other toxic substances and thus is not environmentally friendly. The use of cement or chemicals for soil improvement is also expensive and time-consuming. There is an urgent need to develop new and sustainable construction materials that can reduce the need to use cement or chemicals for geotechnical applications.

Using the latest microbial biotechnology, a new type of construction material, called biocement, has been developed as an alternative to cement or chemicals. Biocement is made by naturally occurring microorganisms at ambient temperature and thus requires much less energy to produce. It is sustainable, because microorganisms are abundant in nature and can be reproduced easily at low cost. The microorganisms that are suitable for making biocement are nonpathogenic and environmentally friendly. Furthermore, unlike the use of cement, soils can be treated without disturbing the ground or the environment, because microorganisms can penetrate and reproduce themselves in soil. Harnessing this natural and inexhaustible resource may result in an entirely new approach to geotechnical or environmental engineering problems and bring enormous economic benefit to the construction industries. The application of microbial technology to construction will also simplify some existing construction processes. For example, the biocement can be in either solid or liquid form. In liquid form, the biogrout has much lower viscosity and can flow like water. Thus, the delivery of biocement into soil is much easier compared with that of cement or chemicals. Furthermore, when cement is used, one has to wait for 28 days for the full strength to be developed, whereas when biocement is used, the reaction time can be much reduced or controlled if required.

The principle of microbial treatment is to use microbially induced precipitation of calcium carbonate or other approaches to produce bonding and cementation in soil so as to increase the strength

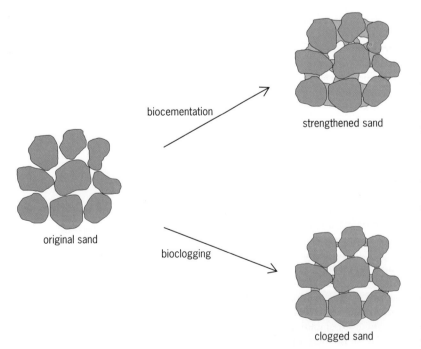

Fig. 1. Schematic of the biocementation and bioclogging processes.

original sand

biocementation

strengthened sand

bioclogging

clogged sand

Fig. 2. Sand column treated using biocement.

and reduce the water conductivity of the soil. A number of studies have been done in recent years. Presently, much of the work is at the experimental stage. However, the scale of treatment has increased rapidly with time and has reached 100 m³ in recent years.

The microbiological processes induce calcium carbonate crystals, other minerals, or slimes, as shown by M. Van der Ruyt and W. van der Zon (2009), L. A. Van Paassen and colleagues (2010), and J. Chu and colleagues (2012). Those minerals or slimes act as cementing agents between soil grains to increase the shear strength of soil and/or to fill in the pores in soil to reduce the water conductivity, as shown in **Fig. 1**. The two processes to increase strength and reduce conductivity have been called biocementation and bioclogging, respectively. The process to deliver the biocement in situ to achieve biocementation or bioclogging is called biogrouting. In the treatment of sandy soil, because the viscosity of biocement grout is low, it is possible to pump it into the ground without mixing. This will enable the construction process to be simplified. The studies so far show that the biocement method is effective in both increasing the shear strength and reducing the water conductivity of soil. Some potential applications of biocement are discussed in the following.

Enhancing shear strength of sand or to make a strong ground. By using the microbially induced calcium carbonate precipitation method, the shear strength of soil can be increased. We know from our childhood experience building sand castles on the beach that dry sand will not stand. However, when dry sand is treated with biocement, it not only can stand as a column, it can also sustain a lot of weight, as shown in **Fig. 2**. When cement or chemicals are used to treat soil, the amount of improvement in the shear strength of the soil depends on the amount of cement or chemical used. Similarly, when biocement is used, the shear strength of the soil is affected by the amount of metal precipitation. One way to measure the shear strength of soil is simply to compress a soil column between two rigid plates, the so-called uniaxial compression test. The shear strength measured by this method is called the uniaxial compressive strength (UCS). In a study by M. Van der Ruyt and W. van der Zon (2009), the UCS of biocement-treated sand was measured for specimens having different calcium-carbonate contents. The results are shown in **Fig. 3**. It can be seen that the UCS strength increases with increasing calcium-carbonate content. The highest UCS obtained was 27 MPa. For normal applications, the UCS strength required is less than 3 MPa. This will only require a calcium content of 100 to 200 kg/m³. To achieve the same UCS strength for sand using cement grouting, the amount of cement required would be between 250 and 300 kg/m³. Because the production of biocement can be cheaper, the overall cost for biocement grouting can be potentially lower.

Mitigation of soil liquefaction during earthquakes by using biogas. Soil liquefaction refers to a phenomenon in which a soil is transformed into a substance that acts like a liquid in response to an external action such as an earthquake. When liquefaction occurs, the ground completely loses its bearing capacity and undergoes large deformation. Soil liquefaction normally occurs in saturated sand deposits during earthquakes. The ground shaking causes the water pressure in the soil—the pore water pressure—to build up. When the pore water pressure has increased to a certain point, soil liquefaction occurs. Soil liquefaction has been one of the major causes of earthquake-related disasters. Liquefaction was

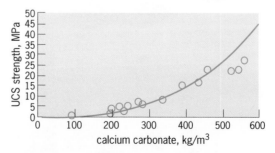

Fig. 3. Unconfined compression strength (UCS) versus calcium carbonate content for biocement-treated sand. (*After M. Van der Ruyt and W. van der Zon, 2009*)

largely responsible for the extensive damage to ports and residential properties in the earthquakes in New Zealand and Japan in 2010 and 2011.

Common methods that can be adopted for mitigation of soil liquefaction include densification and ground modification using cement or chemicals. A new approach that is being developed is called the biogas method. In this method, tiny gas bubbles are generated in situ in saturated sand where liquefaction may occur. When saturated sand is made slightly unsaturated by the inclusion of gas bubbles, the amount of pore-water-pressure generation in the sand under a dynamic load is greatly reduced. According to research, if only about 5% of the water by volume is replaced by gas, the liquefaction resistance of loose sand can be increased by more than 2 times.

However, it is not easy to introduce gas into the ground. Pumping can be used. However, the distribution of gas bubbles introduced by pumping will not be even. Furthermore, the gas pumped into ground tends to be present in the form of aggregated gas pockets rather than individual bubbles. As a result, the gas tends to escape from the ground. One of the most effective ways to introduce tiny gas bubbles in situ is to use microorganisms. This method has the following three advantages over existing methods. (1) Biocement can flow easily in sand, much like water. Gases can be generated easily by bacteria anywhere underground using only a small amount of energy. Thus the biogas method will be much more cost-effective than any other method. The scale of treatment for liquefaction is normally very large, so the potential economic benefit is significant. (2) The gas bubbles generated by bacteria can be distributed more evenly than by other means. This is because the gas bubbles are generated in situ rather than pumped. (3) The gas bubbles generated by bacteria are tiny and less prone to escaping from the ground.

Some model tests using a shake table to generate ground motion were done by our research group. A comparison of ground settlement for a fully saturated sand layer and a sand layer treated with biogas is shown in **Fig. 4**. The settlement is expressed as a settlement ratio, with the settlement for fully satu-

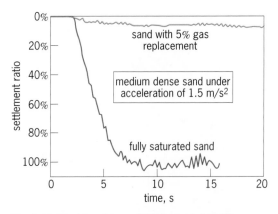

Fig. 4. Comparison of ground settlement induced by ground shaking under an acceleration of 1.5 m/s² for a saturated sand layer and a sand layer with 5% gas replacement.

rated sand being 100%. It can be seen from Fig. 4 that with only 5% gas replacement, the ground settlement generated by ground shaking with an acceleration of 1.5 m/s² can be reduced by more than 90%. Thus, the biogas method is effective in preventing the occurrence of soil liquefaction or reducing the damage caused by liquefaction.

Seepage and erosion control or construction of a water pond in a desert. Biocement can also be used to reduce the water conductivity of sand through the bioclogging mechanism, as shown in Fig. 2. One method that has been developed by our research group is to use urea-reducing bacteria to precipitate a layer of calcium carbonate on top of sand, as shown in **Fig. 5**. This hard layer of crust has a water conductivity of less than 10^{-7} m/s and thus can be used as an impervious layer for water storage. This means we now have a method for building a water pond in the desert. The same method can also be used for erosion control of a beach or riverbank. As the layer of treatment is rather thin, the amount of biogrout used is small. Thus the method can be more economical than conventional methods.

Outlook. The new construction material, biocement, has a number of advantages over cement or chemicals for geotechnical applications. Several

(a)

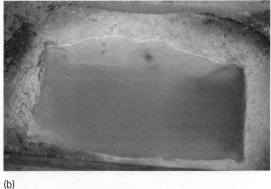

(b)

Fig. 5. Formation of (a) a thin impervious layer on top of sand using the biocement method and (b) a water pond model built using this method in sand.

potential applications exist for ground strengthening, mitigation of liquefaction, and seepage or erosion control. The new microbial technology provides a more cost-effective, sustainable, and environmentally friendly solution to ground improvement. However, a lot of research studies still need to be done before this new approach can be developed into common practice.

For background information *See* BACTERIA; CEMENT; ENGINEERING GEOLOGY; FOUNDATIONS; GROUT; MICROBIOLOGY; SAND; SOIL; SOIL MECHANICS in the McGraw-Hill Encyclopedia of Science & Technology.   Jian Chu; Volodymyr Ivanov; Jia He

Bibliography. J. Chu, V. Stabnikov, and V. Ivanov, Microbially induced calcium carbonate precipitation on surface or in the bulk of soil, *Geomicrobiol. J.*, 29(6):544–549, 2012, DOI:10.1080/01490451.2011. 592929; V. Ivanov and J. Chu, Applications of microorganisms to geotechnical engineering for bioclogging and biocementation of soil in situ, *Rev. Environ. Sci. Biotechnol.*, 7(2):139–153, 2008, DOI:10.1007/s11157-007-9126-3; J. K. Mitchell and J. C. Santamarina, Biological considerations in geotechnical engineering, *J. Geotech. Geoenviron. Eng.*, 131(10):1222–1233, 2005, DOI:10.1061/ (ASCE)1090-0241(2005)131:10(1222); M. Van der Ruyt and W. van der Zon, Biological in situ reinforcement of sand in near-shore areas, *Geotech. Eng.*, 162(1):81–83, 2009, DOI:10.1680/geng.2009.162. 1.81; L. A. van Paassen et al., Quantifying biomediated ground improvement by ureolysis: Large-scale biogrout experiment, *J. Geotech. Geoenviron. Eng.*, 136(12):1721–1728, 2010, DOI:10.1061/(ASCE)GT. 1943-5606.0000382.

Arctic Ocean freshwater balance

As part of the global hydrological cycle, the Arctic Ocean plays a central role by receiving, transforming, storing, and exporting freshwater (FW) from rivers, precipitation, sea-ice melt and growth, and from the Pacific and Atlantic oceans via straits. Changes in the FW balance influence the extent of the sea-ice cover, the surface albedo, the energy balance, the temperature and salinity structure of the water masses, and biological processes in the Arctic Ocean and high-latitude seas. Within the entire Arctic Ocean, the FW at the surface maintains a strong density stratification that prevents release of significant deep-ocean heat to the sea ice and atmosphere. Loss of this FW cap could have grave consequences for the climate, resulting in massive sea-ice melt. Subsequent reestablishment of the FW cap and strong stratification could then result in climate cooling as new sea-ice formation ensues.

Under stable climate conditions, a balance of FW sources and sinks is necessary to maintain unchanging ocean circulation, heat and salt fluxes, and ocean stratification. However, under natural climate variability from seasonal to decadal timescales, the Arctic Ocean can accumulate significant volumes of FW during several seasons or years and then release

this water to the North Atlantic resulting in so-called "great salinity anomalies" that influence ocean circulation and climate conditions in the Northern Hemisphere.

Freshwater balance: sources and sinks. The total FW content of the Arctic Ocean varies from 85,000 to 95,000 km³, depending on the study, and FW is unevenly distributed in the ocean due to influences of wind forcing and FW sources and sinks (**Fig. 1**). In the ocean, FW is a relative term, and is typically considered the salinity anomaly relative to a reference salinity of 34.8 g/kg, which corresponds to the mean salinity of the Arctic Ocean. Change in FW is thus a measure of how much liquid FW was accumulated or lost from the water column bounded by the 34.8 isohaline (constant salinity surface). Low salinity sources that make up the FW pool include river runoff, precipitation, sea-ice melt, and waters of Pacific origin that penetrate the Arctic through the Bering Strait. High salinity sources are saline waters of Atlantic Ocean origin that penetrate the Arctic via the Fram Strait and the Barents Sea. Very salty water (brine) rejected to the ocean during sea-ice formation is an additional source of high salinity.

Liquid FW and solid (sea ice) components leave the Arctic Ocean with the East Greenland current

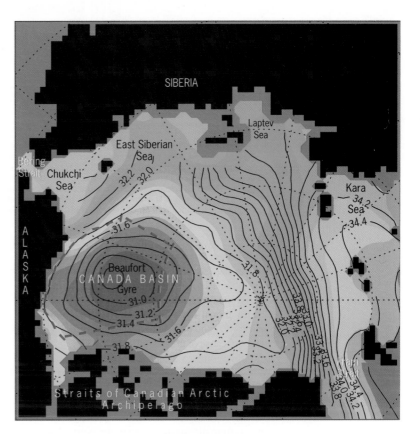

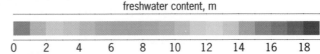

freshwater content, m

Fig. 1. Climatology of FW content in the Arctic Ocean (shown in colors). Solid lines depict mean 1950–1980 salinity at 50 g/kg. FW content is calculated relative to salinity 34.8 on the basis of 1950–1980 data. The Beaufort Gyre Region is bounded by thick dashed blue lines. (*From A. Proshutinsky et al., 2009*)

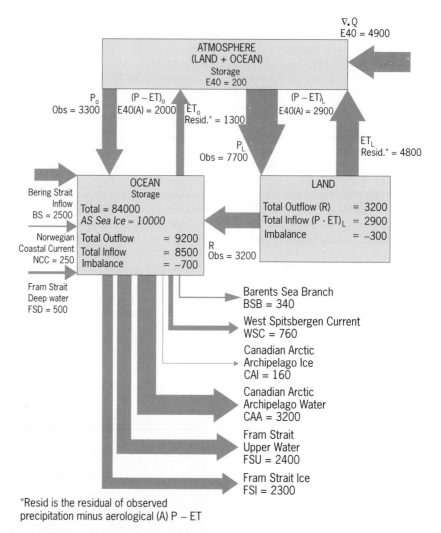

Fig. 2. Annual mean FW budget of the Arctic. The atmospheric box combines the land and ocean domains. The boxes for land and ocean are sized proportional to their areas. All transports are in units of km³ per year. Stores are in km³. The width of the arrows is proportional to the size of the transports. Subscripts L and O denote land and ocean, respectively. E40 (ERA-40) shows results obtained from the European Centre for Medium-Range Weather Forecasts (ECMWF) ERA-40 reanalysis. P and ET depict precipitation and evapotranspiration, respectively. The net precipitation (P-ET) represents water available for runoff (R). ∇Q is the divergence of the horizontal water vapor flux Q integrated from the surface to the top of the atmospheric column. (*From M. C. Serreze et al., 2006; panels are reproduced by permission of American Geophysical Union*)

via the Fram Strait and via the straits of the Canadian Archipelago. A smaller fraction of FW is evaporated (or sublimated) from liquid and solid components within the Arctic Ocean.

The annual cycle of FW and its variability from year to year and decade to decade are difficult to construct because of uncertainties in the various sources and sinks due to a lack of observations (**Fig. 2**). M. C. Serreze and coworkers estimated that the annual mean FW input to the Arctic Ocean is dominated by river discharge (38%), inflow through the Bering Strait (30%), and net precipitation (24%). FW export from the Arctic Ocean to the North Atlantic is dominated by transports through the Canadian Arctic Archipelago (35%) and via the Fram Strait as liquid (26%) and sea ice (25%).

Beaufort Gyre freshwater reservoir. Within the Arctic Ocean, the largest volume of FW stored in the Canadian Basin with its Beaufort Gyre (BG) [Fig. 1] is roughly equal to that stored in all lakes and rivers

of the world and is 10–15 times greater than the annual export of FW (including ice and water) from the Arctic Ocean. Since the 1950s, the BG region was recognized by oceanographers as a unique "reservoir" containing more than 20,000 km³ of FW or approximately half of the FW stored in the Canadian Basin (Fig. 1). Historical observations between 1950 and 1990 showed that the BG FW reservoir was a stable feature of the Arctic Ocean and could be considered as a flywheel of Arctic Ocean circulation. However, recent observations coordinated in the region under the umbrella of the Beaufort Gyre Observing System (BGOS), with support from the National Science Foundation (NSF) Office of Polar Programs Arctic Observing Network (AON), show an unprecedented increase (a gain of more than 25%) of FW in 2003–2011. An important scientific question is whether significant changes in the mechanics of the BG flywheel can be expected in the near future.

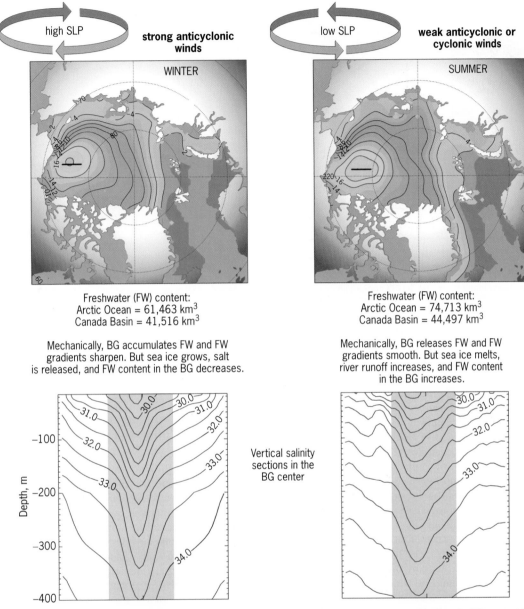

Fig. 3. Conceptual mechanisms of FW accumulation and release in the Beaufort Gyre during an annual cycle. FW content in summer and winter is shown in meters (isolines) calculated relative to salinity 34.8 g/kg. SLP = sea level atmospheric pressure. The bottom panels show salinity distribution along sections in the center of the Beaufort Gyre region. In winter, the wind drives the ice and ocean in an anticyclonic (clockwise) sense so that the Beaufort Gyre accumulates FW mechanically through deformation of the salinity field (Ekman convergence and subsequent downwelling; bottom left panel). In summer, anticyclonic winds are weaker (and may even reverse to be cyclonic). The resultant summer anomaly in Ekman convergence under cyclonic wind forcing releases FW or under weak anticyclonic winds accumulates less FW, thereby relaxing salinity gradients (bottom right panel) and reducing Beaufort Gyre FW content. (*From Proshutinsky et al., 2009; figure is modified by permission of American Geophysical Union*)

The fundamental dynamics governing the BG FW reservoir is that FW is accumulated in the region from different sources by anticyclonic (clockwise) winds dominating over the region as a result of water convergence in the surface ocean (**Fig. 3**). This FW can be released when the anticyclonic forcing weakens or changes sense of rotation to dominantly cyclonic (counterclockwise). FW release from the BG could significantly influence Arctic and global climate. It has been argued that FW release from the Arctic can influence global climate via reduction of the ocean meridional overturning circulation—a release

of only 20% of FW from the BG is enough to cause a salinity anomaly in the North Atlantic with the magnitude of the Great Salinity Anomaly of the 1970s (M. B. Vellinga et al., 2008). In this sense, the BG FW reservoir is "a ticking time bomb" for climate.

Recent BGOS observations and climatological data allow for estimates of the magnitude of FW accumulation and release at seasonal to decadal time scales (**Fig. 4**). The strong FW accumulation trend in the center of the BG (Fig. 4) is linked to an increase in strength of the anticyclonic wind forcing. An atmospheric circulation regime with anticyclonic winds

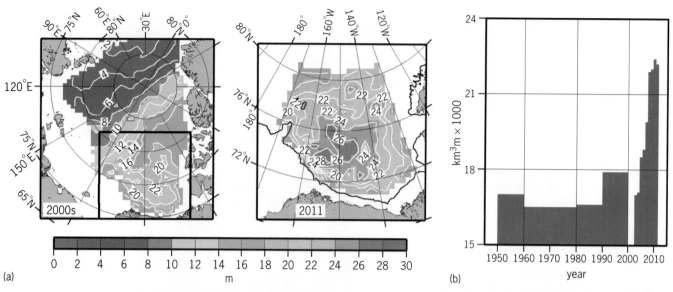

Fig. 4. FW in the BG region. (*a*) FW content (in m) in the Arctic Ocean based on all available ocean measurements from the 2000s. The BG region is delineated by the box. Right panel: FW content (m) in the BG region based on available measurements from 2011 (*b*) Decadal FW content (in thousands of km³) in the BG region before 2000, and annual FW content after 2000. The unprecedented increase of FW content in the BG in 2003–2011 relative to previous decades manifests dramatic changes in the Arctic climate and warns that release of this FW may have significant consequences to Arctic and global climate.

has dominated over the Arctic for at least 14 years, instead of the typical 5–8 year pattern of alternation between anticyclonic and cyclonic regimes. It is to be expected from climatology that the present regime that drives FW accumulation soon will change to one that permits FW release. In fact, the magnitude of BG FW in 2011 is modestly less than 2010, so the change might have already begun.

For background information *see* ARCTIC OCEAN; ATLANTIC OCEAN; HYDROLOGY; PACIFIC OCEAN; SEA ICE; SEAWATER in the McGraw-Hill Encyclopedia of Science & Technology.

Andrey Proshutinsky; Richard Krishfield; Mary-Louise Timmermans; John M. Toole

Bibliography. K. Aagaard and E. C. Carmack, The role of sea ice and freshwater in the Arctic circulation, *J. Geophys. Res.*, 94: 14,485–14,498, 1989, DOI:10.1029/JC094iC10p14485; I. M. Belkin et al., "Great Salinity Anomalies" in the North Atlantic, *Prog. Oceanogr.*, 41(1):1–68, 1998, DOI:10.1016/S0079-6611(98)00015-9; R. R. Dickson et al., The great salinity anomaly in the northern North Atlantic 1968–1982, *Prog. Oceanogr.*, 20(2):103–151, 1988, DOI:10.1016/0079-6611(88)90049-3; A. Proshutinsky and M. Johnson, Two circulation regimes of the wind-driven Arctic Ocean, *J. Geophys. Res.*, 102:12,493–12,514 1997, DOI:10.1029/97JC00738; A. Proshutinsky et al., The Beaufort Gyre Fresh Water Reservoir: State and variability from observations, *J. Geophys. Res.*, 114: C00A10, 25 pp., 2009, DOI:10.1029/2008JC005104; M. C. Serreze et al., The large-scale freshwater cycle of the Arctic, *J. Geophys. Res.*, 111: C11010, 19 pp., 2006, DOI:10.1029/2005JC003424; M. B. Vellinga, B. Dickson, and R. Curry, The changing view on how freshwater impacts the Atlantic Meridional Overturning Circulation. In *Arctic-Subarctic Ocean Fluxes: Defining the Role of the Northern Seas in Climate*, R. R. Dickson, J. Meincke, P. Rhines (eds.), pp. 289–314, Springer, New York, 2008.

Arthropod evolution and phylogeny

It is not an exaggeration to say that life on Earth is predominantly arthropodous. By any criteria, arthropods (members of the phylum Arthropoda) are the dominant metazoans (multicellular animals) on the planet, and this has been true for hundreds of millions of years. This dominance is evident in terms of the number of described species, global biomass, ecological importance, economic impact, and numerous other categories. Arthropods constitute more than 85% of all known species on Earth, and they have the richest fossil record of any animal group. Some typical arthropods are insects, spiders, ticks, and crustaceans. Of course, in additional to the numerous described species, the number of undescribed species of arthropods remains unknown. Arthropods dominate not only in number of species but in number of individual organisms: for example, who could begin to estimate the number of ants on the planet, or the number of copepods in the world's oceans?

Arthropoda. In terms of numbers of described species, the insects rule. Of the estimated 1.8 million described species on Earth, more than half are insects, with the vast majority of those being beetles. Insects are found on all continents and in nearly all habitats, and their diversity in form and function is astounding. However, arthropods also include other extremely significant groups. Foremost among these in terms of species numbers are the incredibly diverse chelicerate groups (including not only spiders, but ticks, mites, scorpions, xiphosurans, opilionids, and many other lesser-known groups) and the crustaceans (often described as the marine equivalent of the insects in terms of ecological and numerical dominance). Apart from insects, chelicerates, and crustaceans, there exists a host of other segmented, joint-legged, chitin-enclosed creatures that are at least distantly related, including tardigrades (water bears) and onychophorans (velvet worms). In terms of marine metazoans, it is also important to mention the extinct trilobites, which constituted another diverse arthropod group that played a major role in the oceans of the world.

Arthropods share several key morphological features that suggest the group is monophyletic, including obvious segmentation, a high degree of tagmatization (specialization of body segments into functional units or regions, such as the head, thorax, and abdomen of insects, or the prosoma and opisthosoma of arachnids), a ventral nerve cord, a more or less open circulatory system, a well-developed chitinous exoskeleton (necessitating growth by ecdysis and all that goes with that process), paired multiarticulate limbs, and (usually) compound paired eyes. They have traditionally been grouped or allied with other cuticle-bearing groups, including tardigrades and onychophorans (and indeed some workers treat these three groups together as the Panarthropoda). Many historical approaches recognized five major arthropodan subphyla: Trilobitomorpha, Crustacea, Hexapoda (the insects and their kin), Myriapoda, and Cheliceriformes. More recent work supports the uniting of crustaceans and insects and the separation of pycnogonids from the chelicerates, resulting in five major extant taxa: Pycnogonida, Euchelicerata, Myriapoda, Crustacea, and Hexapoda; in this scheme, the crustaceans and insects are united as the Pancrustacea (alternatively Tetraconata).

Arthropod relationships. With so much of the Earth's biodiversity contained within the Arthropoda, it might be assumed that the group has been well studied, with all major relationships understood. However, our understanding of arthropod phylogeny and the relationships among (and within) the many groups of arthropods is far from resolved. Salient questions remain concerning the relationships of the various arthropod groups, and the relationships of arthropods to other lineages. Concerning their origin, historically most workers have assumed a relationship to, if not a descent from, the annelids (segmented worms). Are arthropods part of a larger clade (taxonomic group containing a common ancestor and its descendants) that includes all molting animals (extant arthropods, tardigrades, onychophorans, nematode and nematomorph worms, priapulids, priapulans, kinorhynchs, and loriciferans) [referred to as the Ecdysozoa hypothesis], countering the strong morphological and developmental similarities with annelids? Are the arthropods truly monophyletic (having a common descent from a single ancestor), or have the segmented body (and its detailed developmental patterning process), articulated limbs, chitinous exoskeleton (and related modifications), and specific process of molting arisen more than once such that there are several

arthropodous phyla rather than a single phylum? A growing body of evidence suggests that the Crustacea gave rise to the Hexapoda (insects), resulting in the introduction of terms such as Pancrustacea, Tetraconata, and Miracrustacea to accommodate some of the newly recognized clades. Did crustaceans indeed give rise to insects, such that the world's most numerous creatures are actually descendants of a crustacean ancestor? If this is true, from what crustacean group did they arise? How are the major chelicerate groups, and especially the arachnid orders, related to one another? These are some of the seemingly persistent questions about arthropods and their relationships, and many of them have been addressed in recent years with molecular methodology.

Investigations. In a recent and well-supported study, an enormous amount of DNA sequence data was applied to the question of arthropod phylogeny, and fully resolved relationships (with some exceptions) were reported within and among four extant, all-inclusive arthropod clades: Pancrustacea [including the Hexapoda (insects) and Crustacea], Myriapoda, Euchelicerata, and Pycnogonida. The appeal and apparent success of this study were attributable to rather broad taxon sampling (75 arthropod species, representing nearly all major arthropod lineages, plus a diversity of outgroup species that included velvet worms and tardigrades), a very large data matrix (up to 40 kilobase pairs per taxon from 62 protein-coding nuclear genes), and a variety of probability and statistical analyses.

Strong support was shown for the Pancrustacea, for the recognition of some other clades that had been proposed based on morphological evidence (such as the Mandibulata, containing the Myriapoda and the Pancrustacea), and for the monophyly of other traditionally accepted clades [such as the Euchelicerata (Chelicerata minus the pycnogonids), Arachnida, Myriapoda, Branchiopoda, and Malacostraca]. Other previously recognized groupings were clearly rejected, including the formerly recognized crustacean group Maxillopoda, which was proposed earlier (and employed for a surprisingly long time despite mounting evidence against its validity) to encompass such groups as copepods, barnacles, fish lice (branchiurans), mystacocaridans, and other seemingly similar small crustaceans. More contentious findings of the study included the separation of extant crustaceans into three major groups (apart from the insects), referred to as the Oligostraca, Vericrustacea, and Xenocarida, with the latter group Xenocarida (containing the relatively recently described Remipedia and Cephalocarida) being proposed as the sister group to the Hexapoda (insects) [see **illustration**]. The Xenocarida hypothesis is particularly intriguing (although controversial) in that it proposes two groups of completely marine and rather vermiform crustaceans as the sister group to the insects.

Other confirmatory investigations have been carried out as well. Another large effort employing 117 taxa and 129 genes and including more of the basal

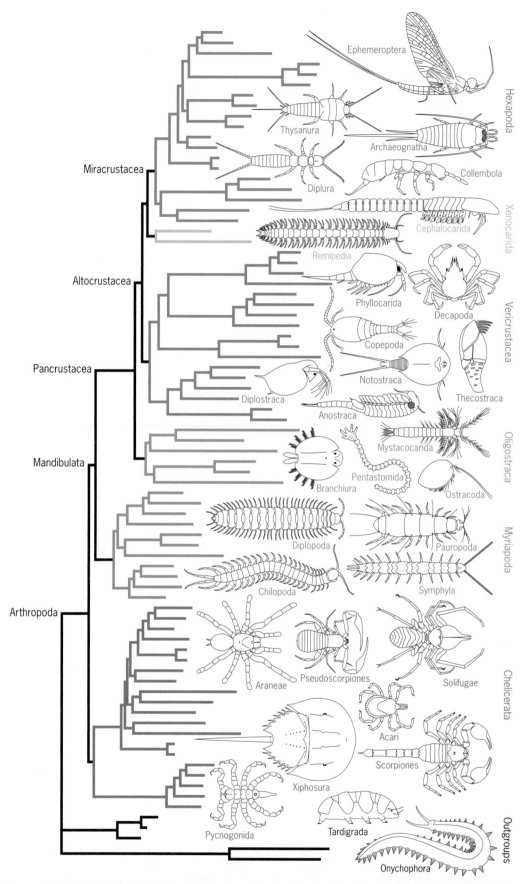

A proposed phylogenetic tree of the Arthropoda based on extensive molecular sequence data (41 kilobases of aligned DNA sequences from 62 single-copy nuclear protein-coding genes, representing 75 different species of arthropods). (*From J. C. Regier et al., Arthropod relationships revealed by phylogenomic analysis of nuclear protein-coding sequences, Nature, 463:1079–1083, 2010; used with permission*)

insect groups and novel expressed sequence tag (EST) data supported many of the aforementioned findings, although it was cautioned that phylogenetic results are remarkably sensitive to methods of analyses. Some investigators reported strong support for the sister-group relationship of Remipedia and Hexapoda. Although these studies were collectively hailed as major steps forward in our understanding of arthropod phylogeny, it would be a mistake to assume that all questions have now been answered. Different genes and different analyses have differed in the clades that they have purported to resolve. Fossils, which provide a window into the sequence of branching and character acquisition in arthropod stem groups, have indicated contradictory results: sometimes they have provided confirmation of molecular-based trees and sometimes they have not.

Conclusions. What do we know at this point? First, based on a combination of morphological, molecular, and fossil evidence, the arthropods are monophyletic. Furthermore, the sister group to the arthropods is very likely the velvet worms (Onychophora); this conclusion is based on both morphological and molecular data. The unity of the Ecdysozoa, which was first proposed based on 18S ribosomal ribonucleic acid (rRNA) sequence data, has been supported by a variety of additional molecular data and is now widely accepted. Still, support for an arthropod–annelid alliance also remains strong. In addition to the well-known morphological similarities between arthropods and annelids, strong evidence from the field of evolutionary development points toward their association. Moreover, components of the same highly complex genetic developmental pathway (the Hedgehog signaling pathway) found in arthropods are present also in annelids, with segmental axial patterning that is remarkably similar (and quite different from that seen in vertebrates), suggesting that the similarity of arthropods and annelids as a result of convergent evolution is unlikely. Within the arthropods, it is almost certain that the Pycnogonida, Euchelicerata, Myriapoda, Pancrustacea (insects plus crustaceans), and Hexapoda are all monophyletic groups. Crustacea is paraphyletic, as it contains the Hexapoda. The clade Pancrustacea (= Tetraconata), uniting the crustaceans and hexapods, is now widely accepted and has been confirmed by characters from neuroanatomy. Within Crustacea, the grouping Mandibulata, uniting the myriapods, crustaceans, and hexapods, also appears quite solid, again based on morphological, molecular, and anatomical data. However, conflicting findings [supporting the grouping Paradoxopda (also referred to as Myriochelata)] persist, although the weight of the evidence seems to favor the Mandibulata hypothesis.

With regard to fossil investigations, recent findings have strongly influenced our understanding of arthropod (and particularly crustacean) evolution, and these findings do not always agree with those of molecular phylogenomics. One salient example (based on a number of molecular studies) concerns the confirmation of the Pentastomida as aberrant crustaceans [related to extant fish lice (Branchiura)]. However, odd, very pentastomid-like fossils are known from the Cambrian, which is long before today's pentastomid hosts appeared on the scene. Are these fossils really pentastomids? If so, what did they parasitize, and why are molecular data consistently supportive of their status as crustaceans? Fossil findings, especially those of Orsten-like fauna from Sweden, China, and other localities, continue to provide new data that cannot be ignored (although the data are yet all too often ignored).

Outlook. In summary, progress concerning the evolution and phylogeny of arthropods is being made. As more is learned about this diverse and challenging group of animals, several historically important hypotheses (for example, Atelocerata, Maxillopoda, and so on) have been discarded, and new ones have been proposed and accepted. However, despite two centuries of study, the arthropods continue to present questions and problems that often seem nowhere close to being resolved.

For background information *see* ANIMAL EVOLUTION; ARTHROPODA; CHELICERATA; CRUSTACEA; ECDYSOZOA; GENOMICS; INSECTA; MYRIAPODA; ONYCHOPHORA; ORGANIC EVOLUTION; PHYLOGENY; PYCNOGONIDA; SPECIATION; TARDIGRADA in the McGraw-Hill Encyclopedia of Science & Technology.

Joel W. Martin

Bibliography. R. C. Brusca and G. J. Brusca, *Invertebrates*, 2d ed., Sinauer Associates, Sunderland, MA, 2003; G. Edgecombe, Arthropod phylogeny: An overview from the perspectives of morphology, molecular data and the fossil record, *Arthropod Struct. Dev.*, 39:74–87, 2010, DOI:10.1016/j.asd.2009.10.002; G. Giribet and G. D. Edgecombe, Reevaluating the arthropod tree of life, *Annu. Rev. Entomol.*, 57:167–186, 2012, DOI:10.1146/annurev-ento-120710100659; J. C. Regier et al., Arthropod relationships revealed by phylogenomic analysis of nuclear protein-coding sequences, *Nature*, 463:1079–1083, 2010, DOI:10.1038/nature08742; C. Tudge, *The Variety of Life*, Oxford University Press, Oxford, U.K., 2000.

Atrial fibrillation

Atrial fibrillation (AF) is a common arrhythmia (any disorder of the heart rate or rhythm) that presently affects approximately 3 million people in the United States (**Fig. 1**). AF is associated with decreased quality of life and increased risk of cognitive decline, dementia, stroke, heart failure (HF), and mortality. The adverse public health impact of AF is also reflected in an estimated annual national cost of $26 billion. This review provides an update on the pathogenesis and management of AF and highlights additional strategies that are needed to prevent and treat AF patients.

Epidemiology of AF. The aging of the population and the accompanying rise in the prevalence of AF have magnified the toll of AF on morbidity and healthcare costs, especially in industrialized countries.

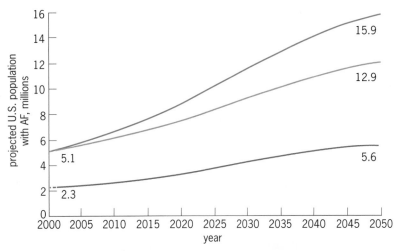

Fig. 1. The estimated prevalence of AF in the United States in the year 2050 ranges from 5.6 million to as high as 15.9 million individuals. The bottom line shows estimates derived from the Anticoagulation and Risk Factors in Atrial Fibrillation Study, comprising cross-sectional data from a large health-maintenance organization. The middle line shows data derived from Olmsted County, Minnesota, during 1980–2000 and projected through 2050 using U.S. Census Bureau population projections. The top line reflects increases in age-adjusted AF incidence similarly extrapolated forward. (*Used with permission and modified from Y. Miyasaka et al., Secular trends in incidence of atrial fibrillation in Olmsted County, Minnesota, 1980 to 2000, and implications on the projections for future prevalence, Circulation, 114:119–125, 2006*)

Over a 15-year period, hospitalizations for AF increased nearly threefold. Established risk factors for AF include older age, smoking, diabetes, valvular heart disease, myocardial infarction, and cardiac surgery. Importantly, an increasing proportion of AF cases are related to HF.

At the end of the twentieth century, the noted cardiologist Eugene Braunwald opined that AF and HF are the two major cardiovascular diseases that continue to burgeon despite advances in cardiovascular medicine. HF and AF frequently coexist, and each condition predisposes to the other. AF affects 30% of all individuals with HF, and the combination carries an ominous prognosis.

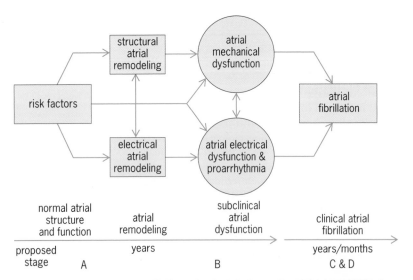

Fig. 2. Model of progression to AF. The pathophysiologic process of atrial remodeling is emphasized, with resultant atrial mechanical and electrical dysfunctions shown as precursors to clinical AF. (*Adapted with permission from R. S. Vasan and D. Levy, The role of hypertension in the pathogenesis of heart failure: A clinical mechanistic overview, Arch. Intern. Med., 156:1789–1796, 1996*)

The epidemiologic similarities between AF and HF are paralleled by clinical and experimental investigations demonstrating significant mechanistic connections between them (**Fig. 2**). HF and AF have overlapping epidemiologic risk factors, and both result from similar mechanisms of myocardial structural remodeling.

Advances in pathogenesis. There have been a number of advances with regard to understanding the pathogenesis of AF.

Genetic basis for AF. The familial nature of AF was reported in the early 1940s, and candidate gene studies identified rare mutations related to AF. Recent genome-wide association studies (GWAS) identified novel genetic loci associated with AF and its risk factors in the community. Presently, three genetic loci (located on chromosomes 4q25, 16q22, and 1q21) have been associated with AF using GWAS (**Fig. 3**). Multiple loci across the genome have been discovered that are associated with established risk factors for AF. However, the causative polymorphisms or pathophysiological mechanisms that relate the GWAS findings to AF and its risk factors are uncertain.

Atrial remodeling and AF. Atrial remodeling, which is an adaptive process describing progressive atrial electrical and structural changes, merits increasing attention in understanding AF's pathogenesis (Fig. 2). Pathologic atrial remodeling often precedes development of AF. Structural remodeling seen in atria of individuals with AF is similar to that seen in myopathic ventricles from individuals with HF [note that the heart is composed of four chambers: two atria (which receive blood returning back to the heart) and two larger ventricles (which pump blood out of the heart)]. Chronic atrial or ventricular pressure overload activates common stress signaling pathways, most notably the renin-angiotensin-aldosterone system (RAAS) of hormones, yielding impaired myocardial vascular growth, myocyte apoptosis (cell death), and interstitial fibrosis. Chronic RAAS activation has been shown to contribute to fibrosis and chamber remodeling in AF experimental studies. Experimental data also reveal that RAAS upregulation influences ion channel function and action potential duration, predisposing to intra-atrial reentry and AF.

Advances in identification of individuals at risk for AF. A recently published risk score for the development of AF showed that established cardiovascular risk factors (C-statistic: 0.76) account for only a part of the AF risk.

Noninvasive measures of atrial remodeling. P-wave indices are electrocardiographic measures that may represent noninvasive markers of electrical remodeling. Echocardiographic left atrial dimension and volume are markers of pathologic atrial remodeling associated with long-term risk for incident and recurrent AF.

Novel AF biomarkers. Biomarkers may provide another means to enhance AF risk prediction and to provide insights into AF pathogenesis. Multiple biomarkers have been examined to assess their diagnostic or prognostic significance in relation to AF. For example, natriuretic peptides (which are involved in regulating sodium and water balance, blood volume, and arterial pressure) were associated with a fourfold higher risk for AF and improved AF risk scores derived in a community-based cohort. The relations between several other circulating biomarkers and AF have been examined, but few have been shown to improve clinical risk prediction instruments. However, a recent study chose 10 candidate AF biomarkers representing major pathophysiologic pathways germane to AF, including inflammation, neurohormonal activation, endothelial dysfunction, oxidative stress, the RAAS, thrombosis, and microvascular damage. In more than 3000 participants, it was found that the 10-biomarker panel was associated with development of new-onset AF ($p < 0.0001$). In stepwise-selection models, B-type natriuretic peptide and C-reactive protein remained associated with AF after multivariate adjustment. However, only B-type natriuretic peptide improved risk stratification beyond established clinical AF risk factors.

Advances in the treatment of AF. There have been a number of advances with regard to the treatment of AF.

Upstream therapies for AF. Despite recent calls for increased emphasis on AF prevention by research panels, few clinical investigations have examined the use of upstream therapies targeting pathophysiologic processes implicated in AF, such as RAAS antagonists or anti-inflammatory agents (for example, statins and steroids). Signals from recent meta-analyses suggest that RAAS inhibitors and statins may reduce the risk of developing AF in selected populations who are at risk for the arrhythmia.

Rate control. Control of the ventricular response rate has long been accepted as a reasonable treatment option for AF patients with minimal or no symptoms. Until recently, however, there were few data to guide rate response goals for patients with AF for whom a rate-control strategy is deemed appropriate. In the RACE II (Rate Control Efficacy in Permanent Atrial

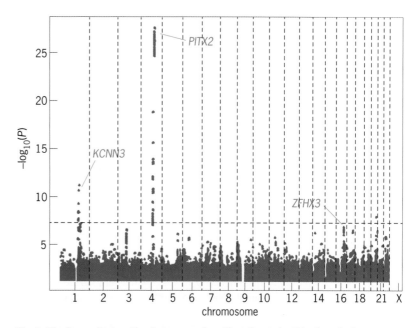

Fig. 3. The figure displays the chromosomal positions for each of the three loci associated with AF. The *x* axis displays the chromosomal position; the *y* axis indicates the statistical significance for the association between the single-nucleotide polymorphism and AF. (*Modified with permission from P. T. Ellinor et al., Common variants in KCNN3 are associated with lone atrial fibrillation, Nat. Genet., 42:240–244, 2010*)

Fibrillation II) study, patients with AF and minimal or no symptoms were randomized to receive strict [resting heart rate: <80 beats per minute (bpm); heart rate: <100 bpm during a 6-minute walk test] versus lenient (resting heart rate: <110 bpm) rate control. Lenient rate control was shown to be noninferior to strict rate control with respect to three-year cumulative incidence of death, hospitalization, thromboembolic events, and bleeding.

Novel anticoagulants. The field of thromboembolic prophylaxis is evolving rapidly (see **table**). In the RE-LY (Randomized Evaluation of Long-Term Anticoagulation Therapy) study, which was a randomized trial involving more than 18,000 patients with nonvalvular AF followed for an average of two years,

Major study (n, comparator)	Agent	Mechanism	Major finding	Specific challenges/ limitations
Major new anticoagulants for prevention of thromboembolic events in patients with AF				
RE-LY (n = 18,113, vs. warfarin)	Dabigatran etexilate, 110/150 mg oral, twice daily	Direct thrombin inhibition	Reduced systemic embolism events in dabigatran (150 mg, twice daily) vs. warfarin	Possible increased risk for myocardial infarction; Increased risk for bleeding, particularly in renal disease; Twice daily
ROCKET-AF (n = 14,264, vs. warfarin)	Rivaroxaban, 20 mg oral, once daily	Factor Xa inhibitor	Rivaroxaban noninferior to warfarin with respect to systemic embolism; Rivaroxaban group had fewer major bleeding events	Higher rates of bleeding in patients with renal impairment
ARISTOTLE (n = 18,201, vs. warfarin)	Apixaban, 5 mg oral, twice daily	Factor Xa inhibitor	Apixaban reduced stroke or systemic embolism vs. warfarin; Lower rates of bleeding in the apixaban group	Twice daily; Not yet approved by the Food and Drug Administration
Limitations of all vs. warfarin				Limited long-term follow-up; No specific antidote for reversal in emergencies or if bleeding; Inability to monitor

investigators found that the anticoagulant dabigatran etexilate given 150 mg twice daily was superior to therapy with the anticoagulant warfarin in the prevention of ischemic stroke (p for superiority < 0.001). Rivaroxaban (another anticoagulant) is a factor Xa inhibitor that has been shown to be more effective than aspirin for stroke prevention in AF patients and also superior to warfarin with respect to systemic thromboembolism and fatal bleeding events in a randomized controlled clinical trial (ROCKET-AF; $p < 0.02$). In contrast, recent studies (ACTIVE-W) have shown that dual-antiplatelet therapy is clearly inferior to warfarin for the prevention of stroke in patients with AF.

Novel antiarrhythmic medications. Recent data showing an increased risk of HF, stroke, and death in dronedarone-treated patients with AF suggest that dronedarone should not be used in this population. Experimental data suggest that ranolazine (an antianginal agent) may reduce the risk of AF in patients at risk for adverse cardiovascular events. Efforts are also in progress to develop atrial-selective agents for the pharmacologic conversion of AF and maintenance of sinus rhythm (SR). Present agents have been limited by risk for QT-prolongation (the QT interval is a section of the electrocardiogram) and torsades-de-pointes [a form of ventricular tachycardia (rapid heart rate)]. Although not yet approved for use in the United States, the efficacy and safety of the atrial-selective agent vernakalant were recently published. In a multicenter study, vernakalant converted recent-onset AF to SR in 51% of treated patients and was well tolerated. Blockade of other atrial-selective potassium channels is being evaluated as a potential means for pharmacologic conversion of AF.

Catheter-ablation of AF. Work by the cardiac electrophysiologist Michel Haïssaguerre and colleagues in the mid-1990s highlighted the importance of pulmonary vein–based triggers in initiating AF, leading to the development of pulmonary vein isolation procedures. Intermediate-term superiority of pulmonary vein isolation in comparison to antiarrhythmic drug therapy for the treatment of selected patients with symptomatic paroxysmal AF has been demonstrated. Pulmonary vein isolation has been deemed a class I treatment for symptomatic patients with AF by the American Heart Association/American College of Cardiology, reflecting the accumulating efficacy and safety data supporting the procedure. The benefit of pulmonary vein ablation with respect to mortality, stroke, and progression from paroxysmal to persistent AF will be examined in the CABANA (Catheter Ablation vs. Antiarrhythmic Drug Therapy for AF) clinical trial. Moreover, the influx of new technologies suggests that ablative therapies will continue to evolve significantly in the near future, with particular emphasis on improving long-term efficacy and reducing the risks from AF ablation. The Medtronic cryoballoon, which has been recently approved for clinical use in the United States, provides an example of what the next generation of technologies for pulmonary vein isolation may look like.

Conclusions. One in four individuals who are 40 years of age and older will develop AF over his or her lifetime. New discoveries have provided avenues for incident and recurrent AF risk prediction, future therapeutic targets, and insights into the epidemiologic similarities between AF and HF. Novel anticoagulants, antiarrhythmic agents, and catheter-based ablation technologies are available and have begun to alter the care of patients with AF. However, there are a number of next steps that must be taken: (1) Better understanding of the epidemiology of AF in developing countries and nonwhite racial groups is needed. (2) Genetic sequencing, epigenomics, transcriptomics, proteomics, and metabolomics are needed to better uncover the genomic and pathophysiologic basis of AF and target interventions. (3) It will be important to determine whether noninvasive measures and biomarkers can improve AF risk reclassification. (4) Strategies for the primary prevention of AF should be developed and tested. (5) It is imperative to determine the safety and efficacy of novel anticoagulants in more generalizable cohorts. (6) Further tests are necessary to evaluate the long-term efficacy of pulmonary vein isolation in patients with paroxysmal AF, its role in patients with persistent and long-standing persistent AF, as well as its risks outside the less-generalizable population enrolled in clinical trials.

For background information *see* CARDIAC ELECTROPHYSIOLOGY; CARDIOVASCULAR SYSTEM; ECHOCARDIOGRAPHY; ELECTRODIAGNOSIS; HEART (VERTEBRATE); HEART DISORDERS; MEDICAL CONTROL SYSTEMS; PATHOLOGY in the McGraw-Hill Encyclopedia of Science & Technology.

David D. McManus; Jared W. Magnani;
Michiel Rienstra; Emelia J. Benjamin

Bibliography. J. P. Bassand, Review of atrial fibrillation outcome trials of oral anticoagulant and antiplatelet agents, *Europace*, 14:312-324, 2012, DOI:10.1093/europace/eur263; E. J. Benjamin et al., Prevention of atrial fibrillation: Report from a National Heart, Lung, and Blood Institute Workshop, *Circulation*, 119:606-618, 2009, DOI:10.1161/CIRCULATIONAHA.108.825380; J. W. Magnani et al., Atrial fibrillation: Current knowledge and future directions in epidemiology and genomics, *Circulation*, 124:1982-1993, 2011, DOI:10.1161/CIRCULATIONAHA.111.039677; R. B. Schnabel et al., Relations of biomarkers of distinct pathophysiological pathways and atrial fibrillation incidence in the community, *Circulation*, 121:200-207, 2010, DOI:10.1161/CIRCULATIONAHA.109.882241; C. T. Tsai et al., Molecular genetics of atrial fibrillation, *J. Am. Coll. Cardiol.*, 52:241-250, 2008, DOI:10.1016/j.jacc.2008.02.072.

Autism and the social brain

Since Leo Kanner's first description of autism in 1943, social difficulties have been recognized as one of the two hallmarks of the condition. Kanner emphasized that the insensitivity to social interaction stood, in some ways, in marked contrast to a significant oversensitivity to the nonsocial world, for example, marked responses to small changes in the nonsocial environment. The observed difficulties in processing social information occur in multiple sensory modalities and are of very early onset. Parents often become concerned about unusual patterns of social interaction and/or lack of social interest in the first year of their child's life, and indeed retrospective reviews of videotapes of first birthday parties of children who were later diagnosed with autism show that these children devoted less attention to the faces and voices of others. In these children, by 2 years of age, deficits in communication, lack of imitation, poor eye contact, and limited joint attention (shared focus) are often striking. A growing body of work suggests that difficulties in these specific social processes probably underlie many subsequent problems in communication, attention, learning, and other aspects of development and that early intervention that focuses on these difficulties is associated with a significantly improved outcome. Although social difficulties persist over time, gains are typically made over the course of development; however, social vulnerabilities remain substantial even when the individual makes significant gains in communication and overall intelligence.

Neurobiological basis of autism. The neurobiological basis of autism was not fully appreciated for several decades. There was early confusion about whether autism might be a form of schizophrenia and about whether experiential factors might contribute to its pathogenesis. As individuals were followed over time, it became clear that there was a markedly increased risk for seizures, with as many as 20% of strictly diagnosed patient cases developing epilepsy (often in adolescence). Similarly, the very strong genetic basis of autism was not appreciated until the first twin studies showed dramatic increases in concordance of identical twins and increased risk, over the population rate of autism, for fraternal twins as well. Various lines of evidence eventually converged, so that autism was recognized as a distinctive disorder. As a result, it was officially recognized in 1980 in the third edition of the *Diagnostic and Statistical Manual of Mental Disorders* (DSM).

Despite the awareness of the importance of social difficulties in autism, research on these difficulties remained minimal for many years. Much of the early psychological research on autism focused on cognitive factors, while traditional neuroscience historically had viewed brain systems in isolation, with little interest in the ways in which social development might relate to brain mechanisms. This situation changed with the development of social neuroscience to study the brain basis of social behavior.

Social difficulties in autism. The earliest work on the nature of social difficulties in autism relied on behavioral ratings, behavioral frequency counts, and simple experimental procedures to investigate social information processes. Although this early work was, in retrospect, rather limited given the nature of the experimental measures then available, it did pro-

vide important clues about differences in the way in which social information was processed. For example, compared to typically developing peers, those with autism have significant difficulties in recognizing faces and emotional expression. Interestingly, those with autism seem not to exhibit a dramatic facial-inversion effect (where faces become more difficult to recognize when presented upside down), which is an effect noted in typically developing infants after approximately 6 months.

Over the past two decades, an increasingly sophisticated body of work on normative infant development (including studies of social attention, attachment, and face and emotion processing) as well as advances in neurobehavioral research methods have accumulated and have served to focus attention on the social brain in typically developing children. Much of this work has focused on the neural processing of social stimuli and the brain regions associated with it. Some of these areas of the brain include the superior temporal sulcus, which is involved in the perception of biological motion; the amygdala, which is involved in the perception of emotion; the orbital frontal cortex, which is important for decision making; and the fusiform gyrus, which is a key brain area necessary for face processing. Refined behavioral measures, such as eye tracking, have revealed significant difficulties in social information processing, particularly in processing dynamic stimuli, and idiosyncratic patterns of viewing social scenes. As first noted in older and more able adults, differences in gaze patterns when viewing socially intense scenes are quite dramatic, with the most able individuals with autism focusing disproportionately on the mouth of a speaker rather than on the rest of the face (where much of the emotional content is conveyed) or, for that matter, on the social-environmental context (for example, failing to observe the facial expression or reaction of nonspeaking participants). This work has been extended to the first years of life and appears to be a pattern of very early onset.

Other studies have also suggested reduced activity in face-related brain regions during perception of faces. Compared to typically developing individuals, those with autism display hypoactivation of the fusiform face-processing area, although these same individuals may activate the same area when viewing nonsocial objects of great personal interest. This lack of specialized processing for faces might be reflected in the lack of a typical facial-inversion effect.

Perception of faces has also been investigated using event-related potentials (ERPs) elicited by presentation of faces. Much of this work has centered on face processing and the N170, which is a large negative ERP component elicited by viewing faces. This methodology, which provides excellent temporal resolution, demonstrates slower processing of faces in children with autism.

Another line of work has used computer-generated simulations and/or point light displays to investigate the perception of biological motion. In these elegant experiments, moving lights represent biological or nonbiological motion. In typical development, very young infants are born with a tendency to focus preferentially on displays that are representative of biological motion, whereas toddlers with autism are more likely to focus on nonsocial movements. In one recent neuroimaging study, it was found that children with autism were less likely to activate a specific brain region (the posterior superior temporal sulcus), and that this lack of activity was directly related to the degree of social dysfunction. This study also revealed differential patterns of activity between unaffected siblings of children with autism and typical controls, suggesting potential compensatory strategies.

This increasing body of research has also led to the development of several conceptual models of social information processing in autism. For example, relative to face processing, one model suggests that the major difficulty lies in brain systems such as the fusiform gyrus, while another approach argues, in essence, that these difficulties may reflect a more basic lack of social interest and orientation. In support of the latter view, infants who have congenital cataracts and have this condition corrected later in life continue to exhibit some subtle difficulties in face perception. These models would potentially have different implications for intervention.

Challenges for the future. Challenges for understanding the social brain in autism arise from several sources. It is clear that autism is a very heterogeneous condition, with considerable variation in symptom patterns and ultimate outcome. It is likely that much of this heterogeneity reflects the complex genetics of the condition. Recent work has focused on a combination of perspectives using complementary approaches to clarify this issue by studying relevant subgroups and subtypes. The observation of potential compensatory strategies in apparently unaffected siblings may provide important clues in this regard. Similarly, the study of what might be viewed as less severe forms of the condition (in the so-called broader autism phenotype) might be helpful in clarifying genetic contributions. The complexity of the latter arises from what appears to be the contribution of many different genes to the ultimate phenotype of autism. Many potential genes have been identified, and their role in brain development is an area of active investigation. The development of more robust animal models will also be of considerable interest. Finally, longitudinal studies of early development are required to determine the extent to which social difficulties and atypicalities in social brain function reflect basic causes of autism versus consequences of developing with the disorder.

For background information *see* AUTISM; BEHAVIOR GENETICS; BRAIN; COGNITION; DEVELOPMENTAL GENETICS; DEVELOPMENTAL PSYCHOLOGY; EMOTION; NERVOUS SYSTEM (VERTEBRATE); NEUROBIOLOGY; PERCEPTION; SOCIOBIOLOGY in the McGraw-Hill Encyclopedia of Science & Technology.

Fred R. Volkmar

Bibliography. G. Dawson, S. Webb, and J. McPartland, Understanding the nature of face processing impairment in autism: Insights from behavioral and electrophysiological studies, *Dev. Neuropsychol.*, 27:403-424, 2005, DOI:10.1207/s15326942dn2703_6; M. D. Kaiser et al., Neural signatures of autism, *Proc. Natl. Acad. Sci. USA*, 107:21223-21228, 2010, DOI:10.1073/pnas.1010412107; J. C. McPartland and K. A. Pelphrey, The implications of social neuroscience for social disability, *J. Autism Dev. Disord.*, 42:1256-1262, 2012; R. T. Schultz et al., Abnormal ventral temporal cortical activity during face discrimination among individuals with autism and Asperger syndrome, *Arch. Gen. Psychiatr.*, 57:331-340, 2000, DOI:10.1.1.43.1035; F. R. Volkmar, Understanding the social brain in autism, *Dev. Psychobiol.*, 53:428-434, 2011, DOI:10.1002/dev.20556.

Bats and wind energy

Humans have harnessed the wind for centuries. In the United States and elsewhere, as a result of the desire for energy independence and renewed interest in renewable energy sources, industrial wind energy facilities have been on the rise. Although wind energy is promoted as "green" energy, bat mortality has been documented at industrial wind energy facilities as a result of turbine operation, raising concerns about the effects on flying wildlife. Common trends in the data include types of bats killed at wind farms and time-of-year bat mortality peaks, and scientists are gaining a better understanding about how bats are killed as a result of wind energy. However, the reason for why bats are getting killed continues to elude researchers. Because of relatively little demographic knowledge of bat populations, as well as recent disease stressors increasing bat mortality, questions persist about the level of bat mortality resulting from wind energy relative to the sustainability of bat populations.

Wind energy and turbines. Wind energy has been the fastest-growing renewable energy source for electricity in the United States. Between 2000 and 2010, wind energy experienced a 17% annual increase. Wind energy is an environmentally friendly energy source because it does not produce air or water pollution, in contrast to fossil fuels.

Industrial wind energy facilities (**Fig. 1**), commonly referred to as wind farms, can contain hundreds of wind turbines that generate electricity as the turbine blades are spun by the wind. Turbines can range in height and wind energy capacity (measured in megawatts; MW): The taller the turbine, the greater is the potential for producing electricity, because taller turbines can support longer turbine blades, and longer turbine blades are able to "catch" more wind compared to shorter turbine blades. Current turbine designs vary in size depending on whether they are situated onshore or offshore. As of 2012, all wind energy development in the United States was onshore. Onshore turbines can be almost 500 ft (152 m) in height [330 ft (100 m) to the nacelle (the enclosure containing the electric

Fig. 1. An industrial wind turbine can reach nearly 500 ft (152 m) into the air and can produce 2.5 MW of electricity. (*Photo courtesy of the National Renewable Energy Laboratory*)

generating equipment in a wind-energy conversion system); and 165-ft (50-m) blade length] and produce 2.5 MW of electricity.

Impact on bats. Concerns about wind energy have been raised because of the effects on flying wildlife. First-generation wind turbines primarily affected birds (specifically raptors), but bat mortality has superseded bird mortality as a result of larger turbine designs and increased wind farm sitings in a diversity of habitats. Twenty-four percent of all North American bat species have been recorded as a mortality at wind energy developments in the United States. Most of the bats that have been killed by wind turbines are migratory bats that roost in trees (**Fig. 2**). These include the Eastern red bat (*Lasiurus borealis*), hoary bat (*Lasiurus cinereus*), and silver-haired bat (*Lasionycteris noctivagans*). However, recent studies have indicated that nonmigratory or short-distance migratory bats may be killed at higher rates than previously thought. These include the little brown bat (*Myotis lucifugus*) and big brown bat (*Eptesicus fuscus*). Regardless of species, turbine specifications, study methodology, or wind facility location, the numbers of estimated bats killed per turbine (corrected mortality) have ranged from 0.1 to 69.6 annually. Mortalities peak during the late summer and early fall, probably because these periods are times of concentrated migration and more bats are flying through the air following the summer breeding season, including inexperienced subadults. Bat mortality has also been associated with nights when wind speeds are relatively low [lower than 18 ft/s (5.5 m/s)], presumably because insects are active and available as a food source for bats.

Because of their secretive and nocturnal lifestyle, bats are poorly understood. For example, little is known about bat numbers (how many bats are there?), their reproduction, their birth and death rates, and the ratio of juveniles to adults. Without this information, it is difficult to determine the health of bat populations and whether the level of mortality reported at wind energy facilities is detrimental to the overall sustainability of bat populations. An additional stressor on cave-dwelling (hibernating) bat species is a disease called white-nose syndrome (WNS). In some parts of the eastern United States,

WNS has killed 100% of bats within a hibernaculum (winter shelter). Thus, a number of questions are raised: How much total mortality can bats endure? Will scientists be aware when the tipping point has been reached at which death rates exceed birth rates for bats? If so, can that trend be reversed with research, management, and education?

Fig. 2. Migratory, tree-roosting bats, including (*a*) Eastern red bat, (*b*) hoary bat, and (*c*) silver-haired bat, are the most common species found as a result of wind turbine–caused mortality. [*Photos courtesy of (a) United States Geological Survey; (b) Paul Cryan/United States Geological Survey; (c) United States National Park Service*]

Fig. 3. Radiograph of a bat killed as a result of collision with a wind turbine. Note the dislocation of the right shoulder and comminuted (shattered/crushed) fracture of the right wing bone.

How bats are killed by wind turbines is relatively well understood. Veterinary diagnostic evidence suggests that bats die as a result of a combination of trauma from colliding with spinning turbine blades (or possibly the monopole that supports the nacelle and blades) and barotrauma (injuries caused by changes in air pressure). Injuries consistent with barotrauma include primarily fatal damage to the internal organs as a result of massive hemorrhaging. Areas of high and low pressure surround spinning turbine blades. As bats fly through the pressure differential, their skeletons are not strong enough to protect their internal anatomy from the pressure drop (unlike birds) and their internal organs hemorrhage. Bat mortality as a result of barotrauma does not require collision with the monopole or turbine blades.

Although it is not possible to definitively assign the cause of death as a result of collision trauma or barotrauma, there are identifiable patterns to the injuries. A recent study investigating the cause of death found that 74% of the bats examined had at least one broken bone, with fractures to the wing bones being the most common. All fractures were to the anterior portion of the body (no fractures were found in the hind limbs), and most of the fractured bones were shattered or crushed into many pieces (comminuted) [**Fig. 3**]. Eighty-one percent ($n = 27$) of all bats necropsied had a pneumothorax (the presence of air or gas in the pleural cavity; collapsed lung) or hemothorax (accumulation of blood in the pleural cavity); 33% ($n = 8$) experienced mild to severe pulmonary hemorrhaging; and 52% ($n = 12$) were found to have hemorrhaging in and around the middle or inner ear. Most bats were found within 90 ft (27 m) of the base of the turbine, with 82% of all bats suffering ≤ 1 broken bone discovered within 30 ft (9 m) of the base of the turbine.

Reasons. Why bats are killed is less well understood. A number of theories have been put forth as to why bats may be "attracted" to wind turbines, but evidence to date does not support any one theory (if any of them at all). Certain bat species roost in trees during the day to rest, and one theory suggests that bats may mistake the monopole supporting the nacelle and turbine blades for a roost tree. An alternative hypothesis suggests that modifications to the landscape to accommodate wind energy facilities may create favorable conditions for insects, thereby providing an abundant and available food source for bats, as well as travel corridors that may cause increased encounters between bats and wind turbines. Some researchers have suggested that bats may be attracted to wind turbines as a result of the sounds emitted by the spinning blades, whereas other researchers have suggested that sounds produced by turbines may disorient bats and make it difficult for bats to avoid a nearby turbine. Based on the limited range of the echolocation of most bats and given the speed with which most industrial-sized turbine blades spin, another team of researchers has suggested that bats simply do not have sufficient time to avoid operating turbines once they are in close proximity.

Possible solutions. Mitigation techniques to reduce bat mortality from wind energy facilities are not well understood and vary in effectiveness. Most often, proper siting of wind energy facilities is suggested to minimize bat fatality. However, siting wind energy facilities where bats are not found is difficult because preconstruction knowledge of bat locations on the landscape is not well understood (and this is true even in cases where preconstruction bat monitoring studies have been carried out). Acoustic deterrents have also been tested, but were deemed ineffective. The mitigation technique that has demonstrated the most promise is curtailment. Curtailment entails altering the cut-in speed at which the turbine blades begin spinning, essentially requiring relatively higher wind speeds to engage the turbine blades. A large-scale curtailment experiment was conducted at an industrial wind farm in Alberta, Canada. A select number of turbines were altered so that either the blades would not turn or they would free-spin (remain near motionless) but not produce energy unless the wind speed was more than 16.5 ft/s (5 m/s). This experiment resulted in a reduction of bat mortality by as much as 60%. However, detailed cost-to-benefit studies have not been conducted to determine the economic impact of this mitigation strategy to a wind energy facility relative to the benefit to bat populations.

For background information *see* AERODYNAMICS; AIR PRESSURE; CHIROPTERA; ECHOLOCATION; ELECTRIC POWER GENERATION; FLIGHT; NEUROBIOLOGY; PHONORECEPTION; TURBINE; WIND POWER in the McGraw-Hill Encyclopedia of Science & Technology.

David Drake

Bibliography. E. F. Baerwald et al., A large-scale mitigation experiment to reduce bat fatalities at wind energy facilities, *J. Wildl. Manage.*, 73:1077–1081, 2009, DOI:10.2193/2008-233; S. M. Grodsky et al., Investigating the causes of death for wind turbine-associated bat fatalities, *J. Mammal.*, 92:917–925, 2011, DOI:http://dx.doi.org/10.1644/10-MAMM-A-404.1; T. H. Kunz et al., Ecological impacts of wind energy development on bats: Questions, research needs, and hypotheses, *Front. Ecol. Environ.*, 5:315–324, 2007, DOI:10.1890/1540-9295(2007)5[315:EIOWED]2.0.CO;2; The Wildlife Society, *Impacts of Wind Energy Facilities on Wildlife and Wildlife Habitat*, The Wildlife Society, Bethesda, MD, 2007.

Bicarbonate physiology and pathophysiology

Bicarbonate (HCO_3^-) is a small inorganic negative ion (anion) that is familiar as a constituent of baking soda (sodium bicarbonate; $NaHCO_3$). Baking soda is a leavening agent: in dough, HCO_3^- dissolved in water reacts with acids (H^+) to form carbon dioxide [CO_2; see Eq. (1)].

$$HCO_3^- + H^+ \rightleftharpoons CO_2 + H_2O \qquad (1)$$

As CO_2 escapes from the solution, bubbles form, causing the dough to rise.

After chloride, bicarbonate is the second most abundant anion in mammalian blood. The ability of HCO_3^- to neutralize H^+ (as in the leavening example) underlies the major physiological role of HCO_3^-: the regulation of [H^+], and therefore of pH, in cells and blood [Eq. (2)].

$$pH = -\log[H^+] \qquad (2)$$

It is critical to maintain pH throughout the body within a narrow range because nearly every physiological process is pH-sensitive. Enzymes of the carbonic anhydrase (CA) family, which are present both inside and on the surface of most cells, catalyze the

---

TABLE 1. Distribution and actions of mammalian HCO_3^- transporters

Family	Gene	Gene product	Location	Action
Slc4	**Group 1. Electroneutral Cl⁻-HCO₃⁻ (anion) exchangers (AEs)**			
	Slc4a1	AE1	Kidney, red blood cells	
	Slc4a2	AE2	Widespread	
	Slc4a3	AE3	Brain, heart	
	Group 2a. Electrogenic Na⁺/HCO₃⁻ cotransporters (NBCes)			
	Slc4a4	NBCe1	Widespread	
	Slc4a5	NBCe2	Brain, liver	
	Group 2b. Electroneutral Na⁺/HCO₃⁻ cotransporters (NBCns)			
	Slc4a7	NBCn1	Widespread	
	Slc4a10	NBCn2	Brain	
	Group 2c. Electroneutral Na⁺-driven Cl⁻-HCO₃⁻ cotransporter (NDCBE)			
	Slc4a8	NDCBE	Brain	
Slc26	**Group 3. Anion (including HCO₃⁻) exchangers**			
	Slc26a3	DRA	Intestines	
	Slc26a4	Pendrin	Inner ear, kidney	
	Slc26a6	CFEX	Widespread	
	Slc26a7	SLC26A7	Kidney	
	Slc26a9	SLC26A9	Lung, stomach	

The action of some Slc26 proteins is controversial. Some members may be electrogenic (that is, carry net charge) and some may act as anion channels.

reaction shown in Eq. (1), ensuring rapid equilibration among HCO_3^-, H^+, and CO_2. Another important role of HCO_3^- in many organs is support of fluid secretion and absorption. HCO_3^- moves into and out of cells via pathways formed by a diverse group of proteins known as HCO_3^- transporters (**Table 1**). Aberrant HCO_3^- transport is a critical factor in diverse disease states, including cancer, diarrhea, epilepsy, heart failure, and hypertension.

Maintenance of blood plasma pH. The physiological pH of blood plasma, and therefore of the fluid that surrounds individual cells, is approximately 7.4. This value is a consequence of $[HCO_3^-]$ being 24 mmol/L and $[CO_2]$ being 1.2 mmol/L, and can be calculated using the Henderson-Hasselbalch equation [Eq. (3)].

$$pH = 6.1 + \log([HCO_3^-]/[CO_2])$$
$$= 6.1 + \log(24/1.2) = 7.4 \qquad (3)$$

Dietary-acid intake and cellular metabolism generate a load of acid and CO_2 that, if unchecked, would alter the equilibrium shown in Eq. (1) and produce a catastrophic fall in plasma pH: that is, an acidosis. The usefulness of HCO_3^- and CO_2 in the blood is illustrated by the following example. If a hypothetical acid load of 1 mmol was added to a liter of solution that had an initial pH of 7.4, but no HCO_3^- or CO_2, the pH of that solution would drop to 3.0 [Eq. (2)]. A comparable fall in plasma pH would be lethal. However, adding 1 mmol of acid to a liter of solution buffered with physiological levels of CO_2/HCO_3^- would cause the pH to drop only to about 7.38 because HCO_3^- would buffer nearly all of the added acid, and in the process produce CO_2, which would evolve into the air, just as in the leavening example [Eq. (4)].

$$pH = 6.1 + \log([HCO_3^-]/[CO_2])$$
$$= 6.1 + \log[(24 - 1)/1.2]$$
$$= 6.1 + \log(23/1.2) = 7.38 \qquad (4)$$

The daily acid load for an average adult human on a typical Western diet is 70 mmol/day. **Figure 1** shows how the body controls $[HCO_3^-]$ and $[CO_2]$, and thus pH. The kidneys maintain blood plasma $[HCO_3^-]$. After renal glomeruli filter the blood plasma and pass it along to renal tubules, kidney cells—predominantly those in the proximal tubules—transport the HCO_3^- back into the blood (preventing its loss in urine) and also generate new HCO_3^- in the blood (replacing HCO_3^- lost to neutralization of dietary and metabolic acids). The contribution

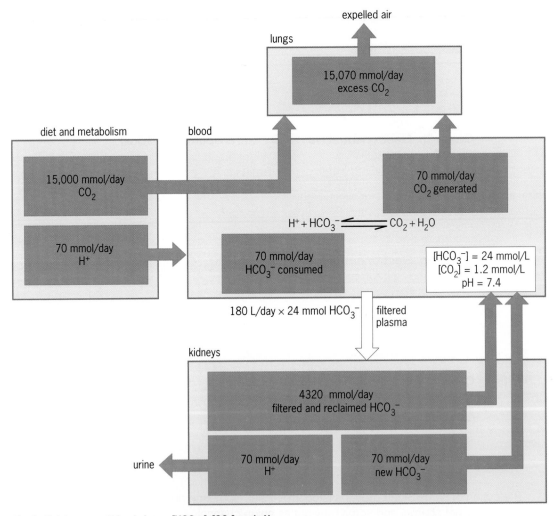

Fig. 1. Maintenance of blood plasma $[HCO_3^-]$, $[CO_2]$, and pH.

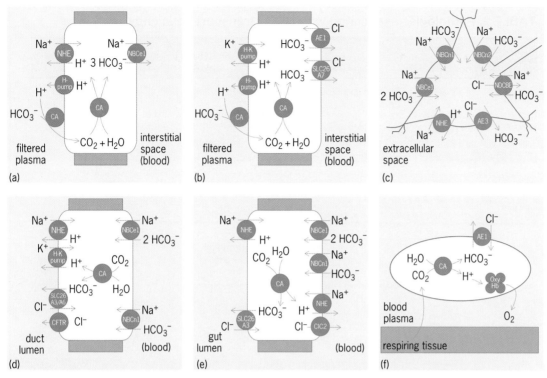

Fig. 2. Cellular mechanisms of HCO_3^- transport. (*a*) In the renal proximal tubule replenishing blood plasma HCO_3^-; AE2 replaces NBCe1 in later proximal tubule segments. (*b*) In the renal collecting duct replenishing blood plasma HCO_3^-. (*c*) In the brain regulating pH_i and pH_s of neurons. (*d*) In salivary glands supporting HCO_3^- and fluid secretion from duct cells. (*e*) In the colon supporting HCO_3^- secretion and fluid absorption by enterocytes. (*f*) In red blood cells promoting gas exchange.

of HCO_3^- transporters in the kidney to maintain plasma $[HCO_3^-]$ is shown in **Fig. 2***a* and *b*. Kidney failure results in a drop in plasma $[HCO_3^-]$, and therefore a drop in pH, known as metabolic acidosis. As one might imagine, genetic defects in any of the HCO_3^- transporters shown in Fig. 2*a* and *b* can result in metabolic acidosis (**Table 2**). For example, individuals with mutations in NBCe1 exhibit low plasma $[HCO_3^-]$—a condition known as proximal renal tubular acidosis—that can be accompanied by signs such as growth retardation, mental retardation, epilepsy, migraines, glaucoma, and paraplegia.

The lungs maintain plasma $[CO_2]$ by excreting the gas in the exhaled alveolar air. It is the partial pressure of CO_2 in the alveolar air (approximately 40 mmHg), together with the solubility coefficient of CO_2 in plasma (0.03 mmol/L per mmHg), that determines the concentration of dissolved CO_2 in arterial blood plasma (that is, $40 \times 0.03 = 1.2$ mmol/L). Lung failure results in a rise in plasma $[CO_2]$, and thus a drop in pH—known as respiratory acidosis. Under these conditions, the kidneys attempt to compensate for the rise in $[CO_2]$ by generating more HCO_3^-, thereby tending to raise pH toward normal [see Eq. (3)]. Conversely, the respiratory system can compensate for the metabolic acidosis (a fall in $[HCO_3^-]$) mentioned in the previous paragraph by increasing the rate of CO_2 excretion, thereby lowering $[CO_2]$ and tending to raise pH toward normal.

Maintenance of pH in and around cells. Alterations in blood plasma pH to a large extent control the pH of the fluid that bathes cells. However, most cells also

have their own mechanisms to regulate the pH of their cytoplasm (intracellular pH; pH_i) and the pH of their extracellular surface (pH_s). Regulation of pH_i is critical to maintain the efficiency of virtually every process in the cell. Regulation of both pH_i and pH_s is critical to maintain the activities of membrane-spanning receptors and transporters, which are simultaneously exposed to the cytoplasm and the cell surface.

Excess intracellular H^+ can be either physically removed or neutralized. When H^+ is abundant inside cells, solute carriers such as Na-H exchangers (NHEs, which exchange incoming Na^+ for outgoing H^+) and H^+ pumps [which use energy from adenosine triphosphate (ATP) hydrolysis to move H^+ out of the cell] actively transport H^+ out of cells (examples of acid-extruding mechanisms). The buildup of extruded H^+ on the cell surface, which would lower pH_s, is minimized by the action of carbonic anhydrases (CAs) that neutralize H^+ with HCO_3^-, thereby lowering HCO_3^- at the cell surface (see example in Fig. 2*a*). At the same time, and often more important, HCO_3^- transporters [typically Na-bicarbonate transporters (NBCs); Table 1 and Fig. 2*a–e*] move HCO_3^- into cells, neutralizing H^+ and thereby maintaining pH_i. The regulation of pH_i in the face of intracellular alkalosis (high plasma alkalinity) is typically achieved by HCO_3^- efflux mediated by anion exchangers (AEs) of the Slc4 family (Table 1; example of acid loading in Fig. 2*a–f*), and in some cell types by NBCs working in an unusual acid-loading mode (see example in Fig. 2*a*). The important roles

TABLE 2. Pathologies associated with HCO₃⁻ transport deficiency

Transporter	Location	Role	Loss-associated diseases and signs
AE1	Kidney	Maintaining blood [HCO$_3^-$] and supporting urinary H$^+$ secretion	Distal renal tubular acidosis, kidney stones
AE2	Bone	Promoting bone resorption	Osteopetrosis
	Liver	Promoting fluid/[HCO$_3^-$] secretion	Biliary cirrhosis
AE3	Brain	Maintaining neuronal pH$_i$	Epilepsy
NBCe1	Brain	Maintaining neuronal and glial pH$_i$/pH$_s$	Mental retardation, epilepsy, migraine
	Eye	Promoting fluid exit from the cornea and maintaining pH$_i$ in various other tissues	Corneal swelling, band keratopathy, glaucoma
	Kidney	Maintaining blood [HCO$_3^-$] and supporting urinary H$^+$ secretion	Proximal renal tubular acidosis
	Intestine	Promoting fluid/[HCO$_3^-$] secretion by the pancreas and duodenum, neutralizing chyme	Pancreatitis, intestinal obstruction
	Teeth	Promoting enamel formation	Poor dentition
NBCe2	Brain	Promoting cerebrospinal fluid secretion	Reduced neuronal excitability
NBCn1	Eye	Unknown	Blindness
	Inner ear	Unknown	Deafness
	Blood vessels	Direct and indirect control of smooth muscle cell contractility by regulation of pH$_i$	Hypertension
NBCn2	Brain	Maintaining neuronal pH$_i$ and composition of cerebrospinal fluid	Autism, reduced neuronal excitability
NDCBE	Brain	Maintaining neuronal pH$_i$	Reduced neuronal excitability
DRA	Intestine	Promoting fluid absorption/[HCO$_3^-$] secretion by the colon	Congenital chloride diarrhea
Pendrin	Inner ear	Secreting [HCO$_3^-$] into endolymph, indirectly maintaining sensory signal transduction	Pendred syndrome (deafness)
SLC26A7	Kidney	Maintaining blood [HCO$_3^-$] and supporting urinary H$^+$ secretion	Distal renal tubular acidosis

played by HCO$_3^-$ transporters in pH regulation for diverse organs are listed in Table 2. In addition, regulation of pH$_i$ and [HCO$_3^-$] per se is also of crucial importance in HCO$_3^-$ secreting cell types (see below).

The pathological consequences associated with the loss of any one transporter that regulates pH$_i$ and pH$_s$ are likely tempered in most cell types by the presence of multiple pH-regulating transporters. However, several pathologies, including hypertension and epilepsy, are linked to the loss of local pH-regulating ability (Table 2). HCO$_3^-$-transport defects are frequently associated with altered neuronal activity because many neurotransmitter-gated ion chan-

nels are exquisitely sensitive to alterations in pH$_i$ and pH$_s$. The pH sensitivity of neuronal excitability is physiologically important because repeated firing tends to acidify neurons, dampening their excitability; this is a negative feedback mechanism that prevents the onset of seizures. In the brain, HCO$_3^-$ transporters of the Slc4 family are responsible for maintaining neuronal pH$_i$—by importing or exporting HCO$_3^-$ (Fig. 2c)—as well as neuronal pH$_s$—by secreting HCO$_3^-$ (as with NBCe1 in glia) or by controlling the secretion and composition of cerebrospinal fluid (as with NBCe2 and NBCn2 in the choroid plexus). Not listed in Table 2 are two pathologies related to aberrant pH$_i$ regulation that are associated with excessive HCO_3^- transport. The first pathology is cancer: Some cancer cells exhibit an elevated NBC activity that allows them to survive within the acidic environment of a tumor. The second pathology is heart failure, which is associated with an overabundance of NBC in heart muscle cells. In these cardiac cells, excessive NBC activity raises intracellular $[Na^+]_i$, which in turn (via the action of a Na-Ca exchanger in the cell membrane) raises intracellular $[Ca^{2+}]$ to toxic levels; then, excessive Ca^{2+} damages mitochondria and promotes cell death.

Support of HCO_3^- secretion. Some organs—typified by the salivary gland—secrete Na^+, Cl^-, and HCO_3^- across their luminal membrane, which is a movement that is accompanied by H_2O moving along its osmotic gradient. HCO_3^--containing saliva is important as a solvent for digestive enzymes, a protectant of tooth enamel from acid attack, and a lubricant for food transit along the esophagus. A HCO_3^--secreting cell from a salivary gland, which exhibits features common to most HCO_3^--secreting cells, is shown in Fig. 2d. Most of the HCO_3^- that is secreted across the apical membrane—via transporters of the Slc26 family (Table 1, group 3) in conjunction with anion-permeable channels such as the cystic fibrosis transmembrane conductance regulator (CFTR)—is generated by intracellular CA. Basolateral NBCs provide a role in support of maintaining pH_i and may also supply HCO_3^- directly for secretion when secretion is maximal. In later parts of the intestine, such as the colon, cells absorb fluid by absorbing Cl^- in exchange for secreted HCO_3^- (Fig. 2e). The important roles played by HCO_3^- transporters in support of HCO_3^- secretion and fluid secretion/absorption for diverse organs are listed in Table 2.

Support of gas exchange in red blood cells. The HCO_3^- transporter AE1 is critical for the functions of red cells. In capillaries, metabolic CO_2 enters red cells, where cytosolic CA catalyzes the conversion of CO_2 and H_2O into HCO_3^- and H^+. The HCO_3^- produced by this reaction is transported out into the blood plasma by AE1. These concerted actions promote gas exchange in the capillaries in two ways: (1) removal of CO_2/HCO_3^- from the red-cell interior maintains a gradient for CO_2 influx into red cells; and (2) the buildup of H^+ inside the red cell causes the O_2 affinity of hemoglobin to fall (the Bohr effect), promoting O_2 release in the capillaries (Fig. 2f). This process is reversed in the alveolar capillaries, where CO_2 efflux from red cells is promoted by the CA-catalyzed combination of HCO_3^- and H^+. Hereditary anemias comprise one important group of pathologies related to loss of AE1 from the red cell, but not associated with defective HCO_3^- transport per se. For example, some individuals with mutations in the gene that encodes AE1 have fragile red cells that break up in circulation. This pathology arises because the AE1 molecule normally acts as a site of attachment between cytoskeletal proteins and the red cell membrane. Loss of these attachment points disrupts the flexible meshwork that allows red cells to deform without fragmenting as they pass through capillaries.

Conclusions. The study of HCO_3^- transport and transporters is a rapidly expanding field with an ever-increasing number of physiological roles and pathologies being ascribed to Slc4 and Slc26 family members. The association of prevalent diseases such as cancer, diarrhea, hypertension, and heart failure with HCO_3^- transport dysfunction means that HCO_3^- transporters may be attractive therapeutic drug targets.

For background information *see* ACID AND BASE; BLOOD; CELL BIOLOGY; ION TRANSPORT; KIDNEY; OSMOREGULATORY MECHANISMS; PATHOLOGY; PH; PH REGULATION (BIOLOGY); SECRETION in the McGraw-Hill Encyclopedia of Science & Technology.
Walter F. Boron; Mark D. Parker

Bibliography. W. F. Boron and E. L. Boulpaep, *Medical Physiology: A Cellular and Molecular Approach*, updated 2d ed., Saunders, Philadelphia, 2012; M. Chesler, Regulation and modulation of pH in the brain, *Physiol. Rev.*, 83:1183–1221, 2003, DOI:10.1152/physrev.00010.2003; E. Cordat and J. R. Casey, Bicarbonate transport in cell physiology and disease, *Biochem. J.*, 417:423–439, 2009, DOI:10.1042/BJ20081634; M. G. Lee et al., Molecular mechanism of pancreatic and salivary gland fluid and HCO_3^- secretion, *Physiol. Rev.*, 92:39–74, 2012, DOI:10.1152/physrev.00011.2011; M. F. Romero, C. M. Fulton, and W. F. Boron, The SLC4 family of HCO_3^- transporters, *Pflug. Arch.*, 447:495–509, 2004, DOI:10.1007/s00424-003-1180-2.

Breakdown of shell closure in helium-10

The study of exotic nuclei at the edges of nuclear stability is one of the most important developments in modern nuclear physics. Unusual forms of nuclear dynamics often arise here. One of the most prominent phenomena encountered is shell breakdown—the deviation from the expected shell structure in these exotic nuclei. On the one hand, in the nuclear shell model, helium-10 (^{10}He) is a "double-magic" nucleus with $Z = 2$ and $N = 8$. On the other hand, it has an enormous neutron excess; its neutron number (N) to proton number (Z) ratio equals 4, which brings it to the edge of nuclear matter asymmetry. Thus, the ^{10}He nucleus is an important system for the development of our understanding of nuclei located far from the beta stability valley and even beyond the neutron and proton drip lines. Here we present new insights into the basic properties of this nucleus, illuminating its shell structure and indicating its strong deviation from the simple shell population picture.

Shell structure in nuclei. For more than 100 years the periodic table of elements has provided a basis for understanding the major chemical laws. The existence of the periodic table is connected with the systematics of quantum-mechanical atomic shells. The nuclide chart, which shows "N versus Z," can be considered an analog of the periodic table in the "realm of nuclei" (**Fig. 1**).

The nuclear shell model is designed to treat systems consisting of particles with semi-integer spin.

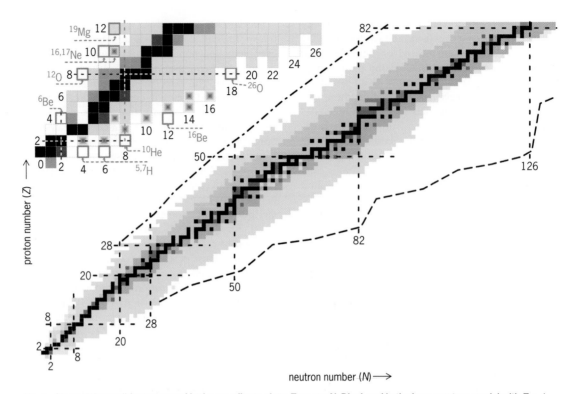

Fig. 1. Chart of the nuclides presented in the coordinate plane *Z* versus *N*. Displayed in the lower part are nuclei with *Z* = 1 − 83. Black squares show stable nuclei that form the stability valley. The areas of nuclei with lifetimes ranging from more than one year to milliseconds are shown in gray with shades varying from darkest to lightest, respectively. The horizontal and vertical broken lines denote the proton and neutron magic numbers. The dash-dotted and long-dashed lines show the expected proton and neutron drip lines. Being produced in some nuclear processes, nuclei situated far on the left (right) side of the proton (neutron) drip line emit protons (neutrons) instantly. In close proximity to these critical zones there are many fragile nuclei observable in the so-called resonance states, signifying the nuclear states characterized by very short lifetimes: 10^{-21}–10^{-19} s is the period which allows a nucleon to traverse a nucleus 10 − 1000 times. The upper left inset presents a close-up of the region of atomic numbers *Z* = 1 − 12. Halo nuclei are marked with shaded crosses. Exotic resonance state nuclei—true 2*p* or true 2*n* emitters—are singled out by green color.

Its essential idea is that the particles are moving in bound orbitals in response to a central force from the remainder of the system. Each orbital has well designated energy, angular momentum, and parity. Electrons, as well as nucleons, since they have a spin of $1/2$, can form such systems. In the Bohr model of the atom, no two electrons occupy the same state, that is, have identical sets of quantum numbers. This principle, attributed to Wolfgang Pauli, results in a finite number of electrons occupying a given energy level, and thus leads to the concept of closed (filled) shells and the noble gases, helium, argon, and so forth.

Basic data (binding energies, nucleon separation energies, spin/parity systematics of excited states, and so forth) indicate that closed shells occur in nuclei at proton and neutron numbers *Z*(*N*) = 2, 8, 20, 28, 50, 82, and *N* = 126 (shown in the nuclide chart in Fig. 1). The shell model treats each nucleon, interacting in the nucleus with the *A* − 1 remaining nucleons, as one moving in a spherical potential well. In accord with the Pauli principle, the lack of vacancies in the nuclear orbitals below the Fermi level results in long free-nucleon paths inside the nucleus. To explain the known specific magic numbers of protons and neutrons, in 1949, M. G. Mayer and, independently, O. Axel, J. H. D. Jensen, and H. E. Suess, proposed a model of independent nucleons confined by a surface-corrected, isotropic harmonic oscillator plus a strong attractive spin-orbit term. This primordial shell model, using a Woods-Saxon potential instead of the harmonic oscillator, describes quite well closed-shell nuclear structures and single-particle and hole states built on them.

Moving away from the stability valley, with *Z* increasingly different from *N*, brings us to a situation of "asymmetric nuclear matter." Sooner or later, with the increased asymmetry, the strong (nuclear) interaction is no longer able to keep the nucleons together. So, the neutron and proton "drip lines" appear (Fig. 1), where the last neutron or proton becomes unbound from the nucleus. Beyond the drip lines, the nuclear systems show up as short-lived resonance states with lifetimes of about 10^{-21} s (around MeV width). Nuclei resonances do not become so broad immediately at the drip lines. The discovery of true two-proton (2*p*) radioactivity of proton drip line nuclei is an important result obtained recently. The search for nuclei on the neutron drip line showing 2*n* radioactivity is now on the waiting list of investigations to be undertaken.

The ^{10}He puzzle. Helium-10 is the second lightest double-magic nucleus, after helium-4. The expectation of the existence of a bound ^{10}He system was discussed for a long time, since it could have had

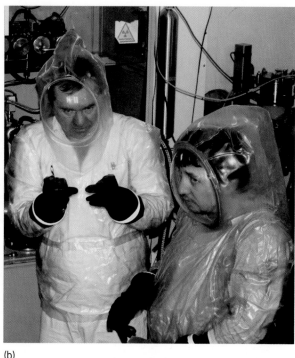

(a)   (b)

Fig. 2. Cryogenic tritium target. (*a*) Refrigerator head unit, capable of cooling down to 26 K (−413°F) the target cell hidden in the thermal screen, which is cooled to 77 K (−321°F; liquid nitrogen temperature). The target cell is a double-chamber unit whose 6-mm-thick internal chamber is filled with tritium gas to 1 bar pressure, and high vacuum is provided by a titanium getter in the small volume between the two chambers. This makes it possible to exclude any leaks of tritium caused by its diffusion through the extremely thin 10-μm stainless steel entrance and exit windows, transparent for particles of interest. (*b*) All operations with the tritium target system are performed by authorized staff in protective suits.

far-reaching consequences. When the nuclear instability of ^{10}He was established, it became important to find out the degree to which it is unbound. This could have an impact on theoretical judgments about the properties of nuclear matter, the nuclear equation of state, and astrophysical nucleosynthesis paths. The last but not the least important point concerns interest in the study of correlations of the ^{10}He decay fragments (^{8}He + n + n). The instability of the two ^{10}He subsystems causes its "true," that is, instantaneous, three-body decay into ^{8}He + n + n. In quantum mechanics, this class of decays is as yet not completely understood. The ^{10}He case is one where ascertaining details of such decays will be an important step for this exciting subject.

Production of ^{10}He. The experimental study of the ^{10}He nucleus is complicated. Promising ways to investigate this nucleus are offered by radioactive nuclear beams available nowadays in several laboratories around the world. In the present work, ^{10}He was produced by the addition of two neutrons to the heaviest particle-stable helium isotope, ^{8}He ($Z = 2$, $N = 6$). The heaviest particle-stable hydrogen isotope, ^{3}H ($Z = 1$, $N = 2$), was used to transfer the two neutrons to ^{8}He, that is, to realize the ^{8}He + ^{3}H → p + ^{10}He reaction. The uniqueness of this nuclear reaction consists in the fact that both the target and the projectile are beta-radioactive nuclei. Because it is a nontrivial task to realize such a nuclear reaction, the attainment of ^{10}He in this way was delayed for a long time.

In this experiment, 172-MeV short-lived (half-life $T_{1/2} = 119$ ms) ^{8}He nuclei were produced in a fragmentation reaction of a boron-11 (^{11}B) nuclear beam obtained from the U400M cyclotron in Dubna (Russia). The in-flight separator ACCULINNA provided isolation and shaping of a beam with an intensity of 10,000 ^{8}He nuclei per second hitting a unique cryogenic tritium target (**Fig 2**). Work with radioactive tritium requires strict adherence to regulations and radiation safety standards. A set of environmentally safe equipment was developed, which made it possible to fill the target cell with tritium gas, evacuate and recover tritium, and perform radiation monitoring along the beam lines and in work rooms. The fulfillment of these requirements was worth the problems entailed, since tritium provides essential advantages, which were used in full in this experiment. At present, equipment for work with tritium gas is available only in very few military laboratories. This part of our work can be seen as an excellent example of the conversion of military technology for application in fundamental science.

Basic properties established for ^{10}He. A remarkable feature of the ^{8}He + ^{3}H → p + ^{10}He reaction is that it is achieved with maximum probability when the recoil protons are emitted in the backward direction, with the ^{10}He partners flying forward and decaying almost immediately into two neutrons and ^{8}He. By detecting the coincident protons and ^{8}He nuclei, one can determine the energy spectrum of ^{10}He nuclei and extract significant information about the angular

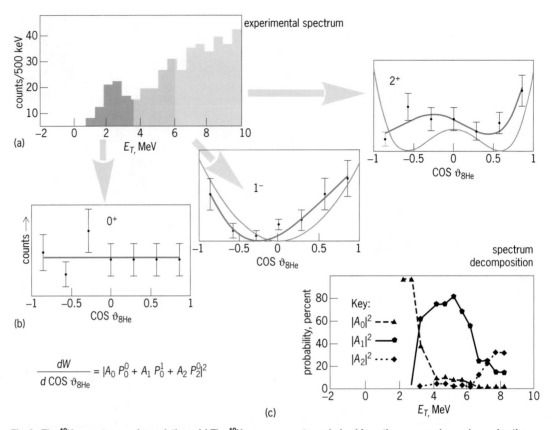

(a)

(b)

$$\frac{dW}{d\cos\vartheta_{8He}} = |A_0\,P_0^0 + A_1\,P_0^1 + A_2\,P_2^0|^2$$

(c)

Fig. 3. The ^{10}He spectrum and correlations. (*a*) The ^{10}He energy spectrum derived from the measured energies and pathways of recoil protons emitted in the ^{8}He + ^{3}H → ^{10}He + p reaction. The total ^{10}He energy E_T is given with reference to the ^{8}He + n decay threshold. (*b*) Angular distributions of ϑ_{8He} in the ^{10}He rest frame for energy ranges indicated by shading. The polar angle ϑ_{8He} is taken with respect to the axis Z parallel to the vector of momentum transferred to ^{10}He in the reaction. In this frame, states with $J^\pi = \{0^+, 1^-, 2^+\}$ in ^{10}He can be related, under certain assumptions, to pure Legendre polynomial distributions $\{P_0, P_1, P_2\}$. Green curves show the fits by coherent sums of these polynomials in the equation shown. Gray curves give the contributions of the isolated Legendre polynomials, showing the importance of the state interference forming the observed correlation patterns. (*c*) Relative contributions of different components to the fit, indicating the $J^\pi = \{0^+, 1^-, 2^+\}$ level ordering in ^{10}He.

and energy correlations of the ^{10}He decay products. In the spectrum obtained (**Fig. 3**), the ground state of ^{10}He is seen as a broad resonance with maximum at $E_T = 2.1 \pm 0.2$ MeV.

The definitely established energy of the ^{10}He ground state with spin-parity $J^\pi = 0^+$ is an important result. Its reliability follows from auspicious features of the two-neutron transfer reactions, making it a trusty tool for the study of nuclei with large neutron excess. The shape and energy position of the ^{10}He ground state verify that, just as in other nuclei with the magic number of eight neutrons, the six neutrons available in ^{10}He above the ^{4}He core take up their positions to complete the $0p$ shell. This is in accord with the recently obtained energy of the ^{9}He ground-state resonance lying at 2 MeV above the ^{9}He → ^{8}He + n decay threshold. Coupling the $S = 1/2$ spin of the odd (unpaired) neutron with the p-shell orbital momentum $l = 1$ gives a spin-parity of the ^{9}He ground-state resonance $J^\pi = 1/2^-$. The two $J^\pi = 1/2^-$ neutrons in ^{10}He are paired, resulting in the ground-state spin-parity $J^\pi = 0^+$. A quite "normal" pairing energy of about 1.9 MeV comes out for the two last (that is, valence) neutrons of ^{10}He.

Correlations and level ordering in ^{10}He. The ground state of ^{10}He is already quite broad, and it can be

expected that the excited states are even broader. Indeed, it can be seen in Fig. 3 that the excitation spectrum of ^{10}He is quite featureless, consisting presumably of broad overlapping structures. However, expressed angular correlation patterns can be formed in transfer reactions in a certain frame, which may allow disentangling of the contributions of states with different J^π. This actually appears to be possible in the case of ^{10}He (Fig. 3). An important finding emerging from the correlation data obtained in this work is that the first excited state in ^{10}He is a 1^- state with very low excitation energy $E^* \sim 2.8$ MeV. In a simple picture of shell population, the 1^- state is formed by a particle-hole excitation where a neutron moves from the $0p$ state (leaving a hole) to a $1s$ state (particle), $0p \rightarrow 1s$, and its energy is directly related to the gap between shells.

The systematics of the lowest 1^- and 2^+ excitations for the $N = 8$ isotones are provided in **Fig. 4**. The 2^+ state demonstrates abrupt variations with shell completion. It can be seen for the 1^- state that the energy gap between the $0p$ and $1s$ orbitals is reduced in ^{10}He by 2–3 times as compared to the isotones that belong to the stability valley, ^{16}O and ^{14}C. The pattern of the ^{12}Be energy levels shown in Fig. 4 demonstrates that the melting (disappearance) of the

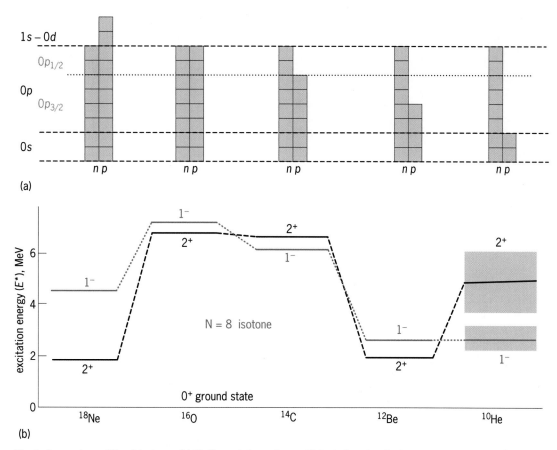

Fig. 4. Comparison of $N = 8$ isotones. (*a*) Shell population scheme. (*b*) Evolution of excitation energy for the first 2^+ and 1^- states. Shaded rectangles indicate the uncertainties of the ^{10}He level positions due to their width.

$N = 8$ shell gap was observed in this nucleus also. However, the established breakdown of the shell closure in the ^{10}He nucleus is quite unexpected: No restoration of the expected shell behavior due to its double-magic nature occurs in this nuclide.

Conclusions. The binding energy of the ^{10}He nucleus is now established. Moreover, ^{10}He is the system with the largest nuclear matter asymmetry, N/Z, for which the excitation spectrum is defined. This spectrum shows anomalous level ordering, providing evidence for the shell structure breakdown in ^{10}He, which is quite unexpected for a double-magic nucleus. The importance of this case is that nuclei with just a few nucleons should become "reference" systems for the expansion of our knowledge further into the as yet unexploited regions of nuclear drip lines. Recently, the first results also promoting this line of research have been obtained about the heavier ground-state true $2n$ emitters ^{13}Li (at GSI, in Darmstadt, Germany), and ^{16}Be and ^{26}O (at the National Superconducting Cyclotron Laboratory, at Michigan State University).

For background information *see* EXOTIC NUCLEI; NUCLEAR SHELL MODEL AND MAGIC NUMBERS; NUCLEAR REACTIONS; NUCLEAR STRUCTURE; RADIOACTIVE BEAMS; TRITIUM in the McGraw-Hill Encyclopedia of Science & Technology.

Gurgen M. Ter-Akopian; Sergey I. Sidorchuk; Leonid V. Grigorenko

Bibliography. B. Jonson, Light dripline nuclei, *Phys. Rep.*, 389:1–59, 2004, DOI:10.1016/j.physrep. 2003.07.004; M. Pfützner et al., Radioactive decays at limits of nuclear stability, *Rev. Mod. Phys.*, 84:567–619, 2012, DOI:10.1103/RevModPhys.84.567; S. I. Sidorchuk et al., Structure of ^{10}He low-lying states uncovered by correlations, *Phys. Rev. Lett.*, 108:202502 (5 pp.), 2012, DOI:10.1103/ PhysRevLett.108.202502; M. Thoennessen, Reaching the limits of nuclear stability, *Rep. Prog. Phys.*, 67:1187–1232, 2004, DOI:10.1088/0034-4885/67/ 7/R04.

Breeding the button mushroom (Agaricus bisporus)

The familiar button mushroom [*Agaricus bisporus* (J. Lange) Imbach (see **illustration**)], originally a forest-dwelling species, has been cultivated continuously for more than three centuries. Today, millions of tons of these mushrooms, with a farm gate value of several billion United States dollars, are produced annually by commercial farms on every continent except Antarctica. White and brown varieties, and both closed-style products ("buttons" and cremini) and open-style products (stuffing cups, flats, and portabellas), are all members of this one species. Several isolated wild populations, as well as collections

Laboratory cultivation of a strain of *Agaricus bisporus* collected from nature in Canada. (*Photo courtesy of Richard W. Kerrigan*)

of commercial cultivars (horticultural varieties), provide a substantial reservoir of genetic diversity. New strains with useful traits are of great commercial interest.

Aspects of breeding: challenges and opportunities. Breeding of new commercial strains of *A. bisporus* was hampered for many years by the opaque nature of sexual reproduction in this species, which in several respects mimics an asexual process. Over a 35-year period, culminating in 1993, it was determined from several lines of evidence that fertile strains each carry two complementary haploid (n) nuclei (thus, they are $n + n$ rather than having the diploid $2n$ state common to plants and animals). The two nuclei ultimately fuse and undergo a normal meiotic process, although with reduced rates of crossing-over, to produce recombined haploid postmeiotic nuclei. These gametic nuclei are then usually nonrandomly reassociated into sexually compatible pairs in binucleate ($n + n$) basidiospores, preserving both fertility and heterozygosity. In this mating-in-the-spore syndrome, called intramixis, the transmission of most or all of the parental genetic material to most offspring allows adaptive genotypes to be maintained (or subtly modified) over multiple sexual generations, thus presenting a superficially "asexual" character while allowing epigenetic and meiotic DNA repair and maintenance functions to operate.

Additionally, outcrossing is not precluded. Haploid offspring are rare in most strains, but they are common in strains carrying the *t* allele at the *BSN* locus, as is seen in members of the population of *A. bisporus* var. *burnettii* in southern California. Haploid offspring readily outcross, facilitating both gene exchange among strains and novel genotypes; to a lesser extent, contact between nonhaploid ($n + n$) strains can sometimes also lead to nuclear exchange and successful mating. Overall, this versatile amphithallic lifestyle maintains a balance between two reproductive objectives with evolutionary consequences: the perpetuation and increased representation of successful parental genotypes, and the generation of novel genotypes from pairs of successful parents.

Key developments. Knowledge of the existence and role of the *BSN* locus in dynamically maintaining or adjusting this balance in natural populations has been exploited to obtain novel breeding stocks from which abundant numbers of formerly rare haploid offspring may now be obtained. This has overcome a primary obstacle to *Agaricus* strain breeding. It is now possible, with large pools of potentially immortal haploid offspring, to screen and select breeding candidates at the haploid (gametic) stage. This is an advantage not traditionally available to plant breeders, although the development of inbred lines (for example, in maize) provides a functionally parallel approach for some plant crop species. However, this creates one of the challenges that mushroom breeders face when working within the legal systems erected to provide Intellectual Property Rights (IPR) protection for plant species: an immortal haploid gametic line does not conform directly to the concept of a plant variety (discussed in the following section).

Other key developments in *Agaricus* breeding in the last few years include the development of whole-genome genetic linkage maps incorporating sequence-characterized markers, and the production of two whole-genome DNA sequences of *A. bisporus*. Two different haploid strains (one from a commercial bisporic European stock; and the other from a wild, tetrasporic Californian stock) were sequenced by an international consortium of scientists supported by the Community Sequencing Program of the U.S. Department of Energy Joint Genome Institute in Walnut Creek, California, which performed the sequence generation and computational assembly of the data. From these two relatively unrelated sequences, many aspects of the organism can be evaluated: how these two strains differ, what they have in common, and how they both compare with other genomes across the fungal kingdom. Subsequent studies are now developing complementary physiological data on these and other strains of *A. bisporus*.

From the breeder's perspective, genome sequences are tools of discovery, diagnostics, and development. Members of gene families that are known to play roles in metabolic processes, for example substrate conversion and nutrient acquisition, have been discovered in the genome sequence, allowing better interpretation of the versatility and ecological niche (plant leaf litter decomposition) of the organism. The discovery of dysfunctional genes [for example, when genes are disrupted by the insertion of a mobile genetic element (a transposon); the genome sequences confirm that many thousands of such elements are present in *A. bisporus*] can explain why a phenotypic trait is anomalously expressed (or not expressed) in a strain. Overall, knowledge of the genotypic inheritance of each offspring can enable the selective breeding of hybrids predicted to be among the most interesting or desirable within the universe of possibilities.

Techniques for genotyping strains (fungi, plants, humans, and so on) have become far more accessible and powerful, and far less expensive, in the past decade. A few years ago, it was the norm to use single-marker methods such as alloenzymes, DNA fragment length markers, and specific polymerase chain reaction (PCR)–amplifiable DNA targets. More recently, multiplex amplification methods incorporating either simple-sequence repeats (which are used in criminology and forensics) or markers with one end of the target anchored in repeated DNA elements have proven much more efficient. In fact, at present, an entire sample genome can be matched (hybridized) to a reference genome on a chip called a microarray, or even sequenced de novo and compared computationally to any number of other reference genomes. It has become practical to survey and compare entire genomes relatively quickly and cheaply.

Some traits, called mendelian, are determined by single genes. In *A. bisporus*, examples of such traits already discovered include mating type, basidial spore number, and cap color. However, other quantitative traits (for example, disease resistances) have a more complex basis and are known to be determined or influenced by multiple genes. When evaluating, or developing, strains with attention to multiple complex traits, it becomes essential to employ a whole-genome approach to genotyping. Additionally, the availability of whole-genome "fingerprints" of strains will have significant impact on strain IPR, not only conferring the ability to identify a particular strain, but also to identify its original sexual parents, and to say whether and precisely how it might have been derived from an original sexual cross.

Strain protection: the IPR landscape. Many fungi, including *A. bisporus*, can be clonally propagated, and this was freely exploited by mushroom spawn purveyors during the historical era when all strains, originally selected from wild populations, belonged in the public domain. With the advent of the first applications for IPR on new button mushroom strains in the 1980s, the situation began to change. It is now generally understood and accepted (although not always embraced) that simple culture cloning of a legally protected *Agaricus* strain does not circumvent IPR protections under any established framework.

Other methods of culture preparation (for example, the preparation of somatic culture selections, or selections of cultures prepared from mushroom tissues, or of cultures prepared from single or multiple spores) may capture subtle variations and could potentially give rise to claims of novelty and inventorship even if derived from a legally protected strain. These methods produce derived strains that exist on a continuum of distinction from the originating culture, from no detectable difference (at one extreme) to slight or moderate behavioral or cultural differences (at the other). The legal question becomes the following: is there some degree of distinction at which claims of novelty could be upheld, circumventing strain IPR? The answer depends primarily upon the legal framework employed.

In the United States, two options for mushroom strain or variety protection exist. The first is the Plant Patent system unique to the United States. Although a U.S. Plant Patent confers the right to control and exploit clonal propagation of a protected variety, the system explicitly recognizes (and exempts) any demonstrable form of novelty, regardless of how obtained. By selecting a relatively trivial somatic cultural variant, or a functionally equivalent inbred descendant, an opportunist can claim novelty and potentially circumvent all rights originally conferred upon the strain's breeder. In addition, use of the protected strain as breeding stock is expressly permitted. This is not a strong form of IPR protection for the mushroom breeder.

A U.S. (or other) Utility Patent provides far stronger protection. In principle, anything that is claimed, and allowed upon examination, is covered. This potentially includes all forms of derived cultures, and all descendent cultures, as long as these elements of the invention have been explicitly claimed and allowed. In practice, however, lacking established precedents, arguments developed in support of effective claims have tended to be complex and lengthy. Internationally, different countries have developed different policies with respect to the allowability of patenting biological material.

Intermediate in the degree of IPR protection afforded is a grant of Plant Breeders' Rights (PBR), which is available in many countries (including those of the European Community, and others, but, for mushrooms, not including the United States or Canada) that are signatories to the International Union for the Protection of New Varieties of Plants (UPOV) Convention. The rights of the breeder are balanced with those of society, primarily by recognizing the "Breeders' Exception" permitting the use of any protected variety as breeding stock for use in future crossed sexual generations. Counterbalancing this is statutory recognition, applied with varying degrees of scope in different countries, that a variety directly derived from an initial (including protected) variety remains covered by any grant of PBR to the original developer, precluding exploitation by opportunistic competitors using commonplace methods of derivation. Such secondary varieties are termed Essentially Derived Varieties (EDVs) and are covered by any original grant of PBR on the initial variety.

The challenge at present is to harmonize the principles of the PBR system, under the UPOV Convention, with the realities of fungal biology, in order to provide effective protection for new mushroom strains. In general, this should be straightforward, following a set of basic principles. To be consistent with broad plant variety protections, fungal EDVs conceptually would include (1) strains derived from an initial strain without an intervening meiotic step; (2) strains wholly derived from a single initial strain or parent; and (3) strains having one or two nuclear genomes that are wholly or preponderantly derived from that of, or those of, an initial strain.

Although such principles seem clear, it is preferable to also explicitly define cases (either by methods employed or by resulting genome characteristics, and ideally by both) to achieve guidelines and practices that are unambiguous from the outset. This is crucial, especially given that definitions of EDVs codified under PBR acts in some territories invoke vague or subjective criteria, such as "retains the essential characteristics . . . " and "does not exhibit any important (as distinct from cosmetic) features that differentiate it . . . ." A fully explicit, uniformly adopted definition of EDVs does not yet exist. For mushrooms, definitions require the most careful consideration in those areas where fungal biology departs from the biology of green plants.

An additional issue for mushroom variety protection concerns the codification of the status of propagated haploid fungal individuals, including unrecombined (nongametic) haploid lines derived somatically from protected $n + n$ varieties. Such questions go to the heart of how the statutory IPR frameworks, particularly those developed to administer Plant Breeders' Rights, will approach fungal biology and mushroom varieties.

Conclusions. As an economic enterprise, the successful breeding of new strains of *A. bisporus* and other mushroom species ultimately hinges upon the ability of the breeder to obtain robust IPR protections that incorporate, at a framework level, knowledge of fungal biology. EDVs of fungi are trivially simple to produce via diverse methods. Explicit definitions of fungal EDVs are essential not only for guidance and practice under the international PBR systems, but are also vital to providing an accepted rationale for corresponding claims under U.S. and other patent law systems. The goal can be envisioned as establishing a harmonious and relatively uniform concept of sanctioned protections for both plant and fungal varieties, or strains, across legal systems, to the extent that this may be possible. The particular value of whole-genome technologies in this undertaking will be to provide unambiguous analysis of the identity and origin of strains, including EDVs and other lines, regardless of how they may have been produced.

For background information *see* AGRICULTURAL SCIENCE (PLANT); BREEDING (PLANT); DNA MICROARRAY; FUNGAL AGRIBIOTECHNOLOGY; FUNGAL BIOTECHNOLOGY; FUNGAL GENETICS; FUNGAL GENOMICS; FUNGI; GENE; GENETIC MAPPING; MUSHROOM; MYCOLOGY; PATENT in the McGraw-Hill Encyclopedia of Science & Technology.

Richard W. Kerrigan

Bibliography. R. W. Kerrigan et al., Meiotic behavior and linkage relationships in the secondarily homothallic fungus *Agaricus bisporus*, *Genetics*, 133:225–236, 1993; R. W. Kerrigan et al., Whole genome sequencing of the cultivated button mushroom *Agaricus bisporus*: History, status and applications, pp. 1–6, in *Proceedings of the 7th International Conference on Mushroom Biology and Mushroom Products (ICMBMP7)*, 2011; A. S. M. Sonnenberg et al., Breeding mushrooms: State of the art, *Acta Edulis Fungi*, 12(suppl.):163–173, 2005.

Bubble-rafting snails

Explaining cases of extraordinary evolution is a fascinating challenge for biologists: fascinating because these cases are so improbable, and challenging because they are historical, and thus cannot be replicated in the laboratory. Instead, evidence from indirect sources, including deoxyribonucleic acid (DNA), anatomy, developmental and life histories, and the fossil record, is necessary to explore how some organisms have adapted to radically different modes of life. In the past several years, for example, research has revealed how early birds and whales adapted to volant (capable of flight) and marine ecologies, respectively, via sequential modifications of ancestral traits.

Janthinids. One such unlikely evolutionary event occurred in open-ocean violet snails in the family Janthinidae. These janthinids are known as violet snails because of the vivid coloration of the shell in the more common genus, *Janthina* (5 species); shells of the rare genus *Recluzia* (2–3 species) are brown. Most snails crawl across surfaces using mucus, which is produced by glands housed in a muscular foot (inspiring the scientific name, gastropod). Janthinids, on the other hand, have evolved a unique means of living at the ocean surface: a raft of mucus bubbles. As the violet snail hangs upside down, suspended at the ocean surface, it produces a special mucus that sets quickly in water, and careful contractions of the foot wrap the mucus around a tiny pocket of air. A bubble is formed. The violet snail repeats this process throughout its adult life, incrementally adding bubbles to a trailing float (**Fig. 1a**). As janthinids have no ability to swim, they must cling to their rafts, drifting passively with oceanic surface currents in an ecosystem called the neuston.

The marine neuston is located at the surface of the Earth's tropical and temperate oceans (an area covering about 40% of the planet). It is rarely studied because of the difficulty in sampling remote locations on a global scale. The base of the neustonic food chain is formed by a mutualism between symbiotic photosynthetic algae (also known as zooxanthellae) and their cnidarian hosts, effectively representing an open-ocean equivalent of a coral reef. The cnidarians, in turn, are preyed upon by janthinids. The cosmopolitan (worldwide) distribution of janthinids means that most beachcombers can observe them, especially during periods of sustained onshore winds, when strandings of the snails are likely. From conservation research, it is known that the approximately 100 species living permanently in the neuston play an important role in open-ocean food webs, representing a major food source for loggerhead turtles, sunfishes, and seabirds.

Marine animals that live without contacting the benthos, or seafloor, are generally considered to have evolved from benthic ancestors. Logically, this makes sense in the case of janthinids: out of more than 50,000 described species of marine snails, fewer than 150 live in the open ocean (the rest are benthic), and only janthinids (7–8 species) build bubble floats.

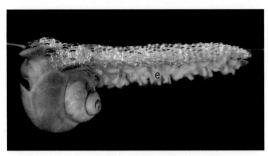

Fig. 1. (*a*) Live bubble-rafting violet snail (*Janthina exigua*) with float and egg capsules (labeled "e") [*photo courtesy of Denis Riek; www.roboastra.com*]. (*b*) Live female wentletrap (*Epifungium ulu*) with egg mass (labeled "e") and dwarf male (indicated by arrow) [*photo courtesy of Adriaan Gittenberger; www.GiMaRIS.com*].

How did the ancestors of modern-day violet snails leave the seafloor and begin rafting by the use of bubbles? In the absence of a living ancestral janthinid, finding the closest living benthic relatives of violet snails could shed light on their evolutionary transition to the neuston.

Finding the benthic sister lineage. Violet snails have been a popular topic of molluscan research since they were first illustrated in the seventeenth century. Before the advent of molecular biology, morphological comparisons between janthinids and other gastropods yielded a list of potential sister families. Based on a global collection of violet snails sampled over five years and representatives of the potential sister families, a multigene molecular DNA analysis (the first to include janthinids) has strongly supported the position that janthinids are a subset of another family of snails, Epitoniidae, commonly known as wentletraps (Fig. 1*b*; also see **Fig. 2**).

Ecologically, violet snails and wentletraps have much in common. Their developmental stages are similar. After a planktotrophic (that is, swimming and feeding) larval stage, individuals metamorphose into juveniles. Next, they develop as protandrous hermaphrodites; in other words, as individuals grow, they pass through a mature male phase before making a final switch to a mature female phase. Although this may seem strange to humans, it is not an uncommon life history among gastropods. Like violet snails, wentletraps eat cnidarians, especially benthic varieties such as corals and sea anemones. In fact,

wentletraps are cnidarian ectoparasites; they live on their prey. Female wentletraps are oviparous (egg-laying). The females build agglutinated masses of mucus and egg capsules, to which the females remain tethered, and often dwarf males are found in association with the female and egg mass complex (Fig. 1*b*). Female janthinids, on the other hand, attach their egg capsules to the underside of the bubble float. An exception is *Janthina janthina*, in which females do not produce egg capsules. Females of *J. janthina* brood larvae inside their gonads and expel live planktotrophic larvae (ovoviviparity).

Two hypotheses for the origins of rafting. The essential janthinid adaptation to a neustonic ecology is the bubble float. Therefore, any hypotheses regarding the origins of janthinid rafting must explain how the float evolved. Two hypotheses have been developed based on morphological and ecological comparisons with extant wentletraps and other marine gastropods. First, the bubble raft may derive from a modified juvenile drogue thread (juvenile drogue hypothesis; note that the term drogue refers to a sea anchor or drag device). Mucus drogue threads are found in juveniles of many other species of benthic snails, and these drogue threads are used for temporary dispersal. The snail remains attached to the drogue, and the end of the thread becomes captured by surface tension. Under this hypothesis, in the ancestral condition, juvenile janthinids would construct floats. Second, the float may represent a modified wentletrap egg mass (egg mass hypothesis). Epitoniid egg masses consist of sand, mucus, and egg capsules. The egg capsules hatch in stages; thus, some egg masses may have a range of capsule stages, including empty husks. In an intertidal species, these husks could trap air, potentially providing temporary buoyancy for both the egg mass and the attached female. Under this hypothesis, the float is a female trait; hence, in the ancestral condition, only females would build floats. Neither hypothesis proposes an immediate historical transition to the neuston. Initial rafting of either juveniles or adult females most likely was temporary, which could have been advantageous in escaping deteriorating local conditions

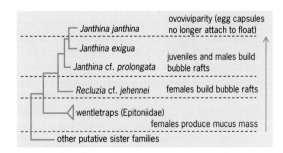

Fig. 2. Evolutionary relationships of janthinids and epitoniids. (*Left*) Reduced evolutionary tree showing the results of a multigene DNA analysis of violet snails (top four species), wentletraps, and other putative benthic sister lineages. Violet snails are a subset of wentletraps. (*Right*) Character states of wentletraps and violet snails showing sequential adaptation of violet snails to a neustonic ecology. The arrow indicates that the character state applies to the particular species and all species listed above it.

or could have provided opportunistic access to incoming neustonic cnidarian prey. In either scenario, the ability to create air bubbles is the final step in adapting to a neustonic ecology.

Juvenile drogue or modified egg mass? To test the two hypotheses, development and life history traits within Janthinidae were examined. Interestingly, the two janthinid genera, *Janthina* and *Recluzia*, each supported a different hypothesis. *Janthina* is consistent with the juvenile drogue hypothesis. Early observations of janthinids after metamorphosis report juveniles attached to a mucus stalk with a clutch of terminal bubbles—a drogue thread. Later in development, juveniles build complete floats. *Recluzia*, on the other hand, is consistent with the egg mass hypothesis. Because *Recluzia* is so rare, no unique ecological information about it had ever been reported, and it was assumed to have the same ecology as *Janthina*. However, a recent study found juvenile *Recluzia* attached to the float of a conspecific (same-species) female, as well as finding further evidence that the association may continue through the male stage of the life cycle of *Recluzia*. In *Recluzia*, float building is a female trait. Not only does this life history resemble that of wentletraps, but it also helps explain how *Recluzia* can persist while remaining rare. Many marine species (for example, barnacles and anglerfish) with low population densities maximize the potential of each reproductive encounter when small males remain associated with larger females. In *Recluzia*, the association is temporary until the small males enter the female stage and can float autonomously.

With one genus supporting each hypothesis, a further question arises: which one (*Janthina* or *Recluzia*) is more ancestral? A morphological comparison of *Janthina*, *Recluzia*, and extant epitoniids showed that *Recluzia* and epitoniids share several character states that are independent of float formation. For example, both *Recluzia* and epitoniids have gravity-sensing organs, called statocysts, which have been lost in *Janthina*. These data support *Recluzia* as the more ancestral janthinid genus. Therefore, the weight of evidence indicates that the bubble rafts of violet snails derive from wentletrap egg masses (egg mass hypothesis).

Adaptation to life at sea. Although an evolutionary reconstruction shows that the janthinid float began as a reproductive structure (an egg mass), the molecular phylogeny shows that the reproductive functions of the float have been subsequently lost in derived violet snail species. This loss is associated with an increase in ecological prevalence (Fig. 2), which may be interpreted as adaptive. In wentletraps, the egg mass functions as an attachment point for dwarf males and as a substrate for egg capsules. In *Recluzia*, the function of a float is added. In oviparous species of *Janthina*, the function of attachment for juveniles and males is lost because they build floats autonomously. Finally, in the ovoviviparous *J. janthina*, the most common species, the float no longer functions as a substrate for egg capsules; it is used only as a float.

Conclusions. Through a sequential modification of float construction techniques, janthinid snails have become increasingly successful at exploiting neustonic resources. As indicated by other cases of ostensibly improbable evolution for new modes of life, this study underscores the importance of data from transitional lineages, either fossil, as in research on birds and whales, or extant, as for *Recluzia* cf. *jehennei*, in reconstructing these compelling evolutionary histories.

For background information *see* ANIMAL EVOLUTION; BUOYANCY; CAENOGASTROPODA; ECOSYSTEM; FOOD WEB; GASTROPODA; MARINE ECOLOGY; MOLLUSCA in the McGraw-Hill Encyclopedia of Science & Technology.   Celia K. C. Churchill; Diarmaid Ó Foighil

Bibliography. C. K. C. Churchill et al., Females floated first in bubble-rafting snails, *Curr. Biol.*, 21:R802–R803, 2011, DOI:10.1016/j.cub.2011.08.011; C. K. C. Churchill, E. E. Strong, and D. Ó Foighil, Hitchhiking juveniles in the rare neustonic gastropod *Recluzia* cf. *jehennei* (Janthinidae), *J. Molluscan Stud.*, 77:441–444, 2011, DOI:10.1093/mollus/eyr020; C. M. Lalli and R. W. Gilmer, *Pelagic Snails: The Biology of Holoplanktonic Gastropod Mollusks*, Stanford University Press, Stanford, CA, 1989; W. F. Ponder et al., Caenogastropoda, in W. F. Ponder and D. R. Lindberg (eds.), *Phylogeny and Evolution of the Mollusca*, University of California Press, Berkeley, 2008.

Cancer cell metabolism

A normal cell and a cancer cell differ in a number of ways. One important difference is seen in their metabolism. Cancer cells have an altered metabolism in comparison to normal cells: Specifically, they rely on the less efficient pathway, called aerobic glycolysis, to produce energy instead of the more efficient pathway, called oxidative phosphorylation (OXPHOS). At first, this might seem counterintuitive, given that cancer cells are fast growing and therefore should require more energy than normal cells. However, the basis for the alteration stems from three basic needs of a cancer cell: energy production via rapid adenosine triphosphate (ATP) generation; increased biosynthesis of macromolecules; and maintenance of cellular redox status (adequate cellular levels of reducing agents such as NADPH or glutathione to inactivate damaging free radicals).

Warburg effect. Normal cells are exposed to a constant supply of nutrients and have control systems that prevent aberrant cell proliferation when there are more nutrients than the cell requires. These control systems usually involve protein growth factors in the extracellular milieu whose levels or activities are enhanced by nutrient levels in the environment. However, cancer cells can overcome the growth-factor requirement for cell proliferation by acquiring mutations in the genes encoding the plasma membrane receptors for the growth factors [such as the epidermal growth factor (EGF) receptor], allowing for continued uptake of nutrients and thus increased cell survival and cell growth even in the absence of the extracellular growth factors. In the 1920s, Otto Warburg observed that cancer cells metabolize the nutrient glucose differently than normal cells do. When normal cells have an abundance of oxygen, they use the oxidative phosphorylation pathway to metabolize glucose to carbon dioxide by oxidation of glycolytic pyruvate; this occurs in the mitochondria. Cancer cells, on the other hand, produce lots of lactate in the presence of oxygen, and this process is called aerobic glycolysis. What this means in terms of energy production (in the form of ATP) is that cancer cells are forced to use glycolysis for ATP synthesis, at a production rate of 2 ATPs per glucose molecule, versus instead of using oxidative phosphorylation, which produces approximately 36 ATPs per glucose molecule. Even under conditions of normal levels of oxygen, 50% of cancer cells will still produce energy via glycolysis, with the rest using the mitochondrial respiratory chain. This means that cancer cells are required to convert most of the incoming glucose to lactate and uptake glucose at a very high rate to meet the demands of energy, biosynthesis, and redox maintenance. However, why the switch? What might be driving the switch is that ATP is generated at a higher rate in glycolysis, and therefore potentially more ATP is made as long as glucose supplies are not limited. Also, glycolysis might be advantageous under hypoxic (low-oxygen) conditions during early tumor development. Finally, glycolysis might allow for a biosynthetic advantage; that is, carbon can be shuttled to make key biomolecules in various pathways.

Metabolic changes in cancer: oncogenes, tumor suppressors, and the tumor microenvironment. When oncogenes (genes that contribute to the conversion of a normal cell into a cancerous cell) are activated or tumor-suppressor genes are inactivated, there is a potential for uncontrolled cell proliferation and tumorigenesis. Likewise, these same oncogenes and tumor suppressors can affect cellular targets, leading to metabolic changes that allow for cancer cell survival. It is these various oncogenes and tumor suppressors that are responsible for the mechanistic changes leading to the metabolic changes in cancer. Only a few of the many oncogenes and tumor-suppressor genes that contribute to these metabolic pathways are mentioned below.

Protein kinase B/Akt (PKB/Akt) is a serine/threonine protein kinase that normally functions to phosphorylate downstream targets that promote cell proliferation, cell survival, glucose uptake, and hexokinase activity. The PKB/Akt pathway plays a role in metabolic conversion by shifting cancer cells toward aerobic glycolysis, through growth-factor stimulation or amplification of certain mutations. It does this in three different ways. First, the gene expression of glucose transporters is induced, which increases glucose usage by the cancer cells. Second, glucose usage is further increased, via PKB/Akt, by inducing the increased expression of hexokinase II (HKII), which phosphorylates glucose to produce glucose-6-phosphate, which is the first step

in glycolysis. Finally, there is induced expression of another glycolytic enzyme, phosphofructokinase, which converts fructose-6-phosphate to fructose-1,6-diphosphate, which is a later step in the glycolytic conversion of glucose to pyruvate. Another protein kinase, mTOR, which functions downstream of Akt and promotes cell growth, is being investigated as a therapeutic target for cancer treatment.

The MYC oncogene encodes a transcription factor that enhances the activity of many genes whose products contribute to the control of growth, proliferation, and apoptosis (programmed cell death). There are different ways by which MYC stimulates glycolysis. Several glycolytic enzymes are activated by overexpression of MYC. These include HKII, glyceraldehyde-3-phosphate dehydrogenase, enolase 1, pyruvate kinase, and lactate dehydrogenase. MYC can also increase production of reactive oxygen species (chemically reactive oxygen-containing molecules, including free radicals), which could damage mitochondrial DNA (mtDNA), impair respiratory-chain activity, and thus allow cancer cells to use aerobic glycolysis.

Mutations in the tumor-suppressor gene, *p53*, are the most commonly observed genetic alteration in human cancer. Overall, p53 plays major roles in maintaining the integrity of the genome, cell-cycle arrest, and apoptosis, as well as metabolism, by regulating glycolysis and mitochondrial respiration. Glycolysis is regulated by p53 through two different targets: TP53-induced glycolysis and apoptosis regulator (TIGAR) and phosphoglycerate mutase (PGM). Loss of p53 and therefore TIGAR expression can lead to an increase in glycolysis by misregulation of a TIGAR target gene, fructose-2,6-bisphosphate, whose activity normally suppresses glycolysis by reducing the level of an intermediate in that metabolic pathway. Likewise, p53 normally suppresses the expression of PGM, which catalyzes the conversion of 3-phosphoglycerate to 2-phosphoglycerate; however, upon p53 loss in cancer cells, PGM's activity is increased and glycolysis is enhanced.

In addition to genetic changes (mutations in oncogenes and tumor suppressors), the microenvironment of the tumor plays a role in changes to the metabolism of cancer cells. For instance, the vasculature of a tumor is structurally and functionally different from normal tissue. This results in variations in oxygenation, pH, and concentrations of glucose that are distributed in the tumor, consequentially contributing to conditions that are stressful for the tumor cell. A stress response elicits the activation of various molecular pathways that allow for an adaptive mechanism to survive under the adverse conditions.

Diagnostic procedures and therapeutic perspectives. A shift in metabolic activity is observed in many (but not all) cancer cells, resulting in high rates of glucose uptake and glycolysis in cancer cells. This increase in glucose usage provides an opportunity to image tumors using a glucose analog, 2-(18F)-fluoro-2-deoxy-D-glucose (FDG), by positron emission tomography (PET). FDG–PET allows for the utilization of these well-defined metabolic abnormalities of malignant tumors for clinical diagnosis. FDG is transported across the cell membrane by the same carrier molecules as glucose and subsequently is phosphorylated by hexokinase. In contrast to glucose, FDG-6-phosphate cannot be further metabolized; furthermore, because it is a highly polar molecule, it cannot diffuse out of the cell and remains trapped intracellularly. Because of this trapping mechanism, there is a steadily increasing FDG concentration in metabolically active cells (potential tumor cells), resulting in a high contrast between tumor cells and normal tissues. In addition to glucose, other important metabolic nutrients (for example, glutamine and glutamate) are currently being tested as new PET agents for cancer. These potential agents could allow another layer of molecular insight with regard to how tumors are functioning and could improve overall detection of tumors that cannot be visualized using FDG–PET. Other methods such as magnetic resonance spectroscopy (MRS) are being evaluated for their usefulness in measuring specific metabolites in various cancers.

Small-molecule inhibitors and other therapeutics that inhibit the processes and enzymes involved in metabolic programming can also have an effect on tumorigenesis by limiting cancer cell bioenergetic flow, stopping growth, inducing apoptosis, and blocking angiogenesis and invasion. These small-molecule inhibitors that could target multiple pathways might be excellent means for investigating their therapy potential. Cancer cells that have activated oncogenes (for example, Ras, Her2, and Akt) or that lack tumor suppressors (for example, TSC1/2, LKB1, and p53) are susceptible to apoptosis under conditions of low glucose or through the use of glycolysis inhibitors. For instance, the HK2 inhibitor, 3-bromopyruvate, induces apoptosis in hepatocellular carcinomas. Similarly, inhibition of pyruvate dehydrogenase kinase 1 (PDK1) can restore activity of pyruvate dehydrogenase (PDH), which would direct pyruvate to the OXPHOS pathway while also triggering apoptosis (by an unknown mechanism) in cancer cells. However, one of the major problems is that there is currently a lack of enzymatic inhibitors that target metabolic steps critical for cancer growth and survival that are specific enough to be used for human therapeutics.

Conclusions. It is now known that abnormal metabolism is a feature of many human diseases, and continual discoveries highlight the importance of metabolism not only in normal biology but in the disease state. However, only by dissecting all the various pathways and understanding how they work together will it be possible to move forward with new advances in the field. Likewise, it is becoming increasingly important that this knowledge be translated into novel diagnostic and therapeutic approaches, particularly in cancer.

For background information *see* BIOLOGICAL OXIDATION; CANCER (MEDICINE); CELL (BIOLOGY); CHEMOTHERAPY AND OTHER ANTINEOPLASTIC DRUGS; ENERGY METABOLISM; METABOLISM; MITOCHONDRIA; ONCOGENES; ONCOLOGY; TUMOR SUPPRESSOR GENES in the McGraw-Hill Encyclopedia of Science & Technology.   Anna M. Puzio-Kuter

Bibliography. R. A. Cairns, I. S. Harris, and T. W. Mak, Regulation of cancer cell metabolism, *Nat. Rev. Cancer*, 11:85–95, 2011, DOI:10.1038/nrc2981; R. G. Jones and C. B. Thompson, Tumor suppressors and cell metabolism: A recipe for cancer growth, *Genes Dev.*, 23:537–548, 2009, DOI:10.1101/gad.1756509; A. M. Puzio-Kuter, The role of p53 in metabolic regulation, *Genes Cancer*, 2:385–391, 2011, DOI:10.1177/1947601911409738; D. A. Tennant, R. V. Durán, and E. Gottlieb, Targeting metabolic transformation for cancer therapy, *Nat. Rev. Cancer*, 10:267–277, 2010, DOI:10.1038/nrc2817; D. A. Tennant et al., Metabolic transformation in cancer, *Carcinogenesis*, 30:1269–1280, 2009, DOI:10.1093/carcin/bgp070.

Cancer stem cells

The cancer stem cell (CSC) concept posits that malignant tumors, like many healthy tissues, can be hierarchically organized at the cellular level and that subpopulations of CSCs, representing the apex of such discernible hierarchies, are exclusively responsible for tumor initiation and propagation. Hierarchical tumor organization denotes the concept that only tumor-initiating CSCs within cancers possess the capacity for generating further tumor-propagating CSC progeny, whereas more differentiated bulk populations of tumor cells with limited replicative potential, also derived from CSCs through asymmetric cell division, do not give rise to CSCs and are not capable of indefinite tumor propagation. According to the CSC model of tumor growth and progression, a CSC is therefore defined as a cancer cell that possesses (1) a capacity for prolonged self-renewal that inexorably drives tumor growth and (2) a capacity for differentiation, that is, the production of heterogeneous lineages of daughter cells that represent the bulk of the tumor mass, but are dispensable for tumor propagation. Consequently, CSCs are thought by proponents of the CSC concept to lie at the root of primary tumor formation and growth and to be also responsible for tumor dissemination, metastasis formation, therapeutic resistance, and posttreatment recurrence. The rapidly expanding interest in CSCs in the field of oncology is based on their exceptional promise as paradigm-shifting novel targets for cancer therapy.

Analysis of cancer stem cells. CSCs can be experimentally identified according to their ability to recapitulate the generation of a continuously growing tumor, and can be isolated or enriched, for example, using antibodies to specific cell-surface proteins preferentially expressed by CSCs. The gold-standard assay in this regard involves transplantation of patient-derived, purified marker-defined candidate CSC populations (or, as controls, tumor bulk populations negative for the candidate CSC marker) into immunodeficient recipient mice capable of accepting human tumor xenografts as a result of a lack of immunological rejection (note that a xenograft is defined as the transplant of an organ, tissue, or cells to an individual of another species). Tumorigenicity rates are then determined, and candidate marker-defined CSC populations (or marker-negative bulk populations) are subsequently reisolated from established heterogeneous primary experimental tumors and regrafted to secondary experimental hosts (and potentially tertiary or higher-order hosts), with tumorigenic capacity of candidate CSC populations in serial tumor xenotransplantation assays demonstrating prolonged self-renewal capacity (that is, the first CSC-defining feature). Additionally, established tumors are analyzed histopathologically to assess whether a candidate CSC population has produced a phenocopy of the cellular heterogeneity displayed by the original patient tumor through differentiation (that is, the second cardinal CSC-defining feature).

In addition to cell sorting–based CSC identification strategies, marker-specific genetic lineage tracing of cancer subpopulations in competitive tumor development models, designed to trace individual cancer cell fates on concurrent xenotransplantation of CSCs and tumor bulk populations, has been used to provide confirmatory evidence for hierarchical tumor organization and to further document CSC-specific phenotype and function. In such lineage tracing experiments, CSCs and tumor bulk populations, each carrying genetically encoded distinct fluorescent labels, are cotransplanted at naturally occurring relative frequencies, with subsequent evaluation of their respective tumorigenic capacity in growing tumors and of their relative ability to give rise to the diverse tumor cell types within developed tumors. Such assays have provided the additional advantage of revealing novel interactions between CSCs and tumor bulk populations, such as CSC fusion with more differentiated tumor cells as a potential mechanism of resistance-associated gene transfer (for example, through epigenetic modifications as a result of nuclear fusion), or CSC secretion of extracellular matrix and growth factors required for efficient tumor initiation and expansion of tumor bulk populations. For example, CSCs in human malignant melanoma

secrete laminin, which acts as a pro-proliferative mitogen (a compound that stimulates cells to undergo mitosis) for melanoma bulk populations. Such assays might also possess the capacity to reveal CSC-activating signals originating from tumor bulk populations or from the CSC niche [for example, vascular endothelial growth factor (VEGF)–induced activation of VEGF receptor 1 (VEGFR1)–expressing CSCs in human malignant melanoma]. Of note, such cellular interactions, which may be operative in naturally occurring cancers, can evade detection when only purified subpopulations of cancer cells are studied.

Cancer subpopulations exhibiting the defining criteria of CSCs have been identified in hematological as well as solid human malignancies, including leukemias, brain cancer, breast cancer, colorectal cancer, head and neck cancer, liver cancer, malignant melanomas, ovarian cancer, and pancreatic cancer. The molecular markers shown to identify CSCs in these human malignancies are summarized in the **table**. Importantly, significant correlations between the abundance of molecularly identified CSCs in clinical tumor tissues and patient parameters such as tumor stage, tumor progression, therapeutic resistance, tumor recurrence, and adverse clinical outcome have been documented in several human malignancies. These findings have provided critical evidence for the diagnostic and therapeutic relevance of the CSC concept.

Tumor initiation and progression in human-to-mouse cancer xenotransplantation models, and as a result determined CSC identity and estimated frequency, can vary significantly depending on the experimental methodology used for identification studies. These observations have given rise to controversies regarding the CSC concept, with some critics questioning the very existence of CSCs and attributing differences in tumor initiation among cancer subpopulations within heterogeneous tumors to xenograft host characteristics alone. However, most recent studies leading to the successful discovery of clinically relevant CSC populations in experimental tumor xenotransplantation assays have carefully addressed the choice of the experimental model system in light of its relevance to naturally occurring human

Molecular markers of cancer stem cells

CSC marker	Human malignancy	Therapeutic target development
ABCB5	Malignant melanoma, hepatocellular carcinoma, colorectal cancer	Preclinical
CD13	Hepatocellular carcinoma	Preclinical
CD19	B-cell malignancies	Clinical
CD20	Malignant melanoma	Clinical
CD24	Pancreatic cancer, lung cancer, bladder cancer	Preclinical
CD34	Hematopoietic malignancies	—
CD44	Breast cancer, head and neck cancer, hepatocellular carcinoma, pancreatic cancer	Preclinical
CD90	Hepatocellular carcinoma	—
CD133	Brain cancer, colorectal cancer, hepatocellular carcinoma, lung cancer, malignant melanoma	Preclinical
CD166	Colorectal cancer	Preclinical
EpCAM/ESA	Colorectal cancer, pancreatic cancer	Clinical

disease. Methodological variations hereby include patient tumor dissociation techniques, tissue site of xenotransplantation, cografting of tumor-promoting growth factors, and observation length for the assessment of tumor formation, all of which may influence the outcome of tumor xenotransplantation assays.

In addition, intrinsic xenograft host characteristics, particularly microenvironmental factors and host immune status, significantly influence CSC phenotype and function. For example, exogenously added extracellular matrix factors normally produced by CSCs in the tumor microenvironment (TME) can markedly enhance tumorigenic potential in human malignant melanoma, and cografting of growth factor–producing human TME-derived fibroblasts can significantly enhance experimental tumorigenic growth of human breast cancer. Tumor host immune status represents a further potential source of experimental variation: Although naturally occurring cancers in human patients typically develop in the context of relatively intact immunity, with some immunogenic cancers capable of inducing potent antitumor immune responses or of developing specific mechanisms of immune evasion, it is well documented that successful establishment of human experimental tumors in xenogeneic animal hosts requires a certain degree of host immunodeficiency in order to facilitate tumor establishment through prevention of xenograft rejection. As a consequence, and furthermore as a result of recently demonstrated novel CSC-specific immunoevasive functions, tumor initiation frequencies and resultant CSC frequency estimates vary depending on host immune status. For example, melanoma stem cells possess the capacity to preferentially inhibit T-cell activation and to support induction of immunological tolerance-conferring regulatory T cells, pointing to specific roles for these CSCs in the evasion of antitumor immunity and in cancer immunotherapeutic resistance. Nevertheless, studies using fully immunodeficient murine xenograft hosts have unequivocally identified critical intrinsic CSC mechanisms of tumorigenic growth that are independent of these immunoevasive CSC properties, leading to confirmation of CSC phenotype and function in several human malignancies. To further refine experimental model systems with regard to their ability to faithfully reflect naturally occurring tumor growth and progression in human patients, humanized xenotransplantation assays, involving grafting of patient-derived tumor cells to chimeric mice already transplanted with syngeneic (genetically identical) immune cells originating from the same patient, are currently being developed. Such assays hold exceptional promise for further enhancing clinical relevance of the CSC concept, based on their ability to model patient-specific mechanisms of syngeneic antitumor immunity or tumor immune evasion.

Comparison of cancer stem cells and normal stem cells. Intriguingly, the CSC-defining functions of prolonged self-renewal and differentiation capacity parallel those of tissue-specific normal stem cells. Additionally, CSCs and normal stem cells share the

properties of relative quiescence and bidirectional interactions with the cellular constituents of the stem cell niche. For example, stem cells induce proliferation of their differentiated progeny through secretion of soluble factors and are reciprocally induced to undergo cell division in response to signals received from the niche. Moreover, CSCs and normal stem cells share immunomodulatory functions conferring immune privilege (the ability to escape immune detection and destruction). These functions, and the associated signaling pathways and cell-matrix and cell-cell interactions, are not only essential for the regulatory balance between self-renewal and differentiation and between quiescence and proliferation of normal stem cells in tissue development, homeostasis, and repair, but they also contribute critically to CSC-driven tumor initiation (for example, self-renewal), malignant progression (for example, formation of vessel-like channels within the tumor, which is a process known as vascular mimicry differentiation), and therapeutic resistance (for example, cellular quiescence and immune evasion). Indeed, they represent a potential Achilles' heel for more effective, CSC-targeted cancer therapy.

Despite the broadly shared properties of CSCs and tissue-specific normal stem cells, normal stem cells do not necessarily represent the cells-of-origin of cancers. Although select murine genetic models of carcinogenesis have revealed that, in the case of intestinal tumors, some could only be induced when particular carcinogenic mutations were introduced into tissue-specific stem cells but not their differentiated progeny, it is well established that cancer can also be induced in differentiated tissue cells on introduction of genetic mutations. These findings indicate that either normal stem cells or differentiated cells may serve as cells-of-origin for clinical cancer development, depending on the cell type targeted for oncogenic transformation. Genetic mutations may hereby serve to activate stem cell-specific self-renewal programs in differentiated cells. Additionally, cell fusion with tissue-specific normal stem cells, and the epithelial–mesenchymal transition (EMT), a developmental pathway by which differentiated adherent epithelial cells with apicobasal polarity and limited migratory and differentiation potential acquire mesenchymal cell–like mobility and developmental plasticity, represent further potential mechanisms for acquisition of CSC properties by mutated differentiated cells. This is supported by observations of the emergence of CSC phenotype on cell fusion in malignant melanoma model systems and by demonstrations of enhanced tumorigenicity of transformed human mammary epithelial cells following induction of the EMT.

The Vogelstein genetic model of carcinogenesis implies that cancer results from gradual accumulation of genetic mutations resulting in malignant transformation of a physiological cell. These genetic alterations eventually lead to the development of a founder cancer cell, which, according to the Weinberg theory of carcinogenesis, acquires self-sufficiency in growth signals, insensitivity to growth inhibitory signals, evasion of programmed cell death (apoptosis), limitless replicative potential, sustained angiogenesis, and capacity for tissue invasion and metastasis. Indeed, accumulative evidence suggests that these hallmarks of cancer are not necessarily equally shared across all malignant cells in heterogeneous tumors, but that they preferentially characterize aggressive subpopulations of cells within the cancer, which may coincide with those subpopulations defined as CSCs. Thus, although CSC generation may proceed according to relatively established mechanisms of carcinogenesis at the cellular level, the novelty of the CSC concept and its profound clinical implications relate to functional hierarchical tumor organization during tumor initiation, progression, resistance, and recurrence, independent of cancer cell-of origin considerations or genetic clonal evolution.

Implications. The existence of CSCs has important implications for the efficacy of current forms of conventional cancer therapy because CSCs can intrinsically express or specifically activate molecular mechanisms that confer increased resistance to chemotherapy or ionizing radiation. Such mechanisms of resistance include cellular quiescence and altered cell cycle checkpoint control, impaired pro-apoptotic pathways, augmented DNA damage repair responses, and reduced drug accumulation resulting from enhanced expression of drug efflux transporters. CSC chemoresistance has been reported in human leukemias, various cancers (for example, brain, breast, colorectal, and pancreatic cancers), and malignant melanomas, and CSC radioresistance has been reported in brain, breast, and colorectal cancers. As a result, conventional chemotherapeutic and radiotherapeutic regimens can fail to eradicate tumor-perpetuating CSC subsets, despite significant killing effects on the bulk population of tumor cells, and CSC persistence might ultimately cause clinical tumor recurrence. Thus, there is an urgent need for consistent eradication of CSC populations for more effective cancer therapy.

Based on the findings that CSCs can be responsible for clinical therapeutic resistance and can be associated with malignant progression in human patients, CSC-targeted treatment strategies represent promising novel approaches aimed at tumor eradication, with anticipated reduced risks of cancer relapse and potential metastasis. Initial proof-of-principle for the potential therapeutic utility of the CSC concept was provided in human melanoma through the demonstration that selective antibody-dependent cell-mediated cytotoxicity (ADCC)–mediated killing of certain CSC subpopulations is sufficient to inhibit experimental tumor growth. This proof-of principle has since been extended to additional solid tumors, including, for example, brain cancer and hepatocellular carcinoma. Several CSC-targeted strategies harbor promise for increasing cancer therapeutic efficacy. These include direct strategies such as CSC killing/ablation through targeting of prospective CSC markers, CSC killing through chemoresistance/radioresistance reversal, disruption of CSC-specific self-renewal pathways, and CSC differentiation therapy (see **illustration**, left panel). Additionally, indirect strategies that target CSC interactions with the CSC niche might serve to inhibit protumorigenic CSC functions for therapeutic benefit—for example, CSC-driven tumor angiogenesis or vasculogenic mimicry, CSC immunomodulatory or immunoevasive functions, or CSC/TME interactions that rely on CSC secretion of extracellular matrix and growth factors (see illustration, right panel). Evidence for the potential therapeutic benefit of all of these approaches has already been provided in preclinical studies.

Importantly, some CSC-targeted approaches are currently also being translated to the clinic. These include, for example, clinical trials of CSC self-renewal inhibition through Hedgehog pathway antagonists in basal cell carcinoma and colorectal cancer, or through Notch pathway antagonists in leukemia and breast cancer. Further ongoing trials involve monoclonal antibody–mediated epithelial cell adhesion molecule (EpCAM) inhibition in breast, colon,

Direct CSC targeting

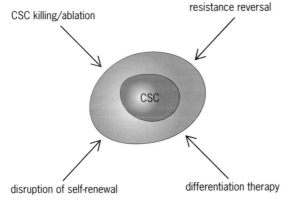

Indirect CSC targeting

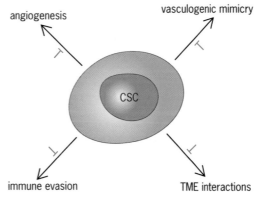

Cancer stem cell (CSC) targeting strategies. Both direct CSC targeting strategies (*left*) and indirect CSC targeting strategies (*right*) have been demonstrated in preclinical model systems to hold promise for cancer therapy (TME = tumor microenvironment).

ovarian, gastric, lung, and gastrointestinal cancers, and clinical use of CD20-targeted antibody-mediated CSC-ablation in malignant melanomas. It is important to recognize that clinical end points of CSC-targeted therapies may differ from more conventional forms of therapy targeted at bulk populations of cancer cells within tumors, with potentially delayed effects of CSC ablation on initial shrinkage of malignant lesions based on the relative rarity of the CSC population, but with anticipated more sustained effects on long-term tumor progression or recurrence, as a result of ablation of the tumor-driving force. Indeed, combination strategies that involve both CSC-targeted treatments and bulk tumor cell killing through conventional forms of therapy might ultimately prove to be most effective for achieving tumor eradication and clinical cures.

Conclusions. In summary, the concept that malignant subpopulations of CSCs, defined by prolonged self-renewal and differentiation capacity, initiate and perpetuate hierarchically organized tumors has been validated experimentally in both hematopoietic and solid human malignancies. Identification of CSCs has led to groundbreaking discoveries unraveling the biological features and functions of these cancer subpopulations in tumorigenesis and cancer progression, and it has allowed identification of novel mechanisms of therapeutic resistance that represent attractive targets for cancer therapy. Experimental proof-of-principle for the potential therapeutic utility of CSC targeting has been established in diverse preclinical tumor models, and the first CSC-targeted strategies are currently being translated to the clinic. These developments shift, for the first time, therapeutic emphasis toward eradication of the most aggressive tumor cell populations within human malignancies. As a result, they hold great promise for significantly improved patient outcomes and, ultimately, achievement of cancer cures.

For background information *see* CANCER (MEDICINE); CELL (BIOLOGY); CELL DIFFERENTIATION; CELL LINEAGE; CHEMOTHERAPY AND OTHER ANTINEOPLASTIC DRUGS; CLINICAL PATHOLOGY; ONCOLOGY; STEM CELLS; TRANSPLANTATION BIOLOGY in the McGraw-Hill Encyclopedia of Science & Technology.   Natasha Y. Frank; Markus H. Frank

Bibliography. N. Barker et al., Crypt stem cells as the cells-of-origin of intestinal cancer, *Nature*, 457:608–611, 2009, DOI:10.1038/nature07602; M. F. Clarke et al., Cancer stem cells—perspectives on current status and future directions: AACR Workshop on cancer stem cells, *Cancer Res.*, 66:9339–9344, 2006, DOI:10.1158/0008-5472.CAN-06-3126; E. R. Fearon and B. Vogelstein, A genetic model for colorectal tumorigenesis, *Cell*, 61:759–767, 1990, DOI:10.1016/0092-8674(90)90186-I; N. Y. Frank, T. Schatton, and M. H. Frank, The therapeutic promise of the cancer stem cell concept, *J. Clin. Invest.*, 120:41–50, 2010, DOI:10.1172/JCI41004; N. Y. Frank et al., VEGFR-1 expressed by malignant melanoma-initiating cells is required for tumor growth, *Cancer Res.*, 71:1474–1485, 2011, DOI:10.1158/0008-5472.CAN-10-1660; S. D. Girouard and G. F. Murphy, Melanoma stem cells: Not rare, but well done, *Lab. Invest.*, 91:647–664, 2011, DOI:10.1038/labinvest.2011.50; D. Hanahan and R. A. Weinberg, The hallmarks of cancer, *Cell*, 100:57–70, 2000, DOI:10.1016/S0092-8674(00)81683-9; S. A. Mani et al., The epithelial-mesenchymal transition generates cells with properties of stem cells, *Cell*, 133:704–715, 2008, DOI:10.1016/j.cell.2008.03.027; T. Reya et al., Stem cells, cancer, and cancer stem cells, *Nature*, 414:105–111, 2001, DOI:10.1038/35102167; T. Schatton et al., Identification of cells initiating human melanomas, *Nature*, 451:345–349, 2008, DOI:10.1038/nature06489; T. Schatton et al., Modulation of T-cell activation by malignant melanoma initiating cells, *Cancer Res.*, 70:697–708, 2010, DOI:10.1158/0008-5472.CAN-09-1952; B. J. Wilson et al., ABCB5 identifies a therapy-refractory tumor cell population in colorectal cancer patients, *Cancer Res.*, 71:5307–5316, 2011, DOI:10.1158/0008-5472.CAN-11-0221; B. B. Zhou et al., Tumour-initiating cells: Challenges and opportunities for anticancer drug discovery, *Nat. Rev. Drug Discov.*, 8:806–823, 2009, DOI:10.1038/nrd2137.

Cancer treatment using magnetic nanoparticles

Patients suffering from any cancer will benefit from early diagnosis and more effective therapeutic modalities. The currently used cancer therapies lack selectivity to target malignant cells, leading to side effects responsible for prolonged and expensive patient recovery or tumor relapse. Thus, advanced drug delivery providing more effective and localized therapeutic modalities is required in order to minimize the harmful toxicity related to the high doses of chemotherapeutic drugs currently administrated. Moreover, it is well known that the efficiency of cancer treatments increases when combining different treatment modalities. In this context, nanomedicine appears as an emerging and multidisciplinary area of knowledge that aims to provide personalized and more efficient tools to detect and remove diseases such as cancer. One of the main challenges faced by nanomedicine is to deliver the cancer treatment at the right place, at the right dose, and at the right moment. In this sense, superparamagnetic iron oxide nanoparticles (SPION) appear to be suitable platforms to act as nanovectors, that is, carriers that selectively transfer infective agents into a targeted body or organ. Because of their small size ($\sim$10 nm), SPION can easily enter into cells with typical sizes of $\sim$10 μm. SPION may simultaneously combine different magnetic, chemical, or biological functionalities for selective cancer cell targeting and removal. Thus, SPION have shown potential for such applications as contrast agents, advanced nanocarriers for selective drug delivery into targeted cells, and intracellular heat sources. Many efforts have been undertaken to functionalize SPION with anticancer agents or antibodies in order to inhibit basic cell functions and to selectively target tumors. At the same time, SPION dissipate magnetic energy into heat when exposed to alternating magnetic fields due to their superparamagnetic properties. This interesting SPION behavior makes it possible to launch localized thermal shock waves into cells, inducing cancer cell death by blocking the cell's DNA repair mechanisms. Thus, SPION would result in a minimally invasive, selective, and efficient therapeutic approach combining different modalities to reinforce the tumor elimination.

SPION synthesis. In order to engineer SPION for cancer treatment, it is necessary to produce SPION with optimal structural (such as a narrow size distribution) and magnetothermal properties. Then, the surfaces of the SPION should be modified to have functional chemical groups, surface charge, high colloidal stability, and biocompatibility. Different physical and chemical methods are currently used to produce SPION with different chemical compositions, sizes, and morphological shapes. However, thermal decomposition of an iron precursor using surfactants and organic solvents is nowadays the method that provides SPION with the best characteristics for biomedical applications: wide size range from 3 to 50 nm, uniform size distribution and morphology (**Fig. 1**), highly crystalline particles, high magnetization values (~70 emu/g), and high magnetically-induced heat power values at accessible alternating magnetic fields (~100 W/g at 100 kHz and 30 mT). At the same time, this chemical route allows surface modification processes to favor the dispersion

of SPION in aqueous media, for example by ligand exchange. Thus, the surface coating provides SPION with hydrophilic characteristics that favor their colloidal stability in water, chemical groups to anchor anticancer agents and antibodies, and, finally, a surface charge that strongly influences their cell uptake. Coated SPION are not individually dispersed in water but form clusters with a given hydrodynamic size. Thus, the cluster colloidal stability and the control of the aggregate size (ideally less than 100 nm) are mandatory requirements for future clinical applications of SPION. Much effort has been invested in the large-scale production of biocompatible SPION with high heat dissipation power values, and a large variety of chemical groups available onto the nanoparticle surface.

SPION functionalization. The multimodal therapeutic approaches for SPION are based on their capabilities to simultaneously play the role of nanocarrier and intracellular heating mediator. Whereas the latter is intrinsically related to the magnetic core of SPION, the former requires a "bottom-up" assembly strategy to transform individual SPION into a platform with customized chemical and biological functionalities. This self-assembling process starts with the particle coating providing surface charge or chemically active groups for "attaching" antibodies or drugs. Two strategies exist for self-assembling drug delivery SPION based on covalent binding or electrostatic adsorption. The choice depends mainly on two criteria: the preservation of the biological

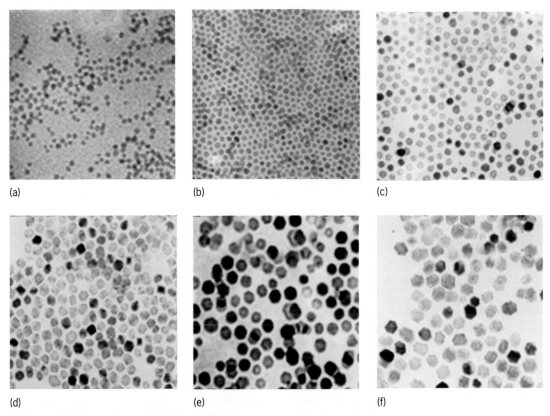

(a)   (b)   (c)

(d)   (e)   (f)

Fig. 1. Representative transmission electron microscopy images (same scale) of SPION synthesized by thermal decomposition with different sizes ranging from 5 to 17 nm. (a) 5 nm. (b) 6 nm. (c) 9 nm. (d) 12 nm. (e) 15 nm. (f) 17 nm. The high SPION size uniformity is notable. (*Image kindly supplied by Dr. Alejandro G. Roca*)

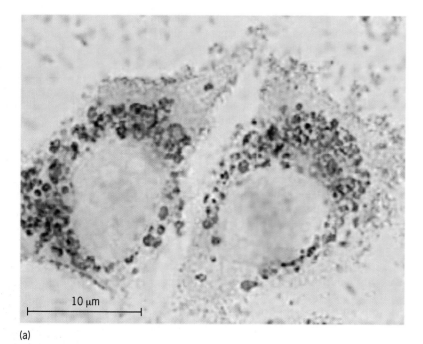

(a)

(b)

Fig. 2. Bright-field optical microscopy images of (*a*) HeLa cells after 24-h incubation with dextran-coated SPION, and (*b*) HeLa cells incubated with 0.1 mg Fe ml-1 aminodextran-coated SPION after cell division. (*Adapted from Á. Villanueva et al., The influence of surface functionalization on the enhanced internalization of magnetic nanoparticles in cancer cells, Nanotechnology, 20:115103, 2009*)

Cell uptake. The SPION engineering toward biological functionalities supplies "virus-inspired capabilities" for acting as nanovectors that overcome biological barriers or deliver therapeutic agents into targeted cells. Recent work has shown that the entry of SPION into cells by endocytosis mechanisms is favored by surface charge (**Fig. 2**). SPION would selectively acquire aspects of these "virus abilities" when entry mechanisms are mediated by membrane receptor interactions via antibodies or peptides, which is of high interest for therapeutic purposes. Iron is a chemical element of life, being present in every living cell and necessary for the production of hemoglobin, myoglobin, and certain enzymes. The essential role of iron favors the degradation and metabolization of iron oxide particles in animal models, leading to a low SPION toxicity which significantly diminishes when particles are coated with polymeric and polysaccharide materials such as polyethylene glycol. Thus, the presence of coated SPION in cells, even at high loads, does not affect the cell morphology and structure or basic functions such as cell division.

Magnetic functionalities. The magnetic properties of SPION provide interesting functionalities for detection and elimination of cancer cells. This is due to their superparamagnetic properties, which include high saturation magnetization values and lack of magnetic order (that is, zero magnetization at zero magnetic field) at physiological temperatures in the absence of magnetic fields under quasistatic conditions, that is, for alternating magnetic field frequencies of the order of millihertz (**Fig. 3**). These characteristics are the main reason for using SPION in biomedical applications because the increase of the particle aggregate size via long-range magnetic interactions is negligible. However, an "opening" of the SPION magnetization cycle (that is, magnetic order) occurs under "dynamic conditions" for

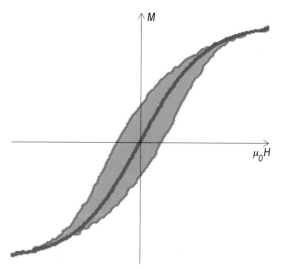

Fig. 3. Schematic representation of magnetization cycles, plotting magnetization (*M*) against magnetic field (*H*), for superparamagnetic systems under quasistatic (blue line) and dynamical (red line) conditions. (μ_0 is the magnetic constant.)

activity of the biomolecule after anchoring onto the SPION surface, and the bond stability under the given physiological conditions. The covalent binding strategy is based on organic chemical engineering requiring the design of linkers with appropriated terminals to anchor anticancer molecules—such as drugs, small interference RNA, peptides, or antibodies—onto iron oxide surfaces. More complex linkers can be engineered for programmed release in order to cast anticancer agents off into cancer cells.

alternating magnetic fields with frequencies of tens of kilohertz. This hysteretic character is related to irreversible magnetization processes observed at high frequencies, leading to magnetic energy dissipation into "heat." The commonly used physical magnitude for evaluating the calorific power of SPION subjected to alternating magnetic fields is the specific absorption rate (SAR), equal to $A \cdot f$, where A is the area of the hysteresis cycle and f is the frequency of the alternating magnetic field. Calorimetric measurements show that the opening of the hysteresis cycles (that is, the SAR value) is strongly influenced by the frequency and amplitude of the alternating magnetic field. The use of SPION as heating mediators to generate "intracellular hyperthermia" above 42°C (107.6°F) has shown significant progress as a therapeutic modality against cancer. This is demonstrated by the start of clinical trials testing SPION as intracellular hyperthermia generators and the development of equipment to generate uniform alternating magnetic fields on a human body by MagForce AG. Very promising SPION biomedical perspectives are related to the successful hyperthermia clinical trials in solid tumors requiring minimally invasive therapeutic modalities such as glioblastoma multiforme. Recent in vivo reports show interesting results regarding significant tumor volume reduction when the natural immune response is stimulated by using SPION as advanced nanocarriers. However, more efforts are needed to assess the synergies when combining different therapeutic modalities, including hyperthermia.

The proof of concept of magnetic nanoparticles for cancer treatment seems to be achieved in the laboratory. It is now time to clarify other important questions regarding the clinical use of SPION, such as the SPION toxicity, the biodistribution, the biodegradation, or the immune response in animal models. The answers to these questions depend on technological progress. Novel devices are needed for quick detection and quantification of SPION in biological samples (tissues, blood, and so forth), or for applying uniform alternating magnetic fields in different targeted organs.

For background information *see* CANCER (MEDICINE); MAGNETISM; MAGNETIZATION; NANOPARTICLES; ONCOLOGY in the McGraw-Hill Encyclopedia of Science & Technology.

Francisco José Teran; María del Puerto Morales; Ángeles Villanueva; Julio Camarero; Rodolfo Miranda

Bibliography. P. Guardia et al., Water-soluble iron oxide nanocubes with high values of specific absorption rate for cancer cell hyperthermia treatment, *ACS Nano.*, 6(4):3080–3091, 2012, DOI:10.1021/nn2048137; K. M. Krishnan, Biomedical nanomagnetics: a spin through new possibilities in imaging, diagnostics and therapy, *IEEE Trans. Magn.*, 46:2523–2558, 2010, DOI:10.1109/TMAG.2010.2046907; K. Maier-Hauff et al., Efficacy and safety of intratumoral thermotherapy using magnetic iron-oxide nanoparticles combined with external beam radiotherapy on patients with recurrent glioblastoma multiforme, *J. Neuro. Oncol.*, 103:317–324, 2011, DOI:10.1007/s11060-010-0389-0; R. Mejías et al., Dimercaptosuccinic acid-coated magnetite nanoparticles for magnetically guided *in vivo* delivery of interferon gamma for cancer immunotherapy, *Biomaterials*, 32:2938–2952, 2011; K. Riehemann et al., Nanomedicine—challenge and perspectives, *Angew. Chem. Int. Ed.*, 48:872–897, 2009, DOI:10.1002/anie.200802585; A. G. Roca et al., Progress in the preparation of magnetic nanoparticles for applications in biomedicine, *J. Phys. D: Appl. Phys.*, 42:224002, 2009, DOI:10.1088/0022-3727/42/22/224002; T. D. Schladt et al., Synthesis and biofunctionalization of magnetic nanoparticles for medical diagnosis and treatment, *Dalton Trans.*, 40:6315–6343, 2011, DOI:10.1039/C0DT00689K.

Charged-particle transmission through insulating capillaries

The interaction of charged particles with the inner walls of electrically insulating capillaries is a topic of considerable interest that was first investigated about 10 years ago and has been studied extensively in the last decade. The studies represent an interdisciplinary area of nanoscience lying at the intersection of atomic physics and materials science, although the emphasis of the work done to date has been on atomic physics. The investigations pose questions of a fundamental nature, as well as offering potential applications in science, medicine, and technology. In addition to the study of interactions that take place between the incident charged particles and the capillary walls, charged-particle transmission in capillaries can be used to produce micrometer- and submicrometer-sized particle beams for use in applications such as controlled surface modification or manipulation of samples, precise sample analysis, and medical diagnosis or treatment, even at the level of a single cell. Applications involving medical uses have been explored quite extensively in recent years. Capillaries might also be used as passive optical elements to focus or collimate charged-particle beams on the micrometer and submicrometer scales in applications in which high-density interaction regions are required, such as the production of crossed or merged beams to be used in studies of interactions with photons.

Guiding. It has been found that slow-moving (as defined later), highly charged ions pass through micrometer- and nanometer-sized insulating capillaries without changing their charge and without losing energy, while faster ions can change their charge and lose energy. The former results were contrary to well-known results for slow ions incident on metallic capillaries, which change their charge state readily. In the case of incident electrons, for which there can be no charge change, energy loss has been observed in some cases, although with somewhat faster speeds, but in other cases little loss occurs. The effect of slow ions traversing electrically insulating capillaries with no charge change and no energy loss is called guiding, a term that refers to the fact that the capillary walls prevent direct collisions and therefore "guide" the ions as they pass through. The term guiding has also been applied to electrons transmitted through capillaries. The large fraction transmitted with no energy loss or change in the charge state indicates that the inner walls of the capillaries collect charges in a self-organizing manner such that electrostatic repulsion inhibits close collisions. This effect, discovered accidentally, has attracted the attention of several groups worldwide, and investigators have focused attention primarily on the fundamental processes involved, with some work on potential applications. Since the discovery of guiding, the properties of charged particles transmitted through capillaries, including the transmission efficiency, incident energy dependence, angular distributions of the transmitted particles, and time evolution, have been studied.

Capillary structures. The capillaries either are cylindrically shaped or have a tapered shape, with diameters of a few to about 100 micrometers (1 μm = 10^{-6} m = 1 millionth of a meter), while smaller capillaries have diameters ranging from tens to a few hundred nanometers (1 nm = 10^{-9} m = 1 billionth of a meter). In typical experiments, the capillary lengths are 100 times longer than these diameters, thereby presenting a long insulating surface for the interactions of charged particles and imposing strict restrictions on how particles move through them. The capillaries have been made of glass in the case of micrometer-sized capillaries, which are large and sturdy enough to be used as self-supporting single capillaries, and of highly ordered arrays of polyethylene terephthalate (PET, or Mylar), polycarbonate (PC), silicon dioxide (SiO_2), and alumina (Al_2O_3) for nanocapillaries, which are too small to be used as single capillaries. Instead, these nanocapillaries are structured in a thin foil of the material, with the many capillaries highly parallel to one another with different densities, and the foil can be mounted to expose the capillaries with their axes parallel to an incident beam of charged particles. The outer surface of the capillaries is covered with a metallic conducting coating, typically gold, to carry away excess charges that are deposited on the surface. The self-supporting micrometer capillaries preclude the possibility of interference from the electric fields of nearby neighboring capillaries; this is unlike the case with nanocapillaries, where there are many nearby capillaries that can affect the field, and thereby the transmission of ions, in a given capillary of interest. The glass sample or the foil sample is then mounted in a goniometer, a precise instrument for manipulation, to rotate the angle and change the position with respect to the incident beam of charged particles.

Charge-up process. In **Fig. 1**, a schematic of the guiding process for positive ions or electrons is shown. Starting with an initially uncharged capillary (Fig. 1a), ions (or electrons) that first strike the surface at a small angle mostly will be "buried" in that wall of the capillary, hence building up the charge on the capillary wall at that point. When an ion strikes the surface, secondary electrons are also produced

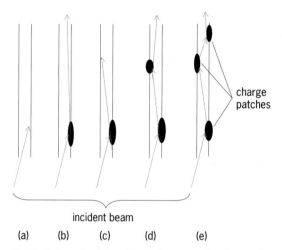

Fig. 1. Schematic of the charge-up process for a beam of ions or electrons entering a capillary. (*a*) Initial charge-up. (*b*) Deflection of the beam to the capillary exit in the opposite direction from the incident beam. (*c*) Additional charge-up, showing deflection resulting in the beam hitting the opposite surface. (*d*) Formation of a second charge patch, again directing the beam to the exit, this time in the same direction as the incident beam. (*e*) Formation of a third patch, resulting in transmission nearly along the capillary axis.

and released; these have the effect of increasing the amount of the positive charge deposited in the patch. For incident electrons, the same process can occur, although electrons are more mobile and the charge is somewhat negated by the production of secondary electrons, with the result that the charge does not build up as quickly. As the charge builds up, it reaches a value where it is sufficient to deflect further incoming particles and prevent them from interacting with the surface at that point. Hence, some of the ions or electrons will be directed toward the capillary exit, causing transmission opposite to the direction in which the incident beam strikes the capillary (Fig. 1b). As the charge continues to grow, the incident beam can be deflected enough to strike the opposite wall instead of exiting (Fig. 1c). However, as the charge on the opposite wall builds up, the beam will again be deflected toward the exit, this time causing transmission in the same direction as the incident beam (Fig. 1d). This process may continue with the formation of a third patch, causing the beam to more nearly approach the capillary axis (Fig. 1e). Because of the angles over which beam transmission has been observed (less than about 10°), at least three charge patches can be formed, eventually producing a stable equilibrium for the transmission along the capillary axis. The third charge patch is likely to be near the capillary exit and thus affects the beam as it leaves the capillary. Charge forming at the capillary exit has been found to affect the angular width of the transmitted ions, with the width increasing as the charge increases.

Experimental setup. To study the properties of charged-particle transmission, the capillary is placed at an angle with respect to the incident beam (**Fig. 2**). Here, ψ represents the tilt angle of the capillary with respect to the beam, and θ is the angle of observation of the transmitted ions or electrons.

Transmission has been observed for tilt angles of about $-10°$ to $+10°$ with respect to the incident beam, with the designation of negative angles being arbitrary to show only the symmetry of the transmission. Two regions of interest may be noted. There is the region of direct transmission, for which the particles have a direct line of sight to the exit, determined by the geometry of the capillary; this is typically $0.5°$ for a capillary with an aspect ratio (length/diameter) equal to about 100. For angles larger than this, the particles have essentially no direct line of sight through the capillary and must rely on an indirect mechanism, such as guiding or scattering, to get through the capillary. When the angular distributions of the transmitted particles are studied, it is found that the centroid of the observation angle coincides with the capillary tilt angle over the range of indirect transmission (**Fig. 3**). This is taken to be the key to guiding by charged particles.

Charged-particle sources, energies, and detection. Ions ranging from a few kiloelectronvolts (1 keV = 10^3 electronvolts) of total energy to MeV/u (1 MeV/u = 10^6 eV/atomic mass unit) energies have been studied, and electron beams with incident energies of 200–1000 eV have also been used. These investigations have been carried out with ion sources that produce slow, highly charged ions, called ECR (electron cyclotron resonance) or EBIT (electron beam ion trap) sources, or with higher-energy accelerator facilities that can produce the desired beams. Electrons have been obtained from filament sources called electron guns. To measure ion transmission, position-sensitive detectors are typically used, while for electron detection, a spectrometer that disperses the final energy of the electrons is utilized. Studies have been done in Europe (Austria, France, Germany, Hungary, the Netherlands, Serbia, and Sweden), in Asia (China and Japan), and in the United States (Kalamazoo, Michigan, and Oak Ridge, Tennessee).

Experimental results. For slow ions with a few kiloelectronvolts of total energy, it is found that the ions transmit with no charge change and no energy loss, whereas higher-energy ions approaching 1 MeV/u do not guide, but instead are transmitted directly or by small-angle scattering and can lose energy. Experimental and theoretical studies have been conducted for slow ($\sim$1–10 keV) positive ions incident on nanocapillary arrays of polymer (PET and PC), SiO_2, and alumina foils, as well as single microcapillaries of straight and tapered glass. Transmission has also been investigated for negative ions with energies of approximately 10 keV incident on alumina nanocapillaries, but no guiding effect was seen.

Time evolution. Time evolution of multiple charge patches for slow positive ions has been observed manifesting itself in spatial variations of about $\pm 1°$ of the transmitted beam as charge-up occurs. Following the variations with angle, the beam approaches the capillary axis as charge equilibrium is reached. Such variations indicate that transmission and guiding of ions through capillaries is a dynamic process that reaches equilibrium with characteristic times that can be quite long and dependent on the current

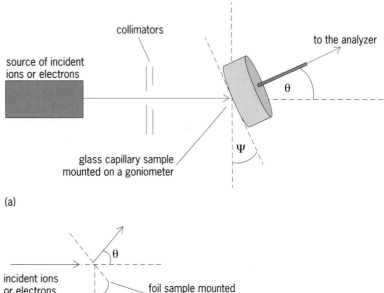

(a)

(b)

Fig. 2. Experimental setups for (*a*) a single glass microcapillary, and (*b*) a foil array of nanocapillaries. The tilt angle ψ and the observation angle θ are shown in each case.

(charge) that is put into the capillary. This formation of multiple charge patches has proven instrumental in understanding the charge-up of the capillaries and the subsequent transmission of particles through them.

Electrons. Electrons, which have been studied far less than positive ions, have been found to transmit less efficiently at total energies up to about 1 keV.

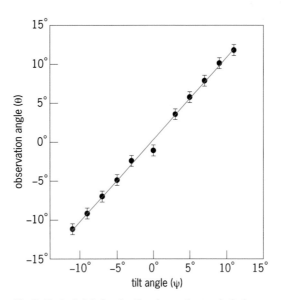

Fig. 3. Typical plot showing the observation angle θ of transmitted particles versus the capillary tilt angle ψ (both defined in Fig. 2). It is seen that the observation angle nearly equals the tilt angle. Any deviations can be attributed to mechanical offsets in the positioning of either angle. The line through the data obtained from a linear fit shows the relationship between the angles. The linear relationship of these angles outside the range (about $\pm0.5°$) where the beam can go directly through the capillary is taken to indicate the presence of particle guiding.

This is attributed to the tendency of their higher mobility to prevent charge-up and secondary electron emission, but they still "guide," although the electrons have been found to lose an appreciable amount of energy—up to 20% of the incident energy in some cases. For electrons of 200–350 eV incident on alumina nanocapillary foils, negligible energy loss was observed for traversing electrons up to tilt angles of about 10°, indicating that the inner capillary walls were charged by electrons in the same manner as for positive ions. Of particular interest are inelastic processes for both electrons and ions, which provide a means of studying the dynamics of charged-particle interactions with insulators through the study of the effect on transmission and guiding.

Secondary electron emission. Also of interest is the importance of secondary electron emission, with energies in the range of 0–50 eV, giving rise to additional positive surface charging for ions and tending to cancel the effect of charging for electrons. While these electrons are often mentioned in reports of transmission in capillaries, there is typically no analysis of them, and they may not even be measured. In any event, secondary electron emission has not been systematically studied to date for either incident electrons or ions traversing capillaries. Certain questions arise with regard to this. For example, where do the secondary electrons go when they are liberated? What kind of electric field do they set up because of their distribution, and how do they modify the existing field? Because of the small dimensions of the capillaries, it cannot be assumed that secondary electrons escape. These questions are important for determining the mechanisms for transmission and guiding.

In summary, this research involves investigation of the fundamental interactions and study of the possible applications of both slow and fast ions and electrons impacting on and traversing along the axes of nanometer- and micrometer-sized electrically insulating capillaries. The mechanisms by which low-velocity, highly charged ions traverse highly ordered structures of insulating nanometer capillaries in a foil as well as single, micrometer-size, self-supporting capillaries have been studied quite extensively for different geometrical configurations and types of insulating materials, and much is known. However, for incident electrons and faster ions, less is known about either the nature of the transmission mechanisms or the characteristics of the transmitted particles. The results of various investigations show that the transmission of electrons is substantially less than that of slow ions. This result points to a fundamental difference between ions and electrons in terms of the ability to deposit and maintain charge. The work also has implications for the production of sources of very small charged-particle beams, and this aspect is being actively studied.

For background information *see* ATOMIC STRUCTURE AND SPECTRA; CHARGED-PARTICLE INTERACTIONS; ION-SOLID INTERACTIONS; ION SOURCES; SECONDARY EMISSION in the McGraw-Hill Encyclopedia of Science & Technology.   John A. Tanis

Bibliography. S. Das et al., Inelastic guiding of electrons in polymer nanocapillaries, *Phys. Rev. A*, 76:042716 (5 pp.), 2007, DOI:10.1103/PhysRevA.76.042716; T. Ikeda et al., Focusing of charged particle beams with various glass-made optics, *J. Phys. Conf. Ser.*, 88:012031 (9 pp.), 2007, DOI:10.1088/1742-6596/88/1/012031; P. Skog, HQ. Zhang, and R. Schuch, Evidence of sequentially formed charged patches guiding ions through nanocapillaries, *Phys. Rev. Lett.*, 101:223202 (4 pp.), 2008, DOI:10.1103/PhysRevLett.101.223202; N. Stolterfoht et al., Transmission of 3 keV Ne^{7+} ions through nanocapillaries etched in polymer foils: Evidence for capillary guiding, *Phys. Rev. Lett.*, 88:133201 (4 pp.), 2002, DOI:10.1103/PhysRevLett.88.133201.

Chemicals from renewable feedstocks

Renewable resources have been used as industrial feedstocks throughout human history, and only in the past two centuries have fossil fuels become the primary sources of carbon-based chemicals. Today, fossil resources are used to supply 96% of the synthetic organic chemicals produced in refineries and chemical production plants. The rapidly increasing population has resulted in a growing demand for chemicals to satisfy the needs of the simultaneously improving living standards. Consequently, one of the most important challenges is sustainable development; that is, how to meet the needs of the present without compromising the ability of future generations to meet their needs. Because the depletion of fossils fuels will occur sooner or later, it is important to develop sustainable processes for the production of carbon-based chemicals. Renewable feedstocks will be the preferred resources for carbon-based chemicals in the future. A renewable feedstock can be defined as a natural resource that can replenish itself in a limited time, preferably within several months, although years, or at maximum a few decades, may be acceptable as well. Biomass is produced in nature by using sunlight to convert carbon dioxide and water into organic compounds, of which 75% is carbohydrate-based with an empirical formula of $C_6H_{12}O_6$ [reaction (1)].

$$6CO_2 + 6H_2O \xrightarrow{h\nu} C_6H_{12}O_6 + 6O_2 \qquad (1)$$

The simplest carbohydrates are the naturally occurring five- and six-carbon monosaccharides, which can be linked together by glycosidic bonds in various combinations to form di-, oligo-, and polysaccharides. The other main components are tannins, resins, fatty acids, and inorganic salts, as microcomponents. Various other substances can be found in biomass such as vitamins, dyes, flavors and aromatic essences, and certain oils and proteins. To use the different components as raw materials or intermediates, appropriate and economical processing technologies should be available in biorefineries by integrating the essential physical, chemical, and biological processes to convert natural raw materials to products such as basic chemicals, intermediates, fine chemicals, and pharmaceuticals.

Although the commercialization of efficient biomass conversion technologies has been slow to emerge, intensive research has begun to identify molecules that could be produced from biomass by efficient processing and conversion technologies. In 2004, a list of potential chemical products was developed that could be produced in a biorefinery, provided the appropriate processes are available or developed. A biorefinery flowchart was developed to show biomass-based chemicals (**Fig. 1**).

Lignocellulose requires pretreatment before converting it into chemicals because of its complex structure, in which cellulose and hemicellulose are encapsulated in lignin by hydrogen and covalent bonds. The particular type of pretreatment depends on the type of lignocelluloses, and it is common for two or more pretreatment processes to be required. Several physical, physicochemical, chemical, and biological methods have been developed for this treatment. In the physical methods, the biomass is mechanically processed to increase the surface area of the material accessible to further treatments via a combination of processes such as chipping, grinding, and milling. Of the physicochemical pretreatment methods, steam explosion is the most widely used, whereby high-pressure steam acts as an acid at high temperature and interacts with the biomass to achieve full or partial hydrolysis of the cellulosic component. In the chemical pretreatment methods, dilute acids, such as sulfuric acid (H_2SO_4) or hydrochloric (HCl) acid, alkali peroxides, organic solvents, or ozone can be used to degrade lignin and hemicellulose.

Biological pretreatments using cellulases also could be considered. The cellulases are usually isolated from fungi (such as *Trichoderma reesei*) and bacteria (such as *Cellulomonas fimi*) and are actually a complex of three different classes of enzymes. This complexity, in combination with the heterogeneous nature of the process (that is, soluble enzymes acting on insoluble substrates), makes cellulases challenging to understand and therefore more difficult to rationally design.

Syngas. Synthesis gas, or synthetic gas (syngas), consists of carbon monoxide and hydrogen in various ratios. The treatment of biomass with air, oxygen, and/or steam to produce gaseous products containing CO, H_2, CO_2, CH_4, and N_2 in various proportions is called gasification. Biomass gasification was practiced in the early decades of the twentieth century and played a major role during World War II, reportedly keeping approximately a million vehicles operational in Europe. The complexity of gasification chemistry is partially due to the varying composition of biomass. For simplicity, biomass can be considered as a pseudocompound $CH_wO_xN_yS_z$, where w, x, y, and z are the ratios of hydrogen, oxygen, nitrogen, and sulfur relative to carbon, respectively. Biomass also contains potassium, sodium, and other alkali metals that can cause slagging

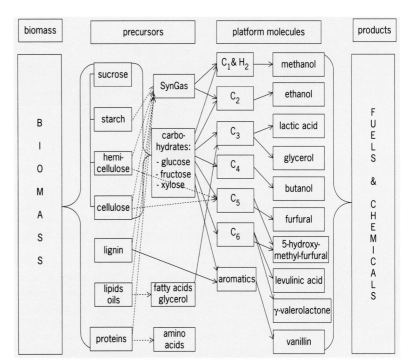

Fig. 1. Biorefinery flowchart. Precursors are the constituents collected and separated from the biomass that are used to produce the platform molecules, which can then be converted into products.

and fouling problems in conventional gasification equipment.

Biomass gasification processes combine physical steps, such as drying, and employ different chemical processes such as pyrolysis, partial oxidation, steam gasification, and reduction. At 100–200°C, water is removed. But at higher temperature in the absence of oxygen, pyrolysis occurs, leading to the formation of gaseous, liquid, and solid products [reaction (2)].

$$CH_wO_xN_yS_z \xrightarrow{\text{heat}} C + \text{tars/oils} + aNH_3 + bN_2 + cH_2S + dH_2O + eH_2 + fCH_4 \quad (2)$$

Because of the complex chemical composition of the biomass, both the carbon and the nitrogen- and sulfur-containing molecules are converted. A portion of the biomass is transformed into condensable hydrocarbon tars, oils, methane, and solid carbon residue or charcoal. Because of the endothermic nature of pyrolysis, energy must be supplied from external heating sources or from exothermic reactions in the reactor. The outlet gas composition of the reactor depends on the composition of the biomass, gasification process, and gasifying agent.

The largest use of syngas is for hydrogen generation via steam-methane reforming. Methanol, dimethyl ether, acetic acid, formaldehyde, methyl *tert*-butyl ether, ethanol, and olefins are also products of syngas-based technologies. Hydrogen and light alkanes, primarily methane, can be produced by aqueous-phase reforming of biomass-derived oxygenates. Industrial hydrogen production is mainly used for ammonia synthesis and petrochemical reactions. Hydrogen can also be used as a reactant in

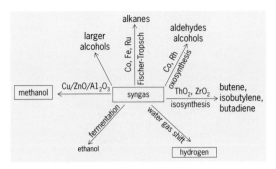

Fig. 2. Products derived from syngas.

biomass conversion strategies. Alkanes and alkenes can be produced from syngas by the Fischer-Tropsch synthesis (**Fig. 2**).

Methanol. A methanol economy has been proposed. Since the technology to produce it from syngas is well established, its availability depends on the conversion of biomass to CO and H_2. A wide variety of catalysts with different compositions of copper and zinc oxides on aluminum-silica supports are commercially available. However, the miscibility of methanol with water combined with its acute toxicity, if ingested, could be a serious issue.

Ethanol. Ethanol is a key liquid produced from biomass, with the largest producers being the United States using corn and Brazil using sugar cane, producing 40 and 25 billion liters in 2009, respectively. The ethanol in the United States is almost entirely produced from cornstarch [reaction (3)]. By contrast,

$$nC_6H_{12}O_6 \xrightarrow{\text{yeast}} 2nC_2H_5OH + 2nCO_2 \quad (3)$$

the main soluble carbohydrate form in sugarcane is the disaccharide sucrose, which can be hydrolyzed and fermented more directly.

The prospects of improving first-generation sugarcane processes are good, with an estimated doubling of production possible using genetic modification of the sugarcane and a 10-fold increase in production from increasing the area cultivated. A potentially more sustainable production method for bioethanol uses cellulose-containing biomass. One of the main differences between current technologies and cellulosic ethanol is the need for the sugars to be released from the lignocellulose. At present, this is very expensive and reducing this expense will be possibly one of the most significant challenges. The ethanol produced from fermentation is in aqueous solution and must be separated by distillation (azeotropic or otherwise), molecular sieves, or membrane pervaporation. Ethanol has been used as an oxygenate in fuels, replacing the methyl *tert*-butyl ether (MTBE) that is being phased out because of groundwater contamination issues.

Lactic acid. Lactic acid can be produced from renewable feedstocks such as corncob, corn stover, wood hydrolyzate, wheat straw hemicellulose, rice and wheat bran, apple pomace, and sugarcane bagasse. Hexoses can be easily fermented by lactic-acid bacteria (LAB), such as *Lactobacillus delbrueckii*. The primary use of lactic acid is in the production of poly(lactic acid) [PLA], a biodegradable plastic (**Fig. 3**). It can be directly polymerized, but the process is more effective when lactic acid is converted to low-molecular-weight prepolymers and then depolymerized to the lactide. A wide range of catalysts have been used for lactide polymerization. The properties of PLA are similar to that of polystyrene and include storage resistance to fatty foods and dairy products equivalent to poly(ethylene terephthalate), excellent barrier properties for flavors and aromas, and good heat sealability. Lactate esters are "green" solvents that can be produced by esterification of lactic acid. Catalytic reduction of lactic acid leads to propylene glycol, and dehydration of lactic acid gives acrylic acid. Biodegradable fibers can be generated from lactic acid, which can be used in biomedical materials or in furniture production.

Glycerol. Glycerol, a triol, is nontoxic, edible, and biodegradable. It could provide important environmental benefits as a new platform chemical. Glycerol is one of the building blocks of fatty-acid esters, or triglycerides, which are important constituents of vegetable and animal fats and oils. The glycerol content of fats and oils varies between 8–14 wt%, depending on the proportion of free fatty acids and on the chain-length distribution of the fatty-acid esters. To obtain glycerol, the oils and fats must be hydrolyzed. Glycerol is also a by-product of biodiesel manufacture. Approximately 75% of the U.S. supply of glycerol is derived from natural sources, and the remaining 25% is produced synthetically.

Glycerol can also be produced via the hydrogenation of cellulose, starch, or sugars, initially producing sorbitol followed by further conversion to mixtures of glycerol and glycols [reaction (4)]. For example, a 15–40 wt% sorbitol solution in water can be

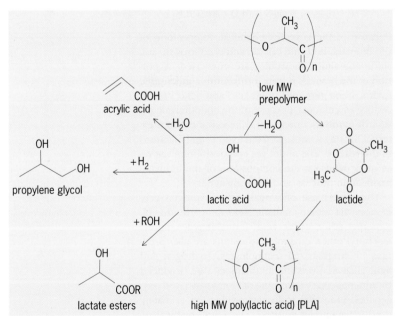

Fig. 3. Products derived from lactic acid.

catalytically hydrocracked to glycerol and glycol products using a nickel catalyst. Ruthenium catalysts have been used to efficiently convert sorbitol to glycerol, although the formation of both ethylene and propylene glycols could also be observed.

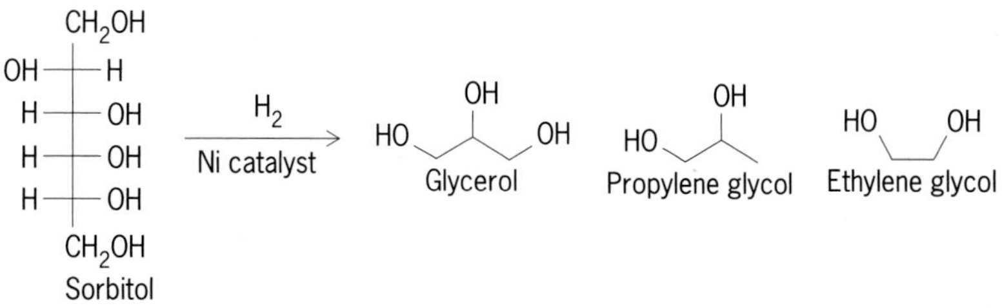

Glycerol offers a very large number of opportunities for chemical production, and it could become one of the more important building blocks in the biorefinery (**Fig. 4**). Glycerol has about 2000 different applications in the food, cosmetic, and pharmaceuticals industries. For example, glycerol is used as a solvent for the preparation of concentrated flavor extracts from natural products such as vanillin or spices. Selective oxidation of glycerol leads to a broad family of derivatives that could serve as new chemical intermediates or as components of new branched polyesters or nylons. By hydrogenolysis, glycerol can be converted to propylene and ethylene glycols, which are mainly used as antifreeze agents but can also be used to make polyesters.

Butanol. Butanol is usually referred to as an advanced biofuel produced by carbohydrate fermentation. Unlike ethanol, butanol is less miscible with water and its boiling point is higher so its processing is slightly different. Since the focus of this review is on chemicals, it will not be discussed further.

Furfural. Furfural is a platform chemical and is the only heterocyclic compound that is produced in large scale (300,000 tonnes/year) by the acid-catalyzed dehydration of pentoses, such as xylose. Sulfuric acid is the most frequently used acid with the highest possible furfural yield being only 70% with the remainder being poorly characterized polymeric by-products known as humins.

Furfural can be used as a solvent or as a starting material for the synthesis of a series of derivatives including furfuryl alcohol, furoic acid, furan, tetrahydrofuran, 2-methyltetrahydrofuran, and related resins (**Fig. 5**).

5-Hydroxymethylfurfural (HMF). 5-Hydroxymethylfurfural can be isolated as a pale yellow solid, and is regarded as being quite reactive. It occurs naturally in foods such as honey, vegetables, coffee, and other beverages in small amounts, and is one of the constituents of aroma in liquors. It can be derived from hexoses by acid-catalyzed dehydration and is a platform molecule for a variety of applications such as lubricants and polymers.

HMF formation has been studied with a range of catalysts in wide variety of solvents such as dimethyl sulfoxide (DMSO), ionic liquids, supercritical fluids, and water. The formation of humins has been a problem as well as other side reactions, including the hydrolysis of HMF to yield an equimolar mixture of levulinic and formic acids. Acetylation, reaction with alcohols, reaction with halogens, or oxidation of hydroxyl group of HMF result in the formation of esters, ethers, halides, or dialdehydes, respectively. The formyl group can also undergo secondary reactions, such as oxidation or reduction, and those lead to the formation of acids or other substituted furans, respectively (**Fig. 6**).

(4)

Levulinic acid. Levulinic acid has been produced from agricultural wastes or cellulose and hemicellulose by the same technology since the 1940s. Initial hydrolysis is achieved with mineral acids (HCl or H_2SO_4) at around 100°C to produce hexoses, with treatment at higher temperatures (200°C) for the

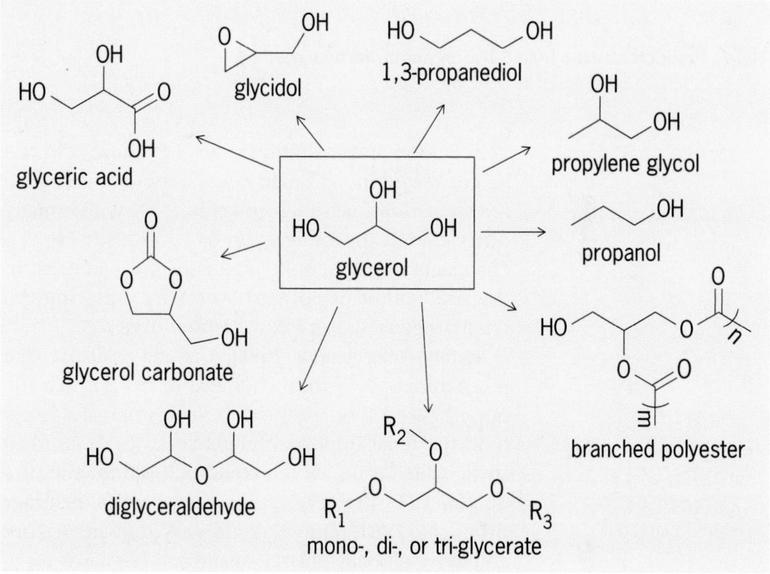

Fig. 4. Some products derived from glycerol.

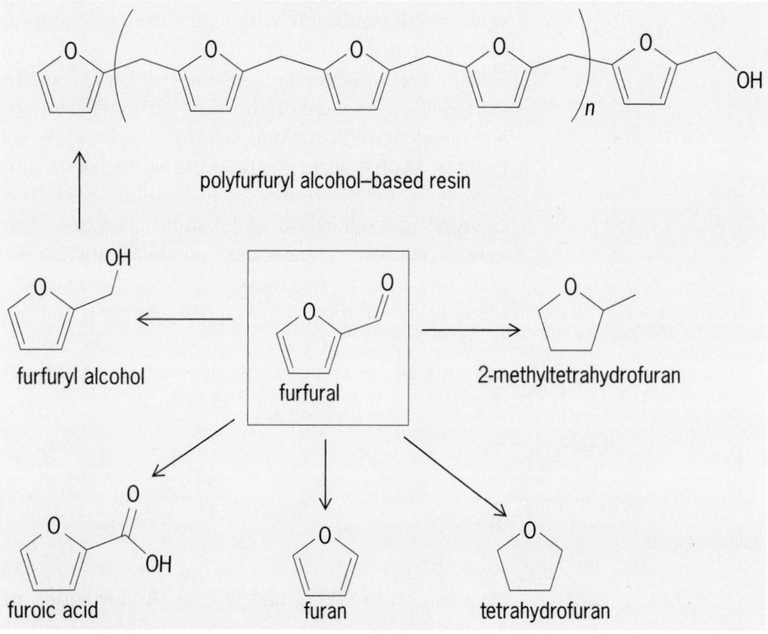

Fig. 5. Selection of products derived from furfural.

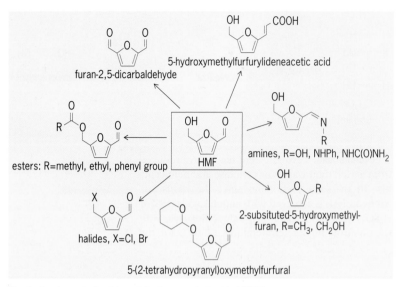

Fig. 6. Products derived from 5-hydroxymethylfurfural (HMF).

dehydration and hydration steps. Levulinic acid can be converted into a broad range of derivatives, such as esters, diols, lactones, or acids, from which many other chemical products can be produced (**Fig. 7**). For example, diphenolic acid is of great interest in the production of polycarbonates by replacing the commonly used, but toxic, bisphenol A.

Gamma-valerolactone (GVL). Gamma-valerolactone is a naturally occurring chemical in fruits and a frequently used food additive. It was proposed to exhibit the most important characteristics of an ideal sustainable liquid, which could be used for the production of both energy and carbon-based consumer products. GVL is renewable and easy and safe to store and move globally in large quantities, has a low melting point (-31°C), a high boiling point (207°C), a high open-cup flash point (96°C), a definitive but acceptable smell for easy recognition of leaks and spills, and it is miscible with water, thereby assisting biodegradation.

The biomass-based synthesis of GVL could start with the dehydration of carbohydrates to 5-hydroxymethyl-2-furfural (HMF), followed by its catalytic hydrolysis to levulinic and formic acids, and completed by the reduction of levulinic acid. The catalytic hydrogenation of levulinic acid has been extensively studied using heterogeneous and homogeneous catalysts. The transfer-hydrogenation of levulinic acid with formic acid to GVL has been demonstrated, making the process hydrogen-independent and readily useable even in remote locations [reaction (5)]. It should be noted that the overhydrogenation of GVL could lead to the formation of 1,4-pentanediol, which can be readily lose water to form 2-methyltetrahydrofuran (2-MeTHF). How-

ever, the formation of even a few hundred ppm of 2-MeTHF could be a safety issue, as it will readily form peroxides under ambient conditions, and any liquid containing more than 80 ppm peroxides is considered hazardous.

GVL can be converted to butenes by its decarboxylation, which can then be oligomerized to form higher olefins. Alkanes can also be produced via the hydrogenation of GVL to 2-MeTHF, followed by the hydrogenolysis of 2-MeTHF. GVL can be converted to other oxygenates, including ethyl 4-ethoxyvalerate and valerate esters (**Fig. 8**).

2-Methyltetrahydrofuran (2-MeTHF). 2-Methyltetrahydrofuran was proposed as a renewable component of the alternative P-fuel (nonpetroleum fuel) and can be synthesized by catalytic hydrogenation of levulinic acid or GVL [reaction (6)]. 2-MeTHF has

$$\text{GVL} \xrightarrow{H_2} \text{OH-OH} \xrightarrow{-H_2O} \text{2-MeTHF} \tag{6}$$

been used as a solvent in the pharmaceutical industry, with full toxicological data being recently collected.

Lignin. Lignin is an amorphous and moderately hydrophobic polymer containing polyaromatic compounds, which provides structural rigidity in plants. It represents 10–30% of biomass and is produced in large amounts as a by-product from the chemical pulping of wood and is perhaps one of the most underused renewable raw materials. The complex structure of lignin makes its extraction from biomass difficult. A widely used process is depolymerization and solubilization in an alkaline-alcohol solution, similar to the Kraft process. The residual lignin can be collected by acid or enzymatic hydrolysis.

Lignin is a by-product of wood processing in which the aim is to remove lignin in three consecutive steps, including depolymerization, breaking the bond between lignin and polysaccharides, and solubilization. Alkaline pulping, known as soda pulping, is the most used method for nonwood plants, but side reactions may occur that prevent further degradation of the lignin polymer. Neutral and sulfite pulping is an important method for manufacturing cellulose derivatives, but during the pulping various types of condensation reactions occur that lead to the formation of a brown color in the lignin. The organosolv method uses alcohols instead of water, leading to chemically altered lignins.

The current main use of lignin is as a fuel to provide heat for pulping and paper manufacturing, but it can also be used directly for heating biomass during gasification. It has been used as a dispersing agent in, for example, oil drilling or as an additive in concrete. Lignin acts as a binding agent for minerals, so it may affect the mechanical properties of soil. Lignin can also function as a binder in composite materials, a filler in polymers, a component in phenol-acetone resins, or a feedstock of carbon fibers. A minor fraction of vanillin is produced from lignosulfonate by

$$\text{levulinin acid} \xrightarrow[\text{H}_2]{\text{HCOOH} \quad \text{CO}_2} \text{OH} \xrightarrow{-H_2O} \text{GVL} \tag{5}$$

alkaline oxidation at high temperature [reaction (7)].

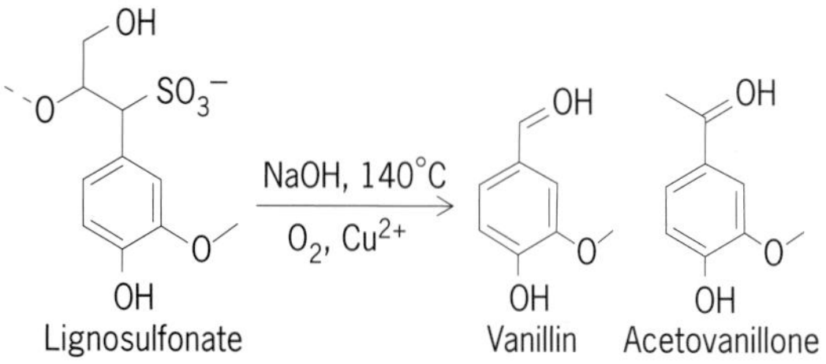

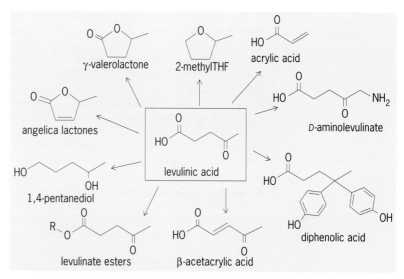

Lignosulfonate — NaOH, 140°C, O_2, Cu^{2+} → Vanillin + Acetovanillone (7)

Proteins. Proteins are present in biomass in variable amounts and may reach average values of 30% in the leaves of some tropical plants. Another important source of proteins is from algae, which have high growth rates, efficiently utilize solar energy, and can grow in conditions that are not favorable for terrestrial biomass. For protein-rich algae, the protein content of the dry matter can be up to 71 wt%.

Protein extraction from dry lignocellulosic feedstocks has only started receiving attention recently, usually with the goal of producing high-quality animal feed. However, a similar process involving preparation of protein concentrates from freshly harvested leaves has been investigated.

Lignocellulosic feedstocks require vigorous destabilization of their structure before undergoing enzymatic procedures. Protein extraction from leaves has been studied for about 50 years, and the original methods involved extraction with water to obtain 50–70% of the protein content. The yield could be increased by using alkaline solutions. Optimization of protein extraction of alfalfa leaves using more advanced techniques resulted in a maximum protein yield of 67%. The protein extracted is actually a mixture of components and has to be further processed to recover the different fractions of protein. In general, microalgae are used as nutritional supplements, rather than the proteins being extracted for other applications.

Biomass-derived proteins can be used as nutrients both for human and animal feeding when applying the water extraction method. However, protein degradation may occur as a consequence of several reactions, which can lead to indigestible components. The presence of nonnutritional components in extracted protein mixtures depends on the original feedstock and includes saponins, tannins, and a number of phytoestrogens.

Essential oils. Essential oils belong to terpenes, which are constructed from isoprene units and classified according to the number of these units, such as monoterpenes (C_{10}), sesquiterpenes (C_{15}), diterpenes (C_{20}), sesterpenes (C_{25}), triterpenes (C_{30}), and tetraterpenes (C_{40}). These oils can be obtained after physical pretreatment (mainly grinding) and steam distillation, extraction, or enfleurage, with each method producing oils with different properties. Widely known terpenes are limonene, menthol, and camphor. Monoterpenes and a few sesquiterpenes also are important economically as perfumes and fragrances. Various essential oils and the terpenes are still used in pharmaceutical preparations,

Fig. 7. Selected products derived from levulinic acid.

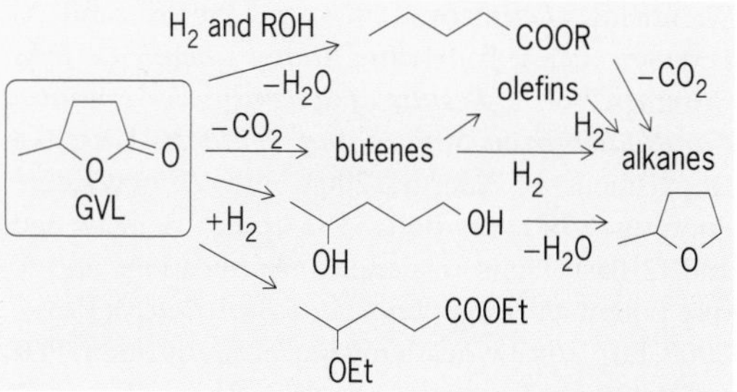

Fig. 8. Gamma-valerolactone (GVL) conversion reactions.

including turpentine, chamomile oil, eucalyptus oil, and camphor. The activity of some plants used in medicine and flavoring is due to their terpene content. Terpenes are also used in the production of synthetic resins.

For background information *see* BIOMASS; CARBOHYDRATE; CELLULOSE; ESSENTIAL OILS; ETHANOL; FURAN; GLYCEROL; GREEN CHEMISTRY; HEMICELLULOSE; LIGNIN; METHANOL; POLYSACCHARIDE; PROTEIN; PYROLYSIS; RENEWABLE RESOURCES; TERPENE; WOOD CHEMICALS in the Encyclopedia of Science & Technology. Edit Cséfalvay; István T. Horváth

Bibliography. A. V. Bridgwater et al., *Fast Pyrolysis of Biomass: A Handbook*, CPL Press, Newbury, U.K., 1999; F. S. Chakar and A. J. Ragauskas, Review of current and future softwood Kraft lignin process chemistry, *Ind. Crop. Prod.*, 20(2):131–141, 2004, DOI:10.1016/j.indcrop.2004.04.016; S. Chiesa and E. Gnansounou, Protein extraction from biomass in a bioethanol refinery—Possible dietary applications: Use as animal feed and potential extension to human consumption, *Bioresource Technol.*, 102(2):427–436, 2011, DOI:10.1016/j.biortech.2010.07.125; A. Corma et al., Chemical routes for the transformation of biomass into chemicals, *Chem. Rev.*, 107(6):2411–2502, 2007, DOI:10.1021/cr050989d; M. Crocker (ed.), *Thermochemical Conversion of Biomass to Liquid Fuels and Chemicals*, Royal Society of Chemistry, Cambridge, U.K., 2010; R. Datta and H. Micheal, Lactic acid: Recent advances in products, processes and technologies—a review, *J. Chem. Technol. Biot.*, 81(7):1119–1129, 2006, DOI:10.1002/jctb.1486; I. T. Horvath et al., γ-valerolactone—a sustainable liquid for energy and carbon-based chemicals, *Green Chem.*, 10(2): 238–242, 2008, DOI:10.1039/B712863K; B. Kamm et al. (eds.), *Biorefineries—Industrial Processes and Products*, Wiley, Hoboken, NJ, 2006; D. L. Klass, *Biomass for Renewable Energy, Fuels and Chemicals*, Academic Press, Waltham, MA, 1998; J. Lewkowski, Synthesis, chemistry and applications of 5-hydroxymethylfurfural and its derivatives, ARKIVOC, 17–54, 2001, http://www.arkat-usa.org/get-file/20028/; H. Mehdi et al., Integration of homogeneous and heterogeneous catalytic processes for a multi-step conversion of biomass: From sucrose to levulinic acid, γ-valerolactone, 1,4-pentanediol, 2-methyl-tetra-hydrofuran, and alkanes, *Top. Catal.*, 48:49–54, 2008, DOI:10.1007/s11244-008-9047-6; K. Weissermel and H.-J. Arpe, *Industrial Organic Chemistry*, 4th ed., Wiley, Weinheim, Germany, 2003; T. Werpy and G. Petersen (eds.), *Top Value Added Chemicals from Biomass*, Vol. 1: *Results of Screening for Potential Candidates from Sugars and Synthesis Gas*, U.S. Department of Energy, 2004, http://www1.eere.energy.gov/biomass/pdfs/35523.pdf; S. Zwenger and C. Basu, Plant terpenoids: Applications and future potentials, *Biotechnol. Mol. Biol. Rev.*, 3(1):1–7, 2008, http://www.academicjournals.org/bmbr/PDF/pdf2008/Feb/Zwenger%20and%20Basu.pdf.

Chemistry through ball milling

Ball milling is one of central techniques of mechanochemical synthesis, that is, chemical synthesis by the action of mechanical force, such as by grinding, stretching, or shearing. Mechanisms underlying chemical reactions in ball milling involve abrasion, shearing, and friction, making ball-milling chemistry one of central areas in tribochemistry. The principal advantages of ball milling over conventional methods of chemical synthesis are (1) enabling chemical reactions in the absence of solvents, therefore resulting in a cleaner synthetic procedure, (2) obtaining products that are difficult or impossible to obtain in the presence of bulk solvents, and (3) expanding the scope of chemical reactions to slightly soluble or inert starting materials.

Although ball milling is sometimes considered as an alternative to simpler mechanochemistry by mortar and pestle, the differences in processing are significant, and the physicochemical properties (such as particle size and morphology) of the products from ball milling and manual grinding are not identical. A significant difference between ball milling and mortar-and-pestle grinding is that the ball-milling reaction is done in a closed vessel, in a controlled environment, and avoids undesirable exposure to the atmosphere. The two most common instruments for ball milling are the vibration mill and the planetary mill. In a vibration mill (also known as shaker or

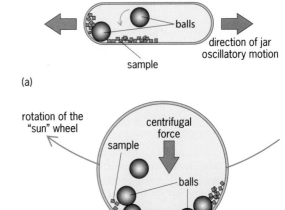

Fig. 1. Schematic representation of the grinding jar and ball motion for (*a*) a vibration mill and (*b*) a planetary mill.

mixer mill), milling is achieved by rapid oscillatory movement (typically 20–50 Hz) of the reaction vessel (milling jar) containing the grinding media, which typically are balls of stainless steel, tungsten carbide, agate, or ceramic (**Fig. 1a**). In a planetary mill, the jar is spun on a central table (the "sun" wheel) and simultaneously rotated around its own axis (Fig. 1*b*). The planetary mill is, in effect, a laboratory version of the industrial tumbler mill in which the gravitational effect is mimicked on a smaller scale by centrifugal force.

The two circular motions result in strong friction and shearing by the grinding media. Laboratory shaker and planetary mills manipulate reaction jars of 5–250 ml in volume and are capable of accommodating samples in the size range of milligrams to several tens of grams. Similar milling instruments at the industrial and pilot-plant scale can handle samples ranging from hundreds of grams to hundreds of kilograms. Another type of laboratory mill is the attritor, which is essentially a stirred ball mill.

Traditional applications. A traditional field of application for ball milling is particle comminution. Ball milling under cryogenic conditions (cryomilling) is used for making powders of biological (animal or plant) tissues, plastics, and rubber materials. Milling is the conventional form of preliminary mineral processing, which provides materials suitable for further treatment in extractive metallurgy. Ball milling can also be used to transform chemicals during mineral processing; for example, in the mechanochemical leaching of antimony (Sb) from tetrahedrite [$(Cu,Fe,Zn,Ag)_{12}Sb_4S_{13}$] concentrates. High-speed milling is an excellent means of mixing metals at the nanoscale. Such mechanical alloying enables mixing of metals that are not readily miscible under conventional melt conditions, the synthesis of metal oxide-metal nanocomposites, and the synthesis of amorphous metals. Different milling procedures are encountered for materials processing in other industries, with one example being liquid granulation in the processing of pharmaceutical solids.

Laboratory techniques and mechanisms. The simplest approach to laboratory ball-milling synthesis is to mill neat (dry) reactants, sometimes with an inert solid diluent. The diluent technique was recently applied for the mechanochemical click polymerization of shock-sensitive organic azides with alkynes. Reactions by neat grinding can be optimized by varying the amount of sample or diluent, the number and the weight of the milling balls, or the time and temperature of milling. Liquid-assisted grinding (LAG), also known as solvent-drop grinding (SDG), uses a small amount of a liquid to enhance reactivity. The stoichiometric amount of added liquid is comparable to or smaller than that of reagents. The liquid accelerates the mechanosynthesis and can enable reactions that do not take place during neat grinding. The liquid will also exhibit chemical effects, by directing the formation of polymorphs or templating hydrogen-bonded or metal-organic frameworks. Switching the grinding liquid in LAG provides a means to modify and optimize mechanochemical reactions, sometimes enabling quantitative reaction yields. The parameter η was introduced for the systematic comparison of neat grinding, LAG, and conventional slurrying and solution syntheses. Expressed in $\mu l/mg$, η is the ratio of the volume of the added liquid to the mass of solid reactants. Whereas an η value of 0 corresponds to neat grinding, an empirical study associated LAG with an η range 0–1, slurrying with an η range 1–10, and solution chemistry with η values larger than 10. The technique of ion- and liquid-assisted grinding (ILAG) uses catalytic amounts of simple salts, in addition to the liquid additive, to facilitate the synthesis of metal-organic materials from metal oxides. The ILAG salt additive can lead to structure templating and/or acid catalysis (with ammonium salts). Another technique for the synthesis of coordination polymers is grinding-annealing, in which a mechanochemical reaction yields a metal-organic coordination complex that polymerizes on thermal treatment.

Inorganic chemistry. Ball milling has been used for a wide range of chemical reactions of inorganic, organic, organometallic, and supramolecular chemistry. The traditional area of application for ball milling is in reactions of metal oxides, including the synthesis of mixed metal oxides (such as perovskites, spinels, ferrites, and garnets), the reduction of metal oxides or their disproportionation (for example, the reduction of magnetite, Fe_3O_4 by aluminum or its disproportionation with oxygen release), and the formation of metal oxides by oxidation, or metal displacement. Equally important are analogous reactions of the chalcogenides (sulfides, selenides, and tellurides) because of their relevance in the mineral industry. A number of such reactions are self-sustainable, such that an exothermic chemical reaction initiated by milling propagates (for the most part) independently. The exothermic nature of many inorganic milling reactions has been used for reaction monitoring, for example, by placing thermocouples on the grinding vessel wall. Another means of monitoring the reaction course is by measuring the change in the pressure or the volume of a gas produced or consumed in the reaction. Inorganic ball-milling reactions can take a different course than expected from normal experience, as exemplified by the reduction of carbon dioxide with metallic gold to form Au_2O_3. Ball milling is a known route to intermetallic compounds, such as silicides, borides, and arsenides.

Organic synthesis. The advent of green chemistry and growing concerns over the use of toxic solvents and energy in chemical processes have fueled the development of organic milling synthesis. A large number of organic reactions have been reported by ball milling, including click reactions (such as the Huisgen addition, urea coupling, and thiourea coupling), [2+4] (Diels-Alder) cycloadditions, Michael additions, hydrogenations and oxidations of carbon-based and nitrogen-based functionalities, and catalytic reactions such as Suzuki or Sonogashira coupling. Synthetic organic mechanochemistry is a rapidly expanding area, and several ball-milling reactions have been demonstrated as being more energy-efficient and less energy-intensive than analogous solution- or microwave-based routes. Ball milling is readily applicable to reactions of asymmetric organocatalysis as well as deracemization.

Metal-organic, organometallic, and coordination chemistry. A number of mechanochemical reactions involving transformations of coordination bonds have been reported, although only a small fraction of these were accomplished by ball milling. Most of these include ligand additions, for example, the addition of a pyridine derivative to a metal halide or acetate to form a coordination polymer. Such reactions have also been observed in organometallic chemistry, for example, the addition of phosphine ligands to metal carbonyls. An important class of metal-organic ball-milling reactions involves ligand replacement, typically of an acetate by another carboxylate ligand. Examples of such reactions involve milling of copper(II) acetate [$Cu(OAc)_2$] with isonicotinic acid (pyridine-4-carboxylic acid) or trimesic acid (benzene-1,3,5-tricarboxylic acid) to form porous metal-organic frameworks with potential applications in gas storage (**Fig. 2a**). Particularly important for green chemistry is the LAG or ILAG conversion of metal oxides or carbonates, as such approaches allow large reductions in energy, solvent, and reaction time, compared to conventional approaches. Oxide-based mechanochemistry has so far been applied for the synthesis of porous metal-organic frameworks based on pillared or zeolitic designs, as well as for the synthesis of pharmaceutical derivatives or metallodrugs (for example, bismuth subsalicylate, the active component of Pepto-Bismol®, Fig. 2b).

Supramolecular chemistry. A number of effects traditionally related to supramolecular chemistry and molecular self-assembly have been recognized in ball-milling reactions. These include the synthesis of a variety of host-guest inclusion compounds and hydrogen- or halogen-bonded cocrystals, the

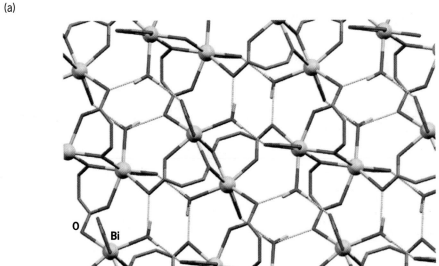

(a)

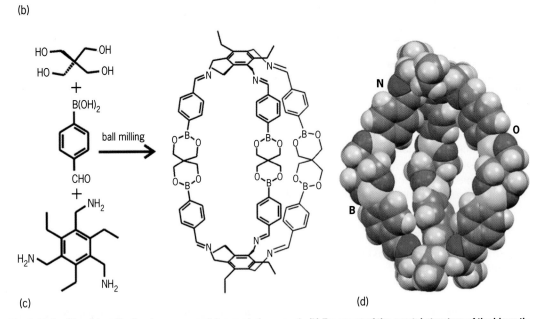

(b)

(c)

(d)

Fig. 2. Ball-milling (*a*) synthesis of a porous metal-organic framework. (*b*) Fragment of the crystal structure of the bismuth disalicylate complex obtained by ball milling. (*c*) Ball-milling synthesis of a cage based on reversible covalent bonds and (*d*) a space-filling representation of the same cage.

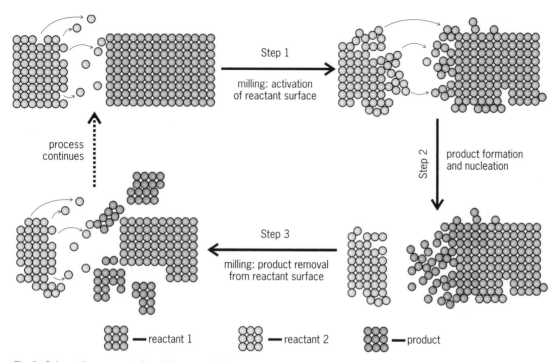

Step 1

milling: activation
of reactant surface

Step 2   product formation
and nucleation

process
continues

Step 3

milling: product removal
from reactant surface

⬢⬢⬢ — reactant 1   ⬢⬢⬢ — reactant 2   ⬢⬢⬢ — product

Fig. 3. Schematic representation of the general three-step mechanism of mechanochemical milling reactions.

formation of covalent or coordination macrocycles and interlocked molecules (rotaxanes and pseudorotaxanes), templating of porous metal-organic frameworks, and the synthesis of cages based on reversible covalent bonds (for example, aldimines or boroxines, Fig. 2c). Reversibility and thermodynamic equilibration of covalent bonds in ball milling were explicitly demonstrated for disulfide metathesis (exchange) reactions. Equilibration behavior resembling Le Chatelier's principle has been observed in hydrogen-bond-forming reactions involving pyrazine and tartaric acids. For chiral solids or molecules undergoing reversible supramolecular or covalent self-assembly, ball milling with low solvent amounts can lead to the spontaneous increase in enantiomeric excess. This effect, which was first recognized by C. Viedma in slurrying a saturated solution of sodium chlorate, has been applied in the spontaneous deracemization of a scalemic (that is, containing a slight enantiomeric excess) aldol compound.

Pharmaceutical form screening. Although the traditional application of milling in pharmaceutical materials technology has been for modifying particle shape and size distributions, an important modern application is in screening for new solid forms of active pharmaceutical materials (APIs). These typically include API polymorphs, pharmaceutical cocrystals, salts, ionic cocrystals, or solvates. Milling has demonstrated superior screening efficiency to a number of alternatives, including screening from melt or solution.

Mechanisms. It is clear that a single reaction mechanism cannot explain the known variety of ball-milling reactions. G. Kaupp has proposed a general mechanistic framework (**Fig. 3**) based on three steps: (1) molecular/atomic diffusion of reactants leading to a chemical reaction, (2) formation of the product phase, and (3) separation of the product phase from the bulk reactant surface. Ball-milling reactions often differ in the mechanism through which step (1) is done. For molecular reactants (for example, organic solids), the mixing is mediated either through gas-phase or surface diffusion by the formation of a liquid eutectic or by the formation of a mobile amorphous phase. As most molecular reactants are low-melting and/or exhibit high vapor pressures, these processes readily take place at or near room temperature and harsh milling is not necessary. For high-melting inorganic solids, the efficiency of reactant mixing is improved by intensive milling, which leads to the formation of transient microscopic areas of high temperature on ball impact (the "hot spot" model) or the formation of highly reactive surface plasma on impact-induced cracking of ionic crystals or infinitely covalent solids (the "magma-plasma" model). A detailed model of inorganic ball-milling reactions has been developed by F. Kh. Urukaev and V. V. Boldyrev.

For background information *see* ASYMMETRIC SYNTHESIS; CLICK CHEMISTRY; COORDINATION CHEMISTRY; COORDINATION COMPLEXES; GREEN CHEMISTRY; GRINDING MILL; INTERMETALLIC COMPOUNDS; LE CHATELIER'S PRINCIPLE; LIGAND; MECHANICAL ALLOYING; ORGANIC SYNTHESIS; SOLID-STATE CHEMISTRY; SUPRAMOLECULAR CHEMISTRY in the McGraw-Hill Encyclopedia of Science & Technology.

Tomislav Friščić

Bibliography. P. Balaž, *Mechanochemistry in Nanoscience and Minerals Engineering*, Springer-Verlag, 2010; T. Friščić, Supramolecular concepts and new techniques in mechanochemistry: Applications to cocrystals, molecules and metal-organic materials, *Chem. Soc. Rev.*, 41:3493–3510, 2012, DOI:10.1039/C2CS15332G; T. Friščić and W. Jones,

Recent advances in understanding the mechanism of cocrystal formation via grinding, *Cryst. Growth Des.*, 9:1621–1637, 2009, DOI:10.1021/cg800764n; S. L. James et al., Mechanochemistry: New and cleaner synthesis, *Chem. Soc. Rev.*, 41:413–447, 2012, DOI:10.1039/C1CS15171A; W. Jones and A. V. Trask, Crystal engineering of organic cocrystals by the solid-state grinding approach, *Top. Curr. Chem.*, 254:41–70, 2005, DOI:10.1007/b100995; G. Kaupp, Reactive milling with metals for environmentally benign sustainable production, *CrystEngComm.*, 13:3108–3121, 2011, DOI:10.1039/C1CE05085K; L. Takacs, Self-sustaining reactions induced by ball milling, *Prog. Mater. Sci.*, 47:355–414, 2002, DOI: 10.1016/S0079-6425(01)00002-0; F. Kh. Urukaev and V. V. Boldyrev, Mechanism and kinetics of mechanochemical processes in comminuting devices: 1. Theory, *Powder Tech.*, 107:93–107, 2000, DOI:10.1016/S0032-5910(99)00175-8.

Cognitive technology

Technology is widely used to augment human capabilities. We can move faster using cars, we can lift heavier weights using cranes, and we can even fly through the air using airplanes. Such technologies have enabled us to surpass human physical limitations, and have transformed what we can achieve physically. Cognitive technology refers to technologies that carry out cognitive operations. Thus, rather than augmenting human physical capacity, these technologies augment mental capacities.

Offloading cognition onto technology. Cognitive technology can increase human mental capacity by enabling to "offload" cognitive operations onto technology. Such offloading reduces cognitive load and thus frees up cognitive resources. For example, using calculators offloads simple mathematical operations to technology. By using calculators humans do not need to spend cognitive resources to do the mathematical operations and cognitive resources are available to do other tasks.

When cognitive technology is used to help humans by enabling to offload cognitive operations, the technology is working in the background. To save time and cognitive resources, humans devolve and delegate cognitive operations that they could do themselves. Such cognitive offloading takes many shapes and forms. One of the first forms of offloading is literacy and other forms of symbolic encoding. It allows humans to increase their memory capacity and free themselves from needing to memorize information that they offloaded.

Offloading is a very basic way that cognitive technology can augment human mental capacity. With the increased sophistication of cognitive technology, they not only support human cognition by offloading, but they are now able to work with humans as cognitive partners.

Cognitive partners through distributed cognition. When cognition is distributed between humans and technology and they work as cognitive partners, the mental operations are divided between the human and technology. Rather than being subservient to the human (when cognition is merely being offloaded), at this level of cognitive technology both human and technology are working together in a cooperative manner.

Many times, distributed cognition entails that the human and technology are performing tasks that are beyond the capacity of the other. Thus, a vital cooperation is formed, whereby each contributes specific and unique capabilities. This means that it is not a matter of mere convenience and efficiency, but that such distributed cognition with cognitive technology is necessary. In forensic investigations, for example, human fingerprint experts do not have the ability to examine millions of fingerprints. Automated fingerprint identification systems (AFIS) have the capacity to examine millions of fingerprints, but they are not able to make conclusive decisions when fingerprints are of low quality, as is often the case in fingerprints found at crime scenes. Therefore, a very effective cooperation exists between human experts and cognitive technology, whereby the technology initially examines millions of fingerprints, selecting the most likely matching candidates and the human reaches a conclusive decision over the small set of potential candidates provided by the technology.

Distributed cognition often relies on sophisticated and expensive technology and is used routinely in highly skilled expert domains, such as in medicine and the military. However, as cognitive technology advances, its use is expanding and is penetrating everyday life. For example, many cars have parking sensors that distribute the cognitive operations. Newer cars include fully automated parking capabilities. These do not work with the driver by assessing distances, but they take over and replace the driver—the higher level of cognitive technology discussed in the next section.

Cognitive technology replaces humans. Cognitive technology has developed so much in recent years that it now not only enable to offload or to distribute cognition, but it can also take over cognitive capacity that was once the sole domain of humans. In some cases, the technology can even replace the human who is no longer needed. For example, Global Positioning System (GPS) and satellite navigation technology has in many cases taken over what humans have done in the past. These technologies tell us in real time where we are and give us directions where to go, with only minimal human involvement. Thus, operations that were almost totally reliant on human cognition and were the sole domain of human activity are now being taken over by cognitive technology.

While offloading requires the technology to work in the shadow of the human, cognitive technology at its highest level leaves the human in the shadow of the technology (if not altogether out of the picture). For example, in aviation, autopilot takes over flying the airplane, leaving the pilot to takeoff and land. The pilot monitors the autopilot and can override it. But as technology advances, pilot interventions are rare and minimal, and future technology is expected to enable total pilot-free flying.

Greater use and reliance on cognitive technology. The creation of cognitive technology and its increasing sophistication and use has brought about the shift from supporting the human (offloading), to cooperative cognitive partnership (distributed cognition), to the technology taking control over some of the cognitive operations altogether. An everyday example to such a shift would be cashiers at retail stores. In the past, cashiers would have to look up the price of sale items, add them up, collect money from the customers, and give them change. At the end of the day, the cashier would add up all the money from the entire day's sales and check the inventory in the store to determine if to order more merchandise. With technological advances, more cognitive operations that were previously done by the human cashier are now performed by the cash register. Initially, the cashier would still look up the prices, but the cash register would add up the cost of the items. Next cash registers calculated how much change to give back as well as the total sales at the end of the day. With further advances, cashiers did not need to look up prices, as the cash register's scanning technology determined the price of items and kept track of inventory and the need to order more merchandise. Today's cash-register technology has advanced so much that it has eliminated the need for a human cashier altogether. Self-checkout enables customers to pay for merchandise without needing a human cashier at all.

As cognitive technology advances, we can expect to see more activities taken over by technology. Activities that are within the sole capabilities of humans will see greater technological involvement, ultimately being totally technologically led with minimal-to-no human participation. The theoretical and practical limits of cognitive technology are unknown. The dangers are discussed in the next section.

Dangers of cognitive technology. We have seen that cognitive technologies are working closely with humans by sharing cognitive operations, and that with their increasing sophistication they can even replace the need for human cognition. Such cognitive technology is clearly a benefit, allowing to increase human capacity and to achieve more. However, are there any dangers in such cognitive technology?

As humans rely more on cognitive technology, not only are they dependent on it, but also the cognitive demands and human abilities change. Human cognition is very flexible and adaptive, and develops to meet environmental demands. Cognitive technology has an effect on the development and capabilities of human cognition. Take, for example, the earlier discussion on the increasing use of GPS and satellite navigation. As we rely on it more, we use it (the technology) rather than our own cognition. With time, our own abilities degrade (for those who grew up in a pre-GPS era), and the new generation that grows up with such technology may never develop the abilities to read maps, navigate, and all the cognitive skills now done by the GPS. Thus, as cognitive technology does more for us, we do less ourselves, and therefore are losing certain cognitive abilities.

That is not necessarily bad, because as we lose certain abilities we may develop new abilities, thus cognitive abilities are changing rather than being lost. Working with cognitive technology may achieve new forms of cognition (such as distributed cognition) and enable us to accomplish things that would never have been possible without the cognitive technology. It is difficult and perhaps impossible to determine, especially in the long run, the effects of cognitive technology, let alone if they are bad or good. Our increased reliance on cognitive technology may be the next step in our evolutionary development, discussed in the next section.

Another example of such a potential danger arises from the use of automated speller checkers in word processing. With the increased sophistication and use of automated spell checking, the need to memorize and know how to spell words comes into question. Even if theoretically it is justified to learn how to spell, the use of automated spell checking changes the cognitive demands and the practical need to learn how to spell well. How important it is to learn how to spell in a cognitive technology age that auto-corrects spelling errors and what are the implications of not learning how to spell are important and complicated questions.

To answer these questions we need to know and understand the long-term effects of using cognitive technology, which we do not know. However, it is clear that cognitive technology will have a profound effect on individual cognition.

Cognitive technology as an evolutionary force. Although we cannot know the specific ways cognitive technology will affect human cognition, we can be certain that it will have such an effect. Brain plasticity and human cognition is very flexible and adaptive to environmental needs. The introduction of cognitive technology makes basic and dramatic changes in the environment in which we operate. Therefore, cognitive technology is a factor that shapes human cognition, and its effects and powers are going to increase the more we use and rely on it.

Even the more basic quantitative offloading can shape human cognition. For example, literacy and other forms of symbolic encoding have changed the demands on memory, and it seems that certain human memory abilities (for example, semantic memory) were superior in preliteracy periods. Such effects are much more pronounced when quantitative effects of efficiency in offloading are replaced with distributed cognition that brings about qualitative effects through new forms of cognition. Cognitive technology acts as an evolutionary force that shapes human cognition. However, its effects on humans do not require millions of years but are within a relatively short time.

The examples discussed show how cognitive technology interacts with and affects human cognition. The distinctions between different modes (for example, offloading and distributed cognition) highlight points in what is a continuum of complex collaborations. The result of these interactions is that cognitive technology not only affects how we think and go about our lives as individuals, but also significantly changes how we communicate, collaborate, and interact with one another. Hence, cognitive technology changes the face of society and the very nature of human cognition.

For background information *see* COGNITION; DECISION ANALYSIS; DECISION THEORY; PSYCHOLOGY in the McGraw-Hill Encyclopedia of Science & Technology.   Itiel Dror

Bibliography. I. E. Dror, Land mines and gold mines in cognitive technologies, in *Cognitive Technologies and the Pragmatics of Cognition*, I. E. Dror (ed.), John Benjamins Publishing, Amsterdam, pp. 1–7, 2007; I. E. Dror and S. Harnad, Offloading cognition onto cognitive technology, pp. 1–23, in I. E. Dror and S. Harnad (eds.), *Cognition Distributed: How Cognitive Technology Extends Our Minds*, 2008; I. E. Dror and J. Mnookin, The use of technology in human expert domains: Challenges and risks arising from the use of automated fingerprint identification systems in forensics, *Law, Probability and Risk*, 9(1):47–67, 2010, DOI:10.1093/lpr/mgp031; I. E. Dror, K. Wertheim, P. Fraser-Mackenzie, and J. Walajtys, The impact of human-technology cooperation and distributed cognition in forensic science: Biasing effects of AFIS contextual information on human experts, *J. Forensic Sci.*, 57(2):343–352, 2012, DOI:10.1111/j.1556-4029.2011.02013.x.

Commercial space activities to support NASA's missions

With the retirement of NASA's space shuttle program, much attention is now focused on the use of commercial space vehicles to reestablish the capability of the United States to transport cargo and crew to and from the *International Space Station* (*ISS*). The policy foundations for this move date back to the Commercial Space Launch Act of 1984, and National Space Policy released by President Ronald Reagan in 1988 (NSDD-293), which encouraged the development of a domestic commercial launch industry and directed NASA and the U.S. Department of Defense to utilize private-sector launch services "to the maximum extent feasible."

President Reagan's support for commercial space transportation was partly a response to an unintended consequence of the space shuttle program, which was envisioned to serve as a much-utilized launch system for NASA, military, and commercial satellites. When the space shuttle failed to achieve the robust launch rates (and consequent low costs) envisioned by NASA, commercial satellite operators turned increasingly to Europe's Arianespace to deploy their payloads. Rocket makers in the United States, like General Dynamics (Atlas), McDonnell Douglas (Delta), and Martin Marietta (Titan), responded by establishing commercial launch divisions to compete with Arianespace and

other foreign launch service providers in Russia and China.

Under the regulatory auspices of the U.S. Department of Transportation (and later the Federal Aviation Administration), these U.S. companies briefly captured a majority of the global market share for commercial launches. It did not take long, however, for the foreign-government-backed competitors to regain their lead with pricing that the U.S. companies could not match. Today, after 2 decades of mergers and acquisitions, U.S. companies such as Boeing, Lockheed Martin, and their joint venture, United Launch Alliance (ULA), serve only a small sliver of the global commercial market, relegating their Atlas and Delta rockets mainly to launching U.S. government payloads.

NASA's path to commercialization. In 2004, NASA's Vision for Space Exploration (VSE) called for the retirement of the space shuttle, and for the agency to rely on commercial vehicles to transport cargo to and from the *ISS*, and for missions beyond Earth orbit. The VSE established a goal for returning humans to the Moon by 2020 in preparation for future missions to Mars and beyond. This goal, and the vehicle architecture devised for it, became known as the Constellation program.

But while VSE sought specifically to rely on commercial cargo launches to the *ISS*, the Constellation architecture included two very expensive rockets (the controversial, potentially duplicative Ares-1 and the heavy-lift Ares-5) that would be funded, in part, by retiring the *ISS* by 2015. This retirement of the *ISS* (necessitated by President George W. Bush's no-growth "pay as you go" budget for NASA) would have limited or eliminated NASA's need for commercial cargo launches.

COTS and CRS. Nevertheless, NASA, in 2006, proceeded with the Commercial Orbital Transportation Services (COTS) program to develop and demonstrate commercial vehicles that could transport cargo (and potentially astronauts) to and from the *ISS*. With about $500 million for initial competitive procurements, and a series of funded and unfunded Space Act Agreement (SAA) contracts, NASA, in 2008, narrowed the field of over a dozen competitors down to two: Orbital Sciences Corporation and Space Exploration Technologies (SpaceX). These two companies received NASA Commercial Resupply Services (CRS) contracts worth about $3.5 billion through 2016. Orbital Sciences would receive $1.9 billion for eight missions, and SpaceX would receive $1.6 billion for 12 missions.

For CRS, Orbital is developing a new Antares launch vehicle (formerly known as Taurus-2), powered by two Soviet-heritage Aerojet AJ26 liquid-fuel first-stage engines. Launching from NASA's Wallops Flight Facility in Virginia, on a Mid-Atlantic Regional Spaceport launch pad, the Antares will lift a Cygnus cargo module that can carry up to 2700 kg (5952 lb) of supplies to the *ISS*. Cygnus will not be able to transfer cargo back to Earth. After delivering supplies to the *ISS*, Cygnus will be jettisoned with trash to burn up upon atmospheric reentry.

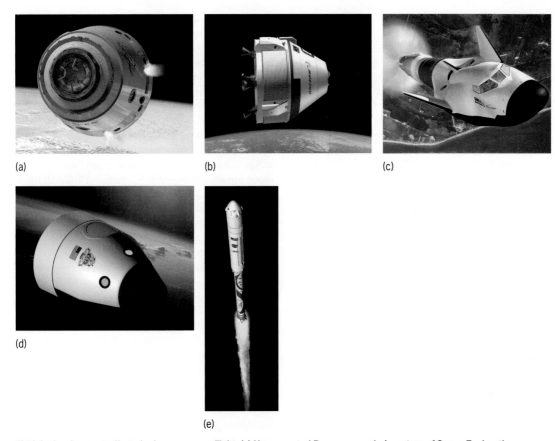

(a)

(b)

(c)

(d)

(e)

Vehicle development efforts for human space flight. (*a*) Human-rated Dragon capsule (*courtesy of Space Exploration Technologies, Inc.*). (*b*) CST-100 capsule (*courtesy of The Boeing Co.*). (c) DreamChaser spaceplane (*courtesy of Sierra Nevada Corp.*). (d) New Shepard capsule (*courtesy of Blue Origin*). (e) Liberty rocket and capsule system (*courtesy of Alliant Techsystems Inc.*).

SpaceX's Falcon-9 rocket uses nine Merlin liquid-fuel first-stage engines, developed in-house by SpaceX. Riding atop the Falcon-9 is the company's multipurpose Dragon capsule, designed to accommodate both cargo and crew (**illus.** *a*). SpaceX has launched the Falcon-9/Dragon from the Cape Canaveral Spaceport, on a launch pad leased from the U.S. Air Force. In addition to carrying up to 6000 kg (13,228 lb) of cargo to the *ISS*, the Dragon capsule can return up to 3000 kg (6614 lb) back to Earth, with a splashdown recovery off the coast of Baja California.

By the time Orbital Sciences and SpaceX were selected for CRS, the Constellation program was falling behind schedule and going over budget. Critics charged that Ares-1 was to blame, requiring NASA to spend billions of dollars to duplicate a capability already available with the Atlas-5 and Delta-4, and that would soon be available with SpaceX's Falcon-9. The following excerpt from the 2004 VSE announcement is at the heart of their concerns: "NASA does not plan to develop new launch vehicle capabilities except where critical NASA needs—such as heavy lift—are not met by commercial or military systems."

In 2009, President Barack Obama formed a blue-ribbon Review of U.S. Human Spaceflight Plans Committee, which concluded that Constellation was substantially behind schedule and would not achieve its exploration goals without many billions of dollars being added to NASA's budget. President Obama cancelled the Constellation program in 2010, and in 2011 put forward an exploration agenda focused initially on human visits to asteroids, extension of the *ISS* to at least 2020, and the accelerated development of commercial cargo and crew capabilities.

CCDev and CCiCap. NASA established a Commercial Crew Development (CCDev) program in 2010, and consolidated the agency's cargo and crew efforts within a new Commercial Crew and Cargo Program Office (C3PO), based at Kennedy Space Center. C3PO and other NASA offices have supported multiple CCDev technology and concept development projects with Alliant Techsystems (ATK), Blue Origin, Boeing, Excalibur Almaz, Paragon Space Development Corporation, Sierra Nevada Corporation, SpaceX, and ULA.

Among these projects are some high-visibility vehicle development efforts for human spaceflight, including the human-rated Dragon capsule by SpaceX, the CST-100 capsule by Boeing, the DreamChaser spaceplane by Sierra Nevada, the New Shepard capsule by Blue Origin, and the Liberty rocket and capsule system by ATK (see illustration). In August 2012, NASA awarded three contracts totalling over a billion dollars to demonstrate these crew transportation capabilities under a Commercial Crew Integrated Capability (CCiCap) program: to Boeing for

CST-100, to SpaceX for Dragon, and to Sierra Nevada for DreamChaser.

Commercial suborbital space flight. Running parallel to NASA's commercial orbital space flight programs, the agency has sought to encourage the development of new commercial suborbital space flight capabilities. Companies like Virgin Galactic, XCOR Aerospace, and Rocketplane Kistler had already begun to develop winged reusable launchers for suborbital space tourism, and other companies like Armadillo Aerospace and Masten Space Systems were developing innovative vertical takeoff and landing suborbital vehicles.

NASA's Commercial Reusable Suborbital Research (CRuSR) program was established in 2010 to fund the use of such vehicles as platforms for innovative research. In 2012, NASA merged the CRuSR program and the Facilitated Access to the Space Environment for Technology (FAST) program into a new Flight Opportunities Program under the agency's Office of the Chief Technologist. This new program competitively secures flight opportunities for research conducted by NASA and universities, providing a new market for space flight companies that previously had focused primarily on space tourism.

Space enterprise zone. One stated goal of NASA's new push for commercial space transportation is to establish Earth orbit as an enterprise zone for private-sector development. By enabling private-sector operations in near-Earth space, NASA hopes it can turn its focus, and its limited resources, toward exploration beyond the current frontiers of human spaceflight.

Already, new companies are emerging to exploit commercial opportunities in near-Earth space and beyond. Planetary Resources Corporation, for example, plans to mine asteroids for valuable minerals and metals, and Stratolaunch Systems is developing a massive new air-launch system. Ultimately, suborbital spaceflight companies like Virgin Galactic and XCOR also aim to provide orbital transportation services, while companies like Bigelow Aerospace and Excalibur Almaz develop new commercial space stations and Moon bases that would expand the market for transportation services beyond NASA's limited requirements.

Outlook. Some would say this is history repeating itself. The SpaceXs, Virgin Galactics, and XCORs of today are not unlike the barnstormers of the early days of aviation. They are pursuing new commercial opportunities in the wake of government-sponsored exploration and technology development, and establishing a new transportation industry that will grow and mature for many decades to come.

The innovation and investment that these commercial players bring to the marketplace has already brought down the cost of space travel. According to a NASA study, SpaceX was able to develop the Falcon-9 for $1.69 billion, less than half of the estimated $3.97 billion that would have been required if NASA or the Air Force were to develop it as a traditional government program. Furthermore, SpaceX's published 2012 price for a Falcon-9 mission is roughly $54 million, less than one-third the price of a comparable Atlas-5 mission (Atlas-5 being more of a traditional government-procured launch system).

Astronauts like Neil Armstrong have lamented the lack of progress seen in human space exploration since Apollo. Futurist Arthur C. Clarke, writing *2001: A Space Odyssey* during the height of the Apollo program, envisioned multiple commercially serviced lunar bases and space stations by the turn of the century. Unfortunately, in our real-world space program, we had to rely principally on NASA to make such major advances, with budgets much constrained since Apollo. Perhaps NASA's growing collaboration with commercial interests will propel humankind's presence in space further and faster.

For background information *see* ASTEROID; ROCKET PROPULSION; SPACE FLIGHT; SPACE SHUTTLE; SPACE STATION; SPACECRAFT STRUCTURE in the McGraw-Hill Encyclopedia of Science & Technology.

Edward Ellegood

Bibliography. National Aeronautics and Space Administration, *The Vision for Space Exploration*, NASA, Washington, D.C., 2004; Review of U.S. Human Spaceflight Plans Committee, *Seeking a Human Spaceflight Program Worthy of a Great Nation*, NASA, Washington, D.C., 2009.

Complex eyes of giant Cambrian predator

The Cambrian Period (542–488 million years ago) is arguably the most important phase in the evolution of complex life. The Cambrian explosion is an evolutionary event that encompasses the rapid proliferation of marine life and the first appearance of most of the major animal lineages (or phyla) familiar to us today. Animals with preservable hard parts (for example, shells and exoskeletons) and animals (termed the bioturbators) that could burrow and mix the sediment became increasingly abundant. This interval also heralds the Cambrian arms race, a phase that amplified ecological complexity, including the advent of predator–prey relationships that completely reshaped food webs within marine ecosystems. In short, the Cambrian explosion is the invention of the modern marine biosphere.

One of the earliest predators to inhabit the Cambrian oceans more than 500 million years ago was the large swimming animal *Anomalocaris* (**Fig. 1**). *Anomalocaris* and its close relatives are generally considered to be the top (or apex) predators of their time. This is based on their anatomy: large body size (some later Ordovician [480-million-year-old] forms were more than 90 cm [36 in.] in length); grasping spinose (spiny) limbs at the front of the head, which were used to grab prey and pass it to a circular mouth with teethlike serrations; and large digestive glands. Moreover, evidence of predation damage to contemporaneous trilobites and also large coprolites (fossilized feces) containing the masticated remains of prey have been attributed to these animals.

Optics of ancient eyes. The stalked eyes of *Anomalocaris* have been known for some time, with discoveries of fully articulated bodies from famous

Fig. 1. Reconstruction of the giant Cambrian predator *Anomalocaris*. (*Illustration by Katrina Kenny*)

Cambrian fossil deposits such as the Burgess Shale in Canada and Chengjiang in China. However, none of the previous occurrences have ever shown the details of the visual surface—for example, lenses.

Recent discoveries from another Cambrian deposit—the Emu Bay Shale on Kangaroo Island, South Australia—that contains exceptionally preserved fossils, including replications of soft tissues such as digestive glands and muscle, have revealed the optical design of some of the oldest arthropod eyes, including those of *Anomalocaris* (**Fig. 2**). Being large and attached to the body by stalks, the eyes of *Anomalocaris* were previously inferred to be of the compound type—the multifaceted variety seen in arthropods such as flies, crabs, and kin—and this has now been confirmed by the Emu Bay Shale fossils. Of particular significance is the size and number of lenses, with each eye up to 3 cm (1.2 in.) in length and containing more than 16,000 lenses, making them among the largest, lens-rich compound eyes to have ever existed. The very large, but considerably younger eurypterids (sea scorpions), such as the 2.5-m-long (8.2-ft-long) *Jaekelopterus rhenaniae* from the Devonian Period (400 million years ago) of Germany, represent some of the rare examples of arthropods with eyes larger than *Anomalocaris*.

The large number of lenses as well as the very acute angle between each lens (less than 1.4°) indicate that a very dense image could be formed, suggesting that *Anomalocaris* would have seen its world with exceptional clarity while hunting in well-lit ocean waters. Only a few arthropods have similar resolution, such as modern predatory dragonflies that can possess up to 28,000 lenses in each eye. The acuity of the stalked eyes of *Anomalocaris* provides additional evidence to support this animal's status as a highly mobile visual predator in the water column, complementing its streamlined profile with flaps along both sides of its body and large tail flukes (fins) that would have allowed for strong swimming abilities. Sharp vision would have provided *Anomalocaris* with a distinct advantage over its other preda-

tory competitors, and certainly over its prey. Many of these other animals either had poor vision in comparison or were completely blind, thus presenting a smorgasbord for *Anomalocaris* to choose from. Moreover, the existence of nektonic (free-swimming) hunters possessing sharp vision within Cambrian marine communities would have placed considerable selective pressures on prey, creating an interaction that would have influenced the arms race that began during this phase in early animal evolution. Prey had to adapt and rapidly evolve defensive armor, such as shells, to counteract the threat of such sophisticated predators; if not, the prey would potentially face extinction.

Evolutionary implications. *Anomalocaris* and its relatives (collectively known as the Radiodonta) have long been problematic with regard to classification because of their unusual body plan. Many paleontologists have considered *Anomalocaris* to be a member or close relative of the arthropods, but there are relatively few anatomical characteristics that unite it with this group. The name arthropod literally means jointed leg, and *Anomalocaris* has one such pair of appendages, its anterior grasping appendages, which

Fig. 2. Complex compound eyes. (*a*) Isolated fossil compound eye with more than 3000 lenses belonging to an unknown arthropod from the early Cambrian Emu Bay Shale on Kangaroo Island, South Australia. Scale bar: 1 mm (0.04 in.); bz = bright zone, an area of the visual field with increased light sensitivity (as a result of larger lenses) and possibly higher acuity. (*b*) The compound eye of a living predatory arthropod, the robberfly *Laphria rufifemorata*, which also possesses a bright zone (bz). Scale bar: 0.5 mm (0.02 in.). (*c*) The stalked eye of *Anomalocaris* from the Emu Bay Shale showing the transition between the visual surface and the stalk (arrows). Scale bar: 2 mm (0.08 in.). (*d*) Enlargement of lenses in the eye of *Anomalocaris*. Scale bar: 0.3 mm (0.012 in.).

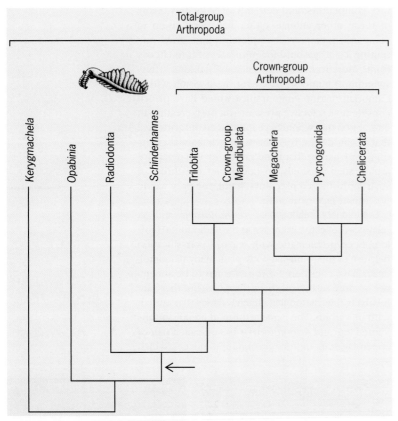

Fig. 3. The evolutionary tree of arthropods showing the basal position of Radiodonta (*Anomalocaris* and kin) relative to the crown-group arthropods [that is, the group consisting of the last common ancestor of living arthropods and all its descendants, including some extinct members (for example, trilobites)]. The arrow indicates the known origin of compound eyes; the five stalked eyes of *Opabinia* from the Burgess Shale may also be compound, but lenses have not been observed in the available fossils. (*Adapted from Fig. 3 of J. R. Paterson et al., Acute vision in the giant Cambrian predator Anomalocaris and the origin of compound eyes, Nature, 480:237–240, 2011*)

The antiquity of the 515-million-year-old eyes of *Anomalocaris* and those of another (unknown) arthropod from the Emu Bay Shale (Fig. 2) suggests that large, sophisticated eyes evolved with astonishing speed (on a geologic timescale), implying that the Cambrian explosion involved rapid innovation at every anatomical scale, from entire body plans to the intricate facets of a compound eye. There is no clear evidence of eyes in any fossil organism found in rocks older than 520 million years of age, despite the considerable diversity of macroscopic life forms in the preceding Ediacaran Period (635–542 million years ago). This lends support to the so-called "light-switch hypothesis," which suggests that the advent of vision was a major trigger for the Cambrian explosion. The introduction of this new sensory organ would have revealed the surrounding environment and its coinhabitants to those who possessed it, allowing the establishment of a new array of ecological interactions, which in turn drove other anatomical innovations.

For background information *see* ANIMAL EVOLUTION; ARTHROPODA; BURGESS SHALE; CAMBRIAN; EYE (INVERTEBRATE); FOSSIL; PALEONTOLOGY; PREDATOR-PREY INTERACTIONS; TRILOBITA; VISION in the McGraw-Hill Encyclopedia of Science & Technology.

John R. Paterson

Bibliography. M. S. Y. Lee et al., Modern optics in exceptionally preserved eyes of Early Cambrian arthropods from Australia, *Nature*, 474:631–634, 2011, DOI:10.1038/nature10097; C. R. Marshall, Explaining the Cambrian "explosion" of animals, *Annu. Rev. Earth Planet. Sci.*, 34:355–384, 2006, DOI:10.1146/annurev.earth.33.031504.103001; A. Parker, *In the Blink of an Eye: The Cause of the Most Dramatic Event in the History of Life*, Perseus Publishing, Cambridge, MA, 2003; J. R. Paterson et al., Acute vision in the giant Cambrian predator *Anomalocaris* and the origin of compound eyes, *Nature*, 480:237–240, 2011, DOI:10.1038/nature10689; J. Vannier and J. Chen, Early Cambrian food chain: New evidence from fossil aggregates in the Maotianshan Shale biota, SW China, *Palaios*, 20:3–26, 2005, DOI:10.2110/palo.2003.p03-40.

have hardened segments separated by membranes, as in arthropods. Although the body of *Anomalocaris* is arthropod-like in being segmented, it does not possess the serially repeated legs characteristic of other arthropods. Thus, the discovery of another definitive arthropod characteristic in *Anomalocaris*, that is, compound eyes, provides unequivocal evidence of this affinity.

The discovery also demonstrates that this type of visual organ appeared and was elaborated very early during arthropod evolution. The position of *Anomalocaris* and other members of Radiodonta on the arthropod evolutionary tree shows that compound eyes originated before other characteristic anatomical structures of the Arthropoda, including a hardened exoskeleton and jointed walking legs (**Fig. 3**).

The processing of visual information received from the large complex eyes of *Anomalocaris* would have required well-developed neurological structures—namely, the optic neuropils [a cluster of neurites (nervous tissue) in the central nervous system] and the protocerebrum (the anteriormost portion of the brain). The most recent common ancestor of modern arthropods is reconstructed as possessing two optic neuropils that transmit to the protocerebrum; thus, it can be assumed that *Anomalocaris* would have possessed the same neural architecture.

Connected vehicles

For as long as vehicles have been driven on roads, they have been essentially independent of each other, and their only connection to the roads has been the physical contact between the tires and the road surface. The only direct connections between vehicles have been intermittent and stationary (for example, jump-starting another vehicle) or unintentional, when drivers failed to avoid crashing into each other. Recent improvements in wireless communication technology have enabled a wide range of virtual connections among vehicles and the roadways, with profound implications for the future of road transportation.

Connected vehicles can make travel safer, faster, more efficient, more convenient, and more entertaining than unconnected vehicles. Travelers in connected vehicles are no longer isolated from the rest of the world or even the rest of the transportation system, but rather can be key parts of a well-integrated and well-coordinated intelligent transportation system. These opportunities have been recognized throughout the industrialized world, leading to substantial research, development, and demonstration programs to prove the benefits in North America, Europe, and Asia.

Many people are already using WiFi communications based on the IEEE 802.11 (wireless) standards to exchange information using their personal computers, tablets, and smart phones. A variation of this type of WiFi, called Dedicated Short Range Communications (DSRC), has been developed specifically for connected vehicle applications to provide highly reliable data transfers, with very short delays, among fast-moving entities. It is currently in the development-and-testing stage, and it is not yet known whether it will be widely deployed. In addition, the rapid advances in cellular communications using third- and fourth-generation (3G and 4G) cellular data transfer have already produced a revolution in smart-phone applications. These same cellular communication technologies are now being used for connected-vehicle applications as well.

Safety improvements. Although road transportation is safer than it has ever been, there are still too many crashes, resulting in high costs in deaths, injuries, and property damage. The U.S. Department of Transportation has estimated that up to 80% of the crashes that are not attributable to excessive alcohol consumption could potentially be avoided by using collision-warning systems based on connected vehicle technology. Many of these crashes occur because drivers fail to notice hazards quickly enough or misjudge the severity of the hazards and therefore do not respond as effectively as they should.

Vehicle manufacturers have already introduced some crash-warning and avoidance systems based on sensors such as radar, laser scanners, or video cameras to detect hazards. These sensors can be expensive, especially if a vehicle were equipped with enough of them to surround it with $360°$ coverage. An alternative to the sensor-based crash-warning and avoidance systems involves equipping each vehicle with an accurate positioning system, such as a Differential Global Positioning System (DGPS) receiver and a DSRC radio system to continually broadcast its position, speed, and heading-angle data to neighboring vehicles. Each vehicle's computer could then determine whether there is a danger of hitting any other vehicle and issue an appropriate warning to the driver using a visible, audible, or haptic alert. These vehicle-to-vehicle safety warnings include forward-collision warning; emergency electronic brake light (warning about braking by a forward vehicle when the driver's front view is blocked by a larger vehicle); blind-spot hazard warning; lane-change warning; overtaking-hazard warning; and crossing-traffic warning.

Safety improvements can also be supported by connections with the roadway infrastructure. Sensors installed on the roadside can detect the motions of vehicles approaching an intersection from all directions, and the data about those vehicles' trajectories can be broadcast by DSRC radio at the intersection to all the vehicles in the vicinity so that each vehicle can determine if it may be on a collision course with another vehicle, even one that is around the corner, where the view is blocked by buildings. The DSRC radio at the intersection can also broadcast the current traffic signal information, so that drivers can be issued an audible in-vehicle alert if they are in danger of violating a red light (or one that is about to turn red).

Some safety improvements can also be achieved by providing drivers with advance knowledge about traffic jams ahead, so that they can be alerted to the need to slow down. A substantial fraction of the rear-end crashes that occur today on limited-access highways are associated with the ends of congestion queues, where drivers approaching at full speed may not recognize that they are approaching slow or stopped traffic. Local information about congested traffic ahead can be communicated to vehicles using 3G or 4G cellular data networks, so it can be presented on in-vehicle displays or as alerts directly to a driver's smart phone.

Traffic flow improvements. One of the greatest challenges to improving traffic flow has been the expense and difficulty of collecting accurate real-time data about the actual traffic conditions throughout the road network. Traffic detection sensors and the communication networks needed to collect data are expensive to install and maintain, which has generally limited their use to the most heavily traveled highways. Connected vehicle technology can overcome this challenge because it enables each vehicle to become its own traffic data sensor. In this concept of vehicles as traffic data "probes," each vehicle periodically records its location and speed and then reports it (anonymously) over a wireless communication link so that the local traffic management center can aggregate the data from all the vehicles to produce a comprehensive picture of traffic conditions. Based on this broad knowledge of traffic conditions, a traffic management center can adjust the timing of traffic signals or freeway ramp meters to minimize delays and queue lengths, as well as respond to unpredictable incidents such as crashes.

Once the traffic data have been collected, they can be reported to the traveling public, who can use this enhanced knowledge to make better-informed decisions about their trips. Many people are already taking advantage of this type of information through websites that they can access by computer or smart phone. The same information can also be communicated directly to in-vehicle navigation systems in connected vehicles, to provide routing updates when a crash or other new incident creates a problem on the originally selected route.

Recent research on Active Traffic Management has shown that traffic jams at highway bottleneck locations can be reduced by using variable speed limits, encouraging traffic to slow gradually as the bottleneck is approached rather than braking abruptly at the last minute when the jammed traffic comes into view. Connected vehicle technology makes it possible to communicate these speed-limit variations directly to vehicles for display to the driver or to feed directly into the cruise control system as the new maximum set speed value (at the driver's discretion, of course) so that it does not require conscious action on the driver's part. *See* ACTIVE TRAFFIC MANAGEMENT.

More dramatic traffic flow improvements will be achievable in the longer term through the use of cooperative adaptive cruise control (CACC) systems. These combine the speed control capability of regular cruise control with forward-looking radar to measure the distance and closing rate to the car ahead and a DSRC communication link between the cars so that the following car can respond quickly and smoothly to speed changes by the lead car. Radar-based adaptive cruise control systems are already commercially available on some cars, where they provide increased driving comfort and convenience but do not help facilitate traffic flow. Adding the connected vehicle capability through DSRC makes it possible for the CACC vehicles to drive closer together in comfort and safety. Field tests have shown favorable driver reactions to the first experimental versions of this technology, and computer simulations show the potential to accommodate about 4000 vehicles per hour per lane on a freeway, which is about twice the current capacity per lane.

Efficiency improvements. Connected vehicles can improve the efficiency of road transportation several different ways by supporting both improved information and improved control. Real-time communication of traffic and transit operational information to vehicles can make it easier for drivers to make the most efficient mode choice for each individual trip. For example, when stuck in unexpected highway congestion, the driver can receive information about a variety of nearby bus or rail transit alternatives, including the availability of parking at park and ride lots and estimated arrival times at the destination, all based on current information. Diverting some of the car drivers that are stuck in traffic to transit will not only help them make their trip more efficient, but also will accelerate the dissipation of the congestion. At a more microscopic level, drivers waste considerable time and energy searching for parking spaces in congested downtown locations. Connected vehicle technology enables them to obtain real-time information about available parking spaces and in some cases to reserve a space so that they can go directly to their destination without extra driving.

Each time a vehicle needs to stop and restart at a traffic signal it wastes significant energy and produces excessive pollutants, compared to steady-speed cruising. Connected vehicle technology enables the traffic signals to communicate their current status and upcoming phase changes to the approach-

ing vehicles so that the vehicles can identify potential glide paths to enable them to coast through intersections without stopping, based on either driver speed control or adaptive cruise control. Initial tests of this approach have shown the potential for significant energy savings in very light traffic conditions, when other vehicles do not interfere with the fuel-efficient glide paths, but the benefits are likely to be reduced in heavier traffic.

Vehicle-to-vehicle communication using DSRC makes it possible for heavy trucks to drive in very close formation as electronically coupled virtual trains, producing significant savings in aerodynamic drag. Experiments have shown that when the trucks are driven with gaps of a few meters, they can reduce their fuel consumption by 5% (for the first truck) to 15% (for some of the following trucks) compared to what they would consume when each truck is driven independently. Ongoing research will be needed to refine all of the technologies needed to ensure the safety of this type of automated driving.

Convenience and entertainment. Entertainment systems are already available for use in the back seats of vehicles, although these normally depend on use of prerecorded media. Connected vehicle technology makes it possible to have real-time downloads of content, as well as web searching and interaction with social media and e-mail applications. The possibilities are almost limitless, except for the challenge of maintaining driver vigilance regarding the driving environment. If a driver is devoting visual or cognitive attention to an entertainment system, he or she is much less likely to be able to respond safely to a hazardous situation on the road, so much concern has been expressed about the necessity of preventing drivers from diverting their attention away from the driving task. There is an ongoing tension between the responsibilities of driving and the lure of entertainment, which has been tilted toward the entertainment side by the surge in smart-phone use.

Future directions. Telecommunications technology has been advancing dramatically in recent years, with product life cycles defined in months from introduction to obsolescence. In contrast, the road vehicle industry deals with product life cycles measured in years, while the roadway infrastructure industry builds its systems to last decades. These three contrasting industries converge and coexist in the connected vehicle space. The balance among them is not clearly defined or stable because of their large differences, making it hard to discern how connected vehicle systems will evolve in the coming years.

The safety-oriented aspects of connected vehicle technology will be decisively influenced by government decisions about the extent to which it should be encouraged or mandated for incorporation in new cars, based on the expected safety benefits. The testing to support those decisions will be done in the period 2012–2013, and assuming that the safety benefits appear substantial, the DSRC technology and connected vehicle safety systems should be entering the automotive market by 2020. Most of the traffic flow, efficiency, and entertainment oriented connected vehicle applications can be implemented using commercially available 3G and 4G cellular communications, so their market penetrations will proceed based on consumer perceptions of their benefits relative to their costs. Growing interest in the general information and entertainment (infotainment) applications is likely to increase interest in automated driving, so that drivers will be able to concentrate on their infotainment at the expense of their driving tasks, without compromising their safety.

For background information *see* DATA COMMUNICATIONS; HIGHWAY ENGINEERING; MOBILE COMMUNICATIONS; RADAR; SATELLITE NAVIGATION SYSTEMS; TRAFFIC-CONTROL SYSTEMS; WIRELESS FIDELITY (WI-FI) in the McGraw-Hill Yearbook of Science & Technology.   Steven E. Shladover

Bibliography. S. Miller et al., *National Evaluation of the SafeTrip-21 Initiative: Combined Final Report*, U.S. DOT Report FHWA-JPO-11-088, March 2011, http://ntl.bts.gov/lib/38000/38500/38510/safetrip_cfr.pdf; W. Najm et al., *Frequency of Target Crashes for IntelliDrive Safety Systems*, U.S. DOT Report No. DOT-HS-811-381, October 2010, http://www.its.dot.gov/research_documents.htm; S. E. Shladover, Three-truck automated platoon testing, *IntelliMotion*, 16(1):7–9, 2010, http://www.path.berkeley.edu/Publications/Intellimotion/Default.htm.

Conservation of archeological wood: the Oseberg find

The Oseberg find is one of the most important archeological finds in Norway and represents the most comprehensive collection of Viking Age wooden objects in the world. The archeological artifacts are displayed at the Viking Ship Museum in Oslo. Unfortunately, the unique objects are threatened by a slow but ongoing deterioration caused by a conservation treatment with alum salts that was done more than 100 years ago. Today, the artifacts of the Oseberg find are in an alarming condition.

The Oseberg find. In 1903, the owner of the farm *Lille Oseberg* came across a ship while digging in a large burial mound on his ground. The farm was situated near Tønsberg, 100 km (62 mi) southwest of Oslo, in the Oslofjord region, which was one of the Vikings' core settlement areas in Norway. The farmer invited the archeologist Gabriel Gustafson, head of the Museum of Antiquities in Kristiania (the former name of Oslo), to the site. He identified the burial mound as a ship grave from the Viking Age. Later, the burial was dated to A.D. 834.

The mound was 40 m (131 ft) in diameter and originally 6.4 m (21 ft) in height. It was excavated in 1904, revealing not only an almost complete Viking ship—the Oseberg ship (**Fig. 1**)—but also two female skeletons and a comprehensive collection of grave goods. Because of the weight of the covering stones and turf [5000 metric tons (5500 tons)], the excavated objects were highly fragmented. The richness of the burial indicates that the buried persons

were probably of a high social standing. Their identity, however, is still unclear.

The grave goods comprise daily-life items as well as richly ornamented ceremonial objects, weaving tools, kitchen equipment, and horse harnesses, which give an impression of Viking life. In particular, the sledges, a wagon (**Fig. 2**), animal head posts, and other items demonstrate the high standard of the Vikings' skills in craftsmanship, especially in wood carving. The carvings portray mostly animal motifs, embedded in geometrical patterns. Remaining particles of pigments allow for the assumption that parts of the objects were originally painted. Apart from the wooden objects, metal objects and a remarkable collection of textiles were also found.

All of the found objects were soaked with groundwater. The water content in this waterlogged wood supported its internal structure, even though significant amounts of the wood polymers were degraded by biological and chemical deterioration processes. If these artifacts were to be dried without prior treatment, they then would shrink irreversibly. Therefore,

Fig. 1. The Oseberg ship.

Fig. 2. One of the ceremonial objects: a richly ornamented wagon.

Fig. 3. The Oseberg objects were mounted together from numerous fragments in the early 1900s.

the water needed to be replaced by a suitable consolidation agent to support the wood structure and avoid shrinkage.

The necessity of such a consolidation treatment depends on the condition of the artifact, that is, the extent of decay. In the case of the Oseberg find, the objects made from oak were less degraded and did not need any consolidation, whereas the ones made from maple and birch were so highly deteriorated that they had to be strengthened to avoid dimensional changes during drying.

The objects made from oak, including the ship, were coated with a mixture of linseed oil and creosote in order to slow down the drying process. Creosote was also widely used as a wood protection agent at this time. After drying, the fragments were mounted together using brass and iron pins and screws. No further conservation treatment was applied to the ship and the other objects made from oak.

There was limited choice in wood consolidation treatments at the beginning of the twentieth century. After a study trip to selected European countries, Gustafson decided to use the alum conservation method to strengthen the highly degraded objects of the Oseberg find.

Alum treatment. One of the earliest methods for conservation of waterlogged wood was developed by C. B. Herbst in the 1850s in Denmark. The idea was to support the degraded wooden structure using alum crystals, recrystallized from a solution that had penetrated the wooden artifact. The method was mainly used in Scandinavia, but it has also been applied to collections in other countries worldwide. The conservation method using alum salts remained in use until the 1950s, when it was re-

placed by the use of polyethylene glycol and artificial resins.

There are several variations of the method known. An excavation report and archival material including a laboratory journal describe how the Oseberg objects were conserved: The artifacts were immersed in a concentrated solution of alum, $KAl(SO_4)_2 \cdot 12H_2O$. The solubility of alum in water decreases considerably with decreasing temperature. Therefore, a hot solution [approximately $90°C$ ($194°F$)] was used that penetrated the wood and replaced the water. After 2–36 h, the objects were removed from the bath and rinsed with water. The recrystallizing alum supported the weakened cellular structure. The objects could now be dried and they would preserve their relative dimensions (that is, there would be minimal dimensional change). After drying, the surfaces were brushed with linseed oil.

As mentioned previously, the objects were highly fragmented. Every single piece was conserved separately. Then, after drying, all fragments were refitted using adhesives, metal pins, screws, and nails (**Fig. 3**). Missing parts were replaced with filler materials and modern wood. The reconstructed objects were coated with a matte lacquering (*mattlakk*). In the 1950s, the objects on display were recoated with a synthetic resin.

Condition of the find. At the beginning of the twenty-first century, conservators observed damage on the alum-treated wooden objects from Oseberg. The wood had become brittle, cracks and voids were observed, and some pieces had fallen apart. The mechanical stability of the wood was similar to that of crisp bread, and the objects were no longer able to carry their own weight. At the same time, similar observations were made for alum-treated wooden artifacts in other Scandinavian collections. Because all objects had been treated with alum, it led to the assumption that the decay was caused by the alum treatment.

X-radiographs show an uneven distribution of alum in the wood. Alum crystals are concentrated in a thin layer on the surface, while the inner part is relatively unconserved. This leads to mechanical tensions, resulting in cracks. Optical examinations as well as X-ray microtomography show that the core is highly deteriorated (**Fig. 4**). The objects are more or less held together by a thin surface layer consisting of the remaining wood, alum crystals, and coatings.

All alum-conserved wooden objects exhibit a high acidity compared to fresh wood or untreated archeological wood. This high acidity is considered to be the main reason for the ongoing decay. Initial chemical analyses using infrared spectroscopy and nuclear magnetic resonance spectroscopy revealed an almost total loss of carbohydrates. Only lignin is left; and even though it is known to be quite stable against acids, this wood component also seems to be affected by acidic deterioration over a 100-year period.

As mentioned previously, the fragments were mounted together using brass and iron pins and screws. There are also original decorative nails

Fig. 4. Synchroton X-ray microtomography shows the grade of deterioration on a microscopic level: (*a*) fresh poplar wood; (*b*) alum-conserved archeological wood. [*Images were taken at the TOMCAT beamline, Swiss Light Source (SLS)*]

present. All these metal pieces are undergoing corrosion processes. This leads not only to a loss of mechanical stability but also to a release of copper, zinc, and iron ions. In addition, iron ions had migrated from the soil into the artifacts during the burial. These metal ions may thus contribute to the decay, as is known from other wooden objects and from paper and parchment conservation. Because these are catalytic processes, only small amounts of metal ions are needed to initiate a deterioration process.

Decomposition of alum. An alum solution at room temperature is slightly acidic, with a pH value between 3.5 and 4, as a result of hydrolysis. Alum-conserved wood, however, exhibits a pH of approximately 1. Where does the acid come from? Reconstruction of the alum treatment following the historical procedure revealed decomposition of alum as the source of the acidity. By heating, insoluble aluminum hydroxide compounds were formed and precipitated, accompanied by the formation of sulfuric acid. The likely reaction is

$$3KAl(SO_4)_2 + 6H_2O \rightarrow KAl_3(SO_4)_2(OH)_{6\downarrow}$$
$$+ K_2SO_{4(aq)} + 3H_2SO_{4(aq)}.$$

Precipitation of $KAl_3(SO_4)_2(OH)_6$ is a slow process; nevertheless, one has to take into account that the original treatment took up to more than 30 hours and that the same bath was used for several treatments. This led to an enrichment of the aluminum hydroxides, resulting in increased acidity.

The use of a hot alum solution for conservation is at least one important source for the acid in alum-conserved objects. There are also indications that, under certain conditions, alum may decompose at room temperature, and metal ions may also play a role in the formation of sulfuric acid in the wood.

Future of the Oseberg find. Successful reconservation treatments of alum-conserved wooden objects have been carried out by the National Museum of Denmark and the Swedish National Heritage Board. The alum salts and the acid were washed out, and the objects were retreated with a modern, more stable consolidator (polyethylene glycol). However, all these artifacts were quite simple and small, without any surface decoration. Furthermore, they were not restored using any additional materials, in contrast to the Oseberg artifacts. As such, a simple washing-out procedure will not be applicable to the Oseberg objects, and a preservation strategy for the Oseberg find has to be designed for its special needs.

A preservation strategy for the Oseberg find has to include neutralization of the acid and strengthening of the remaining wood structure. Moreover, metal ions have to be removed. The application of impregnation agents is challenging because of the varnish layer on the surfaces of the objects and the fact that many objects have been restored to such an extent that undoing these reconstituted sections risks irreversible damage. Moreover, the surface decorations are extremely sensitive with respect to any dimensional change.

Art objects and archeological finds have undergone restoration and conservation treatments for more than 200 years. It is likely that historical conservation treatments and their consequences will be an issue that conservators and conservation scientists will meet with increasing frequency because of the natural aging of the conservation materials used. A successful reconservation of the Oseberg find will not only preserve these unique artifacts for the enjoyment of future generations, but could also be a model case for showing how to deal with past conservation treatments.

For background information *see* ALUM; ARCHEOLOGY; ART CONSERVATION CHEMISTRY; SALT (CHEMISTRY); SHIP DESIGN; SHIPBUILDING; WOOD ANATOMY; WOOD DEGRADATION; WOOD ENGINEERING DESIGN; WOOD PRODUCTS; WOOD PROPERTIES in the McGraw-Hill Encyclopedia of Science & Technology.　　　　Hartmut Kutzke; Susan Braovac

Bibliography. S. Braovac and H. Kutzke, The presence of sulfuric acid in alum-conserved wood—origin and consequences, *J. Cult. Heritage*, in press, 2012, DOI:10.1016/j.culher.2012.02.002; M. Christensen, F. K. Hansen, and H. Kutzke, New materials used for the consolidation of archaeological wood—past attempts, present struggles, and future requirements, *J. Cult. Heritage*, in press, 2012, DOI: 10.1016/j.culher.2012.02.013; A. M. Rosenqvist, The stabilizing of wood found in the Viking ship of Oseberg: Part I, *Stud. Conserv.*, 4:13–21, 1959; A. M. Rosenqvist, The stabilizing of wood found in the Viking ship of Oseberg: Part II, *Stud. Conserv.*, 4:62–72, 1959; A. Unger, A. P. Schniewind, and W. Unger, *Conservation of Wood Artifacts: A Handbook*, Springer-Verlag, Berlin, 2010.

Decay heat

Nuclear reactors produce energy from the irradiation of nuclear fuel by neutrons. Most of the energy comes from splitting of heavy atoms present in nuclear fuel (for example, uranium or plutonium isotopes) into two lighter fragments. This fission process occurs when a heavy isotope like uranium-235 (^{235}U) or plutonium-239 (^{239}Pu) captures a neutron and then splits apart. Additional neutrons are created during the fission process and the kinetic energies from the fission fragments heat the bulk of the nuclear fuel. This heat can be converted into electricity through a cooling process based, for example, on circulating water used to produce steam to drive a turbine generator. A significant consideration is that energy generation in a nuclear power plant does not trigger carbon dioxide (CO_2) release into the Earth's atmosphere, which is one of the main concerns during the operation of conventional coal-burning power stations.

Most of the nuclei created in the fission process are radioactive and decay by emission of electrons, gamma rays, x-rays, and neutrons within a time range from milliseconds to years after the fission. The term "decay heat" refers to the energy from delayed beta and gamma radiation emitted by the fission products, and accounts for about 8% of the total energy produced in a reactor by fission. After a reactor shutdown, decay heat is the dominant source of heating the nuclear fuel and the reactor itself. Accidental loss of cooling with a natural continuation of decay heat release may result in overheating of the fuel and even failure of the fuel integrity. The subsequent release of radioactivity may occur if safety systems are not available or adequate. This was illustrated in the recent Fukushima disaster in Japan.

Call for new studies. Nuclear power, as a CO_2-free energy source, will continue to contribute to energy generation technologies. However, in 2007, even before the Fukushima disaster, the Nuclear Energy Agency (NEA) of the Organization for Economic Cooperation and Development (OECD) called for new studies of decay heat to reduce the uncertainties in computer codes and nuclear data used in the analysis of emergency core cooling system performance. These studies are intended to provide a better understanding of reactor safety and nuclear waste management, and improve the designs of a new generation of nuclear power plants.

Pandemonium effect. Fission products in nuclear fuel often have a large beta decay energy, even exceeding 10 MeV (10^7 eV), and a large excess of neutrons over protons. Therefore, several beta transitions, including so-called allowed Gamow-Teller transformations (no parity change between the parent and daughter nuclear states, maximum spin change of $1\hbar$ unit) and the first-forbidden (FF) ones (parity change, spin change up to $2\hbar$), may compete during the decay process. Many final states are populated in beta decays of neutron-rich nuclei, each with a small beta branching ratio. These weak beta transitions are followed by cascades of gamma transitions, usually passing through a few discrete, low-energy levels. The numerous weak gamma transitions at the top of the cascades are very difficult or impossible to detect with radiation detectors having a low efficiency. The resulting decay scheme usually results in an underestimation of the energy released in the form of gamma radiation, with a corresponding overestimation of the kinetic energy of beta rays (electrons). This effect got the name "Pandemonium effect" after John Hardy and his colleagues, who analyzed the decay of an "unstructured" fictional nucleus named Pandemonium in 1977 (**Fig. 1**). "State of chaos" would be the closest meaning of the word "Pandemonium" in this context (after John Milton's seventeenth-century poem *Paradise Lost*). However, there is an "anti-Pandemonium" method in nuclear spectroscopy. Measurements using a total absorption spectrometer (TAS) having

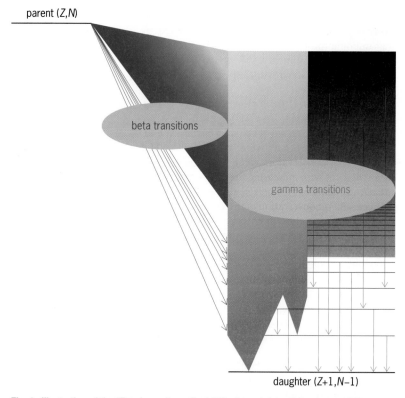

parent (*Z,N*)

beta transitions

gamma transitions

daughter (*Z+1,N−1*)

Fig. 1. Illustration of the "Pandemonium effect." The true picture of the decay of the neutron-rich parent nucleus (*Z* protons and *N* neutrons), with many weak beta transitions and following low-intensity gamma transitions, is given. Evidently, the determination of the true beta strength function, the gamma deexcitation patterns, and their interpretation might suffer from undetected gamma radiation. (*From K. P. Rykaczewski, Conquering nuclear pandemonium, Physics, 3:94, 2010, DOI:10.1103/Physics.3.94*)

(a) (b)

Fig. 2. Modular Total Absorption Spectrometer (MTAS). (*a*) MTAS inside lead shielding structure. (*b*) MTAS ready for measurement of decay heat of fission products at the mass separator on-line to the Tandem Accelerator at Oak Ridge National Laboratory (ORNL).

detection efficiency as close as possible to 100% can result in a true picture of a complex decay.

Studies with TAS devices. A few TAS devices have been constructed in the past from thallium-doped sodium iodide [NaI(Tl)] scintillator material. These spectrometers were attempting to cover a full solid angle around the studied activity. TAS energy resolution is rather poor by today's standards, about 6% full width at half maximum at 1.3 MeV, but the gamma cascades are efficiently summed up inside the detector. The true beta-feeding and decay scheme can be obtained after a careful evaluation of TAS spectra employing a deconvolution procedure.

Responding to the NEA's call for improved decay measurements for nuclides identified as important for decay heat, an international collaboration of nuclear scientists led by Alejandro Algora combined ion-trap techniques with TAS measurements to study several isotopes of refractory elements created in the fission of uranium-238 (^{238}U). The increase in the average energy of gamma radiation release (as compared to previous data) was found to be around or even above 1 MeV for technetium isotopes with mass numbers between 104 and 107, and for molybdenum-105 (^{105}Mo). These new data have helped to solve a long-standing discrepancy in

Y 89	Y 90	Y 91	Y 92	Y 93	Y 94	Y 95
100	64.1 h	58.51 d	3.54 h	10.18 h	18.7 m	10.3 m
Sr 88	Sr 89	Sr 90	Sr 91	Sr 92	Sr 93	Sr 94
82.58	50.53 d	28.6 y	9.54 h	2.66 h	7.5 m	74 s
Rb 87	Rb 88	Rb 89	Rb 90	Rb 91	Rb 92	Rb 93
27.83	17.78 m	15.15 m	2.6 m	58.4 s	4.49 s	5.84 s
Kr 86	Kr 87	Kr 88	Kr 89	Kr 90	Kr 91	Kr 92
17.30	76.3 m	2.84 h	3.18 m	32.32 s	8.57 s	1.84 s
Br 85	Br 86	Br 87	Br 88	Br 89	Br 90	Br 91
2.90 m	55.1 s	55.65 s	16.36 s	4.4 s	1.91 s	0.64 s
Se 84	Se 85	Se 86	Se 87	Se 88	Se 89	Se 90
3.1 m	33 s	14.1 s	5.8 s	1.53 s	0.41 s	>300 ns

(a)

La 137	La 138	La 139	La 140	La 141	La 142	La 143
60 ky	0.09	99.91	1.68 d	3.92 h	92.6 m	14.3 m
Ba 136	Ba 137	Ba 138	Ba 139	Ba 140	Ba 141	Ba 142
7.85	11.33	71.7	83.06	12.75 d	18.27 m	10.7 m
Cs 135	Cs 136	Cs 137	Cs 138	Cs 139	Cs 140	Cs 141
2.3 My	13.16 d	30.17 y	32.2 m	9.27 m	63.7 s	24.94 s
Xe 134	Xe 135	Xe 136	Xe 137	Xe 138	Xe 139	Xe 140
10.44	9.10 h	8.87	3.83 m	14.08 m	39.68 s	13.6 s
I 133	I 134	I 135	I 136	I 137	I 138	I 139
20.8 h	52 m	6.61 h	84 s	24.2 s	6.4 s	2.29 s
Te 132	Te 133	Te 134	Te 135	Te 136	Te 137	Te 138
3.2 d	12.5 m	41.8 m	18.6 s	17.5 s	2.49 s	1.4 s

(b)

Fig. 3. Fission products in the (*a*) mass A∼90 and (*b*) A∼140 regions whose beta decay was studied with MTAS at the HRIBF (ORNL) are marked by the red-bordered squares. The yellow symbols "1" and "2" indicate highest priority for decay heat measurements established by the OECD NEA assessment in 2007. Blue squares indicate nuclides that undergo beta decay, with the half-lives shown; black squares indicate stable or semistable nuclides, with the percentage abundances shown; and the pink square indicates a nuclide that decays by electron capture, with the half-life shown.

the gamma component of the decay heat for ^{239}Pu in the time range from 4 to 3000 s after the fission event.

MTAS measurements. Recently, a new program of decay heat measurements has been initiated at Oak Ridge National Laboratory (ORNL). A Modular Total Absorption Spectrometer (MTAS) has been constructed and commissioned at the Holifield Radioactive Ion Beam Facility (HRIBF; **Fig. 2**). With a mass of nearly 1000 kg (2200 lb), the MTAS array is by far the largest and most efficient TAS ever built. In a full implementation, with MTAS auxiliary detectors and activity transport system, the efficiency for full gamma-energy absorption is at the level of 78–71% within an energy range of 0.5–4 MeV. In addition, the MTAS modular design helps scientists to understand better the measured energy spectra during decay studies of fission products.

The measurements of the decay heat released from neutron-rich nuclei using the MTAS started at the HRIBF in January 2012. These nuclei were produced by bombarding the ^{238}U target with protons from HRIBF Tandem Accelerator to induce fission. The measurements were performed at the mass separator on-line to the Tandem Accelerator. The decays of over twenty fission products, including the seven highest priority decays defined by the NEA decay heat assessment, have been measured (**Fig. 3**). An example of results pointing to deficiencies in currently accepted decay heat data is given in **Fig. 4** for bromine-86 (^{86}Br) decay.

The evaluation and interpretation of decay data collected during the first MTAS campaign should yield the energies and intensities of gamma and electron transitions contributing to the decay heat in nuclear fuel for several nuclei critical to the analysis of decay heat at times of importance to reactor safety. It should also help to verify and develop further the theoretical description of beta-decay patterns for fission products. The reliable modeling of neutron-rich nuclei is needed to understand another process that occurs at high neutron density, namely nucleosynthesis in the rapid neutron capture process in the explosions of hot stars.

For background information *see* BETA PARTICLES; GAMMA-RAY DETECTORS; GAMMA RAYS; NUCLEAR FISSION; NUCLEAR PHYSICS; NUCLEAR POWER; NUCLEAR REACTOR; NUCLEAR SPECTRA; NUCLEAR STRUCTURE; RADIOACTIVITY in the McGraw-Hill Encyclopedia of Science & Technology.   Krzysztof P. Rykaczewski

Bibliography. A. Algora et al., Reactor decay heat in ^{239}Pu: Solving the γ discrepancy in the 4–3000-s cooling period, *Phys. Rev. Lett.*, 105:202501 (4 pp.), 2010, DOI:10.1103/PhysRevLett.105.202501; I. C. Gauld, Validation of ORIGEN-S decay heat predictions for LOCA analysis, in *Proc. of PHYSOR-2006, American Nuclear Society Topical Meeting on Reactor Physics: Advances in Nuclear Analysis and Simulation*, September 10–14, 2006, Vancouver, British Columbia, Canada, American Nuclear Society, La Grange Park, IL, 2006; J. C. Hardy et al., The essential decay of pandemonium: A demonstration of errors in complex beta-decay schemes, *Phys. Lett.*, 71B(2):307–310, 1977, DOI:10.1016/0370-2693(77)90223-4; K. P. Rykaczewski, Conquering nuclear pandemonium, *Physics*, 3:94, 2010, DOI:10.1103/Physics.3.94; Working Party on International Evaluation Co-operation, NEA Nuclear Science Committee, *Assessment of Fission Products Decay Data for Decay Heat Calculations*, NEA/WPEC-25, NEA No. 6284, Nuclear Energy Agency, Organisation for Economic Co-operation and Development, Paris, 2007.

Deep seabed mining

In the last half-century, our understanding of the global mineral resource base has changed drastically. The potential for commercial minerals per unit area in the oceans and seabed now appears to be similar to that on land. Almost three-fourths of the global mineral resources are in or under the sea and are virtually undeveloped. Half of the global seabed minerals are now controlled by coastal nations or small island states within their exclusive economic zones (EEZs), which extend 200 nautical miles (nmi) [370 km] from their shores. The other half are within the areas beyond national jurisdiction, known as the Area, and are controlled by the International Seabed Authority (ISA). The United Nations Convention of the Law of the Sea (UNCLOS), ratified in 1994, has resulted in the largest transfer of control of natural resources in history and in what is probably the most significant development in the history of the modern minerals industry—the exploration of the marine environment and the understanding of the potential for the development of marine minerals. In 1945, the Truman Proclamation on the outer continental shelf provided an initial impetus in the United States to go offshore for minerals other than oil and gas. Now every coastal nation has an EEZ and sovereign rights to its resources. The Area is administered by the ISA as a "common heritage of mankind," and negotiations for mineral recovery from the oceans are made

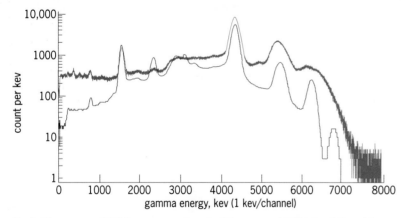

Fig. 4. The measured MTAS energy spectrum (red) is compared to the simulation of the MTAS response (black) to the decay of bromine-86 (^{86}Br). The decay schemes (like the one used for simulation of ^{86}Br MTAS spectrum) are taken from Evaluated Nuclear Structure Data File (ENSDF). This database is kept at Brookhaven National Laboratory. Spectra were normalized to the feeding of the 1.6-MeV excited state in krypton-86 (^{86}Kr). Measured excess of high-energy gamma radiation release is clearly observed. (*A. Fijalkowska et al., private communication, 2012*)

by the countries or entities involved. In the 1960s, the United States led developments in deep seabed mining, based largely on interest at that time in manganese nodules. Now, successful marine mining research and development is encouraged or subsidized by China, Japan, Canada, the United Kingdom, Australia, New Zealand, Papua New Guinea, Russia, Germany, South Africa, Fiji, the Cook Islands, and others. Extensive exploration and improvements in marine mining technology have resulted in the identification of deep seabed resources that were previously undiscovered or beyond economic recovery. These include, besides the ubiquitous manganese nodules, high-cobalt manganese crusts, metalliferous muds, marine phosphorites, seabed massive sulfides (SMS), methane hydrates, deep placer gold, diamonds, deep offshore sands, and precious corals. The greatest challenges to mining are in technology development and the mitigation of environmental issues. Processing of marine ores may be done on the seabed, onboard a surface platform, or on shore, depending on the particular deposit being mined. For the most part, the mining processes will be similar to those used for terrestrial deposits. Although all of the long-term effects of marine mineral resources recovery are not yet fully known, on the basis of current research it appears that the environmental effects of marine mining will, in general, be more benign than for equivalent recovery on land, and it appears highly likely that nondamaging recovery will be sustainable.

Advantages of deep seabed mining. Significant mineral discoveries on land in the twenty-first century have been constrained by increasingly competitive exploration costs in sometimes hostile geographic and political environments. Discoveries are followed by increasingly punitive costs of environmental regulations, land acquisition, government controls, future reclamation needs, and remote-area infrastructure development. Development of deep seabed minerals has the advantage of a mobile infrastructure, vast unexplored areas, potentially high-grade deposits, easy transportation, and not a great deal of competition at this time. Major disadvantages are those of working in the unpredictable climate of the oceans and the need to develop new tools and systems for exploration and mining. Climate issues are being overcome in many instances by working submerged in the calmer waters beneath the oceanic surface, and the needs for new technology are frequently supported by the advances made by industry in space and in deep-water oil and gas recovery. The inclusion of EEZs in a country or state's inventory of mineral resources can have very significant advantages. Alaska's EEZ, for example, covers over twice the area of the state's landmass; Fiji's archipelagic EEZ is over 70 times the total area of the islands, and the EEZ of solitary Johnson Island (U.S.) is 144,000 times greater in area than its land. To be competitive, the raw materials must be recoverable at a cost not greater than the cost of similar terrestrial materials, including transportation to the point of sale. Financing is a major part of the effort, although mineral development on land or sea is very sensitive to mar-

ket price. In today's markets, the comparative costs of recovery may be strongly influenced by secondary environmental or imputed costs.

Deep seabed minerals and mining technology. Deep seabed mining, as used here, is a broad term for the exploration, characterization, and recovery of economic mineral resources, excluding traditional oil and gas, from the marine environment in water depths greater than 200 m (660 ft). Seabed minerals occur in every environment from the surf zone to 6000 m (20,000 ft) in depth, and from the coast to 8000 km (4000 nmi) offshore. The artificial boundaries of EEZs and the Area do not necessarily limit the distribution of the deep seabed mineral resources, which may be found in both locations. Although many mineral occurrences are known, the actual amount of exploration on the deep seabed is still limited, and few operations at this time are beyond the development stage. Despite the fact that every mine is unique, there are only four basic methods of mineral extraction on land or at sea, namely, scraping the surface, excavating a pit or quarry, extracting the mineral through a borehole or other conduit as a fluid, and tunneling underground (**Fig. 1**). The latter method, however, is unlikely to be used for the recovery of minerals from deep seabeds. As with all subsurface mines, deep seabed mining systems must also deal with the transport of the mined materials to a surface facility for further treatment in preparation for sale. This activity may be simple, as in the case of placer materials, or highly complex, as in the case of polymetallic ores.

Manganese nodules. Deep seabed manganese nodules are potato-like concretions of manganese and other metal oxides commonly a few to 10 cm in diameter. They were first discovered during the HMS *Challenger* global oceanographic expedition in 1873 and, although ubiquitous in the deeper areas of the world's oceans around 4000 m (13,000 ft), were regarded at that time as a scientific curiosity. In the 1950s, John L. Mero, a University of California at Berkeley mining engineer, used the reports of the expedition and samples of the nodules stored in the British Museum in London to prepare a groundbreaking study of the contained minerals [oxides of manganese (Mn), iron (Fe), copper (Cu), nickel (Ni), molybdenum (Mo), and other metals], which together would constitute ore-grade material.

The best-known manganese-nodule deposits are in the Pacific in the Clarion-Clipperton fracture zone (CCZ) south and southeast of Hawaii and in the Central Indian Ocean Basin. Interest in the potential exploitation of the nodules generated a great deal of activity among prospective mining consortia in the 1960s and 1970s. Over the years, political and economic factors have prevented them from being exploited. However, technology developed by petroleum companies to recover oil and gas in water depths as great as 4000–5000 m (13,000–16,000 ft), plus the increase in value of Cu, Ni, and Co and the fact that the deposits also contain all of the 17 lanthanide (rare-earth) metals, now is resulting in a reassessment of their economic potential. There

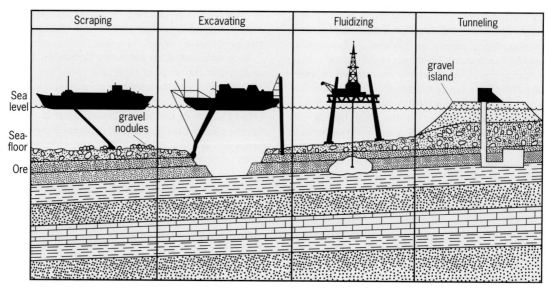

Fig. 1. The four basic methods of mining.

was a reemergence of commercial interest in 2008, when the Authority received two new applications to explore for nodules from private firms sponsored by their respective governments. This resulted in 15-year exploration contracts to Nauru Ocean Resources, Inc., on July 22, 2011, and to Tonga Offshore Mining, Limited, on January 12, 2012.

Production concepts. Surveys have determined that a weight-based cutoff for an economic deposit would exceed 10 kg/m². No regular patterns have been observed in the distribution of nodules in many of the areas surveyed, as both nodule-rich and nodule-depleted patches are found on undulating abyssal plains and on the tops and slopes of ridges. A minable deposit should contain enough nodules to guarantee commercial exploitation of 1.5-4 million tonnes

(metric tons) of nodules for 20-25 years, with metal grades of 1.25-1.5% for Ni, 1.0-1.4% for Cu, 27-30% for Mn, and 0.2-0.25% for Co. The development of mining systems for nodule recovery during the 1960s was intensive and included towed or self-powered mining devices that fed via hydraulic or air lift to a surface platform (**Fig. 2**).

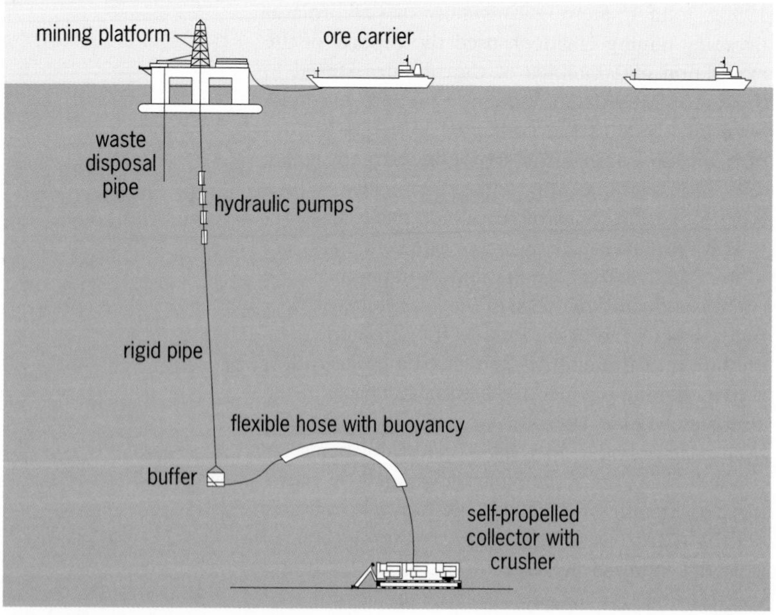

Environmental issues. During the last 30 years or so, following the testing of commercial mining concepts, major research efforts have concentrated on assessing the potential environmental impacts of nodule recovery. Studies included air quality, water quality, geological resources, biological resources, and social and economic resources. An unsettled issue is in the disposal of iron/manganese oxide tailings that comprise over 30% of the volume of the mined ore. Many innovative uses for these materials have been examined and show that they may form a valuable resource in their own right.

High-cobalt crusts. Iron-manganese oxyhydroxide crusts, now termed cobalt crusts, are found on hard rock substrates and are widely distributed throughout the ocean basins on the flanks and tops of sediment-free seamounts, ridges, plateaus, and abyssal hills. They form very slowly in water depths of 400-4000 m (1300-13,000 ft) as pavements or rock encrustations up to 250 mm (10 in.) thick at the rate of 1-10 mm/My. Most Co-rich crusts are commonly found at 800-2500 m (2600-8000 ft) and are potential resources of Co, Ni, platinum (Pt), Mn, titanium (Ti), and rare-earth elements (REEs), with a bulk dry weight of 1.3 g/cm³. The attachment of the crusts to the substrate varies considerably, but the crusts are generally weaker and lighter in weight than the underlying rocks. Certain geographic areas, such as the Cook Islands, which have an archipelagic boundary, may contain both high-cobalt crusts and nodules within their EEZs.

Production concepts. The method of recovery will necessitate the maximum removal of crust with as little substrate as possible to prevent dilution of the ore.

Fig. 2. Basic mining system for deep seabed minerals. (*Courtesy of International Seabed Authority*)

Appropriate systems have been proposed and would follow the flow sequence for deep seabed mining of a bottom-sited miner, a vertical transport system to a surface craft, and a separate or an integrated separation plant, with or without a waste-disposal subsystem. The mining machine would need to be capable of separating the crust layer from the substrate. Various options are available for this, including sharp cutting heads, vibratory heads, or high-pressure water jets. Some separation may be feasible on the seabed. The final design will depend on the environmental setting of the ore.

Environmental issues. As with other undeveloped resources, there are no confirmed environmental problems identified at this time. However, in 2009, the Cook Islands Parliament passed the Seabed Minerals Act to provide the government and people of the Islands, and not outside interests, control over seabed mining. A task force will be assisted and advised by independent professional experts, such as The Pacific Islands Geoscience Commission (SOPAC) and other governmental bodies, before any mining commences in their waters.

Metalliferous muds. Since the first reports of metalliferous deposits in the Red Sea in 1897, more than 40 expeditions have studied these unique deposits. The Red Sea is an example of a seafloor spreading center in the early stages of separation of the Arabian plate from Africa. A spreading rate of about 1 cm/yr has resulted in the formation during the last 4–5 million years of a central rift in the seabed 1500–2000 m (5000–6600 ft) deep with some 25 isolated "Deeps" containing metalliferous muds and brines associated with hydrothermal springs. Potentially economic deposits of Fe (17–29%), Mn (0.1–2%), Cu (0.1–4%), and zinc (Zn) [0.1–10%] sulfides have been identified in certain areas, and estimates of the Atlantis II Deep suggest a tremendous potential of 1.95 million tonnes of Zn, 400,000 tonnes of Cu, 4000 tonnes of silver (Ag), and 60 tonnes of gold (Au). Numerous sites throughout the Pacific have also been shown to contain high concentrations of REEs and yttrium (Y) in deep-sea muds.

Production concepts. Pre-pilot-test mining of the muds in the Atlantis II Deep was done in 1979 from the deep drilling vessel *Sedco 445* and managed by Preussag of Germany. The ship was equipped with a 2200-m (7200-ft) steel drill string with a suction head attached to the end of a standard oil-drilling pipe. The suction head, with a motorized vibrating attachment, broke up the mud and pumped it to the surface. About 15,000 m³ (530,000 ft³) of mud and added brines were recovered for processing by froth flotation. Preussag engineers estimated that commercial recovery of Ag and other metals would require at least 400,000 tonnes of slurry and seawater per day.

Environmental issues. The probable mining and processing of the muds may be expected to add to the risk of pollution. The Saudi-Sudanese Red Sea Commission set out an environmental study program that emphasized the assessment and magnitude of the possible risks resulting from tailings disposal in the Atlantis II Deep over time. The results obtained, so far, indicate that a well-controlled tailings disposal below 1000 m (3300 ft) water depth would keep the environmental impact of such an operation to acceptable dimensions.

Marine phosphorites. Phosphorus is necessary for all life and is one of the three elements [nitrogen (N), phosphorus (P), and potassium (K)] required for plant growth. There are no substitutes for P in agriculture, and it is likely that global land reserves will run out within the next century. Perceived global shortages of P resulted in prices rising from $80/tonne in 2007 to $200/tonne in 2012. There are, however, very large resources of phosphorite (calcium fluorophosphate) in the marine environment on the continental shelves and on seamounts and guyots in water depths to 500 m (1600 ft) or more that have not yet been developed or discovered. The best-known deposits are in New Zealand on the Chatham Rise, where the ore is found at around 400 m (1300 ft) and is of a nature that it may be spread directly on the ground without further treatment, and in Southwest Africa on the Namibian Shelf. Mining or exploration permits have been granted for tracts in each area. The Sandpiper Marine Phosphate Project, owned by Namibian Marine Phosphate (Pty) Ltd (NMP), is located about 50 km (30 mi) offshore and 120 km (75 mi) south-southwest of Walvis Bay. The water depths in the license area range from 180 to 300 m (600 to 1000 ft). NMP was granted a mining license over a defined mineral resource area 25.2 km (16 mi) wide and 115 km (71 mi) long, an area of 2233 km² (862 mi²).

Production concepts. The key operational aspects for the Sandpiper project include a planned recovery of 3 million tonnes of phosphatic sediment by dredging, initially from water depths to 225 m (740 ft) using a trailing suction hopper dredge. The dredged material will be transferred shoreside by pumping from the vessel to a buffer pond, where the oversize will be screened. The resultant slurry will be pumped to a processing plant near Walvis Bay, where it will be deslimed, treated by gravity separation to remove fine shells, and washed and dried. The rock phosphate concentrate will then be transferred to the nearby port of Walvis Bay as the point of sale.

Environmental issues. In January 2012, NMP announced the filing with the relevant Namibian government ministries of a Draft Environmental Impact Assessment (EIA) and a Draft Environmental Management Plan (EMP) report for the marine component of the project. The draft report concluded, among other things, that there are presently no identified issues of environmental significance to preclude the dredging of phosphate-enriched sediments. However, in May 2012, strong opposition from local fisheries caused the operations to be put on hold.

Seabed massive sulfides. Hydrothermal metalliferous sulfides, or seafloor massive sulfides (SMS), associated with tectonic plate boundaries and with values of Au, Ag, lead (Pb), Zn, and other metals were first discovered in 1977 by the U.S. submersible *Alvin* at black smokers emanating from hydrothermal vents

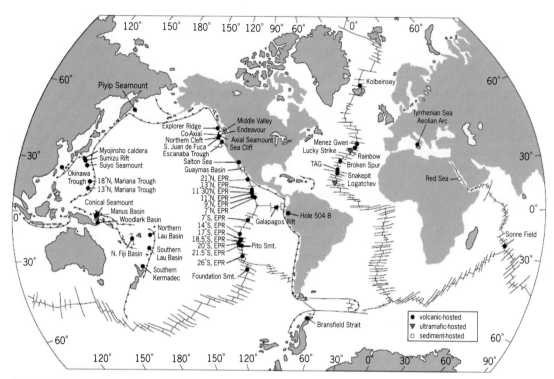

Fig. 3. Global distribution of known hydrothermal deposits of SMS. (*Courtesy of International Seabed Authority*)

in the Galapagos Ridge. This early work preceded discoveries at other spreading centers throughout the world's oceans at water depths from 1000 to 3000 m (3300 to 9800 ft) that have resulted in applications for exploration and mining permits to mining companies within the EEZs of Papua New Guinea (PNG), Tonga, Tuvalu, Fiji, and New Zealand (NZ). In January 2011, the Government of PNG granted the world's first deep-sea mining lease to Nautilus Minerals, Inc., in its territorial waters in the Bismarck Sea for the development of its Solwara 1 project. The lease covers an area of approximately 59 km² (23 mi²) surrounding Solwara 1, 50 km (30 mi) north of Rabaul, where Nautilus intends to mine high-grade Cu and Au deposits at depths of approximately 1600 m (5000 ft). The Solwara 1 SMS system appears to be of a size and grade that is typical of ancient massive sulfide analogs, although base and precious metals are particularly high compared to projects on land, which typically have Cu grade of much less than 1%. Neptune Minerals, also a leading explorer and developer of SMS deposits, has exploration licenses totaling more than 278,000 km² (107,308 mi²) in the territorial waters of NZ, PNG, the Federated States of Micronesia (FSM), and Vanuatu, and licenses pending over 436,000 km² (168,340 mi²) in the territorial waters of New Zealand, Japan, the Commonwealth of Northern Mariana Islands (CNM), Palau, and Italy. Neptune, along with Canyon Offshore, Century Subsea, and Seacore, undertook coring and sea-floor mapping in the company's granted prospecting license (Kermadec PL 39-195), within the EEZ of NZ. Elsewhere, Norwegian scientists reported discovery of the northernmost known hydrothermal vent fields around 71° North on the

Mohns Ridge, a part of the Arctic ridge system (**Fig. 3**).

Production concepts. Nautilus is planning to test production capabilities based on an Indicated Mineral Resource of 870 kt at 6.8% Cu, 4.8 g/t Au, 23 g/t Ag, 0.4% Zn and Inferred Mineral Resources of 1300 kt at 7.5% Cu, 7.2 g/t Au, 37 g/t Ag, 0.8% Zn. Mining operations, as planned, will use large (250-tonne) crawler machines with rock-cutting devices feeding into a pump line to a surface vessel. Initially the ore material will be transported to land for processing, but integration of these operations at sea will undoubtedly follow.

Environmental issues. The planned operation is essentially undersea strip-mining, but it is presently anticipated that the tiny forms of life found on the seafloor at this depth will repopulate an area within 2–3 years. Nautilus claims to be taking a conservative and careful approach. They will be watching the results of their operations and making the needed adjustments.

Methane hydrates. A very significant development has been in the exploration of the vast extent of frozen marine natural-gas hydrates (NGH), estimated by the United States Geological Survey (USGS) to contain more than twice the available energy in the form of methane than is contained in the global resources of all other hydrocarbon fuels. Marine hydrates, often in layers several hundred feet thick, occur in ocean-floor sediments below 300 m (1000 ft) and remain stable a within a range of high pressure and low temperature (P/T). Developments in hydrate research were encouraged or supported in the United States by the Methane Hydrate Research and Development Act of 1996 and include the use of three-dimensional seismic and infrared spectroscopy techniques in exploration and in thermal-exchange extraction methods. The United States, Japan, Canada, India, China, Korea, Norway, Mexico, and New Zealand have all been involved recently in activities dealing with the development of these resources.

Production concepts. Currently, research to develop production technology is based on the three primary concepts of depressurization, thermal stimulation, and chemical stimulation. The depressurization technique reduces the fluid pressure in the porous rocks in contact with the hydrate reservoir. Thermal stimulation increases the temperature of the reservoir above the P/T threshold by in-situ combustion through a close pattern of drill holes or combustion chambers. Chemical simulation entails the injection of carbon dioxide into the formation, where the carbon dioxide is preferentially exchanged for the methane gas. Each of these methods has been demonstrated successfully in the field or laboratory and may be ready to yield field production by 2025.

Environmental issues. The instability of hydrates outside of their stability zone indicates an area where serious research still needs to be done in the potential recovery and transportation of the product.

Deep placer gold. Placer gold deposits have been indicated by side-scan imagery in the deep water of the Gulf of Alaska, where the material has been transported by turbidity currents from the nearshore through an adjacent submarine canyon to water depths of over 2500 m (8200 ft). No attempt has been made at this time to recover the gold or evaluate the deposit.

Production concepts. A seabed miner similar to those used in the mining of the Nome gold deposits may be used, except that the ratio of gold to overburden (about 1:1 million), the weight of the native metal (specific gravity 19 g/cm^3), and the greater distance involved in the vertical transport system may call for separation of the gold, disposal of the waste on the bottom, and some innovative method for bringing the gold to the surface.

Environmental issues. These have not been determined at this time but would be by thorough analysis during the characterization of the deposit.

Offshore diamonds. Diamonds were first located off the Southwest African coast in the 1960s. Because the diamonds offshore had traveled hundreds of miles by river from the interior and had been subjected to heavy surf action on the seabed for thousands of years, the remaining stones were generally of the finest quality. Leases have been awarded along the coasts of Namibia and South Africa.

Production concepts. The diamonds are generally located in cracks and gullies in the bedrock, and early production systems used, among other things, vertical drills like tunnel-boring machines several meters in diameter to penetrate the bedrock, prior to pickup by mechanical means or by suction. In 2007, De Beers Consolidated Mines (DBCM) named its first marine diamond vessel to mine off the South African coast. The ship, *Peace in Africa*, is a $142 million

vessel 176 m (580 ft) long and 28 m (92 ft) wide and is designed to produce an estimated 240,000 carats annually off the Namaqualand coast. The ship is equipped with a large undersea tracked mining vehicle and has a specialized diamond recovery and treatment plant onboard. The company enhanced its crawler technology, used in the offshore Namibian operations, to mine the 8000-km^2 (3100-mi^2) reserve in South Africa at a higher rate. Another feature of the enhanced crawler technology is that it provides a three-dimensional visualization of the seabed and mining operations. The project is exceptional for its grade of 0.1 carats/m^2 and the fact that 95% of diamonds found are expected to be of gem quality. Although the size of the reserve is substantial, only about 0.5% of the area will be mined by the *Peace in Africa* over 30 years. De Beers will consider expanding its fleet of diamond-mining vessels to exploit a greater part of the resource at a greater pace, and will explore the remaining 95% of the diamond reserve on an ongoing basis to extend the project life.

Environmental issues. Although the nature of the seabed habitat and the communities that live in the soft sediment are typically destroyed during the mining process, independent scientific assessments of mining operations on the west coast of southern Africa claim to have demonstrated that natural recovery of the unconsolidated sediment habitats occurs over time. Recovery periods depend on the specific environment and the interactions with other natural processes and vary between 4 and 15 years for the seabed faunal communities in this region.

Deep offshore sands. Coastal erosion has become a serious problem in many parts of the world, and in Pacific Island states, where many of the coastlines are beaches formed from erosion of adjacent coral reefs, measures are being taken to replenish the beaches using offshore sands. During the course of this work it has been noted that there are many and substantial drowned beaches at a depth of about 100 m (330 ft) or more that were formed at a previous still-stands of sealevel. These deposits are currently too deep for standard dredging equipment but are likely resources for the near future.

Precious corals. Small industries for black, bamboo, and red precious corals found on hard substrates in water depths beyond 200 m (600 ft) are maintained in waters off Hawaii, Taiwan, and in the Red Sea. Commonly recovered by dragging bundles of material on which the corals snag. Most modern operations are carried out from small submersibles.

Conclusions. It is apparent that deep seabed mining will provide a high potential for the provision of mineral resources that may be in short supply on a global basis. The present (2012) state of the world economy and the pressures of organizations supporting preservation in preference to development, despite increasing domestic and global demands for resources, indicates the need for a more positive approach to mining than is presently being evinced by authorities in the United States and elsewhere. Continued cooperation with other appropriate entities on a national and international basis is an essential requirement for the development of the global marine mining industry and for the assurance of access in the future to an adequate supply of strategic and critical minerals.

For background information *see* GEOPHYSICAL EXPLORATION; HYDRATE; HYDROTHERMAL VENT; MANGANESE NODULES; MARINE GEOLOGY; MARINE MINING; MINING; ORE AND MINERAL DEPOSITS; PACIFIC ISLANDS; PLACER MINING; PLATE TECTONICS; RARE-EARTH ELEMENTS; RED SEA; UNDERWATER VEHICLES in the McGraw-Hill Encyclopedia of Science & Technology. Michael J. Cruickshank

Bibliography. C. L. Antrim, *What Was Old Is New Again: Economic Potential of Deep Ocean Minerals the Second Time Around*, Center for Leadership in Global Diplomacy, background paper, August 2005; D. S. Cronan, *Handbook of Marine Mineral Deposit*, CRC Press, 1999; M. J. Cruickshank, Marine mining, an area of critical national need, *Mining Eng.*, 63(5):89–93, 2011; M. J. Cruickshank, Technological and environmental considerations in the exploration and exploitation of marine minerals, Ph.D. dissertation, University of Wisconsin, Madison, 1978; M. J. Cruickshank and R. Kincaid, Marine minerals development in the Pacific basin and rim, *Proceedings of the 14th Congress of the Council of Mining and Metallurgical Institutions*, Edinburgh, 2–6 July, pp. 197–206, 1990; M. R. Dobson, Placer deposits in submarine fan channels, *Marine Mining*, 9:495–506, 1990; N. F. Exon et al., Ferromanganese nodules and crusts from the Christmas Island Region, Indian Ocean, *Marine Georesources Geotechnol.*, 20(4):275–297, 2002, DOI:10.1080/03608860290051958; R. H. T. Garnett, Marine placer diamonds, with particular reference to southern Africa, in D. S. Cronan (ed.), *Handbook of Marine Mineral Deposits*, CRC Press, pp. 103–141, 1999; R. H. T. Garnett, Marine placer gold, with particular reference to Nome, Alaska, in D. S. Cronan (ed.), *Handbook of Marine Mineral Deposits*, CRC Press, pp. 67–101, 1999; L. Hannah, *P-ACP Regional legislative and regulatory framework for deep sea minerals exploration and mining*, Secretariat of the Pacific Community, ver. 1, [in draft], January 20, 2012; J. R. Hein et al., Cobalt rich manganese crusts in the Pacific, in D. S. Cronan (ed.), *Handbook of Marine Mineral Deposits*, CRC Press, pp. 239–279, 1999; Y. Kato et al., Deep-sea mud in the Pacific Ocean as a potential resource for rare-earth elements, *Nature Geosci.*, 4:535–539, 2011, DOI:10.1038/ngeo1185; R. Kotlinski and P. Rybár, Technologies of manganese nodules mining, *The 17th Annual General Meeting of The Society of Mining Professors/Societät der Bergbaukunde*, June 17–21, 2006, Košice, Slovak Republic, 2006; S. Kral, Regulations, public awareness still constrain minerals development, *Mining Eng.*, 63(5):94–95, 2011; P. Kulkarni, Beyond unconventional: Unlocking the gas potential of hydrates, *World Oil*, 231(3):21, 2010; J. L. Mero, *The Mineral Resources of the Sea*, Elsevier, New York, 1965; MMS, *Synthesis and Analysis of Existing Information Regarding Environmental Effects of Marine Mining*, U.S. Department of the Interior, Minerals Management Service, Final Report, OCS Study, MMS 93-006, 1993; P. Moore, New wave: Sub-sea mining, *Mining Mag.*, pp. 31–36, September 2010; MTR, South African engineers lead deep sea mining effort, *Marine Technol. Rep.*, November, pp. 38–39, 2007; J. C. Scholten et al., Hydrothermal mineralization in the Red Sea, in D. S. Cronan (ed.), *Handbook of Marine Mineral Deposits*, CRC Press, pp. 369–395, 1999; K. Sherman, The large ecosystem concept, research and management strategy for living marine resources, *Ecol. Appl.*, 14:349–360, 1991; J. C. Wiltshire, Innovations in marine ferromanganese oxide tailings disposal, in D. S. Cronan (ed.), *Handbook of Marine Mineral Deposits*, CRC Press, pp. 281–305, 1999.

Deepest nematodes

The deep subsurface biosphere, extending more than 3 km (1.86 mi) under the Earth's surface, comprises a significant fraction of the global planetary biosphere. Limitations imposed by temperature, energy, oxygen, and space led to the conviction that the deep subsurface was populated exclusively by bacteria and viruses with a total biomass equaling that of the surface. However, no multicellular organisms were believed to be able to survive at these depths. Although many lower invertebrate phyla have a reputation for being able to survive in harsh conditions, the phylum Nematoda (also termed Nemata, comprising the roundworms) is second to none. (Roundworms are not to be confused with earthworms, which belong to the phylum Annelida.) Nematodes have been recovered from the seafloor of deep oceans, hot springs, and acidic seeps (pH 0), and some species recover easily from deep freezing or decades of desiccation. However, the most impressive example of stress resistance in nematodes was demonstrated by the survival of the species *Caenorhabditis elegans* during the 2003 breakup of the space shuttle *Columbia* upon reentry, and the subsequent free fall and impact of the biological container in which the species was contained. In addition to the ability of nematodes to undergo anabiosis (a state of suspended animation induced by desiccation and reversed by the addition of moisture) for extended periods, these organisms continue to metabolize aerobically in hypoxic (oxygen-deficient) environments where the partial pressure of oxygen (O_2) is only 0.4 kPa (2% of the atmospheric value). Therefore, nematodes were regarded as prime candidates to look for in the deep subsurface because they are one of the most successful metazoan phyla on the surface with respect to their abundances, distribution, and physiological tolerances.

The search in the South African deep subsurface led to the discovery of four nematode species recovered at varying depths. One of the species is new to science and has been named *Halicephalobus mephisto*, which is known colloquially as the "devil worm" (see **illustration**).

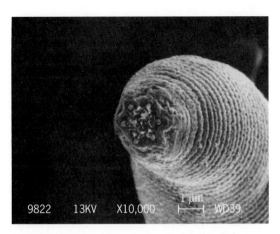

Scanning electron microscope image of the head of *Halicephalobus mephisto*, known as the "devil worm."

Discovery. In 2008-2009, a team from Ghent University (Belgium), University of the Free State (South Africa), and Princeton University (New Jersey) took 22 water samples from boreholes at depths varying from 0.5-3.6 km (0.3-2.24 mi) in the deep underground of South African gold, platinum, and diamond mines. Thousands of liters of mining water and several soil samples were taken as controls to exclude the possibility of contamination as an origin.

The following strategy was used to determine whether any nematode recovered was indeed indigenous and not a recent invader from the surface or a contaminant from mining water used to cool corridors and mine equipment. First, existing filtering equipment and protocols that had been successfully used to collect planktonic microorganisms from thousands of liters of borehole water in mines were adapted for the collection of nematodes. Second, soil samples around boreholes and mining water that might be the origin of contaminants were analyzed for nematodes. Third, the chemical composition and microbial community structure of the borehole water were analyzed to determine whether it was mining water or fracture water. Fourth, concentrations of ^{3}H and ^{14}C were measured to determine the age of the borehole water and whether it represented a recent influx of surface water as a result of mining activities.

The borehole samples resulted in the discovery of four nematode species, including *H. mephisto* at the Beatrix gold mine at depths of 1.3 km (0.8 mi). At depths of 0.9 km (0.56 mi), the Driefontein gold mine yielded two nematode species: *Plectus aquatilis* and a monhysterid specimen that survived but did not reproduce. Another monhysterid nematode (DNA traces only) was discovered at depths of 3.6 km (2.24 mi) in the Tau Tona gold mine; this nematode is the deepest invertebrate found to date, even surpassing the depth [2 km (1.24 mi)] at which a recently discovered annelid was found in a cave in Ukraine. Thus, these discoveries dramatically enlarge the known biosphere for metazoan animals on Earth.

No other metazoans were detected, and samples from two other mines yielded no nematodes. None of the control samples contained the same species, and very few nematodes were actually found in soil samples. Mining water contained no nematodes. Bacterial communities confirmed earlier results indicating that old fracture water was sampled. A ^{14}C analysis indicates that the ages of the nematode-bearing water were 10,000-12,000 years old for the Driefontein mine borehole, 4400-6200 years old for the Beatrix mine borehole, and 2900-5100 years old for the Tau Tona mine borehole. The nematodes, therefore, are not contaminants resulting from mining or from incursion of modern water; instead, they seem to be indigenous to the paleometeoric water (that is, underground water that originally came from the surface and that is either old or mixed with older water coming from even deeper underground).

Description. Nematodes are relatively simple organisms consisting of essentially three tubes: an outer tube (body wall) covered with a strong cuticle; an inner tube (digestive tract); and a third tube (reproductive system) squeezed between the inner and outer tubes. The nematodes collected in the deep underground do not differ from this general morphology and do not exhibit any exotic morphological features that make them stand out from their surface counterparts. However, morphology in nematodes is a poor indicator of evolutionary change. Currently, the genome of *H. mephisto* is in the process of being sequenced, and it remains to be seen to what extent its adaptation to an extreme habitat is reflected at the genetic level. Although nematodes that reach sizes of several meters are known, most are microscopic. *Halicephalobus mephisto* is 0.52-0.56 mm (0.02-0.022 in.) in length and is distinguished from other species of the genus *Halicephalobus* (as well as other nematodes) by its long tail.

Habitat and ecology. The nematodes recovered from the fracture water all feed by ingestion of bacteria and substrate. Typical bacteriophagous nematode behavior consists of grazing on bacteria patches rather than plankton. Based on previous studies on the number of bacteria in the deep subsurface, the ratio of bacteria in patches (5×10^4 bacterial cells per cm^2) to the number of nematodes would be 10^{10}-10^{12} to 1. Based on the weight, bacterivory rates, and respiration/total carbon consumption by nematodes, 10^4 bacterial cells would readily sustain *H. mephisto* for 1 day and 10^{11} bacterial cells would sustain *H. mephisto* for approximately 30,000 years. Therefore, the available known number of bacteria in the deep subsurface is not a limitation on indefinite nematode survival. Oxygen (O_2) is not a limiting factor either. Water samples containing the nematodes yielded O_2 concentrations of 1.3-6.8 kPa. Based on the mass of *H. mephisto*, its maximum metabolic rate would be 7×10^{-8} moles of O_2 per nematode per year. Thus, the estimated O_2 consumption rate could support 1 nematode per liter at the maximum metabolic rate, although 1 nematode per 10^4 liters was actually observed. Moreover, nematodes can maintain aerobic metabolism with O_2 concentrations as little as 4 μM.

The fracture water containing the nematodes ranged from 24 to 48°C (75 to 118°F). However, the latter temperature is above what is generally known for free-living terrestrial nematodes on the surface. Laboratory experiments confirmed that the nematodes recovered could not withstand temperatures much above the temperature of the water in which they were found. Although the ability of nematodes to go into prolonged anabiosis might allow them to survive bouts of excessive temperature, it would render them inactive; moreover, if the stress is too excessive, death would be inevitable. Thus, temperature is a limiting factor for the depth to which nematodes can be found.

Additional observations. The mechanisms by which the nematodes reached the deep subsurface remain as yet unknown. However, based on the genetic and morphological data obtained from the collected nematodes, it is clear that their origin is at the Earth's surface. Furthermore, because O_2 and food are not limiting factors for indefinite survival of nematodes at these depths, it remains to be determined whether, over time, nematodes have evolved unique adaptations to even harsher conditions deeper underground than those currently recorded. Another unanswered question is whether nematodes are the only invertebrates present in the deep subsurface. Although extensive sampling of the fracture water did not reveal any metazoans except nematodes, conditions prevailing at the sampling sites should have been manageable for other invertebrates (for example, annelids).

The discovery of nematodes grazing on bacteria in the deep subsurface indicates a far more complex ecosystem than was previously thought. Considering the high number of nematodes found on the ocean floor and around hydrothermal vents, nematodes should also be found beneath the seafloor, possibly in conjunction with other meiofauna (multicellular organisms with sizes ranging from a few micrometers to a few millimeters) inhabiting the deep [for example, Loricifera reported in anoxic (oxygen-depleted) brines of the L'Atalante basin]. Finally, the ability of multicellular organisms to survive the deep subsurface should be considered in the context of the evolution of eukaryotes on Earth, extinction events, and the search for life on seemingly inhospitable moons and planets (for example, Mars) in our solar system and beyond.

For background information *see* ANIMAL EVOLUTION; BIOSPHERE; ECOSYSTEM; MONHYSTERIDA; NEMATA (NEMATODA); OXYGEN; UNDERGROUND MINING in the McGraw-Hill Encyclopedia of Science & Technology.   Gaetan Borgonie

Bibliography. G. Borgonie et al., Nematoda from the terrestrial deep subsurface of South Africa, *Nature*, 474:79–82, 2011, DOI:10.1038/nature09974; D. Chivian et al., Environmental genomics reveals a single species ecosystem deep within the Earth, *Science*, 322:275–278, 2008, DOI:10.1126/science.1155495; M. Kaufman, *First Contact: Scientific Breakthroughs in the Hunt for Life Beyond Earth*, Simon & Schuster, New York, 2011; D. L. Lee (ed.), *The Biology of Nematodes*, CRC Press, Boca Raton, FL, 2002.

Definitive evidence for new elements 113 and 115

Current research continues to explore the limits of the periodic table of the elements. This article describes new insights into the discovery of elements with 113 and 115 protons. The number of protons, symbol Z, in the nucleus of an atom distinguishes each element. For example, hydrogen has $Z = 1$ and uranium has $Z = 92$.

Searches for new elements. Beginning around 45 years ago, it was theoretically predicted that unknown elements with numbers of neutrons (N) equal to 184, and number of protons either 114, 120, or 126 could form what is called an island of stability. Their half-lives for radioactive decay could be much, much longer than those of elements with 2 to 10 less protons and especially with 8 to 20 less neutrons. The half-lives of nuclei with $N = 184$ and $Z = 104$–108 could be so long that they could even be comparable to the age of our Earth (5×10^9 years), and so exist in nature. These predictions led to significant efforts to search for new elements beyond $Z = 106$, then the limit of the periodic table.

At the Gesellschaft für Schwerionenforschung (GSI) in Germany, scientists bombarded targets of lead and bismuth, the heaviest stable elements in nature, with projectiles up to zinc, $Z = 30$. The projectile energies were kept sufficiently low that following fusion, for example, of iron $Z = 26$ and lead $Z = 82$ to make element 108, only one neutron was evaporated from the compound nucleus. In this way, new elements from 107 to 112 were discovered. These new elements had neutron numbers $N \leq 165$, which are far from 184. Their half-lives became increasingly short as Z increased, with element 112 having a half-life of only 0.7 ms.

Then scientists at the Flerov Laboratory of Nuclear Reactions at the Joint Institute for Nuclear Research in Russia developed a new approach in which more neutron-rich, long-lived radioactive actinide targets from neptunium ($Z = 93$) to californium ($Z = 98$) were bombarded with very neutron-rich, double-magic calcium-48 (^{48}Ca) to make nuclei with neutron numbers 170–177, much closer to 184. (Certain magic numbers of protons and neutrons give special stability to nuclei by forming closed energy shells, much as certain numbers of electrons give closed energy shells to form the noble gases starting with helium. Here $Z = 20$ and $N = 28$ for calcium-48 are both closed shells to make it double magic.) From 1999 to 2005 they reported the discoveries of new elements with $Z = 113$, 114, 115, 116, and 118. In 2010, the new element 117 was discovered by bombarding with calcium-48 the much shorter-lived berkelium-249 (^{249}Bk; 320-day half-life) that was produced in the high-flux reactor at Oak Ridge National Laboratory.

New experiments on elements 113 and 115. In 2011, the reported discoveries of elements 114 and 116 were certified by the International Union of Pure and Applied Chemistry. They were given the names flerovium for 114 and livermorium for 116.

Discoveries of new elements are accepted when there is corroborative evidence from other laboratories or cross bombardments and observations at several excitation energies. For the first time, the discoveries of elements 113 and 115 were reported in 2004 in the reaction ^{243}Am + ^{48}Ca that led to the synthesis of isotopes of element 115 and their subsequent alpha decay, resulting in the discovery of element 113, which was also unknown at the time. At the 248-MeV calcium-48 energy, three decay chains originating from ^{288}115 were registered. At about 5-MeV higher calcium-48 energy, one decay chain of the neighboring isotope ^{287}115 was detected. To provide data to definitely establish the discovery of elements 113 and 115 produced in the reaction of ^{243}Am + ^{48}Ca, this reaction was studied again at the Flerov Laboratory in Russia. This new work was a collaboration between the Flerov Laboratory, Vanderbilt University, Lawrence Livermore National Laboratory, and Oak Ridge National Laboratory.

The new experiment was carried out in two parts, separated by several months. The setup for the experiment is shown in **Fig. 1**. The U400 cyclotron was used to accelerate calcium-48 to five different energies from 240 to 254 million electronvolts (MeV) to study the production of different isotopes of element 115 at the different energies.

The fusion of calcium-48 and the americium-243 is illustrated in **Fig. 2** for what is called a 3n (three-neutron) evaporation channel. After evaporating x numbers of neutrons, the evaporation residue recoiled out of the target to conserve linear momentum. The evaporation residues then passed through a gas-filled magnetic separator, shown in Fig. 1, where they were separated from the undisturbed calcium-48 beam particles and other reaction products.

The superheavy element recoils then passed through two time-of-flight detectors to establish their coming from the separator. Then they were implanted into a semiconductor detector array with 12 vertical position-sensitive strips (Fig. 1). The 12 strips were surrounded by 8 side detectors in an open-ended box to detect alpha particles and fission fragments that escaped detection of their total energy in the strips because of their forward angle of emission. Events detected in this way had larger uncertainties in the alpha energies. To establish that an event in one of the pixels in one of the strips corresponds to an isotope of a new element, the evaporation residue event must be followed by the detection of an alpha particle with an energy and half-life in a range expected from theoretical calculations. This alpha event must likewise occur in the same pixel as the evaporation residue within ≤1 mm. After the first alpha was detected, the calcium-48 beam was switched off and the subsequent alpha emissions and spontaneous fission were observed without background. These subsequent events must all be recorded in their same pixel within ≤1 mm to ensure they form a decay chain.

The first part of the experiment was carried out at four different energies to measure the production probability as a function of energy (excitation

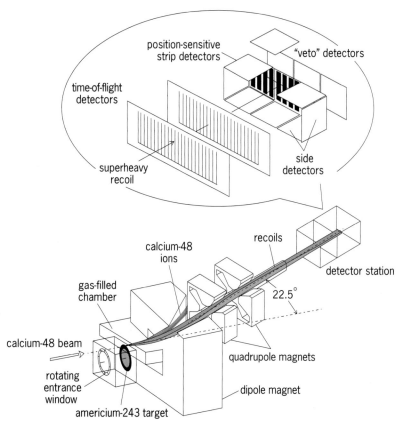

Fig. 1.   Schematic of the Dubna gas-filled recoil separator (consisting of a dipole magnet followed by a doublet of quadrupole magnets). Also shown is the schematic of the detector station at the separator focal plane.

function) for the evaporation of three neutrons from the ^{291}115 compound nucleus to give ^{288}115. In this first part of the experiment, a total of 21 new decay chains of ^{288}115 were observed at four different bombarding energies and one new observed chain was assigned to ^{289}115 at 241 MeV following two-neutron evaporation, as reported in *Physical Review Letters* in January 2012. After submission of these new results, the experiment was continued at 241 MeV. Three additional chains of the type α-α-spontaneous fission following two-neutron evaporation were observed, along with seven additional events of ^{288}115. Following a change in energy to 254 MeV, one new event associated with the four-neutron evaporation channel was observed. The total number of decay chains observed for the

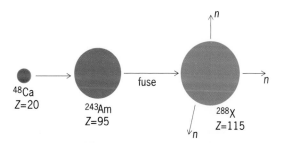

Fig. 2.   An illustration of the reaction of an americium-243 (^{243}Am) target atom with the calcium-48 (^{48}Ca) projectile, with the recoiling element 115 evaporation residue emitting three neutrons. This is called a 3n(3-neutron) evaporation channel. The unnamed element with atomic number $Z =$ 115 is given the symbol X.

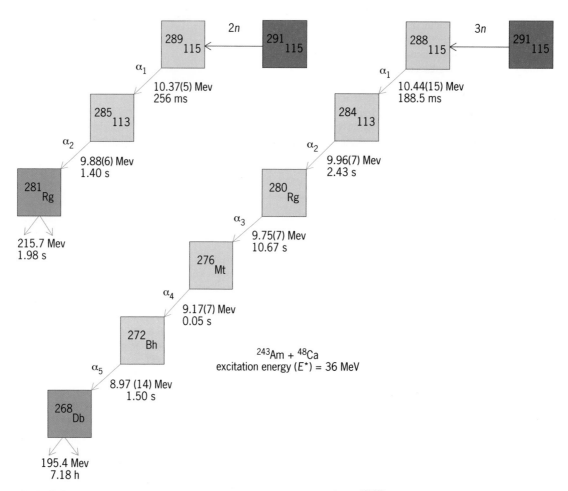

Fig. 3. Examples of the new decay chains of the *2n* and *3n* reaction channels to ^{289,288}115 respectively, from the reaction of americium-243 (^{243}Am) with calcium-48 (^{48}Ca) projectiles. The alpha energies with uncertainties and the times for the nuclei to decay are given next to each decay.

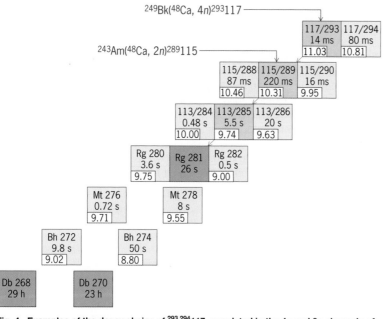

Fig. 4. Examples of the decay chains of ^{293,294}117, populated in the 4n and 3n channels of ^{249}Bk + ^{48}Ca, and of ^{288,289}115, populated in the 3n and 2n channels of ^{243}Am + ^{48}Ca. The alpha energies are given in the white box below the decay times. As indicated by the arrows, the same ^{289}115 is seen in both reactions to provide cross-bombardment checks of the new elements 117, 115, and 113. The alpha energies and decay times for ^{289}115 and ^{285}113 are the averages of the five chains observed in the alpha decay of ^{293}117, which can be compared with those observed directly in the ^{243}Am + ^{48}Ca reaction as illustrated in Fig. 3. The ^{288}115 data in the figure are averages of the three events seen in 2004, which may also be compared with the new data illustrated in Fig. 3.

isotopes of 115 with mass numbers $A = 287$, 288, and 289 are 2, 31, and 4, respectively. Examples of these new 2n and 3n evaporation channel events are shown in **Fig. 3**.

The 28 new chains of ^{288}115 from both parts of the experiment are completely consistent with the first three cases reported in 2004, which are shown in **Fig. 4** for comparison with the example in Fig. 3.

The ^{289}115 decay chains produced in the ^{243}Am + ^{48}Ca reaction illustrated in Fig. 3 are the same as those populated in the alpha decay of ^{293}117 produced in the ^{249}Bk + ^{48}Ca reaction shown in Fig. 4. The good agreement of the new ^{289}115 decay data and the ^{289}115 decay data obtained from the ^{293}117 alpha decay is seen by comparing the data in Figs. 3 and 4. These cross-bombardment data obtained for reactions with two different targets give independent evidence for the discovery of the new element 117 as well as of elements 115 and 113. The peak of the production cross section for the three-neutron channel in the ^{243}Am + ^{48}Ca reaction is the largest of any of the reactions of ^{48}Ca with plutonium to californium targets.

The average lifetimes for ^{289,288}115 are 560 and 250 ms, and for ^{285,284}113 are 7.1 and 1.4 s, respectively. The decay time (only one event was observed) for ^{294}117 with $N = 177$ and lifetime for ^{293}117 with $N = 176$ of 110 and 21 ms, respectively, as shown in Fig. 4, can be compared to 1.3 ms for $^{294}118$ with $N = 176$ and 1 ms for the heaviest element seen via cold fusion in Germany with $Z = 112$ and $N = 165$ (far removed from 184). These new data on elements 113, 115, and 117 show significant increases in the lifetimes with increasing neutron number as they approach $N = 184$. Thus, these lifetime data give strong support for an island of stability around $N = 184$. The new data indicate that we have landed at least on the beaches of the island.

In summary, 31 decay chains of $^{288}115$, two of $^{287}115$, and four of $^{289}115$ have been observed at different energies to give the production probabilities as a function of energy. The $^{289}115$ decays are the same as those following the alpha decay of $^{293}117$ to provide a cross-bombardment check. Together these excitation-function and cross-bombardment data provide definitive evidence for the discovery of the new elements 113, 115, and 117. Their longer lifetimes as the number of neutrons increase give definite support to the prediction of an island of stability around neutron number 184.

For background information *see* ALPHA PARTICLES; AMERICIUM; NUCLEAR FISSION; NUCLEAR REACTION; NUCLEAR SHELL MODEL AND MAGIC NUMBERS; NUCLEAR STRUCTURE; PARTICLE ACCELERATOR; PARTICLE DETECTOR; RADIOACTIVITY; SUPERHEAVY ELEMENTS; TIME-OF-FLIGHT SPECTROMETERS; TRANSURANIUM ELEMENTS in the McGraw-Hill Encyclopedia of Science and Technology.

Yuri Ts. Oganessian; Joseph H. Hamilton;
Vladimir K. Utyonkov

Bibliography. J. H. Hamilton et al., New insights into the discoveries of elements 113, 115 and 117, *Proceedings of the 11th International Conference on Nucleus-Nucleus Collisions, 2012*, Journal of Physics Conference Series (JPCS), forthcoming; Yu. Oganessian, Heaviest nuclei from ^{48}Ca-induced reactions, *J. Phys. G: Nucl. Part. Phys.*, 34:R165–R242, 2007, DOI:10.1088/0954-3899/34/4/R01; Yu. Ts. Oganessian et al., New insights into the ^{243}Am + ^{48}Ca reaction products previously observed in the experiments on elements 113, 115, and 117, *Phys. Rev. Lett.*, 108:022502 (4 pp.), 2012, DOI:10.1103/PhysRevLett.108.022502.

Denisovans

Denisovans (named after the Denisova cave in the Altai Mountains, Siberia, Russia; **Fig. 1**) are hominins that may belong to a previously unknown archaic population or species likely to have originated in Asia during the Middle Pleistocene. If confirmed, this will be the first time that a new *Homo* species has been discovered primarily from genetics, although it is not yet formally described.

Genome and fossil research. The research on the Denisovans began in 2010, when investigators from the Max Planck Institute (Leipzig, Germany), led by Svante Pääbo, retrieved the complete mitochondrial DNA genome from a fifth finger bone tip likely belonging to a juvenile female. (Note that mitochondrial DNA is the circular DNA duplex, generally 5

to 10 copies, contained within a mitochondrion and maternally inherited.) The finger bone was discovered in 2008 by Russian archeologists (from the Institute of Archaeology and Ethnology of Novosibirsk) at the Denisova cave in a stratigraphic layer (labeled 11) dated between 30,000 and 50,000 years ago (the exact age is uncertain, but probably closer to the latter one) and was morphologically undiagnostic. The Denisova cave consists of a large central chamber, measuring approximately 9 m × 11 m, or 99 m^2 (29.5 ft × 36.1 ft; 1065 ft^2), from which several side galleries extend out. It is located near the village of Chorny Anui, about 150 km (93 mi) south of the city of Barnaul (Altai Krai). The name of the cave derives from a hermit, Dionisij (Denis), who inhabited this location during the eighteenth century.

The analysis of the Denisova finger's mitochondrial DNA, published in May 2010, showed that it was a highly divergent genetic lineage that split off from the mitochondrial DNA genomes of modern humans and Neandertals approximately 1 million years ago. This finding suggested the presence in the cave of an unknown archaic hominin, who was dubbed the "X-woman" (because mitochondrial DNA is maternally inherited).

Some months later, in December 2010, further published analysis revealed the complete genome of this Denisova individual, as well as a new mitochondrial DNA genome from a second individual that was similar but not identical to the previous one. With an efficiency (number of endogenous DNA sequences versus exogenous environmental DNA sequences) of 70%, which is unprecedented in nonpermafrost specimens, the Denisova genome was sequenced up to 1.9× coverage with next-generation DNA sequencing methodologies.

Fig. 1. Exterior view of the Denisova cave in the Altai Mountains, Siberia, Russia. (*Courtesy of Bence Viola, Max Planck Institute for Evolutionary Anthropology, Leipzig, Germany; reproduced with permission*)

Different approaches (which investigated modern human sequences in the mitochondrial genome, the residual presence of Y-chromosome sequences, and heterogeneities in nuclear DNA positions among two different genomic libraries) were used to estimate that modern human contamination was only between 0 and <1%.

The analysis of genomic divergence showed that Denisovans were a sister group to Neandertals, from which lineage they diverged approximately 640,000 years ago. Denisovans, Neandertals, and modern humans shared a common ancestor about 800,000 years ago. Thus, it seems that Denisovans and Neandertals had a common ancestry but represent different evolutionary lineages, having probably evolved in parallel in Asia and western Eurasia, respectively.

It was also discovered that Denisovans contributed approximately 4.8% to the genomes of Melanesian individuals from Papua New Guinea and Bougainville Island. Further genetic studies on additional modern Oceanian and Asian populations have indicated that the signals of gene flow from Denisovans could be traced also to Australian aborigines, but not to most Southeast Asian groups. The most plausible explanation for these findings is that the gene flow took place in Southeast Asia itself, from Denisovans into the common ancestors of Australians and New Guineans, but not into the later-arriving populations that gave rise to other Asian groups. If this interpretation is correct, then Denisovans must have occupied an extraordinarily large and ecologically diverse geographic area, from Siberia to tropical Asia.

Additional genetic analysis on the highly polymorphic human leukocyte antigen (HLA) immune system in West Asian groups suggested that some remarkable divergent alleles (alternative forms of a gene), currently present at high frequencies, were possibly introduced by Denisovans into the modern human gene pool through introgression. One possible explanation for the genetic pattern observed is that archaic hominins, having lived in the Eurasian environmental conditions for hundreds of thousands of years, could have possessed immune systems that were better adapted to local pathogens. However, this finding contradicts the previous one, where no Denisovan admixture was detected in modern Asians.

There is an obvious discordance between the Denisovan mitochondrial DNA phylogenetic tree and their population history as deduced from genomic data. This could be explained in different ways: possibly the mitochondrial genome was introduced into the Denisovan gene pool by admixture with an unknown, more archaic hominin; or perhaps the Denisovan population size was so large that it allowed the random survival of very divergent lineages into its mitochondrial gene pool (a phenomenon known as incomplete lineage sorting).

The second Denisovan specimen analyzed is a tooth (**Fig. 2**) found in 2000 in the stratigraphic layer 11.1. It corresponds to a second or third upper molar from a young adult individual (thus representing another individual, different from the one to whom the finger bone belonged) and it displays some archaic

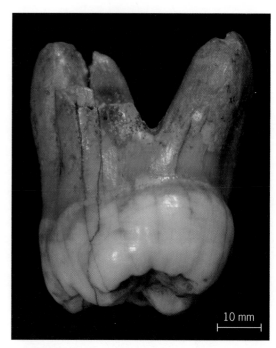

Fig. 2. Denisova molar tooth. (*Courtesy of Bence Viola, Max Planck Institute for Evolutionary Anthropology, Leipzig, Germany; reproduced with permission*)

features, notably a large size, that places it within the range of variation of early *Homo* species such as *H. habilis* or *H. rudolfensis*. The mitochondrial DNA from the Denisova molar differs from that of the finger bone at only two positions. Both Denisovan mitochondrial genomes share a common ancestor only about 7500 years before the assumed age of approximately 50,000 years old for both specimens, thereby suggesting that they belong to the same population. The yet unpublished finding of a second molar tooth at Denisova cave has indicated that their mitochondrial DNA genome is more divergent than the previous ones (and thus corresponds to a third individual). In fact, the Denisovan mitochondrial genomes have more variation than all of the six Neandertal mitochondrial genomes published so far. This again points to a wider geographic range in Denisovans (as well as larger populations and higher genetic diversity) in comparison to Neandertals. Intriguingly, the genetic analysis of a toe bone recently found at Denisova indicates that Neandertals were also present in the cave, probably around 45,000 years ago, after the Denisovans. This, in conjunction with typical Upper Paleolithic stone tools and art objects, including decorative items carved on bone and a stone bracelet, illustrates that Denisova is a unique site where three successive hominins lived (Denisovans, Neandertals, and modern humans).

Outlook. It is not known how the Denisovans would look from a morphological point of view, although it has been suggested that some Middle Pleistocene fossils from China with unknown taxonomical affinities, such as the Dali and Jinniushan fossil skulls, could represent the faces of these extinct hominins. Both of these fossil skulls are morphologically similar and have been described as showing a combination of *H. erectus* and anatomically modern human features. However, it is likely that no direct genetic attribution will be possible because no DNA has been preserved as a result of the age of these remains (approximately 200,000–300,000 years old).

From the scarce information currently available, it thus seems that Denisovans had exceptionally archaic genetic features in both their mitochondrial and nuclear genomes; in addition, they carry archaic morphological features in their teeth. They likely evolved in Asia in parallel with Neandertals that were evolving mainly in Europe and western Asia, and they interbred with the ancestors of modern humans that were moving into Oceania and Melanesia, before their extinction.

For background information *see* DENTAL ANTHROPOLOGY; DEOXYRIBONUCLEIC ACID (DNA); EARLY MODERN HUMANS; FOSSIL; FOSSIL HUMANS; GENETIC MAPPING; HUMAN GENOME; MITOCHONDRIA; MOLECULAR ANTHROPOLOGY; NEANDERTALS; PHYSICAL ANTHROPOLOGY in the McGraw-Hill Encyclopedia of Science & Technology.   Carles Lalueza-Fox

Bibliography. A. Gibbons, Who were the Denisovans?, *Science*, 333:1084–1087, 2011, DOI:10.1126/science.333.6046.1084; J. Krause et al., The complete mitochondrial DNA genome of an unknown hominin from southern Siberia, *Nature*, 464(7290):894–897, 2010, DOI:10.1038/nature08976; C. Lalueza-Fox and M. T. G. Gilbert, Neandertal and Denisovan palaeogenomics, *Curr. Biol.*, 21(24):R1002–R1009, 2011; D. Reich et al., Genetic history of an archaic hominin group from Denisova Cave in Siberia, *Nature*, 468:1053–1060, 2010, DOI:10.1038/nature09710.

Development of drought-tolerant crops through breeding and biotechnology

The greatest challenge faced to meet the needs of food production is the availability of freshwater. Water supplies and subsequent water shortages have contributed to the rise and fall of many civilizations (for example, the Mayan, Anasazi, and Angkor Wat societies), and the challenges associated with adequate food production in the face of insufficient water are no less great now compared with these former civilizations. Crop yields have increased significantly because of the development of modern agricultural practices. As examples, from the 1970s to the 2000s, corn (maize) yields increased by approximately 67% in the United States, cotton yields increased by roughly 188% in India, and wheat yields grew approximately 55% in Mexico. Although these dramatic improvements increased the total global food supply, they have also been punctuated by significant yearly swings in production primarily associated with fluctuations in precipitation during the growing season. In some underdeveloped countries, periodic droughts have resulted in catastrophic famines. An ever-increasing global population and a concomitant decrease in prime agrarian land area create pressure on modern agriculture to increase food production, often with suboptimal water availability.

Drought affects food production by reducing the overall quality and quantity of the crop. Crops respond to drought in a number of ways. Reproductive development is particularly sensitive, so the timing of the stress is important. In maize, maturation of flowers is affected, so synchrony of male and female flowering can be disrupted, leading to fewer pollinations and reduced kernel set when drought occurs at the time of flowering (**Fig. 1**). Water deficiency during the grain-filling period also has a significant detrimental effect on ear growth and development, resulting in yield loss. In addition, drought stress results in reduced stalk strength, so plants in the field may fall over (lodging), making it difficult to harvest the crop. Increased disease is also common in crops that are produced under drought stress, negatively impacting food quality. Heat often occurs in combination with drought, producing further damage.

Fig. 1. Ears from conventional maize plants grown under conditions of full water or conditions with limited water during flowering or during the grain-fill period. The figure illustrates how drought reduces grain yield, and it shows the different responses seen when drought is imposed at different times during development. Grain yields are reported in metric tons per hectare (mt/ha). *Top row*: Flowering stress; 6.3 mt/ha. *Middle row*: Grain-fill stress; 8.8 mt/ha. *Bottom row*: Well watered; 16.6 mt/ha.

Past approaches to mitigate the effects of drought have included irrigation, selection by plant breeders for improved drought tolerance, and agricultural methods that reduce the amount of moisture lost from the soil as a result of evaporation. All of these improvements have challenges associated with them. Irrigation water for agriculture competes with the supply of freshwater that is needed for other purposes, including water used for human consumption. Agricultural practices conserve water, but they may have offsetting complications (for example, increased disease pressure because of plant debris that remains on the soil surface). Drought exacerbates the environmental variability that plant breeders must deal with when trying to select for improved drought tolerance, slowing the rate of gain.

Current research approaches to reduce the impact that drought has on crops can be categorized under the following areas: (1) plant breeding, (2) biotechnology, and (3) agronomic practices. This article will review research to improve drought tolerance of crops through breeding and biotechnology. Most of the examples presented are for maize, as it is one of the world's predominant crops, but many of the principles can be extended to crops in general.

Breeding for improved drought tolerance. Plant breeding has been carried out in crop plants since humans began cultivating crops. Crop plants have been adapted to grow in a wide range of environments through either natural or artificial selection, creating significant genetic diversity, and plant breeders have exploited this diversity to select lines with improved drought tolerance. For example, selective breeding for lines that can tolerate higher plant densities has produced plant populations that averaged 10–25% higher than those typically grown by farmers. These ever-increasing population densities have resulted in a significant increase in yield on a land area basis while maintaining the per-plant yield. Plants grown at high population density need to compete for a limited amount of water; thus, by crowding plants together, maize breeders have selected lines with greater water-use efficiency, so they produce more grain with the available water.

More recently, additional resources have been allocated by breeding programs to directly measure performance under drought stress. These programs have incorporated testing under reduced irrigation conditions in areas such as California and Chile, where little to no natural rainfall occurs during the growing season. They are expected to produce lines that can tolerate significant drought in areas such as the U.S. Western Great Plains, but they also provide protection against occasional droughts that can occur in areas that have, in most years, sufficient rainfall for high yields.

Advances in molecular biology and genomics have resulted in the development of methods that use molecular deoxyribonucleic acid (DNA)–based markers to enhance breeding. Molecular breeding methods using DNA-based markers are now being employed to improve the results of selection under drought. The correlation of molecular markers and

yield values obtained for hybrids that are grown under drought stress conditions enables the identification of segments of chromosomes that may contain genes that provide improved drought tolerance. This information is then used to select lines containing multiple chromosomal segments correlated with improved yield under drought for further testing. This approach helps to identify lines with improved performance under drought stress in a shorter period of time than through more traditional plant breeding methods.

Improving drought tolerance using biotechnology. Biotechnology includes methods to insert a specific gene into a chromosome of the target species using a process called genetic transformation. First, a gene of interest may be identified by testing different candidates in a model species such as the small and rapid-cycling plant *Arabidopsis*. Once a gene is identified that can provide drought tolerance in the model plant, it is introduced into the crop of interest. The genetically engineered crop plant can then be tested under a variety of environments that are similar to what would be found in a farmer's field to identify those that may have potential to become commercial products. These plants are also carefully studied to ensure that they would not introduce risk to human or animal health or to the environment.

Biotechnology improvements associated with drought tolerance may have direct or indirect effects on stress tolerance. Examples of indirect impact include the current development and deployment of insect resistance. Genes selected from *Bacillus thuringiensis* have been inserted into maize. These genes confer resistance to numerous insect pests, including major pests such as the European corn borer (*Ostrinia nubilalis*) and the Western corn rootworm (*Diabrotica virgifera*). These insects can cause damage by chewing through the vascular tissue that moves water throughout the plant or by damaging the root system of the plants. By minimizing or eliminating this damage, plants are better able to withstand drought (**Fig. 2**).

Research is also being conducted to identify any gene (or genes) that can directly improve the performance of maize plants under limited water conditions. Plants have a number of strategies to respond to drought stress. They may escape drought completely by accelerating their flowering time to complete their life cycle before drought occurs. Drought avoidance mechanisms include slowing of growth, as well as reduction of water loss by closing the pores (stomates) on leaves (stomatal closure also reduces photosynthesis) and increasing wax production on the surfaces of leaves. Plants can also develop a degree of true drought tolerance by increasing root growth to improve water acquisition, adjusting cellular solutes to adapt to drier conditions, and producing antioxidants to protect against free radicals that may increase because of reduced photosynthetic efficiency. Plant responses to stress have evolved for survival in natural environments, but crop plants experience artificial conditions; thus, manipulation of the timing or magnitude of these responses may improve plant productivity under cultivation.

One gene that has demonstrated potential for improving drought tolerance comes from a soil bacterium and is called cold shock protein B (CspB). Cold shock proteins rapidly accumulate in bacteria subjected to cold shock and are thought to be important for stimulating growth following stress accumulation and during periods of high metabolic activity. CspB has been shown to bind cellular ribonucleic acid (RNA) and keep it in a single-stranded functional state in the stressed bacteria, thereby promoting protein biosynthesis that is needed for continued growth. The gene was initially tested in the model species *Arabidopsis* and was shown to improve cold tolerance in that species. When tested in rice, the gene promoted tolerance to several stresses, including cold, heat, and drought. Plants with the *CspB* transgene (a gene or genetic material that has been experimentally transferred from one organism to another) were able to continue growing at a faster rate in the presence of these stresses compared to control plants that did not contain the transgene. The gene was then introduced into maize, and transgenic hybrid lines demonstrated improved drought tolerance in the greenhouse and field compared to isogenic lines without the transgene. One line was selected and tested over many locations over several years. Maize hybrids containing this transgene will be tested in farmers' fields in 2012.

Other genes provide benefit to plants under drought stress through other mechanisms, and it is likely that maize can be made even more drought tolerant by the introduction of multiple complementary transgenes. In fact, dozens of genes have been identified that can provide improved drought tolerance to crop plants through a variety of metabolic and developmental pathways. These include improving the ability of the plant to adjust internal solute concentrations to promote growth and avoid injury, and tighter control of the opening and closing of stomates to minimize water loss while maximizing gas exchange for photosynthesis. Therefore, considerable effort has gone into understanding the plant hormones and regulators of gene transcription (transcription factors) that are responsible for coordinating plant responses to stress. An example of a transcription factor that has been shown to improve drought tolerance of maize in the greenhouse and field is the nuclear factor YB gene. Overall, the timing and level of expression of these regulatory factors in relation to the progression of drought conditions and the plant's life cycle often need to be optimized to improve drought tolerance.

Identification of transgenes that improve drought tolerance has primarily involved testing under controlled conditions, usually in a greenhouse, but pots in a greenhouse do not accurately represent field conditions. A current challenge is to more thoroughly characterize the effects of these genes on crop growth and development under field conditions in order to identify those that improve the plant's response to stress under such conditions, and en-

Fig. 2. Commercial maize field located in Cambridge, IL, under drought stress in 2005. Plants on the right contain genes from *Bacillus thuringiensis* (BT) for lepidopteran and coleopteran insect control. Plants on the left are from a non-BT hybrid and were treated with a soil insecticide to control insects.

sure that they have no negative effects on other aspects of growth and development that could reduce yields under optimal growing conditions. Although field testing may seem to be a straightforward process, variability encountered from one field to another, and even within a field, complicates analysis so that a large amount of testing needs to be done to understand the effect of a given gene. This kind of testing is slow and expensive. Therefore, more effort is now also being focused on how to predict field performance more accurately from data generated in controlled environments.

Conclusions. Drought tolerance is a complex trait, and improving crop productivity under water-limiting conditions will require manipulation of multiple genetic pathways. Combining the tools of traditional breeding, molecular breeding, and biotechnology will provide the best opportunity for improving the genetic potential of crops grown with limited water.

For background information *see* AGRICULTURAL SCIENCE (PLANT); AGRICULTURAL SOIL AND CROP PRACTICES; BIOTECHNOLOGY; BREEDING (PLANT); CHROMOSOME; CORN; DROUGHT; GENE; GENETIC ENGINEERING; GENETICALLY ENGINEERED PLANTS; PLANT GROWTH; PLANT METABOLISM; PLANT PHYSIOLOGY; PLANT-WATER RELATIONS in the McGraw-Hill Encyclopedia of Science & Technology.

Jill Deikman; Mark Lawson

Bibliography. B. Barnabas, K. Jager, and A. Feher, The effect of drought and heat stress on reproductive processes in cereals, *Plant Cell Environ.*, 31:11–38, 2008, DOI:10.1111/j.1365-3040.2007. 01727.x; P. Castiglioni et al., Bacterial RNA chaperones confer abiotic stress tolerance in plants and improved grain yield in maize under water-limited conditions, *Plant Physiol.*, 147:446–455, 2008,

DOI:10.1104/pp.108.118828; J. Deikman, M. Petracek, and J. E. Heard, Drought tolerance through biotechnology: Improving translation from the laboratory to farmers' fields, *Curr. Opin. Biotechnol.*, 23:243–250, 2012, DOI:10.1016/j.copbio.2011. 11.003; S. R. Eathington et al., Molecular markers in a commercial breeding program, *Crop Sci.*, 47:S154–S163, 2007, DOI:10.2135/cropsci2007.04. 0015IPBS; G. Head and D. Ward, Insect resistance in corn through biotechnology, pp. 31–40, in A. L. Kriz and B. A. Larkins, eds., *Molecular Genetic Approaches to Maize Improvement*, Springer-Verlag, Berlin/Heidelberg, 2009; J. Su and R. Wu, Stress-inducible synthesis of proline in transgenic rice confers faster growth under stress conditions than that with constitutive synthesis, *Plant Sci.*, 166:941–948, 2004, DOI:10.1016/j.plantsci.2003.12.004.

Distributed propulsion

The main objective of distributed propulsion is to achieve optimum vehicle benefits through integration of aerodynamic, propulsive, structural, and operational elements. The concept could be applied to various vehicle configurations, such as conventional "tube and wing," hybrid-wing-body (HWB), and supersonic aircraft. However, to achieve maximum benefits, it will be necessary to design an aircraft with greater emphasis on propulsion–airframe integration. Among the concepts examined, the turboelectric distributed propulsion (TeDP) system potentially provides a significant enhancement in fuel burn, emissions, community noise reduction, and field length reduction.

Background. Since the introduction of large subsonic jet-powered transport aircraft, the majority of these vehicles have been designed by placing thrust-generating engines in pods either under the wings or on the fuselage to minimize aerodynamic interactions between the airframe and the propulsion system. However, advances in computational and experimental tools, along with new technologies in materials, structures, aircraft controls, and so forth, are enabling a high degree of integration of the airframe and propulsion system in aircraft design.

One of the most promising concepts in addressing current large transport aircraft issues such as fuel burn, noise, and emissions is called "distributed propulsion." The concept is to fully integrate a propulsion system within an airframe such that the aircraft takes full synergistic benefits of coupling airframe aerodynamics and the propulsion thrust stream by distributing the thrust along the wingspan. Some of the concepts are based on the use of distributed jet flaps, distributed small multiple engines, cross-flow fans, or multiple propulsors using an engine core. The engine core in a conventional turbofan engine typically consists of compressors, combustors, and turbines, and provides power to a fan through a common shaft. The latter concept, multiple propulsors per engine core, can be further categorized by the method of how the power is transmitted from the power source to the propulsors. The propulsors can be driven by (1) chain or belt (the Wright Flyer's two propellers were driven by two chains attached to a single engine), (2) gas, either by "cold" discharge from a compressor or "hot" discharge from engine core exhaust, (3) multiple gears and shafts connected to conventional engines, or (4) electric power generated by a turboelectric generator, battery, fuel cell, or other electric power source.

Benefits. The benefits of using distributed propulsion for aircraft could be found in improvement in aircraft performance, noise reduction to the surrounding community, and providing the capability of short takeoff and landing (STOL). Specifically, the following possible benefits of distributed propulsion concepts have been identified through various studies:

1. Reduction in fuel consumption by ingesting thick boundary-layer flow caused by friction between the airframe surface and the air flow just above it, and filling in the wake generated by the airframe with the distributed thrust stream.

2. Spanwise high lift via high-aspect-ratio trailing-edge nozzles for vectored thrust providing powered lift, boundary-layer flow control, or supercirculation around the wing, all of which enable short-takeoff capability.

3. Better integration of the propulsion system with the airframe for reduction in noise to the surrounding community through airframe shielding.

4. Reduction in aircraft propulsion installation weight through inlet/nozzle/wing structure integration.

5. Elimination of aircraft control surfaces through differential and vectoring thrust for pitch, roll, and yaw moments.

6. High production rates and easy replacement of engines, or propulsors, which are small and lightweight.

7. For the multiple fans per single engine core concept, the propulsion configuration provides a very high effective bypass ratio (BPR), the ratio between the air flow rate going into the fans to the air flow rate going into the engine core, enabling low fuel burn and low community noise.

A number of fixed-wing aircraft using distributed propulsion have been proposed and flown to achieve some of the benefits described above. Because what constitutes distributed propulsion for aircraft is not clearly defined—any aircraft with more than one propulsor could be classified as such—the following description can be applied to further reduce the number of possible vehicle configurations: "Distributed propulsion in aircraft application is the spanwise distribution of the propulsive thrust stream such that overall vehicle benefits in terms of aerodynamic, propulsive, structural, and/or other efficiencies are mutually maximized to enhance the vehicle mission." Based on this description, and due to the recent interest in aircraft employing multiple propulsors, the following specific concepts are described below.

Fig. 1. Cruise-efficient STOL vehicle with 12 engines.

gas-driven multifan transport aircraft was conceived and a model was tested for STOL operation. The aircraft used 16 tip-driven fans spread along the top surface near the wing trailing edge (**Fig. 2**). The tip-driven fans were powered by high-pressure discharge air from the low-pressure compressor stages and mounted on a hinged flap to achieve high lift.

Gear-driven multiple fans. A distributed propulsion concept employing a dual fan driven by one engine core on an HWB airframe was studied by NASA. **Figure 3** shows the concept, where an engine core drives two large-diameter fans, via gears and shafts, providing a very high effective bypass ratio. Recently, the Cambridge-MIT Institute's Silent Aircraft Initiative developed the SAX-40 conceptual HWB aircraft. This is a similar concept but with three gear-driven multiple fans per engine core. The purpose of the study was to design an aircraft with noise being

Fig. 2. Gas-driven multiple-fan propulsion system on "tube and wing" configuration.

Multiple engines. Various types of aircraft using multiple propulsors or engines have been proposed and flown. For these aircraft, propulsors such as propellers, turbojets, or turbofans are mounted in front of the wing, at the back of the wing, or within the thick section of the wing. An example is the 1940s YB-49 flying-wing aircraft, which had eight engines completely embedded inside the wing. Recently, a cruise-efficient short-takeoff-and-landing (CESTOL) aircraft (**Fig. 1**) was studied using a hybrid-wing-body (HWB) or blended-wing-body (BWB) transport configuration. This airframe was chosen because of its high cruise efficiency, low noise characteristics, and a large internal volume for integrating an embedded distributed propulsion system. The propulsion system employed 12 small conventional engines partially embedded within the wing structure. They were mounted along the wing upper surface near the trailing edge to enable STOL operation using fan-diverted bypass air to augment the lift.

Gas-driven multiple fans. In the late 1960s, a vertical/short-takeoff-and-landing (V/STOL) air-deflection and modulation (ADAM III) concept was studied for various missions. In this concept, two gas generators and their inlets were installed near the fuselage to provide hot gas to the wing-mounted turbines that drove multiple turbofans through a common shaft. The turbofans and turbines were collocated in the wing section away from the gas generators. The hot gases from the gas generators were routed through long ducts across the wing span to the location where the turbines and fans were installed. The inlets and nozzles for the fans and turbines were also all within the wing structure, away from the gas generators, and provided distributed thrust to the vehicle. Then, in the 1970s, another

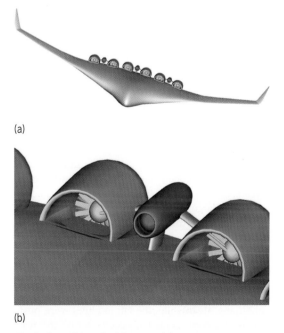

(a)

(b)

Fig. 3. Gear-driven dual-fan propulsion system on hybrid-wing-body (HWB) configuration. (a) Configuration of system on aircraft. (b) Detail of dual fan driven by a single engine core.

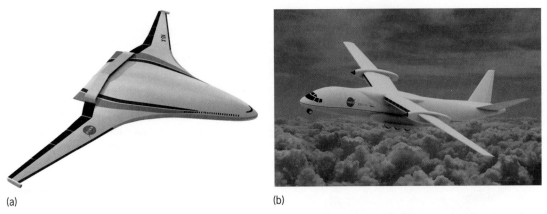

(a)   (b)

Fig. 4. Turboelectric distributed propulsion (TeDP) concept vehicles. (*a*) NASA concept. (*b*) ESAero concept. (*Courtesy of Empirical Systems Aerospace, Inc.*)

the primary design variable addressed, such that the noise would be contained within the perimeter of an urban airport.

Electric-driven multiple fans. To improve performance and to further reduce environmental impacts, a drastic change in the power transmission of a distributed propulsion system for large transport aircraft was proposed and studied on HWB and "tube and wing" configurations. The turboelectric distributed propulsion (TeDP) concept employs a number of superconducting electric motors to drive the distributed fans. The electric power to drive these fans is generated by separately located gas turbine–driven, superconducting electric generators. This arrangement enables the use of many small distributed electric fans. This allows for a very high effective bypass ratio while retaining the superior efficiency of large engine cores, which are physically separated but connected to the fans through superconducting electric power lines. This power transmission method has the desired effect of allowing the power turbine in the electric generator to spin at any desired speed, while the fans spin at their optimum speed. Not only can the speeds of the turbine and fans be different, the use of power inverters between the generators and the fan motors allows the speed ratio to change in flight, giving the effect of a variable-ratio gearbox. In addition, the use of electric power transmission allows a high degree of flexibility in positioning the turboelectric generators and fan modules to best advantage.

A vehicle configuration using this new propulsion concept, dubbed the N3-X, is currently being investigated by NASA (**Fig.** 4*a*). The airframe is based on a large HWB airframe with similar mission characteristics to the Boeing 777 [6000-nautical-mile (11,000-km) range, 103,000-lb (47,000-kg) payload capacity, and a nominal cruise Mach number of 0.84]. In the aircraft configuration examined, the turbogenerators were located at the wing tips, where they would experience undisturbed free-stream flow conditions. Meanwhile, the fan modules were positioned in a continuous fan nacelle across the rear upper fuselage. In this location, they will ingest the thick boundary-layer flow caused by the friction between the airframe upper surface and air flow above

it and fill the wake of aircraft with fan discharge air. These two effects thereby reduce the thrust required by the vehicle.

Recently, a 150-passenger-class "tube and wing" STOL vehicle with TeDP was investigated. A key feature of the concept is the integration of the non-superconducting electric motor–driven fans within the wing (Fig. 4*b*). The inboard wing is separated into top and bottom sections, and the electric fans are completely embedded within the airfoil or wing structure. This feature provides a wing weight benefit through wing bending moment relief because the distributed electric fans and the use of the common nacelle as wing rib structure provide stress relief to the wing structure. In addition, at low speed, thrust vectoring of low-temperature nozzle flow may provide supercirculation around the airfoil to improve lift.

Another key feature of the TeDP concept is the possibility of using liquid hydrogen, both as a cooling fluid for the superconducting system and as fuel for the turboelectric generator engine. Although the studies are still in the preliminary stages, the electric propulsion system did show the potential for a large fuel burn reduction.

For background information *see* AIRCRAFT ENGINE PERFORMANCE; AIRCRAFT PROPULSION; AIRFRAME; BOUNDARY-LAYER FLOW; SHORT TAKEOFF AND LANDING (STOL); TURBINE PROPULSION; TURBOFAN; VERTICAL/SHORT TAKEOFF AND LANDING (V/STOL) in the McGraw-Hill Encyclopedia of Science & Technology.

Louis A. Povinelli; Hyun Dae Kim

Bibliography. E. de la Rosa Blanco, C. A. Hall, and D. Crichton, Challenges in the silent aircraft engine design, AIAA-2007-454, 2007; A. R. Gibson et al., Superconducting electric distributed propulsion structural integration and design in a split-wing regional airliner, AIAA-2011-223, 2011; H. D. Kim, G. V. Brown, and J. L. Felder, Distributed turbo-electric propulsion for hybrid wing body aircraft, in Royal Aeronautical Society, *IPLC 2008: International Powered Lift Conference*, London, July 2008, RAS, London, 2008; H. D. Kim et al., Low noise cruise efficient short take-off and landing transport vehicle study, AIAA-2006-7738, 2006; B. R. Winborn, Jr., The ADAM III V/STOL concept, AIAA-1969-201, 1969.

Effects of global warming on polar bears

The polar bear is the only species of terrestrial mammal so highly adapted to living on drifting pack ice of the Arctic Ocean that its entire life cycle can be fully completed without coming ashore. Only pregnant females need solid ground for setting their maternity dens; however, even these females are known to den on the sea ice in some regions. The polar bear is also a top predator in the Arctic marine environment and a specialized hunter on pagophilic (ice-loving) seals. Because Arctic sea ice is the prime habitat of the polar bear, it is logical to expect that ice disappearance will strongly affect this animal. If global warming continues to progress, perspectives for polar bear survival may look dim; under the worst scenario, the species may disappear. However, to investigate the polar bear's future, it is necessary to consider all factors involved and all opportunities for the species' struggle for survival. Before a judgment can be made, several questions need to be answered: How long will the current global warming continue? Will this warming period be so strong that the sea ice melts completely? Will the Arctic Ocean be ice-free year-round, or only seasonally? How long can polar bears survive ice-free seasons? Can they survive as a species if the ice is gone year-round?

To answer these questions, it is necessary to look at three things: (1) the past, that is, the climate of the Earth during the Pleistocene epoch; (2) the evolutionary history of ancient polar bears; and (3) the present, that is, the behavioral ecology of modern polar bears.

The Middle–Late Pleistocene was an epoch of cyclic glaciations followed by interglacial and ocean transgressions. The Arctic Sea ice cover was formed about 700,000 years ago in the Early Pleistocene. During all Ice Age cycles, the phases of ice-cover degradation were much shorter than the phases of ice growth. During interglacial periods, ice cover disappeared on the continents, but it never completely disappeared from the Arctic Ocean. Cyclic climatic changes and sea-level fluctuations on a lower scale also occurred during the Holocene (the last 10,000 years). However, the predicted scenarios for climate development on Earth are variable and controversial, so their effects on polar bears vary from moderate to extreme.

Since 1990, the behavioral ecology of the polar bear has been studied on Wrangel Island, Russia (**Fig. 1**). Wrangel Island is a large landmass situated north of Chukotka Peninsula, in the middle of a biologically highly productive zone of the continental shelf, and is a key polar bear habitat. It also lies near the Bering Strait. The processes that have been observed on Wrangel Island represent a model of how polar bears respond to ice disappearance. For any animal species, its behavior is a tool to manage its relations with the environment. By studying behavioral ecology, we learn what problems animals are facing under different environmental conditions and how they can solve them.

Fig. 1. Wrangel Island, Russia.

Changes in population processes. The first reaction of polar bears to sea-ice disappearance is their shifting to coastal ecosystems and surviving ice-free seasons on solid ground. This seasonal change of habitats is observed in many areas of the Arctic. In the high Arctic, though, not all bears move to land during summer ice melting. Some of them remain on the main shield of the Arctic pack ice, receding with it to the Central Arctic Basin (CAB). Recent observations have revealed that the numbers of polar bears in the CAB have increased compared to the 1990s. Simultaneously, more ringed seals have been recorded in the CAB as well. Thus, the productivity of marine ecosystems in the high Arctic is currently large enough to support reasonable densities of seals, which in turn provide food for polar bears. All bears observed in the CAB in the summers of 2005 and 2007 ($n = 20$) were in good physical condition. Other observations give clear evidence that summer hunting conditions in the CAB are currently good enough for survival of females and their small cubs.

Polar bears that land on Wrangel Island during ice-free seasons are part of the Chukchi–Alaskan geographical population. Some bears from this population also land on the north coast of Chukotka, and a few land on the northwest coast of Alaska. Thus, every year, this population splits seasonally into four temporarily isolated parts: Wrangel, Chukotka, Alaska, and the CAB. How long this isolation continues depends on the length of the ice-free season. Most bears are stranded on Wrangel Island. The number of bears stuck on Wrangel Island in autumn reached almost 400 in 2007, was approximately 350 in 2010, and was 200–250 in 2009 and 2011. In addition, a few tens of bears landed on the coast of Chukotka, and even fewer landed in Alaska. Overall, the total number of bears from this population surviving ice-free seasons on land probably does not exceed 600–700. As for the CAB, the total number of bears in this population does not seem higher than 1500–1700. In addition, when ice habitats change so dramatically, it is likely that bears from different

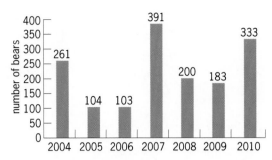

Fig. 2. Number of bears recorded on Wrangel Island during ice-free seasons in 2004–2010.

geographical populations are found in the CAB as well. However, the frequency of polar bear occurrence on Wrangel Island during ice-free seasons was significantly higher in the early 1990s compared to today. During the past 10 years, the total number of bears recorded on the island varied from a maximum of 391 to a minimum of 103 (**Fig. 2**).

Since 2005, the number of females going to Wrangel Island to give birth has not exceeded 60–70. However, in the early 1990s, approximately 350–400 dens were recorded on the island every spring. Cub mortality during the first 1.5 years of life is high as well. During the period from 2004 through 2010, average litter size decreased from the first to the second autumn of cub life by 18% (from 11% to 36% in different years). For the same period, average percent of yearlings versus cubs-of-the-year was reduced by 74.5% and percent of family groups (females with yearlings versus females with cubs-of-the-year) was reduced by 68%.

There are several indications that environmental changes are having a negative impact on the local polar bear population in this region: low numbers of pregnant females going to dens, very high mortality of cubs during their first year of life, regular observations of starving and underfed bears on land and on marginal ice, regular observations of bears with severe wounds, and findings of dead bears on land. There are a number of reasons for these observations: (1) As a result of sea ice melting every year, optimal ice habitats of polar bears are fully excluded from their life cycle during the late summer–autumn seasons. (2) More polar bears are exposed to environmental extremes associated with living on marginal disappearing ice and with swimming in the open sea. (3) Polar bears are forced to live for long periods on land without opportunities for hunting their main prey on the ice, to which they are specialized. As a result, they often are deprived of food for long periods. (4) The polar bear breeding cycle is affected. For example, the timing of pregnant females coming ashore and settling for hibernation is changed, and the timing of emergence from dens in the spring is altered. (5) Because of forced landing and staying on shore for long periods, the frequency of polar bear encounters with humans has increased significantly. As a result, polar bears are more often exposed to shooting, hunting, and poaching.

At the same time, as soon as any improvement in feeding conditions occurs, bears respond immediately, in the course of 1–2 years, by improving their physical condition and litter size. The average litter size of cubs in the autumn was the highest in 2008: 1.81 ($n = 26$). Most probably, this occurred because, in the previous autumn (in 2007), females who went to den were in very good physical condition as a result of very high walrus mortality. Under such circumstances, the stranded bears had plenty of food on the island (**Fig. 3**), and the larger litter size was supported by favorable ice conditions and good hunting opportunities in the spring of 2008.

The proportion of bears coming ashore in good or bad physical condition also varied from year to year (**Fig. 4**). Underfed (skinny) bears in poor body shape were observed in all years on marginal ice and on land. The proportion of underfed bears was the highest in 2002 and 2003, when walruses were not available in coastal ecosystems of the island. The proportion of well-fed (fat) bears increased in years of early and fast mass landing of bears on the island, and in seasons with a high availability of walrus carcasses on the shore. The most reasonable explanation is that early and fast landing reduced the phase of bears living on marginal ice and being exposed to extremes of the open sea. The availability of walrus carcasses also allowed stranded bears to feed well and to gain weight.

Food resources available on land. Polar bears feed typically on walruses, seals, and carcasses of whales, which are available for stranded bears in coastal ecosystems in some seasons. Walruses are the main source of food for stranded polar bears on Wrangel Island. In the absence of ice, walruses haul out on the beach, forming coastal rookeries, where they are available for bears to hunt. Walrus mortality was high in a number of recent years, and polar bears had plenty of food on the beach. In the summer of 2011, an unusually high mortality of young ringed seals on the coast was observed as well. In the absence of walruses or other marine mammals, polar bears will look for any alternative food, including Arctic cod, salmon, molting snow geese, lemmings,

Fig. 3. Bears feeding on a walrus carcass, Wrangel Island, September 2007. (*Photo copyright © Nikita G. Ovsyanikov*)

wolf kills, marine invertebrates (shellfish), carrion of dead birds, and carcasses of reindeer, musk-oxen, and Arctic foxes.

At large food sources, such as walrus rookeries, polar bears gather in numbers and form temporary communities, where bears constantly interact with each other. In such situations, they form a behaviorally structured society: a coastal congregation of bears (**Fig. 5**). In congregations, polar bears show much higher social tolerance to each other than when wandering on the ice. The largest congregation observed on Wrangel Island was 160 bears, including both sexes and all social categories. Bears stay within such congregations as long as food is available at the spot; the length can be 2 or more months, but it is usually a few days or weeks.

Population responds to global warming. In this part of the Arctic, the seasonal ice disappearance causes the local populations of bears to follow determined cycles of changes in spatial distribution, activity patterns, and use of resources. In winter, when the ocean is an optimal living zone (if frozen), polar bears fill the area from land and from the CAB, using one-year-old ice in shallow coastal waters as platforms to hunt. In the summer–autumn season, when the ice melts, bears that were hunting along the edge of the pack ice are cut off from the main pack and, with complete disappearance of marginal ice fields, are forced to land on the shores. Bears that were hunting farther north remain on the main pack and recede with it to the CAB and to the area of the North Pole. Then, as winter returns and the ice expands southward, the cycle repeats itself.

Based on this descriptive model, the same seasonal switch between living strategies has probably helped polar bears to survive periods of interglacial global warming. This strategy will likely be modified in different geographical areas by variations in local conditions. However, if the behavioral reaction of polar bears to various factors can be understood, this may help to explain the drivers of observed processes and to predict potential scenarios for population trends under various conditions. From the current knowledge of polar bear behavior and population processes, some actual and potential reactions of the species to global environmental changes can be outlined:

1. Decreased polar bear numbers; if the trend of warming continues, some populations may disappear because of habitat degradation and increased mortality.

2. Global redistribution of polar bears, including exchanges and mixing among geographical populations, and a seasonal shift of some bears from the continental-shelf zone to the CAB.

3. Increased opportunities for gene flow among some populations.

4. Seasonal splitting of some geographical populations into temporary isolated parts, and population fragmentation in some areas of the Arctic.

5. Extended periods of stranding on land, and seasonal switches to terrestrial lifestyles in coastal and tundra ecosystems.

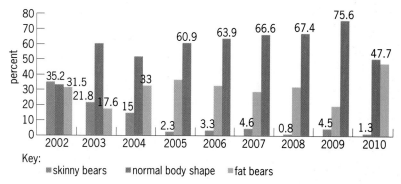

Fig. 4. Percentage of bears in various physical conditions observed on Wrangel Island during ice-free seasons in 2002–2010.

Fig. 5. Part of a bear congregation at a walrus rookery. Bears are feeding on a walrus carcass, Wrangel Island, 2011. (*Photo copyright © Nikita G. Ovsyanikov*)

6. Seasonal switches to alternative food resources and food items, and variability of foraging behavior.

7. Increased maternity denning on drifting pack ice in the CAB (for females that cannot return to land in time for hibernation).

8. Increased sociality during periods of coastal life.

9. Increased nutritional stress in terrestrial habitats and a probability of cannibalism.

10. Increased likelihood for hybridization between polar bears and brown bears.

Perspectives for survival. How unique are the aforementioned processes considering the evolutionary history of the polar bear? In the bear family, the polar bear is the youngest species. As a distinct species, the polar bear appeared on Earth in the mid-Pleistocene, somewhere between 120,000 and 200,000 years ago. However, the most recent genomic analyses of polar, brown, and black bears have revealed that the species is significantly older, originating between 338,000 and 934,000 years ago. In the Pleistocene, polar bears inhabited not only the Arctic basin, but also the coasts of the Northern Atlantic and Baltic Sea basins. For certain periods, they also inhabited terrestrial coastal ecosystems, overlapping with or adjusting to taiga fauna. The polar bear also survived at least four or five major global warming periods during the Pleistocene and two during the Holocene. These global warming periods lasted for thousands of years. Processes observed at Wrangel Island thus help to understand the biological features that are important for the polar bear's survival in a rapidly and dramatically changing environment, and how they managed to survive long global warming periods in the past.

Despite high specialization to living on the Arctic drifting sea ice and to hunting marine mammals, polar bears have retained high ecological and social plasticity. This quality has allowed them to easily switch their life strategies: from a strategy of highly specialized hunters on the sea ice to a strategy of predators–generalists in coastal and tundra ecosystems. Other key characteristics include high cognitive abilities, including a highly developed ability for learning; an ability to consume a wide range of food items; an ability to switch between a lifestyle of a solitary nomadic hunter on the ice to one of a socially tolerant community member; and a capacity to manage complicated interactions with conspecifics (individuals or populations of a single species) by well-developed social behavior. Such adaptabilities enabled the polar bear to survive the previous cyclic climate fluctuations between phases of global cooling (Ice Ages) and global warming (interglacial periods).

Of course, the basis for the adaptabilities of polar bears was inherited from their ancestors, that is, ancient brown bears, which already possessed a full set of features important for successful colonization of the Arctic sea-ice habitats. These bears were used to living in boreal climates and were well insulated. They were good swimmers and capable of crossing large open-water areas. They were ecological generalists with a broad diet, yet they remained skillful hunters capable of killing large prey. They were socially flexible as well, being territorial at times, but gathering for certain periods in large congregations at salmon streams or at whale carcasses cast onto the beach. In addition, they had very high cognitive abilities. Thus, the emergence of the polar bear was not a revolutionary breakthrough; instead, it was a gradual evolutionary fine-tuning of an ancestor predator.

The uniqueness of the current global warming is the fact that, during the entire evolutionary history of the polar bear, there has never been a warming with such high density and activity of humans in the Arctic. Biological features will allow polar bears to survive long periods of global warming, unless sea ice disappears completely from the face of the Earth. Thus, the major current extinction threats for the polar bear are its continuous elimination by humans, the increasing pollution of the Arctic, and other growing disturbances to its habitat.

For background information *see* ADAPTATION (BIOLOGY); ARCTIC OCEAN; BEAR; CLIMATE HISTORY; CLIMATE MODIFICATION; ENDANGERED SPECIES; EXTINCTION (BIOLOGY); GLACIAL HISTORY; GLOBAL CLIMATE CHANGE; GREENHOUSE EFFECT; SEA ICE in the McGraw-Hill Encyclopedia of Science & Technology.
Nikita G. Ovsyanikov

Bibliography. S. C. Amstrup, Polar bear, *Ursus maritimus*, in G. A. Feldhammer, B. C. Thompson, and J. A. Chapman (eds.), *Wild Mammals of North America: Biology, Management, and Conservation*, 2d ed., The Johns Hopkins University Press, Baltimore, MD, 2003; S. C. Amstrup, Polar bears and climate change: Certainties, uncertainties, and hope in a warming world, in R. T. Watson et al. (eds.), *Gyrfalcons and Ptarmigan in a Changing World*, The Peregrine Fund, Boise, ID, 2011; A. Derocher, *Polar Bears: A Complete Guide to Their Biology and Behavior*, The Johns Hopkins University Press, Baltimore, MD, 2012; F. Hailer et al., Nuclear genomic sequences reveal that polar bears are an old and distinct bear lineage, *Science*, 336:344-347, 2012, DOI:10.1126/science.1216424; C. Lindqvist et al., Complete mitochondrial genome of a Pleistocene jawbone unveils the origin of polar bear, *Proc. Natl. Acad. Sci. USA*, 107:5053-5057, 2010, DOI:10.1073/pnas.0914266107; N. G. Ovsyanikov, Current research and conservation of polar bears on Wrangel Island, pp. 167-171, in J. Aars, N. J. Lunn, and A. E. Derocher (eds.), *Polar Bears: Proceedings of the 14th Working Meeting of the IUCN/SSC PBSG, Seattle, Washington, USA*, IUCN, Gland, Switzerland/Cambridge, U.K., 2006; N. G. Ovsyanikov, Polar bear and seals in the Central Arctic Basin: Observations in 2005 and 2007, pp. 451-456, in *Marine Mammals of the Holarctic*, Society for Marine Mammalogy, Kaliningrad, 2010; N. G. Ovsyanikov, Polar bear research on Wrangel Island in 2005-2008 and in the Central Arctic Basin in 2005 and 2007, pp. 171-178, in M. E. Obbard et al. (eds.), *Polar Bears: Proceedings of the 15th Working Meeting of the IUCN/SSC PBSG*, IUCN, Gland, Switzerland/Cambridge, U.K., 2010; S. Schliebe et al., Fall distribution of polar bears along northern Alaska coastal areas and relationship to pack ice position, pp. 558-561, *Marine Mammals of the Holarctic*, Collection of Scientific Papers, St. Petersburg, 2006; I. Stirling, *Polar Bears: The Natural History of a Threatened Species*, Fitzhenry and Whiteside, Markham, Ontario, 2011.

Electrospun polymer nanofibers

Electrospinning is a simple technique that uses a high voltage to draw ultrathin fibers from a polymer solution or melt. The fibers typically have diameters on the scale of several hundred nanometers. Therefore, they are widely known as nanofibers. Because of their small sizes, nanofibers have extremely high surface area-to-volume ratios, and nanofiber-based mats are highly porous, with excellent interconnectivity between the pores. These features, together with the feasibility of alignment, make electrospun nanofibers particularly attractive for applications such as neuroregeneration, wound closure, and filtration.

Electrospinning. A typical setup for electrospinning consists of three major components (**Fig. 1**): a high-voltage power supply, a spinneret (for example,

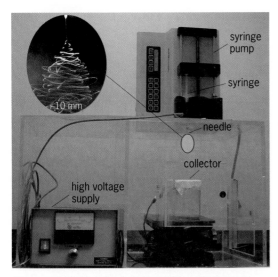

Fig. 1. A typical setup for electrospinning. The inset shows the whipping process of a nanofiber during electrospinning, as captured using a high-speed camera. (*Adapted from J. Xie et al., Electrospun nanofibers for neural tissue engineering, Nanoscale, 2:35–44, 2010, DOI:10.1039/B9NR00243J*)

a hypodermic needle), and a grounded collector (for example, a piece of aluminum foil). When a polymer solution is extruded from the spinneret, it tends to form a spherical droplet confined by surface tension. If a high voltage is applied to the spinneret, positive (or negative) charges will be immobilized on the surface of the droplet. The repulsive interactions between the charges act against the surface tension, causing the droplet to deform into a conical shape commonly known as the Taylor cone. Once the strength of the electric field has surpassed a threshold, a liquid jet will be ejected from the apex of the cone. Before reaching the collector, the liquid jet is continuously stretched in an effort to increase the separation between adjacent charges and thus mitigate the repulsion. A combination of the repulsive interactions among charges on the jet and the electric field force between the jet and the electric field (formed between the spinneret and the collector) drives the liquid jet in a process known as whipping instability (Fig. 1, inset), further stretching and thinning the jet. As a result of the increase in surface area, the solvent in the liquid jet evaporates quickly. In principle, the jet can be constantly reduced in size until it reaches the collector or solidifies to become a fiber. The nanofibers, when collected as a nonwoven mat, appear to be a soft, compliant piece of cloth with a high porosity and a large specific surface area.

Polymers for electrospinning. Polymers are typically used for electrospinning because their long chains can interact effectively with each other to provide adequate viscoelasticity for the solution. During electrospinning, there is a drastic stretching or shear force acting on the liquid jet, which requires the viscoelasticity of the solution to be sufficiently strong so that the liquid jet will not break into short segments. To date, more than 100 different polymers have been successfully electrospun into nanofibers.

These polymers can be generally divided into two groups: natural and synthetic. Examples of natural polymers include collagen, elastin, fibrinogen, silk, DNA, and cellulose, among others. These materials occur in nature and can be purified for electrospinning. Collagen, elastin, and fibrinogen are typically useful for biomedical applications, since they exist in the human body. Examples of synthetic polymers include polystyrene, polyester, nylon (polyamide), polyethylene, and many others. Depending on the application of the nanofibers, different polymers have their own niches. For instance, polystyrene or poly(vinyl alcohol) can be used for fabricating filters, while some polyesters are biodegradable or biocompatible and hence are suitable for biomedical implantations. Since different applications have specific requirements for functionality, having a wide variety of materials to choose from allows one to alter the composition and other properties of electrospun nanofibers to meet the demands.

Control of alignment. Normally, the nanofibers deposited on a collector will adopt a random orientation (**Fig. 2***a*). This can be ascribed to a uniform distribution of charges on the collector, which tend to attract oppositely charged nanofibers from all different directions at roughly the same strength. As a result, the nanofibers will take random orientations, and the structure will be isotropic in the pane of the collector. When an insulating region is introduced in the conductive collector, the continuity of charges will be interrupted, as they will not be able to populate the insulating region. An example of such a collector is a piece of aluminum foil containing a rectangular window in the middle. The charges on the collector can now be divided into two groups across the rectangular window. These two groups of charges exert a uniaxially stretching force on the descending nanofiber, forcing it to orientate perpendicular to the long edges of the rectangular window and be deposited across it. As a result, the nanofibers will be collected as a uniaxially aligned array (Fig. 2*b*).

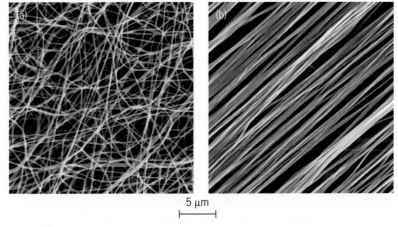

5 μm

Fig. 2. Scanning electron microscopy images showing electrospun nanofibers with different alignments. (*a*) A nonwoven mat of random nanofibers collected on a piece of aluminum foil. (*b*) A uniaxially aligned array of nanofibers deposited across a rectangular insulating gap in the middle of a piece of aluminum foil.

Fig. 3. Fluorescence micrograph revealing the aligned projection of neurites (green). The neurites can replicate the topography of the underlying nanofibers with a uniaxial alignment.

Neuroregeneration. Neuroregeneration refers to the regrowth or repair of nerve tissues after injuries. Such injuries occur to more than 90,000 people every year in the United States alone. A typical neural injury often includes the concussion of several axons, which are also known as nerve fibers. An axon is a long, slender projection of a nerve cell, or neuron, that typically conducts electrical impulses away from the neuron's cell body. When an axon is severed, the end that is still attached to the neuron is referred to as the proximal segment and the other end is called the distal segment. The proximal seg-

ment is the part that will regenerate, and the distal segment is the part that will degenerate.

Scaffolds made of uniaxially aligned nanofibers are particularly useful for neuroregeneration. When an axon tries to regrow, the growth cone at the tip of the proximal segment "sniffs out" the extracellular environment for signals that will instruct the axon to grow toward a specific destination. These signals, or guidance cues, can attract or repel axons. The axon grows when positive cues are presented and retracts when it encounters negative cues. Therefore, building continuous positive cues to guide the growth of an axon from the proximal segment to the distal segment holds the key to successful neuroregeneration. Many techniques have been developed to provide the positive cues (for example, growth factors) for neuroregeneration, but in most cases, these cues are homogeneously distributed around the lesion (damaged site). As a result, the axon may grow along any direction with an equal possibility. To optimize the regeneration, uniaxially aligned nanofibers have been developed to provide directional positive cues for bridging the proximal and distal segments.

The most prominent advantage of using aligned nanofibers is that they can reduce the mismatch between the proximal and distal segments. Many in-vitro studies have shown that the extension of neurites (immature neurons) can be affected by the topography of the nanofibers and consequently will adopt an aligned projection (**Fig. 3**). Since most of the regenerated neurites project parallel to each other, the possibility of intervening among adjacent neurites would be reduced so that the mismatch could be minimized. By folding a sheet of aligned nanofibers into a small conduit with the alignment parallel to the long axis of the tube, a scaffold, known as a neural conduit, can be created for a surgical procedure or an in-vivo study. In a typical in-vivo study, a section of the nerve bundle is dissected and removed to create a lesion, followed by the suturing of a conduit to the nerve gap at both ends to bridge the proximal and distal segments. So far, the longest lesion that has been repaired with reasonable recovery of function by using a neural conduit was around 17 mm.

Wound closure. Unlike neurons, which regenerate through axon sprouting, most wounds heal as a result of cell proliferation and migration. Cells prefer to migrate along the guidance provided by the nanofibers. Aligned nanofibers offer the shortest migratory route by avoiding detours and hence increasing the regeneration rate. In the scenario of wound closure, the injury site is always surrounded by intact tissues. Therefore, nanofibers assembled into a structure that can guide the migration of cells from the healthy periphery inward to the injured center will be very beneficial (**Fig. 4**). Radial alignment provides such a structure. The function of radially aligned nanofibers has been exemplified in the case of dural repair. The dura mater is the outermost of the three layers of meninges that surround the brain and the spinal cord. Its major function is to prevent the leakage of cerebrospinal fluid. Dural repair is often required

Fig. 4. Superimposed fluorescence and scanning electron micrographs showing the migration of cells from the periphery to the center, as guided by a scaffold of radially aligned nanofibers. (*Adapted from J. Xie et al., Radially aligned electrospun nanofibers as dural substitute for wound closure and tissue regeneration applications, ACS Nano, 4:5027–5036, 2010, DOI:10.1021/nn101554u*)

after traumatic injuries or surgery. By applying this new type of scaffold to a defective site in the dura mater, a much faster regeneration is expected.

Radially aligned scaffolds can be fabricated by using a metal ring with a point electrode at the center as the collector. When the point electrode is placed slightly higher than its peripheral ring, most of the descending nanofibers will land first on the central electrode and then connect with different points on the ring collector with an equal possibility. The result will be a scaffold with nanofibers in a radially aligned fashion.

Filtration. Filtration is commonly used for the separation of solids from a fluid (liquid or gas) by interposing a medium through which only the fluid can pass. Historically, filtration media have been based on porous materials made of limestone or stretched Teflon® meshes. Nowadays, a web composed of electrospun nanofibers is often deposited on the medium to further improve its efficiency in removing small particles. There are two primary functions for the web of polymer nanofibers used in a filtration application. First, although traditional media typically have good handling characteristics and favorable economics, they are also associated with relatively poor filtration efficiencies because of the large pore sizes. Since the solids often have a very broad distribution of sizes, most of the small particles still remain in the fluid after filtration, making multiple operations necessary. Because of their fine sizes, electrospun nanofibers are particularly suited for filtering aerosols, or small particles, from air. In most cases, the layer of nanofibers is designed to have a thickness of only several micrometers. Even with such a thin layer, the nanofibers can achieve remarkable filtration efficiency. Therefore, the addition of nanofibers to the surface of a filtration medium can generate a composite structure with excellent handling properties and superior efficiency for many industrial applications. Second, polymeric nanofibers can be used to improve the surface-loading behavior of a typical filtration material. Many filters exhibit depth-loading characteristics when they are exposed to industrial dust environments. Over time, particulate matter becomes deeply embedded in the fibrous structure of the medium, eventually preventing an adequate flow. When a polymeric nanofiber web is applied to the upstream (dirty) side of a filter, the particulate matter is largely caught at the surface of the nanofiber web. This type of surface-loading behavior allows a filter to be cleaned through standard mechanisms such as backward pulsing or shaking.

Outlook. Electrospinning is a remarkably simple and versatile technique that is capable of routinely producing fibers with diameters down to the nanometer scale. The structure of a nanofiber-based assembly can be tailored to accommodate the specific need of the application. With these and other merits, electrospun nanofibers will continue to enjoy a role as a unique class of functional nanomaterials for a broad spectrum of applications.

For background information *see* BIOMEDICAL ENGINEERING; BIORHEOLOGY; CELL MOTILITY; FIL-TRATION; MENINGES; NANOTECHNOLOGY; NERVE; NERVOUS SYSTEM (VERTEBRATE); NEUROBIOLOGY; NEURON; POLYESTER RESINS; POLYMER; REGENERATIVE BIOLOGY; RHEOLOGY; SURFACE TENSION in the McGraw-Hill Encyclopedia of Science & Technology.

Wenying Liu; Ping Lu; Younan Xia

Bibliography. P. Gibson, H. Schreuder-Gibson, and D. Rivin, Transport properties of porous membranes based on electrospun nanofibers, *Colloid Surface A*, 187:469–481, 2001, DOI:10.1016/S0927-7757(01) 00616-1; D. Li and Y. Xia, Electrospinning of nanofibers: Reinventing the wheel?, *Adv. Mater.*, 16:1151–1170, 2004, DOI:10.1002/adma. 200400719; W. Liu, S. Thomopoulos, and Y. Xia, Electrospun nanofibers for regenerative medicine, *Adv. Healthcare Mater.*, 1:10–25, 2012, DOI:10.1002/adhm.201100021; D. H. Reneker and I. Chun, Nanometer diameter fibres of polymer produced by electrospinning, *Nanotechnology*, 7:216–223, 1996, DOI:10.1088/0957-4484/7/3/009.

Evolution of nuclear structure in erbium-158

The response of atomic nuclei to increasing angular momentum (or spin) and excitation energy is one of the most fundamental topics of nuclear structure research and is often studied through high-resolution gamma-ray spectroscopy. Erbium-158 (^{158}Er) is widely considered as a classic nucleus in this field since it exhibits a number of beautiful structural changes as it evolves with increasing excitation energy and angular momentum. At low spin it behaves like a weakly deformed prolate quantum rotor, similar to many other rare-earth nuclei. With increasing angular momentum, it undergoes Coriolis-induced alignments of high-j neutron or proton pairs until a dramatic prolate collective to oblate noncollective transition eventually takes place via the mechanism of band termination. At the highest spins, a spectacular return to collective rotation is observed, and it has been suggested that this is in the form of triaxial strongly deformed structures. This latter suggestion is based on a comparison of energies, spins, and transition quadrupole moments between experiment and theory. This observation confirms the longstanding prediction that such heavy nuclei will possess nonaxial shapes on their path toward fission.

Nuclear structure physics and gamma-ray spectroscopy. Nuclear structure physics concentrates on understanding the structure, behavior, and properties of atomic nuclei, which make up 99.9% of the mass of our everyday world. At present there are 118 known elements and over 3100 known isotopes, of which approximately 270 are stable. The latter are plotted in **Fig. 1**. One of the great quests of nuclear structure physics is to discover the limits of nuclear existence in both proton number Z and neutron number N. Large efforts worldwide are being undertaken to map out the so-called proton and neutron drip lines, which define the limits of nuclear

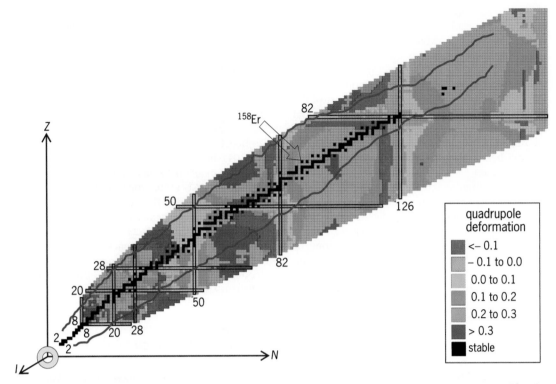

Fig. 1. Chart of nuclei and their calculated ground-state quadrupole deformations as a function of neutron number (*N*) and proton number (*Z*). Magic numbers (2, 8, 20, 28, 50, 82, and 126), which correspond to closed shells, are labeled. The present limits of experimentally known nuclei are between the two thick violet lines above and below the stable nuclei (black squares). A third axis of angular momentum or spin (*I*) is also shown, and it is the evolution of nuclear structure along this axis that this article addresses. The location of the ^{158}Er nucleus is marked.

existence. *See* BREAKDOWN OF SHELL CLOSURE IN HELIUM-10; DEFINITIVE EVIDENCE FOR NEW ELEMENTS 113 AND 115.

Intense studies are also being conducted to understand the pathways far from the line of stability that are taken in the processes that create the heavy elements beyond iron. This plot of *N* and *Z* is similar in many ways to the well-known periodic table of elements in chemistry. For example, certain nuclei exhibit "magic numbers," which represent especially favored or stable numbers of protons and neutrons, akin to the noble gases. These special numbers and their periodicity arise from the underlying quantal shell structure of electrons and nucleons (protons and neutrons) in both atomic and nuclear systems, respectively. The numbers are not the same in the two systems since in nuclei the spin-orbit interaction of the particles' spin and orbital motions plays a dominant role. These numbers (2, 8, 20, 28, 50, 82, and 126) correspond to closed shells, and are indicated in Fig. 1. The colored shading in Fig. 1 indicates the degree of the ground-state deformation or shape of nuclei. It can be seen that as one moves away from the magic numbers the deformation is strongly dependent on the number of "valence" nucleons (protons and neutrons). Again, this is similar to atomic systems, where the number of valence electrons determines an element's chemical properties.

Another great quest in nuclear structure physics is to explore the behavior of nuclei at the limits of angular momentum and excitation energy. This is shown

in Fig. 1 by the third axis, representing angular momentum or spin (*I*). These studies give us deep insight into the many body nuclear problem, and especially the role of the proton and neutron "intruder" orbitals, whose positions move from energies above a magic number orbital to energies below, with increase in deformation, and so play a dominant part in determining the properties (shapes, moments of inertia, pairing correlations, collectivity, and so forth) of nuclei at high spin.

It was as early as 1937 that Niels Bohr and Fritz Kalckar proposed that we could learn about the evolving structure and shape of excited nuclei by detecting their gamma-ray emissions. Again, this is similar to investigating atomic or molecular structure by the radiation emitted when these systems are excited. Gamma-ray spectroscopic studies based on novel detector technologies continue to revolutionize our understanding of the atomic nucleus, revealing an extremely rich system that displays a wealth of static and dynamical facets. For example, the discoveries in ^{158}Er have benefited enormously from the progression of detector technology (**Fig. 2**). The next major step is to move toward the goal of a 4π germanium ball (that is, an instrument with germanium detectors that completely surround a source of gamma rays, so that they cover 4π steradians, the solid angle of the whole sphere), utilizing the mechanism of gamma-ray energy tracking. This will bring yet another revolution to the field of gamma-ray spectroscopy through an increased sensitivity which allows the ability to see very weak, previously unobserved gamma rays, and so will usher in a new era in nuclear structure physics.

Level structures up to band termination. In the field of high-spin nuclear physics, the rare-earth region has always been one of the most favored regions, since nuclei here can accommodate the highest values of angular momentum. In particular, in the erbium-158 nucleus (^{158}Er: $Z = 68$, $N = 90$), numerous fascinating phenomena have been observed with increasing excitation energy and angular momentum (Fig. 2), and it is widely acknowledged as a textbook example of the evolution of nuclear structure. For example, initially its angular momentum is generated by a collective rotation of all the particles together, as described below; then, as the angular momentum increases, this nucleus exhibits Coriolis-induced rotational alignments of both neutron and proton pairs along the yrast line (the line that defines the lowest state in energy for a given angular momentum; Fig. 2 and **Fig. 3**). Erbium-158 was among the first in which backbending, that is, rotational alignment of a pair of high-j (high orbital angular momentum) neutrons (Fig. 3), was discovered (at spin $I \sim 14\ \hbar$, where $\hbar$ is Planck's constant divided by 2π), and it was the first nucleus where a second alignment, of high-j protons, at $I \sim 28\ \hbar$, and a third anomaly in the moment of inertia along the yrast line, at $I \sim 38\ \hbar$, were identified. When spin values reach 40–50$\hbar$, a very different structure becomes most energetically favored (that is, yrast), where this nucleus undergoes a dramatic shape transition from a collective (the case where the rotational axis is perpendicular to the symmetry axis) prolate (American football shape) rotation to noncollective (the case where the angular momentum axis and the symmetry axis are aligned) oblate (discus shape) configurations. The angular momentum of the collective prolate states comes from both collective rotation of the nucleus as a whole and single-particle alignment contributions; whereas the angular momentum of the noncollective oblate states comes completely from the spin alignment of the valence nucleons, and for a particular configuration it has a maximum terminating value where all the valence particles' angular momenta are aligned. Thus, the noncollective oblate band terminates at an energetically favored state where the valence nucleons outside the gadolinium-146 semimagic spherical core (^{146}Gd: $Z = 64$, $N = 82$) all have their spins fully aligned in the same direction. Band termination represents a clear manifestation of mesoscopic physics, since the underlying finite-particle basis of the nuclear angular momentum generation in "classical-like" rotational band structures is revealed. In ^{158}Er, three terminating states, 46^{+}, 48^{-}, and 49^{-}, have been observed (Figs. 2 and 3). Other neighboring nuclei were also found to exhibit similar fully aligned states.

Level structures beyond band termination. It had been a goal for several decades to establish the nature of the states in the rare-earth nuclei well beyond the very favored band-terminating states. Several years ago, a number of weak, individual transitions feeding into the terminating states were

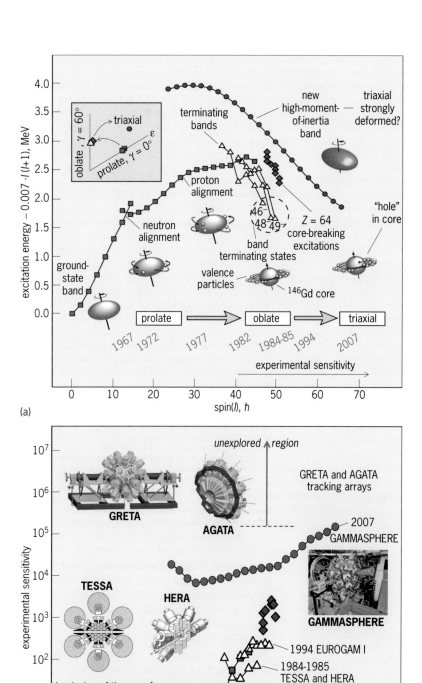

Fig. 2. Nuclear structure and experimental sensitivity as functions of angular momentum (spin) in ^{158}Er. (*a*) Evolution of nuclear structure with excitation energy and spin. The inset illustrates the changing shape of ^{158}Er with increasing spin within the standard (ε, γ) deformation plane (prolate → oblate → triaxial). The parameters ε and γ represent the eccentricity from sphericity and triaxiality, respectively. (*b*) Experimental sensitivity of detection (proportional to the inverse of the observed gamma-ray intensity) is plotted as a function of spin, showing the progression with time of gamma-ray detector technologies that are associated with nuclear structure phenomena in ^{158}Er. HPGe = high-purity germanium. TESSA, HERA, EUROGAM I, GAMMASPHERE, GRETA, and AGATA are names of gamma-ray detector arrays.

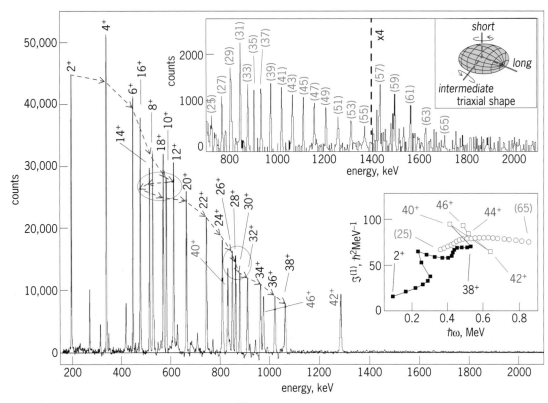

Fig. 3. Spectrum of gamma rays of the yrast band in ^{158}Er that are seen in coincidence with the $44^+ \rightarrow 42^+$ gamma-ray transition. Transitions are all labeled with the states (spin^{parity}) from which they decay. The observed neutron (energy $\sim$ 550 keV, spin $I \sim 14\ \hbar$) and proton (energy $\sim$ 850 keV, $I \sim 28\ \hbar$) rotational alignments are marked in red circles. Upper inset shows coincidence spectrum representative of the most intense collective band at ultrahigh spin (band 1) observed in ^{158}Er. The portion of this spectrum above 1400 keV has been magnified by a factor of 4. The spins assigned are tentative and the parity of the sequence is not known. The triaxial nuclear shape with which the ultrahigh spin bands are associated is schematically illustrated at the upper-right corner. Lower inset shows moments of inertia, $\Im^{(1)}$, as a function of ω, where ω is the rotational frequency, for the yrast sequence in ^{158}Er (collective prolate states: filled squares; noncollective oblate states: open squares; band 1: open circles). Note that the $\Im^{(1)}$ line bends back (toward lower rotational frequency) with increasing spin when the neutron alignment occurs ($\hbar\omega \sim 0.25$ MeV), hence the name "backbending." The 40^+, 42^+, 44^+, and 46^+ states are based on a configuration that terminates at the 46^+ noncollective oblate state. The triaxial collective band shown extends up to a spin of $\sim$65 $\hbar$.

observed in ^{158}Er (and its neighbor, ^{157}Er), but this extended the highest spin by only 1-2 $\hbar$. It has been suggested that the related levels above band termination arise from weakly collective single-particle excitations that break the ^{146}Gd core (Fig. 2). More significantly, in 2007, two rotational structures displaying high dynamic moments of inertia and possessing very low intensities ($\sim 10^{-4}$ of the respective channel intensity) were identified in ^{158}Er (Fig. 3). These structures bypass the well-known band-terminating states and extend over a spin range of $\sim$25 - 65 $\hbar$, marking a spectacular return to collectivity at spins beyond band termination. These sequences have properties, for example moment of inertia values, that are very different from the lower spin states (Figs. 2 and 3). Thus, a new frontier of discrete-line gamma-ray spectroscopy was opened in ^{158}Er toward spin $\sim$ 70 $\hbar$ (the so-called ultrahigh-spin regime).

Initially, based solely on theoretical calculations, it was proposed that the new ultrahigh-spin bands in ^{158}Er were triaxial strongly deformed (TSD) structures. A triaxial nuclear shape has distinct short, intermediate, and long principal axes (Fig. 3). At high spin, collective rotation about the short axis

is expected to be energetically favored over rotation about the intermediate axis (and even more so over the long axis), based on moment of inertia considerations. In a later study using the Doppler-shift attenuation method (DSAM), the transition quadrupole moments (Qt) of the ultrahigh spin bands in ^{158}Er were experimentally determined to be $\sim$9 - 11 eb (electron barn: 1 e = 1.602 $\times$ 10^{-19} coulomb, and 1 barn = 10^{-28} m^2). As the low-spin collective yrast band in ^{158}Er has a measured Qt of $\sim$ 6 eb, this result demonstrates that the ultrahigh spin bands in ^{158}Er are all associated with strongly deformed shapes. However, the measured Qt values appear to be too large for the calculated energetically favored TSD shape (rotating about the short axis and with a calculated Qt $\sim$ 7.4 eb). This puzzling discrepancy has attracted much attention and has motivated more sophisticated theoretical calculations which allow the angular momentum vector to "tilt" between the two minor axes. However, a fully coherent understanding has not yet emerged.

A large number of questions arise from the striking observation of the return of collectivity and unusual nuclear shapes beyond band termination in ^{158}Er. An extended systematic study of ultrahigh spin phenomena in nuclei around ^{158}Er is in progress from which we hope to find answers to the mysteries of the evolution of nuclear structure at the limits of angular momentum and excitation energy.

For background information *see* GAMMA RAYS; NUCLEAR SHELL MODEL AND MAGIC NUMBERS; NUCLEAR STRUCTURE in the McGraw-Hill Encyclopedia of Science & Technology.   Xiaofeng Wang; Mark A. Riley; John Simpson; Edward S. Paul

Bibliography. A. Bohr and B.R. Mottelson, *Nuclear Structure*, vol. 2, Benjamin, New York, 1975; K. Heyde, *Basic Ideas and Concepts in Nuclear Physics*, Institute of Physics, Bristol, U.K., 1999; P. Moller et al., Nuclear ground-state masses and deformations, Atomic Data and Nuclear Data Tables, 59:185–381, 1995, DOI:10.1006/adnd. 1995.1002; S. G. Nilsson and I. Ragnarsson, *Shapes and Shells in Nuclear Structure*, Cambridge University Press, Cambridge, U.K., 1995; E. S. Paul et al., Return of collective rotation in ^{157}Er and ^{158}Er at ultrahigh spin, *Phys. Rev. Lett.*, 98:012501 (4 pp.), 2007, DOI:10.1103/PhysRevLett.98.012501; Y. Shi et al., Self-consistent tilted-axis-cranking study of triaxial strongly deformed bands in ^{158}Er at ultrahigh spin, *Phys. Rev. Lett.*, 108:092501 (5 pp.), 2012, DOI:10. 1103/PhysRevLett.108.092501; J. Simpson et al., Single particle excitations and properties of multiple band terminations near spin 50$\hbar$ in ^{158}Er, *Phys. Lett. B*, 327:187–194, 1994, DOI:10.1016/ 0370-2693(94)90716-1; X. Wang et al., Quadrupole moments of collective structures up to spin $\sim 65\hbar$ in ^{157}Er and ^{158}Er: A challenge for understanding triaxiality in nuclei, *Phys. Lett. B*, 702:127–130, 2011, DOI:10.1016/j.physletb.2011.07.007; F. Yang and J. H. Hamilton, *Modern Atomic and Nuclear Physics*, 2d ed., World Scientific, Singapore, 2010.

Expressed emotion

Schizophrenia is a debilitating psychiatric disorder that affects approximately 1% of the population in the United States. This disorder is characterized by three broad categories of symptoms: (1) positive symptoms (for example, hallucinations and delusions), (2) negative symptoms (for example, restricted affective expression and impoverished thinking), and (3) cognitive dysfunction (for example, deficits in memory and attention).

Scholars have long been interested in understanding the link between family factors and the biological processes that underlie this debilitating disease. Early work in this area focused primarily on questions concerning etiological factors associated with schizophrenia. For example, one early theory speculated that the behavior of the parents [especially that of the mother, who was termed the schizophrenogenic mother] could determine whether their child would develop schizophrenia. At present, such hypotheses regarding the etiological role of family factors with schizophrenia have largely been debunked.

In the 1950s and 1960s, George Brown and colleagues in the United Kingdom began a line of research that represented a paradigmatic shift in the study of family factors and schizophrenia. Although previous research on family factors had examined the question of whether families cause schizophrenia, the research by Brown and colleagues focused on whether aspects of the family environment may shape the course of illness. More specifically, Brown and colleagues were interested in whether individuals with schizophrenia did better or worse depending on aspects of the larger family system in which they lived. To investigate this question, Brown and colleagues completed a series of investigational studies with families caring for a relative with schizophrenia with the goal of identifying certain aspects of the family environment that might influence the course of schizophrenia. It was these studies that ultimately led to the proposal of the concept of expressed emotion.

The construct of expressed emotion (EE) is an umbrella term that subsumes a number of family factors directed by the caregiving relative toward the individual with schizophrenia. These factors include both negative family factors (that is, criticism, emotional overinvolvement, and hostility) and prosocial family factors (that is, positive remarks and warmth). In 1972, Brown and colleagues found that individuals with schizophrenia who were exposed to familial environments characterized by high levels of criticism, emotional overinvolvement, and/or hostility (that is, high EE) were more likely to experience a significant exacerbation (that is, relapse) of psychotic symptoms over the subsequent nine months than individuals with schizophrenia who were exposed to family environments in which the presence of these variables was low (that is, low EE). This finding has since been replicated by numerous studies of individuals with schizophrenia from diverse ethnic backgrounds across multiple continents, and more recent research has extended the study of EE to other psychiatric disorders.

Yet, despite the clear statistical association between EE and a poor course of illness among individuals with schizophrenia, this line of study is not without its critics. In particular, three prominent controversies within the EE literature need to be addressed: (1) because of EE, caregiving relatives are blamed for their ill relative's poor course of illness; (2) the mechanism (or mechanisms) through which EE facilitates negative health outcomes among individuals with schizophrenia is unclear; and (3) cultural variation exists in the relationship between EE and relapse in schizophrenia.

Expressed emotion: trait of family member or family system? Although, the construct of EE was considered a breakthrough in the understanding of schizophrenia, this line of research quickly (and appropriately) drew the ire of family members caring for a relative with schizophrenia. It is important to note that caregiving relatives play a critical, and often thankless, role in the recovery process of individuals with schizophrenia (for example, coordinating medical appointments, providing financial support, and assisting with obtaining/maintaining competitive employment). Not surprisingly, given the herculean effort typically extended by caregiving relatives to support their relatives with schizophrenia, the hypothesis that caregivers are to blame for the ill relative's poor course of illness was not received well by caregiving relatives.

Criticism from family caregivers with regard to EE research highlighted an important and previously unexamined aspect of the EE-relapse association. Whereas early EE research had assumed a specific causal direction in the EE-relapse association (that is, high EE causes worsening of psychotic symptoms), an equally plausible hypothesis was that exposure to greater psychotic symptomatology on the part of the ill relative may elicit more criticism and emotional overinvolvement (that is, high EE) from caregivers. In other words, when confronted with an exacerbation of psychotic symptoms in their relative with schizophrenia, caregivers may respond with greater challenging (that is, criticism) of behaviors that may worsen psychotic symptoms (for example, substance use) and the provision of increased social support (that is, emotional overinvolvement). The goal of such behaviors from caregivers would be to help promote a reduction in the expression of psychotic symptoms by their relative with schizophrenia.

One of the earliest attempts to clarify the direction of the relationship between EE and relapse was completed by Keith Nuechterlein and colleagues. Using structural equation modeling, these investigators examined the longitudinal associations between caregiving relatives' EE and the severity of psychotic symptoms among a group of individuals with a recent onset of schizophrenia. Interestingly, the findings supported both hypothesized directions of the relationship between EE and relapse of psychotic symptoms. More specifically, high EE from caregivers was associated with an increased risk of future symptomatic relapse in their relative with schizophrenia. However, greater symptomatology among individuals with schizophrenia was also predictive of the display of greater criticism and emotional overinvolvement (that is, high EE) on the part of their caregiving relative. In total, these findings suggested a transactional relationship between psychotic symptoms and EE—exposure to greater symptomatology from relatives with schizophrenia elicits more behaviors consistent with high EE from caregiving relatives, which in turn triggers a worsening of symptomatology in the relative with schizophrenia. This relationship is depicted in the **illustration**.

The results of the study by Nuechterlein and colleagues have important implications for the conceptualization of EE. In highlighting the cyclical nature of the relationship between EE and the expression of psychotic symptoms (see illustration), this study suggests that it is inaccurate to view EE as situated solely within caregiving relatives. In other words, because high EE attitudes and behavior emerge as a result of interactions between caregivers and their

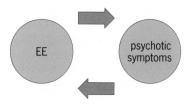

Transactional relationship between caregiving relatives' expressed emotion (EE) and severity of psychotic symptoms in individuals with schizophrenia.

relatives with schizophrenia, EE is not a characteristic of caregivers specifically. Rather, EE is more accurately viewed as a quality of the larger family system. Consequently, previous depictions of EE specifically as a characteristic of caregiving relatives (for example, the frequently used phrase "high-EE caregiver") inappropriately portrayed caregivers as the sole cause for their ill relative's poor course of illness.

What is the mechanism of expressed emotion? The second major controversy within the EE literature involves understanding the mechanism (or mechanisms) through which exposure to high-EE familial environments promotes poor clinical outcomes in individuals with schizophrenia. Specifically, although numerous studies have demonstrated a clear statistical association between EE and the course of schizophrenia, comparatively little attention has been directed toward clarifying the specific mechanism (or mechanisms) through which the components of EE (for example, criticism and emotional overinvolvement) activate the biological mechanisms underlying the expression of psychotic symptoms.

To date, Jill Hooley and colleagues have advanced the most comprehensive theoretical model outlining the mechanisms through which exposure to high-EE familial environments may promote the expression of psychotic symptoms. This model (that is, the stress–diathesis model of EE) suggests that, for individuals with schizophrenia, who by definition have a preexisting disposition (that is, diathesis) for experiencing symptoms of psychosis, exposure to high-EE familial environments is a psychosocial stressor that activates the biological mechanisms underlying the expression of psychotic symptoms. This theory comports with existing research suggesting that (1) individuals with schizophrenia perceive exposure to high-EE familial environments as stressful and (2) exposure to psychosocial stressors is predictive of a worsening of psychotic symptoms among individuals with schizophrenia.

It is important to note, though, that support for the stress–diathesis model of EE is equivocal. For example, if exposure to high-EE familial environments is a psychosocial stressor, one would expect that EE would predict other stress-responsive health outcomes among individuals with schizophrenia. However, a recent study found that EE was not predictive of negative physical health outcomes among individuals with schizophrenia—outcomes that are clearly associated with exposure to psychosocial stressors among people in general. This finding raises significant concerns with regard to whether the stress–diathesis model may fully capture the mechanisms through which EE influences the course of schizophrenia, and it highlights the need for additional research in this area.

Cultural variation of expressed emotion. Scholars have long recognized the critical role of culture in shaping both family structure and practices. As such, the finding that EE predicts relapse among individuals with schizophrenia across diverse ethnic backgrounds and across multiple continents is particularly striking. However, more recent research has suggested that important cultural variation does exist with regard to the EE-relapse association. One of the key findings to date focused on cultural variability with regard to the EE index of warmth.

As noted earlier, EE research has focused largely on negative components of family systems (for example, criticism), and little attention has been directed toward prosocial components (for example, warmth). The impetus for this decision can be traced to the original work by Brown and colleagues, who viewed warmth as a variable that was too complex to be included in the operational definition of high versus low EE. However, in recognition of the importance of studying how positive family functioning may influence the course of schizophrenia, Steven López and colleagues examined the association between family warmth and relapse among Euro-Americans and Mexican-Americans having schizophrenia. The results of this study were compelling: whereas family warmth was not associated with the course of schizophrenia for Euro-Americans, family warmth did predict the course of illness for Mexican-Americans. More specifically, Mexican-Americans having schizophrenia who were exposed to familial environments characterized by greater warmth were less likely to experience a symptomatic relapse. Two possible interpretations were suggested for why family warmth may matter for Mexican-Americans, but not for Euro-Americans. First, given the greater importance of strong family ties in Mexican-American families as compared to Euro-American families, limited familial warmth may be a larger psychosocial stressor for Mexican-Americans with schizophrenia versus Euro-Americans with schizophrenia. Second, as the vast majority of families included in the Mexican-American sample were immigrants from low socioeconomic backgrounds, family warmth may buffer these individuals from the stress associated with living in a foreign and potentially hostile environment. In total, this research has highlighted the important cultural variation that exists with regard to EE.

Conclusions. Despite nearly 60 years of research, the construct of expressed emotion (EE) remains a controversial topic in psychiatric research. The clear predictive association between EE and the course of schizophrenia has led EE to become one of the most studied psychosocial variables in psychiatric research to date. Yet, questions with regard to mechanisms and cultural variation in the EE-relapse association persist and ultimately limit our understanding of this important psychiatric construct.

For background information *see* COGNITION; EMOTION; MENTAL DISORDERS; PSYCHOLOGY; PSYCHOSIS; SCHIZOPHRENIA; STRESS (PSYCHOLOGY) in the McGraw-Hill Encyclopedia of Science & Technology.   Nicholas J. K. Breitborde

Bibliography. N. J. K. Breitborde, S. R. López, and A. Kopelowicz, Expressed emotion and health outcomes among Mexican-Americans with schizophrenia and their caregiving relatives, *J. Nerv. Ment. Dis.*, 198:105–109, 2010, DOI:10.1097/NMD. 0b013e3181cc532d; G. W. Brown, J. L. Birley, and J. K. Wing, Influence of family life on the course of schizophrenic disorders: A replication, *Br. J. Psychiatry*, 121:241–258, 1972, DOI:10.1192/bjp.121. 3.241; J. M. Hooley, Expressed emotion and relapse of psychopathology, *Annu. Rev. Clin. Psychol.*, 3:329–352, 2007, DOI:10.1146/annurev.clinpsy.2. 022305.095236; S. R. López et al., Ethnicity, expressed emotion, attributions, and course of schizophrenia: Family warmth matters, *J. Abnorm. Psychol.*, 113:428–439, 2004, DOI:10.1037/0021-843X.113.3.428; H. I. McCubbin et al., Culture, ethnicity, and the family: Critical factors in childhood chronic illnesses and disabilities, *Pediatrics*, 91:1063–1070, 1993; National Institute of Mental Health, *Schizophrenia*, U.S. Department of Health and Human Services, National Institutes of Health, Bethesda, MD, 2009; K. H. Nuechterlein, K. S. Snyder, and J. Mintz, Paths to relapse: Possible transactional process connecting patient illness onset, expressed emotion, and psychotic relapse, *Br. J. Psychiatry*, 161(suppl. 18):88–96, 1992.

Extinction and the fossil record

The Earth contains an incredible diversity of species that are important for maintaining a functional, healthy world. There are an estimated 8.7 million species living on Earth today (about 1.3 million of them have been documented by scientists), but these species represent only 1% of the species that have lived on the Earth since life originated approximately 3.5 billion years ago. Ninety-nine of every 100 species that have ever lived on Earth are extinct. Extinctions have occurred relatively constantly through time, punctuated by several periods of severe and relatively rapid extinction events. Paleontologists and conservation biologists have realized that these past extinctions can provide much information about the events that are occurring today and the extent of their impact on global biodiversity. By putting modern-day extinctions in the context of the rates and magnitudes of past extinctions, it is possible to understand the types of events that can lead to devastating biodiversity loss.

Past extinctions. Given that 99 of every 100 species that have ever lived on Earth are no longer in existence, what happened to them? Extinctions occur all the time (at least a couple every thousand years or so) and for many different reasons. For example, if a volcano erupts, it may kill a single lizard species that only lived on the slopes of that volcano (an example of a species restricted to a limited range; this species is termed an endemic); likewise, if one species of rodent is consistently outcompeted by another, it may eventually die out. These types of individual extinctions that periodically occur are referred to as background extinctions. However, there are five time periods throughout history that have earned the name "mass extinctions" (although there may be more, depending on how these events are defined). These are periods that stand out from the background as having much faster rates and magnitudes of species losses, resulting in the devastation of species diversity on the planet. The five main mass extinctions recognized by paleontologists are the Ordovician event, the Devonian event, the Permian event, the Triassic event, and the Cretaceous event. Although these events were comparably severe in terms of their impact on biodiversity, the causes of the events and how they proceeded were quite different (see **table**). There have been some suggestions that the human species may be in the process of initiating a sixth mass extinction event today.

To investigate this question, scientists have been endeavoring to quantify exactly what makes a mass extinction. Extinction is typically measured in one of two ways, using extinction rates or extinction magnitudes. Extinction magnitudes describe the absolute loss of biodiversity. For example, if there were 1 million species before an extinction event and 250,000 remained after the event, the magnitude of extinction would be 75%. Extinction rates describe how rapidly extinctions are occurring. For example, the background extinction rate for mammal species since 65 million years ago is about 1.8 extinctions per million species-years. Extinction rates can be described as the number of species that become extinct in a given amount of time. However, because more species are expected to go extinct if the overall number of species is greater, scientists also normalize the rates to the total number of species. In the fossil record, for example, the number of nonpreserved species that were becoming extinct is not always known; however, the proportion of the species examined that went extinct is known. The units of an extinction rate are therefore often reported as the number of extinctions per million species-years. Thus, if 1.8 extinctions occur per million species-years, then 1.8 of every million species would go extinct, on average, per year. This rate is the same as 1.8 extinctions occurring in a group of 1000 species over 1000 years (that is, 1,000,000 species-years). A mass extinction is traditionally defined as having a faster rate of extinctions than the background, resulting in the loss of more than 75% of species, and being global in scope.

Estimating prehistoric extinction rates and magnitudes is challenging because it is necessary to know when each fossil organism appeared in and disappeared from the fossil record. With much hard work and persistence, paleontologists have constructed a number of databases that contain compilations of known fossil specimens that specify how old they are and where they were found. These databases are powerful tools that can be used (1) to determine the first and last known occurrences of each fossil type and (2) to calculate the overall duration of the group. Then, calculations can be made regarding the number of fossil groups at each time period, the difference in numbers of groups between time periods (that is, the magnitude of change), and the rate at which fossil groups became extinct between time periods.

Sixth mass extinction? Now that background and mass extinction magnitudes and rates can be calculated, it is possible to analyze the current biodiversity crisis to determine whether the Earth is currently in the midst of a sixth mass extinction event.

Timing, duration, magnitude, causes, and effects of the five prehistoric mass extinctions, as well as the potential sixth mass extinction*

Mass extinction	Timing	Duration	Magnitude	Causes	Groups with the most extinctions
Ordovician	~443 MYA	3.3–1.9 MYA	86% species	climate change; sea level change; changing ocean chemistry	brachiopods, bryozoans, trilobites, conodonts, and graptolites
Devonian	~359 MYA	29–2 MYA	75% species	climate change; changing ocean chemistry, especially low oxygen levels	brachiopods, trilobites, and reef-building organisms
Permian	~251 MYA	2.8 MYA–160 KYA	96% species	volcanism; global warming; low oxygen levels in ocean; changing ocean chemistry	foraminiferans, trilobites, corals, blastoids, acanthodians, placoderms, pelycosaurs, bryozoans, brachiopods, ammonoids, sharks, bony fish, crinoids, eurypterids, ostracodes, and echinoderms
Triassic	~200 MYA	8.3 MYA–600 KYA	80% species	volcanism; global warming; changing ocean chemistry	conodonts, crurotarsans, therapsids, and amphibians
Cretaceous	~65 MYA	2.5 MYA–1 year	76% species	meteorite; climate change; volcanism; changing ocean chemistry, including low oxygen levels	nonavian dinosaurs, mosasaurs, plesiosaurs, pterosaurs, coccolithophorids, mollusks, and cephalopods
Sixth mass extinction?	Now	10 KYA–500 years	1%–? species	climate change; habitat loss and fragmentation; overharvesting; pollution; low oxygen levels in ocean; changing ocean chemistry	amphibians, coral reefs, large-bodied mammals, large fish, sharks, island species, mollusks, some plants

*MYA, million years ago; KYA, thousand years ago.

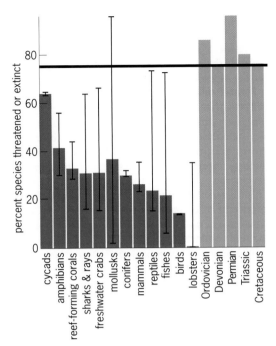

Fig. 1. Extinction magnitudes: The magnitudes of modern and past extinctions. Modern magnitudes are represented as the percent of each group that is recently extinct (*red*) or may soon be extinct (that is, designated by the IUCN as having threatened status; *blue*). Whiskers indicate the lower and upper estimates for the number of threatened species as designated by the IUCN. Past extinction magnitudes (*orange*) represent the estimated percent of total species extinct during each mass extinction. The thick black line is the 75% threshold that defines a mass extinction. (*Modern and threatened species data are from the 2012 IUCN Red List; mass extinction data and the plot are based on A. D. Barnosky et al., Has the Earth's sixth mass extinction already arrived?, Nature, 471:51–57, 2011*)

Unfortunately, it is not possible to directly compare what is happening today to what happened in the past because very different time scales are involved. None of the former mass extinction events occurred overnight. In general, the burst of extinctions that created previous mass extinctions had durations of 160,000 years to as many as 29 million years (the possible exception being the Cretaceous event, in which a huge meteorite struck the Earth and killed the dinosaurs and many other species; however, the rapidity of this extinction event is often disputed; see table). Most of the data that relate to recent extinctions go back only 500 years. If we consider that (1) major climate change is only beginning to occur, (2) the Industrial Revolution only began in the late 1700s, and (3) rapid human expansion only occurred in the last half-century, human impacts have not been around as long as it generally takes to cause a mass extinction. To make modern and historical mass extinctions comparable, it is important to look at the trajectory of human-caused extinctions.

Since 1963, the International Union for Conservation of Nature (IUCN) has been compiling expert opinions regarding how much any given species is in danger of becoming extinct. So far, however, only about 62,000 species have been assessed; this number is approximately 4% of all documented species and less than 1% of all estimated species. Over

the past 500 years, only 0–1% of known species have become extinct, which is far below the magnitude of past mass extinctions (**Fig. 1**). However, because of the very short time frame (in a geological sense) of these extinctions, the rates of modern extinctions are relatively high, and it seems that we are just on the cusp of having a very large number of extinctions occur (**Fig. 2**). Approximately 25% of species in groups that have been thoroughly assessed have been deemed to be threatened.

Using the IUCN data, scientists calculated what extinction rates would be if these threatened species became extinct (Fig. 2). They found that current extinction rates are likely elevated well above background rates, but these rates have not been sustained for long enough to reach the 75% magnitude of extinctions necessary to define a mass extinction (Figs. 1 and 2). Scientists then projected these extinction rates to calculate how long it would take for 75% of vertebrate species to become extinct. If all threatened vertebrates went extinct in the next 100 years, and if that rate of extinction continued, then it was estimated that it would only take 240–540 years to reach the 75% mark. If only the most critically endangered species became extinct in the next 100 years, and if that rate of extinction continued, then it would take as many as 890 to 2270 years to reach the 75% level. Note that these time periods for reaching a mass extinction are much shorter than those estimated for previous mass extinctions,

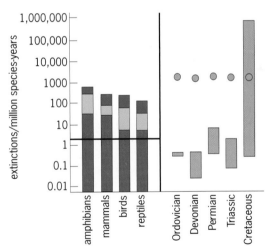

Fig. 2. Extinction rates: The extinction rates for modern and past extinctions. Modern extinction rates are measured in extinctions/million species-years and are based on the extinctions that have occurred over the past 500 years (*red*) or are likely to occur in the near future for species that are listed by the IUCN as critically endangered (*yellow*) or threatened (*blue*). The thick black line is the background extinction rate for North American mammals over the last 65 million years as divided into 500-year time intervals. Note that the background extinction rate for other time periods, that is, when previous mass extinctions occurred, would differ from this black line. Past extinction rates (*orange bars*) are measured in extinctions/million species-years based on the estimated intervals of time over which each mass extinction occurred (see table). *Orange circles* represent past extinction rates if all of the extinctions for each mass extinction event are compressed into a 500-year time interval. (*Plot based on A. D. Barnosky et al., Has the Earth's sixth mass extinction already arrived?, Nature, 471:51–57, 2011*)

indicating that humans have the potential to be a dramatic force on the planet (see table).

Outlook. On a hopeful note, humans have not yet caused the sixth mass extinction. Although current and predicted future extinction rates are well above background rates, not many species have become extinct yet (Figs. 1 and 2). In the near future, though, as climate begins changing even more rapidly, organisms will need to shift their distributions on the landscape to track their habitats (that is, disperse toward areas with better ecological settings). Given that landscapes are fragmented by development and that many places are now unsuitable due to pollution and overharvesting by humans, this will be a significant challenge for many organisms. It will take very careful conservation planning and changes in human behavior to prevent the sixth mass extinction. However, with some drastic changes, it can be done.

For background information *see* ANIMAL EVOLUTION; BIODIVERSITY; EXTINCTION (BIOLOGY); FOSSIL; GEOLOGIC TIME SCALE; GLOBAL CLIMATE CHANGE; MACROEVOLUTION; PALEOECOLOGY; PALEONTOLOGY; TAPHONOMY in the McGraw-Hill Encyclopedia of Science & Technology.　　　Jenny L. McGuire

Bibliography. A. D. Barnosky et al., Has the Earth's sixth mass extinction already arrived?, *Nature*, 471:51–57, 2011, DOI:10.1038/nature09678; J. B. C. Jackson et al., Historical overfishing and the recent collapse of coastal ecosystems, *Science*, 293:629–637, 2001, DOI:10.1126/science.1059199; C. Mora et al., How many species are there on Earth and in the ocean?, *PLoS Biol.*, 9(8):e1001127, 2011, DOI:10.1371/journal.pbio.1001127; M. J. Novacek, *The Biodiversity Crisis: Losing What Counts*, The New Press, New York, 2001; S. Pimm et al., Human impacts on the rates of recent, present, and future bird extinctions, *Proc. Natl. Acad. Sci. USA*, 103:10941–10946, 2006, DOI:10.1073/pnas.0604181103; D. M. Raup and J. J. Sepkoski, Mass extinctions in the marine fossil record, *Science*, 215:1501–1503, 1982, DOI:10.1126/science.215.4539.1501.

Fabrication of flexible polymer solar cells roll-to-roll

The aim of developing carbon-neutral and sustainable energy technologies, such as solar cells, is a race against the clock, as expectations of climate change increase and the reserves of fossil fuels dwindle. Solar cells based on polymers or "plastics" are flexible and allow for mass fabrication in a roll-to-roll (R2R) fashion, much the same as the printing of newspapers. The basic property that allows for this is solubility of the materials in common solvents and even water. Polymers can thus be formulated into liquid inks that can be printed or coated like colors on a printed page. The projected increase in energy demand toward 2050 is around 1 GW a day, which is the equivalent of building a medium-sized nuclear power plant each day. Fabrication of solar cells on such a massive scale will require low costs, a minimal environmental footprint, and high-throughput production. Polymer solar-cell technology can deliver in all categories, but future success depends on the ability of scientists to embrace this paradigm as a rational driving force for further development.

Polymer solar cells at a glance. The polymer solar cell (PSC) has been around since the early 1990s and it is, in essence, very similar to the well-known classical silicon solar cell developed at Bell Labs in the 1950s. Solar photons are absorbed by a semiconducting material when the energy of the photons exceed the energy band gap of the semiconductor. This triggers the excitation of electrons in the semiconductor to an energy state above the band gap, whereby energy is transferred from the photons to the semiconducting material. An excited negatively charged electron leaves behind a charge vacancy that can be regarded as an oppositely charged "hole." Before long, any such promoted electron would collapse back onto the hole because of coulomb interaction and the energy would be "lost." For a solar cell to produce electricity, the electron and the hole must therefore be kept from recombining, forced apart, and collected at different electrodes. This is most often done by combining two types of semiconductors: one that prefers electrons and another that prefers holes. A junction of two such appropriate semiconductors can efficiently separate the electron and the hole. The state-of-the-art PSC combines two types of organic semiconductors in a so-called bulk heterojunction. The preferred hole-conducting material (called p-type, or donor) is a semiconducting polymer, while the electron-preferring part (called n-type, or acceptor) is a small-molecule material based on a C_{60} or C_{70} fullerene (buckyball).

The bulk heterojunction, also referred to as the photoactive layer, owes its success to a finely structured intimate mixture of donor and acceptor materials, which results from the processing of the layer from solution. The layer is formed by solubilizing the donor and acceptor materials in a common solvent and mixing them together as one ink. A film is then cast from this mixture (in ways we will cover in the next section), whereby a nanoscale network of the two materials naturally forms on drying of the film. In this way, the heterojunction is distributed throughout the photoactive layer.

Figure 1*d* presents the archetypical stack of a R2R coated PSC. Central to this is, of course, the photoactive layer. The semiconducting polymers are highly absorbing, and consequently PSCs can be made very thin (on the order of hundreds of nanometers) and still capture most of the available photons. This is why PSCs can be made flexible, as opposed to silicon (Si) solar cells, which are several tens of micrometers thick. The photoactive layer is sandwiched between two other semiconductors, again one is p-type and one is n-type. These layers work like semipermeable membranes, permitting holes to go in one direction and electrons in another, which maximizes the current output from the solar cell. A favorite p-type material is the conductive polymer poly(3,4-ethylenedioxythiophene):poly(styrene

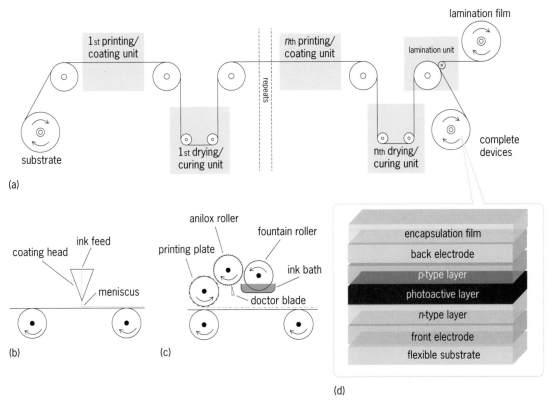

Fig. 1. Basic R2R setup with (*a*) printing/coating units followed by drying/curing unit and a lamination unit at the end. (*b*) A coating unit example showing the principle of slot-die coating. (*c*) A flexographic printing unit. (*d*) A typical R2R fabricated solar-cell stack.

sulfonate) (PEDOT:PSS), while a wide-band-gap oxide, such as zinc oxide (ZnO), is chosen as the *n*-type layer. Electrodes collect the current on each side, one of which has to be transparent or at least semitransparent to allow light to enter the stack. The traditional electrode choice has been indium tin oxide (ITO), but for large-scale production this is a poor choice for several reasons. First, it is a brittle material that is not well-suited for flexible PSC devices, and secondly it is very costly, amounting to around 90% of the total cost of a large-area R2R fabricated device. The total cost comes partly from an energy-intensive fabrication process, but mainly from indium being a very scarce element and thus very expensive. The only present alternative available is to use PEDOT:PSS in conjunction with a silver grid. This, of course, will come at the price of a shadow-loss from the grid. Presently, silver/PEDOT:PSS is the materials combination of choice when it comes to solution-processed electrodes. However, to further reduce the cost of PSCs, it would be beneficial to find a new alternative, preferably a carbon-based material such as graphene.

Roll-to-roll fabrication. So far we have outlined the strong points of PSC technology in terms of its potential for very fast and cheap processing and fabrication, most notably when considering continuous methods as is offered by R2R processing. R2R processing starts with a very long substrate wound on a roll made from a thin and transparent plastic material such as poly(ethylene terephthalate) [PET or polyester]. Such a roll can easily be several 100 m

long and 1 m wide, with a thickness of 10–100 μm. The PET roll is referred to as the web. In an R2R fabrication machine, the web is fed from one end, called the unwinder, to the other end, called the rewinder. The R2R principle is shown in Fig. 1*a*, where the uncoated substrate is unrolled on one end and complete devices are taken up at the opposite end of the machinery on the final roller, hence R2R. As illustrated in Fig. 1*a*, one would have the opportunity of designing the R2R machinery to match any given set of processing requirements. In the case of the stack illustrated in Fig. 1*d*, there are five layers. Each layer would require a specific coating or printing technique, followed by an appropriate drying or curing method. There are quite a few coating and printing techniques, and the ones that so far have proven most relevant are listed in the **table**.

After coating or printing the wet film, the web passes through a subsequent drying or curing section for the formation of a solid film. Methods include simple hot air ovens for the promotion of solvent evaporation and infrared (IR) heating or ultraviolet (UV) curing, if necessary, to induce a chemical hardening of the material. The necessary processing heat is generally low (that is, <140°C, the temperature at which PET starts to deform) and should preferably be as low as 80°C, which would enable exclusive use of solar heating.

Printing versus coating. In R2R processing, one differentiates between printing and coating techniques. Coating, in general, refers to wet film formation through a continuous meniscus (the freestanding

Data for the most relevant R2R printing and coating methods thus far used in PSC fabrication					
Technique	Print/coat, P/C	Speed, m min^{-1}	Ink viscosity, cP[a]	Dimensionality	Resolution, μm (in-plane/height[b])
Flexographic	P	1–1000	1–1000	2	32/100
Gravure	P	1–1000	1–1000	2	25/100
Slot-die	C	1–100	1–100,000	1	1000/0.01
Knife	C	0–100	1–10,000	0	∞/0.01
Flatbed screen	P	0–10	100–100,000	2	100/1000
Rotary screen	P	0–100	100–100,000	2	50/1000
Inkjet	P	0–100	1–1000	2+ (digital master)	25/100

[a]As a reference, at 20°C the viscosity of water is 1 cP and around 10,000 cP for molasses.
[b]Lower bound of resolution given for wet films of inks typically used for PSCs.

film formed by surface tension between the ink reservoir and the substrate, Fig. 1b). In contrast, printing entails the reproduction of a pattern by ink transfer through contact, or, in the special case of inkjet printing, without contact (Fig. 1c). Coating is inherently a low-dimensionality technique, as it only allows for zero- or one-dimensional patterning along the roll. Printing techniques allow for two-dimensional patterning of complex patterns, such as grids or strips, and enables the production of discrete units (solar-cell modules) along the roll, which is why it is necessary to print at least the electrode layers in the solar-cell stack. This does not preclude the use of coating techniques, as these present some unique qualities, such as an unrivaled ability to produce very thin and smooth layers and the possibility of simultaneous multilayer formation, which in the context of PSCs so far have been demonstrated by simultaneous processing of the photoactive layer and the overlying PEDOT:PSS layer using double slot-die coating. Schematic representations of slot-die coating and flexographic printing are shown in Figs. 1b and c, respectively.

R2R from laboratory to industrial scale. Because of the nature of the working mechanisms in PSCs, there is an undeniable interrelation between the performance of a final device and the way it was processed. As a result, it is essential for the successful development of the PSC technology that R2R techniques and equipment are available in the laboratory. For this purpose, laboratory-size and larger equipment is now available.

A medium-sized R2R laboratory setup might include one coating unit and a 2-m-long oven, whereby solar-cell fabrication is done sequentially, such that one layer is coated at the time, the roll is rewound before the next layer is coated, and so on.

The scaling from laboratory-size fabrication to a full industrial-scale production line (Fig. 1a) is not without its challenges. Some techniques simply cannot practically be downscaled too much, as they demand operation at very high speeds or an ink viscosity that is impractical to use for small samples. In such cases, downscaling would simply not allow for probing the right parameters.

State-of-the-art in performance. A solar cell is conveniently benchmarked within three main parameters: (1) power-conversion efficiency [PCE; percentage of the incoming solar light converted

into electric power], (2) lifetime, and (3) process scalability. In this context, PSCs can typically be divided into two categories. Category 1 focuses, above all, on maximizing PCE. The devices are typically very small (<0.1 cm^2 of active area), where the device fabrication and testing is done in an inert atmosphere using generally nonscalable fabrication methods and processes such as spin coating and high-vacuum thermal evaporation. Devices in category 1 have evolved at a fast pace, with a near doubling in PCE over the last five years to above 10%, while there have been close to no reports on either lifetime or scalability of these devices. Category 2 solar cells have somewhat lower PCE in the range of 2–4%, but pride themselves with a focus on lifetime or scalability by using only fabrication processes that are stable and scalable from the substrate to the final encapsulated device. This entails using mostly R2R methods in ambient conditions, while also considering the abundance and toxicity of the materials included in the process. A favorite benchmark in this category is the energy payback time (EPBT) determined through life-cycle assessments (LCAs), which we will touch on later. The final devices are tested with regard to lifetime and stability under different conditions to simulate real use. Category 2 solar cells have seen remarkable progress, with lifetimes now in the range of one to two years, but what is most remarkable is that manufacturing costs per watt-peak (W$_p$) produced have gone down by a factor of 1000 over the last five years. **Figure 2a** compares the learning curves in terms of manufacturing costs per W$_p$ for a mature process based on ITO (Process-One) to an ITO-free process in development (IOne). Current costs are ≈€5 W$_p^{-1}$ and ≈€2.5 W$_p^{-1}$ for the ITO-based and ITO-free processes, respectively.

Environmental footprint. With the compelling need to increase energy production capacity by a GW-a-day, one must carefully assess the environmental footprint of PSC fabrication from cradle to grave, as such a scale of production amounts to a massive investment in terms of energy and materials as well as an enormous amount of waste at the end of usefulness of the PSCs. An initial rule of thumb must be to avoid the use of materials that might be toxic to the people processing or handling the devices, or to the environment. This implies that the use of water as a processing solvent is likely to be mandatory, and chlorinated solvents, which are almost exclusively

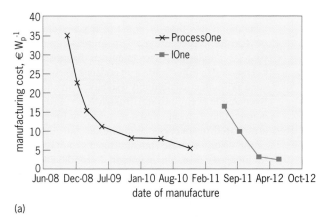

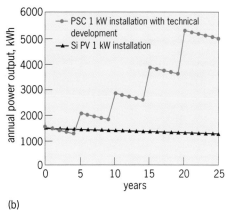

(a)   (b)

Fig. 2. The learning curves for the (*a*) ITO-based ProcessOne and ITO-free IOne PSC manufacturing processes. (*b*) Comparison of the annual power outputs from two solar-cell arrays along a 25-year life span for mature Si photovoltaic (PV) versus PSC technology, with an expected technical development factor in PCE of 1.065/year for PSCs.

used today for processing the photoactive layer, will be avoided. Deeper analysis is done through LCAs, considering everything from the extraction of raw materials through materials handling and fabrication processes as well as waste handling. This identifies materials or processes that should be avoided, such as in the case of ITO, which was discussed earlier. Recyclability is an area that should be explored for recollection and reuse of the metals, as is done for consumer electronics today. However, with the possibility of an all-carbon nontoxic PSC, the prospect of burning the solar cells in a combined heat and power plant (CHP) at the end of their usefulness might be the least wasteful scenario.

Outlook. To produce 1 GW_p-a-day of electricity from solar cells operating at a 10% efficiency would require the printing of 12.5 km^2 of PSCs (with 80% active area) each day, covering the area of Manhattan in less than five days. For comparison, this is only a fraction of the 30,000 km^2 of newspapers that are printed each day worldwide. This could be viewed as a somewhat misleading analogy, although the similarities are many in terms of manufacturing, and it does show that manufacturing at this scale is not farfetched. We are not there yet, as PSCs with lifetimes close to 10 years and PCEs in the range of 10% (referred to as the 10:10 target) are still years away. However, with the current rate of improvement in R2R processing along the lines of the PSC paradigm, it has become obvious that market entry is no longer bound by the 10:10 target. With very cheap solar cells on a roll, one can imagine replacing the solar panels in a large PSC array every 2 to 5 years with the benefits of an increase in performance each time the solar cells are replaced (Fig. 2*b*), and in this way PSC technology could become competitive with other solar-cell technologies sooner than otherwise anticipated. The role of PSCs during the next five years will depend both on scientists and innovative business.

For background information *see* BAND THEORY OF SOLIDS; ELECTRON-HOLE RECOMBINATION; FULLERENE; HOLE STATES IN SOLIDS; INK, PHOTOVOLTAIC EFFECT; PRINTING; SEMICONDUCTOR; SOLAR CELL; SOLAR ENERGY in the McGraw-Hill Encyclopedia of Science & Technology.

Thue T. Larsen-Olsen; Frederik C. Krebs

Bibliography. B. Azzopardi et al., Economic assessment of solar electricity production from organic-based photovoltaic modules in a domestic environment, *Energy Environ. Sci.*, 4:3741-3753, 2011, DOI:10.1039/C1EE01766G; N. Espinosa et al., Solar cells with 1-day energy pay back for the factories of the future, *Energy Environ. Sci.*, 5:5117-5132, 2012, DOI:10.1039/c1ee02728j; M. Jørgensen et al., Stability of polymer solar cells, *Adv. Mater.*, 24:580-612, 2012, DOI:10.1002/adma.201104187; F. C. Krebs et al., The OE-A OPV demonstrator anno domini 2011, *Energy Environ. Sci.*, 4:4116-4123, 2011, DOI:10.1039/c1ee01891d; R. Søndergaard et al., Roll-to-roll fabrication of polymer solar cells, *Mater. Today*, 15:36-49, 2012, DOI:10.1016/S1369-7021(12)70019-6.

Factors related to risk of autism

Autism is a complex neurodevelopmental disorder that is characterized by significant impairment in social interaction and communication, and by restricted, repetitive, or stereotyped behaviors. By definition, impairment is noted early in development, before the age of 3 years. With more than 1% of children affected with this disorder, understanding the factors related to risk of autism is an urgent imperative.

Autism is at least four times more common in males compared to females, and significantly higher rates in the United States have been reported for non-Hispanic white children compared to Hispanic or black children. Although no definitive cause for autism has been identified, the pendulum of public and scientific opinion has swung between two extreme positions: (1) Autism is solely of genetic origin, which led to a decade-long search for an autism gene; or (2) some environmental factor is primarily responsible, ranging from a lack of maternal warmth (the so-called refrigerator mother syndrome) to

<table>
<tr><td>

Factors related to risk of autism
─────────────────────────────────

A. Genetic factors
Male gender
Genetic syndromes
   Fragile X syndrome
   Tuberous sclerosis
   Angelman syndrome
   Rett syndrome
   Gene-deletion syndromes
Genetic variants
   Copy-number variants
   Single-nucleotide variants
   De novo mutations

B. Environmental factors
Advanced parental age
Perinatal factors
   Low birth weight
   Factors related to fetal or neonatal hypoxia
   Parental psychiatric history prior to birth
Other theoretical risk factors
   Prenatal androgen excess
   Deprived early social experience
   Diminished oxytocin

C. "Ruled-out" causes of autism
Vaccines
   Thimerosal-containing vaccines
   Measles–mumps–rubella (MMR) vaccine

</td></tr>
</table>

Control and Prevention reported the rate of autism in children less than or equal to 8 years of age to be 1 in 88, representing a 78% increase compared to 2002. Similar increases in autism diagnoses have been noted in other studies, in contrast to conditions such as intellectual disability that have remained relatively stable over time (see **illustration**). Although it is uncertain to what degree increases in prevalence are related to increased recognition and reporting of the disorder, these statistics have prompted calls for increased research on potentially modifiable environmental factors that may lead to the development of autism.

Genetic risk factors. Over the past 20 years, genetic risk factors for autism have been vigorously sought, in view of sibling and twins studies demonstrating high rates of heritability. In families with one child affected with autism, the recurrence risk (that is, the likelihood that a trait or disorder present in one family member will occur again in another family member) is estimated to be 2–8%. Earlier studies had reported high rates of concordance in monozygotic (identical) twins, who share the same genome, compared with nonidentical twins, who share only the same genetic makeup as regular siblings (72% versus 0%, respectively). These studies gave a calculated heritability rate, or genetic contribution, of approximately 90%. However, a larger, more-recent study, using standardized diagnostic measures of autism, showed a concordance rate of only about 60% in identical twins, and a much higher rate of 21–27% among nonidentical twins. This showed that a larger proportion of the variance (55%) could be explained by shared environmental factors rather than genetic heritability (37%).

environmental toxins and vaccines. Over the past 60 years, debate and controversy have continued to surround the question of etiology in autism (see **table**), with the only consensus being the importance of understanding causality in order to advance diagnosis, treatment, and prevention efforts.

Increasing incidence? Although it was once considered to be a rare condition, dramatic increases in the prevalence rates of autism have emerged over the past decade. Recently, the U.S. Centers for Disease

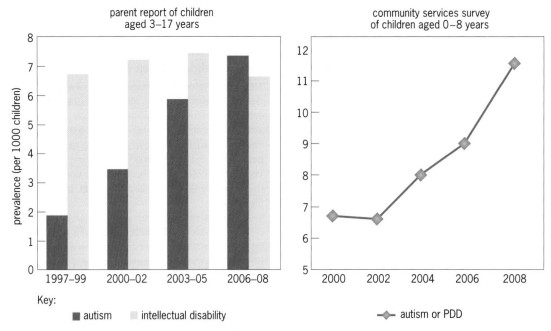

Rising prevalence rates of autism in the United States (1997–2008). PDD: pervasive developmental disorder. (*Left panel: C. A. Boyle et al., Trends in the prevalence of developmental disabilities in US children, 1997–2008, Pediatrics, 127:1034–1042, 2011; right panel: Autism and Developmental Disabilities Monitoring Network Surveillance Year 2008 Principal Investigators, Prevalence of autism spectrum disorders—autism and developmental disabilities monitoring network, 14 sites, United States, 2008, MMWR Surveill. Summ., 61:1–19, 2012*)

It has been estimated that gene-associated syndromes are present in no more than 1–2% of autism cases. Some of the more common gene-associated syndromes are fragile X syndrome, tuberous sclerosis, Angelman syndrome, Rett syndrome, 16 p11 deletion, and 22q deletion, although it has been difficult to link these varying genetic defects with a unifying causal mechanism. With the increasing availability of more sophisticated genome-wide analysis tools, which are able to detect so-called de novo mutations and copy-number variants, hundreds of independent risk loci have been identified. However, these regions have rarely linked a single gene as a risk factor for autism, and they are often associated with risk for a broad range of conditions, including epilepsy, intellectual disability, and schizophrenia. Furthermore, each marker explains only a small percentage of the variance observed in children with autism, providing low levels of sensitivity and specificity. Thus, although autism is thought to be highly heritable, few genetic risk factors have been identified (see table).

The vexed question of vaccines. The environmental factor that has generated the most widespread controversy has been childhood vaccines, including the measles–mumps–rubella (MMR) vaccine and vaccines containing thimerosal (a mercury-based preservative). However, in response to this controversy, multiple large, population-based studies have been conducted in several countries, and all of these studies have convincingly refuted any causal relationship between autism and childhood vaccination. For example, in one Danish study conducted from 1971 to 2000, no differences in the rate of autism were seen between children who had received thimerosal-containing vaccinations and those who had not. In fact, after removal of thimerosal from all vaccines in 1992, the prevalence of autism continued to increase. Another study that examined the potential association between the MMR vaccine and autism was conducted in Canada between 1987 and 1998. Researchers found that the prevalence of autism continued to increase despite decreasing rates of MMR vaccination. After reviewing the worldwide body of evidence, a report from the U.S. Institute of Medicine firmly rejected the possibility of a causal relationship between thimerosal-containing or MMR vaccines and autism.

Advanced parental age. As the search has continued for possible environmental factors associated with autism, multiple studies have shown that advanced parental age may be a factor related to risk, although the mechanism remains unclear. Some have postulated that it could be related to spontaneous genetic alterations in reproductive germ cells that occur with increasing age. However, the most recent study showed no additive risk if both parents were of advanced age, making this hypothesis less likely. Another suggestion was that these parents might have autistic traits themselves, making it less likely for them to find a partner and reproduce early in life, compared with the general population. However, the study also showed that children (from the same family) who were born later (that is, to older parents) had

an increased risk of developing autism. Thus, the association between parental age and autism does not appear to be accounted for by either common genetic inheritance or stable shared environment, but by parent-related factors that increase over time.

Perinatal factors. Studies have also identified numerous perinatal factors associated with the subsequent development of autism. Prenatal risk factors for autism were examined in a meta-analysis of 64 epidemiological studies, and the following statistically significant risk factors were found: maternal gestational bleeding, gestational diabetes, being a first-born infant, prenatal medication use, and maternal birth abroad. Parental psychiatric history prior to birth has also been identified as a potential risk factor.

Another meta-analysis examined the association between perinatal and neonatal factors and autism risk by evaluating 40 independent studies. This meta-analysis found an association with numerous perinatal stressors, including fetal distress, birth injury or trauma, maternal hemorrhage, low birth weight, congenital malformation, and feeding difficulties. It was noted that these associations might reflect consequences of preceding prenatal complications, but there was insufficient evidence to implicate any one perinatal or neonatal factor in the etiology of autism. Nevertheless, many of these factors are associated with fetal or neonatal hypoxia (oxygen deficiency), suggesting that this may contribute overall to the risk of autism.

Prenatal androgen excess. Because of the disproportionate number of males affected by autism and without a clear genetic mechanism, another potential environmental risk factor that has been proposed is prenatal androgen (male sex hormone) exposure. It has been hypothesized by some investigators that autism may be an extreme manifestation of the male brain, as a result of excessive androgen exposure during pregnancy. This theory has been supported by empirical evidence from questionnaire data and behavioral tasks, as well as by a biological marker of prenatal testosterone exposure (the second-to-fourth digit length ratio, with lower mean values in males compared to females).

Social impairment and oxytocin. Social impairment is one of the core features of autism. Social reciprocity and face-processing skills are developed in early childhood, at least partially in response to social stimulation. Several researchers have hypothesized that basic deficits in social perception may underlie all of the other developmental and behavioral abnormalities seen in autism, and that a set of defining experiences early in development may critically affect the establishment of these core social skills. Therefore, it has been proposed that abnormal or restricted social experience may also contribute to impairment in social development in autism.

Face processing is a key factor in the development of social perception. However, it is severely impaired in children with autism. These individuals often have an impaired ability to recognize faces or to fixate on the eyes of others, thus being deprived of crucial social information. Functional magnetic resonance imaging (MRI) studies have demonstrated that individuals with autism have reduced brain activation in face-processing regions when viewing or discriminating faces of adults. Although infants (starting from birth) possess an innate perceptual bias for face-like shapes, the capacity to distinguish facial features develops during the first year of life and is acquired through sensory experience during critical periods of development. It has therefore been proposed that autism may result from deficits in exposure to contingent, responsive facial expressions during a critical period of social development. Studies have shown that children who are exposed to extreme social or sensory deprivation, such as was observed with Romanian orphans in the 1990s, have much higher rates of autistic-like behavior, with the severity of their symptoms positively associated with their age of adoption. Congenitally blind children, who are deprived of visual face experience, likewise have higher rates of autism.

In blinded placebo-controlled trials, oxytocin, which is a neuropeptide associated with social memory and learning, has been shown to enhance direct eye gaze and facial memory in humans. Brain oxytocin receptors also appear to be programmed by early life experience, with decreased expression seen in the brains of animals that received lower levels of contingent maternal care in infancy. Oxytocin deficits in humans have also been implicated in the development of autism, whereas administration of oxytocin (intravenously or intranasally) results in enhanced face perception, improved social responsiveness, and decreased repetitive behavior. Together, these studies suggest that oxytocin may play a role in promoting social perception, and oxytocin deficits may be a risk factor for the development of autism.

Conclusions. Despite the complexity of the question, it is becoming increasingly evident that both genetic and environmental factors play a causal role in the development of autism. The most likely scenario, yet to be adequately tested, is that genetic susceptibility interacts with environmental factors, such as perinatal stress and social experience, to modify gene functioning and the neural processing of social cues across early development, resulting in the behavioral disturbance that is termed autism. As the factors related to risk of autism are more clearly defined, we hope to be better equipped to intervene early and prevent some of the devastating developmental consequences of autism.

For background information *see* AGING; AUTISM; COGNITION; CONGENITAL ANOMALIES; DEVELOPMENTAL GENETICS; DEVELOPMENTAL PSYCHOLOGY; HORMONE; HUMAN GENETICS; NERVOUS SYSTEM (VERTEBRATE); NEUROBIOLOGY; SOCIOBIOLOGY; STRESS (PSYCHOLOGY) in the McGraw-Hill Encyclopedia of Science & Technology.   Lane Strathearn; Eboni Smith

Bibliography. B. S. Abrahams and D. H. Geschwind, Advances in autism genetics: On the threshold of a new neurobiology, *Nat. Rev. Genet.*, 9:341–355, 2008, DOI:10.1038/nrg2346; Autism and Developmental Disabilities Monitoring Network Surveillance Year 2008 Principal Investigators, Prevalence of autism spectrum disorders—autism and developmental disabilities monitoring network, 14 sites, United States, 2008, *MMWR Surveill. Summ.*, 61:1–19, 2012; H. Gardener, D. Spiegelman, and S. L. Buka, Perinatal and neonatal risk factors for autism: A comprehensive meta-analysis, *Pediatrics*, 128:344–355, 2011, DOI:10.1542/peds.2010-1036; J. Hallmayer et al., Genetic heritability and shared environmental factors among twin pairs with autism, *Arch. Gen. Psychiatry*, 68:1095–1102, 2011, DOI:10.1001/archgenpsychiatry.2011.76; L. Strathearn, The elusive etiology of autism: Nature *and* nurture?, *Front. Behav. Neurosci.*, 3:1–3, 2009, DOI:10.3389/neuro.08.011.2009.

False-belief reasoning and bilingualism

Like many life experiences, bilingualism (the ability to speak two languages with equal or nearly equal fluency) affects cognitive development. On the one hand, bilingual children have a smaller vocabulary in each language than their monolingual peers do, and bilingual adults underperform in word-retrieval tasks relative to monolinguals, partly because of the interference of the other language. On the other hand, bilingual language production requires constant monitoring of the target language in order to minimize interference from the competing language. This requires the exercise of executive control, which has been found to strengthen the executive control system of bilinguals.

False-belief tasks. Bilingual children also perform better on false-belief tasks than their monolingual peers. False-belief tasks have been a key test of social cognition development for more than two decades. In the classic Sally–Anne task, children are presented with two puppets, Sally and Anne, who are playing with a toy. When the puppets finish playing, they put the toy into a box, and Anne leaves the scene. While Anne is away, Sally moves the toy to a different box. When Anne comes back, the child is asked where Anne will look for the toy. At around the age of 4, monolingual children typically answer correctly that she will look for the toy in the original container. However, younger monolingual children and some autistic individuals tend to respond according to their own knowledge of the situation, thus failing to show an appreciation of Anne's false belief about the location of the toy. In contrast, bilingual children as young as 3 years of age have shown a precocious success in false-belief tasks.

A recurrent theme in the developmental literature has been whether young children fail traditional false-belief tasks because they are not able to appreciate another person's beliefs or because they are not able to inhibit their own knowledge of the situation. Bilingual children's advantage in traditional false-belief tasks may well be related to both factors. Bilingual children must develop an early sociolinguistic sensitivity to the language knowledge of their interlocutors because they must use their languages accordingly. Even though this fundamental aspect of bilingual children's experience has not been investigated in the context of false-belief reasoning, their awareness that other people do not always speak the same languages as they do may be an early form of appreciating that other people might have a different perspective from their own. Moreover, this early form of perspective taking is combined with an early development of their executive control system, which is necessary in order to focus on the target language and avoid interference from the contextually inappropriate linguistic system. Their advanced executive control would help bilingual children inhibit their own knowledge in false-belief tasks. In general, language learning itself can promote the development of children's "theory of mind" (that is, their understanding of other people's mental states and intentions).

Even though adults perform optimally in standard false-belief tasks, their performance on a variety of judgment tasks can be affected by the "curse of knowledge" (that is, the tendency to be biased by their own knowledge of the situation). It is therefore possible that, if sufficiently fine-grained measures were used, adults might reveal an egocentric bias in their performance on a standard false-belief task. If this is the case, would bilingual adults also have an advantage in perspective taking relative to monolingual adults?

An eye-tracking study on false-belief reasoning in bilingual adults. In a recent study, the false-belief reasoning abilities of adult bilinguals and monolinguals were investigated using a standard false-belief task coupled with an eye-tracking technique. If adult participants first consider the container that they know the toy is in before taking the mistaken protagonist's perspective, then this would imply an egocentric bias. Given bilinguals' sociolinguistic sensitivity to the language background of their interlocutors as well as their enhanced executive control, bilingual adults may well have an advantage in false-belief reasoning analogous to the advantage found in bilingual children.

At Princeton University, 46 undergraduates, including 23 bilinguals and 23 monolinguals, were tested. The bilinguals in this sample had learned a second language (L2) before age 9 and had been using it regularly for 10 years or more. They had learned their L2 in (1) a bilingual household, (2) a monolingual household using a language foreign to their country of residence, or (3) a bilingual school. The bilingual group therefore had found it necessary to switch languages daily or almost daily for a number of years. The L2s were Spanish, Chinese, Hindi, Korean, Russian, Hebrew, French, Farsi, and Japanese. At the time of the study, English was the dominant language of all participants, whether bilingual or monolingual.

The study used a computer version of the classic Sally–Anne task. Two kindergarten characters, Sally and Anne, interact in an animated cartoon. Each child has a favorite toy and a container in which she puts it before going home every day. In the false-belief condition, Anne puts her doll in her basket and goes home. While she is away, Sally moves Anne's doll from the basket to the box. When Anne comes back the next day, participants are asked the following question: "Where will Anne look for her doll?"

It was expected that the adult participants would have no trouble answering the question. However, they were also expected to look at the incorrect container, albeit briefly, before responding correctly. In addition, bilinguals were expected to show less of such egocentric eye movements than monolinguals, and that is exactly what was found: only 26% of monolinguals initially gazed at the correct container, compared with 57% of the bilinguals. There was also a difference in the time that it took participants to fixate on the correct container, with bilinguals showing significantly shorter fixation delays than monolinguals.

Further investigations attempted to ascertain whether the bilingual participants had higher levels of executive control than their monolingual counterparts. One test of executive control is the Simon task, patterned after the game of "Simon Says." In this task, participants press a right-hand key when the word RIGHT appears on a computer screen, and they press a left-hand key when the word LEFT appears. The task is easy when the words RIGHT and LEFT appear on the right and left sides of the screen, respectively, but it is harder when RIGHT and LEFT appear on the left and right sides of the screen, respectively. As expected, bilinguals and monolinguals performed the easy task equally well, but bilinguals were better able to inhibit incorrect responses in the "Simon Says" conflicting situation. Also as expected, participants' performance on the Simon task correlated with their performance on the false-belief task: the less interference they suffered from the wrong response in the Simon task, the less interference they suffered from their own knowledge in the false-belief task. This confirms that both tasks required the use of executive control.

Overall, this study showed that adults suffer from an egocentric bias in false-belief reasoning. Unlike in the case of young children, though, this bias does not affect task performance: adults always answer the false-belief question correctly. However, with the use of an eye-tracking technique, it was found that the majority of adults momentarily consider the egocentric response before correcting this tendency and taking the protagonist's perspective. These results are in line with previous studies revealing an egocentric bias in several different areas of cognition.

Regarding differences between bilinguals and monolinguals, adult bilinguals were observed to suffer less from the curse of knowledge than do monolinguals. This advantage in false-belief reasoning may be maintained from an early age, given that bilingual children outperform monolinguals on false-belief tasks. One of the reasons for this advantage may be the bilinguals' higher level of executive control, which was confirmed by the Simon task. Enhanced executive control would help bilingual participants inhibit their own knowledge of the situation in false-belief tasks, making it easier for them to take the protagonist's perspective. A second factor that may account for the bilinguals' advantage in false-belief reasoning is their early sociolinguistic awareness of their interlocutors' language background. Thus, the need to monitor for the language background of their interlocutors from an early age may also make bilinguals better perspective takers than monolinguals.

Outlook. Future research should determine whether the advantage in false-belief reasoning that has been observed in adult bilinguals relative to monolinguals is maintained from early childhood throughout development. Bilinguals' advantage in executive control tasks has indeed been found in children, young adults, and older adults. In principle, this advantage could help bilinguals maintain an advantage in false-belief reasoning during development. It remains an open question whether the perspective-taking advantage that bilingual children have shown in language-switch situations, for example, is also maintained throughout development and perhaps enhances their false-belief reasoning abilities throughout their life span.

For background information *see* BRAIN; COGNITION; INFORMATION PROCESSING (PSYCHOLOGY); LINGUISTICS; PERCEPTION; PROBLEM SOLVING (PSYCHOLOGY); PSYCHOLINGUISTICS; PSYCHOLOGY in the McGraw-Hill Encyclopedia of Science & Technology.

Paula Rubio-Fernández; Sam Glucksberg

Bibliography. S. Baron-Cohen, A. M. Leslie, and U. Frith, Does the autistic child have a "theory of mind?," *Cognition*, 21:37–46, 1985, DOI: 10.1016/0010-0277(85)90022-8; E. Bialystok, Bilingualism: The good, the bad and the indifferent, *Biling. Lang. Cognit.*, 12:3–11, 2009, DOI:10.1017/S1366728908003477; E. Bialystok and F. I. M. Craik, Cognitive and linguistic processing in the bilingual mind, *Curr. Dir. Psychol. Sci.*, 19:19–23, 2010, DOI:10.1177/0963721409358571; S. Birch and P. Bloom, The curse of knowledge in reasoning about false beliefs, *Psychol. Sci.*, 18:382–386, 2007, DOI:10.1111/j.1467-9280.2007.01909.x; N. Epley, C. K. Morewedge, and B. Keysar, Perspective taking in children and adults: Equivalent egocentrism but differential correction, *J. Exp. Soc. Psychol.*, 40:760–768, 2004, DOI:10.1016/j.jesp.2004.02.002; P. Goetz, The effects of bilingualism on theory of mind development, *Biling. Lang. Cognit.*, 6:1–15, 2003, DOI:10.1017/S1366728903001007; A. M. Kovács, Early bilingualism enhances mechanisms of false-belief reasoning, *Dev. Sci.*, 12:48–54, 2009, DOI:10.1111/j.1467-7687.2008.00742.x; J. E. Pyers and A. Senghas, Language promotes false-belief understanding: Evidence from learners of a new sign language, *Psychol. Sci.*, 20:805–812, 2009, DOI:10.1111/j.1467-9280.2009.02377.x; P. Rubio-Fernández and B. Geurts, How to pass the false-belief task before your 4th birthday, *Psychol. Rev.*, in press, 2012; P. Rubio-Fernández and S. Glucksberg, Reasoning about other people's beliefs: Bilinguals have an advantage, *J. Exp. Psychol. Learn. Mem. Cognit.*, 38:211–217, 2012, DOI:10.1037/a0025162; H. Wimmer and J. Perner, Beliefs about beliefs: Representation and constraining function of wrong beliefs in young children's understanding of deception, *Cognition*, 13:103–128, 1983, DOI:10.1016/0010-0277(83)90004-5.

Flettner rotor ship

The world has known sailing ships for centuries. The thought, romantic to some, conjures images of canvas and the sea, but you don't need a traditional canvas sail to get a propulsive force from the wind—and it's not a new idea! Anton Flettner (1885–1961), a German aviation engineer and director of the Institute of Aerodynamics at Amsterdam, the Netherlands, filed an application for a German patent in 1922 on a ship propelled by vertical circular cylinders rotating in the wind. Then, in 1924, he purchased an old 2000-ton steel-hulled schooner, the *Buckau*, and aided by the now-famous aerodynamicist Ludwig Prandtl, among others, he converted it to the first rotor ship at Kiel, Germany (**Fig. 1**). The ship was 47.5 m (156 ft) long with a beam of 9 m (28.5 ft). Flettner installed two rotating cylinders each 3 m (10 ft) in diameter and 15 m (50 ft) high driven by two 11-kW (15-hp) electric motors with power supplied by a 34-kW (45-hp) diesel generator. The rotors turned at a top speed of 125 revolutions per minute. They were constructed of heavy gauge steel but were still lighter than the masts and rigging they replaced. Performance of the ship was very good. She moved at nearly twice her former speed. She reportedly could tack to 25° to the wind while the original schooner could only tack 45° to the wind. In February 1925, after 62 trial voyages around Kiel loaded with up to 35 metric tons of coal, the ship was put into commercial service hauling lumber from Gdansk, Poland to Leith, Scotland through the North Sea. She handled the stormy weather without trouble. In 1926, the ship was renamed the *Baden Baden*, sailed to New York, and continued her commercial service for several years, mostly in American waters. Eventually she was converted to conventional engine and propeller propulsion as it was deemed to be more efficient and continued service as a motor freighter. She sank in the Caribbean Sea in 1931.

Magnus effect. The principle that created the force necessary to propel the rotor ship is called the Magnus effect after physicist Heinrich Gustav Magnus, who published an explanation of it in 1852 as it applied to the flight of cannonballs. It is also sometimes called the Robins effect after Benjamin Robins, who described the effect in a publication in 1742. This effect is familiar to those involved in many of the ball sports. The common explanation (at least that given to undergraduate aerodynamics students), apparently first presented by Lord Rayleigh in 1877, is

Fig. 1. The Flettner rotor ship in 1924, converted from the schooner *Buckau*.

in terms of inviscid (zero viscosity, ideal fluid) flow with circulation and assumes the circulation in the flow is imposed by the rotating body—a cylinder in the present case. As the flow passes over the cylinder, its rotation and the friction between the air and the cylinder speeds up the flow over the side rotating in the same direction as the flow and slows down the flow over the side rotating in the opposite direction (**Fig. 2**). According to Bernoulli's principle, places in a flow where the velocity is smaller have higher pressure and places where the velocity is greater have a lower pressure with pressure varying with speed squared. The resulting variation of the pressure distribution on the cylinder induced by the circulation in the flow results in a net force perpendicular to the oncoming flow. Because the pressure on the surface of the cylinder is symmetric front and back, this inviscid flow theory predicts no drag force on the body, only the side force, a result that is intuitively unrealistic.

In reality, fluids, including air, do have viscosity. Prandtl, in a 1904 paper, showed that viscosity leads to the formation of a thin "boundary layer" between a solid boundary and a moving fluid in which the flow velocity varies from the velocity of the fluid outside the boundary layer to the velocity of the boundary. The shear stress between layers of fluid in this boundary layer induces a loss of energy in the flow which leads to a phenomenon known as "boundary-layer separation." The boundary-layer flow cannot follow the body all the way around the back side, into what would have been a higher pressure region if the flow could remain attached, but breaks away from the body leaving a region of recirculating flow in the wake (**Fig. 3***a*). The pressure in the wake on the leeward side of the body is not as great as on the windward side leading to a net drag force on the body.

In the case of flow over a rotating cylinder (or sphere), the friction between the moving surface

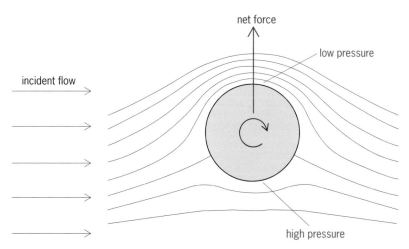

Fig. 2. Inviscid flow about a rotating cylinder.

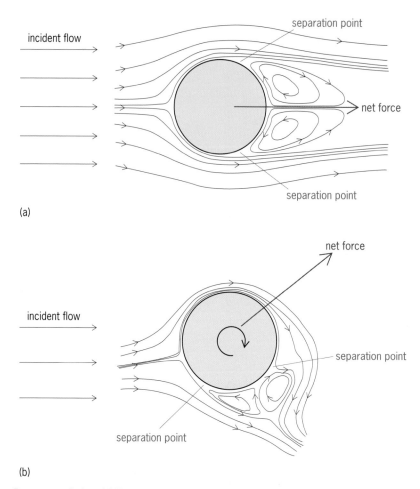

(a)

(b)

Fig. 3. Viscous flows over cylinders. (a) Viscous mean flow over stationary (nonrotating) cylinder showing boundary layer separation and drag force. (b) Viscous flow over rotating cylinder showing modification of boundary layer separation and force on cylinder. (*After H. M. Badr et al., Unsteady flow past a rotating cylinder at Reynolds numbers 10³ and 10⁴, J. Fluid Mech., 220:459–484, 1990, DOI:10.1017/S0022112090003342*)

and the fluid induces circulation in the boundary layer which significantly alters the flow there but is unlikely to have much influence on the outer flow, as would be required to make the inviscid flow expla-

Fig. 4. The rotor ship *Barbara*. (*From http://www.naviearmatori.net/eng/foto-45542-4.html*)

nation of the Magnus effect plausible. What this alteration of the boundary layer flow does accomplish, however, is an asymmetric alteration of the point of flow separation leading to a net force on the body which has a significant component perpendicular to the flow (Fig. 3b). This was recognized by Prandtl in a 1925 publication. On the side rotating in the same direction as the flow, the frictional resistance to the flow—the shear stress—in the boundary layer is reduced leaving more energy in the flow and allowing it to continue farther around the body before separating. On the side moving opposite the flow, the shear stress is increased, removing more energy from the boundary layer flow and causing it to separate from the body farther upstream. The result is an asymmetric wake with an asymmetric pressure distribution on the body, resulting, in a component of the total force on the body normal to the incident flow, and a corresponding reaction in the flow, resulting in a component of velocity in the direction opposite to the force on the body. At top speed, the surface of Flettner's rotors was moving at 19.6 m/s (64.5 ft/sec) or about 38 knots—faster than most of the winds in which it sailed. This allowed for extreme modification of the separation points on the two sides of the cylinder. Because of a phenomenon called "vortex shedding," the flow patterns and resulting forces are oscillatory in nature but are, on average, as previously described.

Other rotor ships. Encouraged by Flettner's success with *Buckau*, the German Admiralty commissioned the 90-m-long (300-ft), 3000-ton, 3-rotor cargo ship *Barbara*, which was launched in 1926 (**Fig. 4**). The *Barbara*'s aluminum rotors were 4 m (13 ft) in diameter, 17 m (56 ft) high, and turned at up to 160 RPM. She could make 19 knots. The *Barbara* transported fruit from Italy to Hamburg under charter until 1929, when she was sold to a new owner who removed the rotors and converted her to conventional propulsion in 1933. *See* ALUMINUM SHIP STRUCTURES.

The worldwide economic depression that began in 1929 seems to have stifled further development in rotor ship technology. However, interest in it has resurfaced whenever oil prices have spiked. In the early 1980s, Lloyd and Henry Bergeson of the Wind Ship Development Corporation experimented with the small 13-m (42-ft) rotor craft *Tracker*. They demonstrated that this screw-propelled craft with a rotor-sail assist could achieve a savings of 30–40% at speeds of 6–7 knots. Once oil prices declined, so did interest in rotor-assisted vessels and development ceased.

Although not actually a rotor ship, Jacques Cousteau's *Alcyone*, launched in 1985, employs the same boundary-layer-altering principle by using suction on the lee side of its "turbosails" (**Fig. 5**). A small fan at the top of 10-m-high (33-ft), fixed, oval sectioned towers draws air through louvers, which can be opened on one side or another to alter the airflow around the tower and produce a force perpendicular to the wind. The 31-m-long (103-ft) turbosail-assisted vehicle is capable of 10.5 knots using a combination of turbosail and conventional propulsion. The *Alcyone* has demonstrated a one-third savings in fuel and exhaust emissions over conventional propulsion alone.

The most recent rotor ship is the *E Ship 1* built for the wind energy company Enercon (**Fig. 6**). Launched in 2009, it is a rotor-sail-assisted vehicle which Enercon uses to transport its wind turbine components worldwide. The 9700-ton ship is 130 m (426 ft) long, with a beam of 22.5 m (74 ft) and a maximum draft of 9.3 m (30.5 ft). She reaches a maximum speed of 17.5 knots using two 3500-kW electric drive motors powered by six diesel generators and the rotor sails. The four rotors placed two forward and two aft are 25 m (82 ft) high and 4 m (13 ft) in diameter. They are driven by a waste heat recovery system using the exhaust gas from the diesel generators (which is also used to cool the ship's interior). Fuel savings are expected to be 30–40%, although no published data seems to be available to date. She has been shipping cargo since August 2010.

For background information *see* BERNOULLI'S THEOREM; BOUNDARY-LAYER FLOW; FLUID MECHANICS; KÁRMÁN VORTEX STREET; SUBSONIC FLIGHT; VISCOS-

Fig. 5. Cousteau's *Alcyone*. (*Courtesy http://www.starthrower.org/products/DDDB/ DDDB_200-249/DDDB_220%20Alcyone%20part%201.htm***)**

Fig. 6. The *E Ship 1*. (*Copyright © Jörn Prestien, MarineTraffic.com***)**

ITY in the McGraw-Hill Encyclopedia of Science & Technology.   Wayne L. Neu

Bibliography. H. M. Badr et al., Unsteady flow past a rotating cylinder at Reynolds numbers 10^3 and 10^4, *J. Fluid Mech.*, 220:459–484, 1990, DOI:10.1017/ S0022112090003342; Y.-M. Chen, Y.-R. Ou, and A. J. Pearlstein, Development of the wake behind a circular cylinder impulsively started into rotary and rectilinear motion, *J. Fluid Mech.*, 253:449–484, 1993, DOI:10.1017/S0022112093001867; Y. T. Chew, M. Cheng, and S. C. Luo, A numerical study of flow past a rotating circular cylinder using a hybrid vortex scheme, *J. Fluid Mech.*, 299:35–71, 1995, DOI:10.1017/S0022112095003417; Lord Rayleigh on the irregular flight of a tennis ball, *Messenger of Mathematics*, 7:14–16, 1877; I. H. Shames, *Fluid Mechanics*, 4th ed., McGraw-Hill, New York, 2002.

Flow patterns in champagne glasses

Fine sparkling wines and champagne are particular in that they are the result of a two-step fermentation process. After completion of the first alcoholic fermentation, some flat champagne wine (base wine) is bottled and then a mixture of yeast and sugar is added. Consequently, a second fermentation starts inside the bottle as the yeast consumes the sugar, producing alcohol and a large amount of carbon dioxide (CO_2). This is the reason why champagne has a high percentage of CO_2 dissolved and the finished champagne wine can be under as much as 5–6 atmospheres of pressure. The gas gushes out in the form of tiny CO_2 bubbles as the bottle is opened.

For consumers and winemakers as well, the role that bubbles usually play in champagne tasting is to awake the sight sense. Indeed, the magic image of champagne is intrinsically linked to the bubbles, which act like "chains of pearls" in the glass of champagne and look like discrete jewelries; they create a cushion of bubbles on the surface. Beyond this first visual aspect, the informed consumer will associate to the bubble behavior one of the main ways to extract flavors; this is because the aroma and bouquet of sparkling wines are CO_2-propelled into the nose and mouth. What is unknown and will be discussed in this article is the consequence of the bubble be-

havior on the dynamics of the champagne inside the glass and consequently on the CO_2-propelling process.

Birth of bubbles and controlled effervescence. The first step is to elucidate how bubbles themselves come into being. Generally speaking, two ways exist, and sometimes coexist, to generate bubble chains in champagne glasses. Natural effervescence depends on a random condition: the presence of tiny cellulose fibers deposited from the air or left over after wiping the glass with a towel, which cling to the glass because of electrostatic forces (**Fig. 1**). These fibers act as nucleation sites in champagne and have recently been identified as tiny microchannels of cylindrical shape. They are made of closely packed microfibrils, themselves consisting of long polymer chains composed mainly of glucose. Each fiber, about 100 μm long, develops an internal gas pocket as the glass is filled. Capillary action tries to pull the fluid inside the microchannel of the fiber, but if the fiber is completely submerged before it can be filled, it will hold onto its trapped air. Such gas trapping is aided when the fibers are long and thin, and when the liquid has a low surface tension and high viscosity. Champagne has a surface tension about 30% less than that of water, and a viscosity about 50% higher. These microfiber gas pockets act as nucleation sites for the formation of bubbles. To aggregate, CO_2 has to push

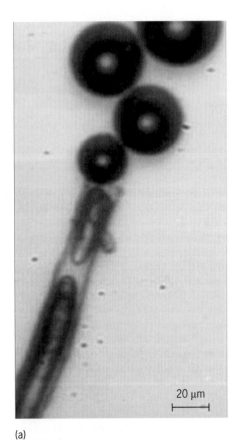

(a)

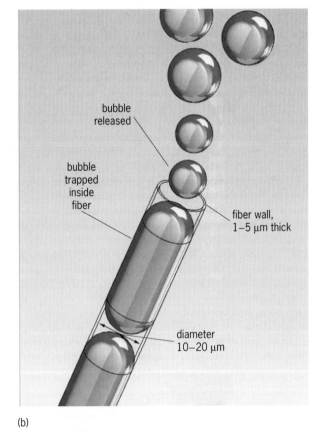

(b)

Fig. 1. Bubbles in sparkling wines do not spring into existence unaided, but require a starting point. These nucleation sites take the form of microscopic cellulose fibers, from the air or a towel used to dry the glass, which trap air pockets as the glass is filled. Carbon dioxide from the wine diffuses into the gas pockets, producing bubbles like clockwork. (*a*) Photomicrograph of microscopic cellulose fiber emitting bubbles. (*b*) Diagram of process.

through liquid molecules held together by van der Waals forces, which it would not have enough energy to do on its own. The gas pockets lower the energy barrier to bubble formation as long as they are above a critical size of 2 μm in radius, because below that size the gas pressure inside the bubble is too high to permit CO_2 to diffuse inside. It should be noted that irregularities in the glass surface itself cannot act as nucleation sites since such imperfections are far too small, unless larger micro scratches are purposely made. Once a bubble grows to a size of 10 to 50 μm, it is buoyant enough to detach from the fiber, and another one forms like clockwork; an average of 30 bubbles per second are released from each fiber.

The bubbles expand from further diffusion of CO_2 into them as they rise, which increases their buoyancy and accelerates their speed of ascent. They usually max out at less than a millimeter in diameter over the course of their 1–5-s travel time up the length of a flute. Because natural nucleation is very random and not easily controllable, another way to generate bubbles is to use a mechanical process that is perfectly reproducible from one filling to the next. This way consists in creating artificial nucleation sites by means, for example, of an impact laser technique applied at the glass bottom surface by glassmakers. With such treatments, glasses are called engraved ones. This engravement method is commonly used by the champagne houses during tasting. Each point of impact has a diameter of about 400 μm. To make effervescence so pleasant-looking to the eye, no less than 20 impacts can be used, the whole being usually circular in shape, producing a regular column of rising bubbles (**Fig. 2**).

Experimental setup. For the present study, glasses engraved at the bottom were used (Fig. 2), which is the most frequently encountered situation, in order to highlight the strong effect played by bubbles on the whole flow dynamics. To observe the flow dynamics, the analysis was based on the laser tomography visualization technique in association with adequate tracers. Filling experiments were carried out at room temperature to avoid damaging condensation on the glass surface. Once poured with champagne, the glass was lighted in its symmetry plane with a 1-mm planar argon laser sheet. Because glasses are circularly engraved at their bottom, the resulting flow exhibited an axisymmetrical behavior, as will be seen further. In this situation, a two-dimensional (2D) examination in the axisymmetry plane could be considered as sufficient to investigate the whole three-dimensional (3D) flow.

Two types of tracers were used. The first one was Rilsan particles having a density ($d = 1.060$), approaching that of champagne ($d = 0.998$). These particles were quasispherical in shape and were neutrally buoyant (75 μm < diameter <150 μm). Rilsan particles were found to be completely neutral with regard to bubble formation. Moreover, they exhibited a high degree of reflectivity when illuminated by a 2-mm-thick laser sheet. The champagne was initially, before pouring, suitably and homogeneously

Fig. 2. To study effervescence in champagne and other sparkling wines, random bubble production must be replaced with controlled creation of bubble streams. The glass bottom is etched with a ring that provides nucleation sites for regular bubble trains. The ring consists of many small impact points from a laser. Glasses etched with a single nucleation point were used in studies to see how a single stream of bubbles would induce motion in the surrounding fluid, and what shape that fluid motion would take.

seeded with Rilsan particles in an attempt to get the flow features from instantaneous velocity fields and deduced streamline patterns, for example. To better highlight the vortical structures and to access the streakline patterns, fluorescent dyes of sulforhodamine B and fluorescein were carefully injected in the lighted plane in complement to the first flow visualization method.

Ring vortex scenario. Flow visualizations have shown that a glass with an engraved circular crown exhibits a steady state of fluid motion reached approximately 30 s after the glass is poured. Because of the high degree of reflectivity of bubbles, one clearly observes the formation of a rising gas column along the vertical glass axis from the treated bottom surface up to the free surface of the beverage. Consequently, a drive process of the surrounding fluid occurs to generate two large contrarotating vortices in the vertical lighted section (**Fig. 3**). These cells are located outside of the rising bubbles close to the wall of the glass in the case of a flute. Because this gas column acts like a continuous swirling-motion generator within the glass, the flow structure exhibits a quasisteady 2D behavior with an axisymmetrical geometry. In this case, the whole domain of the liquid phase is homogeneously mixed.

To complete the trends previously observed, the resulting flow may be observed in an engraved

Fig. 3. Visualization by fluorescent tracers in an engraved flute-glass.

traditional champagne coupe, much wider but shallower than the traditional champagne flute (**Fig. 4**). It can be seen that, as for the previous glass model, the rising CO_2 bubble column causes the main fluid to move inside the glass. Nevertheless, two distinctive steady flow patterns are identified for such a glass shape. One pattern clearly exhibits a 2D axisymmetrical single swirling-ring (annulus) whose cross-section visualization reveals two contrarotating vortices close to the glass axis (Fig. 4*a*). What strongly differs from the champagne flute is that this recirculation flow region does not occupy the whole volume in the glass.

As a consequence, a singular steady flow regime is observed in the external periphery of the glass

which is also axisymmetrical and characterized by a dead zone of no motion. This means that, for a wide-brimmed glass, only about half of the liquid bulk participates in the champagne mixing process.

Relation between swirling motion and mechanisms of adsorption. During the bubble rise, surface-active substances adsorb at the bubble interface, to finally reach and concentrate themselves at the free surface. Moreover, swirling motion induced by the continuously ascending bubble motion contributes to the progressive adsorption of surface-active materials by continuously bringing surface-active materials from the champagne bulk to the air/champagne interface. Therefore, because of the two above-mentioned contributions, the amount of surface-active materials adsorbed at the air/champagne interface progressively increases as time proceeds after champagne is poured into a glass. Actually, because of their amphiphilic structure, some of the various organic compounds found in champagne wines show surface activity, including, for example, alcohols (ethanol, butanol, pentanol, phenyl-2-ethanol, and so forth), some aldehydes (butanal, hexanal, and hexenals), and organic acids (propionic, butyric acid, and so forth). The free surface of champagne is therefore strongly believed to be overconcentrated with regard to surface-active molecules (some of them being potentially aromatic, and thus participating in the global sensorial perception of a sparkling wine).

Numerical modeling. A numerical modeling of flow dynamics induced by the effervescence in a glass of champagne was carried out using the finite volume method by computational fluid dynamics (CFD). The idea of this study was to develop a "universal" numerical modeling allowing the study of bubble-induced flow patterns due to effervescence, whatever the shape of the glass, in order to quantify the role of the glass geometry in the mixing flow phenomena and induced aromas exhalation process. To ensure a continuous and perfectly controlled process of effervescence, glassmakers usually consider circularly engraved glasses (Fig. 2). In such a way, as previously mentioned, the flow structure exhibits a quasisteady 2D behavior (Figs. 3, 4).

(a)

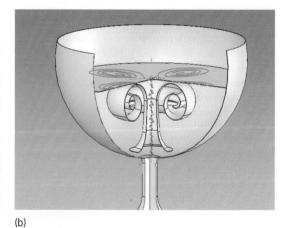

(b)

Fig. 4. Flow in an engraved champagne coupe. (*a*) Visualization by solid tracers. (*b*) Scheme of the fluid motion.

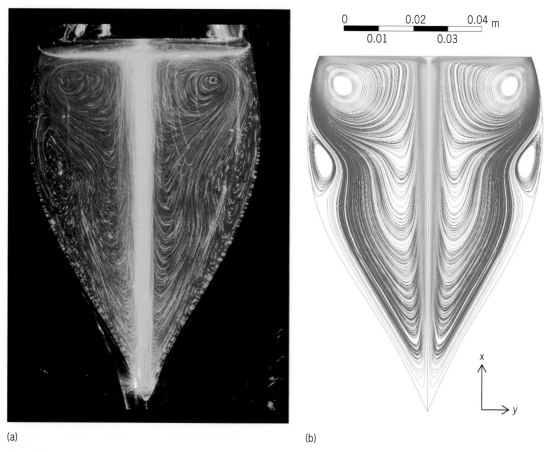

Fig. 5. Flow patterns revealed by (*a*) classical flow visualization and (*b*) streamlines obtained by numerical simulations (CFD).

In this situation, a 2D examination in the axisymmetry plane can be considered as sufficient. Because champagne is a wine in which a gaseous phase as well as a liquid phase are simultaneously present, the flows in a glass of champagne have been simulated numerically with a multiphase model. The results obtained by numerical simulation have been compared with those from experiments using flow visualization techniques to get both qualitative and quantitative viewpoints (Fig. 4). Laser tomography as a qualitative analysis method has been used to visualize the flow patterns and vertical structures induced by the continuous column of ascending bubbles in the reference flute poured with champagne. During the time exposure of a camera, the liquid seeded with solid Rilsan particles and lighted by a planar laser sheet exhibits streamline patterns. Comparison between these experimental streamlines and the numerical ones (**Fig. 5**) shows a good agreement, especially with regard to the location of the vortex cores in the investigated domain. The global flow features are satisfactorily modeled with the CFD-developed code.

Because the kinetics of flavor and gas release also strongly depend on the velocity of the recirculating flows close to the interface, there are strong reasons to believe that these studies provide objective elements and clues to better understand the role of glass shape and engravement conditions in the "olfactive" behavior of champagne and sparkling wines in a glass.

For background information *see* COMPUTATIONAL FLUID DYNAMICS; FERMENTATION; FLUID MECHANICS; NUCLEATION; VORTEX; WINE in the McGraw-Hill Encyclopedia of Science & Technology.

Fabien Beaumont; Gérard Liger-Belair; Guillaume Polidori

Bibliography. G. Polidori et al., Artificial bubble nucleation in engraved champagne glasses, *J. Visual.*, 11(4):279, 2008, DOI:10.1007/BF03182193; G. Polidori et al., Ring vortex scenario in engraved champagne glasses, *J. Visual.*, 12(3):275–282, 2009, DOI:10.1007/BF03181866; G. Polidori et al., Visualization of swirling flows in champagne glasses, *J. Visual.*, 11(3):184, 2008, DOI:10.1007/BF03181703; G. Polidori, P. Jeandet, and G. Liger-Belair, Bubbles and flow patterns in champagne, *Am. Sci.*, 97:294–301, 2009, DOI:10.1511/2009.79.294.

Fluid mechanics of fires

A seemingly simple fire can grow rapidly into an inescapable furious inferno, resulting in a catastrophic fire. It is not unusual for as little as 3 min to elapse from the start of a fire to a full-room inferno, with disastrous consequences. For example, the Rhode Island nightclub fire on the evening of February 20, 2003, began when the band used a "pyrotechnic" display, and the heat was enough to ignite the polyurethane foam behind the stage. Burning spread very rapidly, with a great deal of thick, dark, deadly smoke. The fire in the single-story structure with floor area about $418\,m^2$ ($4500\,ft^2$) became an inferno within 3 min and resulted in the loss of 100 lives. A simple kitchen or room fire can completely destroy a typical 167-m^2 (1800-ft^2) house in less than 1 h.

Structural fire development can be understood more readily when technical knowledge puts the phenomena on a firm scientific footing. Recent extensive study, research, experimentation, and field observations and measurements have led to dramatic developments, and in particular to improvements in the theory of fluid mechanics of fires that assist in understanding and applying scientific information to real-world fire situations.

Basic features. The fluid mechanics of fires is an extremely important topic in loss prevention, safety engineering, and fire protection engineering. Availability of fuel, ignition, fire growth, fire spreading rates, amount and temperature of the smoke layer produced, structural impairment, evolution of toxic gases, and the amount of time available for the safe departure of occupants after the onset of fire are all topics of immense concern. Fire dynamics is the term chosen to represent the topics associated with fire behavior, including ignition, fire development, and fully developed fires. Technical information about fuels, burning rates, fire spread, and flashover and backdraft phenomena is also relevant. Finally, aspects of experimental studies, fire modeling, and fire investigation are also strongly related to, and in fact can be considered an integral part of, the subject of fire dynamics. So-called zone-type and field-type computational fluid dynamic (CFD) modeling approaches to multiroom structural fire modeling are available.

In addition to heat and oxygen insufficiency, fires produce acutely lethal products of combustion, including smoke, carbon monoxide, hydrogen cyanide, acrolein, hydrogen chloride, and nitrogen oxides. These gases are produced in large quantities and driven to remote areas, causing life-threatening situations, as inhalation of fire gases (toxic products of combustion) is the major cause of death in fires. For these reasons, any combination of finishes, combustible building materials, or contents and furnishings that could result in flashover (full-room involvement) in a few minutes represents a severe fire hazard in many types of occupancies. Protection by automatic sprinklers and fire-rated construction separations is often needed or mandated by the relevant code. Computer fire models include their mathematical characterization. One can then assess whether an especially hazardous situation exists, that is, whether a given fire scenario has the potential to develop flashover or full-room involvement.

Initial fire growth. In general, fire behavior is characterized by the following: all fires have a starting place, or "point of origin"; all fires require a source of heat, or "ignition source"; and all fires result from the coming together of heat, fuel, and oxygen, or

"fire cause." The origin and cause cannot always be found by investigators after the fire, although that is their intent, together with determining responsibility for why the fire occurred.

Full-scale furniture calorimeter tests give useful information on the burning rates of many typical household items. The rate of fire growth, the time to reach maximum heat release rate, the value of this maximum, how long the item burns at this maximum, and the rate of decay as the fire burns down are particularly useful to know. Typical upholstered, wood-frame pieces of furniture with fire-retardant polyurethane padding and an olefin cover show the following peak heat release rates and times at which these rates occur for the cases of a chair, a loveseat, and a sofa:

Chair: 2100 kW at 260 s
Loveseat: 2886 kW at 230 s
Sofa: 3120 kW at 215 s

Fitted curves to the data smooth out the "noise" of the experimental measurements and result in smooth curves for the preignition, growth, leveling off, and decay of the heat release rate versus time (**Fig. 1**).

From a practical standpoint, fire spread in compartmentalized buildings is through open or fire-burned-through doors, unenclosed stairways, and nonfirestopped combustible concealed spaces. Some fire codes require a certain separation of spaces within a building, with fire-resistant doors and walls, for example, so that a fire on one side does not penetrate through to the space on the other side quickly. The "fire resistance rating" of the door or wall then comes into play. The sequence of events after the ignition of leaking gas, malfunctioning of an appliance, overheating of cooking oil, or some other event in a room includes what happens immediately after ignition, the plume-ceiling interaction, the ceiling jet–wall interaction, the fully enclosed space with a developed growing upper layer of smoke, and further smoke filling. The smoke and fire development can then continue easily through open doorways to adjacent rooms, which then participate by delivering fresh air and receiving smoke from the initial burn room, and typically more items in the adjacent rooms ignite (**Fig. 2**).

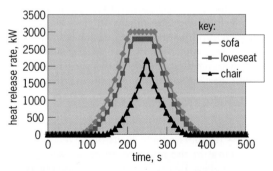

Fig. 1. Heat release rate versus time for typical chair, loveseat, and sofa. (*Data from NIST Experiments F21, F31 and F32*)

Standard fires are called slow, medium, fast, or ultrafast fires depending on their growth rates. They are specified by the t^2 fire growth model, where, after an initial incubation period, the heat release rate $\dot{Q}$ in kilowatts is given by Eq. (1), where α is a

$$\dot{Q} = \alpha t^2 \qquad (1)$$

fire growth coefficient (in kW/s^2). Suggested values for α are:

Slow: $\alpha = 0.002778$ kW/s^2, with fire growth time from 0 to 1 MW in 600 s.

Medium: $\alpha = 0.011111$ kW/s^2, with fire growth time from 0 to 1 MW in 300 s.

Fast: $\alpha = 0.044444$ kW/s^2, with fire growth time from 0 to 1 MW in 150 s.

Ultrafast: $\alpha = 0.177778$ kW/s^2, with fire growth time from 0 to 1 MW in 75 s.

The chair, loveseat, and sofa just described all have ultrafast fire growth rates.

The flame height above the burning location is related to the heat release rate of the burning item and its proximity to a wall or a corner of the enclosure. A standard simple expression for a pool fire (a pan of liquid fuel or a spill of liquid fuel on the floor) or a single item on fire is Eq. (2), where H_f = flame

$$H_f = 0.174(\dot{Q})^{0.4} \qquad (2)$$

height (in meters) and $\dot{Q}$ = heat release rate of the fire (in kilowatts). In Eq. (2), the proximity of a wall or corner may be taken into account via the value of k, where $k = 1$ for a location with no nearby walls, $k = 2$ for a location adjacent to a wall, and $k = 4$ for a location adjacent to a corner. For the same heat release rate, wall proximity increases the flame height as expected, since less air is available around the fire to participate in the burn (**Fig. 3**). Actually, the heat release rate is related to the type of fuel that is burning, its rate of burning per unit surface area, and the area of the spill. To be more precise, then, there are other more realistic flame height theories that also include the equivalent pool diameter in the expression for the flame height. Equation (3) is rec-

$$H_f = 0.23(\dot{Q})^{0.4} - 1.02D \qquad (3)$$

ommended, where D = pool diameter (in meters). Equations (2) and (3) for estimating the flame height above a fire give similar results for fires away from walls, but it must be noted that they are estimates, the flame height fluctuates widely during a fire, experimental data show scatter, and there are cases in which the fire diameter is important for the flame height calculation.

Radiant ignition of nearby items. For many enclosure fires, it is of interest to estimate the radiation transmitted from a burning fuel array to a target fuel positioned some distance from the fire to determine whether secondary ignitions are likely. Considerations of inverse square distance lead to the point-

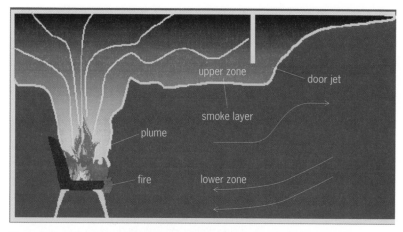

Fig. 2. Schematic of a burning chair, the plume above the fire, hot smoke escaping through the doorway, and fresh air entering to feed oxygen to the fire.

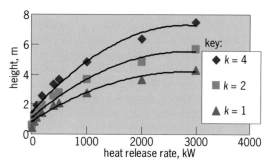

Fig. 3. Flame height versus fire heat release rate ($k = 1$: away from walls; $k = 2$: by a wall; $k = 4$: in a corner). 1 m = 3.28 ft.

source radiation heat flux equation (4), where

$$\dot{q}_o'' = \frac{x_r \dot{Q}}{4\pi r^2} \qquad (4)$$

$\dot{q}_o''$ = incident radiation heat flux on the target (in kW/m^2)

r = distance (radius) from the fire to the target fuel (in meters)

x_r = radiative fraction (the fraction of the total energy released in the fire that is radiated)

$\dot{Q}$ = total heat release rate of the fire (in kJ/s or kilowatts)

Heat flux values can be computed from the heat release of the fire and the distance to the target (see the **table** and **Fig. 4**). At target locations that are very close to the fire (closer than about two fire diameters away from the center of the fire), the point-source approximation becomes less accurate, and the more precise configuration or view factor methodology must be used, whose mathematical treatment is more complicated.

The foregoing discussion has provided information about the burning rate (heat release rate versus time) of a single specified item in the burn room. What happens next? Either the item burns out with little further damage to the surroundings, or one or more nearby items ignite and add fuel to the fire. This can be by direct flame contact (if the second item is sufficiently close) or, more usually, by radiant heat

Heat flux on target, using the point-source approach and a radiative fraction of 0.4*				
Source total heat release ($\dot{Q}$), MW	Distance of target from source			
	1 m	2 m	5 m	10 m
1	31.83	7.96	1.27	0.32
2	63.66	15.92	2.55	0.64
3	95.49	23.87	3.82	0.96
4	127.3	31.83	5.09	1.27
5	159.2	39.79	6.37	1.59
10	318.3	79.58	12.73	3.18

*The heat flux is the quantity ($\dot{q}_o''$) in Eq. (4), in kW/m². 1 m = 3.28 ft.

energy on the surface of the second item becoming sufficiently large. Direct flame contact requires time to pyrolyze (decompose into gaseous form) the fuel and time to heat the gases produced to their ignition temperature. The radiant flux ignition problem is a very complicated issue that depends on many factors. As the fire grows and the radiant energy flux arriving on the second item increases, a simple criterion for ignition of the latter is often used. The minimum radiant heat flux (arriving on the surface of the second item) necessary to ignite the second item is called the characteristic heat flux (CHF). Typical values are:

CHF = 10 kW/m² for easily ignitable items, such as thin curtains or loose newsprint

CHF = 20 kW/m² for normal items, such as upholstered furniture

CHF = 40 kW/m² for difficult to ignite items, such as wood of 13 mm (0.5 in.) or greater thickness

This amount of heat flux represents the minimum amount needed for an item to ignite. The larger the actual incident flux arriving on an ignitable object, the more quickly that object will ignite, provided the incident flux is greater than the characteristic heat flux. The time required for ignition is characterized by an additional parameter called the thermal response parameter (TRP). As an example, as previously noted, an upholstered chair has a peak heat release rate of about 2100 kW, which is associated with the ignition of easy, normal, and difficult to burn items at distances of 2.6, 1.8, and 1.3 m (8.5, 6.0, and

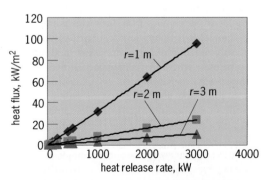

Fig. 4. Radiant heat flux on a target versus fire heat release rate and distance away (*r*). 1 m = 3.28 ft.

4.2 ft), respectively. Methods for calculating the heat flux on a target some distance away and the time for ignition of the object are available.

Flashover. Flashover is characterized by the rapid transition in fire behavior from localized burning of fuel to the involvement of all combustibles in the enclosure. Flashovers entail rapid combustion without explosion, are caused by heat buildup, and occur only in the early (growth) stage of a fire's growth. High radiation heat transfer levels from the original burning item, the flame and plume directly above it, and the hot smoke layer spreading across the ceiling are all considered to be responsible for the heating of the other items in the room, leading to their ignition. Warning signs are heat buildup and "rollover" (small, sporadic flashes of flame that appear near ceiling level or at the top of open doorways or windows of smoke-filled rooms). Factors affecting flashover include room size, ceiling and wall conductivity and flammability, and heat- and smoke-producing quality of room contents.

Flashover is characterized by temperatures over 500°C (932°F) in the upper portions of the room, heat flux of 25 kW/m² [7925 Btu/(h · ft²)] at floor level, near-simultaneous ignition of combustibles not previously ignited, and the filling of almost the entire room volume with smoke and flames. Generally, very high heat release rates occur after flashover, with the result that (subject to oxygen availability) most ignitable items in the room burn, it gets very hot, and the windows break out. Any open window or open doorway permits smoke to escape but also permits more oxygen to be made available and thereby increases the severity of the fire.

Fire size ventilation limit and incomplete combustion. One of the enclosure effects is the availability of oxygen for combustion. If the air in the space, plus that drawn in through openings, plus that blown into the space by HVAC (heating, ventilation, and air-conditioning) systems or other means is insufficient to burn all the combustible products driven from the fuel package, then only the amount of combustion supportable by the available oxygen can take place. This situation is referred to as ventilation-limited burning. When ventilation-limited burning occurs, incomplete combustion takes place, and the combustible products contain unburned fuel. This fuel is not burned in the room, but it often does burn when it combines with air outside the room, and this appears as flame extensions from the room. Also, ventilation-limited burning changes the mass loss rate. The interface between the upper and lower layers of fires nearing flashover and postflashover stages is located near the floor. Then the air mass flow rate $\dot{m}_a$ (in kg/s) into the compartment can be approximated by Eq. (5), where A_O = area of open-

$$\dot{m}_a = 0.5 A_O \sqrt{H_O} \qquad (5)$$

ing (in m²) and H_O = height of opening (in meters). The term $A_O \sqrt{H_O}$ is commonly known as the ventilation factor. The maximum fire size (heat release rate, in MW) supportable by the incoming air is given by

Eq. (6), so that, for example, a door that is 1 m wide

$$\dot{Q}_a = 1.5A_O\sqrt{H_O} \qquad (6)$$

and 2 m high can support a fire whose size is about 4 MW.

Backdraft explosions. A backdraft is an instantaneous explosion of smoke blasting back through a door or window (a combustion explosion). Backdrafts are caused by the introduction of air, and they occur in the early (growth) and late (decay) stages of a fire's growth. Increased occurrence of backdrafts in recent years has resulted from buildings being more tightly sealed and better insulated, increased use of energy-efficient, double-glazed windows, and more synthetic materials being used in furnishings. Backdrafts can occur when large quantities of carbon monoxide (which is itself a fuel) build up as a result of incomplete combustion in ventilation-limited, air-starved fires, and oxygen is then introduced, and when smoldering exists, during the early and late stages of a fire's growth. Warning signs include thick smoke puffing and pushing out of windows and doors; reverse flow of smoke after venting; thick dark brown or black smoke, indicating buildup of unburned gas; and discoloration of window glass. Defensive operations include venting from above to release hot smoke and gases upward; quenching the fire, resulting in lowering its temperature; and flanking the fire (standing aside so as to be out of the way of the direct explosion path).

Computer modeling. Zone fire models solve the conservation equations for distinct zones. Each room is divided into the hot upper layer of smoke coming from the fire and a colder lower layer. The basic assumption is that properties can be approximated throughout the zone by some uniform function. The uniform properties are temperature and smoke and gas concentrations, which are assumed to be exactly the same at every point in a zone. Zone modeling has proved to be a practical method for providing estimates of fire processes in enclosures. The changes in properties such as the size, location, temperature, and optical density of the smoke in each room as time progresses are predicted. One of the best-known zone models is the NIST CFAST.

Potentially greater accuracy in simulating enclosure fires is available via field models or CFD models. The structure is divided into a very large number of subvolumes, and the basic laws of conservation representing mass, momentum, and energy are written as partial differential equations. These represent the variation of the dependent variables (velocity components, density, temperature, and species) as a function of position and time throughout the 3D domain of interest, perhaps one or many rooms, corridors, and stairways within a building. Only over the last 15 years or so has this fundamental differential equation method been applied to fire situations. One of the best-known CFD fire codes is the fire dynamics simulator (FDS) code from NIST.

For background information *see* COMBUSTION; COMPUTATIONAL FLUID DYNAMICS; FIRE; FIRE TECHNOLOGY; FLAME; INVERSE-SQUARE LAW in the McGraw-Hill Encyclopedia of Science & Technology.

David G. Lilley

Bibliography. A. E. Cote and J. L. Linville (eds.), *Fire Protection Handbook*, 20th ed., National Fire Protection Association, Quincy, MA, 2008; B. Karlsson and J. G. Quintiere, *Enclosure Fire Dynamics*, CRC Press, Boca Raton, FL, 2000; D. G. Lilley, Fire dynamics, pp. 467–514 in J. M. Haight (ed.), *Safety Professionals Handbook*, 2d ed., American Society of Safety Engineers, Des Plaines, IL, 2012; K. B. McGrattan et al., *Fire Dynamics Simulator (Version 5), User's Guide*, NIST Special Publication 1019-5, NIST, Gaithersburg, MD, 2008; R. D. Peacock et al., *CFAST—Consolidated Model of Fire Growth and Smoke Transport (Version 6), User's Guide*, NIST Special Publication 1041, NIST, Gaithersburg, MD, 2005; Society of Fire Protection Engineers, *Handbook of Fire Protection Engineering*, 4th ed., SFPE, Boston, MA, 2008.

Fluorination of organic compounds

The introduction of one or multiple fluorine atoms to an organic molecule will profoundly change its physicochemical properties. Accordingly, numerous research fields, such as medicinal chemistry, agrochemistry, and material science, benefit from the use of organofluorine compounds. In 2005, 30–40% of agrochemicals were estimated to contain fluorine, and in 2010, three of the top 10 drugs sold in the United States contained at least one fluorine atom (**Fig. 1**).

The importance of the fluorine atom in agrochemistry and pharmaceutical sciences comes from the fact that fluorine can influence a number of crucial features for the bioactivity of a compound. Because

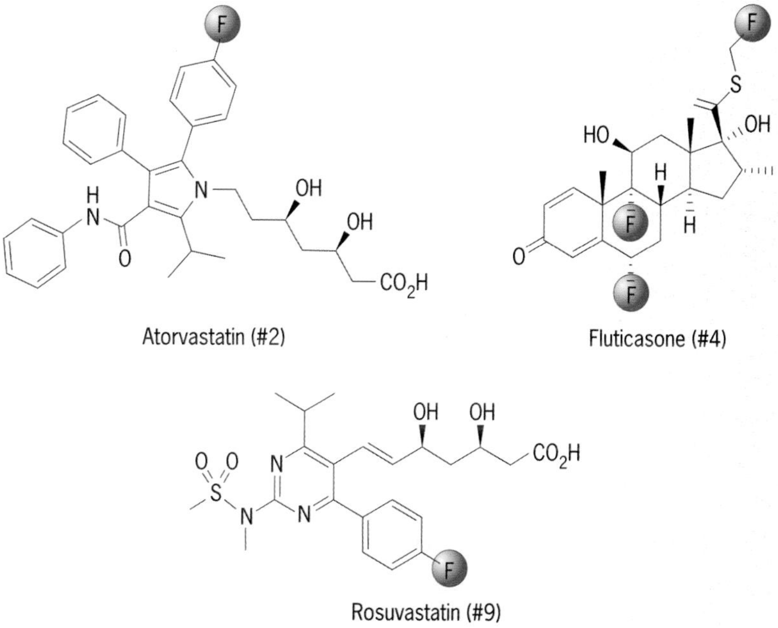

Fig. 1. Three drugs containing fluorine atoms. Their rank in the Top 10 drugs for 2010 by sales in the United States is in parentheses.

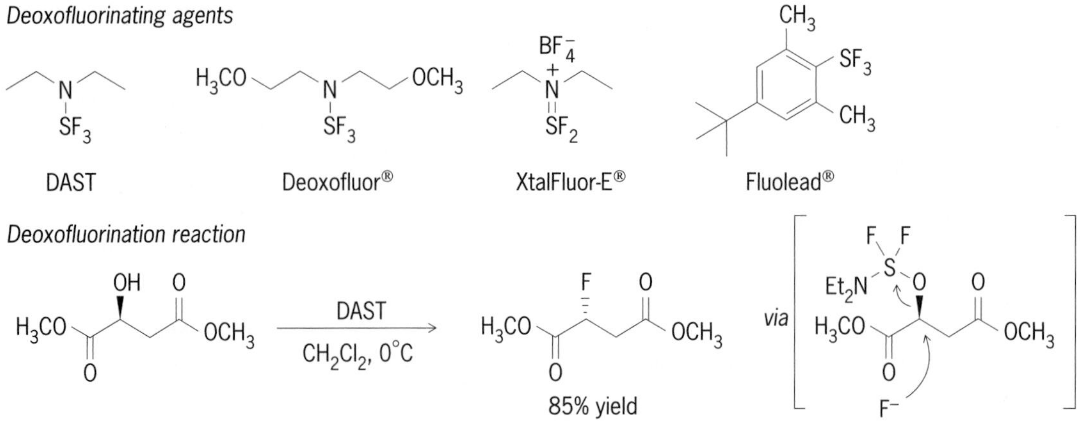

Deoxofluorinating agents

DAST   Deoxofluor®   XtalFluor-E®   Fluolead®

Deoxofluorination reaction

85% yield

Fig. 2. Deoxofluorinating agents and an example of a deoxofluorination reaction.

of its high electronegativity, the highest of all elements on the periodic table, fluorine can affect the acidity or the basicity of a neighboring group such as an amine or a carboxylic acid which, in turn, can influence bioavailability through modulation of the binding affinity and pharmacokinetic properties. The fluorine atom will also modulate the lipophilicity of a molecule, that is, the ability of a molecule to dissolve in lipids or lipid-like compounds. This property is an important factor since a drug may have to cross a cell membrane composed mostly of lipids. The presence of a fluorine atom can also affect the conformation of a molecule, which can directly affect its binding affinity. Also, fluorine atoms will often reduce the rate of elimination of a drug by the body by preventing or slowing down metabolic degradation. In certain drugs, the fluorine atom can even play a crucial role in the mechanism of action.

Fluorine-containing organic molecules are rare in nature. A number of fluoroalkanes, fluoroalkenes, and fluoroaromatics are produced and released through abiogenic pathways, that is, geothermal processes such as volcanoes, biomass fires, and other geological processes. To date, only 18 organofluorine compounds have been discovered to be produced through a biogenic pathway, that is, in living species (with no examples coming from animals or insects). Biosynthesis of some of those fluorinated natural products requires an enzyme capable of affecting fluorination, a fluorinase, which has been recently isolated. In all cases, since natural sources of organofluorine molecules are rare, they cannot serve as a convenient feedstock and, as a consequence,

most fluorine-containing molecules are synthetic and are generated via fluorination of organic compounds.

Synthetic methods for the fluorination of organic compounds. The rest of this article focuses on selected synthetic methods for introducing a single fluorine atom onto organic molecules. The introduction of a fluorine atom can be achieved using three different approaches that vary according to the electronic nature of the fluorine (nucleophilic, electrophilic, and radical). This article is subdivided according to those approaches and includes examples where the fluorine atom is located on an sp^3-hybridized carbon (aliphatic C-F bond) or an sp^2-hybridized carbon (vinylic or aromatic C-F bond). Examples where the fluorine atom is located on an sp-hybridized carbon (acetylenic C-F bond), although known, are rare and of limited use.

Nucleophilic fluorination. In nucleophilic fluorination, fluoride (the anionic form of fluorine) is used as a nucleophile in a substitution reaction on a substrate bearing a leaving group. Typical reagents used for such transformations are metal fluorides (KF being the most commonly used), hydrogen fluoride and its amine-complexes (HF/pyridine), and organic solvent-soluble fluoride sources such as fluoride ammonium salts exemplified by tetrabutylammonium fluoride (abbreviated as TBAF) or fluorosilicate salts such as tetrabutylammonium difluorotriphenylsilicate (TBAT). An example of nucleophilic fluorination using a metal fluoride is shown in reaction (1). In the earlier examples, the

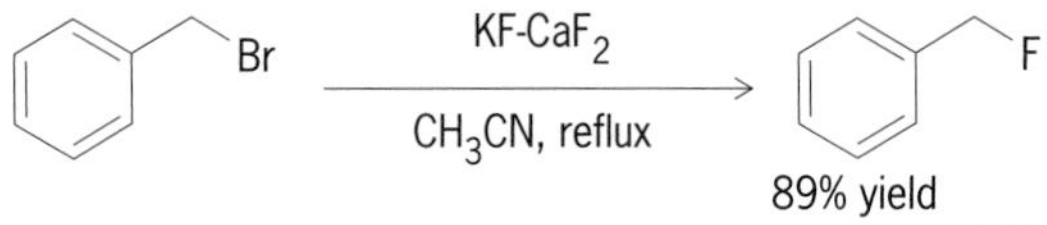

89% yield

(1)

starting molecules already have leaving groups present. A more direct approach is the use of the deoxofluorination reaction. In this case, the reagent activates, in situ, an alcohol (transforming it into a good leaving group), which is then displaced by a fluoride ion. The structures of the reagents typically used for this transformation are presented in **Fig. 2**. With substrates bearing the leaving group on a stereogenic center, an inversion of configuration is generally

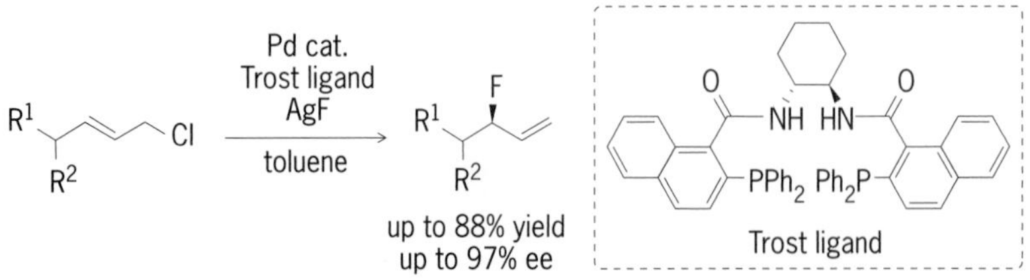

up to 88% yield
up to 97% ee

Trost ligand

Fig. 3. Enantioselective nucleophilic fluorination reactions. Abbreviations: Ph = phenyl, dba = dibenzylideneacetone.

observed since the reaction normally proceeds through a bimolecular nucleophilic substitution reaction (S_N2) as exemplified in Fig. 2.

Generation of enantioenriched fluorinated molecules via nucleophilic fluorination through desymmetrization of meso substrates or from achiral substrates using asymmetric catalysis have been reported. One example using an achiral substrate is shown in **Fig. 3**. In this case, the chiral ligand dictates the absolute stereochemistry.

The preparation of aryl fluorides using nucleophilic fluorination has been traditionally limited to the Halex process, in which halogen atoms located on the aromatic ring serve as leaving groups. However, this method is limited in terms of the substrate scope since the reaction requires an aromatic ring bearing an electron-withdrawing group such as a nitro (**Fig. 4**). A key breakthrough was made when the Pd-catalyzed fluorination of aryl triflates was reported for which this limitation is not present. In this case, the palladium catalyst plays a crucial role in the creation of the C-F bond. Recently, the deoxofluorination of phenols has been described. As for the deoxofluorination of aliphatic alcohols, in-situ activation of the hydroxyl functionality allows the aromatic nucleophilic displacement to occur.

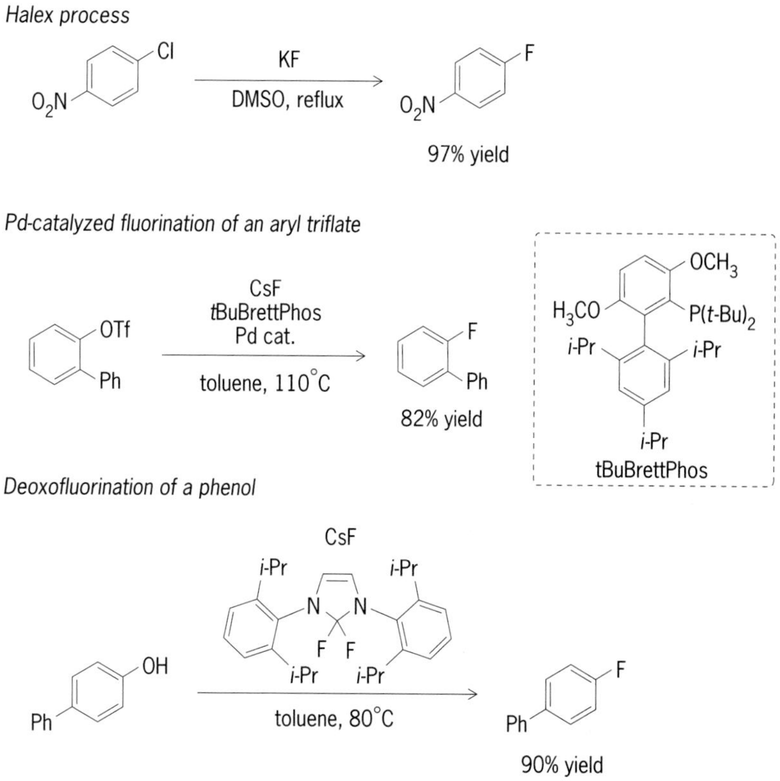

Electrophilic fluorination. When using an electrophilic approach for the fluorination of organic compounds, the role of each partner is reversed. Hence, the substrate will behave as the nucleophile, whereas the fluorine atom will be delivered as an electrophile. Although electrophilic sources of fluorine are often referred to as F^+ sources, they do not actually generate such species since this is energetically highly disfavored. One of the first electrophilic sources of fluorine was elemental fluorine (F_2). However, difficulties associated with its manipulation, its high reactivity, and toxicity have limited its use. The use of commercially available, although expensive, XeF_2 has also been reported. A number of reagents possessing an O-F bond, such as CF_3OF, $CF_2(OF)_2$, CH_3COOF, ClO_3F, and $CsSO_4F$, were also described as electrophilic sources of fluorine with the main drawback that they must be prepared using F_2. In addition, their high reactivity is counter-balanced with generally poor selectivity. In this sense, the development of bench-stable N-F reagents, which are now commercially available at reasonable cost, has completely revolutionized this field. The structures of the three most popular N-F electrophilic sources of fluorine are shown below.

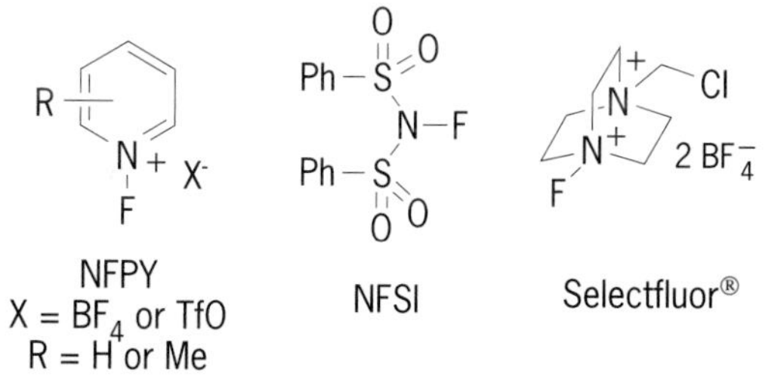

When using this approach to create an sp^3 C-F bond, various carbon-centered nucleophiles can

Halex process

Pd-catalyzed fluorination of an aryl triflate

Deoxofluorination of a phenol

Fig. 4. Examples of aryl fluoride synthesis. Abbreviations: DMSO = dimethylsulfoxide, Tf = SO_2CF_3.

be used, including metal enolates, enols, silyl enol ethers, vinyl acetates, vinyl silanes, imines [reaction (2)], enamines, and alkynes, among others.

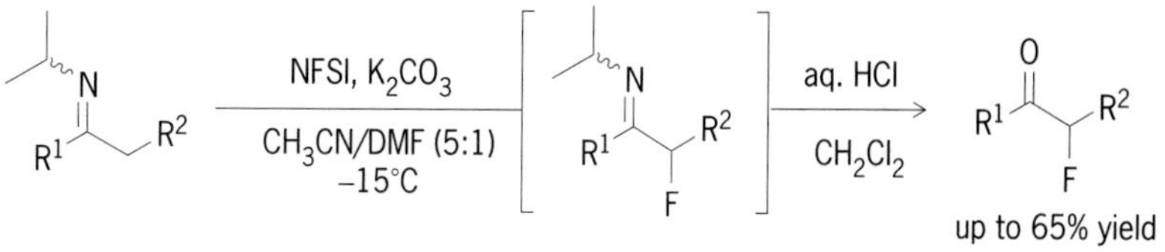

In reaction (2), the newly created stereocenter is racemic.

Two different approaches for the development of asymmetric electrophilic fluorination reactions have been examined: chiral N-F reagents and asymmetric catalysis. In the first case, the use of chiral N-F reagents has been investigated and a number of reagents have been developed. They can be used in a stoichiometric fashion as shown in reaction (3)

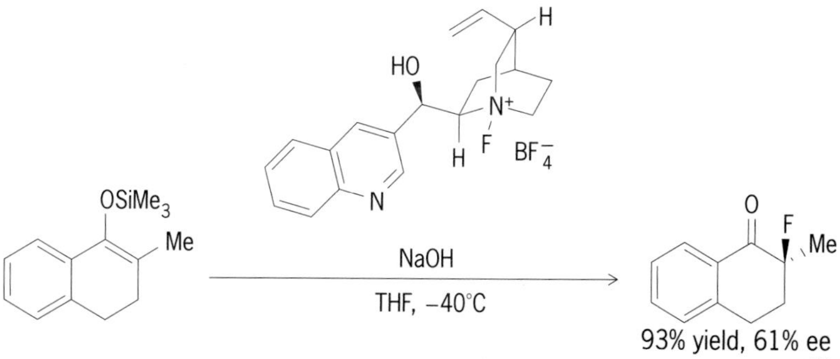

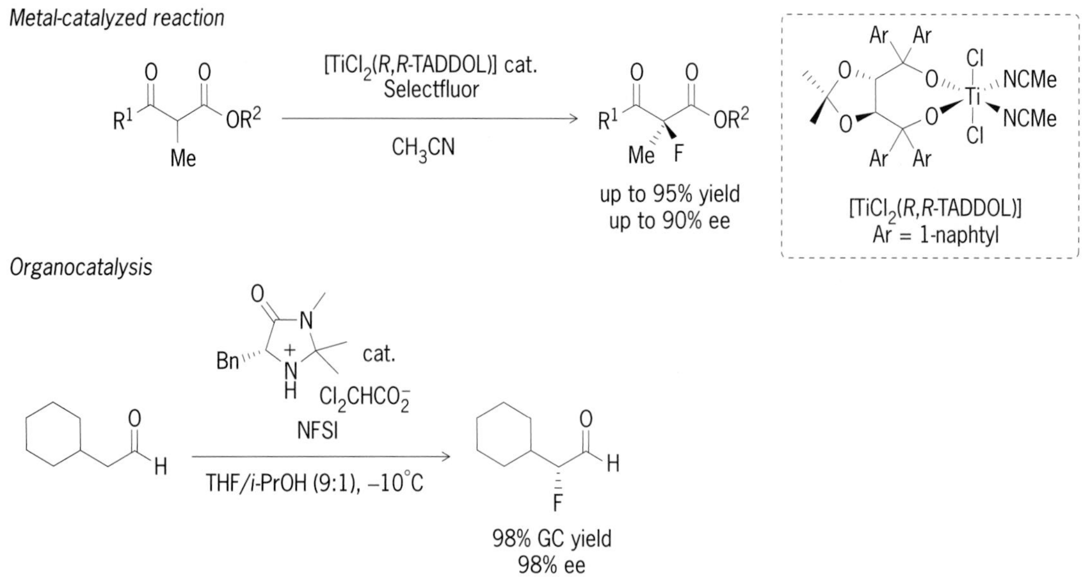

Metal-catalyzed reaction

Organocatalysis

98% GC yield
98% ee

Fig. 5. Examples of catalytic asymmetric electrophilic fluorination. Abbreviation: GC = gas chromatography.

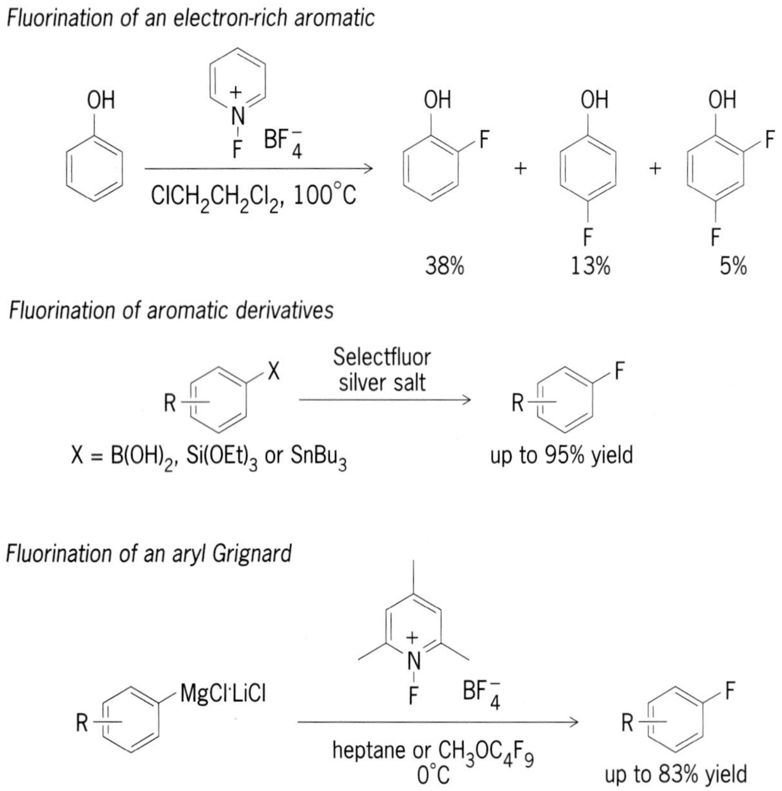

Fluorination of an electron-rich aromatic

38%   13%   5%

Fluorination of aromatic derivatives

X = B(OH)$_2$, Si(OEt)$_3$ or SnBu$_3$ up to 95% yield

Fluorination of an aryl Grignard

up to 83% yield

Fig. 6. Electrophilic fluorination of aromatics.

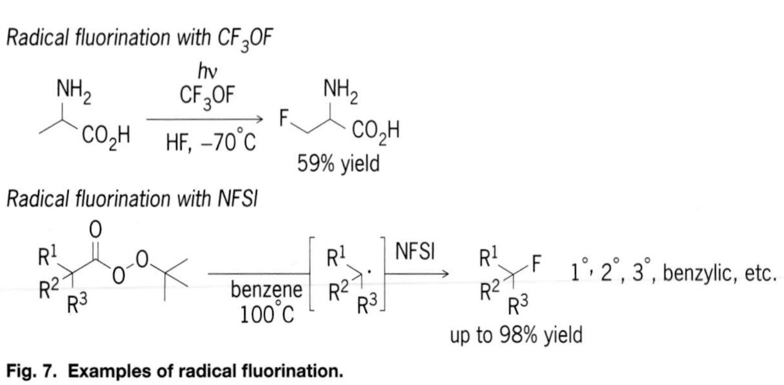

Radical fluorination with CF$_3$OF

59% yield

Radical fluorination with NFSI

1°, 2°, 3°, benzylic, etc.

up to 98% yield

Fig. 7. Examples of radical fluorination.

or in catalytic amounts if a stoichiometric amount of a less reactive achiral electrophilic source of fluorine is used to regenerate the chiral N-F reagent in situ. Alternatively, asymmetric catalysis has been successfully applied for the synthesis of enantioenriched organofluorine compounds bearing an sp^3 C-F bond. **Figure 5** shows an example of titanium-catalyzed enantioselective fluorination and an example of an organocatalytic approach.

For the introduction of a fluorine atom onto an aromatic ring using an electrophilic fluorine source, the nucleophile can either be neutral or anionic (Grignard reagents) as shown in **Fig. 6**. When using electron-rich aromatics as neutral nucleophiles, a mixture of products is often observed since the ring possesses more than one nucleophilic site. The issue is not observed when using other neutral substrates (aryl boronic acids, aryl silanes, or aryl stannanes) or an aryl Grignard, where substitution is only observed at the carbon bearing boron, silicon, tin, or magnesium.

Radical fluorination. Radical fluorination has been studied far less than the other two "traditional" approaches and was mostly limited to F$_2$, XeF$_2$, and CF$_3$OF, until recently. For example, alanine reacts with CF$_3$OF to give fluoroalanine (**Fig. 7**). The difficulty in handling some of those reagents has been one of the limiting factors for further development of this approach. More recently, fluorination of an alkyl radical, generated through thermal decomposition of diacylperoxide, with NFSI has been shown to be possible. In this case, NFSI, a commercially available and bench-stable solid, behaves as a fluorine atom source instead of an electrophilic fluorine source (Fig. 7).

For background information *see* ASYMMETRIC SYNTHESIS; ELECTROPHILIC AND NUCLEOPHILIC REAGENTS; FLUORINE; GRIGNARD REACTION; HALOGENATED HYDROCARBON; HALOGENATION; ORGANIC REACTION MECHANISM; ORGANIC SYNTHESIS; SUBSTITUTION REACTION in the McGraw-Hill Encyclopedia of Science & Technology.   Jean-François Paquin

Bibliography. C. Bobbio and V. Gouverneur, Catalytic asymmetric fluorinations, *Org. Biomol. Chem.*, 4:2065–2075, 2006, DOI:10.1039/b603163c; D. Cahard et al., Fluorine & chirality: How to create a nonracemic stereogenic carbon-fluorine centre?, *Chem. Soc. Rev.*, 39:558–568, 2010, DOI:10.1039/B909566G; T. Hiyama, *Organofluorine Compounds: Chemistry and Applications*, Springer-Verlag, Berlin, Heidelberg, Germany, 2000; S. Purser et al., Fluorine in medicinal chemistry, *Chem. Soc. Rev.*, 37:320–330, 2008, DOI:10.1039/B610213C; K. Uneyama, *Organofluorine Chemistry*, Blackwell Publishing Ltd., Oxford, U.K., 2006.

Foothill abortion

Foothill abortion, or epizootic bovine abortion (EBA), is an economically devastating disease that has a limited geographical distribution. The disease is believed to have existed in California since the 1920s, and it became recognized as a significant deterrent to maximum calf production in California in the early 1950s. Pregnant beef cattle that were grazed in the summer months for the first time on foothill and mountainous terrain were at the greatest risk, aborting fetuses late in gestation or giving birth to weak calves that often succumbed to the infection (**Fig. 1**). The pathology of the disease was characterized in the early 1960s. Affected fetuses often exhibit an extensive ascites (an abnormal accumulation of serous fluid in the abdominal cavity, resulting in an obviously enlarged abdomen), petechial (small, round, red–purple) hemorrhages on mucous membranes (nostrils, nose, and gums), a swollen liver, petechial hemorrhages in the thymus, and enlarged organized lymphoid tissues (including the spleen and lymph nodes) [**Fig. 2**]. Microscopically, the disease is characterized by proliferation of lymphocytes and mononuclear phagocytes, acute vasculitis, and focal-necrotizing lesions that can appear as pyogranulomas in lymph nodes and the spleen. The thymus presents with unique lesions that are characterized by atrophy of cortical thymocytes and infiltration of macrophages; these are pathognomonic (distinctively characteristic) of foothill abortion. In general, the classical and primary diagnosis of foothill abortion is based on the unique gross and microscopic presentation of the fetus, the history of the dam (whether she had been grazed in the foothill or mountainous regions of California in the summer and during pregnancy), and abnormally high levels of immunoglobulins in the blood. Naïve (previously unexposed) dams infected between gestational days 85 and 125 routinely generate fetuses having characteristic gross and histologic lesions. Dams infected late in gestation (generally at more than 150 days) are likely to produce healthy calves, whereas the effect of infection early in pregnancy has not been established.

Economic impact. Today, foothill abortion is recognized in the foothill and mountainous regions of California, Nevada, and Oregon (**Fig. 3a**). Losses are variable, but they can approach 90% when naïve pregnant cattle are grazed on endemic terrain. Animals continuously grazed in enzootic areas develop apparent immunity, thereby reducing (but not eliminating) the fetal losses as cows mature. Losses associated with foothill abortion continue to represent the greatest deterrent to successful calf production in endemic areas. The only control measure currently available is the modification of herd management, in which breeding cycles are altered to minimize exposure of pregnant cattle in the hot summer months to the foothill and mountainous terrain. Such an approach requires fall calving, which often compromises the efficient use of feed for many ranchers.

Disease transmission. Early data on the seasonal distribution of foothill abortion provided compelling evidence that the etiologic agent involved a transmission cycle that included a reservoir host and an arthropod vector. Armed with the information that the natural habitat of a soft-shelled tick, *Ornithodoros coriaceus* Koch (commonly referred to as the Pajaroello tick; Fig. 3b–e), was geographically coincidental with areas in which foothill abortion was enzootic, researchers designed experiments to determine its potential as a vector of the disease. Following the experimental feeding of field-collected *O. coriaceus* ticks on naïve pregnant heifers, foothill abortion was successfully transmitted, establishing this unique tick as a vector, if not the only vector, of foothill abortion.

Tick vector. Historically, the Pajaroello tick (pronounced pa-har-wayo and derived from the Spanish words "paja," meaning straw, and "huello," meaning the undersurface of a hoof) was the subject of considerable folklore. In the early twentieth century, native Mexicans living in the Central California coastal mountains were prone to believe that three tick bites would result in certain death. They feared the Pajaroello ticks more than rattlesnakes, and stories of men losing arms or legs as a result of Pajaroello tick bites were common. Although such folklore appears to be greatly exaggerated, sequential tick bites of humans can result in severe local reactions (suggesting a hypersensitivity), including severe swelling of the involved appendage.

Fig. 1. **Near-term foothill abortions. The aborted calves show typical distended abdomens (resulting from edema) [labeled *a*] and skin lesions [labeled *b*].**

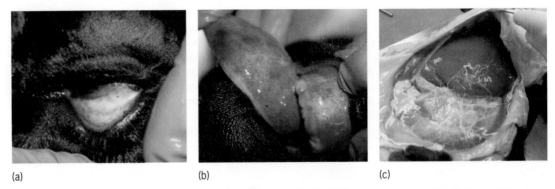

(a)   (b)   (c)

Fig. 2. Classical lesions associated with a case of foothill abortion: (*a*) petechial hemorrhages on eyelid; (*b*) petechial hemorrhages on tongue and gums; (*c*) fluid- and fibrin-filled abdominal cavity, swollen liver, and enlarged spleen.

Ornithodoros coriaceus was first described in 1844 in Mexico by C. L. Koch. The tick is easy to identify because it is the only species of *Ornithodoros* found in the Western Hemisphere that has two pairs of eyes. Other than a single report of the tick in Paraguay, the majority of early collections had been from Mexico and California. Pajaroello ticks differ from hard-bodied ticks because they do not attach to the host. They primarily live in large animal beds (wildlife and cattle) in coarse soils that are mixed with tree litter at elevations ranging from 600 to 8000 ft (183 to 2438 m). They are attracted to "bed-

ded" warm-bodied animals by CO_2 emissions and lacerate the skin with their mouthparts. Adult ticks (nymphs, males, and females) engorge with pooled blood in 5–50 min and drop off the host, whereas the larvae stage may stay attached for 7–9 days. More recently, the tick has been extensively trapped in Nevada and Oregon. It has been suggested that this tick has been slowly moving eastward by taking advantage of migratory mammals and because of the relocation, domestication, and management of livestock by humans. This is supported by the relatively recent recognition of foothill abortion in Nevada and

(a)   (b)

(e)   (d)   (c)

Fig. 3. Habitat and appearance of the Pajaroello tick. (*a*) Typical terrain (sage, bitterbrush, and pinyon pine) inhabited by the Pajaroello tick. (*b*, *c*) Collection of Pajaroello ticks using dry-ice traps (hungry ticks are attracted to CO_2). (*d*) Hungry Pajaroello ticks (desiccated appearance). (*e*) A Pajaroello tick following a blood meal (engorged).

Oregon, which historically had not been recognized as endemic areas for either the tick or the disease.

Disease etiology. Although identification of the vector was an important breakthrough, it did little to advance identification of the organism, and studies directed at identifying the etiologic agent of foothill abortion have been difficult. A variety of microbial suspects were identified as potential etiologic agents, including a member of the psittacosis-lymphogranuloma-venereum (PLV) group, a chlamydial pathogen, an unclassified virus, *Borrelia coriaceae*, and an unclassified spirochete-like organism; all were eventually eliminated as suspects. Extensive efforts over a 50-year period to cultivate the etiologic agent from either the tick vector or diseased fetal necropsy tissue using a variety of synthetic media and lab animal systems were unproductive. Researchers often reflected on the possibility that the pathogen might be cleared from the diseased fetus prior to being aborted; this seemed like a logical hypothesis, given that aborted fetuses had high levels of circulating immunoglobulins and affected tissues were infiltrated with lymphocytes and macrophages.

The ability to transmit foothill abortion by inoculation of pregnant naïve heifers with homogenized necropsy tissues derived from freshly aborted foothill fetuses was first reported in 1983. This observation facilitated the subsequent development of an efficient and reliable system for infecting pregnant animals with the etiologic agent of foothill abortion using cryopreserved thymus homogenates derived from select infected fetuses. The ability to reliably infect pregnant cows with foothill abortion permitted the initiation of studies to determine if the etiologic agent was antibiotic-sensitive; extensive antibiotic treatment at the time of infection successfully eliminated the pathogen, resulting in the birth of healthy calves. Armed with the knowledge that the etiologic agent of foothill abortion was prokaryotic, researchers subjected the cryopreserved infectious tissues to intensive probings using molecular biology. Competitive polymerase chain reaction (PCR) amplification techniques were used to clone and sequence the 16S ribosomal deoxyribonucleic acid (rDNA) gene of a unique and previously undescribed prokaryotic organism. Phylogenetic analysis of the rDNA gene sequence placed the bacterium as a member of the Deltaproteobacteria group and, more specifically, as a myxobacterium. A pathogen-specific PCR was subsequently developed with 100% specificity and 88% sensitivity on fetal necropsy tissues, confirming this bacterium as the etiologic agent of foothill abortion. This PCR was subsequently used to identify the bacterium in the tick vector (*O. coriaceus*), and its preferential residence was found to be in the salivary gland. Prior to the discovery of this unique bacterium, only one mammalian pathogen, *Lawsonia intracellularis* (primarily associated with an enteritis in pigs), had previously been placed within the Deltaproteobacteria group. However, the two pathogens are only distantly related, and the etiologic agent of foothill abortion is currently the only animal pathogen that is recognized to be a myxococcus. Fortunately, though, the phylogenic connection between the bacterial agent of foothill abortion and *L intracellularis* led researchers to employ a modified Steiner silver staining technique to visualize the foothill abortion bacteria microscopically for the first time. Fetal bovine production of foothill abortion–specific immunoglobulins has since been exploited to develop an immunohistochemical (IHC) assay and an immunofluorescence assay for visualization of the bacteria. Disease diagnosis now includes the presence of the bacteria as detected by foothill abortion–specific PCR or IHC techniques.

Animal models. Attempts to advance the study of host-pathogen interactions were compromised by the substantial cost of bovine experiments, the inaccessibility of the developing fetus, and the lack of extensive immunologic probes for components of the bovine immune system. The ability to cryopreserve infectious tissues, which were capable of reliable disease transmission, spurred attempts to develop a more congenial animal model. Initial efforts were directed at pregnant sheep; similarities in the placental structure (the transport of maternal antibodies across the placenta is prevented in both sheep and bovines), a shorter gestation period (five compared to nine months), and the relatively minimal cost made this species an attractive alternative. Unfortunately, disease transmission studies in sheep were unsuccessful; this failure was attributed to the relatively short gestational period of the sheep and the hypothesized slow replication of the bacteria. The failure to propagate the bacteria in the sheep fetus and immunocompetent mice, combined with the immunologic immaturity of the developing bovine fetus, led researchers to test the susceptibility of mice with severe combined immunodeficiency (SCID). SCID mice inoculated with infectious cryopreserved fetal bovine thymus homogenates began to develop a wasting disease and required euthanasia approximately 3 months postinfection. Interestingly, the time from infection to expression of clinical disease in the mice was similar to that in the developing bovine fetus, further suggesting that the agent is slow to replicate. The bacterial pathogen was harvested from the necropsied mouse tissues and passed again in SCID mice with similar results. Then, the second murine passage-derived bacteria were used to successfully infect pregnant cows that subsequently aborted their fetuses with classical foothill abortion pathology and the presence of bacteria in most tissues.

Characterization of the etiologic agent. Recent research efforts have been directed toward characterization of the bacteria and their pathogenesis, but progress has been slow. For example, bacterial purification and characterization have been hindered by the inability to grow the bacteria in culture and by the cell-associated nature of the bacteria. The amount of bacteria in any given fetus is highly variable, and the identification of a fetus that harbors large bacterial loads while also being in good postmortem condition is a relatively rare event. In addition, attempts at identifying bacterial proteins (via mass spectrophotometry) or genes (via PCR) have failed as a result of the presence of contaminating bovine proteins or genes, respectively, and the apparent unique sequences of the bacteria.

Summary. All of the aforementioned advances dealing with foothill abortion have contributed to current efforts toward understanding this unique disease and the associated host-vector-pathogen relationships. Studies directed at the tick vector have identified a highly variable infection rate that ranges between 0 and 50%, which is consistent with the hypothesis that the bacterial agent (not yet given a name) of foothill abortion is not a natural resident of the tick; however, the reservoir of the bacteria remains a mystery. Disease control efforts are in progress, and researchers are currently establishing the safety and efficacy of an unusual live/virulent SCID mouse spleen-derived vaccine for foothill abortion. Immunization of naïve cattle, prior to breeding, has resulted in 100% protection from infection during subsequent pregnancy. Murine-derived bacterial preparations are proving to be easier to work with, and advances in genomic/proteomic characterization are forthcoming, although purification of the bacteria from host cells remains problematic. Bacterial genes are being cloned and expressed for the purpose of establishing improved diagnostics and providing low-cost recombinant vaccines for the near future.

For background information *see* AGRICULTURAL SCIENCE (ANIMAL); ARACHNIDA; BEEF CATTLE PRODUCTION; DISEASE; DISEASE ECOLOGY; EPIDEMIC; EPIDEMIOLOGY; INFECTIOUS DISEASE; IXODIDES; PATHOGEN; POLYMERASE CHAIN REACTION (PCR); VACCINATION in the McGraw-Hill Encyclopedia of Science & Technology. Jeffrey L. Stott

Bibliography. M. L. Anderson et al., Histochemical and immunohistochemical evidence of a bacterium associated with lesions of epizootic bovine abortion, *J. Vet. Diagn. Invest.*, 18:76–80, 2006, DOI:10.1177/104063870601800110; M. T. Blanchard et al., Serial passage of the etiologic agent of epizootic bovine abortion in immunodeficient mice, *Vet. Microbiol.*, 144:177–182, 2010; R. S. Brooks et al., Quantitative duplex TaqMan real-time polymerase chain reaction for the assessment of the etiologic agent of epizootic bovine abortion, *J. Vet. Diagn. Invest.*, 23:1153–1159, 2011, DOI:10.1177/1040638711425573; M. Hall et al., Diagnosis of epizootic bovine abortion and identification of the vector in Nevada, *J. Vet. Diagn. Invest.*, 14:205–210, 2002, DOI:10.1177/104063870201400303; P. C. Kennedy et al., Epizootic bovine abortion: Histogenesis of the fetal lesions, *Am. J. Vet. Res.*, 44:1040–1048, 1983; D. P. King et al., Molecular identification of a novel deltaproteobacterium as the etiologic agent of epizootic bovine abortion (foothill abortion), *J. Clin. Microbiol.*, 43:604–609, 2005, DOI:10.1128/JCM.43.2.604-609.2005; E. T. Schmidtmann et al., Experimental and epizootiologic evidence associating *Ornithodoros coriaceus* Koch with the exposure of cattle to epizootic bovine abortion in California, *J. Med. Entomol.*, 13:292–299, 1976; J. L. Stott et al., Experimental transmission of epizootic bovine abortion (foothill abortion), *Vet. Microbiol.*, 88:161–173, 2002; M. B. Teglas, The geographic distribution of the putative agent of epizootic bovine abortion in the tick vector, *Ornithodoros coriaceus*, *Vet. Parasitol.*, 140:327–333, 2006.

Forensic mycology

Mycology is the study of all kinds of fungi, including mushrooms, mildews, blights, molds, lichens, and yeasts. Until the last few years, the value of fungi to forensic investigation was confined mainly to cases involving poisoning or the illicit use of psychotropic species. Recently, however, they have been useful in providing trace evidence, estimating the postmortem interval, ascertaining the time of deposition, investigating the cause of death, examining events involving hallucinations or poisoning, and biosecurity.

Background. Fungi include species that can only be seen with a microscope as well as those whose spore-producing bodies can weigh several kilograms. The basic body structure consists of a microscopic thread that grows continuously forward. This thread, the hypha, anastomoses with others to form an interconnecting network of tissue called the mycelium. The hypha and mycelium become organized into all the structures found in fungi, including the spore-forming tissues. A sporophore is any structure that produces spores, either asexually or sexually. These can be microscopic or large as in mushrooms and bracket fungi. The largest known living organism is *Armillaria bulbosa* (a honey fungus), found in North America. It produces spores in mushroom-type sporophores, which are produced prolifically, but its mycelium has been reported to occupy an area the size of London.

It was originally thought that fungi were closely related to plants and so, traditionally, were studied by botanists. However, molecular studies have shown that they share more characters with the animals and belong to the same evolutionary group as them. They are heterotrophic and do not carry out photosynthesis. They absorb energy-rich molecules, and other essential nutrients, directly from living organisms or from dead organic material from which organic and inorganic molecules are released through polymeric degradation.

Fungi form an ancient group that was already diverse by 600 million years ago. It is generally accepted that there may be more than 1.5 million species on Earth, at least six times as many as plants. Unlike plants, species of fungi that prove to be new to science are found regularly from all the major fungal groups, and it is unlikely that the actual number will ever be known. It is probably impossible even to make an inventory of species at one location, because many fungi rarely produce sporophores and it requires special techniques to reveal their presence.

Fungal species distributions are not as well studied as those of plants and animals but, except for some molds (which appear to be found in most terrestrial habitats), just like other organisms, fungi are affected by their biogeographical histories and the mass of physicochemical environmental variables prevailing within any habitat. In other words, each fungal species occupies a distinct ecological niche. If they have become adapted to pathogenic or mutualistic relationships with plants and animals, their distributions may mirror those of the other partner(s). Some do not form intimate associations but live as commensals in close proximity to others. However, if they require specific nutrition from the exudates or discarded organic material, the geographical distribution of the donor may influence that of the fungus. Saprotrophic species, which live on any dead or decaying material (and, thus, do not rely on nutritional specificity) will have their geographical ranges affected by their response to the mass of physicochemical variables in the ambient environment.

Trace evidence. A palynomorph is any microscopic organic entity that yields information, including the pollen of flowering plants and conifers, spores of ferns and mosses, and spores of fungi, to name some examples. Fungal spores have proved to be important palynomorphs and have contributed greatly to the resolution of forensic palynological analysis. Although they may be restricted to specific ecological habitats, unlike plants, many (particularly the microfungi) can grow to maturity and produce spores on very small amounts of nutrients, for example, on a smear of blood or glue. They can grow on stone, brick, paving stones, leather, plastics, rubber, paper, and textiles, and their spores can be transferred to anything with which they come in contact. The main source of fungal spores, and their greatest usefulness for forensic investigation, comes from their presence in soils, sediments, and on vegetation.

Some fungi, such as certain species of *Penicillium* and *Aspergillus*, appear to be ubiquitous and are considered to be "weed" species. However, these are usually taxa (groups of organisms) with wide nutritional and ecological requirements. Invariably, they are associated with food spoilage or decomposition of relatively simple organic substrates in soils. Rarely are they species that are associated with living organisms or those that have more specific nutritional requirements, such as wood- or dung-decomposing fungi. The latter have much more restricted distributions and, unlike the weed species, they have specific requirements. Whereas the spores of weed species become distributed widely in the air, the spores of those with narrower requirements generally do not travel more than a few meters away from the source, and often they reach no farther than a few centimeters. This can make them highly significant marker species for a location.

When the footwear or clothing of an offender contacts a palyniferous surface, fungal spores and other palynomorphs are transferred to the material. They can work their way down into the weave of fabric, or become stuck in interstices, and remain in situ for considerable periods. Their length of residence time is unknown, although palynomorphs have been retrieved from fabrics kept in storage after 25 years. Thus, mycological and palynological trace evidence has proved useful in "cold case" studies.

The value of fungal and other palynological trace evidence relies on the whole assemblage of palynomorphs found in a sample. Each organism has specific ecological requirements, and assemblages of species often grow in the same environment, each exploiting its specific niche within the habitat. Thus, it is the whole assemblage (a very large number of specific markers) that gives specificity to a particular place. Additional specificity is conferred by the profile; this is when the individual members of an assemblage are quantified and the relative abundances of the various taxa are estimated. It has been demonstrated many times, through working on criminal investigations, that no two places will yield exactly the same profile of fungal spores or pollen and plant spores. Thus, when the profile retrieved from clothing, footwear, or even a person (for example, hair) is shown to be similar to that of a crime scene, but different from any other claimed to have been visited by that person, evidence of contact is powerful. Evidence is strengthened even more if rare taxa are present in the profile.

In a case of a drug-related murder in East London, U.K., a gunman hid in a cypress hedge and leaned against the trunk of an oak tree. Because it was so shaded by the oak, the plants in the hedge were weak, and were infected by a pathogenic fungus (*Pestalotiopsis funerea*). The spores of the fungus had been dispersed to the foliage of the cypress and the trunk of the oak. They were also present in the leaf litter and debris underfoot. The pollen assemblage obtained from the crime scene, and retrieved from the footwear and getaway car, was shown to be unlike any other from sites claimed to have been visited by the defendants. *Pestalotiopsis funerea* formed part of the profile along with an apparently undescribed species of *Endophragmiella*. The fungal spores provided additional resolution to the palynological profile, and the jury was convinced that there had been contact between the defendants and the crime scene. Many such cases have now benefitted from fungal trace evidence, which provides an independent class of evidence from many forensic markers.

Time since death (postmortem interval). The fungi that infect humans are usually species that are tolerant of body temperature and the body's natural defenses. They range from those that infect the surface of the skin (such as ringworm fungi) to invasive infections such as candidiasis (thrush), and others which form more deep-seated infections in lungs (such as aspergillosis) and other tissues (mycetomas, mycoses). If the immune system is compromised, infections can occur by less specialized, more opportunistic, fungi. These fungi are often associated with spoilage of food and other substances and include species of *Penicillium, Mucor, Aspergillus, Fusarium,* and *Geotrichum*. These are the species that have proved useful for estimating the postmortem interval.

If the growth rates of fungal colonies on, or associated with, human corpses are known, measurement of isolates from corpses may give information on the length of time taken for them to grow and reach the size of the observed colonies on the skin or other pertinent object. Correction factors must be applied to take account of the time for degradation of the immune system, and the lag phase for colonization and establishment. However, there are few data on actual growth rates on dead human tissue, especially considering variations in temperature and humidity. It is necessary, therefore, to carry out experimental trials to establish these parameters de novo in every criminal investigation in which this kind of information is pertinent.

This approach was used successfully in a case for Tayside Police, Scotland, U.K., in which a man had been stabbed to death in his warm flat. The doors and windows were closed and there was no access for flies. The heat also meant that the air was exceedingly dry. Body fluids and blood had spattered over cushions and carpet, and these had developed large numbers of fungal colonies. These colonies were later identified as *Mucor plumbeus*, *Penicillium brevi-compactum*, and *Penicillium citrinum*. After removal of the body, the police set up data loggers, which monitored relative humidity and temperature. The humidity was consistently between 30% and 34% and, as most fungi require a relative humidity of at least 95%, it seemed that drying of the fluids had resulted in inhibition of fungal growth. This was demonstrated by a spurt of renewed growth when a sample of the carpet was rewetted with bovine blood in the laboratory. Samples of several fungi were cultured on a range of media and grown in a range of conditions. Comparison of the sizes of new colonies, both on the carpet and in vitro, with those at the crime scene suggested that the death had occurred five days prior to the discovery of the body. This was consistent with a subsequent admission of guilt.

Time of deposition. Growth characteristics of fungal sporophores (for example, mushrooms) and lichens at crime scenes can give temporal information about changes at a site, as can the sporulation times of various species.

Case example 1. In one notable case, fungal lesions on the still-green leaves of *Rubus fruticosus* (bramble) lying beneath a buried body gave information about the time of burial. In this particular instance, the underside of the leaves had black pustules, which had caused surrounding leaf tissue to redden. The pustules were caused by the sporulation of a common rust fungus, *Phragmidium violaceum*. This fungus produces orange spores in the spring (urediniospores), and dark ones (teliospores) in the late summer and autumn. Combined with evidence from the budding shoots of the bramble plant, and the rust fungal infection, it was possible to ascertain that the body had been buried between the end of September and the beginning of November, in the year prior to discovery. This resulted in the rapid

identification of the deceased and helped obtain a conviction of the offender.

Case example 2. In another case, it was shown that a plastic bag containing a dismembered leg had lain in situ at a site for only about five days, whereas it was assumed by investigators that it had been there for two weeks. The five-day estimate was achieved by observing that a lichen, *Xanthoria parietina*, growing on a twig on the ground had obviously lain beneath the parcel and it had turned green. When growing in full light, the lichen is yellow with orange spore-producing bodies (apothecia). In low light intensities and on the underside of branches, it adopts a gray color with orange apothecia. When kept covered and completely deprived of light, it turns green with faintly orange apothecia. A small-scale experiment, using the lichenized twigs from the crime scene, showed that it required about five days of light deprivation for the green coloration to develop. Thus, the leg was unlikely to have been deposited at the site for more than about five days at most.

Cause of death. Fungal spores, and other microscopic structures, can be identified from gut contents and from food remains. Thus, the consumption of poisonous fungi can be ascertained by direct microscopic examination of gut contents or vomit. Psychotropic fungi can also be identified in the same way, and there have been several cases where flasks of liquid, containers, and even clothing have yielded fungal spores of these fungi.

A young man died several days after consuming a mixed infusion of the roots of two South American vines, even though no other person who had indulged in the event had any effects other than hallucinatory "trips." By microscopic analysis of several sections of the deceased gut (from stomach to ascending colon), it was shown that he had consumed *Cannabis*, magic mushroom (*Psilocybe semilanceata*), and seeds of *Papaver somniferum* (opium poppy). Evidence of the infusion of psychotropic root material was demonstrated by other analytical techniques. It appeared that it was the concoction of the various compounds—derived from plants and fungi—that had been responsible for his death.

Location of corpses and other objects. Much has been written about the potential of using fungi to highlight the presence of burial. Some fungi will often produce sporophores in disturbed ground, while others will not respond by sporophore production until 1–2 years after the perturbation [for example, *Coprinus comatus* (shaggy ink cap) and some *Morchella* species (morels)]. Certain *Hebeloma* species have been suggested as an indicator of buried corpses. This suggestion was based on the fact that the long stalks of some sporophores of these fungi arose in the bodies of dead buried small mammals and extended to mushroom sporophores on the surface. Although there are many records of some of these species, not least in Europe, none has yet been found associated with any buried human remains.

Bioweapons and biosecurity. A few toxin-producing fungi are candidates for use in biological warfare, and there have been attempts at mass culturing to create inocula that can be sprayed onto human populations from the air. Most dangerous is *Fusarium* T2 toxin, the actions of which are measured in minutes, but there have also been attempts to spray carcinogenic mycotoxins, notably aflatoxin. Plant pathogenic fungi can pose major threats to food security if introduced by aerial spraying on staple crops. However, it is accidental introductions that have caused the most serious episodes in recent decades, as in the case of sudden oak (tree) death (*Phytopthora ramorum*) in California, Oregon, and Europe. Biosecurity of borders with respect to fungi has become an increasing concern in agriculture-dependent countries.

Accurate identification of the fungi likely to create these problems is essential, and this is unlikely to be achieved by a generalist employed by forensic science providers.

Conclusion. Mycology has proved to be a most valuable addition to the forensic science armory in a wide range of situations, but to date its potential and application in actual cases has been limited. This is because many investigating officers and forensic science providers have not become aware of, and taken up, the new possibilities it presents.

For background information *see* AFLATOXIN; CRIMINALISTICS; FORENSIC MICROSCOPY; FUNGAL ECOLOGY; FUNGAL INFECTIONS; FUNGI; LICHENS; MUSHROOM; MYCOLOGY; MYCOTOXIN; PALYNOLOGY in the McGraw-Hill Encyclopedia of Science & Technology.   Patricia E. J. Wiltshire

Bibliography. G. S. de Hoog et al., *Atlas of Clinical Fungi*, 2d ed., Centraalbureau voor Schimmelcultures, Utrecht, the Netherlands, 2000; D. L. Hawksworth, The magnitude of fungal diversity: The 1.5 million species estimate revisited, *Mycol. Res.*, 105:1422–1432, 2001, DOI:10.1017/S0953756201004725; D. L. Hawksworth and P. E. J. Wiltshire, Forensic mycology: The use of fungi in criminal investigations, *Forensic Sci. Int.*, 206:1–11, 2011, DOI:10.1016/j.forsciint.2010.06.012; J. M. B. Smith, *Opportunistic Mycoses of Man and Other Animals*, CAB International, Wallingford, U.K., 1989; M. Tibbet and D. O. Carter, Mushroom and taphonomy: The fungi that mark woodland graves, *Mycologist*, 17:20–24, 2003, DOI:10.1017/S0269915X03001150.

Free will and the brain

"[We] do evil by the free choice of our will" (Augustine, *On Free Choice of the Will*, 1.15). These words by Augustine of Hippo—and many similar statements by countless philosophers, theologians, politicians, legal scholars, and academics of many stripes—have formed the basis for Western society's interpretations of human behavior, whether it be sinful or holy, criminal or law-abiding, for nearly two millennia.

However, consider a "normal" individual who, after suffering damage to his prefrontal cortex, behaves in antisocial or even sociopathic ways. Can we still say that this person did evil by the free choice of his will? Furthermore, if damage to the prefrontal cortex causes one person to commit a crime, might it be that every crime, and every behavior then, has an external, deterministic cause? Now, what if the brain damage is healed and the sociopathic behavior stops? Do we excuse the harm that was done previously? With these questions, it is possible to instantly grasp the many social issues that are linked to the age-old notion of free will.

Determinism and the brain. Recent findings in neuroscience have shown that human behavior is a product of an automatic and determined brain. Accordingly, the idea has become prevalent that the brain controls the mind and, as a consequence, there is no free will. The claim is commonly made that there is no essential "you" in your brain that is in charge of your actions.

What is deterministic about our brains? Like all physical entities, our brains must obey the physical laws of the universe. In short, the brain is not unlike a machine, albeit human rather than artificial. Also, this machine is functioning all day, with parallel and distributed systems that have particular decision-making points and centers of integration. The parts of this machine are genetic, innate, and endowed upon us before we are even born.

After centuries of fact-finding and theory busting, the field of neuroscience is well positioned to enter the debate about the existence of free will. Indeed, theory busting is nothing new to neuroscience. For example, the work of intellectual giants, such as the neurobiologist Roger W. Sperry, contradicted the commonly held behaviorist view from the early twentieth century that humans enter the world as a blank slate and that different areas of the brain are undifferentiated and interchangeable (that is, any part of the brain can complete a particular task). Others took this paradigm shift and started to ask questions in new ways, thus revealing new answers.

In spite of how we feel about our own selves as integrated whole entities, the copious evidence from neuroscience has taught us that the brain is a highly specialized system with millions of local processors making important decisions. There is no central command system.

Free will versus personal responsibility. We can address, from a neuroscientific viewpoint, the primary question of this paper: Is human behavior the result of free will? After all, the overwhelming evidence (some of it quite recent) has shown that the brain's overall architecture is indeed under genetic control, with experience playing a critical role in determining the final constraints. For example, training, such as practicing the piano, can rewire synaptic connections and promote dendritic spine formation, and these effects can occur not only in youngsters but in adults. Still, even with this ongoing feedback and rewiring, the fact is the deterministic brain enables the mind. Our brain makes decisions based on experience, genetically determined biases, and mostly without our conscious knowledge. Indeed, when viewing the notion of free will in a modern world, what does the concept even mean? What exactly do we want to be free from? The concept seems to have outlived its usefulness.

If we do not have free will, can we defend the idea of personal responsibility? Yes. A social network can survive, and perhaps thrive, only when the individuals who make up the network accept the rules of the network and accept responsibility for their actions. Personal responsibility is derived from the social layer that we all live in, which exists beyond our own mental layer—which in turn is derived from our brain. Personal responsibility is something that exists in the social layer and not in the brain. It comes out of people and not their brains.

Consider the following: Air-traffic rules were created for the purpose of safe air travel. Now think of how unsafe you would feel if you knew that the air-traffic controllers, and pilots for that matter, were trained exclusively as mechanics. Probably you would not board another airplane because you know that someone who understands how an airplane works mechanically does not necessarily understand the rules by which an airplane must interact safely with other airplanes. In a similar fashion, neuroscientists understand something about the brain, but social norms operate at a different level entirely. Thus, to a neuroscientist, the idea of free will seems nonsensical. However, the idea that personal responsibility, in the context of social rules, is the yardstick by which our actions are judged makes perfect sense.

Responsibility and the law. In a typical criminal justice setting, a defendant is brought to trial so that a jury can determine his or her guilt; this judgment can be inextricably intertwined with whether the defendant's behavior was driven by free will. Consider the insanity defense, which is a claim made by a defendant that he or she is not responsible for his or her behavior as a result of mental illness or handicap. Although it is used rarely, insanity defenses often make news headlines because of the sensationalistic nature of the alleged crimes. It also does not often work. For example, the insanity defense did not help David Berkowitz, New York's "Son of Sam" murderer in the 1970s, who claimed that his neighbor's dog ordered him to kill. On the other hand, John Hinckley, Jr., in a twisted effort to gain the attention of actress Jodie Foster, attempted to assassinate President Ronald Reagan in 1981. He was declared not guilty by reason of insanity and, as of this writing, spends most of his time at a psychiatric facility in Washington, D.C. The insanity defense is unevenly applied in the United States; in fact, it is not even available in some states.

However, if we abandon the concept of free will, then there is no reason to ask juries to determine guilt, innocence, or insanity. Instead, the first phase of a trial should simply determine responsibility. Did the person charged with the crime commit the crime? Yes or no. By this standard, one's mental health does not excuse one's actions because responsibility is based on social norms, and not physiological or mental norms. Of course, once a jury makes the determination of responsibility, there is still the paramount issue of what we do with the responsible party—whether it be release, isolation, rehabilitation, or punishment. These issues need great attention and consideration.

Moving forward without free will. Free will was postulated as an answer to philosophical questions during a time in human history when humans did not have the ability to observe and measure the working brain. Over time, the tools of neuroscience were developed and have become more and more sophisticated. However, these tools are only useful when their use is directed by intellectual leaps and insights that can challenge the status quo. To put it simply, it has taken time for us to even ask the right questions, much less find the right answers, about the means by which the brain and mind interact.

Although some bemoan, and even vigorously challenge, the loss of free will, others find this new outlook refreshing and invigorating. Our understanding of the brain, and how it enables the mind, is still in its infancy. The questions that are now being asked by the geniuses of this generation require the collaboration of psychologists, biologists, anthropologists, computer scientists, engineers, and philosophers. Old barriers are torn down, and antiquated theories are being refashioned or jettisoned altogether. Indeed, many find this world without free will to be liberating.

For background information *see* BRAIN; COGNITION; DECISION ANALYSIS; MOTIVATION; NEUROBIOLOGY; PERSONALITY THEORY; PSYCHOLOGY; SOCIOBIOLOGY in the McGraw-Hill Encyclopedia of Science & Technology.   Michael S. Gazzaniga

Bibliography. Augustine, *On Free Choice of the Will*, translated by T. Williams, Hackett Pub. Co., Indianapolis, 1993; D. S. Bassett and M. S. Gazzaniga, Understanding complexity in the human brain, *Trends Cogn. Sci.*, 15(5):200–209, 2011, DOI:10.1016/j.tics.2011.03.006; M. S. Gazzaniga, Mental life and responsibility in real time with a determined brain, in W. Sinnott-Armstrong (ed.), *Moral Psychology*, vol. 4: *Freedom and Responsibility*, MIT Press, Cambridge, MA, 2013; M. S. Gazzaniga, *The Social Brain*, Basic Books, New York, 1985; M. S. Gazzaniga, *Who's in Charge?: Free Will and the Science of the Brain*, Harper (Ecco), New York/London, 2011.

Frontiers of large cable-stayed bridge construction

Bridges are a physical and philosophical link in civilization, crossing boundaries, overcoming obstacles, conquering difficult terrains, and bringing people together. The quest for building increasingly long spans stems both from necessity and from a sense of achievement in harnessing the forces of nature for human benefit.

The construction of ultra-long-span bridges demands very substantial engineering input throughout the construction planning and execution stages.

The engineering sciences involved draw on a unique combination of complex theoretical underpinning and robust practical experience. Prime examples of the application of engineering sciences when tackling the technical challenges in large bridge construction are found in two projects involving two of the world's longest cable-stayed bridges.

Stonecutters Bridge in Hong Kong is a high-level cable-stayed bridge with a main span (between the towers) of 1018 m across the Rambler Channel at the entrance to the busy Kwai Chung Container Port frequented by the world's largest container vessels. The total length of the bridge, including the back spans from land to the towers, is 1.6 km. The Stonecutters Bridge was opened to traffic on December 20, 2009.

The Sutong Bridge in China crosses the Yangtze River, connecting the cities of Suzhou and Nantong. The 7-span cable-stayed bridge had a record-breaking-long main span of 1088 m and a total length of 2.088 km. The main span closure cable-stayed bridge was successfully completed in June 2007 and the bridge was opened to traffic on June 30, 2008.

Throughout our mission in constructing these two large bridges, innovation has been an integral part of our activities. There were innovations through evolution, in that improvements on existing techniques were made progressively. There also were innovations through revolution, in that new, or even radically new, techniques were developed in the face of a difficult challenge.

The experience gained in the course of harnessing the challenges posed by these two large bridge projects was phenomenal. Working closely with construction contractors AECOM (formerly Maunsell Consultants Asia Ltd.) has been part of the momentum in pushing the frontiers of large bridge construction, and is part of an even greater momentum in striving for the best innovations and further accomplishments in a new era. This article provides a firsthand account of our mission.

Stonecutters Bridge

Stonecutters Bridge in Hong Kong is an ultra-long-span cable-stayed bridge (**Fig. 1**). To enable construction of this magnificent structure, very substantial and complex engineering inputs were required. The contract for the construction of Stonecutters Bridge was awarded to Maeda-Hitachi-Yokogawa-Hsin Chong Joint Venture (JV) in April 2004. AECOM assisted JV in the tender preparation and, upon JV's success in securing the construction contract, AECOM was appointed by JV as consultant for the comprehensive construction engineering services in the construction phase.

Very substantial and complex engineering inputs were involved in enabling the construction of this structure. These construction engineering activities included, among many others, erection analysis; bridge geometry monitoring, control, and adjustment; bridge aerodynamics; wind-tunnel testing; vibration mitigation measures; 60-m-high falsework (temporary shoring towers and supports) systems,

Fig. 1. Construction of the 1018-m-long Stonecutters Bridge.

including precast segmental concrete tower and steel trusses in longitudinal and transverse planes, for the construction of the concrete back spans of the cable-stayed bridge systems; cofferdam design; development of the deck-lifting procedures; navigation simulation; marine traffic management; marine jetty design; temporary traffic management; and geotechnical engineering. This article focuses on selected aspects of our work.

Erection analysis and bridge geometry control. AECOM developed the bridge geometry control strategies and methods for use in a prediction, survey, reanalysis, and possible adjustment cycle during bridge erection (**Fig. 2**). The overall objectives were to ensure that the final target geometry of the completed structure was achieved without unacceptable locked-in stresses and to ensure that the structure had adequate strength and performance at all construction stages.

The framework for bridge geometry control consisted of a coordinated set of activities in the construction planning, fabrication, and erection phases. All activities in these three phases were robustly integrated to support the prediction, survey, reanalysis, and possible adjustment cycle. Erection analyses underpinned many of the activities in the framework and therefore they are described in detail in the following sections.

The erection analyses of the entire bridge structure produced data for geometry control and

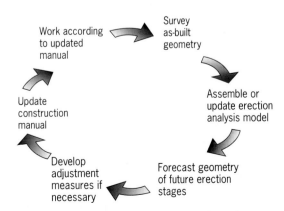

Fig. 2. Bridge geometry control cycle at each erection stage.

Fig. 3. Back span falsework.

identified any special design requirements for temporary works, including falsework, temporary stays or propping, and dynamic stabilization measures such as damping devices. The main outputs from the erection analyses were unstressed lengths of compression and tension members, precamber of flexural members, design loads on temporary works, natural frequencies and mode shapes at intermediate construction stages, and the structural effects of wind loading in the partially erected bridge structure at the erection stages.

Fig. 4. Back span concrete girder construction.

The primary objectives of the global stage-by-stage erection analysis are described below, working through the prediction, survey, reanalysis, and possible adjustment cycle.

1. To establish the stay cable unstressed lengths and structural element prestrains (precambers) required to achieve the target geometry at the end of construction.

2. To establish the stay cable jacking forces.

3. To predict the displacements of the structure at each erection stage for inclusion in the construction manual to enable on-site geometry control.

4. To determine the state of stress in the structure at each erection stage and verify structural adequacy.

5. To use as-built survey data to track the changes in geometry of the bridge and to forecast the geometries in the future erection stages.

6. To identify any corrective actions required to ensure that the target bridge geometry would be achieved.

The principal aspects of the bridge construction that were investigated in erection analysis and bridge geometry control included:

1. Tower and deck cantilever erection cycles

2. Tower alignment control

3. Stress resultants in concrete back spans in the final state

4. Effects of displacements of falsework and temporary foundations on the shape and stress state of the back spans and on the final profile of the main span and towers

5. Effects of creep and shrinkage

6. Effects of temperature and wind on the bridge geometry to correctly interpret the as-built survey data

7. Devising of typhoon procedures to ensure structural integrity

8. Sensitivity analyses on tolerances in input parameters

9. Design of the back span temporary works

10. Design of temporary post-tensioning to concrete cross-girders

11. Main span closure.

Back span falsework and deck construction. At the time of tender preparation the erection of the back span concrete decks were already identified as one of the most significant challenges in constructing the Stonecutters Bridge. The difficult configuration and structural detailing of the deck led to a falsework scheme that emerged as the only viable option. It was one of the most substantial ground-supported falsework systems ever erected, keeping approximately 30,000 tons of superstructure concrete (per back span) supported at a height of about 70 m (**Fig. 3**).

The falsework system was developed by AECOM in close interaction with JV to suit its construction methods and operational means. Safety, efficiency, and constructability were the priorities of the falsework scheme development. The final scheme was modular, had direct load paths, and was highly efficient in the use of materials.

Within each span between permanent piers was a two-bay falsework structure, consisting of three pairs of temporary towers that were braced with steel members and founded on bored piles. In the longitudinal direction, the falsework towers were positioned under the centerlines of the intermediate cross-girders in the deck.

The central portion of each intermediate cross-girder was cast on a birdcage falsework structure that was supported on steel cross-girder trusses that spanned transversely between the temporary towers. Under each intermediate cross-girder in the concrete deck there were four steel trusses, each 5.5 m deep. The trusses were simply supported at their ends on an arrangement of fabricated steel "crown beams." The crown beams were positioned on the top of the temporary towers.

The end portions of each intermediate cross-girder were cast on a birdcage falsework structure that was supported on steel "wing trusses." The wing trusses were 20-m-deep triangular trusses that cantilevered from the temporary towers. The top of each wing truss was supported on the crown beams and the base of each wing truss was clamped to the temporary tower.

The temporary towers consisted of precast segmental concrete blocks with external dimensions of 2×2 m. The majority of the segments were 2 m in height with a hollow core and 250-mm-thick walls. At the connection points with the steel bracing members, the segments were 1 m high and solid. The segments were match-cast with shear keys at the joints. Vertical ducts in the walls of the segments accommodated unbonded high-tensile bars that connected the segments and continued for the full height of the towers. The temporary towers were braced in three orthogonal planes by diagonal steel bracing members.

The temporary towers were supported on single 1.8-m-diameter piles that were founded at depth on bedrock. Each pile had a pile cap that supported the plinth at the base of each tower. Ground beams linked the pile caps to distribute the horizontal forces between the piles and provided rotational resistance to the tops of the piles and bases of the towers.

AECOM created a detailed stage-by-stage finite-element analysis model of the concrete back spans. This model was used to track the changes in structural configuration and loads through the construction process and duly accounted for stiffness contributions from falsework and partially cast concrete members, changes in weight and stiffness during staged casting, and the removal of falsework trusses, as well as the subsequent installation of stay cables and the removal of the falsework tower supports. Furthermore, creep and shrinkage effects as well as the post-tensioning sequence were accurately represented. The complex concrete grillage deck was constructed on the ground-supported falsework. The cross-girders were cast first and subsequently used to support the falsework for construction of the longitudinal girders. Transverse post-tensioning was applied at a number of intermediate stages after cross-girder completion and during longitudinal girder construction. The permanent transverse tendons in the bottom chord had to be augmented by temporary tendons at the top chord in the outer regions of the girders (**Fig. 4**).

The deck was completed first in the center part of the bay before it was then connected to the pier crossheads by "stitch" bays. Before casting these stitch pours, the deck geometry was carefully checked. At this stage, corrections by jacking on the falsework towers would have been possible. This was found unnecessary because the settlement predictions were accurate, thus further economy and speed of construction were achieved.

Deck segment assembly. A number of steps were required to fabricate, assemble, test, paint, and transport the deck segments.

Steel plate fabrication. The steel deck segments for Stonecutters Bridge were formed from thermomechanically controlled process steel, Grade S420M/ML. The plates required a very accurate control process during heating, rolling, and water cooling. This grade of steel is becoming increasingly common in Europe, but it is still relatively unusual in Asia. Consequently, sourcing of the material was difficult and a sufficient quantity was eventually procured from a number of sources in Europe and Japan.

The total weight of deck steel was 33,200 tonnes, with a typical segment weighing 500 tonnes.

Deck plates were fabricated at the workshops of China Railway Shanhaiguan Bridge Group (CRSBG) in Shanhaiguan, Northern China, an advanced facility that has been used for many of the major bridges constructed in China in recent years. At this facility, the steel plates were blasted, primed, and cut. The edges of the plates and associated stiffeners were then beveled and the stiffeners were welded to the deck plates. A typical deck plate weighed 15 tonnes.

For all the welding processes at the fabrication yard, CO_2-gas-shielded arc welding was selected. This form of welding can be adopted for all positions

Fig. 5. Welding of the stiffened panels.

Fig. 6. Deck segment assembly.

(flat, vertical, horizontal, and overhead) and allows for a continuous welding process. **Figure 5** shows the welding of stiffened panels.

From a geometry control viewpoint, the most critical components were the stay cable anchor tubes, which had to be fixed to the segment with an accuracy of 0.1°.

After completion, the deck plates were transported to the next stage in the fabrication process: assembly at Dongguan in the Pearl River Delta.

Deck segment assembly. There were 65 deck segments, each having about 200 components.

Assembly of the deck segments took place on two production lines, which were each capable of working on seven or eight segments at a time. The production lines included a moveable shelter that enabled the assembly to remain in the shade and out

Fig. 7. Heavy lift of the 88-m-long girder units. (*Courtesy of AECOM*)

of the rain (**Fig. 6**). The assembly of deck segments operated on a 60-day cycle for each production run. The deck plates were unloaded at a purpose-built jetty and then prepared in a preassembly area.

Match fabrication and geometry control. The segments were assembled bottom upward on special trestles that were capable of limited adjustment by means of hydraulic jacks. As outlined earlier, the segments were assembled in runs of seven or eight segments, with each segment matched to the adjacent segments to ensure a close fit when finally erected on-site in Hong Kong. The deck segments were carefully assembled with continual checks on alignment and elevation.

When the first three segments of a production run were almost complete, a trial assembly was done. In the trial assembly, the segments were checked for alignment, elevation, segment dimensions, and plate flatness. If acceptable, the excess top and bottom plate material (green) was cut from the segment, leaving the predefined weld gap between segments. Having been matched, adjacent segments were temporarily connected together by bolted splice plates (keeper plates), which would later be used during erection to ensure that the relative geometry on-site was the same as the trial assembly geometry at the assembly yard.

Each typical segment comprised two longitudinal girders connected by a cross-girder. In the case of the back span segments, these components were not welded together at the assembly yard. However, main span segments left the yard completed.

As an integral part of the geometry control, AECOM conducted extensive and varied analyses to determine the data necessary for fabrication and trial assembly of the steel segments. The geometry of the main span deck was predefined by the unstressed shape of the steel deck and the unstressed lengths of the stay cables. The correct unstressed geometry of the steel main deck was achieved by match-assembly of the main deck geometry and was to be reestablished on-site by accurately fitting the new erected segment to the existing cantilever.

Blasting and painting. The existing facilities for blasting and painting the deck segments were significantly enhanced at the assembly yard. All the internal and external surfaces were blasted. The external surfaces were then treated with a traditional coating process of epoxy-rich primer, two coats of epoxy MIO (micaceous iron oxide pigment), and an acrylic-polyurethane topcoat. The internal surfaces have just a coating of 50 micrometers of epoxy-zinc phosphate primer. On completion, the deck segments were silver-grey with a semigloss finish.

Transportation. The painted deck segments were then transported to the storage area by means of two multiwheeled transporters, one for each longitudinal girder. The transporters had synchronized control to ensure a segment was not subject to excessive differential distortion. AECOM conducted detailed finite-element analyses to investigate the feasibility of moving the segments on transporters. The effect on the cross-girder, arising from differential longitudinal movements of the two halves of the transporter system, was examined. The results demonstrated that the cross-girder was capable of resisting the twist induced by the movements in the transporters. Detailed finite-element modeling was also conducted by AECOM to examine the conditions at segment storage. The results led to local stiffening of the welded connections being implemented in conjunction with a rigorous site survey to ensure that the support trestles in the storage yard were at the correct levels prior to the placing of the steel segments.

The segments were moved by the transporters from storage to a dynamically positioned (DP) barge that was grounded at the jetty for easy loading. The segments were then shipped to Hong Kong for erection.

Heavy lift for steel deck erection. The steel segments to be erected in the vicinity of the bridge towers, connecting the back span with the main span, required a different method of erection; that is, they had to be erected over land as it was not possible to lift the segments directly from a barge. A very substantial ground-supported falsework system was originally envisaged, whereby the segments would be lifted from the barge to final deck level. The segments would then be slid along rails at high level to their final position and then welded together. Such a scheme would have had significant impacts on costs. JV therefore reviewed alternative schemes for the heavy lift.

Assembly of the longitudinal girders. The first stage was the construction of a gravity wall jetty, designed by AECOM, adjacent to each tower. This required the removal of about 100 m of seawall on each side of Rambler Channel and the placement of large precast concrete blocks to form the jetty. At the same time, an unloading frame was erected, cantilevering from the tower with temporary stay cables for support.

The steel deck segments were lifted from the barge by the unloading frame and lowered onto carts. The carts were slid along a series of rails by hydraulic jacks until the segments were positioned at ground level below their final position. Once the alignment and elevation relative to each other had been confirmed, the segments were welded together to form the two 88-m-long longitudinal girder units. The cross-girders, at this stage, were placed in storage.

Lifting and sliding the deck. Once the welding of the segments was complete, the 88-m-long longitudinal girder units were lifted using strand jacks mounted on a bracket attached to the tower and a deck-lifting frame cantilevering from the concrete deck. The two longitudinal girders were lifted simultaneously to ensure balance of load between the brackets on the tower. The overall weight of the lift was 4000 tonnes, with a load distribution between tower brackets and deck-lifting frame of about 80:20.

Figure 7 shows the basic arrangement for the heavy lift. Guides were attached to the tower and back span concrete deck falsework system to restrict lateral movement during the lift. Initially the deck

Fig. 8. Deck-lifting operations.

was raised to about 50 m above ground level in a series of 0.5-m strokes of the jacks. Due to the tapering tower form, the longitudinal girder units had to be jacked laterally by 4 m toward one another at that stage.

The lift then continued until the longitudinal girder units were at their final elevation of about 75 m above ground level. At this level, the longitudinal units were jacked again laterally by 2 m inwards and then finally 2 m longitudinally toward the concrete back span deck, leaving a 2-m gap. The decks were secured by ties and props to the concrete deck and by temporary bearings to the tower.

Completion of the heavy lift operation. Following the lift of the longitudinal girder units, extensive surveys were undertaken and the position of the decks was fine-tuned. The cross-girders were lifted, again by strand jacks. Once all five cross-girders had been lifted and the geometry was confirmed, the longitudinal and transverse girder units were welded together, while the concrete stitch was cast and then stressed

Fig. 9. Lifting a main span segment.

Fig. 10. Trial jacking: main span segment mismatch.

Fig. 11. On-site jacking: main span segment mismatch.

between the steel deck and the concrete deck. With the cross-girders welded and the stitch complete, installation of the permanent stay cables commenced.

The heavy lift operation was done two times, for the East and the West of Rambler Channel. On both occasions the lifting and sliding operations were completed successfully within 2 days.

Throughout the planning and the execution of the heavy lift scheme, AECOM conducted rigorous analyses, carried out structural verification of the permanent works, determined necessary strengthening measures, developed detailed geometry–control procedures, and implemented on-the-day back-analysis and control.

Main span deck segment erection. With the heavy lift completed, the focus changed to main span deck erection.

Marine considerations. One of the main constraints to the construction of the Stonecutters Bridge was the need to maintain the flow of shipping in Rambler Channel during the construction of the bridge. Consequently, AECOM assisted JV in developing a number of measures to maintain and control marine traffic during deck-lift. First, a DP barge was used to transport the segments to the lifting location. By reference to GPS satellites the barge was able to automatically position itself and then hold position by means of thrusters, located at the four corners of the barge. Under the terms of the contract, no anchors could be used to secure the barge at the lift location. The second measure was to complete the lift in as short a period as possible. This was achieved by using winches as opposed to strand jacks, giving a lift time of about 40 minutes.

While practical measures were taken in terms of equipment, a number of studies were also undertaken to investigate how to minimize disruption to the port. Current measurements were taken in Rambler Channel, and from them a current atlas was prepared. Using this atlas, the current in Rambler Channel could be predicted for every deck-lift operation. In association with the Hong Kong Pilots Association, ship movements were simulated through the channel during various critical deck-lifting operations. This exercise helped prove that shipping movements did not need to be halted during a deck-lift and also acted as a familiarization exercise for the pilots.

Lifting operations. **Figures 8** and **9** show a typical deck-lift operation. The DP barge was accompanied by four guard boats when moving to the lift location. The lifting gear was lowered to the barge below where it was attached to the segment lifting lugs. Once secured, the lifting frames started to take the load until the segment lifted off. At this point, wooden wedges beneath the segment were immediately removed to avoid rebound and potential resonance between the barge motion and the deck cantilever. The segment was lifted smoothly until it was level with the end of the deck cantilever, where it was secured and the barge and guard boats could be released.

The site connection of a segment in a wide, flexible deck to the tip of a wide deck cantilever under the actions of deck-lifting gantries was investigated by AECOM. The difference in support conditions of the lift-in segment and those of the erected deck were such that their vertical deflection profiles would have a certain degree of mismatch. AECOM's work concluded that connecting the lifted segment to the end of the cantilever required special measures, as the cantilever end would be deformed by the load from the main span-lifting gantries. It was therefore necessary to deform the lifted segment in a similar manner and this was achieved by applying a bowstring prestress system. **Figures 10** and **11** show the arrangement of the external prestressing system designed by AECOM and used in the trial jacking and in the on-site operation, respectively. The system included posts on either side of the cross-girder, two steel sections connecting the posts, and diagonal prestressing bars fixed to the deck plate. Loads were applied by means of hydraulic jacks at the base of the posts, inducing a transverse deformation on the segment. The load applied to the jack was adjusted until the deck plates of the lifted segment matched with the cantilever end. Measurements were also taken across the joint to ensure the weld gap was consistent with the measurements taken at the assembly yard. If the geometry of the lifted segment was satisfactory, the keeper plates were fitted and welding proceeded.

While the lifted segment was being welded, the back span stay cable was installed and stressed. JV and AECOM investigated different welding processes in the search for an optimum solution. The welding process took place in a fixed sequence, with the perimeter weld first. The stiffeners across the erection joint were then welded, the bowstring prestress system removed, and the main span stay cables were installed.

The as-built geometry and its deviation from the design geometry were monitored during erection. When necessary, the unstressed lengths of cable were modified and adjusted to compensate for any local as-built deviation. JV and AECOM continued to monitor and control the geometry of the bridge throughout the erection stages. Using as-built survey data, AECOM continuously back-analyzed and predicted the behavior of the bridge to provide JV with the detailed knowledge to remain in full control.

Wind-tunnel testing. AECOM advised JV on bridge aerodynamics and completed the planning, management, and supervision of the wind-tunnel investigations for the Stonecutters Bridge construction.

Comprehensive wind-tunnel investigations were commissioned to cover conditions arising during erection, including the representation of temporary works and construction plant and equipment, wherever relevant. Tests were also commissioned on the stay cables, including the textured sheathing as a countermeasure to rain-wind induced excitation.

The section model testing verified the stability of the bridge against divergent amplitude response during construction and the efficacy of the guide vanes in mitigating vortex shedding response at the

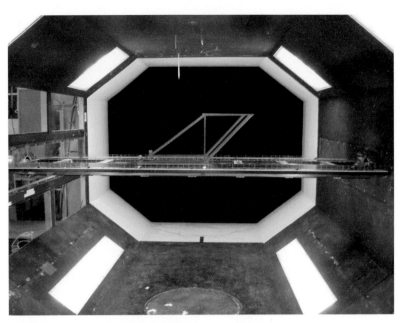

Fig. 12. Section model in the wind tunnel.

erection stages (**Fig. 12**). The guide vanes therefore were installed prior to deck lifting to assist in suppression of vortex shedding during deck cantilever construction. The dynamic tests on the section model also demonstrated the significance of the vertical and torsional aerodynamic damping compared to intrinsic structural damping, and hence the dominance of aerodynamic damping in vertical and torsional buffeting response. The static wind–loading measurements on the section model identified the significance of the temporary handrail system

Fig. 13. Aeroelastic tower model: full-height freestanding condition.

Fig. 14. Aeroelastic bridge model in the wind tunnel.

on the overall drag of the bridge deck. A number of additional tests were done to develop the form and configuration of the temporary safety handrail system, which limited the lateral wind forces on the structure to acceptable levels during construction.

The aeroelastic tower model wind-tunnel investigations verified the aerodynamic performance and structural integrity of the tower during the erection stages, including the full-height freestanding conditions (**Fig. 13**). Damping was effective in mitigating vortex shedding response and such a response was, in general, reduced by the presence of the construction plant and equipment.

In the aeroelastic bridge model testing (**Fig. 14**), the vertical responses recorded were generally less than the a priori analysis, based on the buffeting response observed in the section model testing. The buffeting responses recorded in the aeroelastic bridge model testing also corroborated the observations made in the section model tests, in that the vertical and torsional responses were dominated by aerodynamic damping. The very large aerodynamic

damping of the vertical motion, up to 0.6 log dec at design wind speeds, means that relatively little benefit would be obtained through additional (mechanical) damping. A reduction of only 10% of the resonant contribution to the response in the first vertical model was obtained by adding damping up to 0.15 log dec. These findings meant that the application of damping devices to mitigate buffeting response in these modes would present significant challenges. The aerodynamic damping of lateral response was relatively modest, and additional damping could significantly reduce the dynamic displacements. Reduction by up to 40% was obtained by increasing damping to 0.2 log dec. The effect of the free end of the cantilever appeared only modest, and introducing the end of the adjacent deck cantilever in close proximity had little consistent effect on the critical responses. The aeroelastic bridge model study validated a comprehensive numerical model that was then used for further investigations into the buffeting response of the bridge structure in different erection scenarios. It was found that buffeting effects posed a significant demand on the structure in the cantilever conditions.

Rain-wind induced oscillation tests on the stay cables were done to investigate the effects of a dimpled pattern on the cable sheathing and of increased damping on the dynamic behavior of the stay cables. The dimpled pattern suppressed or alleviated the rain-wind induced vibrations. Increased damping also was effective in suppressing such oscillations. The drag coefficients measured on cables with dimpled surface texture were within the permissible design values.

Sutong Bridge

Until 2012, the 1088-m main span Sutong Bridge in China was the world's longest cable-stayed bridge (**Fig. 15**). The Sutong Bridge project was masterminded and directed by China Jiangsu Province Construction Commanding Department, which has a record of success in projects such as the Jiangyin Yangtze River Highway Bridge, the Runyang Yangtze River Highway Bridge, and now the Sutong Yangtze River Highway Bridge. Contract C3 for the construction of Sutong Bridge was awarded in early 2005 to China Communication Construction Company (CCCC), Second Navigation Engineering Bureau to whom AECOM was a consultant. A fast-track construction program was initiated, such that main span closure was completed in June 2007 and the bridge was opened to traffic in June 2008.

One of the most significant undertakings in the construction of the super-long-span Sutong cable-stayed bridge was construction control (**Fig. 16**). The unique complexity of the Sutong Bridge required specially developed methods and procedures to control bridge geometry and to ensure the safety of the bridge during construction. We will discuss selected aspects of the integrated techniques adopted for Sutong Bridge construction control, with illustrations of the robust principles and practices in the analysis, survey, prediction, correction cycle.

Fig. 15. Panoramic view of Sutong Bridge.

The framework for the bridge geometry control consisted of a coordinated set of activities in the construction planning phase, the fabrication phase, and the erection phase. All activities in these three phases were robustly integrated to support the prediction, survey, reanalysis, and possible adjustment cycle.

Tower erection control. The construction sequence was identical for both the north and the south towers and consisted of construction of tower concrete elements, construction of tower steel elements, application of tower temporary supports, application of tower temporary loads, and application of stay cable loads acting on the tower (**Figs. 17** and **18**).

For the purpose of tower geometry control, a total of 176 erection stages (key events in the tower construction activities) were judged to be of interest, with the last stage being the application of superimposed dead load at the target or reference state. Each of these stages was modeled in the erection analysis.

To achieve the target geometry, all the structural displacements that occurred during the construction stages were taken into account in determining geometrical adjustments for each erection step. The adjustments would consist of overlengths and precamber and preset of the formwork.

Overlengths. As a result of axial shortening and creep and shrinkage effects, the concrete tower shortened during construction. To achieve the target geometry at the reference state, an axial overlength was specified for every concrete lift of the tower. The overlength values were determined from the stage-by-stage erection analysis.

Overlength values used for set-out calculation. For the set-out procedure, the elevation of the top of the concrete lift being set-out was specified as a design elevation, along with the value of the overlength to compensate for the further displacement of the top joint due to additional loads.

Precamber. Owing to the inclination of the lower tower and middle tower legs, the self-weight of the concrete induced deflections of the tower legs in the transverse direction of the bridge. This effect was compensated for by precambering the tower legs. Precamber values were determined from the stage-by-stage erection analysis. The values of precamber and overlength were used to calculate the intermediate expected geometry during the error assessment and correction procedures.

Formwork preset. The deformations of each concrete lift were not only the deflections of the tower legs but also the deflections of the jump-form (climbing-form) scaffolding system. The transverse component of the self-weight of the wet concrete induced deflections in the form and thus the concrete lift geometry followed the deflections of the jump-form system. It was therefore necessary to compensate for this effect by presetting the jump-form system.

Temporary props. The bending of the lower legs was caused by self-weight-induced bending stresses, and these effects were mitigated by installing transverse props (thrusts) between the tower legs. The props were activated by jacking.

Fig. 16. Bridge elevation.

Fig. 17. Tower shaft construction.

Fig. 18. Upper tower stay anchor box installation.

Fig. 19. Back span segment erection.

Fig. 20. Cantilevered segment erection.

Survey data processing. As-built displacements were measured at survey points and tower monitoring points. For the sections in the concrete tower shafts, the as-built displacements at the top of concrete lifts were surveyed upon completion of each concrete lift or after installation of the transverse prop. The expected intermediate geometry of the concrete section was specified at the center of the section. For the stay-anchor boxes, the survey points of the anchor boxes were at the outer corners of the box section. The center of the box section was derived from corner survey data. The expected intermediate

geometry of the anchor box section was specified at the center of the section.

Corrective actions. After calculating the difference between the as-built survey and the expected intermediate values, the corrective measures were made in the set-out for the next cycle to aim at achieving the target geometry of the tower within the allowable tolerances.

Deck and cable erection control. The back span steel segments were prefabricated in nine units and positioned by floating cranes onto the permanent piers and temporary supports (**Fig. 19**). The erection of the cantilevered section of the deck commenced after the erection of three deck units at the tower location (**Fig. 20**), their corresponding cable stressing, and the installation of the temporary fixity onto the tower crossbeam.

Prior to back span closure (**Fig. 21**) it was necessary in the installation cycle to erect the segments at the ends of the double cantilever with the cables in their final positions. Subsequent to back span closure, the back span segments were in position, and single-cantilever construction toward midspan progressed.

The installation sequence for a segment in the cantilevered construction involved lifting the segment from a barge, welding to the existing cantilevered deck, installing the cable stays, and applying first-stage stressing (**Fig. 22**). The lifting gantries were then moved to their forward lifting position, followed by the second-stage and final stressing of the stay cables.

Step 1. Segment lifted from barge.

Step 2. After segment adjusted, matched up, and welded, remove connection between the segment and lifting gantries; first-stage stressing of cables.

Step 3. Move lifting gantries forward; second-stage stressing of cables.

Step 4. Prepare for lifting the next segment.

The typical erection activities consisted of only first- and second-stage stressing. The segment was installed on the unstressed installation geometry, which was defined by its position on the existing structure. The unstressed installation geometry (the geometry excluding the deformation from self weight) was to be the same geometry as had been

Fig. 21. Back span closure.

Fig. 22. Segment erection cycle.

Fig. 23. Main span closure.

finally accepted in the trial assembly. The repetition of the trial assembly geometry was achieved by the connection of the fixing plates that had been welded after final acceptance. The alignment of the newly installed segment was verified on-site by survey.

Cable installation procedure. The stressing of the cable was controlled by the displacement of the socket nut to the prescribed position. This was the primary control parameter. The secondary parameters were cable force and the change of the geometry of the anchor points—displacement due to stressing. The cable force and deck displacement were recorded as a cross-check in the geometry control.

Cable adjustments. The cable length could be changed by adjusting the bearing nuts. This would be necessary for corrections after the detailed erection analysis for fabrication and installation error and after analysis of as-built and as-fabricated positions of the cable anchor points.

Bridge erection was constantly monitored and the future construction was predicted using a three-dimensional (3D) finite-element model of the whole bridge. The erection analysis model correlated to the actual structure through one-to-one correspondence of model and bridge control points. The model provided the target position for each segment installation, together with unstressed lengths for each cable. The model was also used for predictive analysis from any particular erection stage, using the as-built geometry, to predict the final bridge geometry upon completion. This was the prediction, survey, reanalysis, and possible adjustment cycle.

In terms of geometry control of the deck girder installation, the most significant control parameters were the deck unstressed geometry and the cable unstressed lengths. Prior to commencement of the deck segment installation, the cable lengths were updated through analyses based on crucial parameters, including the as-built geometry of the tower, the as-fabricated unstressed geometry of the deck girder, and updated loads' weight densities and stiffness parameters.

Main span closure. At 13:30 hours Beijing time on June 9, 2007, the main span central segment (closure segment "JH") of Sutong Bridge was lifted from a barge (Jin Hong No. 1) in the Yangtze River (**Fig. 23**). All the preparatory work had been meticulously planned and executed, in anticipation of the main span closure operation for the world's longest cable-stayed bridge.

Just over 48 hours later, at around 15:00 hours, June 11, 2007, the welding of the closure segment to the adjacent record-breaking-long deck cantilevers was completed, thereby forming the 8-km link across the Yangtze River between the cities of Suzhou and Nantong.

The Sutong Bridge main span closure method was based on integrating the merits of a traditional Chinese approach using "natural temperature closure" with a typical jacking-back of the two deck cantilevers. The motivation for the combined method was to draw on the strengths of each approach to ensure efficient closure control for the Sutong Bridge. The jacking-apart of deck cantilevers to provide sufficient air gap for fitting the closure segment was achieved by means of stressing and relaxing the

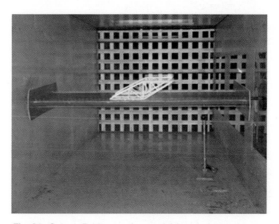

Fig. 24. Sutong Bridge section model for wind-tunnel testing.

Fig. 25. Sutong Bridge aeroelastic bridge model for wind-tunnel testing.

temporary longitudinal diagonal "tie-ropes" at the deck-tower junctions.

The preclosure preparatory work consisted of a number of carefully coordinated activities. Final adjustment and survey were done on the night of June 7, 2007. Survey data was rigorously analyzed to ensure smoothness of the local geometry, including rotational and translational fits as well as to determine the final length of the closure segment.

The closure segment, which was precisely cut according to instructions from the construction control team, was transported to the site in the early morning of June 9, 2007.

The preparation work for the lifting operation was from 10:00 a.m. to 1:00 p.m. on June 9, 2007. At 1:30 p.m., the closure segment was lifted by four deck gantries with four strand jacks. It reached a position immediately below the two cantilever decks at around 15:00 hrs. The precise insertion of segment JH into the closure gap commenced at 6:00 p.m.

Fig. 26. Night view of the completed Sutong Bridge. (*Courtesy China Jiangsu Province Construction Commanding Department*)

when the temperature started to drop below 30°C. Following insertion of Segment JH into the closure gap, fine adjustments to fit the closure segment were completed, involving the progressive launching back of the deck cantilevers toward the closure piece. Welding of perimeter plates commenced at midnight and was completed by 6:00 a.m. of the next day, together with the release of temporary fastenings. The remainder of the time, until completion during the afternoon of June 11, 2007, was taken up by welding of the longitudinal plate stiffeners. The historic closure of the main span of the world's longest cable-stayed bridge was thereby successfully completed (**Figs. 24** and **25**).

Conclusion. In pushing the frontiers of large bridge construction, we have also gained extensive experience in resolving the complex construction engineering involved. Construction engineering for an ultra-long-span cable-stayed bridge was one of the keys to the success of the Stonecutters Bridge and Sutong Bridge projects (**Fig. 26**). The record-breaking-long span and the flexible nature of the bridges demanded robust, versatile, and simple-to-implement methods for geometry control and for ensuring the safety of the structure throughout the erection stages.

[The author wishes to express his deepest gratitude and recognition to Maeda-Hitachi-Yokogawa-Hsing Chong JV, the CCCC Second Navigational Engineering Bureau, the Highways Department of Hong Kong SAR, China Jiangsu Province Construction Commanding Department, the project teams of AECOM, and other parties involved, for their contributions to the achievements described in this article.]

For background information *see* AERODYNAMICS; BRIDGE; CANTILEVER; COFFERDAM; CONSTRUCTION ENGINEERING; PILE FOUNDATION; TEMPORARY STRUCTURES IN CONSTRUCTION; STRUCTURAL ANALYSIS; STRUCTURAL DEFLECTIONS; STRUCTURAL DESIGN; STRUCTURAL MECHANICS; STRUCTURAL STEEL; STRUCTURE (ENGINEERING); SURVEYING; TRUSS; WELDING AND CUTTING OF MATERIALS; WIND TUNNEL in the McGraw-Hill Encyclopedia of Science & Technology.

S. H. Robin Sham

Fungal β-glucans

Humans have been using fungi as medicines for 5000 years or more. β-Glucans (beta-glucans), the only glucans found in fungi, are quite complex and unique in their interaction with the human immune response. Many details remain to be discovered, but some exciting results have been ascertained, with reliable products available in some countries.

Fungi. A fungus (plural fungi) is a member of a large group of eukaryotic organisms that includes microorganisms such as yeasts and molds, as well as the more familiar mushrooms. These organisms are classified as a kingdom, Fungi, separate from plants, animals, and bacteria. Fungal cells have walls that contain chitin, unlike the cell walls of plants (which contain cellulose). Genetic studies have shown that fungi are more closely related to animals than to plants. Fungal biodiversity has been estimated at approximately 1.5 million species, with about 5% of these having been formally classified. Importantly, research has identified compounds produced by fungi that can have an impact on viruses and cancer cells.

β-Glucans. β-Glucans are polysaccharides of D-glucose monomers linked by β-glycosidic bonds (see **illustration**). These heterogeneous polysaccharides of glucose polymers consist of a backbone of β-(1-3)-linked β-D-glucopyranosyl units with β-(1-6)-linked side chains of varying distribution and length. The activities of β-glucans depend on their molecular structure, size, branching frequency, structural modification, conformation, and solubility; the most active forms contain β-(1-3),(1-6) linkages. β-Glucans from many mushrooms have a β-(1-3) backbone with shorter β-(1-6)-linked branches, whereas β-glucan from *Alcaligenes faecalis* contains only β-(1-3)-glucosidic linkages. Other β-glucans have a β-(1-3)-linked backbone with various β-(1-6) glucose substitutions in the backbone residues. Biologically active β-glucans usually have large molecular weights. However, it is unclear whether β-glucans with intermediate or small molecular weights have biological activities, although some of them are active in vivo. The optimal branching frequency is suggested as 0.2 (1 in 5 backbone residues) to 0.33 (1 in 3 backbone residues). For example, although unbranched β-glucan curdlan showed proper biological activity, chemical addition of β-(1-6) glucose residues to the curdlan backbone led to an increase in antitumor activity (because highly branched β-glucans have higher affinity for cognate receptors). Furthermore, soluble β-glucans appear to be stronger immunostimulators than insoluble ones. When insoluble scleroglucan (from *Sclerotium glucanicum*) is modified by sulfation or carboxymethylation, its antitumor activity increases. In addition, orally administered β-glucans may be modified to smaller oligosaccharides in vivo. Thus, the actual β-glucans binding to the immune cell surface receptors in vivo may in fact be these smaller ones.

Fungal β-glucans as immunomodulators. The mycological industry has claimed numerous medicinal benefits of mushrooms, mostly as a result of β-glucans (see **table**). The U.S. Food and Drug Administration has granted GRAS (generally recognized as safe) status to β-glucans.

Together with chitin, the β-glucans are components of fungal cell walls. A high level of biological efficiency has been found in β-1,3-D-glucans and β-1,6-D-glucans isolated from some mushrooms. These polysaccharides increase the number of Th1 lymphocytes, which help protect organisms against allergic reactions. A number of β-glucans [for example, pleuran from oyster mushrooms (*Pleurotus* species) or lentinan from the fruiting bodies of shiitake mushrooms (*Lentinula edodes*)] have shown marked anticarcinogenic activity. β-Glucans also participate in physiological processes related to the metabolism of

Structure of a β-glucan.

fats in the human body, decreasing the total cholesterol content in blood and contributing to reductions in body weight. β-Glucans, generally called biological response modifiers, are now recognized as antitumor and anti-infective drugs; the most popular β-glucan is lentinan. β-Glucans have been shown to protect against infection by bacteria, viruses, and pathogenic microorganisms; they prevent cancer promotion and progression; they have synergistic antitumor effects with monoclonal antibodies and cancer chemotherapeutics; and they promote antibody-dependent cellular cytotoxicity through a biological pathway involved in carcinogenesis. However, β-glucans do not directly affect cancer cells or infectious microorganisms; instead, they exhibit their biological activities through activation of the host's immune system. Macrophages and dendritic cells are the main target cells of β-glucans, although neutrophils, B cells, T cells, and natural killer (NK) cells are also activated. They enhance cytotoxic activity and inflammatory cytokines of primary macrophages and macrophage cell lines, as well as the phenotypic and functional maturation of dendritic cells with significant interleukin-12 (IL-12) production. They also enhance the virus-specific T-cell functions induced by DNA vaccine, acting as a vaccine adjuvant.

Macrophages and dendritic cells have typical cell surface receptors called pattern recognition receptors (PRRs) that detect innately nonself molecules, including pathogen-associated molecular patterns (PAMPs). β-Glucans act as PAMPs and are recognized by PRRs because β-glucans cannot penetrate cell membranes directly as a result of their large molecular sizes. The major PRRs for β-glucans are dectin-1 and the toll-like receptor (TLR). Upon binding with β-glucan, dectin-1 and TLR induce a signaling cascade and activate immune cells. Other receptors, such as complement receptor 3 (CR3), scavenge receptors (SR), and lactosylceramide (LacCer), are also involved. In addition, β-glucans have been shown to support hematopoiesis (the process by which the cellular elements of the blood are formed) suppressed by ionizing radiation or

Cross-index of mushrooms and targeted avenues of research*

	Anti-oxidant support*	Blood pressure support*	Blood sugar support*	Cardiovascular support*	Cholesterol support*	Immune support*	Kidney support*	Liver support*	Respiratory support*	Nerve support*	Sexual health support*	Stress support*
Agaricus brasiliensis (Royal Sun Blazei)				●		●	●					
Cordyceps sinensis (Cordyceps)	●	●	●	●	●	●	●	●	●	●	●	●
Flammulina velutipes (Enokitake)						●						
Fomes fomentarius (Amadou/Ice Man Polypore)						●						
Fomitopsis officinalis (Agarikon)						●						
Ganoderma applanatum (Artist Conk)						●				●		
Ganoderma lucidum (Reishi/Ling Chi)	●	●	●	●	●	●		●	●	●		●
Ganoderma oregonense (Oregon Polypore)					●	●				●	●	
Grifola frondosa (Maitake/Hen of the Woods)		●	●			●					●	●
Hericium erinaceus (Yamabushitake/Lion's Mane)						●					●	
Inonotus obliquus (Chaga)	●		●			●			●			
Lentinula edodes (Shiitake)		●	●		●	●	●	●	●		●	●
Phellinus linteus (Mesima)						●						
Piptoporus betulinus (Birch Polypore)						●						
Polyporus sulphureus (Chicken of the Woods)						●						
Polyporus umbellatus (Zhu Ling)						●			●	●		
Schizophyllum commune (Suehirotake/Split-Gill)						●						
Trametes versicolor (Yunzhi/Turkey Tail)	●					●	●	●				

*These statements have not been evaluated by the U.S. Food and Drug Administration. These products are not intended to diagnose, treat, cure, or prevent any disease.

Source: Courtesy of Paul Stamets/Fungi Perfecti®

cytotoxic anticancer therapy. They also enhance stem cell homing and engraftment.

Specific examples. There have been a number of recent investigations that have studied the effects of diverse medicinal fungal β-glucans on humans. In 2011, a Norwegian group evaluated the effect and safety of a β-glucan from *L. edodes* mycelium, Lentinex®, in healthy, elderly Caucasian subjects in a controlled trial. When given orally, this β-glucan was safe and induced an increase in the number of circulating B cells; in contrast, placebo use had no such effect.

Fractions of *Grifola frondosa* containing β-glucans have been observed to work on several levels against HIV: by direct inhibition of HIV, by stimulation of the natural defense system against HIV, and by reducing vulnerability to opportunistic diseases. Other β-glucan-containing fractions of *G. frondosa* together with dimethyl sulfoxide have shown success in treating AIDS-associated Kaposi sarcoma, a malignant skin tumor.

The active polysaccharides of *Ganoderma lucidum* stimulate cytokine expression, TLR4, and various immune cells (including macrophages, B cells, dendritic cells, and stem cells); in addition, NK-cell-mediated cytotoxicity is enhanced, effectively killing tumor cells. The β-glucans, ganodelan A and B, help release insulin by facilitating the influx of calcium into the pancreas beta cells, lowering elevated blood sugar. Extracts of *G. lucidum* have shown beneficial results with regard to the quality of life in patients having active hepatitis B.

Agaricus blazei (also termed *A. subrufescens*) contains the highest level of β-glucans of any mushroom. They are (1-6),(1-3)-β-D-glucans, (1-6),(1-4)-β-D-glucans, polysaccharide-protein complexes, RNA-protein complexes, and glucomannan. The β-glucans in *A. blazei*, as a result of their low molecular weights, can be absorbed into the body more easily, making them more effective.

A North Korean clinical study investigated 50 patients with diverse malignancies (liver, stomach, lung, colon, larynx, breast, and cervical cancers, as well as lymphoma) treated with *Phellinus linteus*. After 6–12 months, "all patients reported an increase in appetite, general well-being, and reduction (or even control) of pain; the patients with cancer of the stomach gained weight."

Trametes versicolor, which is known as yunzhi, is the mushroom from which the Japanese anticancer drug Krestin® is derived. Its active ingredient is polysaccharide Kureha (PSK), which is a (1-3)-β-glucan bound to a protein. PSK increases white blood cell counts, interferon-alpha, IL-2 production, and delayed-type hypersensitivity reactions. PSK is approximately 62% polysaccharide and 38% protein. The glucan portion of PSK consists of a β-(1-4) main chain and a β-(1-3) side chain, with β-(1-6) side chains. The polypeptide portion is rich in aspartic acid, glutamic acid, and other amino acids and is orally bioavailable. After intratumoral administration, PSK causes local inflammatory responses that result in the nonspecific killing of these abnormal cells. Consequently, local administration of PSK is more efficient than systemic use. PSK exerts tumoricidal activity by inducing T cells that recognize PSK as an antigen and kill tumor cells in an antigen-specific manner. In a number of controlled studies of advanced cancer patients (specifically, patients with lung, breast, stomach, and esophagus cancer), the combination of PSK and typical treatment (surgery, radiotherapy, and chemotherapy) resulted in highly significant extensions of survival (sometimes more than 20 years) when compared to the control treated groups.

Another similar β-glucan is polysaccharide-peptide (PSP), which was first isolated from a cultured deep-layer mycelium of *T. versicolor* in the 1980s. It contains at least four discrete molecules, all of which are true proteoglycans. PSP differs from PSK in its saccharide makeup, lacking fucose and containing arabinose and rhamnose. PSP can be easily delivered by oral route. It has been prescribed successfully for cancer patients to help improve their quality of life, and it has shown positive effects on immune systems before and after surgical treatment, chemotherapy, and radiotherapy.

Compounds in *Hericium erinaceus* (hericenons C–H) are another group of interest. They have been found to encourage the production of nerve growth factor in the brain.

Poria cocos sclerotia are composed mainly of polysaccharides and some triterpenoids. Two of the polysaccharides are poriatin and β-pachyman. Poriatin increases the antitumor effects of some chemotherapeutic agents. Pachyman can be chemically converted to pachymaran, which shows a high degree of antitumor activity.

The medicinal effects (including a significant dose-dependent hypoglycemic effect) of *Tremella fuciformis* are attributed mostly to acidic glucuronoxylomannans from the fruiting bodies. Three heteroglycans (T1a, T1b, and T1c) have been isolated that induce human white blood cells to produce IL-1, IL-6, and tumor necrosis factor, which have significant immunostimulating potentials. They also improve immunodeficiency, including that induced by AIDS, stress, or aging. Most of these effects come from long-term consumption of the fungus.

Lastly, a group of Slovakian pulmonologists investigated the effect of pleuran, an insoluble β-(1-3/1-6) glucan from oyster mushrooms (*Pleurotus ostreatus*), on selected cellular immune responses and incidence of upper respiratory tract infection (URTI) symptoms in athletes. Fifty athletes were randomized to pleuran (Imunoglukan®) or placebo supplementation for three months. Pleuran significantly reduced the incidence of URTI symptoms and increased the number of circulating NK cells.

Future directions. Before most patients can benefit from a series of fungal β-glucans, further explorations, identifications, and demonstrations are necessary. Many research groups, predominantly in Asia and Europe, are actively exploring the diverse beneficial properties of these molecules and complexes, based on traditional usage and animal studies. However, financial support is hard to acquire, the majority of products are difficult to patent, and the market is often flooded with dubious "miracle cures" that so far benefit unscrupulous manufacturers.

For background information *see* CANCER (MEDICINE); CELLULAR IMMUNOLOGY; CHEMOTHERAPY AND OTHER ANTINEOPLASTIC DRUGS; FUNGAL BIOTECHNOLOGY; FUNGI; GLUCOSE; IMMUNOLOGY; IMMUNOTHERAPY; MUSHROOM; MYCOLOGY; POLYSACCHARIDE in the McGraw-Hill Encyclopedia of Science & Technology.   Georges M. Halpern

Bibliography. B. Chan and G. M. Halpern, *The Yin and Yang of Cancer*, Square One Publishers, Garden City Park, NY, 2007; P. C. K. Cheung (ed.), *Mushrooms as Functional Foods*, Wiley, Hoboken, NJ, 2008; G. M. Halpern, *Healing Mushrooms*, Square One Publishers, Garden City Park, NY, 2007; H. S. Kim et al., Stimulatory effect of β-glucans on immune cells, *Immune Network*, 11:191–195, 2011, DOI:10.4110/in.2011.11.4.191; S. P. Wasser, Current findings, future trends, and unsolved problems in studies of medicinal mushrooms, *Appl. Microbiol. Biotechnol.*, 89:1323–1332, 2011, DOI:10.1007/s00253-010-3067-4.

Fungal zoospores in aquatic ecosystems

Fungi represent one of the last frontiers of the undiscovered biodiversity and the related functions that challenge aquatic microbial ecology today. The number of fungi present on Earth has been estimated to be approximately 1.5 million species, of which about 97,000 have so far been identified, corresponding mostly to species thriving in moist soils, lotic (actively moving water) systems, mangroves, and wetlands, or to economically interesting pathogens of humans, animals, and plants. Recent environmental ribosomal deoxyribonucleic acid (rDNA) surveys of microbial eukaryotes have unveiled a large reservoir of unexpected fungal diversity in pelagic (open water) systems, emphasizing their ecological potentials for ecosystem functioning, and have opened new perspectives in the context of food-web dynamics. Typical pelagic fungi are members of the phylum Chytridiomycota, which occupies the basal branch of the kingdom Fungi. These true fungi usually produce small zoospores that typically have a single, posteriorly directed flagellum. The so-called zoosporic fungi, or chytrids, are universally present in the world's pelagic ecosystems, where they play major roles, primarily as parasites and saprotrophs (utilizers of dead organic matter). These roles, however, remain mostly cryptic in classical microscopy studies because chytrids are small in size and lack conspicuous morphological features, a situation that makes them hardly distinguishable from many flagellated protists (for example, sessile choanoflagellates or bicosoecids). Previously, the modes of nutrition for all heterotrophic flagellates (which ingest organic compounds for their nutrition) in the plankton were thought to be restricted to bacterivory, but zoosporic fungi are not bacterivores (that is, bacterial feeders). It is now clearly evident that (1) not all heterotrophic flagellates thriving in pelagic systems are either protists or bacterivores, and (2) parasitism and saprophytism from fungal flagellates might represent important potential functions in these ecosystems.

Ecological conceptualization of the chytrid life cycle.
Chytrid species have an interesting life cycle in the context of the pelagic realm, with the two main stages (that is, sporangium and zoospore) having different effects on food-web dynamics. In the typical life cycle, a free-living zoospore encysts (becomes enclosed in a cyst) in the host and expands intracellularly as a tubular rhizoid, that is, the nutrient-conveying system for the formation of fruit bodies (the infective sporangium) from which propagules (motile zoospores) are released into the environment. The hosts of parasitic chytrids in aquatic systems are highly diverse, including prokaryotic organisms (for example, cyanobacteria) and eukaryotic phytoplankton, protists, invertebrates (for example, larvae of insects, rotifers, nematodes, and crustaceans such as copepods, ostracods, and cladocerans), flowering plants, and fungi. On the other hand, chytridiomycosis epidemics are known to produce massive amounts of zoospores, which are now known to be valuable food sources for zooplankton. The two main development stages of chytrids thus highlight two overlooked ecological potentials in food-web dynamics: (1) parasitic predation of host populations, most of which are inedible (that is, unexploited by grazers), and (2) the subsequent trophic (feeding) link via the release of suitable zoospore food for zooplankton (**Fig. 1**). In addition, it has been shown recently that chytrid parasitism of cells within the filaments of cyanobacteria during bloom events can result in a mechanical fragmentation of the inedible filaments into shorter-size edible filaments, thereby enhancing the contribution of fungal parasites to the bloom's decline (**Fig. 2**).

Fungal zoospores are valuable food sources in natural waters. The cytoplasm of chytrids contains storage carbohydrates (including glycogen), storage proteins, a wide range of fatty acids, phospholipids, sterols, and other lipids. When chytrids reproduce, most of the cytoplasm is converted into zoospores, which swim away to colonize new substrates or infect new hosts. Lipids are considered to be high-energy compounds, and some are important for energy storage. Indeed, lipids are present mainly in the form of endogenous reserves, often as membrane-bound vesicles called lipid globules, which can be observed easily in the cytoplasm of fungal zoospores by light and electron microscopes. The sizes and numbers of lipid globules within the zoospores vary, and their ultrastructure is complex. The chemical composition of the lipids, including fatty acids and sterols, has been characterized for a number of genera of zoosporic fungi. These endogenous reserves are consumed during the motile phase of the zoospores. They presumably provide energy for the movement of flagella during the motile phase (which can last for up to several hours), as well as for the attachment and germination of zoospores on the appropriate substrates or hosts (Fig. 2). In addition, many zoosporic fungi can grow in the laboratory on minimal synthetic media containing one carbon source, such as cellulose, xylan, starch, or chitin, along with salts containing nitrate, sulfate, and phosphate. This establishes fungi as potential competitors of bacteria and primary producers for essential minerals (Fig. 2).

There are other significant functions for the high-energy compounds found in fungal zoospores, especially as food resources for zooplankton and probably for many other consumers in aquatic ecosystems. Fungal spores and hyphae in general are known to be eaten by a large number of different consumers in both aquatic and soil ecosystems, including a variety of mycophagous (fungus-eating) protozoa (such as amebae and flagellates), detritivores (consumers of dead organic matter), grazers (such as filter-feeding zooplankton), and benthic (bottom-dwelling) suspension feeders. Because most of these consumers can only discriminate between food resources by size, it is expected that zoospores as well as hyphae and nonmotile spores would be eaten by many of these consumers; however, published records are lacking. Fungal zoospores are well within the range

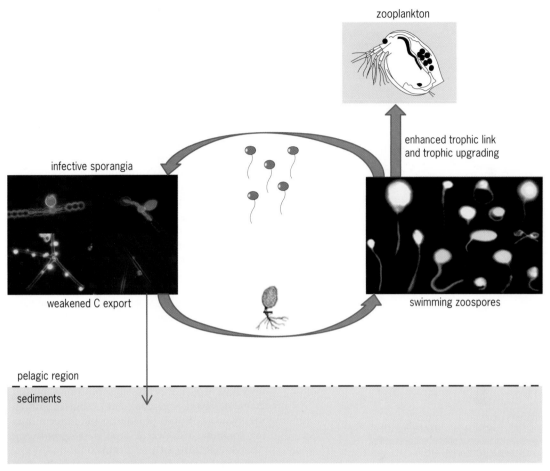

zooplankton

enhanced trophic link
and trophic upgrading

infective sporangia

weakened C export

swimming zoospores

pelagic region

sediments

Fig. 1. Ecological conceptualization of the chytrid parasite life cycle. Swimming zoospores are valuable food sources and exhibit endogenous reserves of essential lipids for the growth of zooplankton. When preferentially hitting large-sized and inedible hosts (for example, cyanobacteria and diatoms), zoospores develop into infective sporangia that produce other propagules. Because parasites kill their hosts, the blooms of large-sized inedible phytoplankton may not totally represent trophic bottlenecks toward the sediments.

of a good particle size (2–3 μm in diameter) for zooplankton feeding behavior and, consequently, when fed upon, matter is transferred to higher trophic levels in the food chain. For example, zoospores [prior to growing into a mature thallus (body)] are efficiently grazed by crustacean zooplankton such as *Daphnia* species. Thus, zoospores may provide organic compounds containing nitrogen, phosphorus, and sulfur; mineral ions; and vitamins to grazing zooplankton (Fig. 2).

Most interestingly, zoospores are a particularly good food source because of their nutritional qualities. Presumably, many consumers must obtain at least some essential nutrients from their food sources because these compounds cannot be produced de novo. One example is found in the cladoceran *Daphnia*. Recent research has shown that zoospores of the parasitic chytrid, *Zygorhizidium*, are quite rich in polyunsaturated fatty acids (PUFAs) and cholesterol, which are essential nutrients for the growth of *Daphnia*. These zoospores are found to facilitate the trophic transfer from the inedible large diatom hosts (*Asterionella* species) to species of *Daphnia*. In addition, PUFAs and cholesterol are known to promote growth and reproduction in other crustaceans

as well. This phenomenon, known as the trophic upgrading concept, is of significant importance in aquatic food webs because it highlights not only the quantity but also the quality of the matter being transferred via fungal zoospores (Fig. 2).

Impact on carbon flow and on food web properties and topology. Given that food webs are central to ecological concepts, it is important to establish the role of parasites in the structure and function of food webs. The application of mathematical tools, such as models, is useful and allows trophic network representations through carbon flows. Recently, the first such model for chytrids parasitizing phytoplankton, based on freshwater lake data, has been analyzed. Model results support recent theories on the probable effect of parasites on food-web function. In Lake Pavin (Massif Central, France), during spring, when inedible diatoms are the dominant primary producers, the epidemic growth of chytrids significantly reduces the sedimentation loss of algal carbon to the detritus pool through the production of grazer-exploitable zoospores. This contributes to longer carbon path lengths, higher levels of activity and specialization, and lower recycling indices. Thus, these food-web properties highlight a significant role for chytrids

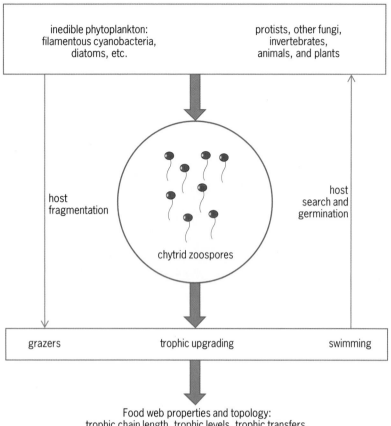

Fig. 2. The hosts for chytrid parasitic production of zoospores in the water column are numerous and diverse within animals and plants that colonize the water column. Zoospores can swim and hit uninfected hosts in the water column, or can be grazed and upgrade the diet and growth of zooplankton. These activities establish zoosporic fungi as an important ecological driving force in the food-web dynamics of aquatic ecosystems, where various properties and the overall topology of the trophic network can be affected by parasites in many complex ways.

in the stability of aquatic ecosystems (Fig. 2). This agrees well with the theory that parasites should lead to a stabilized trophic network because they increase the connectance (that is, the number of observed links divided by the number of possible links) as well as the related numbers of trophic links, species richness, trophic levels, and trophic chain lengths of the food web. However, the potential effects of parasites on food-web stability are a complex issue with regard to community ecology.

Conclusions. The prevalence of zoosporic parasites and saprobes (saprotrophs) in aquatic ecosystems represents a challenge to our view of the trophic modes of heterotrophic flagellates. At present, this view is restricted to classical protistan modes along the evolutionary continuum between autotrophy (the capacity to synthesize food from inorganic substances) and phagotrophy (the process by which unicellular organisms derive their food by engulfing and digesting other cells). Zoosporic parasites and saprobes, in addition to being able to resist adverse conditions and use different sources of carbon and nutrients, can affect plankton food-web

functions and ecosystem properties and topology, including stability and trophic transfer efficiency. For example, by providing upgraded food sources for zooplankton, fungal parasitism can change the balance between autotrophy and heterotrophy in pelagic systems. Perhaps, a new paradigm-shift point in the development of aquatic microbial ecology is being approached. Furthermore, parasitic lifestyle is generally highly subtle and, for example, it can control competition by dominant species for resources, thereby promoting species coexistence and diversity. Parasites can also form long-lived associations with hosts, reducing their fitness for survival, or allowing infected hosts to remain strong competitors, although few models exist for microbial fungus–host systems.

For background information *see* ECOLOGICAL COMMUNITIES; ECOSYSTEM; FOOD WEB; FRESHWATER ECOSYSTEM; FUNGAL ECOLOGY; FUNGI; LIPID; MYCOLOGY; PARASITOLOGY; PHYTOPLANKTON; PROTOZOA; ZOOPLANKTON in the McGraw-Hill Encyclopedia of Science & Technology.   Télesphore Sime-Ngando

Bibliography. C. Gachon et al., Algal diseases: Spotlight on a black box, *Trends Plant Sci.*, 15:633–640, 2010, DOI:10.1016/j.tplants.2010.08.005; F. H. Gleason et al., Fungal zoospores are valuable food resources in aquatic ecosystems, *Inoculum*, 60:1–3, 2009; M. Jobard, S. Rasconi, and T. Sime-Ngando, Diversity and functions of microscopic fungi: A missing component in pelagic food webs, *Aquat. Sci.*, 72:255–268, 2010, DOI:10.1007/s00027-010-0133-z; S. Rasconi, M. Jobard, and T. Sime-Ngando, Parasitic fungi of phytoplankton: Ecological roles and implications for microbial food webs, *Aquat. Microb. Ecol.*, 62:123–137, 2011, DOI:10.3354/ame01448; T. Sime-Ngando and N. Niquil (eds.), *Disregarded Microbial Diversity and Ecological Potentials in Aquatic Systems*, Springer, New York, 2011.

Fuzzy databases

Traditional relational databases (DBs) are not designed, implemented, and populated to store and retrieve information using data that can be considered uncertain or imprecise. Imprecise data frequently occurs when we refer to extant documents where the best we can do is to approximately date them. For example, the Dead Sea scrolls are generally dated between the years 150 BC and AD70. In this case, it is clear that some uncertainty exists when we consider the exact date when these documents were written. In other cases, the degree of uncertainty may not be clear. Consider, for example, a group of people whose height is being measured. By current universal standard, we can safely assume that a person who is 6 ft (183 cm) tall can be considered a tall individual. However, if somebody is only 5 ft 11 in. (180 cm), can we say that this person is not tall? Here the distinction between being tall and not being tall is not that clear. To address issues concerning data that is uncertain or imprecise, it is necessary to consider a new type of databases called fuzzy databases (FDBs). Fuzzy databases are called "fuzzy" because their theoretical formalization is based on fuzzy logic (FL) and fuzzy set theory (FST). The former deal primarily with reasoning that is approximate rather than fixed. That is, in fuzzy logic, there is a gradation of values between something being absolutely false (generally indicated with a value of 0) and something being absolutely true (generally indicated with a value of 1). For this reason, fuzzy logic is said to be a many-valued logic as opposed to traditional logic for which there are only two possible values: true or false. Fuzzy logic, in turn, is based on the concept of fuzzy set theory in which the notion of membership may range between the values 0 (not belonging to the set) and 1 (belonging to the set). For example, considering the previous examples of individuals and their heights we can say that a person who is 6 ft 4 in. (193 cm) belongs to the set of tall people. This person has a membership value of 1 because he or she clearly belongs to the set of tall people. A person who is 5 ft 11 in. (180 cm) may have a 0.7 membership value in the set of tall people. Therefore, whenever we consider the concept of fuzzy database (FDBs) it is necessary to view it as including and being based on FL and FST. These two concepts are present at different levels during the operation of a fuzzy database, particularly at the representation and storage of imprecise or vague data and during the access of the data stored in an FDB through queries (questions posed to the database) that may contain imprecise data. FL and FST also are present during the entire life cycle of the FDB. Since their inception, the concepts of FL and FST have been widely applied in science and technology fields. In computer science, the concept has been gaining relevance because of the growing and more demanding needs for applications that require that data be manipulated in a more human-like rather than machine-like manner. Applications of this type are common in medical and business environments.

Motivation. Consider the situation where we would like to find some information about a particular person. We may not know the exact value of the person's age, but we may know that she is young, or that she is between 20 and 25 years old, or that she is older than some other person whose age we really do know. Within the database itself, the range of possible values considering the age attribute may vary from total ignorance—typically represented in an FBD with a NULL value—to a particular and precise value that can establish that this person is 22 years old. There are also some other situations where we would like to make more flexible queries to the database. For example, a person may query a database to look for a hotel that is not expensive and is located close to a particular conference site. In a classical relational database, the user would be forced to clarify the terms "not expensive" and "close" probably by providing a range of values that seem acceptable. For example, does not expensive mean a hotel where rooms are let for under $50 per

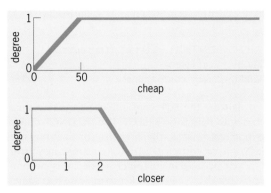

Membership function for cheap and closer fuzzy sets.

night? Likewise, does close mean that the hotel is less than 2 mi (3.2 km) from the conference site? Let us further assume that there is a hotel that costs $51 per night and is located at 2.1 mi (3.4 km) from the conference site. If this were indeed the case, this hotel would be excluded from the results produced by the database. To make this point more relevant, we may assume that this is the only hotel within a 3-mi (5-km) radius that is "close" to satisfying the user request and yet the user will never know about it. A fuzzy representation of this situation can be a trapezoidal representation of expensive and near fuzzy sets (see **illustration**). The problem of rigidity or lack of flexibility in classical relational databases introduces two problems: one is the exclusion of responses on the edge (limit) of the search criteria (not completed responses), as in the aforementioned case, and the other is, the nondiscrimination of answers. For example, a particular query may retrieve so many rows that it may not be easy to discriminate the best possible answer.

Applications. As indicated before, traditional information systems and, in particular, traditional relational databases are designed to answer queries that are written without any degree of ambiguity. Trying to pose queries to a relational database where the query may contain imprecise natural language may be impossible, if not cumbersome, at best. To accommodate these types of queries, it is necessary to use an FDB management system. FBD applications are very diverse and may be used in business and management areas such as merit-based employee promotions, organizational performance assessments, shopping preferences, travel planning and accommodation for events and tourism, decision-aid systems, and medical diagnostic systems, just to mention a few. Systems using FDBs are said to be close to human reasoning because they allow the use of imprecise terms such as young, high, very close, and so on. At the implementation level, such systems include the representation and querying of imprecise or fuzzy data, facilitating the storage and recovery of these data. Fuzzy systems query the FDB using an extension of the traditional structured query language (SQL) called SQLf (or SQL fuzzy). Queries using SQLf commands can be issued directly to the system or embedded in programs that access the database. A basic query block in SQLf may look like example (1).

```
SELECT[DISTINCT] <list-of-attributes>
FROM <relations>
WHERE <fuzzy_conditions>                    (1)
```

Following the SQL convention, all queries must start with the clause SELECT. The DISTINCT clause is optional, that is, the user may or may not use it. We have written DISTINCT enclosed in square brackets because in computer science, it is customary to use this type of bracket to indicate that something is optional. However, when writing the actual query, the square brackets are left out. The DISTINCT clause, when used, eliminates duplicated rows from the answer generated by the query. Following either the clause SELECT or DISTINCT, the user names the attributes of interest, which is indicated in the SQLf block by <list-of-attributes>. When writing the names of the attributes, these attributes must be separated by commas. If there are two or more attributes with the same name in any of the relations (tables), it is necessary to qualify the attribute by preceding it with the name of the table to which it belongs. In this case, the table name and the attributed are separated by a period. The WHERE clause is used to define the conditions that need to be satisfied by the rows of the table or tables that intervene in the query. The process of obtaining the rows that satisfy a particular query may be viewed, for explanation purposes, as a three-step process. First, the database system forms the Cartesian product of the relations specified in the FROM clause that satisfy the fuzzy conditions established in the WHERE clause. Second, after the Cartesian product has been formed and within the set of rows that satisfy the conditions stated in the WHERE clause, the database only displays the attributes listed after the SELECT clause. As an illustration of some queries in SQLf, let us assume that we are working with the tables (2) shown below. As is customary for databases when referring

```
Participant (last_name, first_name, age)
Accommodations (hotel_name, guest)
Hotel (name, address)                        (2)
```

to a table, the name of the table is followed by its attributes separated by commas. A query to search for young participants in hotels very close to UCLA follows the tables (3).

```
SELECT Participant.last_name,
    Participant.first_name
FROM Participants, Hotels, Accommodation
WHERE Participant.age = young AND Hotel.
    address = very close AND
Accommodation.host =
    Participant.name AND
Accommodation.hotel = Hotels.name;          (3)
```

This query joins (brings together) information from three different tables as indicated by the table names following the FROM clause.

In the preceding query, we have qualified, on purpose (but not technically required), the name of attributes to make explicit the table to which they belong. Within the WHERE clause, we use the fuzzy terms "young" and "close." The term "very" is a fuzzy modifier.

The following query (4) illustrates how queries

```
SELECT Hotel.name
  FROM Hotels
  WHERE Hotel.address = close AND
    Atleast3
  SELECT * FROM Participants,
    Accommodation
  WHERE (Accommodation.guest =
    Participant.last_name) AND
  (Accommodation.hotel_name =
    Hotel.name)
    AND (Participant.age = young)           (4)
```

can be "nested," that is, how a query can contain another query. Here the query intends to search for hotels closer to UCLA with at least three young participants accommodated.

In query (4), we have also qualified the attributes by preceding them with the name of the table to which they belong. The asterisk in the nested query is used to retrieve all attributes from the tables indicated in its corresponding FROM clause. The asterisk is used for convenience because it allows us to state implicitly the name of all attributes. For sake of clarity, we have enclosed all conditions in parentheses. From the syntactical point of view of the query, it is not required to do so. Within the WHERE clause, we use the fuzzy terms "young" and "close" and the fuzzy quantifier "Atleast3."

As query (4) illustrates, the use of fuzzy logic opens a world of possibilities for retrieving information from an FDB by introducing the use of linguistic hedges such as very, slightly, and approximately, just to mention a few. These fuzzy terms allows users to issue queries such as list all hotels very near to the conference site at an approximate distance of 1 km (0.6 mi), or list all hotels whose price is slightly over $50. Another type of query includes the use of linguistic quantifiers such as most, some, and any. The use of these quantifiers brings some degree of flexibility to the classic existential and universal quantifiers; for example, list any hotel where the majority of guests are young or list all hotels that meet most of the following criteria: low price, closer to the conference, and good services.

Architecture. All FDBs currently developed or under consideration intend to resolve the lack of flexibility of classical relational databases when confronted with imprecise information. However, to introduce a new technology, it is necessary to take into account the best way to incorporate it and its associated costs. One way to introduce a new technology is to develop a new system from the ground up. Another approach is to consider existing systems and extend or modify them to accommodate a new set of features. It is evident that each alternative has its advantages and disadvantages, such as development time and the appropriate adaptation of these extensions. Developing a system from the ground up obviously offers the most flexibility because it can be laid out as envisioned. However, this approach is also prone to run into unforeseeable circumstances and exceed allotted budgets. For these reasons and generally to be able to work within budget allocations, it is preferable to make extensions to existing systems that have been tested and accepted. There are three basic forms of achieving these extensions. We will refer to them respectively as loose, medium, and tight coupling architectures. A loose coupling architecture occurs when the component in charge of handling data according to the extensions under consideration resides in a layer outside the core or engine of the database management system. The integration with the current system and users is achieved via interfaces. A medium coupling architecture implements specific extensions by augmenting the database functionality by using stored procedures or user-defined functions. A tight coupling architecture occurs when the database engine or core is modified to accommodate the new extensions. Currently, efforts are directed to extend both the SQL language and the database engine. The three architecture couplings have been implemented using existing relational databases. Experimental results show that implementation of tightly coupled architectures is very efficient in terms of performance, but its portability is very poor. The loose couple approach is very portable, but its performance is less efficient.

In general, fuzzy databases represent an important technological advance because they allow users to have greater flexibility in data representation and in expressing queries. Although most of the implementations are still in the academic realm, the results are very promising. The future beneficiaries of this new technology extend from businesses and governments to home users.

For background information *see* DATABASE MANAGEMENT SYSTEM; DATA STRUCTURE; DATA WAREHOUSE; FUZZY SETS AND SYSTEMS in the McGraw-Hill Encyclopedia of Science & Technology.

Ana Aguilera; Ramon A. Mata-Toledo

Bibliography. J. Galindo, A. Urrutia, and M. Piattini, *Fuzzy Database Modeling, Design and Implementation*, Idea Group Publishing, Hershey, PA, 2006; Z. Ma, *Fuzzy Database Modeling of Imprecise and Uncertain Engineering Information*, Springer, New York, 2006; L. Tineo, *Una contribución a la Interrogación Flexible de Bases de Datos: Evaluación de Consultas Cuantificadas Difusas*, Ph.D. Thesis, Simón Bolívar University, Caracas, Venezuela, 2005; L. Yan and Z. Ma (eds.), *Advanced Database Query Systems: Techniques, Applications and Technologies*, IGI Global, Hershey, PA, 2011; L. A. Zadeh, Fuzzy sets, *Inform. Control*, 8(3):338–353, 1965, DOI:10.1016/S0019-9958(65)90241-X.

Gamma-ray bursts

Gamma-ray bursts are short cosmic blasts of very-high-energy electromagnetic radiation (gamma radiation) that are recorded at an average of about once per day by detectors placed above the Earth's atmosphere. At distances of billions of light-years, the energy emitted in a gamma-ray burst (GRB) is more than a billion billion (10^{18}) times the energy emitted each second by the Sun. The mechanism for the origin of gamma-ray bursts remains a focus of current research. Gamma-ray bursts longer than 2 s appear to be associated with the supernova explosions of massive stars in distant galaxies. Recent evidence suggests that shorter gamma-ray bursts are likely the result of merging neutron stars or black holes, although other explanations are also possible. There are indications that both types of gamma-ray bursts lead to the birth of black holes.

History of observations. In 1963, the United States and other nations signed the Nuclear Test Ban Treaty, which forbade the testing of nuclear weapons in or above the atmosphere. To ensure compliance with this treaty, the United States launched the *Vela* series of Earth-orbiting satellites to look for violations that would produce blasts of gamma radiation. No violations of the Nuclear Test Ban Treaty were ever reported. However, in 1973, Ray W. Klebesadel, Ian B. Strong, and Roy A. Olson announced the discovery of the first gamma-ray bursts and provided evidence of their cosmic origin based on data from the *Vela* satellites.

During the next 20 years, an international array of satellites slowly built up a catalog of information about the brightness and energies of the gamma rays in scores of gamma-ray bursts. Some of these satellites were located in the solar system at great distances from Earth, whereas others were in near-Earth orbits. The combined data from this Interplanetary Network (IPN) allowed the positions of some of the bursts to be located with greater precision. However, until the notable event that occurred on March 5, 1979, the identification of the gamma-ray bursts with familiar objects that radiate at, for example, visible wavelengths of light, proved elusive.

March 1979 event. On March 5, 1979, at least nine satellites in the Interplanetary Network were swamped with a blast of gamma radiation, 100 times stronger than any burst detected previously. Because of the large number of detections converging from different angles, the Interplanetary Network was able to locate the burst precisely. Surprisingly, it emanated from a supernova remnant in our neighboring galaxy, the Large Magellanic Cloud, at a distance of about 150,000 light-years. A sinusoidal signal was seen in the gamma-ray light as the intensity decayed, followed hours later by another, weaker burst, with many more bursts occurring over the next month. Studies of gamma-ray bursts for the next decade assumed that the March 5, 1979, event was a prototype for all gamma-ray bursts, and many scientists searched for sinusoidal oscillations and repeated

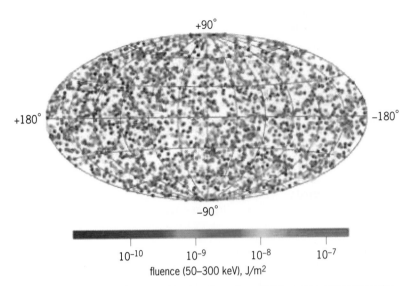

Fig. 1. Equal-area projection map of the locations of over 2700 gamma-ray bursts on the sky as measured by the BATSE detector on the *Compton Gamma-Ray Observatory*. Different shades indicate the total energy emitted (fluence) for each burst. The plane of the Milky Way Galaxy spans the Equator. (*Courtesy of NASA/CGRO/BATSE Team*)

weaker bursts at the positions of seemingly similar events. However, it is now known that the March 5, 1979, event was not a "classic gamma-ray burst" but the defining member of a new class of cosmic explosions—the soft gamma repeaters (SGRs). These seemingly similar detonations result from starquakes on the surfaces of rotating neutron stars that possess extremely strong magnetic fields, with strengths up to 10^{11} tesla. SGRs are now also known as "magnetars" due to the extreme fields present in these isolated neutron stars.

CGRO observations. The *Compton Gamma-Ray Observatory* (*CGRO*), launched into Earth orbit by the National Aeronautics and Space Administration (NASA) on April 5, 1991, had a complement of four gamma-ray–detecting instruments, including one designed specifically to study gamma-ray bursts: the Burst And Transient Source Experiment (BATSE). During its 9-year lifetime, BATSE detected and located over 2700 gamma-ray bursts; **Fig. 1** shows their positions and fluxes on a map of the sky. There is no obvious clustering of the bursts near the plane of the Milky Way Galaxy (horizontally across the center of the figure) or anywhere else. This result surprised almost all astronomers, as it had been widely believed that the gamma-ray bursts must arise from within the Milky Way Galaxy.

Two possible interpretations were advanced to explain the lack of any pattern in the gamma-ray burst skymap: Either the bursts were very close to the solar system, or they were very far away, at cosmic distances. In the latter case, however, the energies required to power the bursts were so large that scientists had trouble explaining the cause of the bursts. The evidence that gamma-ray bursts were truly located across the universe did not appear until 1997.

Discovery of afterglows and distances. On February 28, 1997, a gamma-ray burst was observed by detectors on board *BeppoSAX*, an Italian–Dutch satellite. Eight

hours later, after maneuvering the satellite to use a different set of detectors, the *BeppoSAX* observing team discovered an x-ray source that was located at the position of the initial blast of gamma rays. A second observation 3 days later showed that the source was fading—they had discovered the first x-ray "afterglow" of a gamma-ray burst, radiation emitted by the cooling embers of the much hotter initial explosion. The *BeppoSAX* team then alerted observers at ground-based telescopes, leading to the discovery of the first visible-light afterglow. On May 8, 1997, the *BeppoSAX* team located another x-ray afterglow, and quick work resulted in the first determination of a distance to a gamma-ray burst—about 7×10^9 light-years—providing the first real evidence that gamma-ray bursts are located at cosmic distances.

The distances to the gamma-ray bursts cannot be obtained solely by using gamma-ray data, but result primarily from the measurement of red-shifted spectral lines in the visible-light spectrum of the afterglow. (Spectral lines are red-shifted due to the expansion of the universe. As the universe expands, light waves traveling great distances are stretched by the expansion. Longer-wavelength light is "redder," hence the term redshift. With a few additional assumptions about the expansion history of the universe, a measured redshift can then be converted into a distance.) As of 2012, distances had been measured to over 170 gamma-ray bursts; the farthest, detected on April 29, 2009, by NASA's *Swift* satellite, is at a distance of about 13.02×10^9 light-years (redshift of 8.1). The average distance to the long bursts is 10.6 billion light-years (or a redshift of 2.2), whereas the short bursts are considerably closer (average redshift about 0.5, or a light travel time of 5 billion years). For comparison, the most distant galaxy seen so far is located at about 13.07×10^9 light-years (at a slightly higher redshift of 8.6). Evidence has accumulated that bursts with afterglows are located in regions of distant galaxies that are actively forming stars.

Observations with burst alert systems. As BATSE accumulated observations of gamma-ray bursts, ground-based observers became eager to help unravel the mystery of the bursts' origins. To facilitate these efforts, in 1993 Scott Barthelmy created a system for distributing information about the bursts, including crude positions, to interested observers. Originally called BACODINE (BAtse COordinates DIstribution NEtwork), the system used telephones, pagers, and eventually the Internet to notify interested astronomers of burst arrivals within seconds. In 1998, the system was renamed the Gamma-Ray Burst Coordinates Network (GCN). It now sends out burst alert notices from several satellites, and also provides a repository for the ground- and satellite-based observers to post their scientific results. In many cases the burst alerts are transmitted directly to the control systems for robotic telescopes linked through the Internet. The telescopes then autonomously repoint to observe the sky locations of the bursts.

The Robotic Optical Transient Search Experiment (ROTSE) was the first autonomous telescope to be connected to the Internet-driven burst alert system.

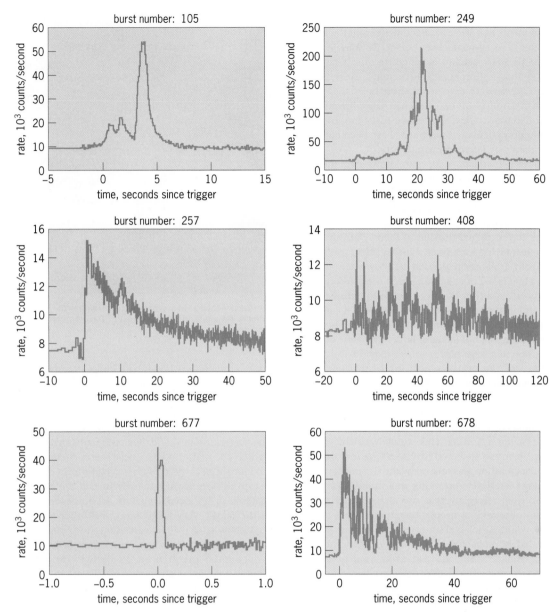

Fig. 2. Time histories of various gamma-ray bursts detected by BATSE on the *Compton Gamma-Ray Observatory*, showing a broad range of time scales. (*Courtesy of NASA/CGRO/BATSE Team*)

After more than 5 years of monitoring the burst alert notices, on January 23, 1999, ROTSE quickly repointed to acquire a series of images beginning 22 s after the initiation of a burst reported by BATSE. It was the first burst to be "caught in the act." ROTSE's images of the rapidly fading "prompt" optical emission (rather than the afterglow) were so bright (ninth magnitude) that the optical flash could have been seen by an observer using binoculars. It was realized that, with a sufficiently quick response, it would be possible to observe optical light that was emitted during the burst itself, as well as the afterglows. Following the launch in November 2004 of NASA's *Swift* satellite, the hunt for the origin of gamma-ray bursts intensified as many additional ground-based observers joined the burst chase. One of the first *Swift* gamma-ray bursts (on January 17, 2005) was observed using onboard x-ray, ultraviolet,

and optical telescopes while flaring activity continued for hundreds of seconds. *Swift*'s rapid response has allowed many ground-based robotic telescopes to study bursts in their early phases, as well as compiling detailed observations of afterglow emission across the electromagnetic spectrum.

Observational properties. The time histories of the brightnesses of gamma-ray bursts show a remarkable range of variation (**Fig. 2**). Some bursts last only a few milliseconds, whereas others show multiple peaks that last for minutes. These widely varying characteristics pose a difficult challenge to scientists seeking to understand the origin of the bursts.

There are two classes of gamma-ray bursts that are distinguishable primarily by their durations and spectral characteristics: short bursts, with durations less than 2 s and more energetic (harder) spectra; and bursts lasting longer than 2 s, with less energetic (softer) spectra. Before 2005, the afterglow measurements that were obtained were from the longer, softer class of gamma-ray bursts. On May 9, 2005, NASA's *Swift* satellite detected the first x-ray afterglow from a short burst, and on July 7, 2005, NASA's *HETE-2* satellite detected a short burst that led to the first optical afterglow, seen in data collected with ground-based telescopes. Since that time, the x-ray afterglows of many short bursts have been detected, primarily by *Swift*'s x-ray telescope.

Possible origins. With the knowledge that the longer-duration gamma-ray bursts are being created at vast distances, in star-forming regions of galaxies, speculation has increased that they are somehow related to supernova explosions of massive stars that have reached the end of their relatively short, nuclear-fueled lives. Such massive stars form preferentially in denser regions in galaxies where there is an abundance of material to seed the star-formation process, and they burn very brightly, exhausting all their fuel within a few million years. When a massive star exhausts its nuclear fuel, it will collapse inwardly while flinging out its outer layers in a supernova explosion. If a large piece, greater than 3 solar masses (6×10^{30} kg or 1.3×10^{31} lb), of the inner core remains intact, it can collapse directly into a black hole.

There is direct evidence that a subset of long gamma-ray bursts is associated with supernovae. On March 29, 2003, an extremely bright gamma-ray burst was detected by NASA's *HETE-2* satellite, and its position was relayed by the Gamma-Ray Burst Coordinates Network. This well-studied burst produced an afterglow that was detected by visible-light observers at ground-based telescopes around the world. A detailed study of the afterglow light revealed the characteristic signature of a supernova, located only 2×10^9 light-years away—one of the closest gamma-ray bursts ever detected. Further studies of this supernova proved conclusively that it was related to the gamma-ray burst, providing indisputable proof that the two events were connected. On February 18, 2006, *Swift* detected a gamma-ray burst precursor to a supernova, which occurred later that week, as astronomers worldwide trained their telescopes on the amazing event. This burst, at a distance of 440 million light-years, was the second-closest burst on record, offering another clue to the gamma-ray burst–supernova connection. As of 2012, there are now six GRBs accompanied by well-observed supernovae.

In the supernova scenario—named the "collapsar" or "hypernova" model—an extremely massive star, greater than 40 times the mass of the Sun (2×10^{32} kg or 4×10^{32} lb), explodes, and then its inner core collapses to form a black hole. The newly born black hole then accretes some of the swirling and collapsing matter to form jets. These jets carry matter and energy outward, plowing through the collapsing layers of matter, and creating internal shocks that produce gamma rays (**Fig. 3**). The interaction of the jets with the collapsing material produces a wide variety of different gamma-ray time histories,

Fig. 3. Conception of a hypernova explosion with gamma-ray emitting jets. (*Courtesy of Dave Ambrecht, General Dynamics***)**

forming gamma-ray bursts with different numbers of peaks and durations—depending on the amount and distribution of the matter that is being shocked. When the jets impact the more tenuous matter blown off earlier by the star as it evolved, additional shocks create x-rays, and then visible and finally radio light as the matter cools. These lower-energy emissions are seen as the afterglow, which varies depending on the density and temperature of the ambient material in the nearby vicinity. Radio observations at the Jansky Very Large Array and other facilities have shown that the radio emissions from gamma-ray bursts are highly collimated into beams or jets, thus reducing the energy needed to power the bursts. (Because the gamma rays that we see are in narrow jets aimed in our direction, rather than emitted uniformly in all directions, the energy inferred in the gamma-ray burst is lowered by a factor of around 100.)

The first important clues to the origin of the shorter gamma-ray bursts—those with durations less than 2 s—were revealed in 2005 with the discovery of afterglows by *Swift* and *HETE-2*. The initial observations were consistent with the orbital decay of two compact objects in a binary system: two neutron stars, black holes, or perhaps one of each. Einstein's theory of gravity (general relativity) predicts that binary orbits will decay by the emission of gravitational radiation. Several binary pulsar (neutron star) systems are now known (including the original system, discovered by Russell Hulse and Joseph Taylor). As the two objects finally gravitate together, their death spiral will lead to the formation of a black hole, accompanied by a relatively small disk of debris. The formation of jets from the accretion of this disk onto the rapidly spinning black hole can release a tremendous surge of gamma rays (**Fig. 4**). In this case, there is little nearby material to create afterglow emission, and the gamma rays can escape the stellar envelope

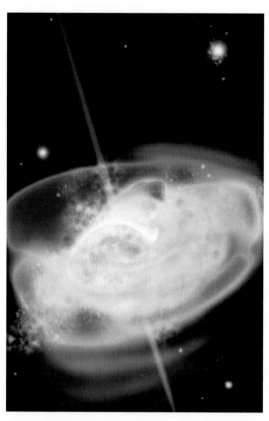

Fig. 4. Conception of merging neutron stars. (*Courtesy of Aurore Simonnet, Sonoma State University*)

more quickly, because not much obscuring matter is present. Hence the gamma-ray bursts that are formed in the merger scenario are much shorter, and should not be accompanied by long-lived afterglows.

Continued observations, primarily with the *Swift* satellite during the past 7 years, have now demonstrated conclusively that short gamma-ray bursts are cosmological in origin and are not associated with supernovae. The hosts of short gamma-ray bursts are both star-forming and elliptical galaxies, and the locations of the gamma-ray bursts within the galaxies seem to trace the optical light distribution, and hence, an older stellar population, rather than the ultraviolet light distribution, which is associated with younger stars. The short gamma-ray bursts are therefore clearly not being formed in the same manner as the long gamma-ray bursts. As of 2012, however, the nature of the progenitors of short gamma-ray bursts remains an unsolved problem: The observations are consistent with the compact object merger (that is, neutron star–neutron star, or black hole–neutron star) scenario, yet other possibilities cannot be ruled out. These alternative formation scenarios include white dwarf mergers, magnetar creation, or either white dwarf or neutron star accretion-driven collapse. In particular, there is increasing evidence that a subset of the short gamma-ray bursts may be related to superflares from distant magnetars, similar to the blasts seen on March 5, 1979 (discussed above) and December 27, 2004 (see below).

Important recent observations of individual GRBs. Observations with NASA's *Swift* satellite have pro-

vided many examples of unusual bursts, which have provided important keys to our present-day understanding of burst energetics, beaming, and progenitors. The brightest burst seen from Earth was GRB 080319B, located in the constellation Boötes, at a distance of 7.5 billion light-years. This burst, dubbed the "naked-eye" burst, had a peak visual magnitude of 5.3, and could have been seen without a telescope by Earth-bound observers. Scientists concluded that one of its particle jets appears to have been aimed squarely at Earth. In June 2008, astronomers combining data from *Swift*, the W. M. Keck Observatory in Hawaii, and other facilities identified the first gas molecules in the host galaxy of a gamma-ray burst, GRB 080607 in the constellation Coma Berenices, providing a glimpse of star formation in a galaxy 11 billion light-years distant, and at a time one-sixth the present age of the universe. On January 2, 2009, magnetic fields were first directly measured in a gamma-ray burst's afterglow by a specialized camera on a telescope operated by astronomers from the Liverpool John Moores University in the United Kingdom.

NASA's *Fermi Gamma-Ray Space Telescope* has observed several record-setting gamma-ray bursts, including GRB 080916C, the burst with the greatest total energy, equivalent to 9000 typical supernovae (**Fig. 5**). GRB 090510 displayed the fastest observed motions, with ejected matter moving at 99.99995% of light speed. And the highest-energy gamma ray yet seen from a burst, 33.4×10^9 electronvolts (about 13×10^9 times the energy of visible light), was detected by *Fermi* from GRB 090902B.

Very high energy gamma-ray emission has now been seen by *Fermi*'s Large Area Telescope (LAT) in over 30 gamma-ray bursts. In all cases, the very high energy (over 100 MeV) gamma rays are delayed with respect to the initial prompt emission (at 1–2 MeV). For four of the brightest LAT bursts, these high-energy emissions persisted for up to 2000 s. On May 10, 2009, *Fermi*'s LAT observed the short GRB 090510, which had a very sharp initial peak. After traveling for 7.3×10^9 years, the observed (less than 1 s) time delay in arrival times between photons differing in energy by a factor of a million was used to set limits on the constancy of the speed of light to one part in 10^{17}. A slower light speed at very high energies is a prediction of some theories of quantum gravity, which have now been ruled out.

Other phenomena. Although the vast majority of events that trigger gamma-ray-burst telescopes are either short or long gamma-ray bursts, there have also been some recent discoveries of other events that were later seen to represent entirely different phenomena. For example, on August 22, 2008, NASA's *Swift* satellite as well as several other orbiting instruments reported multiple blasts of radiation from a new soft gamma repeater, SGR J1550-5418. This new magnetar was also studied extensively by scientists using data from the European Space Agency's *XMM-Newton* and *International Gamma-Ray Astrophysics Laboratory* (*INTEGRAL*) satellites. Unlike GRBs, most magnetars are located in our Milky Way

(a)

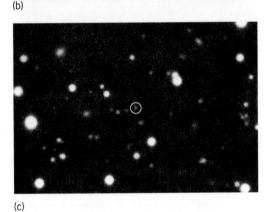

(b)

(c)

Fig. 5. Images of GRB 080916C, which exploded on September 16, 2008. (*a*) Burst as seen by the *Fermi* Large Area Telescope (LAT). Colored dots represent gamma rays of different energies. Visible light has energy between about 2 and 3 electronvolts (eV). The blue dots represent lower-energy gamma rays (less than 10^8 eV); green, moderate energies (10^8–10^9 eV); and red, the highest energies (more than 10^9 eV). The center of the burst has the highest energies, and appears light blue or white due to the superposition of several photons with complementary false colors. A video at www.nasa.gov/mov/314162main_GRB080916C_LAT_600.mov compresses about 8 min of *Fermi*LAT observations of GRB 080916C into 6 s (*NASA/DOE/Fermi LAT Collaboration*). (*b*) Superposition of *Swift*'s UVOT and XRT images. The x-ray afterglow is indicated by various shades of yellow and orange to reflect the intensity of the source. Ultraviolet colors are shown as red (v filter), green (b filter), and blue (u filter; *NASA/Swift/Stefan Immler*). (*c*) On September 17, 31.7 h after GRB 080916C exploded, the Gamma-Ray Burst Optical/Near-Infrared Detector (GROND) on the 2.2-m Max Planck Telescope at the European Southern Observatory, La Silla, Chile, began acquiring images of the blast's fading afterglow (circled; *MPE/GROND*).

Galaxy, and thus are relatively close by Earth, at distances of tens of thousands (rather than billions) of light-years. On December 27, 2004, a giant flare from SGR 1806-20 was so intense that it measurably affected Earth's upper atmosphere from a distance of 50,000 light-years. And the unusual 28-min-long "Christmas burst" on December 25, 2010, has been interpreted either as a distant merger of a neutron star with the core of a red giant companion, as viewed through the giant's cocoon, or, alternatively, as a relatively nearby tidal disruption of a large comet-like object triggering a crash of debris onto a neutron star.

Most recently, on March 28, 2011, NASA's *Swift, Hubble Space Telescope*, and *Chandra X-Ray Observatory* teamed up to study one of the most puzzling cosmic blasts yet observed. Although the initial blast in the constellation Draco resembled a gamma-ray burst (and thus was dubbed Swift J1644+57), lingering high-energy emission and repeated flaring that continued for more than 6 months indicated that something unusual was happening. Radio observations of the galaxy revealed that a central source had appeared in the same location as the x-ray flares. Radio outflows then began to expand from the galaxy's central massive black hole, as a Sun-like star was shredded and captured by the black hole. Infalling material continues to be channeled outward in relativistic jets, one of which is aimed toward Earth, producing a very luminous x-ray source.

Present and future studies. In mid-2012, nine satellites were observing gamma-ray bursts and participating in the third Interplanetary Network: NASA's *RHESSI, Mars Odyssey, MESSENGER, Swift*, and *Fermi* spacecraft; NASA's *Wind* experiment on the Russian *Konus* mission; the Japanese *Suzaku* satellite; the Italian Space Agency's *AGILE* mission; and ESA's *INTEGRAL* mission. Active areas of gamma-ray burst research include studies to pin down the progenitors of the short gamma-ray bursts, high-energy (GeV and TeV) observations to understand the origin of the very-high-energy emissions seen in some gamma-ray bursts, as well as continued studies of unusual events such as Swift J1644+57. A very exciting future development (expected sometime after 2015) would be the direct detection of gravitational waves by gravitational-wave observatories such as LIGO (Laser Interferometer Gravitational-Wave Observatory) in conjunction with a short gamma-ray burst. This widely anticipated discovery would provide definitive confirmation of the merger scenario for short gamma-ray bursts.

For background information *see* BLACK HOLE; COSMOLOGY; GALAXY, EXTERNAL; GAMMA RAYS; GRAVITATIONAL RADIATION; LIGO (LASER INTERFEROMETER GRAVITATIONAL-WAVE OBSERVATORY); NEUTRON STAR; PULSAR; RADIO TELESCOPE; REDSHIFT; RELATIVITY; SUPERNOVA in the McGraw-Hill Encyclopedia of Science & Technology.   Lynn Cominsky

Bibliography. E. Berger, The environments of short-duration gamma-ray bursts and implications for their progenitors, *New Astron. Rev.*, 55:1–22, 2011,

DOI:10.1016/j.newar.2010.10.001; J. S. Bloom, *What Are Gamma-Ray Bursts?*, Princeton University Press, Princeton, NJ, 2011; N. Gehrels, E. Ramirez-Ruiz, and D. B. Fox, Gamma-ray bursts in the *Swift* Era, *Annu. Rev. Astron. Astrophys.*, 47:567–617, 2009, DOI:10.1146/annurev.astro.46.060407.145147; J. E. McEnery, J. L. Racusin, and N. Gehrels (eds.), *Gamma-Ray Bursts 2010*, AIP Conference Proceedings, v. 1358, 2011; G. Schilling, *Flash! The Hunt for the Biggest Explosions in the Universe*, Cambridge University Press, Cambridge, U.K., 2002; G. Vedrenne and J.-L. Atteia, *Gamma-Ray Bursts: The Brightest Explosions in the Universe*, Springer/Praxis, Berlin/Chichester, U.K., 2009.

Gas shielding technology for welding

Welding is a common method for joining two metallic materials together with high structural integrity. When joints need to be leak-tight, lightweight, or free of contaminant-trapping seams or surface asperities, welding tends to be specified.

There are many welding techniques, each with its own advantages and disadvantages, including forge welding, gas tungsten arc welding, friction stir welding, and laser beam welding, to name a few. The objective of all these techniques is a structural joint that meets the requirements of a particular component or assembly. A key practice in producing high-quality welds is the use of a shielding gas.

Metallic bonds, or joints, are produced when metals are put in intimate contact. In the solid-state "blacksmith welding" process, now called forge welding (FOW), the site to be joined is pounded into intimate contact. The surfaces to be joined usually need to be heated to make it easier to deform the metal. The surfaces are sprinkled with a flux to remove the surface oxides and given a concave shape so that contamination can be squeezed out of the joint as the surfaces are pounded together; otherwise, the contaminants would be trapped in the joint and would weaken the weld.

Fusion welding is done at higher temperatures and with an exposed molten metal surface, resulting in a greater tendency for atmospheric interaction to produce weld contamination. In fusion welding processes, a heat source that is sufficiently intense to cause local melting is applied to the seam to be welded; ideally, the molten metal simply flows together and solidifies as a joint. Bulk weld metal contamination as well as surface contamination is encountered in fusion welding processes. Contamination is typically prevented by shielding the hottest, most reaction-prone surfaces.

References to gas shielding appeared soon after the introduction of the carbon arc welding (CAW) process in the early 1880s. Although the CAW process is ostensibly unshielded, the carbon arc is not entirely inert. Carbon from the electrode reacts with the atmosphere to produce a shielding effect (as carbon does when reducing iron ore in blast furnaces), but it can also carburize steels to the point of producing brittle welds under suitable conditions.

In the late 1880s, the consumable metal electrode processes came into use. This was the familiar "stick" welding process, known today as shielded metal arc welding (SMAW). The word "shielded" is used because coatings on the welding electrode decompose to produce a molten slag that flows over and shields the molten weld puddle or emits a shielding gas. Coatings for welding electrodes emerged in the early 1900s. However, the shielding provided by coated electrodes is not adequate for avoiding contamination in aluminum and magnesium.

Gas tungsten arc welding (GTAW). In 1941, the Heliarc process, using an inert (helium) shielding gas, was invented by Russell Meredith at Northrop Aircraft Company for producing magnesium airframes. The patent was bought by the Linde Division of the Union Carbide Corporation, which marketed a "Heliarc" welding torch as well as a torch that used cheaper argon gas to replace helium as the inert shielding gas. This was the origin of the gas tungsten arc welding (GTAW) process, informally known as TIG (that is, tungsten inert gas) welding.

In the GTAW process, a tungsten electrode is used. The electrode can withstand high temperatures and melts at around 3380°C. It is not as reactive as carbon. The tungsten electrode is surrounded by a nozzle that emits an inert shielding gas. Atmospheric contaminant gases are swept away by the shield-gas flow, as seen in **Fig. 1**. The higher the shield-gas flow rate, the lower the contaminant concentration that reaches the hot metal at the arc contact site.

The shield-gas column may be distorted by the motion of the welding torch, and this can reduce the thickness of the gas barrier between the arc contact and the contaminating atmosphere. If this effect is not too severe, it can be compensated for by increasing the shield-gas flow rate. At very high flow rates, however, the gas-column flow can become turbulent, entraining contaminating atmospheric gases, and the shielding effect is lost. If the weld metal is sufficiently reactive to be damaged at temperatures outside the gas column, additional shielding may be required. A hot root surface (the opposite side of the workpiece from the welding torch) may be

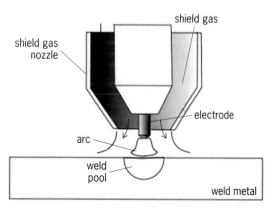

Fig. 1. Gas tungsten arc welding (GTAW) torch nozzle.

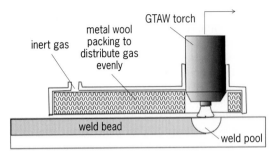

inert gas
metal wool packing to distribute gas evenly
GTAW torch
weld bead
weld pool

Fig. 2. A trailing shield. For the flat welding position, the shield gas (for example, argon) should be heavier than air.

protected by a "backup purge." This backup purge will flood the hot length of root pass with an inert or nonreactive gas emitted from a box or similar fixture with a slot in the upper surface close to the weld root. A hot crown surface (on the side of the welding torch) may be protected using a "trailing shield," a box following the torch and emitting inert or nonreactive gas to flood the hot metal trailing the torch (**Fig. 2**).

Shield-gas arc effects. In the GTAW process, the arc is produced within the shield gas, and the characteristics of the shield gas determine the characteristics of the arc, such as its intensity and stability. The sustained electrical discharge of the arc is generated by collisions of electrons with gas atoms or molecules. The gas in this conducting condition is called a plasma. The collisions must knock out enough new free electrons to balance the electrons absorbed by the ionized atoms for equilibrium to be maintained. The electrons get their collision energy from the electric field (the voltage gradient within the discharge), which accelerates the electrons between collisions. The amount of energy accumulated between collisions is determined by the structure of the gas atoms/molecules.

The gases commonly used in the GTAW process are helium and argon. The smaller and tighter outer arrangement of electrons in helium atoms and the larger and looser outer arrangement of electrons in argon atoms result in different ionization potentials (that is, the energy required to detach an outer electron). To initiate an arc during the interval between collisions, the electron must be accelerated across a potential of approximately 24.5 V for helium and 15.8 V for argon, the ionization potentials of helium and argon, respectively. Once the arc is initiated, the collisions heat up the atoms/molecules of the gas, and the kinetic energy of the atoms/molecules contributes to the ionization process, reducing the requirement for the electric field.

In SMAW, an arc is initiated by bringing the electrode very close to the workpiece to allow a sufficient voltage drop across the small mean free path of a free electron. This may be initiated by the liberation of a stray photon and the subsequent collision. Once the arc is started, it can be extended to a controllable length, which may be only a few millimeters.

In GTAW, an arc is typically initiated at a distance from the workpiece surface such that the tungsten is

not in danger of being contaminated by a surface interaction. A high-frequency generator in series with the arc circuit can generate free electrons in the gas and initiate the arc.

Helium requires greater power from the weld generator and delivers more heat to the workpiece surface than argon because of its higher ionization potential and its tendency to lose more heat as a result of its higher thermal conductivity. Thus, helium is less stable and is more prone to cause fluctuations of the arc than argon.

Unreactive but not necessarily inert gases, such as hydrogen, nitrogen, or carbon dioxide, can also be used as shield gases under appropriate conditions. The shielding emissions from carbon electrodes or electrode coatings fall into this category. Molecular, as opposed to monatomic, inert gases can dissociate in the arc and recombine at the workpiece to give off heat and produce a hotter effective arc.

Arcs are not symmetrical. In the direct current electrode negative (DCEN) operational mode, the main current carriers in the arc—the small, fast-moving electrons—are emitted at the cathodic electrode. Here, they absorb heat and are conducted to the anodic workpiece, where they give up heat. DCEN, often referred to as "straight polarity," is often the preferred mode of operation because more heat is delivered to the workpiece, where it is wanted, rather than at the electrode, where it is not desired.

In the direct current electrode positive (DCEP) or "reverse-polarity" mode, the electrons are emitted at the cathodic workpiece and conducted to the anodic electrode. In this mode, a curious effect was observed on the workpiece surface. A high-speed movie of an aluminum workpiece surface revealed an array of tiny flashes on the surface near the arc, reminiscent of a landscape subjected to a bombing raid. The flashes may be interpreted as local dielectric breakdowns of a thin layer of surface oxide charged by the settlement of positive ions from the arc on the oxide surface. This blasting away of surface oxides is the origin of the well-known reverse-polarity cleaning effect. With metals that have accumulated a tenacious oxide layer, the optimal heating of the DCEN mode is sometimes sacrificed for the cleaning effect of the DCEP mode.

It is possible to mix the optimal heating mode and the optimal cleaning mode by using an oscillating-current waveform. An alternating-current (ac) mode is a simple mixing scheme. More complex square-wave forms are also available.

Gas metal arc welding (GMAW). Filler metal is often added to a GTAW pool in the form of a wire, which may be handheld or fed automatically. If a consumable wire electrode replaces the nonconsumable GTAW electrode itself, the GTAW process becomes the gas metal arc welding process. In the GMAW process, the effect of the shield gas on metal transfer from the electrode to the weld pool must also be considered. Whereas the GTAW filler wire can simply be immersed in the weld pool, if there is to be a GMAW arc, the end of the consumable wire electrode must be separated from the weld pool. The metal may transfer in large drops (globular transfer) or as a fine spray (spray transfer).

Globular transfer of metal may form occasional metal bridges between the electrode and the weld pool. The short-circuiting bridges heat rapidly and explode, with an abundance of metal spatter. Other mechanisms that cause spatter are electromagnetic forces, trapped thermal energy in surface indentations of bubbles, and pool impacts from larger drops.

Variable-polarity plasma arc welding (VPPAW). Shielding does not address every form of contamination. For example, it does not remove surface oxides or other surface contamination, such as residual oils, that is already present on the weld-metal surfaces. Oily matter tends to decompose in the arc, yielding hydrogen. More hydrogen can dissolve in molten aluminum than in solid aluminum. Thus, an aluminum alloy weld pool that has sufficient dissolved hydrogen contamination precipitates bubbles of hydrogen (porosity) on the solidifying surface.

Porosity may be addressed through surface preparation, such as by a solvent wipe, white-glove handling, and surface scraping. Even so, a certain amount of porosity may have to be ground out and rewelded. Additionally, the weld process could be designed to allow for easy escape of porosity.

Porosity can also be flushed from the weld. The plasma arc welding (PAW) process has flushing capability. In 1957, Robert M. Gage of the Linde Division of Union Carbide was issued U.S. Patent Number 2,806,124 for the "Arc Torch and Process." This is usually regarded as the birth of PAW, although nonwelding plasma arc torches already existed.

The arc or plasma torch emits a jet of plasma similar to the flame of a gas welding torch, except that it is much hotter. An arc is initiated between a tungsten electrode and a water-cooled constricting nozzle surrounding the electrode. As the gas heats up, it expands and jets out of the constricting nozzle as a plasma with enough force to press into the molten metal pool where the jet impinges on the workpiece. If conditions are suitable, the jet will penetrate through the workpiece and emerge out the back side. This mode of operation is called "keyholing." In keyholing, the plasma jet entrains and flushes out gas impurities that might cause porosity. Meanwhile, the extremely hot molten metal surrounding the plasma jet is flooded with shield gas emitted from a secondary nozzle surrounding the PAW plasma-constricting nozzle, just as the GTAW shield-gas nozzle surrounds the tungsten electrode. The PAW torch nozzle is shown in **Fig. 3**.

The PAW process is adequate for welding steels, but aluminum alloys are covered with a tenacious thin oxide layer. In the PAW keyholing mode, the molten metal on the forward surface of the keyhole tends to migrate and merge into a unified pool behind the keyhole. The backflow occurs because the surface tension of the molten metal increases as the metal cools and pulls the molten metal with it. As the plasma jet advances and melts more metal at

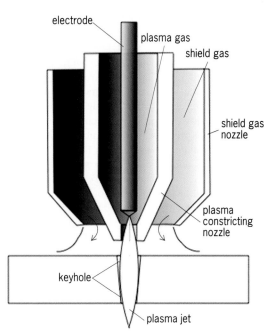

Fig. 3. Plasma arc welding (PAW) torch nozzle. The water-cooling channels of the plasma-constricting nozzle are not shown.

the keyhole surface, surface-tension-gradient-driven lateral flow removes it and feeds it to the weld pool at the back of the keyhole. In aluminum alloys, the tenacious oxide layer can be imagined as plastic wrap covering the molten metal surface, disturbing the backflow from the melting surface and preventing consolidation of the two lateral streams. Instead of a uniform-quality weld, a rough and "blobby" structure will emerge in the wake of the plasma jet.

Similar to that for GTAW discussed earlier, a remedy for the oxide problem is to incorporate enough reverse-polarity cleaning effect in the weld process to blast away the surface oxide during the consolidation of the molten weld pool. The variable-polarity plasma arc welding (VPPAW) process incorporates a power supply that does this. The polarity-reversing waveform for the current gives the process a characteristic buzzing sound.

Friction stir welding (FSW). Gas shielding, plasma flushing, and reverse-polarity cleaning as part of the VPPAW process have not proved adequate in preventing all forms of contamination. Newer aluminum alloys containing lithium can interact with nitrogen from the air at temperatures of approximately 360°C, well below that of welding, to form aluminum-lithium nitride. Lithium nitride is well known for storing large quantities of hydrogen. Such alloys may show secondary porosity that does not appear in the first weld pass, unlike porosity caused by simple supersaturation of the melt. Instead, it will emerge in response to heating by a second weld pass.

In spite of attempts to use heated titanium getters (reactive materials that absorb gases) to reduce contaminants in the weld gas, producing ductile welds in the solid-state process proved difficult until friction stir welding was introduced.

In the FSW process, a rotating pin stirs the weld seam surfaces together, while a shoulder caps the weld and prevents the weld metal from flowing up around the pin. If the shoulder were not present to prevent metal upflow, a trench would be left in the wake of the pin. The metal adjacent to the pin softens with the heat, but no gross melting occurs. However, in some circumstances, low-melting phases present in small volume fractions may melt. The axial compression force is large, and the weld metal is under considerable pressure. Welds generally involve a single pass. Friction stir welds in aluminum-lithium alloys do not exhibit the tendency for low ductility exhibited by fusion welds in that metal. If a friction stir weld is exposed to the heat of a fusion weld pass, it may exhibit secondary porosity.

Current FSW practice does not generally use gas shielding (unlike fusion welding), although the surfaces to be joined may be brushed and solvent cleaned. The large shear imposed upon the seam as it enters the deformation field around the tool expands the seam surface to expose a great deal of clean surface for welding. Although large oxide chunks or other hard particles could embrittle the surface, FSW is not especially sensitive to seam contamination. As FSW practice is extended to higher-melting metals, protection of hot metal surfaces may require shielding in the form of a trailing shield and back purge or an inert atmosphere box with a sealed moving surface incorporating the tool.

Laser beam welding (LBW). LBW is a high-power-density beam welding process. The power density is so high that when the fine focused laser beam impinges on a metal surface, the metal vaporizes (**Fig. 4**).

In LBW, the tendency of a metal surface to reflect light makes the power coupling of a laser beam to a metal surface complex. A laser beam can be attached to a robotic arm to perform elaborate welding operations automatically in a production environment, and recent advances have produced handheld laser welding torches.

Like the GTAW and VPPAW processes, LBW raises the temperature of exposed metal to a level at which the atmosphere would react with it detrimentally. To prevent this, a shield-gas nozzle emits a flow of inert gas around the welding laser beam.

Lasers impose unique shield-gas challenges that are not encountered during some of the traditional welding techniques. Laser welding systems may include the previous-generation systems, such as CO_2 and Nd:YAG (neodymium-yttrium aluminum garnet), as well as some of the newer fiber (doped optical fiber) laser systems or hybrid systems. An incorrect shielding gas or turbulent flow can cause plasma formation or secondary plumes. Selection of the shielding gas and laser-welding parameters are critical for clean and defect-free laser welds.

Many laser systems use gas shielding systems similar to those of traditional GTAW, including distribution cups and bodies, gas lenses, diffusion screens, and nozzles. These assemblies are often modified to accommodate the optic path, and it is good practice to run a gas flow test. Flow tests are a simple visual tool to understand the varying flow characteristics through the shielding assembly and see whether turbulent flows that could adversely affect the weld are present. Backup purges or trailing shields might also be appropriate for some situations. The interaction of secondary flow or trailing shielding gas should also be investigated in a flow test prior to implementing a new gas system for laser welding.

Unlike the case with the GTAW and VPPAW processes, evaporated metal from LBW may be ionized to create a plasma that absorbs and attenuates the laser beam. This has been observed in handheld laser welding with Nd:YAG and fiber laser systems. The pulse frequency, peak power, and interaction of shielding gas can cause plasma formation that makes it difficult for the operator to observe the weld and can also cause damage to the workpiece or hardware. A lower pulsing rate in a quasi-continuous wave operation can help minimize the plasma formation. Additionally, helium has replaced argon in most handheld laser-welding operations, which helps to minimize plasma formation because of its higher ionization potential, as discussed earlier. In some instances, a cross-current of gas or plasma suppression jet is required to remove the plasma.

For background information *see* ALUMINUM ALLOYS; ARC DISCHARGE; ARC WELDING; ELECTRIC FIELD; GRAIN BOUNDARIES; IONIZATION POTENTIAL; LASER WELDING; MEAN FREE PATH; PLASMA (PHYSICS); WELDING AND CUTTING OF MATERIALS in the McGraw-Hill Encyclopedia of Science & Technology.

Arthur C. Nunes, Jr.; Paul R. Gradl

Bibliography. P. M. Bhadha, How weld hose materials affect shielding gas quality, *Weld J.*, 78(7):35–40, 1999; G. Bjorkman, Backside shielding device for aluminum-lithium VPPA welding, in *Aluminum-Lithium Alloys for Aerospace Applications Workshop*, edited by B. N. Bhat, T. T. Bales, and E. J. Vesely, NASA Conf. Publ. 3287, pp. 229–236, 1994; L. P. Connor (ed.), *Welding Handbook*, 8th ed., vol. 1: *Welding Technology*, American Welding Society, Miami, 1987; R. Green, How to avoid gas-related weld flaws, *Weld J.*, 90(7):44–46, 2011; P. Gradl

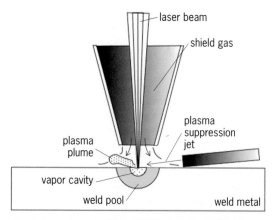

Fig. 4. Laser beam welding (LBW) operation. Many different configurations have been used to shield the weld and suppress plasma obstruction of the beam.

and K. Baker, Improved Assembly for Gas Shielding During Welding or Brazing *NASA Tech Briefs*, p. 21, June 2009, http://ntrs.nasa.gov/archive/nasa/casi.ntrs.nasa.gov/20090022331_2009022387.pdf; A. F. Manz, Ways to limit spatter, *Weld J.*, 89(9):86–87, 2010; A. C. Nunes, Jr. et al., Variable polarity plasma arc welding on the space shuttle external tank, *Weld J.*, 63(9):27–35, 1984; R. L. OBrien (ed), *Welding Handbook*, 8th ed, vol. 2: *Welding Processes*, American Welding Society, Miami, 1991; S. Yang, B. Carlson, and R. Kovacevic, Laser welding of high-strength galvanized steels in a gap-free lap joint configuration under different shielding configurations, *Weld J.*, 90(1):8-s–18-s, 2011.

Genetic exchange among bacteria

The transfer of genes is traditionally thought to occur in a vertical fashion, with genes being passed from parent cell to daughter cell or from parent to offspring through either sexual or asexual reproduction. In contrast, horizontal gene transfer (HGT), which is also known as lateral gene transfer, is the passing of genetic material from one organism to another organism, either related or unrelated, through nonreproductive mechanisms. HGT can occur in all organisms from bacteria to humans. Genes transferred through HGT can be harmful, neutral, or beneficial to the organism. Through natural selection, the harmful genes will be lost, and the beneficial genes will remain to be passed horizontally along from donor to recipient.

The classic example for HGT was observed by Frederick Griffith in 1928. Griffith demonstrated that a nonvirulent strain of *Streptococcus pneumoniae* could turn virulent when exposed to a heat-killed virulent strain. Later, it was confirmed that the transfer of naked deoxyribonucleic acid (DNA) from the dead bacteria was responsible for the resulting virulence. A more recent example was the discovery that marine bacterial genes involved in algal polysaccharide degradation were transferred to the gut bacteria of Japanese individuals who consumed seaweed as part of their daily diet. Of great concern to the medical community, HGT has now also been shown to play a significant role in antibiotic resistance in pathogenic bacteria, which can lead to the creation of "superbugs."

HGT mechanisms. In general, HGT is more common between closely related individuals because of similarities in genomic structure and genetic machinery. Small gene fragments to whole chromosomes can be transferred with no apparent decline in success; however, in divergent organisms, shorter sections of genetic material have a higher occurrence of successful transfer. Naturally occurring mechanisms for HGT include conjugation, transformation, transduction, and endosymbiont gene transfer (**Fig. 1**). Conjugation requires the physical contact of two bacteria, in which DNA is exchanged through a portal created by the bacteria themselves. Transformation, which is the mechanism that occurred in Griffith's

experiment, is the process by which bacteria pick up naked DNA from the surrounding environment. Transduction is a term used to describe HGT by a virus. This occurs in bacteria by bacteriophages and in eukaryotic cells by several types of viruses, most notably retroviruses. Endosymbiont gene transfer refers to the relocation of endosymbiont genes into the nucleus of the host. This has been well characterized for mitochondria and plastids. Artificial means of gene transfer have been developed for medical and scientific purposes; these methods include gene therapy and genetic engineering of laboratory animals. Artificial HGT in eukaryotic cells is called transfection, which can involve the use of electricity or chemicals to cajole the eukaryotic cell to take up the naked DNA.

HGT webs: the evolutionary tree. The exchange of genes through HGT has complicated our understanding of evolutionary relatedness (phylogeny). Because of the bias of HGT occurring among genetically similar organisms, the frequency of genes crossing at the species, genus, or family level is greater in comparison to the frequency of genes crossing between domains and phyla. This bias results in phylogenetic patterns that are similar to those seen with vertical inheritance, making it difficult to distinguish between connections based on shared ancestry and those created by HGT. To add further complexity, interspecies HGT can cause (1) closely related species to look more distantly related and (2) distantly related species to look more phylogenetically related. Therefore, HGT transforms the evolutionary tree of life into a web or net of life (**Fig. 2**).

The human microbiome. In the human body, there are approximately 10 bacteria for each human cell. These bacteria can live in distinct anatomical niches, with each niche having a defined bacterial species profile (**Fig. 3**). The microbiota (microbial flora) is conserved, being passed from parent to child. The mother's microbiota is passed to the child through direct contact with the mother during birth and breast feeding; thus, the baby has a microbial population that is similar to that found in the mother's vagina and milk. It is known that the child also receives the father's microbiota, but the mechanism (or mechanisms) is not understood.

Bacteria within a given niche work together to provide essential functions (for example, metabolism, reproduction, and defense) at each anatomical site. Alterations to the composition of bacteria at a given site can affect these functions. It is commonly thought that the microbiomes (all microbes and microbial genes within a biotic community) of adults are highly resilient. However, if the bacterial community experiences repeated perturbation, it will not be able to recover and the microbe could be lost from that anatomical site. One common perturbation is antibiotic use. Selective loss of a key microbe as a result of an antibiotic would lead to the loss of that microbe's function within the niche, as well as the overgrowth of competitor microbes, which could further alter the functions within the anatomical niche. Genetic alterations of a microbe via either

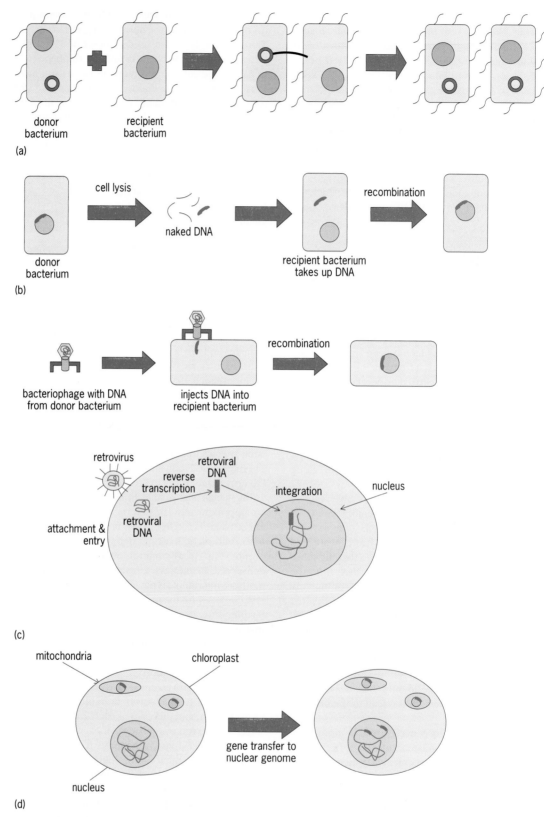

Fig. 1. Horizontal gene transfer (HGT) can occur through at least four defined mechanisms. In bacteria, HGT can occur via conjugation (*a*), transformation (*b*), and transduction (*c*, upper panel). In eukaryotic cells, naturally occurring HGT can occur through transduction (*c*, lower panel) and endosymbiont gene transfer (*d*). Unrelated genes are incorporated into the recipient genome via nonhomologous recombination, whereas related genes may be incorporated via homologous recombination.

mutation or HGT could also serve as a perturbation to the niche.

Network of gene exchange in the human microbiome. To gain a better understanding of HGT within the human microbiome, Eric Alm and colleagues at the Massachusetts Institute of Technology analyzed DNA sequences of 1183 different bacteria from numerous people living all over the world. Specifically, they analyzed the contribution of phylogeny (species relatedness), geography (the location of the people in their study), and ecology (the niche in the human microbiome) to the rate of HGT in bacteria. The investigators were surprised by two findings: (1) the sheer number of recent transfer events in their bacteria cohort, which numbered 10,770 (87% of these were unique genes); and (2) the ecological niche was the single most important contributor to HGT in the human microbiome. As previously mentioned and in theory, phylogeny should have been a larger contributor. However, it was further demonstrated that selective pressure in a given niche forced a greater amount of HGT of genes, contributing to organism survival; these genes included antibiotic resistance genes. For example, analysis of bacterial isolates that cause meningitis revealed a greater number of HGT events of genes that are known to promote disease progression, such as hemolysins, adhesions, and antibiotic resistance genes. This exchange of virulence and antibiotic resistance genes between unrelated bacterial species has prompted great concern in the medical community.

Antibiotic resistance. The discovery of penicillin by Alexander Fleming in 1928 and the subsequent production and routine use of antibiotics since the 1940s have revolutionized human health. Pathogenic bacteria that cause epidemic illnesses such as bubonic plague (*Yersinia pestis*) and cholera (*Vibrio cholerae*) could now be successfully eliminated. However, the overuse and misuse of antibiotics have led to antibiotic resistance in pathogenic bacteria. Bacteria can acquire antibiotic resistance through random mutations or by acquiring a resistant gene through HGT. Because HGT can facilitate the transfer of a functional resistance gene, the rate of bacteria gaining antibiotic resistance is accelerated through this mechanism, thereby leading the way for the acquisition of multiple resistance genes.

HGT builds superbugs. Pathogenic bacteria possessing multiple antibiotic resistance genes (sometimes referred to as multidrug resistance) are called superbugs. Superbugs are collectively one of the greatest threats to human health in that they are extremely difficult, if not impossible, to treat, because most known antibiotics do not kill them. The work carried out on HGT and the human microbiome has provided insight into a disturbing future for the creation and spread of superbugs. Again, HGT in the microbiome is driven primarily by direct contact in a given anatomical niche. Therefore, genes are being passed between unrelated species of bacteria. In addition, recent HGT events have been detected in people living on different continents. If the HGT event is the acquisition

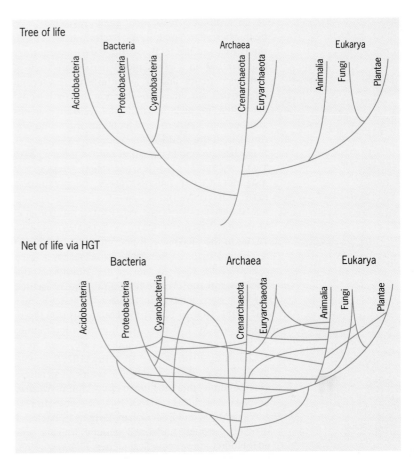

Fig. 2. Horizontal gene transfer (HGT) transforms the tree of life into a web of life. The traditional evolutionary tree (phylogenetic tree) is based on genetic similarities stemming from a common ancestor. HGT allows for the passing of genes to unrelated species, creating a network.

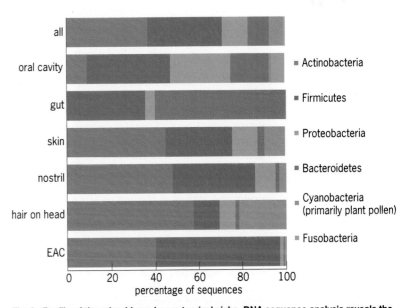

Fig. 3. Profile of the microbiome by anatomical niche. DNA sequence analysis reveals the relative abundances of the six most abundant bacterial phyla in the human body. The phyla shown represent 99.6% of the sequences sampled. Cyanobacteria were related primarily to plant chloroplast sequences (that is, pollen) derived from environmental exposures. EAC = external auditory canal. (*From E. K. Costello et al., Bacterial community variation in human body habitats across space and time, Science, 326:1694–1697, 2009; reprinted with permission from AAAS*)

of an antibiotic resistance gene, resistance can be transferred to a pathogenic bacterium, which can then be passed all over the world. If we then consider that bacteria in a given niche harbor an array of resistance genes and if a pathogenic bacterium infects that site, it is possible for the pathogenic bacterium to pick up multiple resistance genes and become a superbug, which can spread rapidly throughout the world.

For background information *see* ANTIBIOTIC; BACTERIA; BACTERIAL GENETICS; BACTERIAL PHYSIOLOGY AND METABOLISM; BACTERIOLOGY; DEOXYRIBONUCLEIC ACID (DNA); DRUG RESISTANCE; GENE; GENETICS; MICROBIAL ECOLOGY; PATHOGEN; PHYLOGENY; TRANSDUCTION (BACTERIA); TRANSFORMATION (BACTERIA) in the McGraw-Hill Encyclopedia of Science & Technology.   Rebekah L. Waikel; Audrey J. Baute

Bibliography. C. P. Andam et al., Multilevel populations and the evolution of antibiotic resistance through horizontal gene transfer, *FEMS*, 35:756–767, 2011, DOI:10.1111/j.1574-6976.2011.00274.x; I. Cho and M. J. Blaser, The human microbiome: At the interface of health and disease, *Nat. Rev.*, 13:260–270, 2012, DOI:10.1038/nrg3182; S. Smillie et al., Ecology drives a global network of gene exchange connecting the human microbiome, *Nature*, 480:241–244, 2011, DOI:10.1038/nature10571; J. Wiedenbeck and F. M. Cohan, Origins of bacterial diversity through horizontal genetic transfer and adaptation to new ecological niches, *FEMS*, 35:957–976, 2011, DOI:10.1111/j.1574-6976.2011.00292.x.

Genomics of depression

Depression has been characterized recently as "a heterogeneous disorder with a highly variable course, an inconsistent response to treatment, and no established mechanism." This alarming characterization for the disease that is the leading worldwide cause of years lost as a result of disability potentially reflects a fundamental flaw in previous experimental approaches that have focused on few candidate biological systems to understand the pathology of the illness. Recent genome-wide or "genomic" approaches depart from these traditional investigations because they rely on the unbiased knowledge of the whole genome to identify changes in the structure or function of genes in association with the disease.

Major depressive disorder. According to the *Diagnostic and Statistical Manual of Mental Disorders* (Fourth Edition, DSM-IV), major depressive disorder (denoted as depression) is diagnosed by a variable set of five symptoms for a continuous period of two weeks. Depressed mood or reduced interest in activities that were previously enjoyable (anhedonia) represent core symptoms of depression, which are often accompanied by cognitive (attention and concentration), physiological (weight, locomotor, and sleep-pattern changes), and frequent high-anxiety symptoms. Yet, the drugs used to treat depressed patients have not fundamentally changed since their discovery by chance more than 50 years

ago. These drugs mostly target modulatory neurotransmitter systems, such as the monoamines (serotonin, norepinephrine, and dopamine), but robust evidence for primary deficits in these systems in depression is still missing. One underlying reason for the lack of breakthrough in novel drug targets may be that the variable symptom-based diagnostic criteria for depression do not correspond to a common set of disrupted biological systems (as, for example, in the way that unchecked cell growth identifies cancer). Many researchers and funding institutions, including the National Institute of Mental Health (NIMH), are now convinced that focusing on symptom dimensions and their associated neural network may be a more suitable approach to unravel mechanisms of complex brain diseases. For example, the core symptomatology of depression points to a critical deficit in affect regulation, specifically a high propensity for low mood and a reduced ability to experience positive emotions. Affect regulation is subserved by a specific set of brain areas, or neural network, which includes cortical (prefrontal and cingulate cortex) and subcortical (amygdala, ventral striatum, and hypothalamus) brain regions. Moreover, a targeted "endophenotype" approach (for example, focused on affect dysregulation and its associated neural network) may be more successful than targeting the whole depressive syndrome; and results may have consequences for treating mood-related symptoms in subjects having not only depression, but also other major mental illnesses, with an altered mood component.

Genomic approach. The investigation of genes in complex psychiatric disorders relies on a comprehensive approach that takes into account nested biological scales, whereas mechanisms of disease begin with altered gene function mediating changes in basic cellular functions, subcellular signal integration, cellular circuitry, and the neural network, together contributing to the expression of associated symptom dimensions within the broader syndrome. The systematic analysis of genes, or genomic approach, is based on the unbiased knowledge of the whole genome and seeks to identify changes in the structure or function of genes in association with the disease. For structural changes, investigators look for the presence of DNA variants, such as single nucleotide polymorphisms, sequence repeats, or copy number variants, which are present in affected individuals, but not present in control subjects. The rationale for this approach is that those structural variants may affect the function of nearby genes either during development or in adulthood, thereby increasing the vulnerability of the subject for developing the psychiatric illness. Although these studies have been successful in diseases with more straightforward Mendelian inheritance patterns, results from genome-wide association studies (GWAS) for structural variants in depression have been disappointing. Currently, no single finding has withstood the hurdle of genome-wide statistical rigor and of robust replication in independent cohorts. Explanations for these shortcomings invoke a lack of statistical power as a result of the large number of variants investigated (in the millions), the engagement of too many genes with very small effects, the clustering of patients on variable diagnostic criteria, the association with the whole syndrome rather than endophenotypes, and the strong environmental component of depression. These issues will be addressed in the coming years using larger cohorts (including tens of thousands of subjects) and more refined criteria for disease or endophenotype classification.

A second type of genomic approach seeks to identify changes in the function of genes in association with the disease. The rationale is that, regardless of their origin, changes in gene function mediate disrupted biological functions and behavioral output in a bottom-up integrated fashion, and may thus point to biological mechanisms underlying the illness pathology. At the basis of functional genomic studies is the gene or DNA microarray technology. DNA microarrays monitor the expression and relative changes in multiple gene transcripts or messenger RNAs, which are assumed a priori to represent changes in the function or activity of those genes. Because information is obtained for thousands of genes in a simultaneous and unbiased way, DNA microarrays provide the equivalent of functional snapshots of genes, cells, and brain tissue at a particular moment in time.

Several groups have now applied this approach using brain samples from postmortem subjects with depression and compared results to control subjects. Supporting the original unbiased premise, results have largely not confirmed the presence of deficits in neuromodulatory monoaminergic systems, but instead point toward deficits in glutamatergic and gamma-amino butyric acid (GABA) neurons (the major excitatory and inhibitory cells of the central nervous system) and in astrocytes and oligodendrocytes (two types of glial cells providing support to neurons). More specifically, the recycling of released glutamate by astrocytes is a necessary step for maintaining adequate pools of stored neurotransmitters in neurons, and key components of this pathway appear deregulated in depression, potentially leading to reduced glutamatergic availability and reduced signaling. However, because postmortem measures rely on gene transcript and protein levels rather than on the functions of these molecules, the exact extent of changes in neurotransmitter recycling is not known.

At the same time, reduced expression of markers for GABA neurons has been reported, potentially providing a counterbalancing reduced inhibition for reduced glutamatergic excitation. Excitatory pyramidal neurons are under the negative control of several types of inhibitory neurons, and human postmortem gene expression studies report selective deficits in genes coding for markers of GABA neurons targeting the dendritic compartment of pyramidal cells. These inhibitory neurons are characterized by the expression of small neuropeptides, such as somatostatin, cortistatin, and neuropeptide Y, which are all three downregulated by 20–40% in depressed patients compared to nondepressed controls. In contrast, markers of inhibitory neurons that target the pyramidal cell body and axon (expressing parvalbumin or cholecystokinin, for instance) appear mostly undisturbed. These findings were observed in the context of reduced expression of brain-derived neurotrophic factor (BDNF), a neuropeptide essential for maintaining neuronal function and plasticity, and previously implicated in depression and in the action of antidepressant medication. In subjects with depression, BDNF was reduced by 20–30% compared to control subjects, similar to the neuropeptide effect sizes. Notably, BDNF is necessary for maintaining the specific expression of the three markers of dendritic inhibition.

Therefore, considering that dendritic inhibition modulates incoming information, whereas other cell body or axonal inputs mostly regulate the output of pyramidal cells, the combined interpretation of these findings has been a putative low BDNF- and GABA-mediated capacity to integrate incoming information, in turn leading to dysregulated affect, because these findings were consistently observed across brain regions involved in affect regulation. Other findings emerging from unbiased genomic investigation of depression include consistent decreases in the amygdala and nearby cortical regions of multiple markers of oligodendrocytes, that is, the glial cells responsible for generating the insulating sheaths surrounding the axonal projections of neurons, hence potentially further contributing to dysregulated information processing relating to emotionally salient and affect-related stimuli.

Conclusions. Together, these recent findings provide exciting new leads, which can be translated into basic science studies aimed at investigating the precise molecular mechanisms engaged in the illness and to test whether they are causal to the behavioral phenotype. For example, genetic manipulations in a mouse model have now demonstrated that altered BDNF function is sufficient to partially recreate the dysregulated gene profile that is observed in the brains of subjects with depression. Thus, considering the current limitations of these approaches, which include various technical and statistical issues, small sample sizes, disease heterogeneity, and the presence of multiple clinical and demographic parameters, it is surprising and encouraging that robust leads into biological mechanisms of brain dysfunction in depression have already emerged. Genomic studies are still in their early phases. New technical developments, including large-scale direct sequencing of gene transcripts, larger subject cohorts for both GWAS and microarray approaches, and novel analytical perspectives such as the new field of gene coregulation networks, are poised to rapidly increase the scope and sensitivity of genomic studies. Finally, another trend emerging from genomic studies across psychiatric and neurological disorders is that of partial overlaps in molecular pathologies (for example, reduced BDNF and somatostatin), although in different molecular and biochemical contexts. Therefore, it is thought that systematic genomic analyses of brain dysfunction across syndromes, in parallel with similar studies in blood, may lead to a novel system for disease classification based on biological markers, which will augment the current symptom-based diagnostic of depression and other major mental illnesses.

For background information *see* AFFECTIVE DISORDERS; BRAIN; DNA MICROARRAY; GENE; GENETICS; GENOMICS; HUMAN GENETICS; NEUROBIOLOGY; SYNAPTIC TRANSMISSION in the McGraw-Hill Encyclopedia of Science & Technology.   Etienne Sibille

Bibliography. R. H. Belmaker and G. Agam, Major depressive disorder, *N. Engl. J. Med.*, 358:55–68, 2008, DOI:10.1056/NEJMra073096; P. V. Choudary et al., Altered cortical glutamatergic and GABAergic signal transmission with glial involvement in depression, *Proc. Natl. Acad. Sci. USA*, 102:15653–15658, 2005, DOI:10.1073/pnas.0507901102; J. P. Guilloux et al., Molecular evidence for BDNF- and GABA-related dysfunctions in the amygdala of female subjects with major depression, *Mol. Psychiatry* (epub), 2011, DOI:10.1038/mp.2011.113; R. Plomin and O. S. Davis, The future of genetics in psychology and psychiatry: Microarrays, genome-wide association, and non-coding RNA, *J. Child Psychol. Psychiatry*, 50:63–71, 2009, DOI:10.1111/j.1469-7610.2008.01978.x; A. Tripp et al., Reduced somatostatin in subgenual anterior cingulate cortex in major depression, *Neurobiol. Dis.*, 42:116–124, 2011, DOI:10.1016/j.nbd.2011.01.014; G. W. Valentine and G. Sanacora, Targeting glial physiology and glutamate cycling in the treatment of depression, *Biochem. Pharmacol.*, 78:431–439, 2009, DOI:10.1016/j.bcp.2009.04.008.

Geoengineering proposals

Solar radiation management
Spraying SO_2 into the stratosphere to enhance cloud albedo
Spraying engineering nanoparticles into the stratosphere to enhance cloud albedo
Placing reflective mirrors, discs, or particles in Earth orbit
Painting roofs and other structures with reflective material
Placing solar reflectors in the desert

Carbon dioxide removal
Using physical and chemical processes to remove and store CO_2
Enhancing CO_2 removal and storage by terrestrial plants
Burial of biomass
Reforestation and afforestation (creating new forests)
Fertilizing the oceans to stimulate CO_2 removal by plankton

Source: D. B. Resnik and D. A. Vallero, Geoengineering: An idea whose time has come?, *J. Earth. Sci. Climate Change*, 2011, S1.

Geoengineering: enhancing cloud albedo

Engineers must appropriately scale solutions to problems. Since changes in climate are global in scale, the problems associated with these changes must be addressed at very large scales. Engineers also must select from approaches to a problem, ranging from prevention to remedies. To date, proposals regarding climate change have been preventive, that is, to reduce emissions of greenhouse gases. Remedies may also be employed. These large-scale remedies are the domain of geoengineering.

Global-scale interventions have yet to be taken to address climate change. However, intentional large-scale interventions are being considered. These are known as geoengineering. Currently, removal of greenhouse gases and management of solar radiation are the two prominent geoengineering solutions being considered (see **table**). Most climate-change models indicate that the buildup of carbon dioxide (CO_2) is the principal driver for warming. Removal of CO_2 includes methods for extracting the gas from the atmosphere and storing or sequestering it. Solar radiation management does not attempt to address the underlying causes of climate change; rather, it attempts to counteract increases in greenhouse gases by blocking solar radiation or increasing the reflectivity of clouds or the Earth's surface.

Sulfate particles. Scientists make hypotheses based on their observations of natural phenomena. For example, the 1991 Mount Pinatubo eruption emitted 10 million metric tons of sulfur dioxide (SO_2) into the stratosphere, which was converted to sulfate particles. Sulfate particles in the atmosphere increase cloud albedo, which is the reflectivity of solar radiation into space. Noting this, Nobel Prize-winning geochemist Paul Crutzen hypothesized that spraying sulfur dioxide into the stratosphere would increase cloud albedo and could result in mean cooling of the Earth by 0.5°C. According to Crutzen, sulfate particles will last one or two years longer if SO_2 is sprayed into the stratosphere rather than the troposphere, and have a greater impact on cloud albedo. To generate this amount of cooling, about 2 million metric tons of SO_2 would need to be sprayed annually. This represents 3.6% of the 55 million metric tons of SO_2 emitted into the atmosphere each year from the burning of fossil fuels.

From a thermodynamic and climatic perspective, this type of geoengineering appears to be achievable. However, the atmosphere is quite complex. Changing one variable, SO_2 concentration, will affect other variables. For example, spraying SO_2 into the stratosphere may adversely affect human health and the environment if it interferes with the ozone layer or other chemical balances. In the troposphere, SO_2 is converted into sulfuric acid (H_2SO_4), producing acidic precipitation that damages ecosystems and threatens sensitive plant and animal species.

Also, inhalation of SO_2 contributes to respiratory problems, such as airway constriction and asthma exacerbation. As such, most nations regulate the emissions of SO_2. Short-term exposure is associated with increased visits to emergency departments and hospitalization for respiratory problems, especially among young children and asthmatics. So, it may be difficult to justify increasing the global mean concentration of SO_2 while decreasing local concentrations.

Other problems can result from inaction. That is, solving only the cooling problem does nothing about the increasing concentrations of atmospheric CO_2, which can be associated with other problems. For example, when CO_2 dissolves in seawater, it changes the ionic strength and other chemical properties of the water, which increases the net acidity of oceans and other surface waters. The oceans have increased in acidity by 30% since 1750, due to increases in anthropogenic carbon dioxide in the atmosphere. This trend may decrease the availability of calcium carbonate, threatening species that form shells from this compound, such as mollusks, corals, and some types of plankton. A reduction in these species could have wide-ranging effects on other marine species and ecosystems, because many organisms feed on mollusks or plankton or depend on coral reefs for shelter.

Increasing cloud albedo may also lead to environmental problems. Even though the global temperatures may be stabilized, the reflectivity increase could affect precipitation patterns, tropical storm activity, temperature distribution, and wind. Also, much of the radiation that is reflected by sulfate particles strikes the Earth as diffuse light. This increases the whiteness of the daytime sky and may reduce the efficiency of plant photosynthesis and solar power.

Further, the complexity of the atmosphere means that the expected results will be highly uncertain, whatever actions are taken. Thus, spraying too much SO_2 into the stratosphere could lead to excessive cooling, droughts, floods, or other meteorological events.

Nanoparticles. Photophoretic forces occur when there is a temperature differential between an aerosol and the surrounding gas (see **illustration**). This process can enhance albedo. One recently proposed approach is to spray disc-shaped engineered nanoparticles composed of layers of aluminum oxide (Al_2O_3), metallic aluminum, and barium titanate ($BaTiO_3$) into the stratosphere. The aluminum layer provides high solar-band reflectivity with high transparency to outgoing thermal infrared radiation. This produces large mass-specific cooling. The Al_2O_3 layer reduces the rate of oxidation of the aluminum surface. The $BaTiO_3$ layers thickness is determined by the electrostatic torque from the atmospheric electric field so as to orient the disc to optimize levitation, overcoming gravity. That is, the disc is analogous to a colloidal suspension within a mixture of stratospheric gases. Nanoparticles' low mass and large relative diameters could take advantage of photophoretic and electromagnetic forces to levitate above the stratosphere.

The relative low reactivity and resistance to oxidation substantially increase the time that such nanoparticles may remain suspended in the atmosphere, compared to the sulfate particles generated from spraying SO_2. Their specific reflective properties could also be controlled through engineering and design, so that they would produce correct amount of diffuse light. Also, because the nanoparticles would be above the stratosphere, they would be less likely to interfere with ozone chemistry and, unlike sulfate, would not produce acid rain.

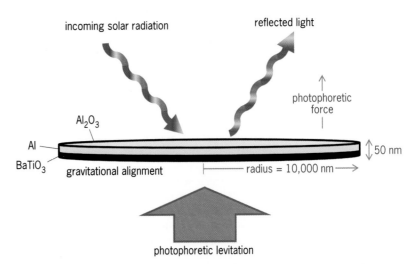

Photophoretic levitation using a nanoscale composite disc with an aluminum oxide upper layer barium titanate bottom layer. (*Based on information provided by D. W. Keith, Photophoretic levitation of engineered aerosols for geoengineering, Proc. Natl. Acad. Sci. USA, 107:16428–16431, 2010, DOI:10.1073/pnas.1009519107*)

The efficacy of nanoparticles in reducing atmospheric temperatures is much less certain than that of SO_2. The eruption of Mt. Pinatubo indicates the cooling potential of SO_2, but no analogous process for nanoparticles has been observed in the troposphere.

Stabilizing global temperatures at higher CO_2 levels would not address the problem of ocean acidification, and it might affect precipitation patterns, temperature distribution, tropical storms, and winds. The risks may be even more uncertain than those for SO_2.

Little is known about the direct and indirect risks from nanoparticles. Indeed, there could be significant environmental and public health risks of spraying nanoparticles into the stratosphere, which are not well understood at this point.

Weighing benefits and risks. Ideally, any decision to implement a geoengineering proposal should be based on a thorough understanding of the benefits and risks. Proposals should only be initiated when there is sufficient evidence that the benefits outweigh the risks, and serious harms can be prevented or avoided. However, we currently lack a thorough understanding of the benefits and risks of most geoengineering approaches, because of their scale and complexity. Such large-scale engineering efforts have no direct precedents and may require a high degree of international cooperation. Large uncertainties about their potential success and potential effects on human health and the environment may keep such large projects from occurring, because people may not want to take risks that are difficult to predict or manage.

Because the benefits and risks of most geoengineering proposals are uncertain at this point, a precautionary approach is warranted. Smaller-scale, lower-risk projects should be implemented before larger and riskier ones.

Arguably, there is less controversy associated with greenhouse gas removal than with enhancing albedo, due to a better understanding of the benefits and risks of greenhouse gas removal. Indeed, carbon sequestration projects are underway, although these are also controversial as their scale increases, due to questions about the efficiency of storing CO_2 in geological formations and potential escape of gases from deep ocean storage, as well as concerns about costs.

Traditional cost–benefit analyses have not been conducted at the global scale. Even for relatively small-scale projects, these analyses often cannot quantify social scientific factors, such as political and economic variables. Thus, even though reducing greenhouse gas emissions is preferable to either removing them or addressing the climatic changes brought about by their increased concentrations in the troposphere, the benefits of prevention may well be underestimated. Notably, three of the world's largest greenhouse gas emitters (China, India, and the United States) did not ratify the Kyoto Protocol on climate change, principally because of concerns about its effect on their economies. Although negotiations continue on a new climate-change treaty, it is not known whether the international community will be able to reach an agreement to reduce greenhouse gas emissions. Barring a significant reduction in worldwide greenhouse gas emissions, geoengineering proposals will likely continue to receive serious consideration.

[Acknowledgment: This research is the work product of an employee or group of employees of the National Institute of Environmental Health Sciences (NIEHS), National Institutes of Health (NIH), however, the statements, opinions or conclusions contained therein do not necessarily represent the statements, opinions or conclusions of NIEHS, NIH or the United States government.]

For background information *see* ACID RAIN; AEROSOL; ALBEDO; ATMOSPHERE; ATMOSPHERIC CHEMISTRY; GLOBAL CLIMATE CHANGE; GREENHOUSE EFFECT; NANOPARTICLES; STRATOSPHERE in the McGraw-Hill Encyclopedia of Science & Technology.

Daniel A. Vallero; David B. Resnik

Bibliography. D. Bodansky, May we engineer the climate? *Climate Change*, 33:309–321, 1996, DOI:10.1007/BF00142579; P. Crutzen, Albedo enhancement by stratospheric sulfur injections: A contribution to resolve a policy dilemma?, *Climate Change*, 77:211–220, 2006, DOI:10.1007/s10584-006-9101-y; A. Dessler and E. Parson, *The Science and Politics of Global Climate Change*, Cambridge University Press, Cambridge, U.K., 2006; K. Elliott, Geoengineering and the precautionary principle, *Int. J. Appl. Phil.*, 24:237–253, 2010; V. J. Fabry, B. A. Seibel, and R. A. Feely, Impacts of ocean acidification on marine fauna and ecosystem processes, *ICES J. Mar. Sci.*, 65:414–432, 2008, DOI:10.1093/icesjms/fsn048; D. Jamieson, Intentional climate change, *Climate Change*, 33:323–336, 1996, DOI:10.1007/BF00142580; J. T. Keihl, Geoengineering climate change: Treating the symptom over the cause?, *Climate Change*, 77:227–278, 2006, DOI:10.1007/s10584-006-9132-4; D. W. Keith, Geoengineering the climate: History and prospect, *Annu. Rev. Energy Environ.*, 25:245–284, 2000, DOI:10.1146/annurev.energy.25.1.245; D. W. Keith, Photophoretic levitation of engineered aerosols for geoengineering, *Proc. Natl. Acad. Sci. USA*, 107:16,428–16,431, 2010, DOI:10.1073/pnas.1009519107; D. W. Keith, E. Parson, and E. G. Morgan, Research on global sun block needed now, *Nature*, 463:426–427, 2010, DOI:10.1038/463426a; E. Kintisch, Asilomar 2 takes small steps toward rules for geoengineering, *Science*, 328:22–23, 2010, DOI:10.1126/science.328.5974.22; S. H. Schneider, Geoengineering: Could—or should—we do it?, *Climate Change*, 33:291–302, 1996, DOI:10.1007/BF00142577; The Royal Society, *Geoengineering the Climate: Science, Governance, and Uncertainty*, The Royal Society, London, U.K., 2009; J. Tollefson, Geoengineering faces ban, *Nature*, 468:13–14, 2010, DOI:10.1038/468013a; J. Tollefson, The sceptic meets his match, *Nature*, 475:440–441, 2011, DOI:10.1038/475440a; D. A. Vallero, *Environmental Biotechnology: A Biosystems Approach*, Elsevier Academic Press, Amsterdam, 2009; T. M. Wigley, A combined mitigation/geoengineering approach to climate stabilization, *Science*, 314:452–454, 2006, DOI:10.1126/science.1131728; L. F. Wiley and L. O. Gostin, The international response to climate change: An agenda for global health, *JAMA*, 302:1218–1220, 2009, DOI:10.1001/jama.2009.1381.

Hedgehog signaling proteins

An understanding of the molecular programs that control the formation of specific cell types in complex multicellular organisms is essential for therapeutic efforts focused on improving tissue regeneration and disabling cancerous cell growth. The success in characterizing these programs in the last few decades has been largely built on genetic results using the embryonic development of model organisms, such as fruit flies and mice, to identify genes that control cell differentiation. The creation of specific cell types, such as muscle, fat, and blood cells, is a coordinated process dependent on different cellular precursors communicating cell differentiation cues to one another. Perhaps not surprisingly then, many of these molecular programs are controlled by secreted proteins (proteins released from cells) that transmit intercellular signals between neighboring or distant cells. The secreted Hedgehog (Hh) signaling molecule is essential for embryonic patterning and the formation of the limbs and central nervous system. Frequently, as a consequence of genetic mutations, the cellular responses controlled by the Hh protein can be exploited to promote cancerous cell growth. Based on this understanding, chemicals that inhibit these cellular responses have been developed for the treatment of certain forms of skin and brain cancers. This brief overview of the Hh pathway will describe its inception from efforts to understand the molecular basis of development and its remarkable journey to therapeutic relevance in cancer.

Building the Hh pathway using fruit flies. In 1995, the geneticists Christiane Nüsslein-Volhard and Eric Wieschaus were awarded the Nobel Prize in Physiology or Medicine for devising an experimental approach to identify genes essential for the control of embryonic development. Their strategy was to apply a chemical that induces random mutations in deoxyribonucleic acid (DNA) to fruit fly embryos and then to identify genes important for cell fate decision making based on the associated developmental defect. The names given to these genes often describe the appearance of the associated mutant animals. For example, loss of the Hh gene distorts the normal larval pattern of alternating naked and denticle-covered segments such that only denticle-covered segments remain, thereby yielding a hedgehog-like appearance (**Fig. 1**). By collecting genes that function in the same aspect of development and then using genetic techniques to understand their relationship, a signal transduction pathway emerges that describes a group of genes that coordinately control a biological process. Even as geneticists were characterizing the genes mutated in the fruit fly studies, new DNA sequencing techniques that enabled rapid gene identification revealed the presence of these insect genes in mammalian genomes. In other words, fruit flies and humans rely on the same cellular programs to acquire different cell types during development. Using the results from the fruit fly studies as a road map, geneticists were able to identify similar genes in mammals and evaluate their function by engineering mice that lack these genes (so-called gene knockout mice).

Atypical molecular switches gate Hh pathway activity. An understanding of the molecular mechanisms that support signaling at the cellular level is a first step toward realizing how a signaling system is regulated at the organism level and how it might be chemically manipulated for therapeutic goals. Hh protein production is dependent on a self-cleavage mechanism that concomitantly results in the attachment of a cholesterol molecule to the protein (**Fig. 2a**). This unusual modification, which has only been observed in the Hh protein, limits the ability of Hh to diffuse such that a gradient of the protein emanating from the Hh-producing cells can be formed across a field of responding cells. In this fashion, responding cells can then receive a different dose of Hh protein and adopt a corresponding cell fate outcome. This phenomenon, for example, is essential for Hh-controlled determination of digit number in the limbs.

Activation of the Hh pathway begins with the engagement of Hh to its receptor Patched (Ptch), which is a protein that is threaded through the cell membrane. Inhibition of Ptch function on binding to Hh somehow results in the activation of Smoothened (Smo), which is another membrane-associated protein that is essential for the subsequent biochemical events leading to the expression of cell regulatory genes controlled by Hh. Ptch is related to a family of proteins that are presumed to have the ability to transport small molecules across the cell membrane

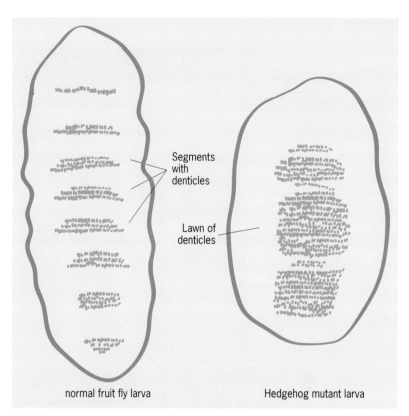

Fig. 1. A developmental defect in fruit flies provides a robust readout for genetic afflictions that alter Hh signaling: normal fruit fly larval segmentation (*left*) and the segmentation pattern in animals with defective Hh signaling (*right*).

or to "flip" lipids between the two lipid layers that constitute most membranes found in a cell. Coupled with observations that Ptch can regulate Smo in a substoichiometric fashion, a model has emerged that entails Ptch indirectly regulating Smo activity by gating its access to a small molecule produced by cells such as a lipid (Fig. 2b). Support for this model has also been buttressed by observations that Smo can be influenced by a variety of cholesterol-like molecules known as oxysterols via direct engagement. The influence of cholesterol on Hh pathway activity suggests that cholesterol metabolism at the organism level can potentially influence the ability of cells to make appropriate cell fate decisions. Because reactivation of the Hh pathway is essential for some aspects of wound healing, it is conceivable that an individual's cholesterol levels could impact bodily repair and the aging process.

Cellular tower and Hh signal transduction. In 2003, using a forward genetic screening strategy in mice similar to one successfully applied in fruit flies, the research team of Kathryn V. Anderson at the Sloan-Kettering Institute in New York identified a role for a long-neglected cellular structure known as the primary cilium in the formation of the central nervous system. On further analysis, the cause of an associated developmental defect in this formation could be traced back to compromised Hh signaling. It is now understood that the primary cilium functions as a signaling hub where Hh pathway components can accumulate and engage in meaningful interactions

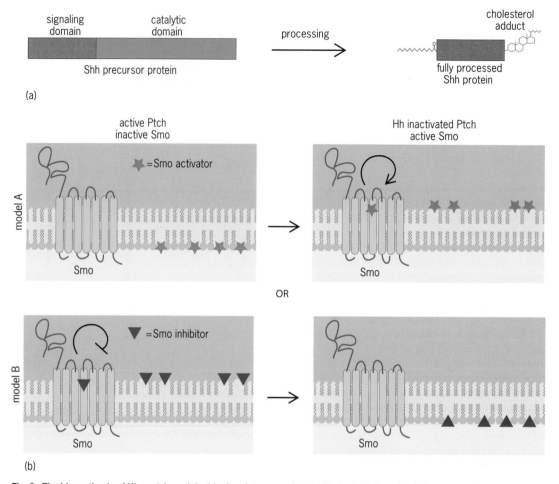

Fig. 2. The biosynthesis of Hh protein and the biochemistry associated with the initiation of cellular response to Hh.
(*a*) Biosynthesis of Hh protein. The Hh precursor protein is autocatalytically cleaved in a reaction that results in the addition of a cholesterol molecule to the mature signaling molecule. A fatty acid (palmitate) is also added before the Hh protein is released from the cell as a fully processed protein. (*b*) Indirect regulation of Smo by Ptch via a mysterious cellular chemical. Ptch may function as a lipid flippase that transports lipids from one leaflet of the lipid bilayer to the other. This molecule may be either an activator (model A) or inhibitor (model B) of Smo. Ptch inactivation by Hh may therefore gate Smo activity by regulating the access of Smo to this molecule.

(**Fig. 3***a*). This discovery is also remarkable for other reasons. First, there is no evidence for a similar role of the primary cilium in fruit flies, suggesting an evolutionary divergence in how Hh signaling occurs in mammals. Second, the primary cilium in many ways is thought of as a vestige of unicellular ancestors that used the structure for motility and to communicate with one another. Finally, genes thought to function in the primary cilium have been implicated in a number of diseases known as ciliopathies, including polycystic kidney disease and retinal degeneration. Thus, by linking Hh signaling to this structure, the basis of these diseases will need to be reevaluated for therapeutic possibilities premised on modulating the Hh pathway.

Management of cancer with Hh pathway inhibitors.
Armed with an in-depth understanding of the molecular basis for Hh signaling from decades of investigation, a number of cancer types have since been linked to deviant activation of the Hh pathway. These cancers typically involve mutations that render Ptch inactive or that force Smo into a constitutively active state (Fig. 3*b*). The range of cancers that can be in-

duced by misactivation of Hh pathway activity can be observed in patients with Gorlin syndrome, which is a tumor syndrome associated with loss of Ptch activity and basal cell carcinoma, medulloblastoma, and rhabdomyosarcoma. In sporadic forms of these cancers, the Hh pathway is also frequently misactivated.

In the late 1950s, a group of sheepherders in Idaho had noticed an unusual number of lambs born with cyclopia (a single eye) after a drought had forced ewes to be moved to higher pastures for feeding. The tall corn lily plant (*Veratrum californicum*) was singled out as the source of the mutagen that had been dubbed cyclopamine. Approximately two decades later, the research group of Philip A. Beachy, who had identified the gene sequence for the fruit fly Hh, generated knockout mice lacking Sonic Hedgehog (Shh), which is one of three Hh genes identified in mammals. Remarkably, these mice also developed cyclopia, suggesting that cyclopamine would somehow disrupt Hh signaling. Indeed, cyclopamine was subsequently shown to directly target the Smo protein. With the realization that Smo was druggable (targeted by a drug), several groups (including the

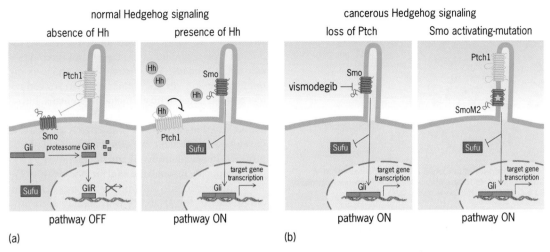

Fig. 3. Normal and cancerous Hh signaling. (*a*) Normal Hh signaling. In the absence of Hh protein (*left*), the Hh receptor Ptch is found in the primary cilium, a cellular antennae-like structure, where it inhibits the activity of Smo. In this state, the DNA-binding protein Gli is inactive as a result of inhibition by the Suppressor of fused (Sufu) protein and proteolytic processing by a large enzyme complex known as the proteasome, which typically destroys proteins. The resulting GliR protein from proteolytic cleavage enters the nucleus and inhibits the expression of genes controlled by Gli. In the presence of Hh (*right*), the Ptch receptor departs from the primary cilium, and the Smo molecule moves into the primary cilium. Once there, Smo activates the Gli protein by inhibiting Sufu and blocking the proteolysis of Gli. The intact Gli molecule then is able to enter the nucleus and induce the expression of Hh-regulated genes. (*b*) Cancerous Hh signaling. A mutation in Ptch resulting in loss of its activity (*left*) induces Smo to accumulate in the primary cilium and to continuously activate Gli. The Smo inhibitor vismodegib (also known as GDC-0449 and Erivedge™) blocks this form of deviant Hh pathway activity. Less frequently, an activating mutation in Smo (*right*; the SmoM2 mutation) similarly gives rise to a continuous Hh pathway response.

Beachy team) isolated synthetic chemicals that were more potent, more specific, and easier to obtain than cyclopamine. One of these molecules termed GDC-0449 (vismodegib) has been approved by the U.S. Food and Drug Administration for treating metastatic basal cell carcinoma.

Perspective. Galvanized by the success of the Hh inhibitor, efforts to target other signal transduction pathways with similar roles in cell fate regulation have gained momentum. The shared dependence of tissue renewal and cancers on these pathways suggests that cancer-initiating cells have properties that are akin to normal stem cells that reside in adult tissues. Whereas this hypothesis continues to be debated, there is general consensus that the study of these pathways and the clinical evaluation of agents that modulate their activity are essential to both regenerative medicine and anticancer therapy.

For background information *see* CANCER (MEDICINE); CELL BIOLOGY; CELL DIFFERENTIATION; CELL FATE DETERMINATION; CILIA AND FLAGELLA; DEVELOPMENTAL BIOLOGY; DEVELOPMENTAL GENETICS; EMBRYOLOGY; FATE MAPS (EMBRYOLOGY); GENE; PROTEIN; SIGNAL TRANSDUCTION in the McGraw-Hill Encyclopedia of Science & Technology.

Xiaofeng Wu; Lawrence Lum

Bibliography. D. Huangfu et al., Hedgehog signalling in the mouse requires intraflagellar transport proteins, *Nature*, 426:83–87, 2003, DOI:10.1038/nature02061; L. Jacob and L. Lum, Deconstructing the Hedgehog pathway in development and disease, *Science*, 318:66–68, 2007, DOI:10.1126/science.1147314; J. J. Lee et al., Secretion and localized transcription suggest a role in positional signaling for products of the segmentation gene *hedgehog*, *Cell*, 71:33–50, 1992, DOI:10.1016/0092-8674(92)90264-D; C. Nüsslein-Volhard and E. Wieschaus, Mutations affecting segment number and polarity in *Drosophila*, *Nature*, 287:795–801, 1980, DOI:10.1038/287795a0; J. Taipale et al., Patched acts catalytically to suppress the activity of Smoothened, *Nature*, 418:892–896, 2002, DOI:10.1038/nature00989.

H5N1 virus (bird flu) controversy

Two groups of researchers created a genetically modified version of bird flu virus that could easily spread among ferrets (the animal model for influenza research) in the laboratory. Their research was driven by the need to understand how influenza viruses can mutate into more lethal strains before they cause pandemics that could kill millions of people. These scientists want to publish their research. However, the announcement of the studies at a conference created controversy among experts, public health officials, and the general community. It started a universal debate over scientific openness versus the fear that terrorists or rogue states might use the influenza "cookbook" to create a global doomsday influenza pandemic.

Background. Influenza is not a new disease. Historians have described outbreaks of this highly contagious respiratory illness as early as 400 BC. The word influenza was derived from the Medieval Latin word *influentia* because it was believed that diseases were caused by bad heavenly fluid or influence. Influenza was used as a general term. In English, diseases correlated with most respiratory

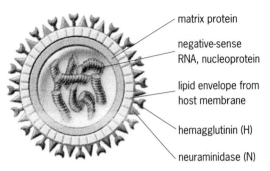

Fig. 1. Structure of influenza A virus. (*Courtesy of M. K. Cowan and K. P. Talaro, Microbiology: A Systems Approach, 2d ed., McGraw-Hill, New York, 2008*)

tract infections became known as influenza, or more commonly the modern flu.

Influenza viruses can cause severe respiratory illness, resulting in hospitalizations and deaths. They are responsible for at least 10 global pandemics over the past three centuries, including three pandemics in the twentieth century alone. The most notorious influenza pandemic is the 1918 Spanish flu pandemic. This pandemic killed more people than acquired immunodeficiency syndrome (AIDS) has killed in 25 years. In fact, it killed more people than the plagues of the Middle Ages killed in a century. Estimated deaths range from 20–50 million, including 675,000 people in the United States. In 2009, there was a pandemic swine influenza that ended up being relatively mild. However, it killed about 1 in 10,000 people who became ill from the virus and claimed about 14,000 lives in the world.

Influenza viruses and their continuing threat. Influenza is caused by three different types of influenza viruses. These are designated influenza type A, influenza type B, and influenza type C. All three types can infect humans. The focus of this article is on influenza type A viruses. Influenza A viruses can infect a broad range of hosts, including humans, birds, pigs, cattle, horses, and marine mammals (for example, dolphins and seals). Influenza A viruses are enveloped and contain two different spikes of glycoproteins, hemagglutinin (H) and neuraminidase (N), which protrude through the viral membrane (**Fig. 1**). They are divided into subtypes or variants based on differences in the H and N viral surface glycoproteins. The N and H spikes undergo variation in their structures. The different subtypes are assigned numbers, for example, H1, H2, H3, N1, N2, and N3. There are 17 different subtypes of H spikes and 9 different subtypes of N spikes, allowing for 153 possible combinations.

All known subtypes of influenza A viruses infect wild waterfowl (for example, ducks and geese), constituting the natural reservoir. Ducks or geese infected with influenza viruses are asymptomatic. Wild waterfowl shed guano (excrement) laden with influenza viruses. It has been found that nearly 100% of chickens infected with bird influenza viruses die. Because influenza is highly contagious and deadly in chickens (and other domesticated poultry, such as turkeys), effective measures to control these outbreaks involve culling (killing) the entire flock of chickens so that the outbreak does not spread to neighboring farms. Only viruses with H1, H2, H3,

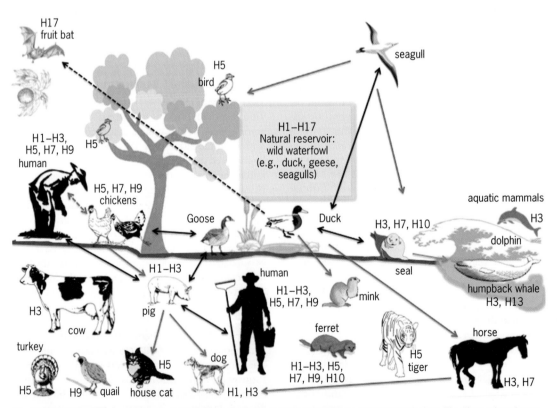

Fig. 2. Habitat of influenza A viruses. Different species harbor different subtypes of influenza A virus. The illustration shows cross-species transmission of influenza A viruses through wild-bird reservoirs.

H5, H7, and H9 subtypes combined with N1 or N2 are known to infect humans. Both human and avian influenza viruses readily infect pigs (**Fig. 2**).

Influenza is a wily virus. The H and N influenza spikes continually undergo two types of changes, known as antigenic shift and antigenic drift. Antigenic shift is a major, abrupt change in the H or N spikes and results from the recombination of genetic material from different viral strains, creating a new influenza virus strain (**Fig. 3**). For example, if a pig were to be coincidentally infected by both a human strain and a bird strain, a new hybrid strain might develop that could then infect a human; the pig served as a blender, somewhat like a mixer, creating a new hybrid virus. Flu pandemics commonly originate in China, where millions of pigs, birds, and people live in close quarters, allowing for new combinations of strains and enormous opportunity for virus leaps among species. Antigenic drift, conversely, is a minor change in the H and N spikes occurring over a period of years.

The outcome of both antigenic shift and antigenic drift is that the virus can evade the antibody defense mechanisms of the host by changing the surface N and H spikes of the influenza virus. Because these changes continually occur, the viral strains that make up flu vaccines need to be adjusted from year to year. Antibodies and immunity against a previous year's viral strains are effective only to the extent that some of these strains may, coincidentally, be the same. If the changes in the virus did not occur, then a flu shot would provide long-term, possibly lifetime, immunity.

Avian or bird flu. In May 1997, a 3-year-old boy in Hong Kong became ill with influenza-like symptoms and died 12 days later. A diagnostic laboratory isolated an influenza A strain that contained an H subtype that could not be identified by any available reagents. In August 1997, it was confirmed that the strain was a highly pathogenic H5N1 avian influenza. This was the first documented case of an H5 subtype of avian origin to infect humans. This jump or species leap of an avian strain directly to humans (bypassing the pig intermediary) had never happened before. The Hong Kong H5N1 strain was also responsible for sudden die-offs of chickens on three rural farms in March 1997.

Experts believe the H5N1 virus in Hong Kong spread from migrating shorebirds to ducks through fecal contamination of water. The virus was then transmitted to chickens and later was established in live bird markets. During transmission among different species, the virus became very pathogenic in chickens and was occasionally transmitted to humans from chickens in the markets. There were 18 human cases of the H5N1 influenza in Hong Kong in 1997. Six of the individuals died, resulting in a 33% mortality rate; this was even more deadly than the 1918 Spanish influenza, which had an estimated mortality rate of 3% of the world's population.

Since 1997, sporadic outbreaks of H5N1 continue to plague eastern Asia and Russia, causing huge economic losses within the poultry industry. Most

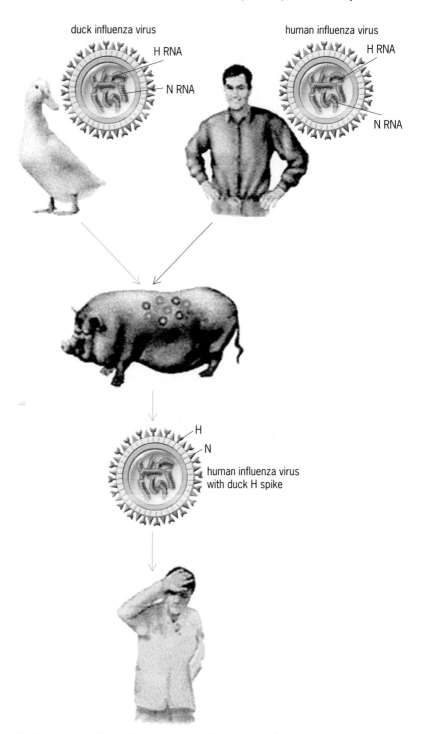

Fig. 3. Antigenic shift of influenza viruses (H: hemagglutinin; N: neuraminidase). (*Courtesy of M. K. Cowan and K. P. Talaro, Microbiology: A Systems Approach, 2d ed., McGraw-Hill, New York, 2008*).

human cases of H5N1 are believed to occur from direct contact with infected poultry. H5N1 is not highly contagious among humans; however, it is highly lethal in humans. From January 2003 to April 2012, there were 602 human cases of H5N1 influenza. Alarmingly, out of these cases, 355 people died (59% mortality rate; **Fig. 4**). Given the fact that influenza A viruses change so rapidly, tracking their antigenic changes in birds, pigs, and other animals might seem impossible. Experts continue to forewarn that we

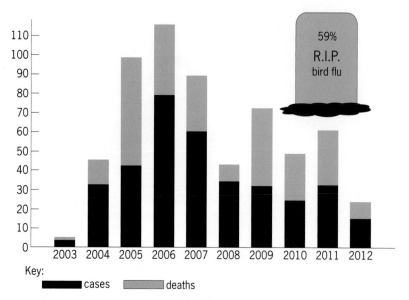

Key:

■ cases   ▨ deaths

Fig. 4. Since 2003, 602 confirmed human cases of H5N1 bird flu have led to 355 deaths. (*Data from World Health Organization Global Influenza Programme as of April 12, 2012*)

must prepare for an unpredictable viral pandemic with catastrophic potential.

Doomsday virus or no big deal. In September 2011, virologists made an alarming announcement to scientists attending the European Scientific Working Group on Influenza Conference held in Malta. Two research teams genetically manipulated an H5N1 influenza virus into a version of virus that could easily pass from ferret to ferret by airborne transmission. The collaborative research was led by Ron A. M. Fouchier from the Erasmus Medical Center in the Netherlands and Yoshihiro Kawaoka from the University of Wisconsin–Madison and the University of Tokyo.

Although experts at the meeting acknowledged that this type of work helps influenza investigators to "prepare for the unpredictable," the National Science Advisory Board for Biosecurity (NSABB) reviewed the findings and placed an embargo on the manuscripts for publication describing the research. The lab-made H5N1 viruses were kept under high security, but there were concerns that publishing the recipe for these viruses in scientific journals could fuel the efforts of terrorists who would try to create a biological weapon or "Armageddon virus." Even more troubling, there was concern that the Fouchier/Kawaoka viruses could escape the laboratory, triggering an influenza pandemic with millions of deaths. Overall, the findings ignited intense public debates and criticism in the media on the benefits and potential harm of this type of research. However, just before the end of 2011, the NSABB advised that the manuscripts could be published as long as the methods to create the viruses would be too vague for potential terrorists. This decision put the burden of ethics on the shoulders of the editors of *Nature* and *Science*.

In January 2012, a letter signed by more than 30 senior influenza researchers was published in

Nature and *Science* declaring a 2-month moratorium on H5N1 research that focused on the airborne transmissibility of the H5N1 bird flu strain among ferrets. The intentions of Fouchier and Kawaoka were to carry out research that would further the ability of scientists to prepare aggressive measures in the event that a mutated H5N1 virus showed up in the wild. The lab-made H5N1 virus provided a model to test the effectiveness of H5N1 vaccines and antivirals against the new H5N1 influenza strain.

In March 2012, Fouchier announced updated research results that differed from the results originally presented in Malta. He clarified that the mutant flu viruses failed to spread among ferrets 100% of the time. He concluded that the lab viruses infecting the ferrets do not spread as well as seasonal influenza viruses or pandemic viruses. A second set of experiments demonstrated that only 1 of 8 ferrets infected with the mutant lab H5N1 viruses developed severe influenza and that none died. In addition, if extremely high doses of the mutant viruses were administered directly into the lungs of six ferrets, all six died; however, through airborne transmission, it was affirmed that the mutant lab H5N1 viruses did not cause severe disease in the ferrets. Anthony Fauci, director of the National Institutes of Allergies and Infectious Diseases, recommended that this updated information should be included in revised manuscripts submitted to *Science* and *Nature*.

Communicating H5N1 research and biosecurity. Journal editors have faced biosecurity issues in the past. An example of research that burdened publishers occurred as early as 2001. Scientists genetically modified a mousepox virus to be 100% fatal in mice by introduction of a single gene, resulting in the virus being superlethal. Mousepox is not known to hurt or infect humans. The work was published purely as a scientific paper at a time when there was no form of biosecurity review.

In 2005, a paper submitted by Stanford University researchers titled "Analyzing a Bioterror Attack on the Food Supply: The Case of Botulinum Toxin in Milk" was published. The researchers developed a mathematical model to predict human deaths if botulinum toxin contaminated thousands of gallons of milk in the United States. In October 2011, the entire reconstructed genetic code of the ancient *Yersinia pestis* bacterium was published, comparing its sequence to modern *Y. pestis* genomes. *Yersinia pestis* caused Europe's fourteenth-century plague (also referred to as the Black Death). The ancient bacterial DNA was isolated from the teeth and skeletons of plague victims buried in cemeteries in East Smithfield of London.

We are thus at a crossroads in which well-intentioned scientific research has the potential to be misused for nefarious purposes. This can be referred to as a dual-use dilemma. When physicists observed atomic fission in 1938, they thought their research may have beneficial applications in medicine and energy production; however, it could also lead to the creation of a devastating atomic bomb. Some of the same discoveries that lead to advancements in medicine and public health can be adapted to the development of weapons of mass destruction.

Dual-use dilemma. In February 2012 at a meeting assembled by the United Nations, experts agreed that full disclosure of the information needed to create the mutant H5N1 lab strains was preferable over a censored version of the research of mutant flu viruses. In addition, editors of both journals (*Science* and *Nature*) have been open to more discussions by influenza researchers, policy experts, and scientists outside of the immediate field before the manuscripts are published. Francis Collins, director of the National Institutes of Health, issued a statement in April 2012 about the flu research: "This information has clear value to national and international public health preparedness efforts and must be shared with those who are poised to realize the benefits of this research."

However, publication of the bird flu studies faced another obstacle. The Netherlands government blocked Fouchier's publication because of export technology rules. Export control laws limit the international shipment of technologies that could be used for dual-use (in this case, the mutant flu viruses could be used as bioweapons). At the time of this writing, the export control restrictions have not been lifted, putting Fouchier's manuscript in limbo.

For background information *see* ANIMAL VIRUS; EPIDEMIC; EPIDEMIOLOGY; FERRET; INFECTIOUS DISEASE; INFLUENZA; PUBLIC HEALTH; VACCINATION; VIRULENCE; VIRUS; ZOONOSES in the McGraw-Hill Encyclopedia of Science & Technology.   Teri Shors

Bibliography. D. Butler, Fears grow over lab-bred flu, *Nature*, 480:421–422, 2011, DOI:10.1038/480421a; D. Butler, Flu surveillance lacking, *Nature*, 483:520–522, 2012, DOI:10.1038/483520a; J. Cohen and D. Malakoff, On second thought, flu papers get go-ahead, *Science*, 336:19–20, 2012, DOI:10.1126/science.336.6077.19; M. Imai et al., Experimental adaptation of an influenza H5 HA confers respiratory droplet transmission to a reassortant H5 HA/H1N1 virus in ferrets, *Nature*, 486:420–428, 2012, DOI:10.1038/nature10831; D. Malakoff, Flu controversy spurs research moratorium, *Science*, 335:387–389, 2012, DOI:10.1126/science.335.6067.387; M. Specter, The deadliest virus, p. 32, *The New Yorker*, March 12, 2012.

Higgs boson detection at the LHC

On July 4, 2012, two collider experiments, CMS (Compact Muon Solenoid) and ATLAS (a toroidal LHC apparatus), at the Large Hadron Collider (LHC) simultaneously announced the discovery of a particle with mass 125 GeV (about 130 times heavier than the proton), which is consistent with the standard-model Higgs boson. The Higgs boson is the final particle to complete the picture of the standard model. It was proposed more than 40 years ago by a number of physicists, including P. W. Higgs, for whom the particle was named. The Higgs boson is the remnant of the so-called Higgs mechanism, which

was designed to give masses to fermions and gauge bosons. The discovery of the Higgs boson can further consolidate our understanding of the origin of mass and electroweak symmetry breaking. This is perhaps the most significant discovery of modern particle physics. It will have influence on all theoretical and experimental research in the future.

Standard model and the Higgs mechanism. The standard model of particle physics consists of a set of matter particles or fermions called leptons and quarks, and a set of force carriers responsible for the electromagnetic, weak, and strong interactions. Leptons consisting of charged leptons and neutrinos come in three flavors—electron, muon, and tau. Quarks, however, come in six flavors—up, down, strange, charm, bottom, and top. They increase in mass from a few thousandths of a gigaelectronvolt (GeV) to more than 100 GeV. A schematic diagram is shown in **Fig. 1**.

So far we are certain of a theory, which is called the standard model, based on a gauge symmetry $SU(2) \times U(1)$, from all the previous experimental data and the discovery of the W and Z bosons in 1983. The immediate problem confronting such a theory is that the gauge symmetry requires all the particles to be massless; this is a problem because we know that, in reality, each particle, except for the photon, has a mass. Thus we come to a deadlock: We have a beautiful theory that can unify weak and electromagnetic interactions but fails to explain mass. Higgs and others proposed the so-called Higgs mechanism that can generate the masses of fermions and gauge bosons. To understand the Higgs mechanism, we can imagine the following.

Suppose the universe started out with gauge symmetry, and every particle in it was massless. Theorists create a potential in the universe by introducing a Higgs field ϕ, which spreads across the whole universe. The idea is simple, similar to what we learn about the electric or gravitational potential. A test

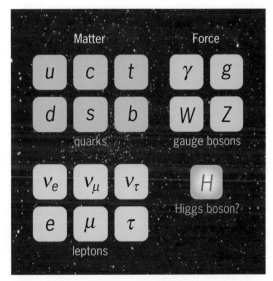

Fig. 1. Structure of the standard model of particles and forces.

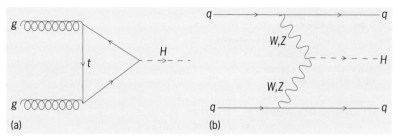

Fig. 2. Feynman diagrams for (*a*) gluon fusion and (*b*) vector-boson fusion (VBF).

charge always wants to go to where the potential is the lowest—the most stable place. Imagine that at the beginning, this potential has a minimum at the origin of the Higgs field, $\phi = 0$, which respects the gauge symmetry. Then, because of some more fundamental reasons, the potential changes and develops another lower minimum away from the origin. So the next thing that happens is that everything will go to the new minimum, because it is more stable. The whole universe will run to the new minimum— we can say that it is like a bubble sweeping across the whole universe. At the new minimum, the original gauge symmetry is not respected anymore, and thus we have breaking of the original gauge symmetry, a phenomenon called electroweak symmetry breaking. As a consequence, almost every particle receives a mass, thus solving the problem of the origin of mass.

The Higgs boson is a remnant of this picture. The finding of this Higgs boson is strong evidence that what has been described in this picture does happen. In the foregoing description, we mentioned that there are some more fundamental reasons why the potential develops a new minimum away from the origin. The standard model does not provide any such reason, other than as an ad hoc condition. Indeed, the fundamental question of how the potential develops a new minimum and thus triggers electroweak symmetry breaking is an active research area. For example, supersymmetry, a symmetry framework between bosons and fermions, provides a dynamical mechanism that triggers the breaking of the electroweak symmetry. Current experiments are also actively searching for the evidence of supersymmetry.

Higgs boson production and decay. At the LHC, the most dominant production of the Higgs boson comes from gluon fusion, and the next dominant by vector-boson fusion (VBF). The Feynman diagrams are shown in **Fig. 2**. Gluons residing inside the proton can annihilate, via a loop of top quarks, into the Higgs boson; while the W or Z bosons annihilate into the Higgs boson in VBF. The most important difference between gluon fusion and VBF is the two accompanying jets in VBF, as shown in Fig. 2. Experiments can develop a set of selection cuts to identify the jets, called jet-tag, to distinguish between the two types of fusion. Gluon fusion depends on the particles running in the loop, thus allowing any other particles from any new theory to run in the loop;

while VBF is a clean probe of electroweak symmetry breaking.

The mass and decay modes of the Higgs boson and contaminations from other experimental backgrounds determine the detection method for the Higgs boson. Experiments have no presumed mass range for the Higgs boson, and so they have to look everywhere allowed by theory and previous searches, namely, from 115 GeV up to 1000 GeV, although there are hints from experiments at the Large Electron Positron (LEP) Storage Ring (which was closed down in 2000 to make way for the LHC) that the Higgs mass should be less than 180 GeV. The decay branching ratios of the Higgs boson are shown in **Fig. 3**. The curve $\gamma\gamma$ down near the bottom of the figure turns out to be the most important mode for masses between 115 and 130 GeV. The branching ratio is about 2×10^{-3}, that is, this mode happens twice in a thousand decays, while the other decay modes either suffer from large backgrounds or the detection is not clean enough. **Figure 4** is a computer-aided picture of the decay $H \rightarrow \gamma\gamma$; this is called diphoton production. The two green lines denote the two photons, for which the energy and direction can be recorded by the detector. The energy-momentum four-vector of the photons can be added to obtain the mass of their parent particle—invariant mass $m_{\gamma\gamma}$. The CMS and ATLAS groups then plotted all the events with two photons versus the invariant mass $m_{\gamma\gamma}$, and surprisingly they found a bump near 125–126 GeV (**Fig. 5**). One of the experiments, CMS, carefully estimated the statistical significance of such a bump and found the probability that the background can fluctuate to the bump is about 10^{-5} (corresponding to a 4σ deviation in Gaussian distribution), which is very small but still cannot be claimed as a discovery. Particle physicists are very careful about claiming a discovery as they need to establish a probability as small as 10^{-7} (about 5σ). The ATLAS experiment found something similar.

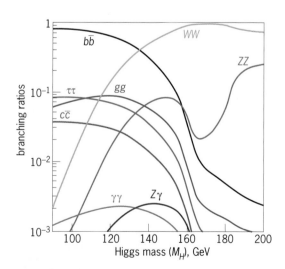

Fig. 3. Decay branching ratios of the Higgs boson, plotted as functions of the Higgs mass (M_H). (*Courtesy of LHC Higgs Cross Section Working Group, https://twiki.cern.ch/twiki/bin/view/ LHCPhysics/CrossSections*)

What else could they do? They also looked for other decay modes of the Higgs boson. The second best channel is $H \rightarrow ZZ^* \rightarrow llll$, where the final state consists of four charged leptons, including $e^+e^-e^+e^-$, $\mu^+\mu^-\mu^+\mu^-$, and $e^+e^-\mu^+\mu^-$. The Z boson is the neutral weak gauge boson. The electrons and muons are very clean objects in the detector and therefore their momenta can be precisely measured, so that the mass of their parent particle can be reconstructed as m_{4l}. Again, experiments then plotted all the events with four charged leptons versus the invariant mass m_{4l}, and amazingly they also found a bump near 125–126 GeV. The CMS experiment carefully estimated the statistical significance of the bump and found that the probability for the background to fluctuate to the bump is about 10^{-3} (about 3σ). ATLAS also found a similar bump. Since the bump from the two-photon final state coincided with the bump from the four-charged-lepton final state, the statistical significance could be added around 125–126 GeV, which summed up to about the 5σ level (corresponding to a probability of 10^{-7}). The ATLAS results had similar significance. A discovery was then claimed by both CMS and ATLAS experiments.

What is next? The discovery of the Higgs boson marks the beginning of a new era in particle physics. Theoretical arguments tell us that the standard model with a Higgs boson cannot be the final theory, because such a Higgs boson with a mass of 125 GeV will receive a huge quantum correction as large as the grand unification scale. Some forms of new physics have to appear around the teraelectronvolt (TeV) scale. Supersymmetry will then be a strong candidate because it predicts a Higgs boson with mass below 130 GeV. Theorists have already begun to limit the parameter space of various supersymmetry models or extended Higgs models. *See* NATURAL SUPERSYMMETRY.

Conversely, since the observed rate of diphoton production is somewhat above the expectation of the standard model Higgs boson, while the other modes are slightly below it, although most of them still have large statistical uncertainties, some other types of scalar particles have been proposed as speculations, such as the radion field of the Randall-Sundrum model and the fermiophobic Higgs boson. These models can have a scalar boson whose mass and decay modes are similar to those of the current data, thus mimicking the standard model Higgs boson. We have to find a method to distinguish these models. The best way to do so would be to use VBF. From the Feynman diagram in **Fig. 2b**, VBF directly probes the couplings of the scalar boson to the W and Z bosons, which are linked to electroweak symmetry breaking. Using dijet tagging techniques, the VBF can be singled out. Comparisons of various proposed models can be seen in **Fig. 6**. All these models can give rates comparable to the standard model Higgs boson in the gluon-fusion process, but it is clear that the predictions in VBF are very different. The radion can give only a very small rate in VBF, while the fermiophobic (FP) model will give a very large rate, and the other models predict a range.

Fig. 4. An event showing the decay $H \rightarrow \gamma\gamma$. *(Courtesy of J. Incandela, Status of the CMS SM Higgs Search, CERN Webcast seminar, July 4, 2012, https://indico.cern.ch/ conferenceDisplay.py?confld=197461)*

Testing the production rates in VBF can help distinguish various models and identify the nature of the newly discovered boson.

Conclusions. The Higgs boson detection is not the end of particle physics, even though the picture of the standard model is completed, but the beginning

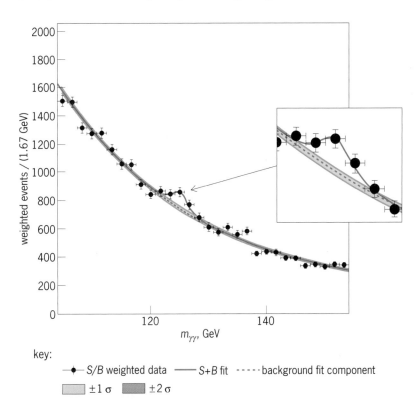

Fig. 5. Event rate of two-photon final state versus the invariant mass $m_{\gamma\gamma}$. Here, S stands for the signal from the Higgs boson and B stands for the continuum background. The broken line is a smooth fit to the background. The $\pm 1\sigma$ and $\pm 2\sigma$ bands represent 1 and 2 standard deviations from the background fit. A clear bump, above statistical fluctuation, is seen at around 125 GeV. *(Courtesy of J. Incandela, Status of the CMS SM Higgs Search, CERN Webcast seminar, July 4, 2012, https://indico.cern.ch/conferenceDisplay.py? confld=197461)*

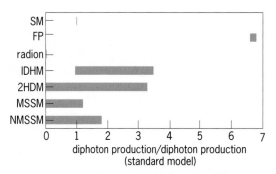

Fig. 6. Ratios of diphoton production rates in various models relative to the standard model in vector-boson fusion. Here, SM, FP, IDHM, 2HDM, MSSM, and NMSSM denote standard model, fermiophobic, inert doublet Higgs model, two Higgs doublet model, minimal supersymmetric standard model, and next-to-minimal supersymmetric standard model, respectively. (*Courtesy of J. Chang et al., Distinguishing various models of the 125 GeV boson in vector boson fusion, in press, http://arxiv.org/abs/1206.5853*)

of a new era. Properties of the discovered boson will be investigated in full strength with both experimental and theoretical efforts. The directions of a TeV-scale theory will be more transparent. We are now one step closer to the theory of everything.

For background information *see* ELECTROWEAK INTERACTION; GAUGE THEORY; GLUONS; HIGGS BOSON; INTERMEDIATE VECTOR BOSON; PARTICLE ACCELERATOR; QUARKS; STANDARD MODEL; SUPERSYMMETRY; SYMMETRY BREAKING in the McGraw-Hill Encyclopedia of Science & Technology.   Kingman Cheung

Bibliography. J. Chang et al., Distinguishing various models of the 125 GeV boson in vector boson fusion, in press; K. Cheung and T. C. Yuan, Could the excess seen at 124–126 GeV be due to the Randall-Sundrum radion?, *Phys. Rev. Lett.*, 108:141602 (5 pp.), 2012, DOI:10.1103/PhysRevLett.108.141602.

High-power diode lasers

Over the last decade, high-power diode lasers have experienced significant performance and reliability improvements, creating new opportunities in various fields of science and technology. Today, high-power, broad-area diode laser chips and bars are well-established laser sources for a variety of applications, including telecommunications, materials processing, medical applications, and solid-state laser

pumping. These compact lasers can provide high optical power, high efficiency, and long lifetime at low cost. These improvements have been made possible by optimal design, improved epitaxial growth, and mature fabrication techniques. This article highlights the recent progress in the design, fabrication, and performance of high-power diode laser chips and bars.

Design and development. A Fabry-Perot edge-emitting diode laser is a forward biased *P-I-N* heterojunction in which the *I*-(intrinsic) section consists of direct-band-gap quantum-well gain material and a low-loss waveguide. The **illustration** is a schematic of a typical broad-area Fabry-Perot laser.

A key figure of merit for high-power diode lasers is their power conversion efficiency, also known as wall-plug efficiency (η_{wp}). It is defined as the ratio of emitted optical power (P_{opt}) to injected electrical power (P_{elec}). The power conversion efficiency of diode lasers is the highest among all types of light sources. In order to maximize power conversion efficiency, one has to reduce the internal optical loss and to maximize differential quantum efficiency, defined as the ratio of the number of emitted photons to the number of injected charge carriers.

High-power diode lasers are extremely efficient in converting electrical energy into laser output, with conversion efficiency better than 70%. However, the excess power not converted into light is converted into heat. As the laser operates at higher electrical power, a fraction of the injected electrical power is dissipated in the form of heat, which in turn causes a rise in the temperature of the active region, causing degradation of the threshold current and efficiency. The rise in the active region temperature is given by the equation below, where Z_T is the thermal impedance of the laser (its resistance to the

$$\Delta T = Z_T \cdot (P_{\mathrm{elec}} - P_{\mathrm{opt}}) = Z_T \cdot P_{\mathrm{elec}} \cdot (1 - \eta_{\mathrm{wp}})$$

flow of heat). For high-power operation, this thermal impedance should be minimized. This objective is achieved by bringing the active region very close to the heatsink through *p*-side-down mounting of the laser. The use of quantum-well and strained quantum-well structures has been found to improve temperature characteristics of the lasers. The thermal resistances of the solder layer and of the heatsink itself must also be taken into account.

Two factors limit the output power of high-power Fabry-Perot lasers: thermal rollover and catastrophic optical mirror damage (COMD). Thermal rollover, which is the saturation of output power due to laser heating, is related to the degradation of the threshold current density and conversion efficiency due to the heating of the device under continuous-wave (CW) operation. This happens when nonradiative recombination becomes significant, causing major heat generation in the active region and degradation of the laser performance. Reducing internal optical loss and maximizing conversion efficiency can help to eliminate thermal rollover. Recent progress has resulted in internal losses smaller than 0.6 cm^{-1}. With

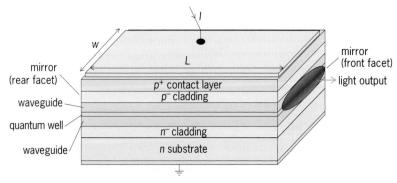

Schematic of a high-power Fabry-Perot laser.

the availability of high-quality n-type substrates and high-precision, defect-free crystal growth, the performance of high-power diode lasers has significantly improved to the point that the thermal rollover is not the main limiting factor.

While the progress in design and growth has helped to practically eliminate thermal rollover, COMD has become the limiting cause of the achievable output power. COMD degradation of the diode facets is caused by facet heating due to nonradiative surface recombination of carriers. Facet heating results in a reduced band-gap energy of the material, which consequently increases the absorption coefficient at the lasing wavelength, causing the rise in temperature to exceed the melting point of the laser crystal. Pushing the COMD threshold to higher-power densities requires a reliable facet passivation technique, which involves removing any nonradiative centers from the facets and then protecting the facets from further oxidations or other impurities by depositing (in general) a lossless protective dielectric material. The maximum CW output power achievable in diode lasers depends on the quality of facet passivation. Several passivation techniques have been developed targeting removal of surface oxides and dangling bond imperfections, and minimizing optical absorption at the facets. Devices with improved passivation have demonstrated COMD thresholds as high as 30 MW/cm^{-2}.

Recent progress. Outstanding performance and reliability of high-power diode lasers have been achieved by optimization in every aspect of laser development, including growth, processing, passivation, and packaging. High-precision, defect-free epitaxial structures with accurately controlled layer thicknesses and compositions are grown by molecular beam epitaxy (MBE) or metal-organic chemical vapor deposition (MOCVD) in multiwafer reactors. Gallium arsenide (GaAs)-based laser structures covering the 780–1000-nm wavelength range have proven to have the highest power and performance among the semiconductor lasers. Strained quantum-well active regions are used in providing higher optical gain and improved temperature characteristics. The growth of aluminum-free structures with simpler passivation techniques has pushed the COMD threshold to higher powers.

High-power Fabry-Perot lasers are cleaved into individual chips or bars with a cavity length in the range of 1–3 mm. The reflectivity of the mirrors is then controlled by passivation and application of dielectric coatings to the front and rear facets. Typical values of mirror reflectivity are $R_1 < 10\%$ for the front facet and $R_2 > 95\%$ for the rear facet. Efficient heat dissipation from the active region is achieved by bringing the active layer close to a heatsink. To this end, the chips are mounted p-side down on cooled heatsinks (such as copper or CuW) using high-temperature solders. High-temperature solders such as AuSn provide high thermal conductivity combined with long-term reliability.

Diode lasers operating in the 780–980-nm wavelength range with conversion efficiencies in excess of 65% are commercially available. Single-emitter InGaAs/GaAs/AlGaAs or InGaAs/GaAs/InGaP lasers at wavelengths of approximately 980 nm, with CW output powers in excess of 20 W, have been reported for an $\sim$100-μm emitting aperture. Commercially available fiber-coupled single-emitter 980-nm lasers can provide over 10 W of CW output power. There are three main material choices for 808-nm devices (which are ideal for pumping solid-state materials such as Nd:YAG): tensile-strained GaAsP, compressively strained InGaAsP, or compressively strained AlInGaAs. Single-emitter output power in excess of 10 W is achieved at 808 nm. Two types of active layers, aluminum-based AlGa(In)As or aluminum-free Ga(In)AsP, are used for 760–790 nm lasers. Broad-area lasers with output powers of over 7 W have been achieved without saturation.

In a single-emitter laser with a COMD threshold of approximately 20 MW/cm^2 and an emitting aperture of 100 μm, the maximum achievable power before catastrophic degradation can reach just over 20 W. Power scaling by increasing the emitting aperture toward the millimeter regime causes unwanted amplified spontaneous emission and parasitic lasing in the lateral direction, drastically degrading the performance of the laser. In order to achieve usable outputs in the hundreds of watts and kilowatts, laser bars and stacks have been developed. In a laser bar, N broad-area single emitters ($N = 7$–30 of them), each with an emitting aperture of approximately 100 μm, are placed in a one-dimensional array, providing a power scaling of almost N times that of single-emitter lasers. An important characteristic of a diode bar is the fill factor, defined as the ratio of the pumped area to the total area of the bar. In order to increase the maximum power, high-fill-factor bars with long cavities and high power conversion efficiency are needed. However, as the fill factor increases, heat dissipation becomes even more important. As a result, thermal rollover is presently the main limitation on the maximum achievable power in diode bars.

In the past few years, significant progress has been reported in the maximum achievable power. The minimizing of operating voltage and internal loss, in combination with the use of high-efficiency, long-cavity lasers, has been critical in achieving higher power. Peak optical power in the range of 400–500 W has been demonstrated from a high-fill-factor, 1-cm-wide laser bar. On the path to kilowatt laser bars, highly efficient, two-sided cooling of the bars with 5-mm cavity length has provided maximum power of over 800 W from a 77%-fill-factor array at 25°C (77°F). Laser bars with a CW output power of 80–100 W in the 800–980-nm wavelength range are commercially available. These bars have a typical fill factor of 20–50% and power conversion efficiencies better than 60%. For kilowatt output powers, laser bars are stacked together. Commercially available stacks can deliver output powers in excess of 4 kW.

In addition to high CW output power, several applications require wavelength stability, narrow linewidth, and high beam quality over a wide range of temperature. In broad-area devices, beam filamentation (loss of coherence) is the main effect that limits the device brightness and beam quality. To increase brightness by inhibiting multi-transverse-mode operation for increased coherence, tapered unstable resonators have been used to maintain transverse-mode selectivity. Similarly, in Fabry-Perot diode lasers, as the injection current increases, more longitudinal modes can reach threshold, resulting in broad spectral linewidth emission and wavelength shift. In order to narrow the spectral linewidth, wavelength-selective mirrors (such as gratings) are typically used. Distributed feedback and distributed Bragg reflector lasers are used to generate spectrally narrow linewidth emission with less temperature sensitivity. For spectral linewidth stabilization, volume Bragg gratings (a wavelength-selective periodic structure formed in a transparent material) have also been deployed.

For background information *see* CHEMICAL VAPOR DEPOSITION; CRYSTAL GROWTH; LASER; QUANTIZED ELECTRONIC STRUCTURE (QUEST); SEMICONDUCTOR DIODE; SEMICONDUCTOR HETEROSTRUCTURES in the McGraw-Hill Encyclopedia of Science & Technology.
Mahmoud Fallahi

Bibliography. L. Bao et al., Performance and reliability of high power 7xx nm laser diodes, *Proc. SPIE*, 7953:79531B-1–12, 2011, DOI:10.1117/12.875842; D. Botez, Design considerations and analytical approximations for high continuous wave power, broad-waveguide diode lasers, *Appl. Phys. Lett.*, 74(21):3102–3104, 1999, DOI:10.1063/1.124075; P. Crump et al., Advances in spatial and spectral brightness in 800–1100 nm GaAs-based high power broad area lasers, *Proc. SPIE*, 7483:74830B1-10, 2009, DOI:10.1117/12.829617; C. H. Henry et al., Catastrophic damage of $Al_xGa_{1-x}As$ double-heterostructure laser material, *J. Appl. Phys.*, 50(5):3721–3732, 1979, DOI:10.1063/1.326278; M. Ziegler et al., Surface recombination and facet heating in high-power diode lasers, *Appl. Phys. Lett.*, 92:203506 (3 pp.), 2008, DOI:10.1063/1.2932145.

Independent system operator

An independent system operator (ISO) is the transmission operator of, typically, a large-area transmission system owned by more than one transmission owner and financially independent of all owners, producers, buyers, transmitters, loads, and other participants. The purpose is to permit all producers, buyers (loads and intermediaries), and other users to use the transmission system on an equal basis. That is, any generator that wants to connect to the transmission system, whether or not associated with the transmission owner, is treated in the same way. Anyone who wants to schedule use of the transmission system (such as a party that wishes to schedule a flow across a system to be delivered remotely) must be permitted to do so on an equal basis with the loads served by that transmission owner.

Background. Beginning as early as the 1970s, various countries and governments began considering the deregulation of various industries that had previously been considered as necessarily regulated by government agencies. Initial drivers were the new entrants into the industry who wanted access to the large infrastructure supporting operations. For the telephone communications system, these new entrants were manufacturers or providers of telecommunications equipment or small telephone systems that wanted equal access to the larger monopoly or quasi-monopoly system. Their initial efforts began in the 1960s. For airline deregulation, the drivers were smaller airlines who wanted access to airport landing slots (times and gates) as well as passenger routes and existing airlines, who felt they could provide better service and lower prices under a competitive model. Those efforts were largely complete by 1978.

With the deregulation of the telephone and airline industries and the earlier deregulation of the railroad transportation industry, government regulators began to consider whether the electric power industry would not also benefit from some kind of deregulation. As with earlier industry deregulation, the drivers were independent producers who wanted access to captive utility customers and customers who felt they were subsidizing other customers through the regulated rates they were paying.

Deregulating the electric industry. Various governments and countries had different ownership models, and deregulation of electric power industries took different paths in different countries. This article concentrates on the United States and Canada (the regulatory discussion is for the United States only), though the concept of an ISO came to be adopted in many parts of the world. In the United States, the industry was primarily served (in terms of load) by vertically integrated (owning and operating generation, transmission, and distribution facilities) investor-owned utilities. Significant geographic areas were (and are) served by government- and customer-owned systems, but serving a smaller total load. In the United States, investor-owned utilities were typically regulated by a state regulatory commission, part of the state government. The Federal Energy Regulatory Commission (FERC) was charged with overseeing the interstate operations of the electric system.

The U.S. Congress passed the Energy Policy Act of 1992 largely to address energy efficiency, but also requiring state regulators to assure that their regulated utilities did not have competitive advantages over smaller companies. This resulted in various actions by utilities and especially independent generators and aggregators. The level of interutility transactions rapidly increased, and various parties complained that access to the transmission system was not comparable for all parties. Seeing a need to regulate interstate commerce, because transmission lines cross state boundaries and transactions of energy routinely cross state boundaries, the FERC, through its normal rules proposal process, issued Order 888, *Promoting Wholesale Competition*

Through Open Access Non-discriminatory Transmission Services by Public Utilities, in April 1996.

The FERC defines an ISO. Order 888 is the primary deregulation FERC order for the U.S. electric utility industry, and it discussed ISOs and provided a list of ISO principles. In providing guidance for ISOs, the FERC noted that some industry members had been considering setting up ISOs as a way of promoting competition and open access to the transmission system. Order 888 listed these ISO principles, with discussion for each:

1. The ISO's governance should be structured in a fair and nondiscriminatory manner.

2. An ISO and its employees should have no financial interest in the economic performance of any power market participants.

3. An ISO should provide open access to the transmission system and all services under its control at nonpancaked rates pursuant to a single, unbundled, grid-wide tariff that applies to all eligible users in a nondiscriminatory manner. ("Pancaked rates" refers to the practice in place before the FERC issued Order 888, whereby each small transmission zone through which a transmission transaction was scheduled added its own transmission charge, thus "stacking up" or "pancaking" the total transmission charge. "Unbundled tariff" references the solution to pancaked rates.)

4. An ISO should have the primary responsibility in ensuring short-term reliability of grid operations.

5. An ISO should have control over the operation of interconnected transmission facilities within its region.

6. An ISO should identify constraints on the system and be able to take operational actions to relieve those constraints within the trading rules established by the governing body.

7. The ISO should have appropriate incentives for efficient management and administration and should procure the services needed for such management and administration in an open competitive market.

8. An ISO's transmission and ancillary services pricing policies should promote the efficient use of and investment in generation, transmission, and consumption.

9. An ISO should make transmission system information publicly available on a timely basis via an electronic information network consistent with the Commission's requirements.

10. An ISO should develop mechanisms to coordinate with neighboring control areas.

11. An ISO should establish an Alternative Dispute Resolution (ADR) process to resolve disputes in the first instance.

The FERC noted that several entities were considering establishing an ISO, and encouraged their formation. Order 888 specifically addressed the operations of what it called "tight pools," and required four named "tight pools" to submit restructuring documents by the end of 1996, noting that several already had been considering restructuring before the Order was issued. The reason for requiring restructuring of those pools was that their transmission owner members essentially shared use of the coordinated transmission system on an equal basis. Those four "tight pools" were

1. New York Power Pool (NYPP), which became New York ISO (NY-ISO).

2. New England Power Pool (NEPOOL), which became ISO New England (ISO-NE).

3. Pennsylvania-New Jersey-Maryland Interconnection (PJM), which became PJM Interconnection.

4. Michigan Electric Coordinated Systems (MECS), which became an independent transmission company (ITC).

While it was a difficult process for those involved, the first three of these set up organizations that were eventually determined by the FERC to be ISOs. MECS took a different route, and set up an ITC by transferring the transmission assets of the utilities to a separate entity. Various methods were used to isolate the original transmission owners from the ITC. For example, one company transferred its transmission assets into a separate transmission company and transferred the stock in the transmission company to the stockholders of the utility. The ITC owns and operates the transmission system independent of the generators and loads connected to its system. In many respects, it meets the FERC principles for an ISO.

Due to the prior operations of NYPP, NEPOOL, and PJM, the three pools that became ISOs, they also became the market operators for their ISOs. While the model of the ISO also being the market operator is fairly common in ISO formation and operations, it is not required by Order 888. By operating the market, the ISO has the immediate authority to dispatch and direct the operation of generators for economics and reliability, control ramp rates for transactions, and provide reserves for contingencies. [Ramp rates are the permitted total changes in transaction levels over a short time period (ramp period), typically five or ten minutes, centered on the hour or half-hour.] There are advantages to being both ISO and market operator.

ISOs today. As ISOs came into being under FERC Order 888, other associations of transmission owners and generators saw benefits to being in an ISO. In North America, as of 2012, there are 10 ISO/RTO organizations (see **illustration**). [A regional transmission organization (RTO) is different from an ISO only in scope—it serves a larger area.] ISOs have considerable variation in their organizations. Due to the forced schedule of the three initial ISOs in what was a very contentious environment, they tend to have a high level of stakeholder activity and independent boards of directors. The California ISO also came into being under pressure from regulators in the state. Other ISOs in North America developed more naturally from the expectation of benefits by multiple parties. There is a tendency for the stakeholders in those ISOs to defer more to the ISO organization for the initiation of projects or changes. For example, in the PJM Interconnection, PJM will plan the transmission system enhancements to assure that all reliability issues are addressed. Others are expected

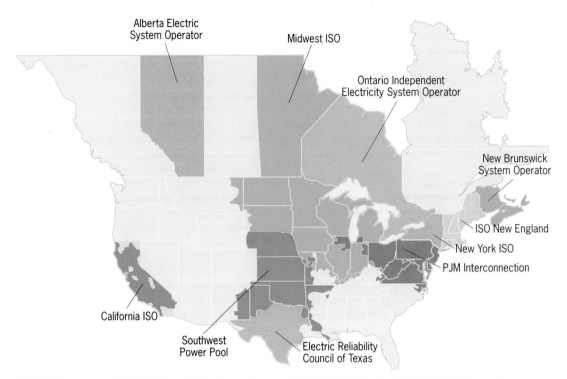

ISO/RTO Council map. (*ISO/RTO Council, http://www.isorto.org/site/c.jhKQIZPBImE/b.2604471/k.B14E/Map.htm*)

to submit plans or proposals for enhancements to increase efficiency or decrease congestion. The stakeholders in MISO (Midwest ISO) expect it to address both reliability and efficiency with its transmission expansion proposals.

ISOs are in place in many parts of the world. When the United Kingdom privatized the electric systems of England, Wales, and Scotland, National Grid became the owner and operator of the transmission system for England and Wales. A separate system is set up in Scotland and does not function as an ISO. The European Union has promoted ISOs as a way of encouraging competition, open access, and freedom for end users to choose. Just as not all areas of North America are served by ISOs, not all areas of Europe are served by ISOs. The regulatory approach in the European Union was to "unbundle" transmission operations from generation ownership. The term generally used in Europe is Transmission System Operator (TSO). The model is different from the U.S. ISO model. European TSOs not only operate the transmission system but also own (or lease) and maintain the system. In general, the goal of the European Union is to have a fully integrated European Electric System with full access, even if operated by different TSOs.

There are transmission operation organizations in South America, Asia, and Australia/New Zealand that operate in line with the principles defined by the U.S. FERC, but that would not be considered ISOs.

For background information *see* ELECTRIC DISTRIBUTION SYSTEMS; ELECTRIC POWER GENERATION; ELECTRIC POWER SYSTEMS; ELECTRIC POWER TRANSMISSION in the McGraw-Hill Encyclopedia of Science & Technology.   Richard Kafka

Bibliography. B. Arizu, W. H. Dunn, Jr., and B. Tenenbaum, *Transmission System Operators—Lessons from the Frontlines*, Energy and Mining Sector Board, World Bank Group, Washington, D.C., 2002.

Indoor navigation for first responders

On December 3, 1999, a tragedy occurred in Worcester, Massachusetts, at a fire in a large aging brick warehouse, the former Worcester Cold Storage and Warehouse Co. building. What at first seemed like a routine response situation by the Worcester Fire Department turned out to be a deadly incident resulting in six firefighters losing their lives when they became disoriented and lost in the complex structure.

This event had a major impact on the Worcester Polytechnic Institute (WPI), located just a few blocks from the warehouse. It motivated members of the Electrical and Computer Engineering Department to develop technology that would help prevent such a tragedy in the future. If the firefighters had had better knowledge of where to look for their fallen companions, they might all have been able to get out safely. The goal of the research conducted at WPI and reported here is the creation of a location system for personnel such as firefighters that requires no preexisting infrastructure and that can perform accurately in a hazardous, near-zero-visibility, radio-frequency- (RF-) propagation–challenged indoor fire environment and in the context of the extreme physical activities of firefighters. Note, too, that for a variety of reasons, the system enables the incident manager to track and locate each responder, but those responders will not have that location information. This is different from the situation with indoor-based civil location services, where individuals need to know where they are with respect to the location of services or stores that they wish to use. These services tend to rely on preexisting infrastructure.

There are many RF-based technologies available, such as RFID and Wi-Fi fingerprinting, that provide rudimentary location information, but these systems also require a prior mapping survey or permanent instrumentation of the building to be effective. In the case of the warehouse fire described above, there was no power in the building, no floor plans, and no infrastructure in place—that is, a situation like that found in nearly every fire ground, according to the fire service. Since these conditions are typical of what is found at an incident site, whether it is a residential or a commercial structure, a feasible and effective system must also work without any assumptions regarding what is known about the building and without any infrastructure in place.

In the system to be described, an incident commander would be constantly updated with the locations of his/her personnel. Since every minute is critical in the case of a fire incident, the system must be able to be put into operation quickly, without requiring any calibration, manual configuration, or complicated setup.

There is much debate on the required level of accuracy for a first responder location and tracking system, and ideas vary depending upon whether the system is targeted for firefighters, law enforcement, or other users. Most first responders indicate that providing the correct floor level ranks first in importance, followed by identifying the responder's position on the floor. The system should also record position estimates over time so that the paths taken by personnel are available in case it is necessary to direct personnel back the way they came or to send a rescuer to them. Other features of a fully deployed system may include a wireless data channel for sending information such as environmental and physiological conditions of the individuals being tracked.

Many approaches have been explored and solutions proposed to address the indoor location problem. The existing Global Positioning System (GPS), for example, uses RF electromagnetic waves to perform position estimation, but it suffers from poor accuracy indoors in its standard deployment. Other approaches include inertial navigation; machine vision; signals of opportunity, such as television signals; and ultrasound-based systems. These systems sometimes use a single-sensor technology, but recently there has been a trend toward combining many different sensors with some kind of data fusion to produce an integrated result.

The system described here uses RF signals for position estimation with a deployable architecture and much more complex signal fusion and solution processing than that found in GPS or in typical range- or signal-strength–based systems using ad hoc or dynamically distributed mesh networks. One of the chief advantages of an RF-based approach is the lack of error growth over time (drift), which is a prevalent problem with inertial systems and one that is significantly exacerbated by the high physical impact nature of many firefighter activities. Another problem that may be experienced by GPS or inertial-based systems is long acquisition or calibration times required after the systems are powered on or the periodic calibration steps required during the use of an inertial-only system. With the specific RF system being developed, the position solution at a given location will be just as accurate many hours after deployment as it is immediately after deployment. Furthermore, there is no need to have any outdoor receivers in place and operating prior to the first responders entering the building and prosecuting their mission, unlike the case for some inertial and some ad hoc radio-based schemes. The first responder community is quite adamant about needing a system that can be deployed as time permits and not as a precursor to beginning lifesaving activities. Further, any system must be small, lightweight, and not too expensive for widespread adoption.

RF-based indoor position estimation. The top-level architecture of the system described here is inverted compared to the GPS system architecture in as much as each person to be tracked and located inside a building wears a low-cost lightweight transmitter device with an antenna to generate a signal, and the "satellite" units outside the building act as receivers. The receiving units, by exchanging data and working in concert, then process the received signals to determine the location of the transmitter in three dimensions. A driving motivation behind the design decisions of the system architecture was to ensure a final deployable system that would be as inexpensive as possible. This meant that the units carried by each of the personnel to be tracked should be as simple as possible in construction, while the smaller number of receiving stations could have more processing capability.

The system concept is illustrated in **Fig. 1**a. As shown in Fig, 1b, the current implementation aggregates multiple receivers (four) into a unit that also provides the data channel necessary for remote system control and transmission of the captured RF signal waveforms to the central processor and display unit (not shown in the figure).

At the core of any radio location scheme is the exploitation of the effects of radio propagation, either the attenuation or the time delay as a function of distance or some combination of both, as described in the following sections.

Time of arrival–based location estimation. The simplest form of propagation delay–based RF location is based on the concept that the range between devices can be estimated by the round-trip time of a signal exchange. Let c be the speed of light in our environment. Electromagnetic waves at radio frequencies in air travel at very close to the speed of light in a vacuum, 299,792,458 m/s. Obviously range, r, can be determined from the relationship $r = ct$, where t is the one-way propagation delay.

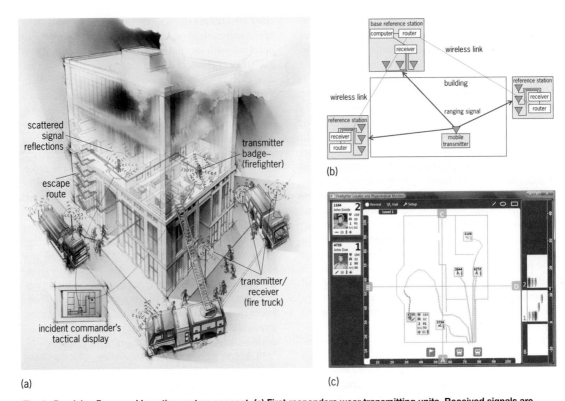

(a)

(b)

(c)

Fig. 1. Precision Personnel Location system concept. (*a*) **First responders wear transmitting units. Received signals are picked up by reference units that compute the 3D locations of all the individuals for display in real time on the incident commander's display.** (*b*) **The mobile transmitter broadcasts a multicarrier wideband (MC-WB) signal, which is received by multiple receiver stations. Data are sent over a wireless data link to the base reference station to estimate and display the mobile transmitter location.** (*c*) **Example of an incident commander's display.**

Thus, a time of arrival (TOA) method applies information about the range from the transmit antenna to several receive antennas outside the building to compute the position of the transmitter by means of various techniques known generally as multilateralization. The simplest form of TOA multilateralization follows from the geometric intuition that, in the ideal case, the solution must reside at the intersection of circles with radii equal to the direct path distances centered at the respective receiver antenna sites. More sophisticated forms of multilateralization involve the use of an overdetermined set of radii and an error minimization process to estimate the transmitter location. The process of multilateralization of course assumes that the locations of the receive antennas are known. Thus, when the system is deployed, a procedure must be executed to determine the positions of these receive antennas. This may be performed manually by surveying with measuring devices, but in an emergency situation, this is not practical. Thus, the locations of the receive antennas should be determined automatically by the system, using a radio positioning approach similar to the way the personnel-attached transmitters themselves are located, but obviously based upon a bootstrap operation. Regarding placement of the antennas, in the firefighter application, receive antennas would most likely be fixed to fire trucks, embedded in ladders placed on the exterior of the building, or manually deployed in the vicinity of the accessible sides of a building.

Time difference of arrival–based location estimation. In practice, there are reasons to avoid the symmetric transceiver architecture required to determine the absolute distance from a transmitter to a receiver via round-trip estimation in favor of an asymmetric one. In the latter case, the mobile unit may be implemented as a lower-complexity and lower-cost transmit-only device. Since the internal clock on such a transmitter cannot be synchronized with the receiver clocks, one-way propagation time is no longer accessible; instead, only differences in the ranges from this transmitter to the receivers can be determined. Determination of location from this range difference information is known as the time difference of arrival (TDOA) approach and uses a form of multilateralization that in the error-free case can be conducted by finding the intersection of hyperbolic arcs determined by the range-to-transmitter differences associated with pairs of receiving antennas.

Multipath problem. The greatest challenge pertaining to indoor location with radio frequency electromagnetic waves is that posed by the complexity of the radio propagation environments involved. Radio waves are reflected by metal objects and discontinuities in the index of refraction (such as at the interface of air and a masonry wall), which are plentiful in indoor environments. Even a simple wood-frame single-family residential dwelling is a torture chamber for radio propagation thanks to in-wall electric wiring, plumbing, appliances, furniture, fireplaces, flues, and ducting. As a result, the signals ultimately

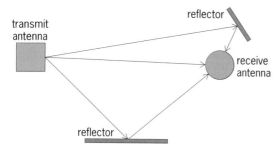

Fig. 2. Receiver captures attenuated direct path signals as well as reflected signals, a phenomenon known as multipath.

received by an RF location system are a combination of the so-called direct path signal and reflected signals, called multipath, as depicted in **Fig. 2**.

Multipath is a dominant contributor to the failure of GPS systems to function accurately indoors. Development of a successful RF indoor location system must find a way to take the multipath problem into account and mitigate its effects. Notably, the direct path is always the shortest path, since it travels directly from the transmit antenna to the receive antenna and hence has the smallest delay. However, owing to shared spectrum bandwidth limitations these signals often overlap. Furthermore, the direct path signal may he highly attenuated or missing. Thus a robust solution involves disentangling these components.

Multisignal fusion location estimation. Because of the issues inherent with the TOA, TDOA, and other techniques using distance estimation followed by multilateralization approaches, and the policy and performance problems surrounding the use of ultrawideband signals, the early efforts to develop the WPI Precision Personnel Location system considered alternative location algorithms and narrower-bandwidth ranging signals. The current system adopts a multicarrier wideband (MC-WB) signal. This signal consists of a sum of unmodulated sinusoids, typically but not necessarily evenly spaced in frequency. The signal is wideband, yet spectrally friendly in that each carrier is unmodulated and so may occupy guard bands between existing services. Also, the processing means that have been adopted allow for gaps to be introduced into the otherwise contiguous carrier set to permit it to coexist with other services.

To address the special problems introduced by severe multipath, and as a result of lessons learned from both theoretical and experimental investigations, signal processing at the receivers implements an algorithm called singular-value array reconciliation tomography (σART, pronounced "SART"). This novel algorithm obtains a position estimate directly with received data from all of the receive antennas, rather than by determining the individual ranges from the transmitter to each receive antenna independently followed by a multilateralization solution of the source position.

Figure 3 displays the outcome of the σART algorithm to determine the location of a firefighter in a

house. The figure can be considered as a contour map indicating the level of likelihood that the transmitter is located at each point, with the deepest-red point indicating the peak and hence the best estimate of the location of the transmitter worn by the firefighter.

Conclusions. This article describes the use of RF-based approaches to implement a location and tracking system suitable for use in the difficult area of indoor location. The challenges of implementing an indoor location and tracking system are many, and the additional considerations imposed by certain user requirements such as those for first responders make many technologies that look promising in the laboratory environment impractical in the real world.

However, if the problems of multipath can be overcome, then RF-based positioning offers the promise of simple, inexpensive transmitter-based systems that do not require time-consuming initialization or calibration, while at the same time providing drift-free solutions. The article addresses how traditional signal processing techniques that attempt to work with individual range-based estimates fail in the indoor environment, but novel signal processing approaches that use all the received data directly to provide a position estimate have experimentally been shown to work to the extent of providing an accuracy on the level of that identified as necessary by first responders.

Acknowledgments. This work has been supported and funded by a variety of organizations, including the Department of Justice. The findings reported here include many contributions from research

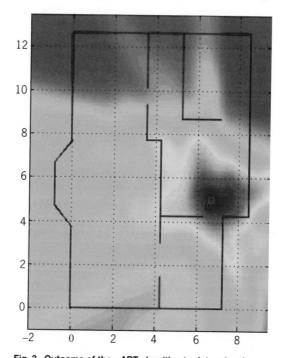

Fig. 3. Outcome of the σART algorithm to determine the location of a firefighter in a house. Location of transmitter (indicated by X) is at the point of the highest peak in the σART scan metric.

students at WPI over the last few years. The work is continuing, with the goal of providing the firefighting service and other first responders with a low-cost, reliable indoor location and tracking system to help prevent the occurrence of a tragedy similar to that which happened at the Worcester Cold Storage Warehouse.

For background information *see* ELECTRONIC NAVIGATION SYSTEMS; SATELLITE NAVIGATION SYSTEMS in the McGraw-Hill Encyclopedia of Science & Technology.   R. James Duckworth; David Cyganski

Bibliography. A. Bensky, *Wireless Positioning Technologies and Applications*, Artech House, Boston, 2008; J. Coyne, D. Cyganski, and R. J. Duckworth, FPGA-based co-processor for singular value array reconciliation tomography, in *16th Annual IEEE Symposium on Field-Programmable Custom Computing Machines*, 2008; E. D. Kaplan and C. J. Hegarty (eds.), *Understanding GPS: Principles and Applications*, 2d ed., Artech House, Boston, 2006; K. Pahlavan, X. Li, and J. Makela, Indoor geolocation science and technology, *IEEE Comm. Mag.* 40(2):112–118, February 2002, DOI:10.1109/35.983917.

Inflammasomes

Inflammasomes are intracellular, multiprotein complexes that, on activation, recruit and activate caspase-1, the aspartic acid protease (an enzyme that digests proteins) responsible for cleavage of the inactive procytokine forms of interleukin-1-beta (IL-1β) and interleukin-18 (IL-18) into their proinflammatory biologically active forms. Both acute inflammation in response to injury or infection and chronic inflammation present in various diseases involve IL-1β and IL-18. Thus, inflammasomes represent a critical mechanism in these processes. Whereas there are approximately 20 cytoplasmic proteins potentially able to form inflammasomes, only a few have been extensively studied, including NLRP3, NLRC4, and AIM2. Although many cell types express the sensor proteins required for inflammasome assembly and activity, the study of inflammasomes has so far focused largely on cells of monocytic origin. Consequently, additional biological roles for inflammasomes may emerge.

Background. Prior to 2002, although the IL-1β-converting properties of caspase-1 were known, the molecular basis for caspase-1 activation was unclear. Early studies that sought to identify proteins with caspase recruitment (CARD) domains that might be involved in apoptosis (programmed cell death) led to the discovery of DEFCAP/NALP1 (now known as NLRP1), which contains a C-terminal CARD domain. NALP1 was first shown in the laboratory of Jürg Tschopp to mediate the assembly of a high-molecular-weight complex comprising NALP1, ASC, and caspase-1. Formation of this complex preceded and was necessary for IL-1β cleavage and secretion; thus, the complex was dubbed the "inflammasome." Around the same time, a large family of approximately 23 proteins was described with structural

features similar to DEFCAP/NALP1, which is today known as the NLR (nucleotide-binding and leucine-rich repeats) family because of the shared central nucleotide-binding domain (NBD) and C-terminal leucine-rich repeats (LRRs). NLR proteins can be divided into five subfamilies based on differences in their N-terminal domains: those with an acidic transcriptional activation domain (NLRA), a baculovirus apoptosis inhibitory repeat (NLRB), an unclassified domain (NLRX), a pyrin domain (NLRP), and a CARD domain (NLRC). NLR proteins are involved in adaptive as well as innate immunity, and some are important for embryonic development. Most NLRs are cytoplasmic proteins and are thought to persist in an inactive conformation until activated by specific stimuli. Members of the NLRP and NLRC subfamilies are known to form, or are thought capable of forming, an inflammasome complex.

Activation. Inactive NLR monomers in the cell cytoplasm undergo a conformational change in response to infection or exposure to various environmental factors. Activated NLR monomers multimerize using their NBD and LRRs, leading to the recruitment and activation of caspase-1. Caspase-1 in turn cleaves IL-1β and IL-18, allowing their secretion. Assembly of the inflammasome complex requires pyrin and CARD domains, which are members of the larger death effector domain (DeD) family that exhibit the six α-helical bundle structure typical of the death domain fold found in members of the tumor necrosis factor (TNF) receptor family. These domains mediate protein interactions by permitting homotypic domain associations (for example, pyrin domains interact with other pyrin domains), enabling both NLRPs and NLRCs to assemble an inflammasome (see **illustration**). NLRP proteins recruit the CARD-containing caspase-1 protein indirectly through first recruiting the adaptor ASC (also known as PyCARD) through the interaction of their pyrin domains. NLRP association with ASC leads to caspase-1 recruitment via the interaction of their CARD domains. NLRC proteins are believed to recruit caspase-1 through the direct interaction of the NLRC and caspase-1 CARD domains. The various interactions between NBDs, LRRs, and the recruited ASC and/or caspase-1 proteins are thought to result in a heptameric complex. A still higher order compact "speck" structure (approximately 1 micron in size) that forms adjacent to the cell nucleus has been observed for some inflammasomes, whereas another results in a more filamentous arrangement near the plasma membrane.

Because of their structural similarity to the NBD-LRR class of disease resistance proteins in plants, it has been generally thought that inflammasomes result from NLR recognition of specific microbial products. The Toll-like receptors (TLRs), which constitute another class of pathogen receptors, use their LRRs to bind microbial products ranging from deoxyribonucleic acid (DNA) to lipids. Thus, LRRs within the NLRs are believed to be the site of recognition, although this has not been well established experimentally. However, some microbial products have been detected in the speck structure, suggesting a

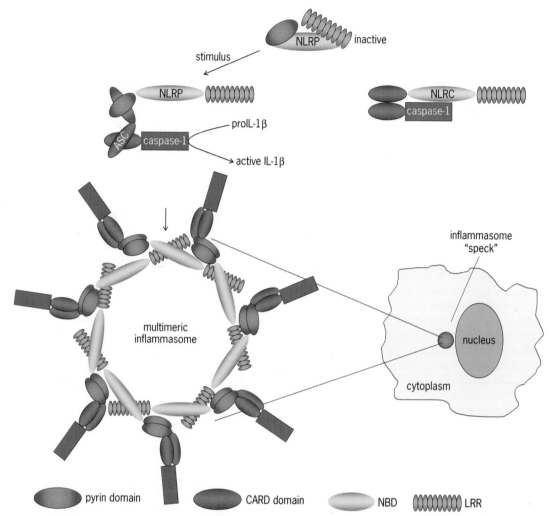

Schematic of inflammasome activation. The inflammasome is assembled through the interaction of pyrin (NLRP) and CARD (NLRC) domains to yield a large macromolecular complex. The various steps are shown for the NLRP arrangement. A similar stimulus-induced activation of NLRC likely leads to an inflammasome structure similar to the NLRP inflammasome, but without the ASC adaptor protein.

more direct interaction. Because the number of proteins potentially able to trigger an inflammasome is large, the specific details of their activation will likely be unique.

Prototypic inflammasomes. The best-studied (and therefore prototypical) inflammasomes are those formed by activation of NLRP3, NLRC4, and AIM2. Whereas NLRP3 recruits caspase-1 indirectly via the adaptor ASC, NLRC4 is thought to recruit caspase-1 directly. Although not an NLR, AIM2 (absent in melanoma 2) also recruits ASC through a pyrin domain interaction followed by caspase-1 recruitment. AIM2 likely represents a family (the PyHIN family) of non-NLR proteins having the potential for inflammasome formation.

The NLRP3 inflammasome is activated following cellular exposure to a variety of molecules associated with pathogens [for example, bacterial ribonucleic acid (RNA), peptidoglycan components, and lipopolysaccharides], molecules derived from the host (for example, monosodium urate crystals, cholesterol, amyloid aggregates, oxidized low-density lipoproteins, and mitochondrial DNA),

microbial toxins (for example, nigericin), or environmental toxins (asbestos, silica, alum, and particles made from polystyrene, silicon, or titanium). Consequently, many common pathogens activate the NLRP3 inflammasome. The precise mechanism for NLRP3 activation is not known. However, it is thought that the various NLRP3 agonists ultimately trigger the release of mitochondrial DNA, which (when oxidized) stimulates formation of the NLRP3 inflammasome. Although the NLRP3 inflammasome responds to these multiple stimuli, NLRC4 appears to be specific for the bacterial protein flagellin responsible for the motility of some bacteria (for example, *Salmonella*). AIM2, conversely, specifically binds double-stranded DNA (via its HIN domain). AIM2 is activated in response to a number of double-stranded DNA viruses and some intracellular bacteria.

Disease connections. Autosomal dominant mutations in NLRP3 are believed to be causal in a set of autoinflammatory conditions affecting humans. These syndromes include familial cold urticaria (FCU), Muckle-Wells syndrome (MWS), and neonatal-onset multisystem inflammatory disease (NOMID). Together, these syndromes are also known as the cryopyrinopathies (cryopyrin was an early name for NLRP3). The cryopyrinopathies represent a spectrum of inflammatory manifestations ranging from a cold-induced rash with arthralgia (FCU) to a chronic inflammatory syndrome with associated bone deformities and inflammatory damage to multiple systems including joints, sight, and hearing (NOMID). The autoactivation of NLRP3 in these syndromes is thought to be a consequence of conformational changes resulting from specific mutations within the nucleotide-binding domain as opposed to exposure to specific agonists.

In addition, a variety of mouse models of human disease have implicated inflammasomes as a likely important element in type II diabetes and atherosclerosis. Obesity can lead to metabolic syndrome (prediabetes) and type II diabetes. Palmitic acid and ceramides from fat also activate the NLRP3 inflammasome. The resulting IL-1β is linked to declining function of pancreatic islet cells that produce insulin. Cholesterol crystal activation of the NLRP3 inflammasome contributes to early atherosclerotic plaque formation, thereby promoting atherosclerosis.

Despite interesting connections to the above pathologies, the NLRP3 inflammasome also helps to provide protection from influenza infection and is responsive to a wide range of human pathogens, including *Staphylococcus aureus* and *Listeria monocytogenes*.

Inflammasome regulation. Inflammasome activation is controlled at multiple levels. An initial signal is typically required for abundant expression of certain NLRs, the adaptor ASC, caspase-1, and the uncleaved, procytokine form of IL-1β. A second signal is provided by pathogens, host-derived molecules, or environmental molecules that can trigger the assembly of the inflammasome complex. Whether the known inflammasome agonists act directly or indirectly on NLRs is not clear, but both mechanisms are probable. A third level of control is evidenced by proteins that can prevent or disrupt inflammasome assembly. These include proteins with solitary pyrin or CARD domains, that is, the so-called pyrin-only proteins (POPs) and CARD-only proteins (COPs). Another NLR protein, NLRP10/PyNOD, can also interfere with inflammasome assembly/activation. Although NLRP10 is present in mice and humans, the POPs and COPs are limited to humans and their closest primate relatives. POPs have also been described in viruses such as the myxoma virus, which is a rabbit pathogen. In the case of myxoma virus, the viral POP interferes with both inflammasome activation and other innate immune signals; in doing so, an effective antiviral response is not mounted and the animal succumbs to the infection. Inflammasome inhibition is likely to be a highly successful pathogen strategy for circumventing early host immune responses and establishing infection.

For background information *see* APOPTOSIS; ARTERIOSCLEROSIS; CELL (BIOLOGY); CELLULAR IMMUNOLOGY; CYTOKINE; DIABETES; ENZYME; IMMUNOLOGY; INFECTION; PATHOGEN in the McGraw-Hill Encyclopedia of Science & Technology.   Jonathan A. Harton

Bibliography. A. K. Abbas, A. H. Lichtman, and S. Pillai, *Cellular and Molecular Immunology*, 7th ed., Elsevier, Philadelphia, 2012; B. K. Davis, H. Wen, and J. P. Ting, The inflammasome NLRs in immunity, inflammation, and associated diseases, *Annu. Rev. Immunol.*, 29:707–735, 2011, DOI:10.1146/annurev-immunol-031210-101405; O. Gross et al., The inflammasome: An integrated view, *Immunol. Rev.*, 243:136–151, 2011, DOI:10.1111/j.1600-065X.2011.01046.x; K. Murphy, *Janeway's Immunobiology*, 8th ed., Garland Science, New York, 2011; D. L. Longo et al., *Harrison's Principles of Internal Medicine*, 18th ed., McGraw-Hill, New York, 2012; A. Rubartelli et al., Interplay between redox status and inflammasome activation, *Trends Immunol.*, 32:559–566, 2011, DOI:10.1016/j.it.2011.08.005; T. Strowig et al., Inflammasomes in health and disease, *Nature*, 481:278–286, 2012, DOI:10.1038/nature10759; J. Tschopp, F. Martinon, and K. Burns, NALPs: A novel protein family involved in inflammation, *Nat. Rev. Mol. Cell. Biol.*, 4:95–104, 2003, DOI:10.1038/nrm1019.

Intelligent microgrids

Microgrids are electrically and geographically small electric power systems capable of operating connected to or islanded from a larger terrestrial grid. Electrically small refers to the amount of installed generation capacity (typically less than 50 MW) and its voltage level (kV or less). Geographically small refers to the spatial dimensions of a microgrid, which can range from a personal office to an entire residential community.

There are a variety of reasons for the establishment of microgrids. For example, a facility with a very large set of dc loads may find it more reliable and efficient to establish a dc microgrid to service those loads rather than serving each independently from an ac source. Another reason for a microgrid may be the difficulty in finding appropriate locations for large power plants or transmission grids like those that exist today. A microgrid can facilitate either putting the appropriately sized generation near loads or locating loads closer to existing generation plants, reducing the need for more extensive infrastructure and possibly enhancing energy security.

By far the most flexible trait of microgrids is their islanding capability, which allows them to separate from their larger (parent) power system and convert to an independent "micro" power system. This flexible trait is also an inflexible exigency that requires microgrids to be energy independent in order to support their loads for an intended period—from hours to months—depending on the microgrid type. Although microgrids can be as small as a single generator connected to a load, today they are commonly incorporated into facilities such as data centers, office buildings, hospitals, submarines, ships, university campuses, and military installations.

Attributes of intelligent microgrids. Microgrids are considered "intelligent" when their operation is computer controlled so they can adapt to changing conditions autonomously. This requires significant intelligent control, sensors, and automation. Examples of intelligent control include fault mitigation and automatic topology reconfiguration. Faults are detrimental and unintentional interruptions in electrical supply. Intelligent microgrids not only detect and isolate faults on their own, but also reroute power from an alternate source to maintain service continuity. Topology reconfiguration can be used to reduce distribution losses, but also to reduce vulnerability to external attacks. These control actions are automated, but may also need occasional human intervention. For example, outages that cannot be fixed automatically must be repaired by a maintenance crew.

The distribution of electric power in grids is changing. An emerging trait is the penetration of renewable energy at distribution-level voltages (typically below 35 kV). One important factor stimulating this change is distributed generation such as solar panels. This type of generation produces desirable bidirectional power flows in a microgrid, but also provides challenges and new incentives to the design of electrical power distribution.

Due to the islanding capability of microgrids, an important aspect in electric power distribution is a high level of redundancy. Microgrids near 50 MW can have between two and 10 turbine generators of various sizes to provide power. This redundancy permits choosing a correct set of generators to efficiently power and support large dynamic loads.

The distribution system may be configured so that important loads (such as hospitals and defense systems) can be supplied by either of two paths. This is critical to achieve maximum functionality during unintended outages. In addition, this reconfiguration provides opportunities for routine use of smart optimal control. The same control system, if well designed, can provide a power-system configuration that provides maximum efficiency during routine operation and maximum flexibility during emergencies.

Another attribute of intelligent microgrids is that the power system does not have to be designed to supply the maximum of all possible loads. Under a "fiat" control, a microgrid operator can dictate which loads are served to extend survivability. This decision is normally in accordance with a prior determination of which loads are critical for each period of time.

While the breadth of topics on intelligent microgrids is large, the topics that follow highlight common automation and self-healing aspects of intelligent microgrids.

Dynamic balancing. Microgrids are power systems with finite inertia, which makes them susceptible to destabilization and overloads following disturbances. ("Finite inertia" is jargon to indicate that islanded power systems have limited generation and significantly less ability to respond gracefully to abrupt changes in load compared to interconnected terrestrial power systems.)

Dynamic balancing is an intelligent control strategy that ensures generation output and controllable-load demand match, while satisfying operational constraints in real time. (The real-time decision time step for finite inertia systems is commonly 10–100 ms.)

Microgrids can incorporate a large number of intermittent energy resources, such as wind and solar. In this configuration, generation capacity varies according to both the weather and the time of day. Moreover, plug-in electric vehicles (PEVs) are proliferating. When this occurs in an isolated microgrid, charging events risk causing frequency and voltage oscillations.

In addition to these potential sources of destabilization, the transition of a microgrid from grid-connected mode to islanded mode may overload or underload a microgrid's generators. Thus, dynamic balancing is a critical aspect of intelligent microgrids to match generation and load in a system with renewable energy intermittency, high penetration of PEVs, and finite inertia.

When sudden load or generation changes occur in microgrids, dynamic balancing regulates the setpoints of generators or loads to reduce the mismatch during the transient state. This mitigates frequency and voltage oscillations in the system. However, combustion generation units, such as diesel generators and microturbine generators, cannot respond quickly to sudden changes in load. A delayed response can result in significant frequency and voltage oscillations. In this situation, energy storage or additional controllable loads are used to compensate for the generation and load changes in microgrids due to their response times. Energy storage devices commonly used for dynamic balancing include batteries, ultracapacitors, and flywheels. Controllable loads typically used for dynamic balancing include service loads such as washing machines, dryers, air conditioners, heaters, and other large loads (including PEVs).

Short-circuit fault protection. Short circuits are unwanted contacts between an energized conductor and a return path (such as ground or neutral wire). This condition can produce large currents in a system, be a source of fire, and damage costly equipment. Microgrids operating in islanded mode have their sources synchronously coupled by electronic power converters, which can limit short-circuit fault currents.

Microgrids with ac and dc power distribution can be supplied from different types of power sources located apart from each another (such as solar arrays, fuel cells, and wind turbines). Moreover, energy storage units (such as batteries and supercapacitors) can also be connected to the distribution grid through electronic power converters. Although these sources can, in principle, source short-circuit fault currents, their power electronic converters limit the fault current when compared to power systems having only machines as their sources. Electronic power converters limit fault current by reducing the time their internal switches permit power to flow to the circuit, thus reducing available power, or by simply turning off to reduce the power to zero. The merit in this controlled throttling mechanism (not part of all-machine-based power systems) is that current-limiting functions and high interrupting speeds can be achieved.

While microgrids fed from multiple sources provide higher reliability, it can be difficult to isolate the faulted part of the system. However, the tradeoff is that controllable converters can limit fault currents by changing their output setpoints. These considerations bring about innovative fault protection methods to eliminate disruptive currents and provide rapid system reconfiguration.

Notwithstanding the presence of ground returns, terrestrial microgrids depend on a ground (that is, "earth") path as a return conductor for fault currents. These solidly grounded systems dominate for safety reasons and result in relatively high line-to-ground fault currents which must be interrupted quickly to limit damage to equipment.

Nonetheless, there are certain microgrids that are meant to operate in islanded mode most of the time. Such microgrids can be designed to (intelligently) operate with a single line-to-ground fault, where "ground" in this context refers to an "accidental" common structure (for example, a chassis) of the power system. This type of operation requires fully ungrounded or high-resistance ground systems, which pose significant challenges with respect to the grounding design and the identification of fault locations.

Load reenergization. Reclosers are protective devices that automatically open and reclose a line when a disturbance is sensed. In the case of high-impedance faults on distribution lines (for example, foreign objects in contact with conductors), a reclosing action often eliminates the source of the fault.

Reclosers operate in a "single shot" mode across a live and a dead line. If the fault is cleared, the recloser retains its closed position. If not, additional attempts follow until, if the fault does not clear in a preset number of tries, the recloser latches to its open position. This operation, while useful, reduces system reliability and degrades power apparatus due to the high inrush currents pulled by motors and transformers. (This degradation may also affect domestic electronics such as garage-door openers, televisions, and residential furnaces.)

For microgrids in general, switching transients due to asynchronous reclosing on motor residual flux spur the failure of protective devices and often cause nuisance trips resulting in a failure of the restoration process. Associated with this conventional reclosing action is a delayed voltage-level recovery that complicates the restoration process and may lead to undervoltage load shedding. A surrogate example is the restoration of a downstream deenergized load section. Reclosing action from a live to a dead wire poses issues such as large inrush currents, downstream motor-torque oscillations, speed droops (that is, reductions), and low-voltage conditions until the system reaches equilibrium.

Intelligent microgrids favor a more adaptive restoration process such as "soft reclosing." This approach involves load-side energy storage equipped with a ramp-up inverter tied to the distribution lines. Upon detection of grid outage, the inverter starts its operation in islanded mode and supplies the deenergized network portion with a voltage and frequency ramp. Before the ramping action settles, the inverter's output voltage and frequency are matched to a grid voltage-and-frequency reference signal that is available via a remote communication link. After the ramping settles, the energy storage system supports the load for a certain length of time, but expects the utility to recover or supplementary power sources within the microgrid to come online before the stored energy is fully consumed.

When the grid recovers from the disturbance, reconnection of the disconnected network (powered from the stored energy source) and the grid takes place. During this live-to-live wire reconnection, the aforementioned issues are not present because the grid voltage is synchronously stepped onto an energized load section. This soft reclosing approach increases service quality and preserves proper functioning of the power apparatus.

This load restoration solution applies to microgrids of different sizes. For microgrids limited to small areas, the opportunities for this technology are favorable due to the reduced transmission distances of the communication link (that is, grid reference and coordination signals). However, challenges arise in microgrids that have redundant power sources where more complex coordination is required between multiple generation and load-side energy storage units.

Communication and control. Intelligent microgrids commonly rely on a centralized controller to assume overall supervisory control and energy management responsibilities. This controller communicates command action to other dispersed controllers local to diverse load and distributed generation points. Data collected from local points is also communicated back to the central controller and processed, and commands are issued across the communications network. This control and communication infrastructure is commonly used to maintain system stability, balance load and generation, or achieve energy management goals as initiated by the operator or initiated automatically by the system.

Due to the small spatial dimensions of islanded microgrids, as compared to terrestrial transmission systems, controller communication propagation delays are not typically a significant issue; however, the overall communication latency does require examination for proper real-time control and energy management. The latency includes delays in communication nodes, delays due to congestion, and propagation delays.

The timing associated with communication is particularly important in islanded mode and during the transition between grid and islanded modes, where the decision-time windows are much shorter than when there is grid support for maintaining stability and voltage-and-frequency regulation. (It is commonly assumed that large terrestrial grids have infinite decision time.) Since the impact of communication issues in microgrids can be minor (depending on the microgrid type), intelligent microgrid design, operation, and real-time control rely on integrated control strategies considering the reliability and redundancy of the measurement and communications infrastructure.

Certain intelligent microgrids have more stringent measurement and communication requirements than others. (A driver of this stringency is vital loads such as defense systems.) In such cases, reconfiguration in response to a fault, or predictive reconfiguration in response to anticipated damage or loss of generation, requires synchronized measurements, millisecond decision making, and communication to local actuators.

Computer simulation plays an important role in anticipating system behavior. These forecasts, however, are commonly performed on desktop computers not set up to receive field measurements. This limitation is well known, and is being overcome by allowing computer simulations to run concurrently with a power system in real time (that is, with simulation speeds synchronized with incoming data) and—optionally—to give operators the ability to run simulations faster than real time. Running simulations faster than real time is a highly sought, highly challenging simulation scheme that allows operators to "fast forward" a simulation in time and play "what if" scenarios to predict system behavior. While such simulation tools can be indispensible for reliable microgrid operations, the computational resources required are proportional to the microgrid size being monitored, the detail desired, and the rate at which the data comes in. Supercomputing centers are strong candidates for this task.

Autonomous operation. Autonomous operation in microgrids combines passive and active control schemes to attain high levels of reliability and security. This operating approach involves little or no human intervention as it performs several aspects of the control and operation locally by using communicationless controls for power balancing and regulation of voltage and frequency. There are also microgrids that use coordinated and remotely-controlled schemes. There is a design tradeoff, however, to establish a balance between centralized control schemes versus distributed ones. Regardless of where the balance is, the possibility of a communication system failure must be accounted for when assessing power-line-loss contingencies.

Voltage- or frequency-droop-based control schemes are commonly used as part of the power management system of a microgrid. Autonomous droop-based controls integrate a wide range of generation technologies that are geographically dispersed and may connect to or disconnect from the microgrid at any time during islanded operation.

A noticeable advantage of droop control is that it enables power sharing among various sources without the need for a fast and widespread communication infrastructure. Hence, communication requirements vary among the various devices in a microgrid. In such cases, these requirements may become a secondary issue in the overall access to supervisory control information as part of a microgrid's energy management function. This trait expands the horizon of applicable communication methods to also include low-bandwidth and intermittent communication schemes based on satellite or radio frequency media.

Certain microgrids are divided into predefined operating zones that are autonomously independent, but may operate in a coordinated manner. Each zone is locally controlled and protected against system disturbances. The local zones may be defined according to the power quality and reliability or established as part of the protection methodology to provide proper coverage and fast fault detection and clearing. These types of microgrids are conventionally designed in a centralized fashion to achieve a high level of security and dependability. Any change in the system operation or energy requirements is determined by the control room and communicated to the power sources. In these environments, communication systems play a critical role in automation, control, and remote status monitoring.

Power system protection has also benefited from the advent of modern-day communication schemes. Traditionally, monitoring equipment relied on a pilot wire to communicate across service zones. Modern designs are moving away from this copper wire approach as digital communication signals can be more easily transferred and shared by many protective devices. This implies the use of Internet-based communication to relay information to and from monitoring equipment that can be geographically distributed over a wide service area. The availability of information across a wider area allows autonomous and coordinated decisions to isolate problems and continue service where interruption can be avoided. The application of this communication is becoming especially important in light of microgrids with renewable energy sources, storage units, and decentralized control methods.

Summary and conclusions. Microgrids are of growing technical significance. The constraints found over different microgrids, the appropriate focus on extreme levels of reliability, and the need to be efficient over a wide range of operating conditions show that intelligent microgrids lead the development over traditional (that is, older, nonintelligent) microgrids.

The wide range of electrical and geographical microgrids presents commonalities as well as challenges in their design and operation. Large microgrids can be expanded since their physical footprints are of less concern. In smaller, confined microgrids, sizing may be fixed *a priori*.

Most microgrids are geographically confined and feature relatively short communications distances.

Although this inherently reduces the communication network's propagation delays, the overall latency requirements remain stringent due to the finite inertia, fast dynamics, and proliferation of power electronics controls in microgrids. (Wireless communication is a good option for certain microgrids, but it may not be suitable for other microgrid types.)

Protection, stability, and power electronics are important to the future of all microgrids. Different microgrids have unique attributes that will likely lead to somewhat different solutions. The possibility of cross-fertilization in these areas is large, however. Because microgrids can operate connected to a larger power system, it is essential that protection schemes warrant reliable and safe operation by using predefined setting groups, advanced settings computed online, and operational adaptation of the settings of relays or reclosers.

Modern microgrids include a variety of power sources, renewable sources, and energy storage systems. These elements interface with the distribution bus through controllable electronic power converters in dc and hybrid ac-dc systems. This controlled multisource system configuration introduces higher reliability and survivability, and brings about new options and challenges in terms of protection against short-circuit faults.

Maintaining reliability of supply during normal and emergency conditions (such as a natural disaster) requires an integrated approach to design and reliability analyses where the intertwined physical and cyber aspects are simulated and studied concurrently. From an operational view, both the electrical and the control systems can benefit from faster-than-real-time modeling and simulations that enable predictive decision making and control in anticipation of a material event, such as sudden variability in supply or unanticipated system failures.

Advances in communication technology provide secure and economically viable media choices for the control and protection design of microgrids. Experience gained from pilot-based (that is, classic) protection schemes is invaluable in the selection of operating modes and protection zones to achieve autonomy and visibility in the overall power and energy management of the system. Passive protection may indeed provide the ultimate backup, but over the next decade more embedded processing power will be presented to provide additional operational intelligence and efficiencies.

For background information *see* COMPUTER; CONTROL SYSTEMS; ELECTRIC DISTRIBUTION SYSTEMS; ELECTRIC POWER SYSTEMS; ELECTRIC PROTECTIVE DEVICES; ELECTRIC VEHICLE; ENERGY STORAGE; SIMULATION; SOLAR ENERGY; WIND POWER in the McGraw-Hill Encyclopedia of Science & Technology.

Fabian M. Uriarte; Robert E. Hebner

Bibliography. S. P. Chowdhury and P. Crossley, *Microgrids and Active Distribution Networks*, The Institution of Engineering and Technology, Stevenage, Hertfordshire, U.K., 2009; R. H. Lasseter, MicroGrids, in *Proc. 2002 IEEE Power Engineering Society Winter Meeting*, New York, pp. 305–308, 2002, DOI:10.1109/PESW.2002.985003; R. H. Lasseter and P. Paigi, Microgrid: A conceptual solution, in *Proc. 2004 IEEE 35th Annual Power Electronics Specialists Conference, PESC 04*, Aachen, Germany, pp. 4285–4290, 2004, DOI:10.1109/PESC.2004.1354758.

Invasive species during the Late Devonian biodiversity crisis

The Late Devonian biodiversity crisis occurred approximately 385–370 million years ago and ranks as one of the five largest biodiversity crises in Earth's history. Although the Late Devonian is often referred to as a "mass extinction" event, recent analyses indicate that extinction rates were not statistically elevated above the typical or "background" extinction rate in geologic time (**Fig. 1**). Instead, the primary cause of biodiversity loss was reduced formation of new species during this interval. Consequently, attempts to explain the causes behind the dramatic biodiversity loss and ecological reorganization must examine the evolutionary processes that control speciation, that is, the formation of new species. Explanations that focus solely on mechanisms that promote extinction, such as asteroid impacts, climate change, or oceanic overturn, cannot satisfactorily explain the observed biodiversity patterns. New studies focused on speciation in Late Devonian marine organisms of Laurentia (the ancient continent that included present-day North America, Greenland, and part of Western Europe) have demonstrated that a series of rapid marine transgressions (pulses of sea-level rise) facilitated the geographic expansion of invasive, generalist species and eliminated the primary way that new species typically form. In short, the Late Devonian biodiversity crisis is primarily attributable to invasive species, which is a problem that is also present in modern ecosystems.

Late Devonian overview. The Late Devonian biodiversity crisis occurred over a span of several million years, with the most severe effects related to the end of the Frasnian stage of the Devonian (Fig. 1). Biodiversity loss was substantial. Approximately 20% of marine families and 50% of marine genera became extinct during this interval. Reef communities, particularly the framework-building stromatoporoid sponges and the rugose and tabulate corals, were the most severely affected. During the Middle Devonian (precrisis) interval, tropical reefs occupied ten times the lateral extent of modern reef ecosystems. However, the Late Devonian biodiversity crisis was so severe that marine ecosystems suffered a fundamental collapse and reorganization, causing vast reef tracts of the Middle Devonian to disappear from Earth's oceans for almost 100 million years. Other common constituents of the marine community, including brachiopods, trilobites, ammonites, and conodonts, also experienced dramatic turnover (changes in faunal composition) at the family and ordinal levels.

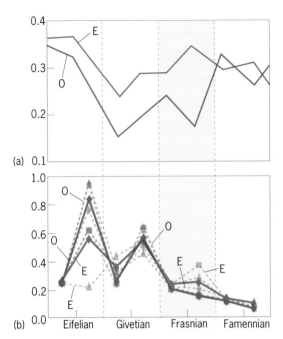

Fig. 1. Speciation/origination and extinction rates before, during, and after the crisis interval. The crisis interval (Frasnian stage) is shaded. (*a*) Proportion of generic extinction (E, red line) or origination (O, blue line) per interval. (*b*) Instantaneous rates of species extinction (E, red lines) and speciation (O, blue lines) for two brachiopod genera [*Schizophoria* (diamonds on dashed line) and *Floweria* (squares on dashed line)], one bivalve genus [*Leiopteria* (triangles on dashed line)], and all three clades combined (diamonds on solid line). [*Modified from A. L. Stigall, Speciation collapse and invasive species dynamics during the Late Devonian "Mass Extinction," GSA Today, 22(1):4–9, 2012*]

The transition from the Middle to the Late Devonian is also marked by increased geographic ranges of species and reduced endemicity (uniqueness of faunal assemblages) between tectonic basins. During this time, the highly endemic Middle Devonian faunas were replaced by a cosmopolitan (worldwide) Late Devonian fauna. This biogeographic shift was caused by the pervasive expansion of geographic ranges of individual species within shallow marine environments of Laurentia associated with transgressions (**Fig. 2**). The expansion of geographic ranges and the transition from an endemic to a cosmopolitan biota in the Late Devonian have been documented in many taxa, including rugose corals, brachiopods, foraminifera, fishes, conodonts, trilobites, and land plants.

These individual invasion events aggregated to cause the overall loss in endemism. Fundamentally, for a species to successfully invade a new tectonic basin, organisms of that species must be able to survive in both the original and new basins as well as the corridor linking the basins. Typically, this requires organisms to survive environmental conditions that differ from those within the ancestral range of the species. Consequently, the most successful invader species are those with broad ecological tolerances (that is, they are ecological generalists), which can exploit a wide array of environmental

conditions. The relationship between generalist ecology and invasion success is apparent in modern invasive species, and it appears to hold for ancient invaders as well. Therefore, the ultimate result of a large number of species invasions during the Late Devonian was to introduce an array of new generalist taxa to each tectonic basin. Indeed, the same set of taxa established themselves in many basins, which promoted the development of the cosmopolitan fauna of the Late Devonian. Reduced speciation was a secondary effect of this faunal homogenization (see below).

Characterizing speciation depression. To understand why speciation rates were depressed during the Late Devonian, it is first critical to ascertain how they were depressed. Speciation is the process by which a population becomes genetically isolated from the parent population and thereby forms a new evolutionary unit in which member organisms reproduce only with other organisms within the unit (that is, a species). This process has a strong temporal and geographic component; that is, speciation happens in a particular place to a particular ancestral–descendant population pair. The reproductive isolation required for speciation typically arises as a by-product of a physical separation of ancestral and descendant populations in geographic space, which is referred to as allopatric speciation. Allopatric speciation occurs via two pathways: vicariance and dispersal. During vicariant speciation, the ancestral population becomes divided geographically by the creation of a new barrier, such as a mountain range, within the ancestral geographic range. Speciation by dispersal occurs when a small population of the ancestral species actively crosses an existing barrier and becomes geographically isolated. Vicariance is passive, whereas dispersal is an active process, and they generate different biogeographic patterns

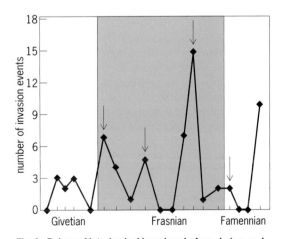

Fig. 2. Pulses of interbasinal invasions before, during, and after the crisis interval. The crisis interval (Frasnian stage) is shaded. Arrows indicate episodes of sea-level rise. (*Modified from A. L. Rode and B. S. Lieberman, Using GIS to unlock the interactions between biogeography, environment, and evolution in Middle and Late Devonian brachiopods and bivalves, Palaeogeogr. Palaeoclimatol. Palaeoecol., 211:345–359, 2004*)

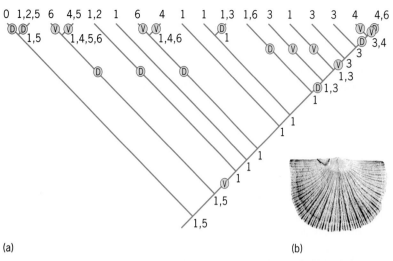

(a)   (b)

Fig. 3. Phylogenetic hypothesis for the brachiopod genus *Floweria*, with biogeographic states substituted for terminal taxa and mapped onto the ancestral nodes. (*a*) Inferred episodes of speciation by vicariance (V) and dispersal (D) are indicated. Biogeographic areas: 0, Europe; 1, Northern Appalachian Basin; 2, Southern Appalachian Basin; 3, Michigan Basin; 4, Iowa/Illinois Basin; 5, Missouri; 6, Western United States. (*b*) *Floweria chemungensis* is a representative species. [*Modified from A. L. Stigall Rode, Systematic revision of the Devonian brachiopods Schizophoria (Schizophoria) and "Schuchertella" from North America, J. Syst. Palaeontol., 3(2):133–167, 2005*]

when viewed in an evolutionary framework (**Fig. 3**). Species-level hypotheses of ancestor–descendant relationships, therefore, provide the context in which to identify the style of speciation during the history of clades (taxonomic groups containing a common ancestor and its descendants).

Analysis of speciation style in four groups of shallow marine species (an order of predatory crustaceans, two genera of brachiopods, and one bivalve genus) from Laurentia revealed that speciation by vicariance was abnormally low during the Late Devonian, particularly during the crisis interval. Overall, vicariance comprised approximately 30% of speciation events compared to 70% by dispersal. This level of vicariance contrasts starkly with the distribution of speciation modes among faunas at other times in Earth's history. Of modern species that have been examined, approximately 75% arose through vicariance. Similarly, vicariance is more common than dispersal throughout much of the Paleozoic Era (Cambrian, Ordovician, and Early Devonian trilobites exhibit approximately 55% vicariance). Therefore, the reason that Late Devonian species rates were reduced is because vicariant speciation, typically the dominant mode of species formation, was stopped.

Causes of speciation depression. As described previously, the Late Devonian was characterized by frequent and widespread invasions of species into new tectonic basins that were facilitated by periodic transgressive events. These invasions introduced species with generalist ecologies into new communities, where they commonly thrived. The combination of general expansion of geographic ranges and the prevalence of generalist species was a fatal combination for vicariant speciation. First, vicariance requires division of ancestral species ranges into smaller geographic areas; this step is difficult to achieve during a biogeographic regime (the state

or organization of an ecosystem) characterized by geographic expansion. Second, following range division, each incipient species has a relatively small population size. These newly established populations would have been in competition with the invasive generalist species. In ecological experiments with modern species, organisms with generalist ecologies frequently are more successful at acquiring resources (such as food or space) than specialist taxa with limited populations. Consequently, most incipient species that did form via vicariance are more likely to go extinct quickly (and hence leave no fossil record) than to proliferate and become established. This sequence of events ensures that few new species would form via vicariance, and it also results in preferential extinction of specialist taxa following the invasion of immigrant generalist taxa into tectonic basins. Indeed, the preferential extinction of specialist taxa (characterized by limited geographic ranges) is a hallmark of the Late Devonian biodiversity crisis.

Impact of invasive species. Ultimately, the dramatic speciation loss and slightly elevated extinction rates that characterize the Late Devonian biodiversity crisis are attributable to biogeographic and macroevolutionary changes that resulted from the rampant species invasions of the Frasnian stage. Rises in sea level created marine links between previously isolated tectonic basins that provided a pathway for the invasion of generalist species into biogeographic areas outside their ancestral ranges. Once established in these new basins, they competed with native taxa, resulting in the preferential extinction of species with small geographic ranges (interpreted to be ecological specialists) during the crisis interval. The presence of invasive generalist species also impeded the formation of new species by vicariant speciation. Because vicariance is the dominant pathway by which speciation normally occurs, this reduced the overall speciation rate substantially—in fact, to zero in many clades (for example, as seen in Fig. 1). Importantly, human-transported invasive species are rampant throughout ecosystems in the modern world. Thus, the Late Devonian biodiversity crisis may offer a lesson about invasive impacts and species prioritization for conservation managers today.

For background information *see* ANIMAL EVOLUTION; BIODIVERSITY; BIOGEOGRAPHY; DEVONIAN; ECOLOGICAL COMMUNITIES; ECOLOGICAL SUCCESSION; EXTINCTION (BIOLOGY); INVASION ECOLOGY; PALEOECOLOGY; PALEOGEOGRAPHY; PALEONTOLOGY; POPULATION DISPERSAL; SPECIATION; SPECIES CONCEPT in the McGraw-Hill Encyclopedia of Science & Technology.   Alycia L. Stigall

Bibliography. G. R. McGhee, Jr., *The Late Devonian Mass Extinction*, Columbia University Press, New York, 1996; A. L. Stigall, Invasive species and biodiversity crises: Testing the link in the Late Devonian, *PLoS ONE*, 5(12):e15584, 2010, DOI:10.1371/journal.pone.0015584; A. L. Stigall, Speciation collapse and invasive species dynamics during the Late Devonian "Mass Extinction," *GSA Today*, 22(1):4–9, 2012, DOI:10.1130/G128A.1.

KAP1 protein

The protein KAP1 (also called TRIM28) is expressed at low to moderate levels in most normal tissues. However, it is upregulated (that is, it has increased expression) in many human cancers, and it has been suggested that inhibition of KAP1 function may be a rational approach for cancer treatment. KAP1 is involved in regulating several cellular pathways, and it is critical to define the normal role of KAP1 to understand how targeting KAP1 might affect tumor growth and/or response to chemotherapeutic agents. Studies in which the levels of KAP1 have been artificially manipulated have suggested that KAP1 plays different roles in different cells. For example, KAP1 has been shown to be essential both for maintaining the pluripotency (that is, the capacity to generate any cell type in the body) of embryonic stem cells and for allowing these stem cells to properly differentiate. Similarly, studies of adult cells have indicated that KAP1 can either promote or inhibit cell differentiation, depending on the cell type. Insight into how KAP1 can have such diverse functions has come from identifying its protein interaction partners. This article focuses on three KAP1-containing complexes that are proposed to inhibit transcription [the process by which deoxyribonucleic acid (DNA) is transcribed into ribonucleic acid (RNA), which is then translated into the proteins that function to create the different types of body cells], facilitate DNA repair (the process by which the cell identifies and corrects damaged regions of the genome), and inhibit apoptosis (a programmed cell death pathway that controls various aspects of normal development and helps rid the body of unwanted or damaged cells).

KAP1 and transcriptional silencing. KAP1 is highly modular in structure, having several domains that interact with proteins associated with modification of chromatin (the DNA–protein complex in chromosomes) and gene regulation. The N-terminus of KAP1 contains a high-affinity protein interaction domain that is necessary for interaction of KAP1 with KRAB-ZNFs (a family of site-specific DNA-binding transcription factors). Each KRAB-ZNF has at least one KRAB domain (the part of the protein that interacts with KAP1) and 3–30 zinc fingers, which are small DNA-contacting protein modules having a folded structure that is stabilized by binding to zinc. The central region of KAP1 interacts with HP1 family members. HP1 proteins are termed "epigenetic readers" because they can interact with specific epigenetic marks, such as histone H3, which is trimethylated on lysine 9 (H3K9me3). The C-terminus of KAP1 contains a domain that functions to recruit and activate SETDB1 (an epigenetic writer), which is the enzyme that creates the H3K9me3 epigenetic mark. KAP1 can also interact with the NURD histone deacetylase complex via its C-terminal region. The ability of the NURD complex to remove an acetyl mark from H3K9 may enhance SETDB1's ability to methylate this lysine residue of histone H3. Thus, locking KAP1 onto the genome is accomplished in

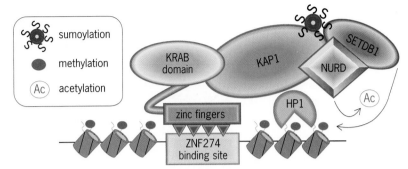

Fig. 1. The KAP1 complex involved in gene silencing. Locking KAP1 onto the genome is accomplished in four steps: a KRAB-ZNF (with at least one KRAB domain and 3–30 zinc fingers) binds to the genomic DNA; KAP1 binds to the KRAB-ZNF; SETDB1 and NURD bind to KAP1, creating H3K9me3; and HP1 binds to KAP1 and to H3K9me3, providing a second contact point between KAP1 and the genomic DNA. Chromatin marked by H3K9me3 is highly condensed and is called heterochromatin. The condensed nature of heterochromatin is unfavorable for gene expression and thus KAP1 complexes containing a KRAB-ZNF, SETDB1, NURD, and HP1 have been implicated in transcriptional silencing.

four steps: (1) a KRAB-ZNF binds to the genomic DNA; (2) KAP1 binds to the KRAB-ZNF; (3) SETDB1 and NURD bind to KAP1, creating H3K9me3; and (4) HP1 binds to KAP1 and to H3K9me3, providing a second contact point between KAP1 and the genomic DNA (**Fig. 1**). Because HP1 can form dimers, this may favor recruitment of a second KAP1 complex to the adjacent genomic region; then, reiterative creation of H3K9me3 by the recruited SETDB1 followed by binding and dimerization of HP1 may allow spreading of the chromatin region marked by H3K9me3.

Chromatin marked by H3K9me3 is highly condensed and is called heterochromatin. The condensed nature of heterochromatin is unfavorable for gene expression and thus KAP1 complexes containing a KRAB-ZNF, SETDB1, NURD, and HP1 have been implicated in transcriptional silencing. Because KAP1 itself is not a DNA-binding protein but is brought to the DNA via interaction with KRAB-ZNFs, it is the KRAB-ZNFs in a given cell type that will dictate the genomic localization of KAP1 and hence determine which genes in the cell are silenced by KAP1. The localization of KAP1 to different target sites depending on which KRAB-ZNFs are expressed in a given cell may also explain the apparent dichotomy of the role of KAP1 in differentiation. Whether KAP1 promotes or inhibits differentiation may depend on which KRAB-ZNFs are expressed in a particular tissue.

KAP1 and repair of damaged DNA. KAP1 has also been implicated in the repair of damaged DNA. Normally, KAP1 is sumoylated [that is, undergoes attachment of SUMO (small ubiquitin-like modifier) proteins] on several amino acids in the C-terminal region of the protein. This increases the activity of SETDB1, helping to lock the KAP1/SETDB1 complex at sites of H3K9me3. However, if there is DNA damage, KAP1 is phosphorylated by a damage-induced kinase called ATM, causing a switch between the sumoylated and phosphorylated forms of KAP1. This causes reduced activity of SETDB1, which leads to a reduction of H3K9me3 levels and an unlocking of the HP1-mediated genomic clamp, allowing a

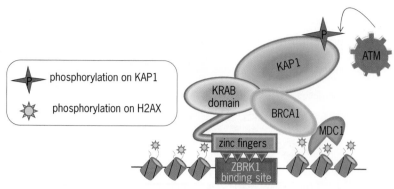

Fig. 2. The KAP1 complex involved in DNA repair. If there is DNA damage, KAP1 is phosphorylated by a damage-induced kinase called ATM, causing a switch between the sumoylated and phosphorylated forms of KAP1. This eventually leads to a relocalization of KAP1 to DNA damage foci. The relocalization process is initiated when the histone variant H2AX is phosphorylated in response to the induction of DNA double-stranded breaks. Note that only a subset of the proteins that localize to sites of DNA damage are depicted.

relocalization of KAP1 to DNA damage foci. The relocalization process is initiated when the histone variant H2AX is phosphorylated in response to the induction of DNA double-stranded breaks. A protein called MDC1 binds to phosphorylated histone H2AX at or near sites of DNA double-strand breaks and recruits a protein called BRCA1 that can also interact with ZBRK1 (a KRAB-ZNF). Thus, a working model for the DNA damage-induced relocalization of KAP1 includes the following steps: (1) phosphorylated H2AX is bound by MDC1; (2) BRCA1 then binds to MDC1; (3) ZBRK1 interacts with BRCA1; and (4) KAP1 is recruited by interaction with ZBRK1 via the KAP1 N-terminal domain (**Fig. 2**). It has been proposed that KAP1 may facilitate a local decondensation of chromatin, allowing access of DNA repair proteins. Although it is not yet clear how this is achieved, perhaps phosphorylation weakens interaction of KAP1 with the NURD histone deacetylation complex, allowing a more acetylated (and thus more accessible) chromatin structure. A return to the sumoylated form of KAP1 mediated by the protein phosphatase 1b is thought to assist in reactivating SETDB1, increasing levels of H3K9me3, and stabilizing KAP1 at its normal genomic sites.

KAP1 and apoptosis. Apoptosis is a process of programmed cell death that can remove damaged cells from the body. KAP1 is involved in inhibition of apoptosis via regulation of the levels of the apoptosis-promoting p53 protein, which is a DNA-binding transcription factor that regulates genes encoding proteins that mediate cell suicide. The level of cellular p53 is kept low by the action of a protein called MDM2, which ubiquitinates p53, leading to degradation of the p53 protein [note that ubiquitination is the conjugation of ubiquitin (a 76-amino-acid protein) to cellular proteins; this process regulates a broad range of eukaryotic cell functions and in particular targets proteins for degradation]. KAP1 can cooperate with MDM2 to enhance p53 degradation. KAP1 (and the NURD complex) is brought to the p53-containing complex by interaction of the N-terminus of KAP1 with MDM2. The reduction in p53 activity by KAP1 probably results from the fact that it can be acetylated on the same residues that can be ubiquitinated. However, these events are mutually exclusive; actions that cause a loss of acetylation favor ubiquitination. When KAP1 engages the p53 complex, this results in deacetylation of p53 by the NURD complex (histone deacetylases can target proteins other than histones) followed by increased ubiquitination of p53 by MDM2 and destruction of p53 (**Fig. 3**). Reduced p53 levels will result in reduced transcription of p53 target genes. In addition, because KAP1 is a transcriptional repressor, it is possible that the interaction with KAP1 converts the p53 transcriptional activator into a repressor of its target genes.

Should KAP1 be targeted for inhibition in tumors? As indicated previously, KAP1 can promote the degradation of p53, reducing apoptosis. If only the role of KAP1 in apoptosis is considered, then perhaps it would be sensible to attempt to inactivate KAP1 in tumors because the ability of tumor cells to evade apoptosis can play a significant role in failure of many therapeutic regimens. An increase of apoptosis in tumor cells as a result of KAP1 depletion may reverse this trend. However, the role of KAP1 in transcription must also be considered. Cell growth is a balance between the actions of growth-promoting genes (oncogenes) and growth-inhibiting genes (tumor suppressors), and relative expression of these two classes of genes is critical in determining phenotypes of normal and cancer cells. As mentioned previously, KAP1 is brought to the DNA via a KRAB-ZNF. For example, KAP1 may be recruited to the genome by a KRAB-ZNF that is inappropriately upregulated in the tumor, perhaps resulting in the silencing of a tumor suppressor gene that is normally turned on in that cell type. If so, this also might suggest that targeting KAP1 in a tumor would be a good approach. However, KAP1 may also be required for suppression of an oncogene, with this normal silencing being dysregulated in the tumor because of overexpression of other proteins. For example, MAGE proteins (a family of proteins that are highly expressed in various cancers) have been shown to bind to the N terminus of KAP1 at the interaction site for KRAB-ZNFs. Therefore, it is

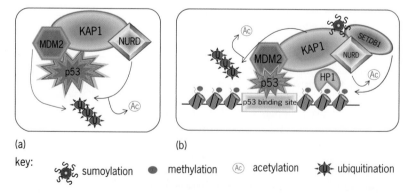

Fig. 3. The KAP1 complex involved in apoptosis. KAP1 can lead to degradation of p53 (*a*), possibly turning any remaining p53 into a transcriptional repressor (*b*). The level of cellular p53 is kept low by the action of a protein called MDM2, which ubiquitinates p53, leading to degradation of the p53 protein. KAP1 cooperates with MDM2 to enhance p53 degradation, thereby inhibiting apoptosis.

possible that the tumor-specific MAGE proteins compete with ZNFs for binding to KAP1, resulting in loss of genomic recruitment of KAP1 and perhaps allowing expression of normally silenced oncogenes. Under these circumstances, targeting KAP1 in the tumor may not produce the desired antitumorigenic effects.

Conclusions. KAP1 can influence several critical cellular processes, including transcriptional silencing, response to DNA damage, and regulation of apoptosis. Reduction of KAP1 levels may influence these processes in ways that are beneficial in some circumstances, but not in others. It should be kept in mind that deficiencies in KAP1 may also have additional effects not discussed herein. For example, KAP1 is expressed at high levels in the brain, and studies in mice have shown that deficiency of KAP1 can lead to stress and anxiety symptoms. Rather than inactivating all functions of KAP1, it may be better to have a clear understanding of the role that KAP1 plays in the particular cell type of interest and then to target specific KAP1-interacting proteins, such as an inappropriately expressed KRAB-ZNF if reduced gene silencing is the goal or an inappropriately expressed MAGE protein if increased gene silencing is the goal.

For background information *see* APOPTOSIS; CANCER (MEDICINE); CHEMOTHERAPY AND OTHER ANTINEOPLASTIC DRUGS; CHROMOSOME; DEOXYRIBONUCLEIC ACID (DNA); DNA METHYLATION; DNA REPAIR; GENE SILENCING; HISTONE; ONCOGENES; PROTEIN; TRANSCRIPTION; TUMOR SUPPRESSOR GENES in the McGraw-Hill Encyclopedia of Science & Technology.

Peggy J. Farnham

Bibliography. A. J. Bannister and T. Kouzarides, Regulation of chromatin by histone modifications, *Cell Res.*, 21(3):381–395, 2011, DOI:10.1038/cr.2011.22; C. A. Brady and L. D. Attardi, p53 at a glance, *J. Cell Sci.*, 123(part 15):2527–2532, 2010, DOI:10.1242/jcs.064501; K. L. Cann and G. Dellaire, Heterochromatin and the DNA damage response: The need to relax, *Biochem. Cell Biol.*, 89(1):45–60, 2011, DOI:10.1139/O10–113; S. Iyengar and P. J. Farnham, KAP1: An enigmatic master regulator of the genome, *J. Biol. Chem.*, 286:26267–26276, 2011, DOI:10.1074/jbc.R111.252569; R. Urrutia, KRAB-containing zinc-finger repressor proteins, *Genome Biol.*, 4(10):231, 2003, DOI:10.1186/gb-2003-4-10-231.

Kepler mission

Kepler is the tenth mission launched under NASA's Discovery Program and NASA's first mission dedicated to the search for planets beyond our solar system, with the goal of determining the frequency of Earth-sized planets within the habitable zones of Sun-like stars. Extrasolar planets are detected using the transit method, in which the passage of a planet across the disk of a star produces minute (10^{-2}–10^{-5}) reductions in the amount of light from that star. *Kepler* monitors the brightness of stars using a space-based wide-field telescope coupled to a precision photometer operating in a single broad optical bandpass. Approximately 160,000 main-sequence stars make up the primary observing program. Several thousand additional sources are normally observed during each observing season; these sources comprise the asteroseismology, guest observer, and other science targets. Using data from its first 16 months of operation, *Kepler* has identified more than 2300 exoplanet candidates, validated on the order of 40 single- and multiple-planet systems, discovered the first circumbinary planets, discovered a super-Earth in the habitable zone of a Sun-like star, and made fundamental advances in the fields of stellar structure and variability.

Exoplanets. Since the seventeenth century, when the nature of stars and our solar system was uncovered, astronomers have speculated about other planets. Driven by the Copernican principle, the enormous numbers of Sun-like stars observed in the nearby Milky Way, and early concepts of planetary system formation, scientists posited that planets are common and habitable planets likely. But until the mid-1990s, all claims of possible exoplanets remained unconfirmed. This lack of verified exoplanets was due to the immense technological challenge of identifying an object located adjacent to its host star with an optical brightness contrast of less than 10^{-6}.

Since the mid-1990s, the new field of exoplanets has blossomed. Advances in high-precision spectrophotometry enabled the first planets to be detected using the Doppler technique, in which extremely small shifts in stellar spectral lines are observed as planets orbit their parent stars. Once extrasolar planets were confirmed, astronomers also began developing the transit method, where the brightness of a star decreases by a small amount when an orbiting planet passes across the star's disk, analogous to the transit of Venus viewed from Earth on June 5–6, 2012. The amount of dimming is a function of the relative sizes of the star and the planet and, with sufficient precision, can identify Earth-sized planets orbiting Sun-like stars. Other methods (for example, microlensing, transit-timing variations, and direct imaging at optical and infrared wavelengths) have identified small numbers of exoplanets, but as of 2012, the Doppler and transit methods have produced the vast majority of known or candidate exoplanets. *See* 2012 TRANSIT OF VENUS.

Each technique offers advantages and limitations. The Doppler method is strongly biased toward identifying massive planets in close orbits. At present, Earth-sized planets cannot be detected with this method. Transit detections require highly precise photometry, constrained by the ever-varying atmosphere, which limits Earth-based signals to a few parts in a thousand (>0.1%). For a more complete census of exoplanets, including detection of Earth-sized planets, a space-based observatory is required to provide the precision photometry.

Kepler science operations. *Kepler* was proposed to NASA's Discovery Program and selected as its

Fig. 1. Depiction of the *Kepler* spacecraft. The solar panel is to the right, above the high-gain antenna. The telescope, enclosed in the gold-colored thermal blanket, forms the spine of the spacecraft; the imaging photometer is located in the base section. (*NASA*)

lar interest is the identification of Earth-sized planets orbiting in the habitable zone of Sun-like stars. The habitable zone is defined as an annular region around a star in which a rocky planet with a sufficient atmosphere could contain liquid water on its surface. Planets with surface water permit the possibility of life.

Kepler (**Fig. 1**) was launched on March 7, 2009, into an Earth-trailing heliocentric orbit. During its prime mission, the spacecraft continuously monitors one location on the sky, a 115 $\square^\circ$ (square degrees) field of view (FOV) in the Cygnus region, centered at $\alpha = 19h\ 22m\ 40s$, $\delta = 44°30'00''$ (**Fig. 2**). A large FOV is required to monitor a statistically significant number of FGKM dwarf stars, as the fraction of these stars expected to show transits is $\leq 1\%$. With 42 CCDs containing about 95 megapixels, the largest photometer yet orbited, *Kepler* can detect changes in brightness to well under 0.01%. An Earth-sized planet transiting a solar analog produces a dip of 84 ppm (0.0084%). As a consequence of the large FOV, *Kepler*'s nominal spatial resolution is $\sim 6''$; each pixel spans $\sim 4'' \times 4''$. Continuous observing lasts for approximately one month, followed by a one-day gap, during which time the spacecraft is oriented toward Earth for data downlink. Because of bandwidth and storage limitations, only photometry of preselected stars is recorded, not of the entire field. Photons are collected using two exposure times, termed short- and long-cadence modes. Short cadences last 58.8 s; long cadences last 1766 s. The long-cadence list is constrained to 170,000 targets or 5.44 million pixels. The short-cadence mode is limited to 512 targets or 43,520 pixels.

Kepler's exoplanet program monitors stars with a range of brightnesses of $9 \leq Kp \leq 16$, where *Kp* refers to an object's Kepler magnitude, which is its source brightness as observed using the unusually broad bandpass of the *Kepler* photometer (400–900 nm). For normal stars, *Kp* is approximately equivalent to a star's brightness as seen through a red filter. *Kepler* is capable of observing objects outside of this brightness range with some loss of precision. Objects as faint as 19th magnitude are on the list.

Kepler science operations are centered at NASA's Ames Research Center, which selects targets, plans observations, and operates the data processing and calibration pipeline. Spacecraft operations are executed by Ball Aerospace at the Mission Operations Center in Boulder, Colorado. Mission data are archived for community access at both the Milkulski Archive for Space Telescopes and NASA's Exoplanet Archive.

Finding the right stars. The search for terrestrial planets begins with an appropriate list of stars to be monitored for transits. The *Kepler* Project developed the target list by executing an extensive ground-based observing program of the entire FOV. Calibrated photometry in several bands was used to classify stars by correlating observed colors with parameter grids generated from synthetic stellar spectra. The output of the Stellar Classification Program is the *Kepler Input Catalog* (the KIC), a tabulation

10th mission in 2001. The scientific goal is to describe the diversity of planetary systems by monitoring large numbers of stars for transits, and then to utilize the detection statistics to estimate the distribution of planet sizes, orbital parameters, host-star properties, and multiple-planet systems. Of particu-

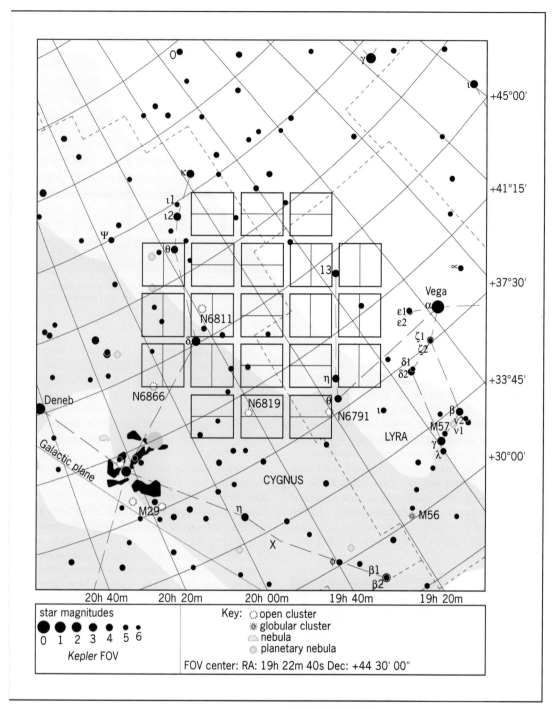

Fig. 2. Location of the *Kepler* field of view (FOV) on the sky. Each rectangle marks the field of one of 42 CCD detector pairs, each with 2200 × 1024 pixels. The center of the FOV lies approximately 12° from the galactic plane. (*NASA*)

of ~4.4 million sources within the FOV. Both the science team and other *Kepler* observers "mine" the KIC to identify objects of interest to their science programs.

Photometry pipeline: pixels to light curves to planets. The *Kepler* Processing Pipeline, a series of software modules coded in MATLAB, converts raw photometry into calibrated temporal sequences of source brightness, termed light curves, then searches for transit signatures within these light curves. A number of steps are involved: (1) Calibrate each pixel by correcting for detector properties. (2) Assign time stamps, estimate and remove various background signals, and construct light curves by summing a pixel set for each source that maximizes the signal-to-noise ratio. (3) Perform systematic error corrections to mitigate a number of instrumental signatures, including positional drift, differential velocity aberration, focus changes after pointing maneuvers, and cosmic rays. (4) Search the calibrated light curves for potential planet transit signals and perform a series of validation tests. The output of the

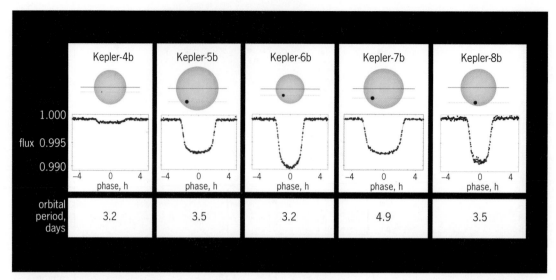

Fig. 3. Light curves of *Kepler*'s first five planet discoveries, showing the decrease in the monitored star's brightness during the transit. The orbital inclination can be estimated by the duration of the transit, which depends on how close the planet comes to the center of the stellar disk. (*NASA*)

validation process is a list of planet candidates, for which three transit signals are required. Two transits permit a period estimate, which is validated by observation of a third (predicted) transit. Examples of planet transits are shown in **Fig. 3**.

Since transits of planets in the habitable zone of Sun-like stars would occur about once a year and require three transits for verification, a minimum of three years is needed to identify and validate Earth-size planets orbiting Sun-like stars. Multiple-planet systems present major challenges in untangling the transit signals from individual exoplanets, yet *Kepler* has uncovered a number of such systems, including the spectacular 6-planet system Kepler-11 (**Fig. 4**).

Revealing exoplanets. In a broad sense, *Kepler* provides two types of information about exoplanets. The first is the primary science: statistics on exoplanets, the frequency distribution of planet sizes, orbits, and host-star properties. The second is the discoveries of individual exoplanet and multiplanet systems. During the first three years of operation, each analysis has produced surprises.

Exoplanet statistics. In February 2012, the Kepler team announced results from the first 16 months of data collection. The total number of viable candidate planets stands at more than 2300, with 20% of the host stars showing evidence for multiple planets. Within the multiplanet systems, short-period ($P \leq 10$ days) giant planets appear rare when compared to those in single-planet systems. The new data increased the relative number of smaller planets and of those in longer-period orbits (the latter expected as the mission progresses). In **Fig. 5**, statistics for the planet sizes in the February 2012 data release are displayed. Detected planets are predominately "super-Earth–" or Neptune-sized. Increasing numbers of small planets are being identified as the pipeline and detection statistics improve. **Figure 6** presents a portrait of the *Kepler* planets and their host stars.

New and strange worlds. Perhaps garnering the most public excitement has been *Kepler*'s discovery of truly strange new worlds. Here are a few examples.

Kepler-11 is a densely packed system with six confirmed planets. Five of the six have orbits smaller than Mercury's, whereas the sixth lies in an orbit a little smaller than Venus's (Fig. 4). All of the planets orbiting Kepler-11 are larger than Earth (with radii between 2 and 4.5 Earth radii), comparable in size to Uranus and Neptune.

Kepler-16b, announced in September 2011, is the first planet proven to orbit two stars, and is termed a circumbinary planet (**Fig. 7**). From Earth's perspective, the two stars regularly eclipse each other. The

Fig. 4. Artist's conception of the 6-planet Kepler-11 system. This view shows the simultaneous transit of three of the planets as observed by NASA's *Kepler* spacecraft. (*NASA*)

planet also transits each star, permitting measurement of the size, density, and mass of the planet. The fact that the orbits of the stars and the planet are aligned to within a degree suggests that the planet formed within the same protostellar disk as the stars. This system has been compared to the fictional *Star Wars* planet Tatooine.

Kepler-22b is the first planet confirmed to orbit within the habitable zone of its host star. This planet has a radius of approximately 2.4 Earth radii, classifying it as a "super-Earth." With an orbital period of 290 days, Kepler-22b lies in the middle of the habitable zone of a star very similar to the Sun.

Kepler-36, announced in June 2012, shows two planets orbiting very close to each other, the closest planet pair of any system currently known. One is a rocky planet with a radius of 1.5 Earth radii in a 14-day orbit; the other is probably a gas giant with a radius of 3.7 Earth radii in a 16-day orbit. At their closest approach, these two planets are separated by only about 5 times the Earth-Moon distance.

Kepler's amazing data, combined with expanding ground-based Doppler spectroscopy campaigns, have now made possible the field of comparative exoplanetology, the systematic study of planetary systems. Discovery of an Earth-sized planet in the habitable zone of a Sun-like star is just a matter of time.

Astrophysics. In addition to its planet-finding mission, *Kepler* results are also being used to rewrite textbooks on variable phenomena in both stellar and extragalactic astronomy. While 99% of observed stars do not show transits, *Kepler*'s precise multiyear monitoring provides a unique data set quantifying source variability for normal stars, known variable sources, and active nuclei in galaxies. *Kepler*'s impact on asteroseismology, stellar pulsational modes, activity cycles, and eclipsing and contact binaries will rival its impact on exoplanet science. Asteroseismology concerns the analysis of acoustic oscillations in stars excited by surface turbulence and convection, which permits astronomers to probe stellar interiors in ways that are not otherwise possible. Asteroseismology also helps inform exoplanet characterization because absolute sizes for transiting exoplanets require knowledge of the host-star radii, which can be determined to accuracies approaching a few percent via these methods. Likewise, *Kepler*'s unique combination of precision and continuous monitoring opens a new window on the strong variability seen in the nuclei of galaxies hosting supermassive black holes. *Kepler*'s impact on extragalactic science is just beginning. In its first 3 years of operation, *Kepler*'s "other" science has also resulted in the unexpected, including detection of oscillations probing the cores of red giant stars, binary stars in highly elliptical orbits, and unusual hot, compact transiting companions.

Perhaps *Kepler*'s most important contribution to stellar astrophysics will be the investigation of stellar activity cycles. These cycles used to be monitored with spectrophotometry of specific lines. *Kepler*'s capabilities have opened up a whole new window

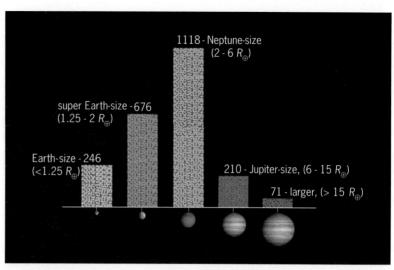

Fig. 5. Summary of the February 2012 *Kepler* planet candidates catalog release, showing the size distribution of detected planets (with radii expressed in terms of the Earth's radius, $R_\oplus$). (*NASA*)

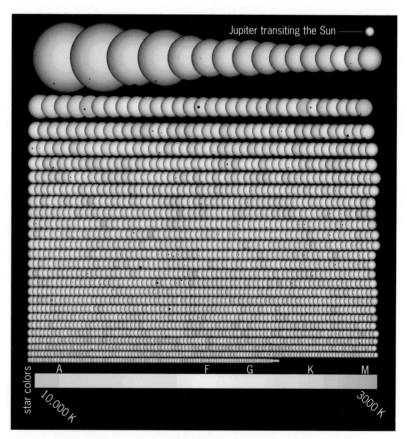

Fig. 6. In this illustration, created by Jason Rowe of the Kepler Science Team, all of *Kepler*'s planet candidates as of December 5, 2011, are shown in transit with their parent stars, ordered by size from top left to bottom right. Simulated stellar disks and the silhouettes of transiting planets are all shown at the same relative scale. (*NASA*)

on this subject. The combination of photometric precision, continuous monitoring, and a long temporal baseline is producing a unique data set to track stellar variability over many time scales. Assuming a mission duration of, say, 6 years, Kepler will be able to monitor the activity of on the order of 100,000 stars, primarily main-sequence stars in the

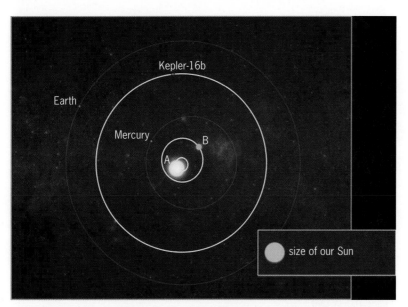

Fig. 7. Illustration of the Kepler-16 system, showing the planet Kepler-16b and the orbits of its twin stars (A and B). For reference, the orbits of Mercury and Earth are shown in blue. This planet orbits at a distance comparable to that of Venus in our own solar system. The size of the Sun's disk at lower right is scaled to the disks of the binary stars and not to the orbits of the planets. (*NASA*)

solar neighborhood. *Kepler* can, in essence, "see" the integrated optical-band variations from the entire star, of which the chromospheric emission lines are just a part. Only a space-based photometer can observe down to the level needed to monitor small brightness variations accurately.

Already, initial analysis of noise levels in *Kepler* observations has detected a larger intrinsic stellar noise signal than had been projected prior to launch. This noise excess implies that stars are more variable at low levels than previously understood and also hinders confirmation of small planet transits.

Extended mission. Prior to launch, *Kepler*'s primary mission duration was set at 3.5 years. In spring 2012, as part of NASA's biennial review of operating missions, the project submitted a proposal for an extension. Given *Kepler*'s steady stream of discoveries and the need to overcome the apparent noise from stellar activity to identify Earth-sized planets, an extension was approved. *Kepler* may operate through fiscal year 2016. We can expect continued announcements of new science from *Kepler*'s gaze for a few more years.

For background information *see* ASTRONOMICAL IMAGING; EXTRASOLAR PLANETS; HELIOSEISMOLOGY; LIGHT CURVES; MAGNITUDE (ASTRONOMY); SPACE PROBE; STAR; TRANSIT (ASTRONOMY) in the McGraw-Hill Encyclopedia of Science & Technology.

Michael N. Fanelli

Bibliography. C. A. Haswell, *Transiting Exoplanets*, Cambridge University Press, Cambridge, U.K., 2010; J. W. Mason (ed.), *Exoplanets: Detection, Formation, Properties, Habitability*, Springer/Praxis, Berlin/Chichester, U.K., 2008; C. Petit, Stellar oddballs, *Sci. News*, 179(12):18, June 4, 2011, DOI:10.1002/scin.5591791219; S. Seager, *Exoplanets*, University of Arizona Press, Tucson, 2010.

Limitations on increasing cellular system data rates

Cellular wireless data rates are fundamentally limited by available bandwidth. Noncellular systems are able to use higher frequencies, but are currently limited to line of sight. Improvements in algorithms and electronics are allowing systems to achieve performance that is coming closer to theoretical limits. However, demand for data over wireless networks is growing rapidly. Unlike wired and fiber networks, which can arbitrarily increase capacity by adding new cables, the shared aspect of the radio spectrum limits its growth because of the potential for interference. Cellular architectures address this, but there is increasing complexity and cost of deploying and operating a network with increasingly smaller cells. Although hard to predict, it appears that the growth in demand for cellular wireless data may outpace the advances necessary to meet this need.

Dominance of wireless communications. Wireless communications has become a dominant communication medium, with recent growth driven primarily by cellular communications. Starting with telegraph and broadcast radio in the early twentieth century, today the wireless spectrum is being used by a variety of systems. Some of the major categories are: satellite, including satellite broadcast, satellite communications, and the Global Positioning System (GPS); broadcast radio and television; maritime and aerospace communications and navigation; unlicensed personal radio, including WiFi, Bluetooth, Zigbee, and cordless phones; VHF/UHF analog radios [family radio service (FRS), general mobile radio service (GMRS), trunked mobile radio (TMR), and so forth]; military radios; cellular; and microwave point-to-point.

This article will focus on limitations on digital data rates for cellular communications, but many aspects will be applicable to other digital wireless systems.

Propagation environment. A number of factors affect wireless data rates. Free-space path loss can be defined as "loss in signal strength of an electromagnetic wave that would result from a line-of-sight path through free space, with no obstacles nearby to cause reflection or diffraction." Signal attenuation caused by free-space path loss is given by Eq. (1),

$$S = P_t \left(\frac{\lambda}{4\pi d} \right) G_t G_r \qquad (1)$$

where

P_t = transmitted power
S = signal at receiver
d = distance from emitter
λ = wavelength of radio energy
G_t = transmit antenna gain
G_r = receive antenna gain
$d \gg \lambda$

An isotropic antenna radiates or captures energy equally in all directions from the source. An isotropic antenna has $G_t = G_r = 1$.

Equation (1) represents the idealized best case. In the real world, additional factors determine signal level at the receiver. For a specific system where parameters such as transmit power, frequency, and antenna gains are fixed, the combination of these factors leads to a loss model defined by Eq. (2).

$$S = k_s / d^m \qquad (2)$$

The factors that are not dependent on distance, such as frequency, are included in the system constant, k_s. Others, for example atmospheric absorption, are dependent on distance. The combination of distance-dependent factors are included in the d^m factor, where, typically, $2 < m < 6$. The value for m varies based on the environment and can be arrived at empirically using experimental measurements. In an example environment where $m = 3$, the signal attenuates with distance much faster than the free-space model.

While the ideal $1/d^2$ expression is exact and based on a physical phenomenon, the $1/d^m$ expression is an approximate model for a number of distance-dependent attenuation effects.

Taking into account noise and channel capacity, it can be shown that the relationship between bit rate and distance for two systems A and B, with all other system factors the same, is given by Eq. (3),

$$\left(\frac{d_A}{d_B} \right)^m = \frac{r_{bB}}{r_{bA}} \qquad (3)$$

where

d_A = distance between transmitter and receiver for system A

d_B = distance between transmitter and receiver for system B

r_{bA} = bit rate of system A

r_{bB} = bit rate of system B

The model is statistical. In the example given by the **illustration**, a point (r,d) on the curve represents 90% of the area at distance d covered by a cellular base station that can achieve a bit rate r. A lower percentage coverage would shift the curve to the right (corresponding to a higher bit rate).

While advanced wireless techniques can improve data rates or reduce power requirements, they typically affect the constant k_s (shift the curve horizontally) and do not change the slope of the $1/d^m$ characteristic. As such, advanced techniques such as MIMO can offer incremental improvements, but not address the fundamental relationship caused by distance. (MIMO is short for multiple input, multiple output. It takes advantage of spatial multiplexing: the ability to transmit multiple information streams in the same bandwidth through the use of multiple transmitting and receiving antennas. MIMO takes advantage of multipaths by processing the different combinations of the transmitting signal at the receiving antennas to resolve the original transmit streams.)

Propagation loss due to environmental factors is a fundamental limitation to data rates. Since signals are attenuated so rapidly with distance in typical environments, increasing transmit power to increase data rate has diminishing returns, especially for battery-powered terminals. For example, each doubling of the effective radiated power (ERP) increases the signal level by 3 dB. In contrast, challenges in urban and suburban environments, such as penetrating 3 in. (76 mm) of lumber (-2.8 dB at 900 MHz), 8 in. (203 mm) of masonry block (-12 dB at 900 MHz), or 3.5 in. (89 mm) of reinforced concrete (-27 dB at 900 MHz), easily eat away at those gains.

Available bandwidth. The demand for higher-speed communications necessitates wider frequency band allocations for a given number of users. Since traditional cell bands have already been allocated and each provider has their fixed allocation of the spectrum, the solution to increasing system capacity generally comes down to one of two solutions: using the larger bandwidths available at higher frequencies or decreasing cell size.

The most desirable frequencies for these types of radio-frequency (RF) links are those below 10 GHz because of the negligible effects of atmospheric absorption and rainfall loss. Above 10 GHz, the effect of O_2, H_2O, and other molecules absorbing RF energy as it propagates drastically increases the attenuation of the signal.

There are issues with utilization of additional frequency bands at higher frequencies. RF links below 10 GHz are minimally affected by atmospheric absorption and rainfall loss. Beyond 10 GHz, the absorption effect of O_2, H_2O, and other molecules becomes more and more significant. While there are advantages of higher frequencies (such as their smaller wavelength, resultant reduction in component size, and ability to perform beam–forming to achieve very narrow beam widths that efficiently direct energy), there are a number of implementation issues that have yet to be addressed in the high-volume, low-cost terminals used for cellular communication. While it is likely that advanced techniques and technology will improve our ability to use higher frequencies for cellular communication, the spectrum limitations will continue.

Since data-rate demands and numbers of users are increasing, and bandwidth is limited, it is important to make efficient use of available bandwidth. Early designers of mobile telephony recognized this, and came up with the cellular concept. It has these fundamental features:

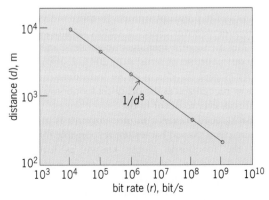

Example of the relationship between bit rate and distance.

1. Multiple transmitters are distributed over the service area, each providing coverage to a small portion (cell).

2. An individual base station is allocated a subset of the channels (frequency bands) available to the entire system.

3. Neighboring base stations are assigned different subsets of channels to minimize interference.

4. The available channels are reused across the service area, at a distance that minimizes interference.

5. To increase capacity, the number/density of base stations may be increased and transmit power decreased, resulting in smaller cells providing additional capacity without increased spectrum allocation.

Modern cellular systems provide higher capacity through smaller cells. The terms "microcell," "picocell," and "femtocell" refer to smaller base stations and antennas that are easier to deploy than traditional cells. Advances in technology continue to reduce size, power consumption, and cost of all types of base stations. Networks using smaller cells are more complex to design and operate, as the larger numbers of cells have more issues with coverage and interference, and require more complex backhaul networks to connect them to other network elements.

Advances in network planning (including self-configuring networks) and base–station technology will continue to drive the use of smaller cells.

Long-Term Evolution. Long-Term Evolution (LTE) is a mobile telephony standard that is being driven by increasing demand for data rates within the finite limitations of spectrum availability. Standards participants with decades of experience in cellular systems have developed innovative techniques to get the most out of the available spectrum. Here are some attributes of LTE that address spectrum efficiency:

1. Multiple-antenna MIMOs at terminal and base stations that can achieve higher bandwidth efficiency than conventional modulations.

2. Different spectrum bandwidths are allowed, from 1.4 to 20 MHz, without changing fundamental system parameters or equipment design.

3. Frequency reuse and interference coordination between cells.

4. Time domain duplex and half-duplex frequency domain duplex modes enable flexible allocation of available spectrum.

5. Voice over Internet protocol (IP) on an all-IP network allows more efficient use of bandwidth for voice traffic.

6. Dynamic, flexible allocation of individual traffic channels in the time and frequency domain to allow for the most efficient allocation of available bandwidth where it is needed.

7. Orthogonal frequency-division multiplexing (OFDM) modulation, fast packet scheduling, and link adaptation that make the most efficient use of the radio channel, to maximize throughput and minimize errors in the presence of the impairment known as fading.

8. Support for different categories of terminal equipment to address differing throughput needs.

9. Ability to utilize noncontiguous spectrum (in LTE-Advanced). *See* LONG-TERM EVOLUTION (LTE).

Implementation. As compared to terminals, base stations are less constrained in size, power, and cost; there are many fewer of them used in a network; and they have fixed locations. Cellular systems have been designed to put a greater capability, power, size, and weight burden on the base station.

Advances in electronics and communications algorithms have allowed us to come closer and closer to theoretical limits for capacity over wireless channels. In the base station, there are fewer constraints to implementing these advances. Advances in electronic technology are increasing processing capacity and radio performance. Advanced antenna designs make use of algorithms to steer and narrow the antenna beams to use the spectrum more efficiently.

Terminals have more limitations. Battery capacity limits the performance and power of terminal radios. Size constraints limit the flexibility in antenna steering and directionality.

For background information *see* INVERSE-SQUARE LAW; MOBILE COMMUNICATIONS; MULTIPLEXING AND MULTIPLE ACCESS; RADIO SPECTRUM ALLOCATION; RADIO-WAVE PROPAGATION; VOICE OVER IP in the McGraw-Hill Encyclopedia of Science & Technology.

Michael D. Rauchwerk; Joseph Battaglia

Bibliography. J. D. Kraus and R. J. Marhefka, *Antennas for All Applications*, 3d ed., McGraw-Hill, New York, 2002; T. S. Rappaport, *Wireless Communications: Principles and Practice*, 2d ed., Prentice Hall, Upper Saddle River, NJ, 2002; S. Sesia, I. Toufik, and M. Baker, *LTE—The UMTS Long–Term Evolution: From Theory to Practice*, 2d ed., Wiley, Chichester, U.K., 2011.

Listeriosis outbreak

The bacterial pathogen *Listeria monocytogenes* is a virulent microbe that is responsible for causing foodborne illness worldwide. *Listeria monocytogenes* is found in many locations in the environment, including soil and water, as well as in the intestinal tracts of many animals. This microbe is one of the most dangerous of the pathogens transmitted through food, causing severe disease in immune-compromised patients and in pregnant women. In these individuals, *L. monocytogenes* causes a variety of illnesses, including meningitis, meningoencephalitis, and spontaneous abortion of fetuses. In 2011, *L. monocytogenes* caused the deaths of 30 individuals in what is now recognized as the deadliest foodborne illness outbreak ever to occur in the United States.

Background. *Listeria monocytogenes* is a Gram-positive rod-shaped bacterium responsible for causing the foodborne illness known as listeriosis (**Fig. 1**). *Listeria monocytogenes* is non-spore-forming and also has the ability to grow well at cooler temperatures, making it a psychrotroph. This means that it can continue to grow on foods even when

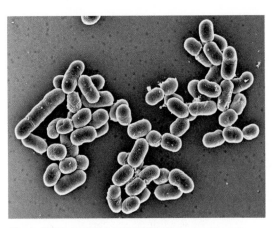

Fig. 1. Electron micrograph of *Listeria monocytogenes*. (*Courtesy of M. K. Cowan, Microbiology: A Systems Approach, 3d ed., McGraw-Hill, New York, 2012*)

they are stored in a refrigerator. *Listeria* is also a halophile, capable of growing at increased concentrations of salt.

This species is known to produce a number of virulence factors. The most prominent of these factors is listeriolysin O, which is a protein that destroys host-cell membranes. *Listeria monocytogenes* is a facultative intracellular pathogen that, once taken up by a host cell, commandeers the host-cell internal structures to facilitate its survival and infection. This mechanism, known as actin polymerization, is rare among infectious microbes and in fact was first discovered to occur with this pathogen.

The infection cycle of *L. monocytogenes* begins when this species triggers its own internalization by host cells in the intestinal epithelium. Once inside the host cell, *Listeria* escapes the host-cell vacuole using listeriolysin O. After reaching the cytoplasm, the bacteria then employs a protein known as ActA to bind a host-cell complex known as Arp2/3, which in turn activates actin nucleation at one end of the bacterial rod. This has the effect of pushing the bacterium in one direction through the host cell, causing a protrusion of the cell membrane, which then extends into the neighboring cell (**Fig. 2**). After the bacterium enters the new cell, it escapes from the host-cell vacuole, and the cycle begins again. This use of the host-cell actin system allows the bacterium to completely evade many of the immune defenses of the host organism.

Listeriosis. *Listeria monocytogenes* causes a wide range of disease symptoms, known collectively as listeriosis, in susceptible patients. These symptoms include septicemia (blood infection), meningitis, meningoencephalitis, and cervical infections in pregnant women, which may lead to spontaneous abortion of the fetus or stillbirth. Pregnant women usually suffer from a relatively mild, flulike illness; however, this infection can cause life-threatening illness in the neonate, as *L. monocytogenes* is one of relatively few bacterial species that can cross the placenta to infect the fetus. This results in miscarriage or stillbirth, depending on the trimester during which the mother is exposed to the pathogen. Babies who sur-

vive the initial *Listeria* infection of the mother may develop a blood infection known as granulomatosis infantiseptica, in which pus-filled granulomas form and are distributed through the entire body. Other infants may develop meningitis or meningoencephalitis after birth. In addition to pregnant women, persons at higher risk for developing listeriosis include older adults and people with weakened immune systems. Individuals in these categories who experience flulike symptoms within 2 months after eating contaminated food should seek medical care, and they should inform their physician about any consumption of the contaminated food. Listeriosis is treated using antibiotics, typically ampicillin alone or in combination with gentamicin.

Foodborne outbreaks resulting from *Listeria* have been transmitted through a number of different foods. The first outbreak in which *Listeria* was identified as a foodborne pathogen occurred in 1981 in Nova Scotia, Canada, when 41 individuals were diagnosed with listeriosis because of the consumption of contaminated coleslaw. The cabbage used in making the coleslaw had been fertilized with raw sheep manure, which contained *Listeria*. The largest outbreak in the United States prior to the 2011 outbreak happened in 1985, when *Listeria*-contaminated cheese caused 142 cases of listeriosis, leading to 28 deaths and 20 miscarriages. At the time, doctors at the Los Angeles County–University of South California Medical Center began noticing that pregnant Hispanic

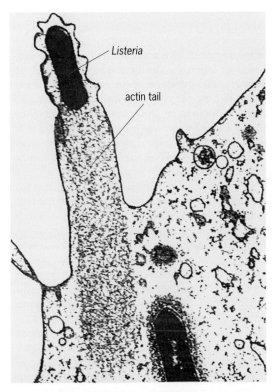

Fig. 2. Representation of the cell infectious process by *Listeria monocytogenes*. Note that the bacterium polymerizes the cellular actin to propel itself through the cell and into the neighboring cell. (*Courtesy of J. M. Willey, L. M. Sherwood, and C. J. Woolverton, eds., Prescott's Microbiology, 8th ed., McGraw-Hill, New York, 2011*)

Number of Persons

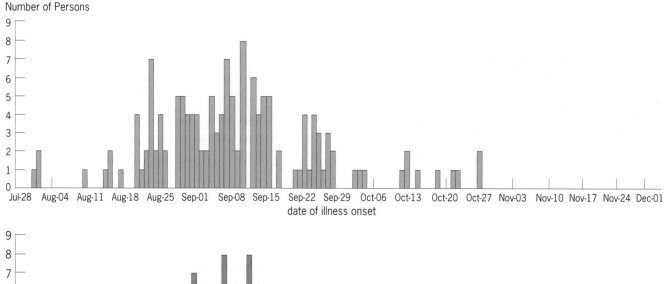

Fig. 3. Multistate outbreak of listeriosis linked to whole cantaloupes from Colorado. *Top:* Number of ill persons by date of illness. *Bottom:* Number of ill persons by date of clinical specimen collection. (*Courtesy of Centers for Disease Control and Prevention*)

women were ill with a disease that was later diagnosed as listeriosis. During a 6-month period, 58 mother–infant pairs were identified with the disease, and 96% of the pairs were Hispanic. Epidemiologists therefore focused on foods consumed primarily by Hispanics, and they soon identified a Mexican-style soft cheese as the culprit. Despite much investigation, the exact source of the *Listeria* could not be determined. It was later found that unsanitary conditions existed at the manufacturing plant, that the pasteurization techniques were faulty, and that milk from the dairy could not be ruled out as the source for the bacterial contamination.

Since the California outbreak of 1985, foods have been monitored for the presence of *Listeria*. If *Listeria* is identified as being present, the food types are recalled in an attempt to minimize potential foodborne outbreaks. For example, in April 2012 alone, recalls of foods having potential *Listeria* contamination included alfalfa sprouts, deli-style sandwiches, Tomme d'Or specialty cheeses, Quebec sausages, and ready-to-eat pizza calzones. This incredible variety of potential sources for disease underlines the ability of this species to contaminate and grow in many different food types.

Contaminated cantaloupe and the outbreak of 2011. In 2011, the first inkling that there was an outbreak of listeriosis in the United States came at the end

of July, when the first patient was identified with the disease (**Fig. 3**). By the end of this outbreak, 146 persons with *Listeria* infection were reported to the Centers for Disease Control and Prevention, and the outbreak was spread across 28 states. When epidemiologists looked for the source of the contamination, they were able to track it to cantaloupes grown on a single farm in the southeastern region of Colorado. The victims ranged in age from infancy to 96 years, with most cases occurring in individuals over 60. Seven of the illnesses occurred in pregnant women or their children; in one case, the illness caused a miscarriage. In total, 30 deaths were reported, spread across 12 states where the contaminated cantaloupes had been sold. Of the 140 patients who were able to provide information on what they ate, 131 (94%) confirmed that they had eaten cantaloupe within the month before the onset of illness.

Investigative steps. Steps in a foodborne outbreak investigation require the initial detection of a possible outbreak, followed by defining and locating cases of the disease. Next, a hypothesis is generated by interviewing patients and their respective families about recent food consumption and potential sources of the pathogen involved. Based on this hypothesis, testing then begins by analyzing other patient histories and examining food samples using microbiological techniques. At this point, an association may be possibly identified between the food and the pathogen. If an association is identified, epidemiologists work to locate the origin of the food itself to contain the outbreak. If an association is not identified, the epidemiologists continue to attempt to identify the source of the outbreak, which may not ever be found.

Conclusions. Approximately 800 cases of listeriosis are reported each year in the United States, with three or four confirmed outbreaks occurring. Foods that have been reported are usually meats and cheeses, including hot dogs, deli meats, and soft cheeses. Produce is not often reported, although cases do occur. For example, cantaloupes were responsible for the outbreak in 2011, sprouts were responsible for an outbreak in 2009, and cabbage was responsible for the first identified outbreak, which occurred in 1981. As a preventive measure, individuals at high risk of developing listeriosis, including pregnant women and AIDS patients, should avoid certain foods such as deli meats and soft cheeses.

For background information *see* BACTERIA; BACTERIAL GROWTH; BACTERIOLOGY; CANTALOUPE; FOOD MICROBIOLOGY; FOOD POISONING; FOOD SCIENCE; LISTERIOSIS; MEDICAL BACTERIOLOGY; PATHOGEN; VIRULENCE in the McGraw-Hill Encyclopedia of Science & Technology.   Marcia M. Pierce

Bibliography. P. Cossart and M. Lecuit, Interactions of *Listeria monocytogenes* with mammalian cells during entry and actin-based movement: Bacterial factors, cellular ligands and signaling, *EMBO J.*, 17:3797–3806, 1998, DOI:10.1093/emboj/17.14.3797; M. K. Cowan, *Microbiology: A Systems Approach*, 3d ed., McGraw-Hill, New York, 2012; J. Czajka and C. A. Batt, Verification of causal relationships between *Listeria monocytogenes* isolates implicated in food-borne outbreaks of listeriosis by randomly amplified polymorphic DNA patterns, *J. Clin. Microbiol.*, 32:1280–1287, 1994; G. Franciosa et al., Characterization of *Listeria monocytogenes* strains involved in invasive and noninvasive listeriosis outbreaks by PCR-based fingerprinting techniques, *Appl. Environ. Microbiol.*, 67:1793–1799, 2001, DOI:10.1128/?AEM.67.4.1793-1799.2001; P. R. Murray, K. S. Rosenthal, and M. A. Pfaller, *Medical Microbiology*, 6th ed., Mosby, St. Louis, 2009; E. Nester, D. Anderson, and C. E. Roberts, Jr., *Microbiology: A Human Perspective*, 7th ed., McGraw-Hill, New York, 2012.

Long-Term Evolution (LTE)

LTE is the leading global broadband mobile radio technology, and a step in the series of mobile radio technologies developed by the Third-Generation Partnership Project (3GPP) following in the heritage of GSM (Global System for Mobile Communications) and UMTS (Universal Mobile Telecommunications System).

The first version of the LTE specifications, known as Release 8, supports peak transmitted data rates of up to 300 Mbps in the downlink (that is, from base station to mobile terminal) and 75 Mbps in the uplink. Later versions have increased these rates to 3 and 1.5 Gbps respectively, and 3GPP continues to introduce further enhancements. It is designated a "fourth Generation" (4G) mobile communications standard by the International Telecommunication Union (ITU).

Development of LTE. The 3GPP consortium which developed LTE is a worldwide collaboration of standards development organizations, with nearly 400 member companies from across the telecommunications industry. A key feature of 3GPP is the active involvement of network operators as well as equipment vendors, resulting in a uniquely practical and flexible standard with a strong emphasis on providing robust performance even under challenging mobility scenarios.

3GPP commenced the work to design LTE in 2004, and the first specifications were released in 2007. LTE was designed to support a wide range of applications, including voice, video, file transfer, and gaming, to name but a few.

LTE is based on a new air interface compared to previous generations of mobile communication systems. Its main features are: flat IP network architecture; orthogonal multiple access, derived from multicarrier technology; advanced multiantenna technology; a fully packet-switched radio interface; and common design for operation in paired and unpaired spectrum.

Each of these, and some other key aspects, is outlined below.

Flat IP network architecture. The LTE radio access network consists of a single type of node, the eNodeB (eNB), which incorporates both base station and radio resource management functionalities. This helps to ensure that LTE has a very low latency, both for packet transmission and for connection setup.

The eNodeBs are interconnected by a standardized interface known as the X2 interface, across which both control signaling and data can be transferred (**Fig. 1**). The connection between the eNodeB and the core network [known in LTE as the Evolved Packet Core (EPC)] is provided by the S1 interface to a Serving Gateway (S-GW) or Mobility Management Entity (MME) node. As in UMTS, the mobile terminals are known as User Equipment (UE).

Multicarrier multiple access. In place of the code-division multiple access (CDMA) scheme used in UMTS, LTE uses orthogonal multicarrier multiple access schemes: orthogonal frequency-division multiple access (OFDMA) in the downlink, and single-carrier frequency-division multiple access (SC-FDMA) in the uplink.

OFDMA breaks down the wideband transmitted signal into a large number of narrowband subcarriers. These are closely spaced such that they are orthogonal to each other in the frequency domain, resulting in a high spectral efficiency. In LTE, different groups of subcarriers can, broadly speaking, be allocated to transmissions for different users, in units

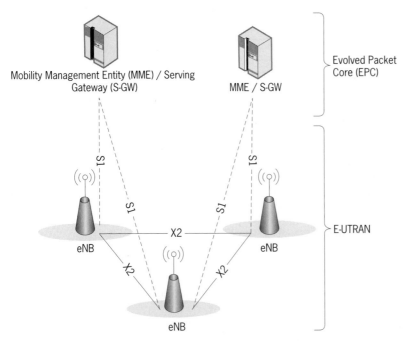

Fig. 1. LTE system architecture. (*Reproduced by permission of and copyright © 3GPP*)

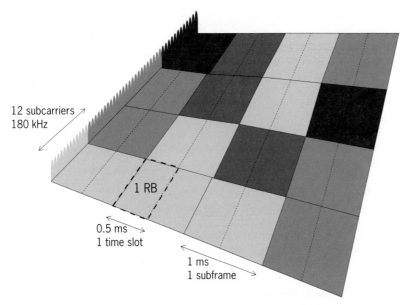

Fig. 2. Transmission resource structure of LTE. Different shades represent transmissions for different users.

known as resource blocks (RBs). One RB consists of 12 subcarriers for a period of 0.5 ms (**Fig. 2**).

Control signaling is usually transmitted in the first few OFDMA symbols of each subframe, using the whole system bandwidth in order to maximize frequency diversity.

Since OFDMA uses multiple subcarriers in parallel, the symbol rate on each subcarrier is low compared to the total combined data rate. This means that the symbol duration is long, so that the delay-spread which arises from multipath propagation can be contained within a guard period occupying only a small portion of each symbol's duration. In order to maintain orthogonality of the subcarriers, the guard period is generated as a cyclic prefix (CP) by repeat-

ing some samples from the end of each symbol at the beginning, as shown in **Fig. 3**.

The subcarrier spacing is chosen as a trade-off between resilience against delay-spread and resilience against frequency shifts such as those that arise from Doppler shifts. A small subcarrier spacing enables a large number of subcarriers to be used in a given spectrum allocation, thereby enabling a long CP to be used without it representing a high overhead as a proportion of the symbol duration. However, a small subcarrier spacing is more sensitive to inter-carrier interference (ICI). As a compromise between these factors, LTE uses a fixed subcarrier spacing of 15 kHz, regardless of the system bandwidth.

By using different numbers of subcarriers, LTE can operate in a wide range of different bandwidths, from 1.4 MHz to 20 MHz.

An OFDMA receiver can be implemented as a low-complexity Fast Fourier Transform (FFT), which makes OFDMA ideally suited to downlink transmissions, where the cost of the receiver in the mobile terminals is crucial. However, the peak-to-average power ratio (PAPR) of a transmitted OFDMA signal is high, which makes it less suitable for the uplink. Therefore the LTE uplink uses single-carrier frequency-division multiple access (SC-FDMA), which is a variation of OFDMA, but with an initial precoding stage using a discrete Fourier transform (DFT). The DFT precoding results in a single-carrier waveform which exhibits a significantly lower PAPR than OFDMA.

The use of SC-FDMA for the LTE uplink enables there to be a high degree of commonality between the downlink and uplink signal structures, including the 15-kHz subcarrier spacing.

Multiantenna technology. LTE incorporates several techniques to exploit the benefits of multiple antennas.

Downlink transmit diversity is provided for up to four antennas at the eNodeB. With two transmit antennas at the eNodeB, a Space-Frequency Block Code (SFBC) is used. With four transmit antennas at the eNodeB, the antennas are treated in pairs, with each pair using independent SFBC encoding in a frequency-switched transmit diversity (FSTD) scheme.

In the uplink, transmit diversity is supported by means of antenna switching, whereby the UE changes the antenna from which it transmits from time to time.

Receive diversity is also used in LTE to provide additional robustness against multipath fading. The basic specified LTE performance requirements assume that all UEs have at least two receive antennas from which the signals are combined using a technique like maximal ratio combining (MRC).

LTE also incorporates spatial multiplexing transmission modes which make use of multiple antennas at both ends of the radio link to transmit multiple parallel data streams using the same time and frequency resources when the radio channel characteristics are suitable. These "MIMO" (multiple-input, multiple-output) modes are mainly "closed-loop," whereby the UE uses the eNodeB's broadcast reference signals to derive an estimate of the downlink channel response in order to feed back a recommendation of a preferred precoding matrix (selected from a specified codebook of matrices) to be applied at the eNodeB to maximize the supportable data rate via as many spatial "layers" as the radio channel can support.

The first release of LTE provides for up to four spatial layers to be transmitted simultaneously to a UE (although only the highest category of UEs has the capability to support more than two layers). In suitable propagation conditions, this provides the possibility for the peak data rate to be increased by a factor of up to four compared to single-layer transmission.

Open-loop beamforming transmission is also supported by LTE, whereby no explicit feedback is sent from the UEs to guide the transmitter precoding at the eNodeB; instead, it is assumed that the eNodeB may have a correlated array of antennas (for example, eight closely-spaced elements) and that the uplink and downlink channels are therefore spatially correlated. The eNodeB can then estimate the angle of arrival of uplink transmissions and form a transmission beam in the same direction. The precoding that is required in this case is not constrained by a finite codebook, and UE-specific reference signals are therefore precoded with the same precoding vector as is used for the data and transmitted together with the data to provide the UE with the phase reference for demodulating the data.

Packet-switched radio interface. LTE benefits from packet switching at the radio interface by means of dynamic scheduling.

Dynamic packet-based scheduling in the frequency domain is facilitated by the multicarrier transmission schemes in uplink and downlink which enable different groups of subcarriers to be allocated to different UEs depending on the radio channel conditions in different parts of the carrier bandwidth. These allocations can be changed as often as every 1 ms. The modulation and coding scheme (MCS) for each transmission can also be varied dynamically depending on radio channel conditions, in conjunction with fast hybrid ARQ (automatic repeat request) with incremental redundancy.

One exception to this dynamic scheduling in LTE is known as "semi-persistent scheduling," whereby a particular set of subcarriers can be allocated to a UE at regular intervals with a fixed MCS. This is particularly suitable for services like voice over IP (VoIP).

FDD and TDD operation. LTE is designed to operate in both paired and unpaired spectrum, in frequency-division duplex (FDD) and time-division duplex (TDD), respectively. These two modes use the same basic multiple access technologies and parameters, with the main difference being that for TDD operation LTE can be configured with a variety of different ratios of uplink timeslots to downlink timeslots. One example is shown in **Fig. 4**. A short guard interval is provided at the switch-point between downlink and uplink, to provide a nonzero

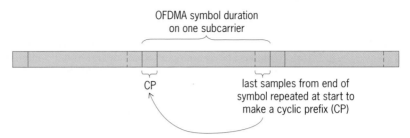

Fig. 3. **OFDMA guard period.**

switching time for the equipment and to absorb propagation delays. The uplink-downlink timeslot configuration can be set to match traffic patterns or to match interference conditions on neighboring carrier frequencies.

Multimedia broadcast/multicast services (MBMS). LTE provides highly spectrally efficient support for broadcast services by means of a mode of operation known as multimedia broadcast single frequency network (MBSFN). In this mode, multiple base stations transmit broadcast data simultaneously, such that a UE will receive multiple versions of the same signal with different delays due to the different propagation delays from the different cells. Provided the base stations are sufficiently tightly synchronized for their signals to arrive at the UEs within the CP at the start of each OFDM symbol, there will be no inter-symbol interference (ISI), which makes the MBSFN transmission appear like a transmission from a single large cell. This gives a significant increase in the received signal to interference-plus-noise ratio (SINR), since inter-cell interference is translated into useful signal energy for the data reception.

Interference management. LTE is designed to operate with a frequency reuse factor of unity. Cell-specific scrambling is used to whiten inter-cell interference. However, neighboring cells can also cooperate to avoid scheduling transmissions for cell-edge users in the same RBs in adjacent cells, and some standardized signaling is provided to support this.

For the downlink, signaling can be exchanged between eNodeBs over the X2 interface so that one eNodeB can inform the neighboring eNodeBs whether it is planning to keep the transmit power for each group of 12 subcarriers below a certain limit. This enables the neighboring cells to take into account the expected level of interference on each group of subcarriers when scheduling transmissions to UEs in their own cells, for example, avoiding scheduling transmissions to cell-edge UEs on those subcarriers.

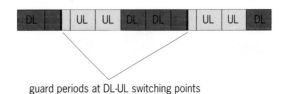

guard periods at DL-UL switching points

Fig. 4. **Example of a TDD timeslot configuration in LTE.** UL = uplink; DL = downlink.

For the uplink, eNodeBs can exchange signals to indicate whether they have detected high levels of interference on certain groups of subcarriers, or to inform neighboring eNodeBs of an eNodeB's intention to schedule uplink transmissions by cell-edge UEs on certain groups of subcarriers, and therefore that high interference might occur on those subcarriers. Neighboring cells may then take this information into consideration in scheduling their own users.

In the uplink, a further important tool for managing inter-cell interference is "fractional power control," whereby the eNodeB can control the degree to which each UE compensates for the path loss when setting its uplink transmission power. This can be used to maximize system capacity by trading off fairness for cell-edge UEs against the inter-cell interference generated toward other cells.

Self-optimization. LTE provides tools to enable some aspects of the network configuration to be optimized automatically. These tools are collectively known as "self-optimizing networks" (SON). They may include: coverage and capacity optimization; energy-saving features; automatic configuration of physical cell ID; optimization of mobility and handover, including for example detection of handovers that are too early, too late, or not to the most appropriate cell; load balancing; and automatic identification of the neighbor cells for each cell. Automatic optimization of such features can reduce the effort required in manual planning and optimization of LTE networks.

Enhancements to LTE: LTE-Advanced. Subsequent to the first version of LTE, 3GPP continues to develop LTE by the addition of new features and enhancements. The Release 9 version was finalized in 2010, and Release 10 in 2011.

Release 10 was a particularly important step in the development of LTE, resulting in LTE being known from then on as LTE-Advanced. This release was designed to fulfill the requirements of the ITU (International Telecommunication Union) for the system to be designated an IMT-Advanced (International Mobile Telecommunications-Advanced) system. In Release 10, the supported peak data rates are increased to 3 Gbps and 1.5 Gbps in the downlink and uplink, respectively.

LTE-Advanced includes carrier aggregation, whereby multiple carriers at different frequencies are used in conjunction with each other, up to a

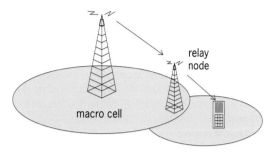

Fig. 6. Operation of a relay node.

total bandwidth of 100 MHz (**Fig. 5**). This enables network operators to make more effective use of diverse spectrum allocations, as well as multiplying the available data rates by the number of aggregated carriers.

LTE-Advanced also features enhancements to the multiple antenna transmission schemes. LTE-Advanced supports up to eight layers in the downlink and four in the uplink.

Relay nodes are also introduced in LTE-Advanced, operating like eNodeBs with a wireless backhaul connection to another eNodeB using the same spectrum as is used to communicate with the UEs (**Fig. 6**). Such relays may be used for filling coverage gaps, extending the coverage of rural cells, and potentially in some cases for capacity enhancement.

Future versions of LTE will include support for coordinated multipoint operation (CoMP), with increased cooperation between eNodeBs, and other enhancements to deliver further improvements in areas such as spectral efficiency and energy efficiency.

For background information *see* MOBILE COMMUNICATIONS; MODULATION; MULTIPLEXING AND MULTIPLE ACCESS; PACKET SWITCHING; VOICE OVER IP in the McGraw-Hill Encyclopedia of Science & Technology. Matthew Baker

Bibliography. S. Sesia, I. Toufik, and M. Baker (eds.), *LTE—The UMTS Long Term Evolution: From Theory to Practice*, 2d ed., Wiley, Chichester, U.K., 2011; C. Zhang, S. L. Ariyavistakul, and M. Tao, LTE-Advanced and 4G Wireless Communications—Part 1, *IEEE Comm. Mag.*, 50(2):102–103, February 2012, DOI:10.1109/MCOM.2012.6146488; C. Zhang, S. L. Ariyavistakul, and M. Tao, LTE-Advanced and 4G Wireless Communications—Part 2, *IEEE Comm. Mag.*, 50(6):26, June 2012, DOI:10.1109/MCOM.2012.6211482.

Main-group multiple bonds

The valence orbitals of the main-group elements (groups 1, 2, and 13–18) are fully occupied or have empty *s* and *p* orbitals, while those of the transition metals are dominated by the presence of *d* orbitals, in addition to *s* and *p* orbitals. It is well accepted that the main-group elements usually have relatively strong covalent bonds using *s* and *p* orbitals compared with the transition metals, which form highly ionic chemical bonds. As a result, compounds of the

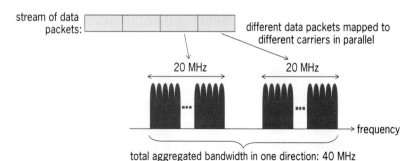

Fig. 5. Example of carrier aggregation.

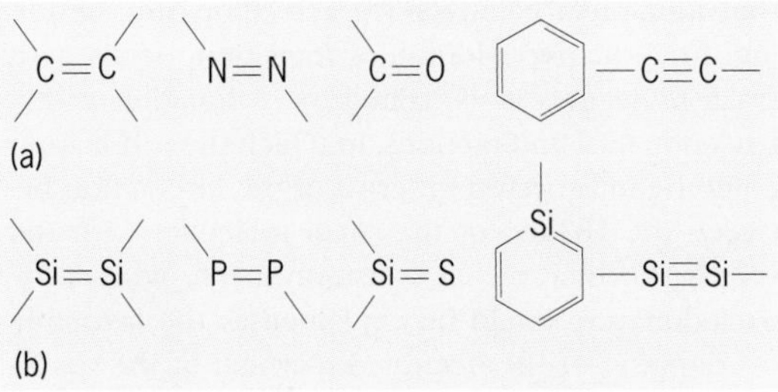

Fig. 1. Multiple-bond compounds (*a*) of the second-row elements; these are very stable and play important roles in organic chemistry. (*b*) Multiple-bond compounds of the heavier elements; these are highly reactive and difficult to synthesize and isolate.

main-group elements show relatively low reactivity compared to transition metals, and very few examples are known of the activation of small molecules (such as those bearing H—H, C—H, or C—O bonds) with compounds of the main-group elements, in contrast to the rich chemistry of the transition metals.

Multiple-bond compounds of the heavier main-group elements. Organic chemistry is based on the first- and second-row elements, such as carbon, hydrogen, oxygen, and nitrogen. Although some heavier elements are now recognized as being indispensable for naturally occurring biological systems, organic compounds containing heavier main-group elements are still rare, and their intrinsic chemistry was not fully understood until recently (**Fig. 1**). In the case of the lighter main-group elements, multiple-bond compounds, such as olefins, azo compounds, aromatic species, nitriles, acetylenes (alkynes), and so on, are readily available as stable compounds and are known to play very important roles because of their versatile reactivity and unique character in functional materials. However, the heavier congeners of such multiple-bond compounds are highly reactive and were reported to be not accessible as stable compounds under ambient conditions until the late 1970s.

This classical double-bond rule was broken by the first synthesis and isolation of stable Si=C, Si=Si, and P=P double-bond species after the 1970s by taking advantage of the steric protection afforded by bulky substituents, the so-called kinetic stabilization method (**Fig. 2**). Inspired by this pioneering work, a number of multiple-bond compounds of the heavier main-group elements have been synthesized as stable species and fully characterized to reveal their unique properties, which are sometimes quite different from those of their lighter-element analogs. Presently, synthesis and isolation of heavy olefins (double-bond compounds of the heavier group 14 elements), heavy azo compounds (double-bond compounds of the heavier group 15 elements), and heavy ketones (double-bond compounds of the heavier group 14 and 16 elements, except for the case of lead-containing heavy ketones) have been accomplished. In addition to these, aromatic compounds containing heavier group 14 elements (heavy aromatics) have been synthesized and isolated as stable species.

Steric protection, using bulky substituents, prevents oligomerization (polymerization) and/or intermolecular addition reactions. Steric protection, which causes the least perturbation from electronic and coordinative influences, is of great importance in view of its superiority in the synthesis and isolation of stable double-bond species. Thus, the multiple-bond systems of the heavier main-group elements are now no longer imaginary species but a practically available class of compounds, provided that we have appropriate steric protecting groups and synthetic methods. As a result of competitive and aggressive studies on these multiple-bond systems of the heavier main-group elements, we are now starting to think about how to use the heavier main-group elements for developing new reactions and functions.

Heavy acetylenes. Since the first reports of Sn=Sn double-bond species by M. F. Lappert and coworkers in 1973 and Si=Si double-bond species by R. West and coworkers in 1981, a variety of heavy ethylenes (doubly bonded compounds of the heavier group 14 elements, $R_2M=MR_2$, where M = Si, Ge, Sn, and Pb) have been synthesized and isolated as stable compounds with unique properties. In addition, the synthesis and isolation of heavy acetylenes (RM≡MR, where M = Si, Ge, Sn, Pb) were recently reported and have attracted much attention.

Acetylenes, carbon-carbon triple-bond compounds, are known to have linear structures with one C—C σ bond and two equivalent π bonds perpendicular to the σ bond. In contrast, the heavy acetylenes were found to have a bent geometry as a stable structure, consisting of distorted σ and π bonds in addition to one π bond (**Fig. 3***a*–*d*). Such unique characters are most likely interpreted in terms of the stabilization afforded by the mixing of π and σ* orbitals due to the weak π bond of the longer M≡M bond, as compared with those of the acetylenes. The most important feature of the heavy acetylenes is that one of their π bonds is weaker than the other, and hence they have a higher HOMO (highest occupied molecular orbital) and a lower LUMO (lowest unoccupied molecular orbital), as compared with those of C≡C bonds

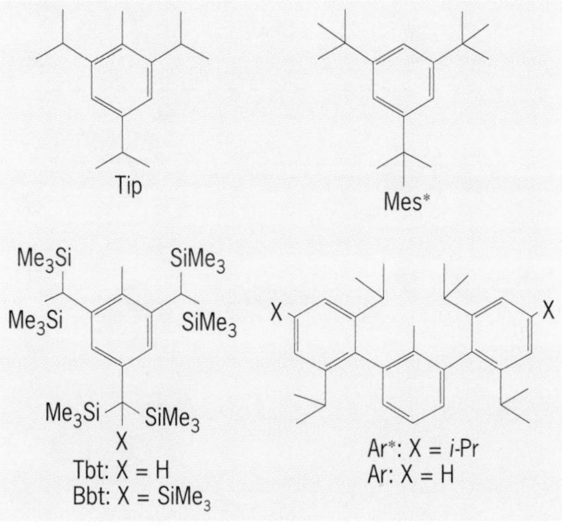

Fig. 2. Examples of steric protecting groups.

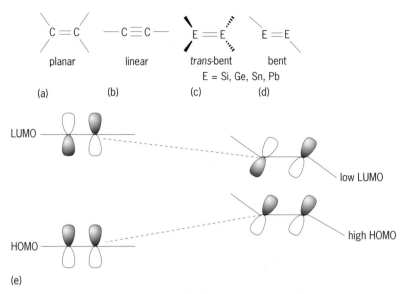

(a) planar

(b) linear

(c) *trans*-bent

(d) bent

E = Si, Ge, Sn, Pb

LUMO

HOMO

low LUMO

high HOMO

(e)

Fig. 3. Characteristics of acetylenes and heavy ethylenes. (*a*) Ethylene. (*b*) Acetylene. (*c*) Heavy ethylene. (*d*) Heavy acetylene. (*e*) HOMO-LUMO energy gaps for acetylene (*left*) and heavy acetylene (*right*).

of the acetylenes. That is, heavy acetylenes should have considerably smaller HOMO-LUMO energy gaps (Fig. 3*e* and **Fig. 4**).

Activation of small molecules with heavy acetylenes. It is well known that transition-metal complexes readily undergo oxidative addition toward the reactive chemical bonding of small molecules, such as H—H, C—X (where X = halogen, etc.), and C≡C bonds, resulting in the formation of a reactive transition-metal complex with a higher oxidation state. Subsequent transmetallation, followed by reductive elimination, enables them to function as

transition-metal catalysts. The activation process during the oxidative addition of transition metals with small molecules is explained by σ donation and π back-donation interactions, in which the σ donation could be interpreted in terms of the interaction between the HOMO of the small molecule with the vacant *d* orbital of the transition metal, whereas π back-donation could be explained as the favorable overlapping of the occupied *d*-orbital of the transition metal with the vacant antibonding orbital of the coordinated small molecule.

Based on a similar concept, the HOMOs and LUMOs of heavy acetylenes can make effective π back-donation and σ donation interactions favorably in phase with the antibonding and bonding orbitals of a small molecule, respectively, because of their higher HOMO and lower LUMO levels. As in the case of a transition metal, the occupied π orbital of a heavy acetylene can effectively overlap with the antibonding orbital of a small molecule as a π back-donation, making the chemical bond weaker. In other words, the molecular orbitals of heavy acetylenes have suitable character to activate small molecules, and it is naturally anticipated that heavy acetylenes will work as active catalysts as well as transition-metal catalysts.

Recent studies on the reactivity of disilyne (**Fig. 5***a*), digermyne (Fig. 5*b*), and distannyne (Fig. 5*c*) revealed that all of them react readily with ethylene gas to give the corresponding ethylene adducts (Fig. 5*d–f*), as the result of activation of ethylene molecules. As shown in **Fig. 6**, these reactions are reasonably interpreted in terms of the initial σ donation of ethylene π bonds toward the LUMO of the heavy acetylene associated with π-back

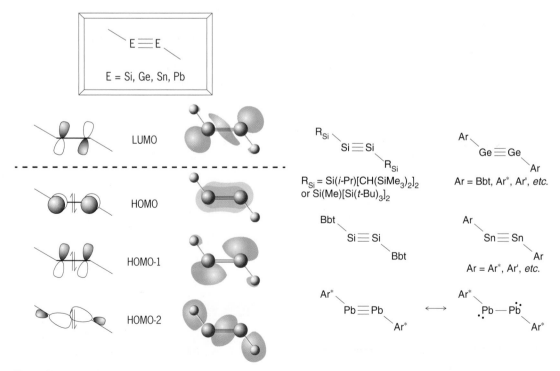

Fig. 4. Heavy acetylenes with bent structures.

donation interactions between the HOMOs of heavy acetylenes and ethylene π^* orbitals followed by the corresponding intramolecular rearrangements. Interestingly, the ethylene adduct (Fig. 5f) of distannyne (Fig. 5c) is reported to undergo regeneration of distannyne via the reversible elimination of ethylene by some external stimulation, such as evacuation or heating.

While hydrogenation reactions using molecular hydrogen are playing important roles in various research areas, it is commonly accepted that transition-metal catalysts are indispensable for the activation of strong H—H bonding. In this context, P. P. Power and coworkers examined the reaction of digermyne with molecular hydrogen and found that digermyne readily underwent hydrogenation at room temperature to give the corresponding hydrogenated products, such as $Ar'(H)Ge{=}Ge(H)Ar'$, $AR'H_2Ge{-}GeH_2Ar'$, and $Ar'GeH_3$. The reaction of distannyne with molecular hydrogen was also reported to give the corresponding hydrogenated tin products via the scission of H—H bonding. The driving force for these hydrogenation reactions of the heavy acetylenes digermyne and distannyne should be the effective coordination of the H—H bonding σ electrons toward the low LUMO of heavy acetylenes in a favorable phase, which is very similar to the initial step of hydrogenation reactions using molecular hydrogen along with a transition-metal catalyst. The silicon—silicon triple bond of a disilyne, $R_{Si}Si{\equiv}SiR_{Si}$ (where $R_{Si} = Si(i\text{-}Pr)[CH(SiMe_3)_2]_2$) is also known to undergo insertion toward the NH bond of amines (R_2NH), giving the corresponding adducts, $[R_{Si}(R_2N)Si{=}Si(H)R_{Si}]$; that is, the disilyne can work as an activator toward the NH bond of amines. In addition, it has already been reported that heavy acetylenes can activate small molecules such as butadienes, acetylenes, and C=O bond-containing compounds.

As can be seen from the above examples, it can be concluded that heavy acetylenes can activate the strong bonds in small molecules. This is most likely because heavy acetylenes have high HOMO and low LUMO energy levels as well as the appropriate orbital circumstances for σ donation and π back-donation. These new findings indicate strongly that the activation of small molecules, which has been believed to be successful only in the cases of transition-metal-catalyzed systems, can now be achieved with reactive low-coordinated systems of the heavier main-group elements (Fig. 6).

Outlook. While it is well known that heavier main-group elements can form the corresponding high-coordinate species because of their large covalent radii relative to those of second-row elements, recent remarkable developments in the chemistry of low-coordinated species of heavier main-group elements, such as multiple-bond compounds, now enables us to synthesize and isolate them as stable compounds. It is evident that multiple-bond compounds of the heavier main-group elements have unique and interesting reactivities in the activation of small molecules (which was only realized by taking advantage of

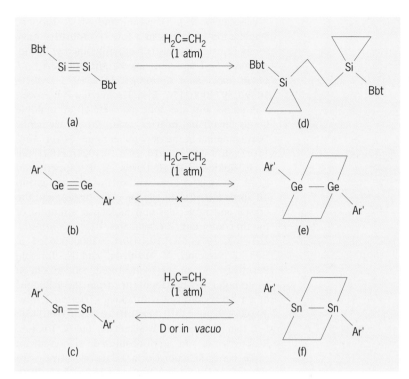

Fig. 5. Activation of ethylene using heavy triple-bonded compounds.

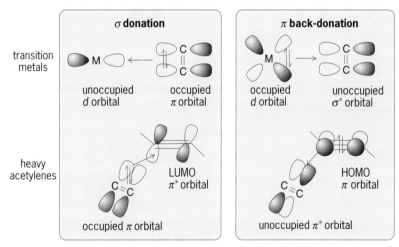

Fig. 6. Examples of σ donation and π back-donation for transition metals and heavy acetylenes.

transition-metal complexes) because of their intrinsic character of high HOMO and low LUMO levels. Although the development of useful catalytic systems has not been achieved yet, it is naturally expected that the heavier main-group elements will make new and great contributions to fields in which transition metals have played important roles. The future of main-group chemistry is truly promising, as the time is right for using a much larger variety of elements in catalytic chemistry and materials science.

For background information *see* ALKYNE; CATALYSIS; CHEMICAL BONDING; MOLECULAR ORBITAL THEORY; PERIODIC TABLE; TRANSITION ELEMENTS in the McGraw-Hill Encyclopedia of Science & Technology.

Norihiro Tokitoh

Bibliography. P. J. Davidson and M. F. Lappert, Stabilisation of metals in a low co-ordinative environment using the bis(trimethylsilyl)methyl ligand; coloured SnII and PbII alkyls, M[CH(SiMe₃)₂]₂, *J. Chem. Soc., Chem. Commun.*, 317a, 1973, DOI:10.1039/C3973000317A; R. C. Fischer and P. P. Power, π-Bonding and the lone pair effect in multiple bonds involving heavier main group elements: Developments in the new millennium, *Chem. Rev.*, 110(7):3877–3923, 2010, DOI:10.1021/cr100133q; D. E. Goldberg, D. H. Harris, M. F. Lappert, and K. M. Thomas, A new synthesis of divalent group 4B alkyls M[CH(SiMe₃)₂]₂ (M = Ge or Sn), and the crystal and molecular and molecular strcuture of the tin compound, *J. Chem. Soc., Chem Commun.*, 261–262, 1976, DOI:10.1039/C39760000261; J. S. Han, T. Sasamori, Y. Mizuhata, and N. Tokitoh, Reactivity of an aryl-substituted silicon–silicon triple bond: 1,2-disilabenzenes from the reactions of a 1,2-diaryldisilyne with alkynes, *Dalton Trans.*, 39:9238–9240, 2010, DOI:10.1039/C0DT00115E; J. S. Han, T. Sasamori, Y. Mizuhata, and N. Tokitoh, Reactivity of an aryl-substituted silicon-silicon triple bond: Reactions of a 1,2-diaryldisilyne with alkenes, *J. Am. Chem. Soc.*, 132:2546–2547, 2010, DOI:10.1021/ja9108566; R. Kinjo et al., Reactivity of a disilyne RSi≡SiR (R = Si^iPr[CH(SiMe₃)₂]₂) toward π-bonds: Stereospecific addition and a new route to an isolable 1,2-disilabenzene, *J. Am. Chem. Soc.*, 129:7766–7767, 2007, DOI:10.1021/ja072759h; V. Y. Lee and A. Sekiguchi, *Organometallic Compounds of Low-Coordinate Si, Ge, Sn, Pb*, Wiley, 2010; Y. Mizuhata, T. Sasamori, and N. Tokitoh, Stable heavier carbene analogues, *Chem. Rev.*, 109(8):3479–3511, 2009, DOI:10.1021/cr900093s; Y. Peng et al., Addition of H₂ to distannynes under ambient conditions, *Chem. Commun.*, 6042–6044, 2008, DOI:10.1039/B813442A; Y. Peng et al., Reversible reactions of ethylene with distannynes under ambient conditions, *Science*, 325(5948):1668–1670, 2009, DOI:10.1126/science.1176443; Y. Peng et al., Substituent effects in ditetrel alkyne analogues: Multiple vs. single bonded isomers, *Chem. Sci.*, 1:461–468, 2010, DOI:10.1039/C0SC00240B; P. P. Power, Main-group elements as transition metals, *Nature*, 463:171–177, 2010, DOI:10.1038/nature08634; T. Sasamori and N. Tokitoh, Group 14 multiple bonding, in *Encyclopedia of Inorganic Chemistry*, 2d ed., R. B. King (ed.), John Wiley & Sons, Chichester, 1698–1740, 2005, DOI:10.1002/0470862106.ia301; A. Sekiguchi, R. Kinjo, and M. Ichinohe, A stable compound containing a silicon-silicon triple bond, *Science*, 305:1755–1757, 2004, DOI:10.1126/science.1102209; G. H. Spikes, J. C. Fettinger, and P. P. Power, Facile activation of dihydrogen by an unsaturated heavier main group compound, *J. Am. Chem. Soc.*, 127:12232–12233, 2005, DOI:10.1021/ja053247a; K. Takeuchi, M. Ikoshi, M. Ichinohe, and A. Sekiguchi, Addition of amines and hydroborane to the disilyne RSi[triple bond]SiR (R = Si^iPr[CH(SiMe₃)₂]₂) giving amino- and boryl-substituted disilenes, *J. Am. Chem. Soc.*, 132:930–931, 2010, DOI:10.1021/ja910157h; R. West, M. J. Fink, and J. Michl, Tetramesityldisilene, a stable compound containing a silicon-silicon double bond, *Science*, 214(4527):1343–1344, 1981, DOI:10.1126/science.214.4527.1343.

Measles outbreak

Measles is a highly infectious and communicable disease that is caused by the measles virus, which is a member of the genus *Morbillivirus*. After the measles vaccine was first introduced in the United States in the 1960s, the incidence of measles began to decrease. Despite the effectiveness of this vaccine, the number of cases has been increasing in recent years, culminating in a severe outbreak of 223 cases in 2011. This outbreak is believed to have resulted from parents failing to have their children vaccinated, making it easy for the virus to spread among the vulnerable population.

Background. Measles is an acute disease that is most often contracted in childhood. It is caused by the measles virus, which is a single-stranded, negative-sense enveloped ribonucleic acid (RNA) virus of the genus *Morbillivirus*; this genus is classified within the virus family Paramyxoviridae. Measles is highly contagious, infecting 90% of exposed individuals within unvaccinated populations. Infected individuals transmit the virus through coughing and sneezing, releasing the virus in aerosolized secretions.

Measles virus is believed to have evolved from the rinderpest virus in the 11th or 12th century. Rinderpest was a viral infectious disease of cattle and other ungulates, resulting in high mortality in affected animals. The rinderpest virus was targeted by the Food and Agriculture Organization of the United Nations in a vaccination campaign that lasted decades, culminating in the full eradication of the disease in 2011. Unlike rinderpest, measles continues to cause disease worldwide, and the number of cases is rising in the United States.

Measles is typically diagnosed through its primary symptoms: a fever lasting up to four days, cold symptoms, cough, and conjunctivitis that makes the eyes appear red, swollen, and weepy. Koplik's spots, which are small white lesions on the cheek inside the mouth, may appear and can be used in diagnosis. These spots disappear quickly, typically within a single day, so they are not always useful in identifying measles cases. Several days after the initial fever begins, a maculopapular rash begins to develop, starting on the head and spreading to the rest of the body (**Fig. 1**). Maculopapular rashes are characterized by flat, red areas with small swollen bumps on the skin; these rashes are also referred to as erythematous because of the redness that occurs. The rash can last for up to eight days after it first appears; infected individuals are contagious for 2–4 days before the rash appears and for up to 5 days after its appearance, allowing the individuals to transmit the virus for as many as 9 days.

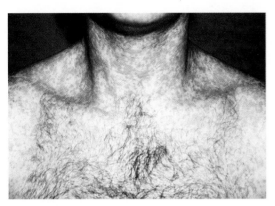

Fig. 1. Measles rash. (*Photo by Kenneth E. Greer; courtesy of M. K. Cowan, Microbiology: A Systems Approach, 3d ed., McGraw-Hill, New York, 2012*)

A significant issue concerning measles is its ability to compromise the immune system of the patient. Individuals who are recovering from this disease often suffer setbacks in the form of complications, which range from relatively mild symptoms, such as diarrhea, to severe and sometimes fatal infections, including bacterial pneumonias and acute encephalitis. These complications occur more often in patients with poor nutrition, including impoverished children in developing countries, and in patients with underlying immune compromise; for example, in patients having acquired immune deficiency syndrome (AIDS), the mortality can be as high as 30%.

Vaccination. Before the measles vaccine was made available in 1963, virtually every person in the United States got measles by the age of 20. Since then, there has been a 99% reduction in the incidence of measles. However, measles is still being "imported" from other countries, typically as a result of unvaccinated individuals who travel into measles-endemic regions.

The measles vaccine was made possible by the work of Thomas C. Peebles, who developed a method of cultivating the virus that he isolated from patients having the disease. After Peebles cultivated the virus in sufficient quantity, John Franklin Enders produced a measles vaccine that was tested on children in New York City and in Nigeria. This vaccine trial proved successful, and the vaccine was rushed into use in 1963. At this time, nearly twice as many children died from infection with measles compared to those infected with polio; however, polio received much greater attention from the public because of the perception of the damage caused by the polio virus, leading to permanent neurological damage or paralysis in many children and adults.

When the measles vaccine was first introduced, one dose was considered sufficient to protect individuals against the disease. Maurice Hilleman developed the combination vaccine MMR, which produces immunity to measles, mumps, and rubella viruses in a single vaccine. At the time that it was put into use, only one dose was recommended, which protected approximately 90–95% of children who were vaccinated. However, in 1989, the Amer-

ican Academy of Family Physicians, the American Academy of Pediatrics, and the Centers for Disease Control and Prevention's Advisory Committee on Immunization Practices recommended that a booster dose be given, increasing the protection level to 99.7% in children. In fact, in the first 20 years since the vaccine was initially introduced, it has prevented 52 million cases, 5200 deaths, and 17,400 cases of mental retardation, and it has been amazingly cost-effective, achieving a net savings of $5.1 billion (**Fig. 2**). However, rates of vaccination have subsequently decreased among some groups, leading to the current situation in which many children are no longer protected against this disease.

Current status. In 2011, 223 cases of measles occurred in the United States. Most of these cases were clustered in 17 different outbreaks that resulted from exposure of the measles virus to vacationing individuals who had traveled to other areas of the world where measles outbreaks were occurring, or they arose because foreign visitors carried the disease with them to the United States. Approximately half of the cases originated in Western Europe, where 37,000 measles cases were reported; the majority of these cases were from France, Italy, and Spain.

Of the cases in the United States, 141 occurred in individuals who were eligible for the MMR vaccine but had not received it. According to the Centers for Disease Control's Morbidity and Mortality Weekly Report (MMWR), "vaccine-eligible patients were defined as United States residents who (1) were unvaccinated or had unknown vaccination status, (2) did not have any contraindications for vaccination, and (3) were either born after 1957 and aged ≥12 months without previous documentation of presumptive evidence of immunity to measles or aged 6–11 months with recent history of international travel."

Of the 141 vaccine-eligible patients, nine (6%) were infants in the age range of 6–11 months who had recently traveled internationally with their families. Fourteen patients (10%) were aged 12–15 months, which is the age recommended for

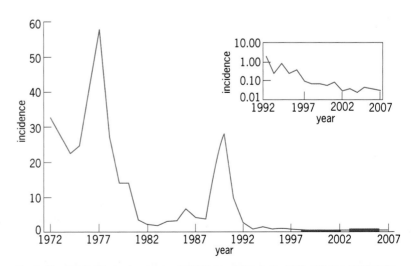

Fig. 2. Measles incidence by year per 100,000 population in the United States, 1972–2007. For the inset graph, the y-axis is given in log scale. (*Courtesy of Centers for Disease Control and Prevention; http://www.cdc.gov/mmwr/preview/mmwrhtml/mm5653a1.htm*)

receiving the first dose of the MMR vaccine. Finally, 66 (47%) were aged 16 months through 19 years. Of these 66 patients, 50 had not been vaccinated because of a philosophic, religious, or personal objection.

The fact that so many individuals were at risk for the disease because of their lack of vaccination is a significant threat to public health in the United States. This trend originated from parents fearing that the vaccine could cause autism. In 1998, British researcher Andrew Wakefield suggested in a published report that the MMR vaccine triggered autism. This study was determined to be fraudulent in 2011, and the article was retracted by the journal. However, its influence remains strong despite widespread press coverage and discussion. Many parents of autistic children still believe that autism is directly linked to vaccines and their preservatives, including the mercury-based thimerosal that was the original topic of the study by Wakefield. Thimerosal was removed from vaccines following the study in 1998 to try to reassure parents that the vaccines were safe. Despite this removal and the debunking of the study by Wakefield, many parents continue to fear that vaccination could cause autism in their children.

Conclusions. Most people think of measles as a childhood disease that is relatively benign. The reality is that this disease is incredibly contagious, spreading rapidly in susceptible populations, and it can be fatal in 3 out of every 1000 patients despite medical treatment. During the 2011 outbreak, approximately one-third of the patients required hospitalization, although there were no fatalities. As of June 2012, only 37 cases of measles had been reported in the United States, hopefully indicating that the number of cases will be significantly less than the spike that occurred in 2011. At least one dose of the MMR vaccine among eligible adults is recommended by the Centers for Disease Control. This recommendation is increased to two doses in high-risk populations, which include any person traveling abroad, college students, and health-care personnel.

For background information *see* BIOLOGICALS; EPIDEMIC; IMMUNITY; IMMUNOLOGY; INFECTIOUS DISEASE; MEASLES; PUBLIC HEALTH; RINDERPEST; VACCINATION; VIRUS in the McGraw-Hill Encyclopedia of Science & Technology.   Marcia M. Pierce

Bibliography. H. D. Clifford et al., CD46 measles virus receptor polymorphisms influence receptor protein expression and primary measles vaccine responses in naive Australian children, *Clin. Vaccine Immunol.*, 19:704–710, 2012, DOI:10.1128/CVI.05652-11; M. K. Cowan, *Microbiology: A Systems Approach*, 3d ed., McGraw-Hill, New York, 2012; Y. Furuse, A. Suzuki, and H. Oshitani, Origin of measles virus: Divergence from rinderpest virus between the 11th and 12th centuries, *Virol. J.*, 7:52, 2010, DOI:10.1186/1743-422X-7-52; N. Komune et al., No evidence for an association between persistent measles virus infection and otosclerosis among patients with otosclerosis in Japan, *J. Clin. Microbiol.*, 50:626–632, 2012, DOI:10.1128/JCM.06163-11; P. R. Murray, K. S. Rosenthal, and M. A. Pfaller, *Medical Microbiology*, 6th ed., Mosby, St. Louis, 2009; E. Nester, D. Anderson, and C. E. Roberts, Jr., *Microbiology: A Human Perspective*, 7th ed., McGraw-Hill, New York, 2012.

Microtomography

Computer tomography (CT) is a nondestructive imaging technique to reconstruct an object's internal structure, based on a series of projection images recorded from different directions by using some type of penetrating radiation, usually x-rays. Originally developed by Godfrey Hounsfield in 1971, medical CAT (computer-aided tomography) scanners quickly became a standard diagnostic tool, and their use in other fields soon became widespread. The development of microfocus x-ray sources and 2D digital detectors led to a new family of CT scanners, the micro-CT scanners. In its most basic implementation, a micro-CT scanner consists of a fixed source and detector setup in between which a sample can rotate, in contrast to medical scanners where source and detector rotate around the patient (**Fig. 1**). Micro-CT has undergone very intensive development over the last few decades, and is quickly becoming a standard microscopy technique in many areas of science and industry. In recent years, the spatial resolution of micro-CT systems has dropped below 1 μm (spurring the term nano-CT). Micro-CT is also performed at synchrotrons, which offer more specialized possibilities, but this work will not be covered in this article.

Tomography. The underlying principle of tomography is based on the work of Johann Radon, who described in 1917 the mathematics for reconstructing a function from a set of line integrals (or more practically "projections") of that function. Any probe that interacts with an object in a way that can be described by line integrals along paths through the object is suitable for tomographic reconstruction. The most common probes are x-rays because of their high diagnostic value and because they travel along straight lines (in first-order approximation). Other forms of tomography include neutron tomography, acoustic tomography, positron emission tomography (PET), and single-photon-emission computed tomography (SPECT).

As an x-ray beam travels through a sample, the x-ray photons can interact with the material inside, causing the original intensity to drop. This is described by the attenuation law of Lambert-Beer, given by the equation below, which forms the basis of

$$I = I_0 \cdot e^{-\int \mu(x) \cdot dx}$$

x-ray tomography. Here, I_0 is the original beam intensity; I is the intensity of the beam after traversing the sample; and $\mu(x)$ is the local attenuation coefficient (expressed in cm^{-1}), which is proportional to the mass density ρ (g/cm^3) and the mass attenuation coefficient μ/ρ (cm^2/g), which depends on the material composition and the x-ray energy. In the typical energy range relevant for x-ray microtomography (10–450 keV), the two main interaction processes contributing to this quantity μ are the photoelectric effect and Compton scattering.

Reconstruction methods. Two types of reconstruction algorithms are commonly used in micro-CT, analytical methods and iterative methods. Analytical methods are based on the Fourier slice theorem, which describes how a distribution can be retrieved based on its projections. In the case of parallel beam acquisition (or by extension, fan beam geometry) this leads to the very intuitive filtered back-projection (FBP) method. Most micro-CT scanners with 2D detectors use cone-beam geometry. Although the central slice can be reconstructed "exactly," slices outside the central plane cannot, resulting in so-called cone-beam artifacts, which can be considered as crosstalk between slices away from the central plane. Nonetheless, the practicality of 2D detectors together with the efficiency of the popular FDK algorithm (introduced by L. A. Feldkamp, L. C. Davis, and J. W. Kress in 1984) for cone beam geometry meant a breakthrough in the use of micro-CT, and cone beam geometry is still the most common implementation in use today. For critical applications, helical acquisition, which provides "exact" reconstructions, has become used more often, despite its more complex acquisition and reconstruction algorithms.

On the other hand, there are the iterative reconstruction methods. By simulating a projection of an intermediate solution and comparing this simulation to the real acquired projection, the intermediate solution can be adjusted accordingly in an iterative way. The advantage of this approach is that one can include knowledge about the imaging process in the forward projection step. For example, in the case of samples composed of a single material, the number of possible solutions is drastically reduced. This principle is referred to as discrete tomography.

Typical scanner implementations. X-ray radiation used for micro-CT is typically generated with an x-ray tube. These produce a so-called bremsstrahlung spectrum composed of a broad continuum stretching from zero to the acceleration voltage, and some emission lines characteristic of the target material. Typical acceleration voltages range from a few kiloelectronvolts up to about 450 keV. To detect the x-ray image, 2D flat-panel detectors are most common. They are made of a scintillator layer [Gadox, cesium iodide (CsI), or other materials] coupled to a photosensitive detector [charge-coupled device (CCD), complementary metal-oxide-semiconductor (CMOS), or amorphous silicon (a-Si)]. Direct detection detectors, which convert the x-ray photon directly into an electrical charge, are less common. 1D line detectors are still used in high-energy CT scanners (energies greater than 200 keV) in order to provide sufficiently high detection efficiency and reduce scatter issues.

The most common implementation of a micro-CT scanner is a closed cabinet which can range in size from a small desktop machine to a room-size cabi-

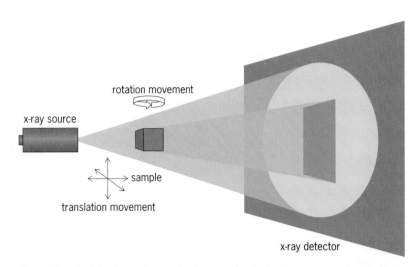

Fig. 1. The principle of cone-beam microtomography. A microfocus x-ray source projects a conical beam onto a pixilated 2D digital detector to record the shadow image cast by the sample.

net. Second, there are small-animal scanners, aimed at preclinical research on laboratory animals. As in medical scanners, the tube and detector rotate on a gantry so that the animal can remain stationary. These scanners are sometimes combined with other modalities such as PET or SPECT in order to merge anatomical data (seen with CT) with functional data (seen with PET/SPECT). Finally, there are custom open-type systems, mostly in research environments, constructed inside radiation-proofed rooms (**Fig. 2**). They can offer more operational freedom and allow for peripheral equipment for more demanding experiments.

Applications. Most CT applications require visual assessment by using 3D volume rendering. This generates 3D images of the object under study and allows for various manipulations such as rotating, changing transparency, making virtual cuts, and false coloring. This is useful to investigate the presence of

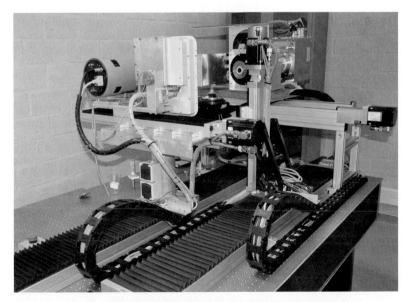

Fig. 2. An example of a custom-made research scanner with two sources and two detectors, capable of scanning large objects up to 40 cm (16 in.) in diameter and small objects with resolutions down to 400 nm. (*Courtesy UGCT, Ghent University*)

Fig. 3. Example of a 3D rendering of an aluminum foam, and a morphological analysis of the pore sizes represented in false color.

cracks or inclusions, to visualize shapes and structures, and so forth. A next analysis step is often the quantification of specific parameters such as size distributions (for example, in foams; **Fig. 3**), porosity mapping (for example, in petrology), network connectivity (for example, in particle filters), and quantitative density profiling (for example, in osteoporosis studies). This step requires dedicated 3D morphological analysis tools. CT data can also be used to extract surface or volume meshes for finite element modeling, or to compare real production samples with their computer-aided design (CAD). Finally, the 3D data can be physically printed in 3D for rapid prototyping purposes. Very recently there have been major developments towards metrology, in-line tomography at production lines, and 4D tomography or dynamic tomography of time-dependent phenomena such as failure under pressure or collapse of foams.

Limitations. Micro-CT has some limitations. First, it follows from theoretical and practical arguments that the relative resolution one can obtain in a given sample is of the order of $1/N$ times the sample size, with N the number of pixels on the detector in one dimension. Adequate sampling of a volume of N^3 voxels requires that the number of projections to acquire is of the same order as N. In most micro-CT applications, N ranges from a few hundred to about 2000 pixels. A volume of 2000^3 voxels therefore requires 8GB of memory space (in 8-bit representation). The time for reconstruction also increases rapidly with N since the numerical complexity of reconstruction is proportional to N^4. As a result, micro-CT requires lots of computing power, fast throughput, and large amounts of storage.

Second, micro-CT scanners rely on magnification to overcome the intrinsic detector resolution. The

focal spot size of the source is then a direct measure for the spatial resolution that can be obtained. Increasing spatial resolution thus requires a smaller focal spot and longer measurement times, since the x-ray flux is roughly proportional to the focal spot size. Microfocus sources typically have a minimal focal spot size of a few micrometers, but some reach 300 nm or even less. Typical scan times can range from a few minutes up to a few hours, and also depend on the desired relative resolution, required signal to noise ratio, and so forth.

A major disadvantage of x-ray tubes compared to synchrotrons is the polychromaticity of the x-ray beam. Most reconstruction methods are based on the idealized exponential attenuation law for monochromatic radiation. The strong energy dependency of the attenuation coefficient causes low-energy x-rays to be attenuated more than high-energy x-rays, an effect known as beam hardening. This results in typical reconstruction artifacts such as cupping (seemingly higher attenuation values in the outside areas of a sample) and streaks. To counteract this, one can tune the x-ray spectrum by adding filters or applying software corrections.

Phase contrast. In micro-CT, the spatial resolution of the detector can be enough to resolve the effects caused by coherent scattering or refraction inside the sample. This so called in-line or propagation phase contrast manifests itself as an edge enhancement of features in the sample, which can be useful to improve contrast in low-attenuation samples, but it can also cause reconstruction artifacts that can lead to misinterpretation. The x-ray optics used at synchrotrons have also been applied at micro-CT setups to manipulate the x-ray propagation in such a way that the phase contrast contribution can be separated from the absorption contrast, allowing reconstruction of the phase-related part of the complex refractive index.

New detector technologies. Existing detectors are used mostly in the integrating mode, but photon-counting detectors that effectively detect and process individual x-ray photon interactions are being developed. Adding energy discrimination capabilities allows separation of x-rays according to their energy, offering promising new ways to perform spectral CT for improved material identification and higher quantitative accuracy, or simply to better address beam-hardening artifacts.

For background information *see* ABSORPTION OF ELECTROMAGNETIC RADIATION; COMPUTERIZED TOMOGRAPHY; MEDICAL IMAGING; MICRORADIOGRAPHY; PHASE-CONTRAST MICROSCOPE; RADIOGRAPHY; SYNCHROTRON RADIATION; VOLUMETRIC DISPLAYS; X-RAY TUBE in the McGraw-Hill Encyclopedia of Science & Technology.   Manuel Dierick

Bibliography. J. C. Elliott and S. D. Dover, X-ray microtomography, *J.Microsc.*, 126:211–213, 1982, DOI:10.1111/j.1365-2818.1982.tb00376.x;   G. T. Herman, *Image Reconstruction from Projections: The Fundamentals of Computerized Tomography*, Academic Press, New York, 1980; A. C. Kak and M. Slaney, *Principles of Computerized Tomographic*

Imaging, IEEE Press, New York, 1988; J. Radon and P. C. Parks (translator), On the determination of functions from their integral values along certain manifolds, *IEEE Trans. Med. Imag.*, 5(4):170–176, 1986, DOI:10.1109/TMI.1986.4307775; S. W. Wilkins et al., Phase-contrast imaging using polychromatic hard x-rays, *Nature*, 384:335–338, 1996, DOI:10.1038/384335a0.

Mineral evolution

Mineral evolution frames the science of mineralogy in the context of Earth's 4.5-billion-year geological history. The central premise of mineral evolution is that Earth's near-surface distribution of minerals has changed significantly as a consequence of physical, chemical, and biological processes. Earth's mineral evolution can be divided into three eras—(1) the era of planetary formation, (2) the era of crust and mantle reworking, and (3) the era of biologically mediated mineralogy—and further subdivided into 10 stages (see **table**). A surprising result of this approach is that two-thirds of the more than 4500 known species of minerals on Earth are the consequence of biological activity, and thus most of Earth's mineralogical richness would not occur on a nonliving planet or moon.

Minerals, which are defined as crystalline compounds of well-defined structure and composition, could not have formed immediately after the Big Bang because the principal elements were gaseous hydrogen and helium. The earliest minerals in the cosmos had to await the formation of large stars and the generation of carbon and other elements heavier than helium through nuclear fusion reactions. Diamond, the earliest mineral species to appear, crystallized when that first generation of large stars exploded into supernovas. As the expanding, carbon-rich portions of the hot stellar envelope cooled to below 4400 K, nanocrystals of diamond started to grow. Graphite, another crystalline form of pure carbon, was probably second (it forms at about 4000 K). Perhaps a dozen microscopic mineral phases appeared altogether—what have been called the "ur-minerals." For tens of millions of years they were the only crystals in the universe.

Era of planetary formation. Planets are the engines of mineral evolution. The first burst of mineralogical novelty occurred in star-forming nebulas, where a new-formed star baked a flat, rotating disk of dust and gas. During this first stage of mineral diversification, about 60 different mineral species crystallized, mostly from molten droplets or "chondrules" that clumped together to form the most primitive chondrite meteorites. Among the new minerals to appear were the first iron-nickel metal phases, sulfides, phosphides, silicates, and oxides that are found in the least-altered chondrite meteorites. Chondrites clumped together by gravity into planetessimals, which grew large enough to partially melt, differentiate, and experience alteration by heat, water, and collisions with other bodies (**Fig. 1**). These processes increased the diversity of minerals to about 250 different species, which have fallen to Earth continuously throughout its history. These 250 minerals represent the starting point of all planets and moons.

Once Earth formed, three processes have led to mineral diversification: (1) the progressive separation and concentration of different chemical elements from their relatively uniform original distribution in the dust that formed the solar system; (2) a greater range in combinations of environmental conditions, including pressure and temperature; and (3) the influence of living cells in producing and sustaining far-from-equilibrium conditions (such as the oxygen-rich atmosphere), as well as their ability to catalyze mineral-forming reactions that would not occur in the absence of life. The extent to which these three processes occur at or near a

The three eras and ten stages of mineral evolution on Earth		
Era and Stage	Age, billions of years	~Total number of species
Pre-Earth "Ur-minerals"	>4.6	12
The Era of Planetary Formation (>4.55 billion years ago)		
1. Unaltered chondrite meteorites	>4.56	60
2. Achondrite meteorites	>4.56 to 4.55	250
The Era of Crust and Mantle Reworking (4.55 to 2.5 billion years ago)		
3. Evolution of the igneous rocks	4.55 to 4.0	350 to 500*
4. Granite and pegmatite formation	4.0 to 3.0	1000
5. Global plate tectonics	>3.0	1500
The Era of Biologically Mediated Mineralogy (>2.5 billion years ago to the present)		
6. The anoxic living world	3.9 to 2.4	1500
7. The Great Oxidation Event	2.4 to 1.85	>2000
8. The intermediate sulfide ocean	1.85 to 0.85	>4000
9. Snowball Earth events	0.85 to 0.542	>4000
10. Skeletal biomineralization	0.542 to the present	4300+

*Number depends on volatile content.

Fig. 1. The first era of mineral evolution is represented in meteorites, which reveal such mineral-forming processes as partial melting and differentiation, alteration by water, and impact processes. This Imilac pallasite meteorite represents the separation of silvery iron-nickel metal from golden magnesium silicate minerals. (*Copyright* © *Meteorites Australia http://www.meteorites.com.au/*)

planet's surface determines the degree of its mineral evolution.

Era of crust and mantle reworking. The mineral evolution of Earth's crust occurred as a result of varied geochemical and petrologic processes, such as volcanism, the release of the volatiles that formed the oceans and atmosphere, cooling and crystallization of magmas, metamorphism under temperature and pressure, plate tectonics, and associated episodes of fluid-rock interactions. These processes changed the face of Earth, generating the first continents and ultimately leading to an estimated 1500 different mineral species. This second era of mineral evolution can be divided into three stages.

Every terrestrial planet and moon experiences the mineral-forming igneous processes of stage 3, which were outlined in Norman Bowen's classic 1928 book, *The Evolution of the Igneous Rocks.* Bowen described how a sequence of different minerals forms whenever a molten rock such as basalt cools and crystallizes. The first crystals to grow from the melt have a composition quite different from the bulk magma. A key process is the settling of denser crystals or floating of lighter crystals, which results in a continuously changing composition of the remaining hot liquid, and thus a cooling sequence of mineral species. Even on water-poor worlds such as Mercury or the Moon, magma cooling and crystallization can yield as many as 350 different mineral species. And if water and other volatiles are abundant then the mineralogical diversity is greatly enhanced by the formation of hydroxides, hydrates, carbonates, and evaporite minerals, leading to as many as 500 different mineral species. Mars, which appears to have had large oceans early in its history, has probably progressed this far in its mineral evolution.

The first crust of every terrestrial world is the dense, dark volcanic rock called basalt. Stage 4 of mineral evolution occurs on planets that have sufficient inner heat to partially re-melt that basaltic crust to form less dense, silica-rich rocks of the granite family. Mineral diversity arises by cycles of partial melting and the concentration of rare elements to form as many as 500 distinctive minerals of the elements lithium, beryllium, boron, niobium, uranium, and a dozen other rare elements. All of these elements have been present since the time of the ur-minerals, but they occur in concentrations much too low to make their own minerals. It may take a billion years or more to achieve the required concentrations of these elements in rock bodies called complex pegmatites. It is possible that Venus reached this kind of stage 4 mineralization, but it is probable that neither Mars nor Mercury are large enough to form significant volumes of granite, much less produce mineral-rich pegmatites.

Plate tectonics, which drives stage 5 of mineral evolution, may be unique to Earth in our solar system. Subduction—the process in which a segment of Earth's crust plunges down into the mantle—causes water-rich, chemically diverse materials of the crust to partially melt and thus concentrate rare elements on a vast scale of millions of cubic kilometers. These buoyant mineral-rich fluids then return to the surface through volcanism, with huge associated ore bodies. For example, massive sulfide deposits associated with subduction volcanism contain more than 150 new sulfosalt minerals. Additionally, dozens of mineral species appeared at the Earth's surface for the first time when deeply subducted wedges of relatively low-density rocks popped back to the surface and carried a wealth of high-pressure minerals, including jadeite and diamond.

A total of perhaps 1500 different mineral species may have formed during the processes of crust and mantle reworking that characterize stages 3 through 5 of Earth's mineral evolution. However, that total is only about a third of the more than 4500 known minerals. Another process, unique to Earth, contributes to this diversity.

Era of biologically mediated mineralogy. Earth differs from all other planets and moons in the solar system because it is a living world. Life has transformed the oceans, the atmosphere, and the rocks and minerals. Indeed, it appears that fully two-thirds of all known mineral species on Earth are a consequence of the biosphere.

Earth's earliest life, microbes that survived in an anoxic environment, had little effect on mineralogical diversity. Stage 6 of mineral evolution is characterized by localized biologically mediated rock formations, including extensive formations of iron minerals and local carbonate reefs, but life contributed little to expand the number or distribution of minerals.

The biological innovation of oxygenic photosynthesis and the rise of an oxygen-rich atmosphere, Stage 7 of mineral evolution, changed that. The "Great Oxidation Event" (GOE), when atmospheric oxygen rose to perhaps 1% of modern levels approximately 2.4–2.2 billion years ago, began to transform Earth's surface mineralogy. More than 2500 of the known mineral species are oxidized and hydrated weathering products of other minerals, and those species are unlikely to have developed in an environment lacking oxygen.

Stage 8, which spanned the period from about 1.85 billion to 850 million years ago, is a time when Earth's

Fig. 2. The third era of mineral evolution saw the dramatic influence of life on mineralization. The trilobite *Hoplolichas* from the Ordovician of Russia displays an elaborately sculpted calcitic shell. (*Robert Hazen collection*)

near-surface environment and biosphere appears to have changed little. During this period the ocean's interface between an oxygen-rich surface layer and oxygen-poor depths gradually got deeper, but few new mineral species are thought to have appeared.

The ninth stage of mineral evolution saw a 300 million year interval when at least three global glaciation events occurred—times commonly referred to as "snowball Earth" episodes, because some models suggest that Earth was frozen from the poles to the Equator. Geologists debate whether ice completely covered the planet, but the mineral ice became the dominant surface species for periods in excess of 10 million years. Ice could not prevent volcanoes from breaking through the frozen surface and adding carbon dioxide to the atmosphere, which led to intervals of rapid greenhouse warming and retreating ice, accompanied by rapid increases in the generation of clay minerals.

The tenth and final stage of mineral evolution, which began approximately 540 million years ago, was marked by the innovation of bioskeletons made of carbonate, phosphate, and silicate minerals (**Fig. 2**). At the beginning of this stage, Earth's land surface was mostly barren rock. The rise of land plants about 400 million years ago significantly altered the appearance of the Earth's surface and led to rapid production of soils, including a dramatic increase in the rate of clay mineral production.

Implications of mineral evolution. The recognition that Earth has undergone dramatic changes in its near-surface mineralogy through more than 4.5 billion years provides a new perspective on mineral sciences. Earth scientists have long known that geology is at heart an historical science. However, mineral collections are traditionally organized on chemical and structural measures. Specimen labels record names, chemical formulas, localities, and crystal information, yet rarely is the age of mineral specimens recorded.

The mineral evolution approach allows each planet and moon in the solar system to be placed in a broader mineralogical context, while providing a conceptual framework for searching for extrater-

restrial life. By framing mineralogy as an historical narrative that is entwined with the epic processes of planet formation, volcanism, plate tectonics, and the origin and evolution of life, mineralogy assumes a more central role in the Earth sciences.

For background information *see* ASTROBIOLOGY; ATMOSPHERE, EVOLUTION OF; BASALT; BIOSPHERE; CARBONATE MINERALS; CLAY; MINERALS; COSMOCHEMISTRY; DIAMOND; GRANITE; GRAPHITE; IGNEOUS ROCKS; JADEITE; MAGMA; MARS; METEORITE; MINERAL; MINERALOGY; OXIDE AND HYDROXIDE MINERALS; PEGMATITE; PLANET; PLATE TECTONICS; SILICATE; SOLAR SYSTEM; SUBDUCTION ZONES; ULTRAHIGH-PRESSURE METAMORPHISM; VOLCANO in the McGraw-Hill Encyclopedia of Science & Technology.

Robert M. Hazen

Bibliography. N. L. Bowen, *The Evolution of the Igneous Rocks*, Princeton University Press, Princeton, NJ, 1928; R. M. Hazen, The evolution of minerals, *Sci. Amer.*, 303(3):58–65, 2010; R. M. Hazen, *The Story of Earth: The First 4.5 Billion Years from Stardust to Living Planet*, Viking, NY, 2012; R. M. Hazen et al., Mineral evolution, *Am. Mineral.*, 93:1693–1720, 2008, DOI:10.2138/am.2008.2955; A. H. Knoll, D. E. Canfield, and K. O. Konhauser (eds.), *Fundamentals of Geobiology*, Wiley-Blackwell, Oxford, U.K., 2012.

Morphometrics in paleontology

Morphometrics is the quantitative analysis of morphology. In paleontology, it is used to study evolution, systematics, and functional morphology. Although morphometrics can be as simple as distinguishing fossil samples based on shell heights, the term is usually applied to multivariate analysis involving tens or hundreds of variables. Morphometrics is most commonly used to differentiate among fossil species. However, it is also used to study rates of evolution, to distinguish between directional change and stasis, to determine the effects of environment on morphology, to measure modularity and morphological integration, to assess changes in morphological disparity, and to reconstruct phylogenetic history. Until the 1980s, most morphometric variables were trait measurements, such as the lengths and widths of bones or other structures. Recently, however, geometric morphometrics, which uses Cartesian coordinates (one of the most useful systems of coordinates for locating points in a given space by means of numerical quantities specified with respect to some frame of reference) of landmark or outline points instead of measurements, has come to predominate paleontology.

Methods. Morphometrics based on trait measurements usually results in a multivariate ordination (a representation of objects with respect to one or more coordinate axes) that is used to assess similarity and difference in form. Measurements are normalized, usually by log transformation, and standardized to have means of zero and unit variances. Principal components analysis (PCA) is a common ordination technique that simply rotates the objects to their

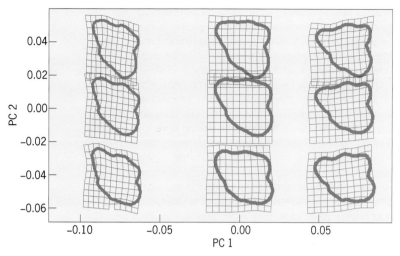

Morphospace of marmot molars. The axes of the graph represent the major axes of variation in a sample of teeth. The variation in shape within the morphospace is shown by models of semilandmarks (tinted area) and thin plate spline deformation grids.

major axes based on the correlation or covariance matrix of the normalized variables. The first few principal component axes show most of the variation in the sample. Visual inspection or statistical tests are used to determine whether groups are distinguishable based on their morphology. Because all linear measurements taken on a fossil are usually correlated with the fossil's size, the first principal component is usually a size axis. For some applications, size is a desirable property; however, for others, size may be a nuisance parameter that can be removed by simply discarding the first principal component or by factoring it out. Two closely related ordination techniques, canonical variate analysis (CVA) and discriminant function analysis (DFA), are used to find the linear combination of variables that best differentiate known groups. These methods can be used to classify an unknown fossil to a group based on the set of trait measurements. Multivariate regression can be used in morphometrics to determine the relationship of morphological shape (and size) to continuous variables, and multivariate analysis of variance (MANOVA) can be used to determine the relationship to categorical variables.

Geometric morphometric methods (GMM) are based on Cartesian coordinates of points placed on morphological features. Points that by themselves represent a structure, such as the apex of a shell, the junction between bones, or the position of a tooth root, are called landmarks, whereas series of points placed along a curve or surface, such as a shell aperture or a bone margin, are called semilandmarks. Regardless of their type, these points can be two dimensional (2D) or three dimensional (3D). Advantages of GMM are that size can be mathematically removed from the analysis and that results can be visualized as deformations of the shape of the fossil rather than as simple scatter plots or tables of numbers. The most common way to remove size, rotation, and translation from the coordinates is by Procrustes superimposition, in which the points of

each object are scaled to unit size, their centroids superimposed, and their coordinates rotated about the centroid until the sum of squared distances between them is minimized. Standard multivariate ordinations and statistics can then be applied to the superimposed shapes, allowing for reduction of statistical degrees of freedom that are lost by the superimposition. When applied to GMM data, principal components space is often called a "shape space" because every point in it is defined by a particular arrangement of landmarks or semilandmarks. The shapes associated with the points can be modeled to show transformations from one part of shape space to another, including modeling of evolutionary change in shape (see **illustration**). The ordination and multivariate statistical analyses described above can all be applied to GMM data.

Several non-Procrustes alternative GMM techniques exist. Fourier analysis is a method applied to closed curves and uses Fourier decomposition to describe shape variation in terms of sine and cosine waves, or harmonics. Fourier analysis predates other GMM methods by approximately 20 years. Eigenshape analysis is another method applied to curves; it describes curves as angular functions (also known as Zahn and Roskies functions). Eigensurface analysis is used for 3D surfaces represented by 3D semilandmarks, and spherical harmonics is a Fourier-based technique for surface data.

Procrustes methods are useful for analyzing similarities and differences in overall shape, but they are largely incapable of studying variation in landmark positions within a shape because the superimposition technique removes the objects from frames of reference in the real world. Procrustes-based morphometrics can indicate that two shapes differ in terms of displacements of landmarks (or semilandmarks) relative to one another, but they cannot determine specifically which landmarks are varying. Euclidean distance matrix analysis (EDMA) is a method specifically designed to study variation at points within a shape. EDMA does not use superimposition, but instead characterizes shape by a matrix of interlandmark distances. Analysis of this matrix identifies which landmarks are the most variable and in what way that they vary relative to the other landmarks in the shape. EDMA has been criticized because it lacks some of the strengths of Procrustes-based analyses, particularly visualization; however, for its primary purpose, EDMA is superior to Procrustes.

Phylogenetic history. Most paleontological morphometric studies involve several samples occurring at different stratigraphic levels or comprising several different taxa. These samples are not statistically independent of one another because they share a phylogenetic history, which must often be taken into account as part of the morphometric analysis. Whether phylogenetic patterns will predominate morphometric ordinations depends on the rate of shape evolution, the depth of common ancestry, and the amount of shape homoplasy (correspondence between organs or structures in different organisms acquired as a result of evolutionary convergence or of parallel evolution) in the group being considered. Paleontologists have vociferously disagreed about whether morphometric data are or are not "phylogenetic," but nearly all morphometric data have a phylogenetic history regardless of whether it is obvious in the results. Several comparative techniques exist for ascertaining the strength of the phylogenetic signal in quantitative data, and related techniques can be used either to map morphometric data onto a phylogenetic tree or to correct for the effects of phylogeny in statistical analyses. Some of the most powerful approaches to morphometrics use phylogenetic structure to assess long-term patterns of evolution, morphological adaptation, and the colonization of new areas of morphospace (a representation of the possible form, shape, or structure of an organism) by one or more clades. Analysis of morphological disparity, or the differences among contemporaneous members of a clade, often accompanies studies of taxonomic diversity through time.

For background information *see* ANIMAL EVOLUTION; BIOMETRICS; COORDINATE SYSTEMS; FACTOR ANALYSIS; FOSSIL; MATHEMATICAL BIOLOGY; PALEONTOLOGY; PHYLOGENY; SKELETAL SYSTEM; STATISTICS; SYMMORPHOSIS in the McGraw-Hill Encyclopedia of Science & Technology.   P. David Polly

Bibliography. J. Claude, *Morphometrics with R*, Springer, New York, 2008; I. L. Dryden and K. V. Mardia, *Statistical Shape Analysis*, Wiley, New York, 1993; Ø. Hammer and D. Harper, *Paleontological Data Analysis*, Blackwell, Oxford, U.K., 2006; R. Reyment, *Multidimensional Paleontology*, Pergamon Press, New York, 1991; M. L. Zelditch et al., *Geometric Morphometrics for Biologists: A Primer*, Elsevier Academic Press, London, 2004.

Multivesicular bodies: biogenesis and function

Multivesicular bodies (MVBs) or multivesicular endosomes are synonymous with late endosomes. Transport vesicles derived from the plasma membrane fuse to form early endosomes, which in turn undergo fission and fusion events to mature into late endosomes that eventually fuse with lysosomes (see **illustration**). The distinguishing characteristic of MVBs is that they contain vesicles and other membranes within their lumen [intraluminal vesicles (ILVs)], which are contained within a single limiting membrane bilayer. ILVs begin accumulating in early endosomes and continue to accumulate until the late endosome stage. Fusion of late endosomes with lysosomes allows for the degradation of ILVs and their contents. Alternatively, in some cases, MVBs may fuse with the plasma membrane to allow secretion of ILVs into the extracellular milieu. MVBs can also serve as intermediates in the formation of more specialized lysosome-related organelles, such as melanosomes or lytic granules, where components enveloped into ILVs are used to form some of the intraluminal constituents of these organelles. A variety of trafficking pathways can contribute to the formation of ILVs, re-

flecting the fact that MVBs and the ILVs within them fulfill a variety of cellular roles.

Importance of MVBs and role of ubiquitin. The biogenesis of ILVs within MVBs solves an important topological problem for cells by providing a way to deliver integral membrane proteins (which contain not only luminally exposed domains, but also cytosolic domains and membrane-spanning segments) completely into the endosomal lumen. This allows the MVB pathway to completely degrade entire membrane proteins once the MVBs fuse with lysosomes and deliver ILVs to the interior of the lysosome. Without this pathway, delivering membrane proteins to only the limiting membrane of endosomes and lysosomes would limit access of lysosomal hydrolases to only the luminal side of membrane proteins, leaving the membrane-spanning segments and the cytosolic domains intact. This distinction is important when considering particular processes such as the function and degradation of cell-surface growth-factor receptor kinases. Receptor tyrosine kinases, including the receptors for epidermal growth factor (EGFR) and platelet-derived growth factor (PDGFR), are delivered and degraded in lysosomes as part of a program to attenuate their activity once they are activated by ligands. These receptor kinases assemble a signaling complex on their cytosolic side, and cessation of signaling activity can only be accomplished once the active cytosolic domain of the receptor kinase is sequestered within the ILVs of MVBs.

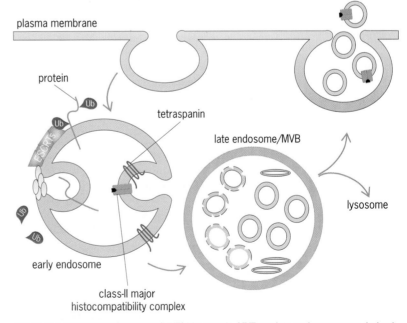

Schematic of multivesicular body (MVB) biogenesis. MVBs or late endosomes are derived from early endosomes. The MVB lumen contains membranes and vesicles that confer the characteristic appearance and properties of the MVBs. The MVB lumen contains vesicle-holding proteins that are targeted for degradation in lysosomes. These proteins are initially sorted into intraluminal vesicles (ILVs) by becoming ubiquitinated (Ub = ubiquitin) and then becoming recognized and sorted by the so-called ESCRT apparatus. Once MVBs fuse with lysosomes, the ILVs and their protein constituents are degraded. MVBs also may contain other internal vesicles, which house proteins such as tetraspanins or the class-II major histocompatibility complex. These internal vesicles can be secreted as exosomes once MVBs fuse with the cell surface. MVBs and lysosomes also contain luminal membranes enriched in the lipid LBPA (lysobisphosphatidic acid).

MVBs play a central role in degrading integral membrane proteins in lysosomes. Membrane proteins are incorporated into ILVs that bud from the limiting membrane of endosomes. This process is conserved in all eukaryotic cells and serves as the major way that cell-surface membrane proteins are ultimately downregulated and degraded. The main sorting signal that allows membrane proteins to be sorted into ILVs is ubiquitin, which is a 76-amino-acid protein that is covalently attached to lysine residues within targeted proteins. Ubiquitin can be attached by a whole host of ligases, with each having specificity for different substrate proteins. Ubiquitin can also be removed from proteins by deubiquitinating enzymes that hydrolyze the peptidelike bond between ubiquitin and the substrate lysine to which it is attached. This provides a versatile mechanism to direct proteins for degradation along the MVB pathway because it allows the cell to designate any membrane protein for degradation simply by conjugating it with a common sorting signal. This also provides a highly regulatable system because ubiquitin can be added to or removed from proteins by a whole host of ubiquitin ligases and deubiquitinating enzymes, respectively, with each probably having their own repertoire of substrate proteins and regulatory mechanisms dictating their activity.

Ubiquitin acts as a self-contained sorting signal by its ability to bind to sorting machinery located on the surface of endosomes. Collectively, this machinery encompasses the ESCRTs (endosomal sorting complexes required for transport), which comprise interacting complexes responsible for gathering ubiquitinated membrane proteins and forming the ILVs that will ultimately incorporate the ubiquitinated membrane proteins. The ESCRTs are found within endosomal subdomains near where ILVs are eventually formed. These subdomains also contain phosphatidylinositol 3-phosphate (a phosphorylated lipid) and clathrin, which interact with ESCRTs and help define a patch on endosomes for cargo sorting and ILV production. Collectively, the ESCRTs are thought to act sequentially, with ESCRT-0 and ESCRT-I playing major roles in recognizing ubiquitinated cargo, and they also serve to recruit remaining ESCRTs to the endosomal surface. ESCRT-0 and ESCRT-I have a number of subunits that bind ubiquitin via so-called ubiquitin-binding domains, thus allowing them to recognize and gather ubiquitinated membrane proteins. ESCRT-I recruits ESCRT-II and ESCRT-III, which together can deform the membrane to help bud vesicles from the limiting membrane and complete the fission process that ultimately separates the forming ILV from the limiting membrane. Deformation of the membrane to allow for vesicle budding is driven in part by the ability of ESCRTs to bind lipids, such as phosphatidylinositol 3-phosphate, which is enriched on endosomes. Fission of ILVs, whereby they separate from the limiting membrane to reside as free-floating vesicles within the endosomal lumen, is accomplished by ESCRT-III, which forms membrane-associated polymers thought to constrict the membrane to allow

for vesicle fission. ESCRT-III also recruits deubiquitinating enzymes that remove the ubiquitin from the cargo, allowing ubiquitin to be recycled and avoid degradation in the lysosome. The ESCRTs, in particular the assembly of the ESCRT-III polymeric structure, is antagonized by Vps4, which is a multimeric protein that hydrolyzes adenosine triphosphate (ATP) to provide the energy for dismantling of the ESCRT-III polymer. The precise mechanisms responsible for ILV formation are not entirely known. Importantly, the process of formation of ILVs is fundamentally different from that of other transport vesicles, including clathrin-coated vesicles at the plasma membrane or COPI-coated vesicles at the Golgi. These latter vesicles are formed by assembling cytosolic proteins into polymers that bud vesicles into the cytosol, thus resulting in vesicles coated with the very proteins used in their formation. MVB ILVs are also formed by cytosolic proteins, but they bud into the endosomal lumen, away from the cytosol, and the proteins responsible for ILV formation are not incorporated into the final ILV and largely remain on the limiting endosomal membrane.

Exosomes. The aforementioned ESCRT apparatus plays an essential role in sorting ubiquitinated proteins into ILVs for their eventual degradation in lysosomes. However, multivesicular endosomes can still be observed in the absence of key ESCRT components, demonstrating that multiple pathways exist for generating endosomal ILVs and membranes. Exosomes constitute one particular class of ILVs, and they are secreted upon fusion of MVBs with the plasma membrane. Exosomes can interact with and sometimes fuse with the plasma membrane of neighboring cells. Exosomes can contain not only proteins, such as antigen-presenting major histocompatibility complexes, but also messenger RNAs (mRNAs) and microRNAs, all of which could potentially regulate a variety of processes of cells that encounter exosomes.

Exosomes are enriched in cholesterol, sphingolipids, and tetraspanins (a family of topologically similar small integral membrane proteins with four membrane-spanning segments). Both in vivo studies as well as in vitro studies with isolated liposomes indicate that these components work together to form a class of ILVs. It is not clear yet what types of sorting signals are recognized to incorporate cargo proteins into these ILVs; however, the process does appear to be independent of ubiquitin. In addition, it is not yet clear how mRNAs or microRNAs are captured into exosomes. How MVBs are directed to the plasma membrane to secrete their contents rather than fuse with lysosomes and thus degrade their contents is not yet clear, but it relies in part on certain endosomal Rab GTPases (guanosine 5′-triphosphate–binding proteins) that direct trafficking of other endosomal compartments.

Other pathways involved. Another pathway that contributes to the generation of multivesicular endosomes is the autophagy pathway. Endosomes fuse with autophagosomes, thereby allowing autophagosome contents eventual access to the lysosomal lumen for degradation. These hybrid organelles have also been termed amphisomes to highlight their dual origin in both endocytic and autophagic pathways. Autophagosomes are characterized by a double-membrane structure that grows around cytosolic components (for example, cytosol, organelles, and protein aggregates). During the later stages of formation, autophagosomes acquire fusogenic membrane proteins on their outer membrane that direct fusion with late endosomes and lysosomes. Fusion allows the outer membranes of the autophagosomes to become contiguous with the limiting membranes of the MVBs, while the inner membranes of autophagosomes are delivered to the endosomal lumen, providing an additional source of intraluminal membranes that characterize MVBs. Typically, intraluminal membranes from autophagosomes are fated for eventual degradation upon MVB fusion with lysosomes. However, it has also been shown that some autophagocytosed proteins can be subsequently secreted, indicating that this could occur by fusion of MVBs/amphisomes with the plasma membrane.

Another source of intraluminal membranes within MVBs is the lipid BMP/LBPA [bis(monoacylglycero) phosphate/lysobisphosphatidic acid]. Intraluminal membranes containing LBPA are distinct from those containing cholesterol and tetraspanins or ubiquitinated membrane proteins sorted by the ESCRT machinery. LBPA-containing intraluminal membranes are observed in mature late endosomes and lysosomes, indicating that they form in the later stages of the endocytic pathway. Although the formation of these membranes does not require the central ESCRT machinery, formation is controlled by an ESCRT-associated protein called ALIX, which binds LBPA and plays a key role in generating LBPA-containing internal membranes. LBPA internal membranes appear to be able to fuse back with the limiting membrane of MVBs, which may be important for the ability of viruses to escape endosomes and gain access to the cytosol for replication.

The MVB biogenesis pathways described above proceed after the formation of early endosomes. However, at least one additional way to create MVBs/endosomes involves their generation right at the step of internalization from the plasma membrane. This MVB pathway is generated by the process of macroendocytosis, which generates large endosomes with ILVs generated from fragments of the plasma membrane. One of the key proteins mediating macroendocytosis is Pincher/EHD4, which is an ATPase that mediates uptake of large areas of ruffled plasma membrane in a manner independent of other internalization routes such as clathrin- or caveolin-dependent endocytosis. These types of multivesicular endosomes provide important functions in neurons, serving as a relatively stable organelle from which internalized nerve growth-factor receptors can continue to signal from intracellular locales. In addition, these MVBs can be transported to other subcellular locations within neurons to promote processes in a highly spatially regulated way.

Lysosome-related organelles. Finally, MVBs can serve as intermediates that give rise to a number of lysosome-related organelles. Lysosome-related organelles (LROs) represent a broad class of specialized organelles that share characteristics with typical lysosomes. LROs include cytotoxic granules secreted by T cells and natural killer cells, melanosomes that store melanin pigment, and lamellar bodies that store surfactant in type II lung cells. Many LROs remain connected with the endocytic system, but their identity is conferred by cell-specific expression of specialized cargo and sorting machinery. Not all LROs have intraluminal membranes. However, for some LROs, internal membranes may provide an important mechanism to complete their development or function. One example is melanosomes, which construct a proteinaceous scaffold within their lumen that stabilizes the melanin pigment. The scaffold proteins themselves start as integral membrane proteins, which are initially delivered to the melanosome lumen by ILVs that form from the limiting membrane.

For background information *see* AUTOPHAGY; CELL (BIOLOGY); CELL MEMBRANES; CELL ORGANIZATION; ENDOCYTOSIS; LYSOSOME; LYSOSOME-RELATED ORGANELLES; UBIQUITINATION in the McGraw-Hill Encyclopedia of Science & Technology.   Robert C. Piper

Bibliography. J. Huotari and A. Helenius, Endosome maturation, *EMBO J.*, 30:3481–3500, 2011, DOI:10.1038/emboj.2011.286; R. C. Piper and D. J. Katzmann, Biogenesis and function of multivesicular bodies, *Annu. Rev. Cell Dev. Biol.*, 23:519–547, 2007, DOI:10.1146/annurev.cellbio.23.090506.123319; S. B. Shields and R. C. Piper, How ubiquitin functions with ESCRTs, *Traffic*, 12:1306–1317, 2011, DOI:10.1111/j.1600-0854.2011.01242.x; M. Simons and G. Raposo, Exosomes—vesicular carriers for intercellular communication, *Curr. Opin. Cell Biol.*, 21:575–581, 2009, DOI:10.1016/j.ceb.2009.03.007.

MYB transcription factors in plants

Over the past few decades, significant efforts have been invested to understand how specific facets of plant growth, development, and metabolism are regulated. Such studies have highlighted the importance of the strict regulation of gene expression, and the critical roles played by a certain class of regulatory proteins called transcription factors (TFs). Among these transcriptional regulators, the large and functionally diverse MYB superfamily has been shown to play a preponderant role. The maize COLORED1 (C1) MYB domain protein was the first transcription factor identified in plants and was found to be required for the synthesis of anthocyanin pigments in corn kernels, giving a specific bluish-black color to the grain. Following on from this work, the functions of MYB proteins have been investigated in numerous plant species belonging to different clades: annual dicotyledons (for example, *Arabidopsis*, petunia, or snapdragon), perennial dicotyledons (for example, apple, grapevine, or poplar), monocotyledons (for example, maize, rice, or wheat), and

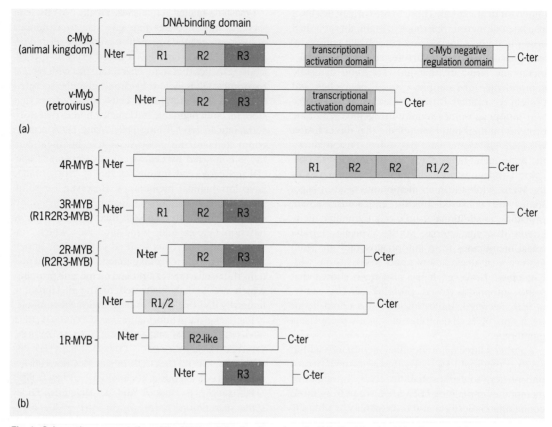

Fig. 1. Schematic representation of the MYB protein structures found within the (*a*) animal kingdom and (*b*) plant kingdom.

gymnosperms (for example, loblolly pine or white spruce). These studies have contributed to deciphering the structure, evolution, and specificity of the MYB superfamily. Furthermore, they have also

MYB repeat helix-turn-helix (HTH) structure

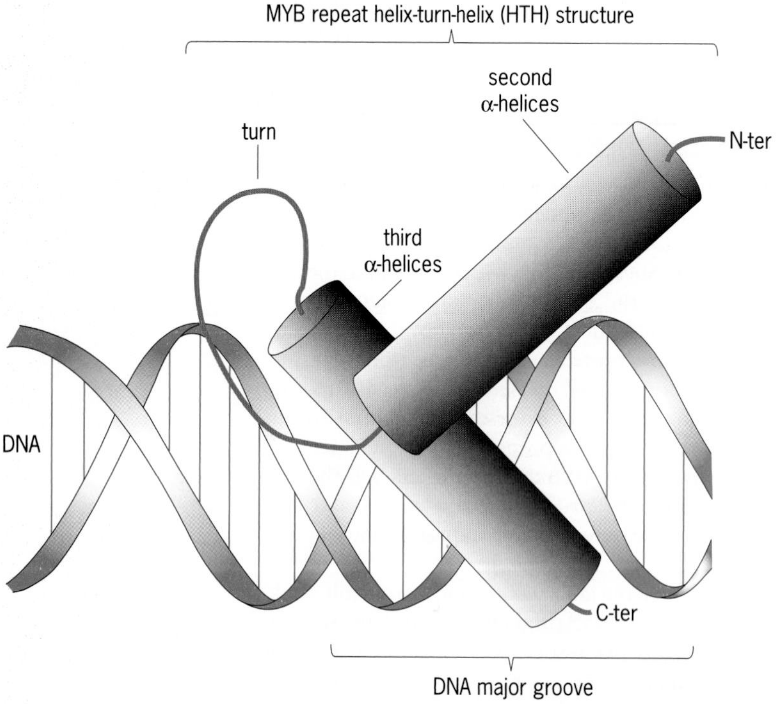

Fig. 2. Schematic representation of the MYB helix-turn-helix (HTH) domain bound to DNA. The third helix (DNA-recognition helix) intercalates in the major DNA groove and makes direct contact with target DNA sequences.

permitted elucidation of the specific biological roles played by numerous MYBs throughout the plant life cycle. Because of this accumulated knowledge, the MYB superfamily is one of the best-characterized classes of transcription factors in plants.

Protein structure. MYB proteins are present in all eukaryotes and are characterized by a highly conserved DNA-binding domain (DBD) generally found at the N-terminus side of the protein: the MYB domain (**Fig. 1**). This domain was initially discovered in v-Myb, which is the oncogenic component of the avian myeloblastosis retrovirus (from which the MYB acronym derives) that causes acute leukemia. v-Myb is a truncated version of c-Myb, a central regulator of proliferation and differentiation of hematopoietic cells, from which it is derived. In contrast, the C-terminal region of MYB proteins is extremely variable and usually functions as a regulatory domain. In plants, the MYB domain generally contains up to four imperfect amino acid sequence repeats (R) of approximately 50 amino acids. These repeats have been named R1, R2, or R3 on the basis of their similarities with the three repeats found in the reference MYB, c-Myb. Each repeat contains three α-helices, with the second and the third forming a helix-turn-helix (HTH) structure that binds to the major groove of target DNA sequences (**Fig. 2**). The HTH contains three evenly spaced tryptophan residues that form a hydrophobic core, which plays a role in the sequence-specific binding of DNA.

Evolution and classification. Compared to the animal kingdom, the plant MYB superfamily is greatly expanded and encodes one of the largest groups of transcription factors; this is exemplified in the model dicotyledonous plant *Arabidopsis*, in which 10% of the identified transcription factors belong to the MYB superfamily. Depending on the number of adjacent repeats present in the DBD, the majority of the MYB proteins can be divided into four different classes. The smallest class corresponds to the 4R-MYB group, represented in diverse plant genomes such as moss, rice, grape, or *Arabidopsis* by a unique gene. 4R-MYB group members contain four R1- and/or R2-like repeats and, to date, the precise role of the proteins belonging to this group has not been determined. The prototypic R1R2R3-MYB proteins (3R-MYBs) are encoded in higher plant genomes by five genes, which are involved in the control of the cell cycle. Interestingly, R1R2R3-MYB and R2R3-MYB proteins are evolutionarily closely related. Two opposing models suggest that these two classes were derived from each other, by the loss or gain of the R1 repeat. The largest class of MYB transcription factors is composed of the R2R3-MYB group, which typically contains more than 100 members in plants. In *Arabidopsis* and rice, 64% and 59% of the *MYB* genes encode R2R3-MYB proteins, respectively. The expanded size of this MYB group is in agreement with the observation that they regulate plant-specific processes, and this suggests a role in generating phenotypic diversity in the plant kingdom. R2R3-MYB transcription factors have a modular structure with a highly conserved DBD located in the N-terminal region, and a variable C-terminal part containing either an activation or repression domain. Based on DBD conservation and C-terminal amino acid signatures, R2R3-MYB proteins have been divided into different subgroups. Some subgroups are species specific (for example, *Arabidopsis* or poplar), indicating that their members have acquired specialized functions. Interestingly, R2R3-MYBs belonging to the same subgroup in different plant species share broadly conserved functions, whereas paralogous genes (genes that arose by duplication and later diverged in sequence or location from the parent gene) from one plant species can regulate the same process, but in different cell types, or act redundantly. The 1R-MYB group includes different subclasses of MYB factors that contain a single or partial MYB-repeat. The R3-type MYB subgroup encodes MYBs involved in the control of cellular morphogenesis (for example, root-hair patterning) and secondary metabolism (for example, anthocyanin biosynthesis). Interestingly, R3-MYBs often act as inhibitors by competing with R2R3-MYBs from which they are likely to have evolved. The second subgroup is composed of MYB factors that contain the evolutionarily older R1/2 repeat, proposed as the ancestor of the R1 and R2 repeats. The characterized R1/2-MYBs were found to encode core components of the circadian clock. The 1R-MYB proteins from the GARP class regulate diverse functions such as organ morphogenesis, chloroplast development, and the response to phosphate starvation. The RAD and the SMH (single MYB histone)

subgroups possess a similar SANT/MYB-like domain (which contains an R2-like repeat), but they diverge in their C terminal. RAD proteins regulate flower and fruit development, whereas SMH proteins have telomeric DNA-binding properties. The last 1R-MYB subgroup encodes the IBPs (indicator binding proteins), which are telomeric DNA-binding proteins thought to be required for telomere maintenance.

Functional diversity. Plant growth and development involve the tight regulation of thousands of genes, at both the spatial and temporal levels. Such control is achieved by diverse regulatory networks, in which MYB proteins play key roles. Among all the MYBs expressed in plants, the R2R3-MYB family shows the greatest diversity of function in plant-specific processes, designating this group as the most specific to this lineage. For instance, biosynthesis of various primary or secondary metabolites present in diverse plant species (for example, *Arabidopsis*, maize, grape, poplar, or pine trees) is regulated by several R2R3-MYBs. This includes the flavonoid compounds, flavonols, anthocyanins, and proanthocyanidins, involved in ultraviolet filtration, floral pigmentation, or embryo protection, respectively, as well as some major components of the plant secondary cell wall (for example, lignin, cellulose, and xylan). In contrast, the biosynthesis of some species-specific metabolites, such as glucosinolates or sesquiterpenes (defense molecules against pathogens or predators mostly found in Brassicales or gymnosperm species, respectively), can also be regulated by specialized R2R3-MYBs. Extensive studies, mainly carried out on *Arabidopsis*, have shown that several R2R3-MYBs are also involved in the determination of cell fate and identity. This includes the control of trichome and stomatal development in shoots, cell shape in petals, and root-hair patterning. In addition, they play a critical role during the reproductive phase because they control both the differentiation of synergid cells during female gametophyte development and the formation of the outer integument of the seed coat. Key R2R3-MYB proteins regulate various crucial steps of plant development, including hypocotyl elongation, axillary meristem initiation, and lateral organ separation during the vegetative phase, as well as anther and pollen growth during flower maturation. In numerous species (for example, *Arabidopsis*, tobacco, tomato, and maize), R2R3-MYBs were also shown to regulate shoot morphogenesis and leaf patterning. As autotrophic sessile organisms, plants are subjected to various unfavorable environmental conditions that can be grouped into two categories: biotic stresses (for example, pathogens, insects, and herbivores) and abiotic stresses (for example, drought, cold, wounding, and nutriment starvation). Plants respond to these environmental constraints by modifying the expression of large sets of genes, which will in turn induce a number of metabolic, physiologic, and developmental changes. The involvement of R2R3-MYBs in both types of stresses is well documented in numerous plant species, including crops (wheat, maize, and rice). For example, in

Arabidopsis and grape, the limitation of water loss by stomatal closure in response to drought is controlled by some R2R3-MYBs. R2R3-MYB induction of the hypersensitive cell death program in response to pathogens is another well-documented example.

Regulatory networks. MYB activity can be significantly modulated by various regulatory mechanisms. These include posttranscriptional regulation by microRNAs and *trans*-acting, silencing RNAs (ta-siRNAs); posttranslational modification through protein–protein interaction (with MYBs or other regulators); protein modification (for example, phosphorylation, oxidation, or *S*-nitrosylation); and protein conjugation (for example, ubiquitination or sumoylation). Moreover, MYB gene expression can be modulated by diverse types of transcription factors (for example, MADS or bHLH), including MYB proteins. Finally, the genes targeted by MYB proteins encode both structural proteins (for example, enzymes) and regulatory proteins (for example, homeodomain or MYB proteins), indicating that MYBs function at multiple levels in regulatory networks.

Prospects. The next challenge in the understanding of the MYB superfamily will be to determine all the functions played by all the protein members. Such knowledge will be valuable for modeling the regulatory networks by which the MYB proteins control plant growth and development, and this can be used as a tool for the targeted improvement of crop species through classical breeding or transgenesis.

For background information *see* CELL (BIOLOGY); CELL BIOLOGY; GENE; PLANT DEVELOPMENT; PLANT GROWTH; PLANT METABOLISM; PLANT PHYSIOLOGY; PROTEIN; TRANSCRIPTION in the McGraw-Hill Encyclopedia of Science & Technology.   Christian Dubos

Bibliography. F. Bedon et al., Subgroup 4 R2R3-MYBs in conifer trees: Gene family expansion and contribution to the isoprenoid- and flavonoid-oriented responses, *J. Exp. Bot.*, 61:3847–3864, 2010, DOI:10.1093/jxb/erq196; C. Dubos et al., MYB transcription factors in *Arabidopsis*, *Trends Plant Sci.*, 15:573–581, 2010, DOI:10.1016/j.tplants.2010.06.005; C. Dubos et al., MYBL2 is a new regulator of flavonoid biosynthesis in *Arabidopsis thaliana*, *Plant J.*, 55:940–953, 2008, DOI:10.1111/j.1365-313X.2008.03564.x; M. B. Prouse and M. M. Campbell, The interaction between MYB proteins and their target DNA binding sites, *Biochim. Biophys. Acta*, 1819:67–77, 2012, DOI:10.1016/j.bbagrm.2011.10.010; O. Wilkins et al., Expansion and diversification of the *Populus* R2R3-MYB family of transcription factors, *Plant Physiol.*, 149:981–993, 2008, DOI:10.1104/pp.108.132795.

Nanocomposites in aeronautics

The commercial application of nanocomposites in the aerospace industry has commenced and is due to become a major factor in reducing cost of ownership, improving performance and safety, increasing environmental sustainability, and reducing environmental impact.

Principles of nanocomposites. A composite comprises two (or more) distinct material components; in the case of nanocomposites, it comprises a dispersion of nanoparticles (or nanofillers) within a matrix such as polymer, ceramic, metal, glass, or rubber. Nanoparticles can be used as the primary filler or can provide a sublevel of reinforcement in a macro-composite material. Such multiscale reinforcement is seen in many natural systems, such as bone, which contains nanoplatelets of apatite. The importance of this nano-level reinforcement has been shown to be significant, since if the apatite crystals grow, the bone loses its toughness and becomes brittle, a condition known as osteoporosis.

Nanofillers are generally described as having at least one dimension less than 100 nm (1 nanometer is 1 millionth of 1 mm) and can range in shape from discrete particles, platelets, rods, fibers, or nanotubes to continuous fibers produced by a range of processes (**Table 1**). Nanoparticles can exhibit very different mechanical, chemical, optical, and thermodynamic properties from those of the equivalent larger particles or bulk material. One important factor is the high surface area offered by nanoparticles—for example, 10-nm-diameter spherical particles give a theoretical surface area of 30 m^2/g. In the case of nanofibers, the surface area/volume ratio can be 1000 times greater than that for a conventional fiber such as carbon fiber.

The earliest engineering polymer nanocomposites were developed by Toyota Research in the 1970s by adding clay to nylon-6. Clay minerals have a nanolayered structure that can be split into nanoplatelets during processing. Early work showed that nanoparticles could influence the polymer glass transition temperature, affect crystallization behavior, and impart macro properties such as improved modulus (**Table 2**).

Key factors in understanding how to achieve the full potential of nanocomposites have been in the areas of (1) characterization—development of high-resolution imaging (for example, scanning tunneling electron microscopy), interfacial measurement chemical analysis (for example, Raman spectroscopy), and characterization (for example, Raman spectroscopy); (2) production—novel manufacturing methods for producing nanoparticles and fibers; and (3) processing—novel nanocomposite processing techniques.

Nanocomposites manufacturing. A key issue learned during the development of macrocomposites was the importance of correct processing. Manufacturing of nanocomposites was originally approached in the same way as for microfillers, that is, blend in the filler and produce the component. However, the end results were either inconsistent or deleterious. It is now known that the inconsistency is due to incorrect processing, and that additional manufacturing steps such as nanoparticle functionalization and dispersion (see **illustration**) are usually required.

Functionalization or compatibilization prepares the nanofiller for chemical bonding with the matrix.

TABLE 1. Examples of different manufacturing methods for making nanofillers

Production method	Examples
Ball milling	Clays (montmorillonite, boehmite, talc), calcium carbonate, alumina
Colloid chemistry	Silicates
Physical vapor deposition (PVD)	Metal oxides, metals
Chemical vapor deposition (CVD)	Carbon (carbon nanotubes, graphene, fullerene, continuous nanofiber)
Hydrothermal synthesis	Metal oxides, many inorganics
Flame spray pyrolysis	Al_2O_3, CeO_2, ZnO, TiO_2
Laser ablation in liquid	Metals, oxides
Chemical extraction	Nanocellulose
Electrospinning	Polymer nanofibers

TABLE 2. Main categories of engineering nanofillers with typical applications

Nanofiller	Example of use
Filler: Mineral For example, montmorillonite clay, boehmite, alumina (synthetic) nanofillers	Increase in mechanical performance, barrier properties
Filler: Metal For example, silver, gold, copper	Increase in electrical conductivity, antimicrobial properties
Filler: Inorganic For example, graphene, silica, cellulose, polyhedral oligomeric silsesquioxane; fullerenes	Increase in mechanical performance, conductivity, and wear resistance; low friction; fire retardancy properties
Filler: Nanofiber For example, carbon nanotubes (CNTs), carbon nanothread	Improved mechanical and electrical performance; increased electrical conductivity of polymer (~1% CNT replaces 10% carbon black), thermal conductivity, and energy (as in batteries)

This stage could, for example, involve adding a capping agent after particle production.

Dispersion is also critical to final performance. Correct dispersion prevents agglomeration and achieves a uniform dispersion of nanofiller. Poor dispersion of the nanofiller can lead to (1) agglomeration of nanoparticles, reducing performance, (2) phase separation in polymers, and (3) poor interfacial strength (adhesion to matrix). Dispersion techniques include ultra-high-shear mixing and ultrasonics. The most recent techniques may combine dispersion and functionalization roles—for example, physicochemical techniques for producing core-shell nanoparticle fillers that are ready for direct incorporation into a polymer master batch.

Alignment of fillers such as carbon nanotubes within a nanocomposite is an important new area, and techniques such as application of an electrical or magnetic field during curing are showing promise.

Aerospace applications. Construction materials in primary and secondary aircraft structures are subject to years of testing, and although nanocomposites will feature in future aircraft, current examples of commercial use are rare. The Lockheed Martin F35 Lightning II fighter is the first mass-produced aircraft to feature CNT-reinforced polymer (CNRP), which is being used in wingtip fairings and one hundred other non-load-bearing applications.

Environmental protection is an important driver. In order to provide protection from lightning strikes, composite aircraft wings currently have conductive bronze mesh embedded, which is parasitic weight. Lightning strikes generate a rapid energy pulse with temperatures in thousands of degrees Celsius, which can degrade the resin system and create a pressure wave. Thermosetting resins such as epoxies are being enhanced with carbon nanotubes for carbon-fiber-reinforced polymer (CFRP) construction to provide an intrinsically conductive system while eliminating parasitic weight. The structure could also provide electrical bonding and grounding for onboard electronics.

Components used in jet engines or rocket motors are subject to high temperature, thermal oxidation, thermal stresses, and, for moving parts, mechanical stress. Novel ceramic thermal barrier coatings (TBCs) with nanoscale features or containing nanofillers are being developed. These typically comprise combinations of a base layer, such as partially stabilized zirconia (ZrO_2), alloyed with a mixture of stabilizing oxides, such as yttria and ceria

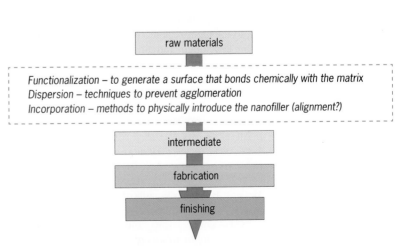

Steps in nanocomposite manufacture.

(Y_2O_3, CeO_2). Techniques such as flame spraying or electron-beam physical vapor deposition (PVD) are typically used to deposit these coatings. The use of nanocomposite TBCs is achieving increased adhesion, reduced oxygen permeability, and improved thermal shock resistance.

Metallic corrosion in water and aqueous environments, often combined with fatigue loading (corrosion fatigue), is a major problem, particularly in aging aircraft. The phasing out of chromate primers, cadmium, and chromium, which have excellent performance, has stimulated the search for replacement materials, and nanofillers offer a route to achieve environmentally friendly products.

Tires and sealing components such as gaskets may need to withstand temperature, pressure, gas permeation, and aggressive liquids such as hydraulic fluid. Nanofillers such as the molecular silica polyhedral oligomeric silsesquioxane (POSS) are being incorporated into functionalized rubber as crosslinking agents to radically improve mechanical performance. In high-temperature vulcanized rubber, POSS nanofillers have been found to increase resistance to thermal degradation.

Inside the cabin, fire retardancy is an important aspect of cabin interior structures, and nanofillers such as modified clay have been used to replace less effective or more toxic fillers. Recently, electrochromic coatings on windows that can darken at the flick of a switch have been used to replace blinds.

"Smart" structures have been investigated for many years, and nanomaterials are finally bringing these concepts to fruition. Self-sensing composite structures can now incorporate microelectromechanical systems (MEMS) devices with wireless data transmission, powered by an energy-harvesting device. Energy sources currently being exploited include nanoscale-layer or nanocomposite fibers to harvest vibration or strain (piezo), thermal (thermoelectric), and solar (organic photovoltaic) energy. Self-repair using programmed release of embedded resin systems is now closer to realization. Nano-enabled harvesting and high-performance batteries also promise realization of "active" structures, such as morphing flight surfaces that tailor their shape based on a feedback signal, just as in nature birds alter their wing configuration according to the flight regime. Novel anti-icing systems are being developed, and ultrasmooth nanocoatings are being applied to passenger jet external surfaces to reduce air friction, leading to a 2% reduction in fuel consumption.

For background information *see* ADAPTIVE WINGS; CARBON NANOTUBES; CHEMICAL VAPOR DEPOSITION; COMPOSITE MATERIALS; FULLERENE; GRAPHENE; MICRO-ELECTRO-MECHANICAL SYSTEMS (MEMS); NANOSTRUCTURE; NANOTECHNOLOGY; RAMAN EFFECT; SCANNING TUNNELING MICROSCOPE in the McGraw-Hill Encyclopedia of Science & Technology.

Martin Kemp

Bibliography. Z. Guo and L. Tan, *Fundamentals and Applications of Nanomaterials*, Artech House, Norwood, MA, 2009; F. Hussain et al., Review article: Polymer-matrix nanocomposites, processing, manufacturing, and application: An overview, *J. Compos. Mater.*, 40:1511–1575, 2006, DOI:10.1177/0021998306067321; J. R. Potts et al., Graphene-based polymer nanocomposites, *Polymer*, 52(1):5–25, 2011, DOI:10.1016/j.polymer.2010.11.042; S. Thomas and R. Stephen (eds.), *Rubber Nanocomposites: Preparation, Properties & Applications*, Wiley, Singapore, 2010.

Nano-electro-mechanical systems (NEMS)

Nano-electro-mechanical systems (NEMS) represent an important new class of device which has a growing range of applications, from tests of basic quantum mechanics through nanoscale metrology to a vast number of sensors. These promise a variety of applications, from single-phonon detection to ultrasensors for physical parameters such as mass, force, charge, spin, and chemical specificity. Single-molecule biosensing, information storage and processing technologies, and nanoscale refrigerators can be expected in the longer term.

Of course, miniature mechanical systems have a long history of utility (one may recall the Antikythera mechanism, pocket watches, tuning forks, and the Babbage difference engine) but have been eclipsed over the past century or so by the rise to dominance of electronic machines. In recent years this situation has begun to change, and mechanical systems are again growing in popularity, with applications across the fields of switches, mirrors, sensors (especially biosensing), logic, and even quantum mechanical systems. In common with most electronic circuits, sensors, and devices, these mechanical systems are also undergoing rapid reduction in size, somewhat analogous to the Moore's law situation in microelectronics.

The mechanical counterrevolution began by coopting the techniques of the microelectronics industry to augment mechanical performance. It has been accelerated by the ability to fabricate and integrate microscale electromechanical systems (MEMS), especially small cantilevers and membranes. These systems, with dimensions from around 1 mm to 10 micrometers (μm), have found very wide industrial applications as accelerometers (in every "smart" phone and airbag), radio-frequency switches, and medical sensors. Seemingly inevitably, the next stage in development has involved further reductions in the physical scale of the mechanical systems. Thus NEMS are devices that exhibit both electrical and mechanical behavior at the nanoscale.

Problems and opportunities. The ability to make ever-smaller systems brings not only advantages but also serious challenges, especially relating to the techniques for driving and detecting the movement of NEMS. As the physical scale of a mechanical system is reduced, the time required for execution of a significant movement is reduced. Thus nanoscale switches should operate faster than their micro-equivalents.

The natural flexural frequencies of a mechanical beam depend on the length L, thickness d in the direction of flexure, material from which it is constructed, and any tension that is deliberately applied or exists in the structure on account of the fabrication method. In the absence of applied tension the flexural frequencies vary as d/L^2, and for a material with Young's modulus E and density ρ, the lowest mode of resonance for a beam clamped at both ends is f_0, given by the equation

$$f_0 \approx \left(\frac{E}{\rho}\right)^{1/2} \left(\frac{d}{L^2}\right)$$

We may consider a multiwall carbon nanotube of diameter 10 nm and length 0.3 μm. A carbon nanotube is the stiffest material known, with $E \sim 1$ TPa and a low density of $\sim$2000 kg/m^3. Substitution in the above equation leads to a lowest flexural mode of around 2 GHz. High-frequency resonators will almost inevitably also lead to rather stiff response, so the displacement for a given applied force will be small. This means that special techniques will be required to read out these picometer displacements, and some of these are described below.

NEMS resonators with high frequency and Q. Frequency is the most precisely measurable physical parameter. In the world of sensitive measurement, it is often advantageous to convert a small change in some parameter to a small frequency shift in a resonator, particularly if the resonator has very low loss, or, in the language of electronics, a high quality (Q) factor. So NEMS resonators are characterized by small size, high frequency of operation, and high Q factors. The only disadvantage of a high Q is that the available input bandwidth for a signal to be detected extends only to f_0/Q, where f_0 is the basic resonant frequency of the NEMS resonator mode. To realize the high Q value, it is necessary to limit the coupling of vibrational energy from the NEMS resonator to its surroundings. The damping provided by air at atmospheric pressure is sufficient to limit the Q factor to around a few hundred, and immersion in a liquid, as is often required for biological measurements, provides even stronger damping. So, for the most sensitive operation, a NEMS device will, if possible, be operated in modest vacuum, typically less than 10^{-4} mbar (10^{-2} Pa). Similarly, there can be losses at the anchorage points of the beam or plate. These can be minimized by mechanical impedance mismatches as well as by ensuring that the glue or weld is sufficiently rigid to minimize losses.

Materials and fabrication. One key issue for NEMS resonators is to optimize the quality factor. This both reduces the influence of noise and makes easier the detection of the resonator's mechanical motion. The range of materials employed is vast and depends on the specific requirements. Silicon and silicon on insulator (SOI) are widely employed, as are silicon nitride (Si_3N_4) and silicon carbide (SiC), which can both be prepared in ultra-thin layers. Piezoactive materials, such as aluminum nitride (AlN) and zinc oxide (ZnO), also have much to

offer for direct electromechanical activation. Carbon materials are of growing importance both for ultraminiaturization and for their unique properties. Carbon nanotubes have been used as quasi-1-D mechanical resonators and, on account of their high stiffness, very high resonant frequencies (up to the gigahertz range) have been achieved. Graphene, a single atomic layer of hexagonal carbon atoms, is also highly suitable for 2-D membrane resonators.

Thin-film deposition. Typically, a thin film of the desired NEMS material is deposited on a suitable substrate and then is patterned using lithography based on conventional optical or electron-beam methods, the latter if submicrometer features are required. In addition, milling using focused ion beam methods has also proved useful in the deep-sub–micrometer size regime.

Free-standing NEMS and etch release methods. Following patterning of the beam or plate design, it must be released from its substrate to allow for flexural mechanical motion. A wide variety of release processes have been attempted, including back-etching, using some end-point detection method, undercutting using an anisotropic plasma etch, or use of an intermediate sacrificial layer which can be removed by an etchant that does not attack the material of the NEMS resonator itself. In the case of reduced dimension, carbon single tubes can be positioned using a micromanipulator mounted in a scanning electron microscope. Graphene films have been transferred by floating onto a liquid surface and then depositing the film onto a suitably patterned substrate. In some cases, a combination of thin-film growth method and substrate choice can give rise to high stress in a NEMS element, and this can be an advantage in not only increasing the resonant frequency for a given dimension but also increasing the quality factor. **Figure 1** shows a schematic of the components involved in a generic thin-film NEMS fabrication process.

Driving and detection of NEMS motion. In a resonant NEMS device it is necessary to drive the beam or sheet at one of its mechanical flexural resonances. A wide variety of methods have been employed which are appropriate to different scales of length and frequency. Probably the most widely used is piezoelectric response.

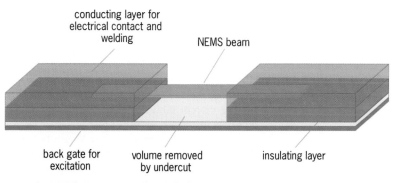

Fig. 1. **Schematic of a generic double-clamped cantilever, showing typical features and fabrication steps.**

Piezoelectric excitation. Here the NEMS movement may be produced by piezoelectric activation. The NEMS element (cantilever or membrane) can be made of a piezo-active material, such as lead zirconate titanate (PZT) or AlN, and a voltage applied across it to generate displacement. Alternatively, a separate piezoelectric element can be mechanically coupled to the NEMS element to drive the NEMS motion indirectly. The advantage is simplicity, but disadvantages are that the available range of piezoelectric materials is small, they tend to have moderate mechanical losses and therefore are not suitable for high-Q resonators, and there is a danger of exciting unwanted movement in other parts of the NEMS supporting structure, which complicates the readout process. Piezoelectric detection can also be realized effectively at the nanoscale, especially using piezoresistive NEMS resonators, which can show relatively large resistance changes as a result of strain induced by displacement from equilibrium configuration.

Capacitive driving and detection. Perhaps the most widely used technique for both drive and readout of the NEMS element motion is through capacitive coupling between the element and some closely spaced electrodes. Applying an alternating voltage between the NEMS (which must be conducting) and one or more electrodes gives a time-varying electrostatic force. The same arrangement can also be used to detect motion if the electrodes are connected to a sensitive transconductance amplifier which can measure the phase change of the impedance as the drive frequency is swept through a mechanical resonance.

Microwave driving and detection. A more recent development, particularly attractive as the frequency of NEMS resonators increases toward the gigahertz range, involves microwave detection. This is mainly an extension of the capacitive methods described above, except that high-Q microwave resonators are available, especially at cryogenic temperatures, where superconductivity can be employed. This means that high excitation fields can be achieved with modest input power. Microwave amplifiers with noise figures close to thermal noise limits are also available and can be used to read out the NEMS displacement.

Lorentz force excitation. Using the Lorentz force, an excitation method can be applied that does not require ultra-fine lithography and that can apply the same variable force over a number of spatially distributed, electrically conducting cantilevers. A static direct-current (dc) magnetic field of flux density B is applied to the cantilevers, orthogonal to the excitation direction. A radio-frequency current flowing in a cantilever will produce a similarly time-varying force that is orthogonal to both the magnetic field and the current flow. Although the maximum current flow decreases with the cross-sectional area of the conductor, it increases as the length becomes shorter. Thus, this form of excitation is not as size-dependent as the others previously discussed.

Optical readout. Scanning probe microscopes such as atomic force microscopes (AFMs) frequently use optical detection to sense the very small ($\sim$1-nm) movement of a cantilever as it is scanned over a surface to sense the varying force between the surface and the tip of the cantilever. An optical beam is reflected from the surface of the cantilever onto a split photodiode, and any movement is sensed by the change in voltage difference from the diode components. This optical lever technique can be made extremely sensitive; however, as the dimensions of the cantilever approach that of the wavelength of light, the system becomes less sensitive. It is for this reason that NEMS require other detection methods. Excitation can also be carried out using amplitude-modulated radiation pressure from a light beam. The same method may produce low-frequency excitation through opto-thermal effects such as the bending of bilayer materials due to differential thermal expansion, but this is rarely useful at the higher frequencies of NEMS devices. Optical detection in which the NEMS element forms one part of a Fabry-Pérot interferometer provides another approach which is of comparable sensitivity to the optical lever, and much more compact.

Nonlinear behavior and the Duffing equation. The simple harmonic motion which provides a good approximation for the movement of a cantilever under the influence of a force will break down at some force level when nonlinear effects become significant. These can arise from increased strain within the material, arising from significant deflection from the unstressed form (generally applicable to double-clamped beams), or from the effects of non-negligible inertia (applicable to singly clamped resonators for which the previous nonlinear effect is absent). The onset of nonlinearity leads to an effective cubic term in the force-extension response, and this leads to the motion being described by the Duffing equation. One useful result of this is that the response to a swept drive frequency is hysteretic; that is, the response to an increasing frequency does not follow the same behavior as a decreasing frequency over the same range. This means that such a nonlinear resonator can be used as a "latching memory" element.

Sideband cooling. An interesting recent development is the use of sideband cooling methods, which were originally developed for laser beam cooling of single atoms or ions, for macroscopic cantilevers such as NEMS resonators. The basic principle applies to the NEMS resonator a quasi-static electromagnetic force from a resonant source which is strongly coupled to the NEMS resonator (**Fig. 2**). This could be an optical resonator or even a microwave resonator. The effect on the cantilever will be to shift its resonant frequency a little, because the spring constant of the cantilever will be altered (increased, in general). However, there is also a back-action effect on the electromagnetic resonator. There will be a frequency-dependent phase shift between the movement of the cantilever and the response of the electromagnetic force on the cantilever. If the power applied to the electromagnetic resonator is exactly at its resonant frequency, there will be no phase

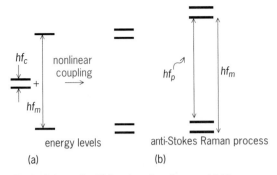

Fig. 2. Schematic sideband cooling diagram. (*a*) Microwave photons (frequency f_p) are incident on the coupled system, consisting of a microwave coaxial resonator (resonance frequency f_m) capacitatively coupled to a mechanical resonator (resonance frequency f_c, where f_c is much smaller than f_m). Nonlinear coupling leads to mixing between the two resonators, so energy can be exchanged, provided it is conserved: $hf_p = h(f_m \pm f_c)$, where h is Planck's constant. (*b*) The negative sign applies if f_p is less than f_m (that is, red detuned), and leads to energy extraction from the cantilever. This is an anti-Stokes Raman process and reduces the thermal fluctuation amplitude, giving cooling. If f_p is greater than f_m (the signal is blue detuned), then the cantilever resonator is heated in the analog of a Stokes transition. (The diagram does not show separately the red and blue detuning cases.)

lag between movement of the cantilever and the resulting force. However, if the electromagnetic excitation frequency is below the resonant frequency, there will be a phase lag between movement and the force response, whereas a positive detuning will lead to a phase advance. The effect of this phase shift means that the force arising from cantilever motion will appear to be proportional to the rate of change of the cantilever position, that is, its velocity. Thus, the applied detuned electromagnetic force looks like an additional velocity-dependent force. For negative detuning the force opposes the cantilever motion and slows it down, whereas for positive detuning it is in the same sense as the cantilever motion and so speeds it up. The former case is equivalent to cooling, whereas the latter corresponds to heating. In each case the cantilever Q is also changed. For negative detuning the cantilever Q is reduced, whereas for positive detuning the Q is increased. The cooling can be observed in the total thermally excited noise appearing at the mechanical resonant frequency, which will depend on the electromagnetic drive signal as well as detuning. The entire process can be seen as analogous to an optical Raman process (Fig. 2). The "cooling" applies only to the mechanical motion of this flexural resonance and not to the lattice of the mechanical resonator. Nevertheless, this form of noise reduction promises a wide range of precise measurement applications. It has proved possible to reduce the thermal noise power by four orders of magnitude in this way, reducing the apparent temperature from 300 K to tens of millikelvins (mK).

Applications. A major breakthrough in condensed-matter physics was achieved in 2010 when A. N. Cleland and his colleagues were able to demonstrate thermodynamic cooling of a mechanical resonator

to its ground state. This experiment involved a drum resonator made of a sandwich of thin-film piezoelectric AlN between layers of aluminum, having a resonant frequency f of around 6 GHz (**Fig. 3***a*). This resonator was cooled to a temperature $T = 25$ mK, at which temperature the mechanical mode should have contained on average zero phonons ($k_B T \ll hf$, where k_B is the Boltzmann constant and h is Planck's constant). To demonstrate the achievement of cooling to the ground state, the resonator was coupled to a superconducting qubit (Fig. 3*b*), and the combined system was manipulated by a sequence of magnetic field and electromagnetic pulses which allowed, among other tricks, a single phonon to be launched into the resonator and then retrieved. The importance of this experiment is the demonstration that such a mesoscopic mechanical system, consisting of ~10^{20} atoms, obeys the fundamental rules of quantum mechanics, behaving as a single quantum object which can be put in a superposition with other quantum objects.

Logic. As previously noted, NEMS systems have the ability to demonstrate nonlinear mechanical resonance and therefore bistability, and this is the basis of a logic system in principle. NEMS logic gates are practically attractive because they require very low power levels for operation. It has been estimated that NEMS logic can be comparable in size and speed to current CMOS circuits. In the longer term, combining cryogenic cooling to the ground state with logic operations could allow mechanical quantum computing to be envisaged.

General sensing. The ability to measure very small shifts in the resonance frequency of high-Q mechanical resonators is the basis of exquisitely sensitive

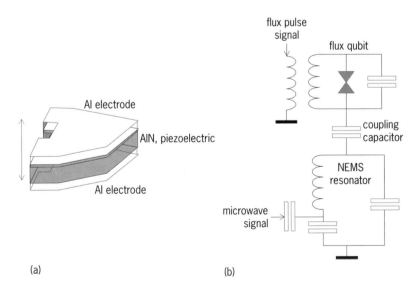

Fig. 3. Experiment to demonstrate quantum ground state of a mechanical resonator and coupling to a superconducting qubit. (*a*) Outline of mechanical resonator consisting of a piezoelectric layer (aluminum nitride, AlN) between aluminum (Al) electrodes. A dilatational mode was excited by a high-frequency voltage across the electrodes, with motion as indicated by the double arrow. (*b*) Schematic of lumped electrical circuit, showing coupling between NEMS resonator and a flux qubit consisting of a Josephson junction and capacitor. The qubit is tuned by application of a magnetic flux signal and the mechanical resonator is interrogated by transmission of a microwave signal. (*After A. D. O'Connell et al., Quantum ground state and single-phonon control of a mechanical resonator, Nature, 464:697–703, 2010*)

mass detection which has been demonstrated with NEMS. As a result of the combination of high Q and low mass, the addition of a small mass to a NEMS resonator (such as a captured macromolecule) will shift the resonant frequency by a measurable amount. In this way, single-atom detection has been demonstrated. This opens the way for unprecedented sensitivity in mass spectrometry. Similarly exceptional performance has been shown by NEMS sensors for force (tens of attonewtons), displacement (subfemtometers), and charge (10^{-24} C). These should not be seen as hard limits to what may be achieved, because further reductions in size and increases in frequency and Q of the resonators will almost certainly lead to further improvements in all these areas.

Biosensing. As was previously mentioned, the requirement for many biological measurements to be made in vivo or at least in a simulation of a living environment, meaning room temperature and a liquid medium, seems to be at odds with what NEMS sensors have to offer. Some clever developments have demonstrated ways in which these limitations may be circumvented. The inevitable fluid damping can be almost eliminated from a NEMS resonator if the fluid is contained within the resonator. For example, a microfluidic channel may be patterned in a NEMS cantilever and fluid made to flow through it while sensitive detection is carried out. If there are biologically active processes occurring in the medium, the changes can be detected through force or mass changes, yet the internal fluid damping will be orders of magnitude less than if the cantilever were immersed. Thus, the sensitivity advantages of high-Q operation can be retained. Medical diagnostics, DNA sequencing, and drug screening have been proposed as areas where the intriguing potential of NEMS could be applied in the future. The general impression is that the potential for this revived but miniaturized ancient technology is enormous and the landscape of future applications is only now being explored, let alone mapped out.

For background information *see* ANHARMONIC OSCILLATOR; ATOMIC FORCE MICROSCOPY; CARBON NANOTUBES; GRAPHENE; HYSTERESIS; INTERFEROMETRY; LASER COOLING; MAGNETISM; MICRO-ELECTROMECHANICAL SYSTEMS (MEMS); MICROFLUIDICS; MICROLITHOGRAPHY; PIEZOELECTRICITY; Q (ELECTRICITY); QUANTUM COMPUTATION; QUANTUM MECHANICS; RAMAN EFFECT in the McGraw-Hill Encyclopedia of Science & Technology.

Ling Hao; John Gallop

Bibliography. T. P. Burg et al., Weighing of biomolecules, single cells and single nanoparticles in fluid, *Nature*, 446:1066–1069, 2007, DOI:10.1038/nature05741; J. Chaste et al., A nanomechanical mass sensor with yoctogram resolution, *Nature Nanotechnol.*, 7:301–304, 2012, DOI:10.1038/nnano.2012.42; K. L. Ekinci and M. L. Roukes, Nanoelectromechanical systems, *Rev. Sci. Instrum.*, 76:061101 (12 pp.), 2005, DOI:10.1063/1.1927327g; L. Hao, J. C. Gallop, and D. C. Cox, Excitation, detection and passive cooling of a micromechanical cantilever using near-field of a microwave resonator, *Appl. Phys. Lett.*, 95:113501 (3 pp.), 2009, DOI:10.1063/1.3224912; A. D. O'Connell et al., Quantum ground state and single-phonon control of a mechanical resonator, *Nature*, 464:697–703, 2010, DOI:10.1038/nature08967.

Natural supersymmetry

The fundamental laws of physics as we know them are encapsulated in a structure known as the standard model. The standard model invokes three generations of quarks and leptons as the fundamental constituents of matter. The strong nuclear interaction describes how quarks interact with each other via the exchange of "force" particles known as gluons in a subtheory known as quantum chromodynamics (QCD). The strong nuclear force is responsible for the binding of quarks into protons, neutrons, pi mesons, and other strongly interacting particles; through subsidiary interactions, it is also responsible for binding protons and neutrons together to make the atomic nucleus. In the standard model, the weak nuclear force, which is responsible for nuclear decay and for the fusion reactions that power the Sun, is mediated by the exchange of the massive vector bosons, W and Z. Within the standard model, the weak interaction is unified with the familiar electromagnetic force (which governs the interaction of charged particles via the exchange of photons).

Higgs boson. The entire structure of quarks and leptons interacting through strong, weak, and electromagnetic forces requires the mathematical structure of nonabelian gauge symmetry to render the theory calculable (renormalizable). A problem arises in that gauge symmetry requires the force carriers to be massless, in contradiction to weak interaction phenomenology, since the short-range weak force requires massive vector bosons. A way forward was suggested by Peter Higgs and others in the 1960s: Invoke, in addition, a scalar (Higgs) field whose ground state does not respect the gauge symmetry. The original gauge symmetry is "spontaneously broken"; the force carriers (gauge bosons) become massive, hiding the original symmetry, while the underlying quantum field theory (QFT) still enjoys the mathematical benefits of the original gauge symmetry. A consequence of spontaneous symmetry breaking is that a physical scalar (spinless) boson, the Higgs boson, remains. As of 2010, all matter states in the standard model had been verified to exist with the exception of the Higgs. Recent data from the European Organization for Nuclear Research (CERN) Large Hadron Collider (LHC), running proton-on-proton collisions at 7×10^{12} electronvolts (7 TeV) in 2011 and 8 TeV in 2012, yields evidence for the existence of the Higgs boson with mass around 125×10^9 eV (125 GeV). *See* HIGGS BOSON DETECTION AT THE LHC.

Supersymmetry and dark matter. While the discovery of the Higgs boson is a great triumph, undoubtedly worthy of a Nobel prize, it comes with its own set of theoretical baggage. Spin-$\frac{1}{2}$ particles such as quarks and leptons are protected from the worst divergences of QFT by chiral symmetry, while spin-1 particles are protected by gauge symmetries. The hapless spin-0 Higgs field is unprotected, which means quantum corrections push its mass far beyond 125 GeV, to the highest energy scales occurring in the theory—possibly the scale of grand unification or string compactification, some 14–16 orders of magnitude higher than it ought to be. This pathology of spinless particles has led many to question whether such states may really exist in nature. However, in the mid-1970s, a new quantum spacetime symmetry, known as supersymmetry (SUSY), emerged, which ties fermions to bosons, so that the spin-0 bosons enjoy the same protective symmetry as their spin-$\frac{1}{2}$ partners, thus stabilizing the Higgs boson mass. The introduction of SUSY then implies the existence of spin-0 partners of all quarks and leptons, the so-called squarks and sleptons. Also, there should exist spin-$\frac{1}{2}$ partners of all the gauge bosons, the so-called gluinos, charginos, and neutralinos. While these whimsically named new states of matter might be construed as figments of theorists' overly active imaginations, there seem to be real hints from data that the "sparticles" do indeed exist with masses in the teraelectronvolt (TeV) range. Most notable of these hints is that the sparticles contribute to the evolution of the strength of the three forces of the standard model in such a way that they unify at an energy scale around 10^{16} GeV, as predicted by simple grand unified theories (GUTs). In addition, it is now known that the bulk of matter in the universe is composed of some unknown slow-moving massive quasistable neutral particles. The standard model includes no such dark-matter candidate, while SUSY theories contain several possibilities.

LHC search for sparticles. Along with searching for Higgs particles, one of the main goals of the LHC is to search for the sparticles that are expected from the minimal supersymmetrized standard model (MSSM). The most popular SUSY model, known as minimal supergravity (mSUGRA) or constrained MSSM (CMSSM), when matched to a variety of low-energy data sets, seemed to predict superparticles within the early reach of the LHC. The strongly interacting sparticles, the gluinos and squarks, could then be produced at observable rates at the LHC. Once produced, they were expected to decay through a cascade into particle plus sparticle pairs until the lightest SUSY particle was reached, which is usually assumed to be absolutely stable, and which may in fact be one of the candidate dark-matter particles. The classic signature for SUSY was then the presence of many jets plus large missing transverse energy apparently carried off by the dark-matter particles. Alas, after the initial data sets were analyzed in early 2012, no evidence for sparticles was found, leaving at least some SUSY proponents scratching their heads, while the skeptics looked on in amusement.

Need for an alternative SUSY model. A third class of physicists remained hopeful that SUSY exists, but were skeptical of the particular manifestation, that is, the paradigm mSUGRA/CMSSM model. This model makes several simplifying assumptions, including the assumption that all scalar particles receive a common, or universal, mass contribution around the GUT scale. This "universality" assumption provides one method for suppressing unwanted SUSY contributions to known processes that violate flavor and CP symmetries at levels beyond observation. However, in generic supergravity models the universal mass assumption is not well motivated. In addition, the mSUGRA/CMSSM model has required more and more fine tuning of parameters as the LHC search limits on sparticle masses have increased. Models without such ugly fine tunings are said to be "natural."

Sparticle spectrum of natural SUSY. If one goes back to examine the structure of the scalar field potential energy that gives rise to the weak scale in the standard model, and in particular to the Z boson mass, $M(Z) \sim 91.2$ GeV, then one can disentangle the various SUSY contributions that ought to conspire in a natural way, without large cancellations (with the consequence that no single contribution should be very large), to yield the measured value of $M(Z)$. One key element is that the mass of higgsinos, the superpartners of the Higgs bosons, enters directly into the scalar potential with no suppression, and so ought not be much larger than $M(Z)$, say less than 150–200 GeV. Other sparticle masses enter the scalar potential only through higher orders in perturbation theory; their contributions are suppressed by powers of their coupling to the Higgs fields but multiplied by their mass squared. If the coupling is small, then the contributing sparticle masses can be much larger. The largest of these couplings, known as Yukawa couplings, occur for the third-generation quark superpartners, the top and bottom squarks. These sparticles are then expected to have masses bounded by about 1 TeV. Gluinos contribute through higher orders to these squark masses, and so they too should be bounded, in this case by about 3 TeV. However, the first- and second-generation squarks and sleptons have only tiny Yukawa couplings; thus, they can be much heavier, in the 10–50-TeV range. This can be regarded as a net positive in that such large masses suppress (via decoupling) the expected anomalous flavor- and CP-violating processes. Thus, naturalness, along with a decoupling solution to the SUSY flavor and CP problems, favors a sparticle mass spectrum very different from the previous paradigm mSUGRA/CMSSM model. The resulting construct, dubbed natural SUSY, anticipates a spectrum of very heavy first- and second-generation squarks, well beyond reach of the LHC. The spectrum would include intermediate-scale gluinos and third-generation squarks, which may or may not be accessible to LHC searches. The rather light higgsinos predicted by natural SUSY in the 100–200 GeV range could be produced at large rates at the LHC, but the small energy release from their distinctive decays means they would be hard to pick out from known standard model processes. The upshot is that requiring naturalness in the model parameters implies that SUSY will be much harder

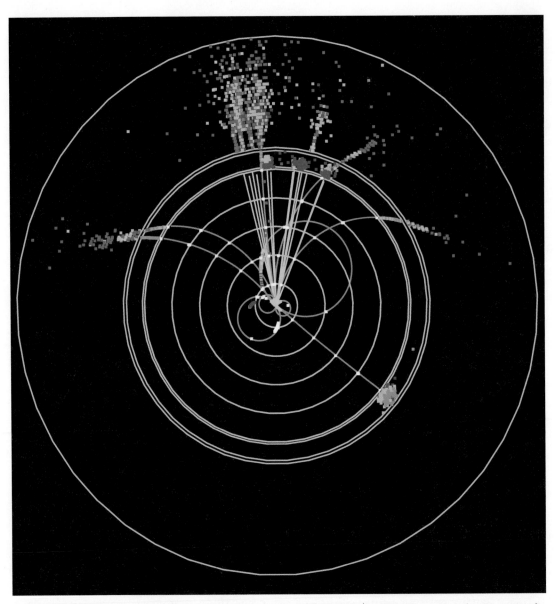

Simulated production of supersymmetric dark matter in an electron-positron (e^-e^+) collision at a center-of-mass energy of 500–1000 GeV at the proposed International Linear Collider (ILC).

to discover at the LHC than the previous paradigm mSUGRA/CMSSM model.

Dark matter in natural SUSY. As LHC limits on sparticle masses increased, the standard calculation of the dark matter (in this case weakly interacting massive particles, or WIMPs) relic density in the mSUGRA/CMSSM model almost always predicted a vast overabundance, in contradiction to measured values. For natural SUSY, the lightest SUSY particle is a higgsino, which annihilates strongly in the early universe, leading to a standard underabundance. This can be regarded again as a net positive, in that it leaves room for production of axions as well, which seem necessary if one is to solve a nagging conundrum in quantum chromodynamics called the strong CP problem.

Proposed International Linear Collider. The global particle physics community is contemplating a long-term plan for future facilities. One of these options,

the International Linear Collider (ILC) would collide electrons (e^-) against positrons (e^+) at center-of-mass energies of 500–1000 GeV. Such a machine could easily pick out signals from light higgsinos, which are a hallmark of natural SUSY (see **illustration**). In fact, such a machine would be a higgsino factory in addition to a Higgs factory, and would make a definitive test of what is emerging as the new paradigm model for weak-scale SUSY.

For background information *see* DARK MATTER; ELEMENTARY PARTICLE; GRAND UNIFICATION THEORIES; HIGGS BOSON; INTERMEDIATE VECTOR BOSON; PARTICLE ACCELERATOR; STANDARD MODEL; SUPERSYMMETRY; SYMMETRY BREAKING; WEAKLY INTERACTING MASSIVE PARTICLE (WIMP) in the McGraw-Hill Encyclopedia of Science & Technology.   Howard Baer

Bibliography. H. Baer et al., Natural supersymmetry: LHC, dark matter and ILC searches, *J. High Energ. Phys.*, 2012(5):109 (25 pp.), 2012,

DOI:10.1007/JHEP05(2012)109; H. Baer, V. Barger, and P. Huang, Hidden SUSY at the LHC: The light higgsino-world scenario and the role of a lepton collider, *J. High Energ. Phys.*, 2011(11):031 (21 pp.), 2011, DOI:10.1007/JHEP11(2011)031; C. Brust et al., SUSY, the third generation and the LHC, *J. High Energ. Phys.*, 2012(3):103 (42 pp.), 2012, DOI:10. 1007/JHEP03(2012)103; R. Essig et al., Heavy flavor simplified models at the LHC, *J. High Energ. Phys.*, 2012(1):074 (33 pp.), 2012, DOI:10.1007/ JHEP01(2012)074.

Near-field scanning microwave microscope (NSMM)

Electromagnetic waves in the microwave frequency range are an essential tool for the investigation of material and device properties across a broad range of applications. Examples of materials of interest include ferroelectric materials, ferromagnetic materials, superconductors, semiconductors, graphene, carbon nanotubes, fullerenes, and life-science materials, among others. These materials are studied at multiple scales, including bulk, thin film, and ideally down to the scale of individual molecules. Examples of devices of interest include: electron charge- and spin-based nanoelectronics, bio-inspired devices, superconducting devices, and magneto-resistive devices, among many others.

The field of atomic-scale microscopy was revolutionized during the 1980s by the inventions of the scanning tunneling microscope (STM) and the atomic force microscope (AFM). Since then, a number of scanning probe techniques have been developed, including Kelvin force microscopy, piezo-force microscopy, scanning capacitance microscopy, ferromagnetic-resonance force microscopy, and near-field scanning optical microscopy, to name but a few. This article describes the near-field scanning microwave microscope (NSMM), which is optimized for material and device characterization in the microwave frequency range.

General characteristics. Measurement of the interaction of electromagnetic waves with matter is one of the most successful ways to characterize material properties. In general this is done in the laboratory setting, either in free space or inside a resonant cavity, by scattering well-characterized electromagnetic waves from a material of interest. The measured scattered waves carry information about the material, provided that the interaction of electromagnetic waves and the material is well understood. By integrating this approach with a scanning probe microscope, the interaction is confined to a smaller volume, allowing for spatially resolved measurements. Specifically in the case of the NSMM, the near-field components of the scattered electromagnetic wave are detected in the conventional microwave region (300 MHz–300 GHz) of the spectrum.

A wide variety of condensed-matter and biological systems benefit from metrological tools that can characterize complex, multicomponent systems

with micrometer- to nanometer-scale resolution across a broad frequency range. The NSMM provides local, nondestructive measurements of the complex conductivity, complex permittivity, and complex permeability at microwave frequencies under a variety of external conditions. Variable external conditions include: electric field, magnetic field, temperature, pressure, illumination, and ambient atmosphere. Provided that the NSMM is calibrated and the interaction of the near-field microwaves with the sample under test is modeled accurately, quantitative information can be extracted from measured signals.

Near-field configuration. In the near-field configuration, a probe, either in the form of a cantilever or a sharpened wire, is brought near a sample at a distance that is very small compared to the wavelength of the applied microwave signal. The microwave signal interacts via the probe with the sample. For this measurement system, (1) the field distribution in the near field can be considered as "quasistatic" with spatial extent similar to the static fields, but with a simple time variation; (2) the spatial resolution does not depend on the wavelength of the microwave signal but

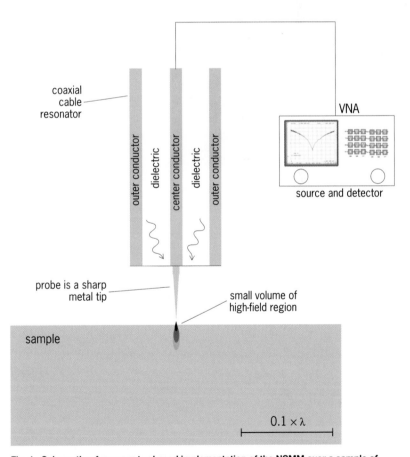

Fig. 1. Schematic of a resonator-based implementation of the NSMM over a sample of interest. For a 10 GHz signal, the scale-bar is roughly 10% of a typical wavelength ($\lambda \sim 3$ cm). The resonator is made out of a coaxial microwave cable with a sharp metal tip attached to one end. The cable connects to a source and a detector at the other end. A vector network analyzer (VNA) can serve as both the source and detector. The microwave signal arrives at the tip (shown by wiggly arrows), illuminating the sample. Although the field penetrates to the skin depth (shown by a gray, droplet-shaped region), with the help of the sharp tip, high concentration of fields is achieved in a small volume (shown by a black, droplet-shaped region), which allows for spatial resolution in the nanometer range (much smaller than the wavelength).

rather on the height of the probe above the sample and the probe dimensions; and (3) the fields are localized, generating local microwave currents in the sample, and as a result the reactive energy storage and power dissipation are concentrated in a small volume (**Fig. 1**).

In the far field, any measurement will consist of an average over the volume of the cubic wavelength or the sample size, whichever is smaller. The large spatial extent of the fields in the far field restricts the resolution to the Abbè limit ($\lambda/2$, where λ is the wavelength). By operating in the near field, the NSMM circumvents this limitation. The field distribution in the resonator depends on the electromagnetic boundary

conditions that are perturbed in the NSMM by the interaction of the resonator with the sample. To achieve condition (2) above, the coupling element is made to be a sharp tapering needle, where the diameter of the taper can be reduced to nanometer range. This enables the probe resolution to be at least 10^5 times better than the wavelength of the signal, and it also improves the signal-to-noise ratio.

Resonator-based implementation. A typical NSMM consists of a resonant circuit (resonator) that is connected to a microwave source, and a detector, which is typically a commercial vector network analyzer (VNA; Fig. 1). This design approach is compatible with feedback systems that control tip sample height and allow for integration of NSMM with other microscopies such as the STM (as shown in **Fig. 2a**) and the AFM. The STM-style feedback keeps the probe at a nominal height of 1 nm above the sample, while the AFM keeps the probe either at nanometer distance or in contact with the sample with a force of a few nanonewtons. In addition, these designs are broadband, so they are also suitable for local spectroscopic measurements.

Integration with STM and AFM. The integration of the NSMM with either the STM or the AFM is the state-of-the-art mode of operation in industrial and research laboratories (Fig. 2a). These two modes of operation have made the NSMM a very promising tool for further research in nanoscale electronics, spintronics, and bioelectronics. In addition, these two modes enable simultaneous collection of topographic data, which increases the amount of extracted information from the sample. Because one can apply a variable voltage at the tip of the sample in STM- or AFM-based techniques, one has the desired capability of making bias-dependent measurements.

Measurement procedure. The basic idea of measurement with the NSMM is to measure the microwave reflection coefficient (S_{11}) at a chosen frequency. The frequency is selected by finding the minimum of the S_{11}-versus-frequency curve in the absence of the sample. Later, the sample and probe are brought together and the curve is measured again (marked as "with sample" in Fig. 2a). Due to the tip-to-sample interaction, the frequency of the minimum shifts by Δf (Fig. 2a) and the quality factor of the resonator changes. This frequency shift and change in the quality factor are recorded by the detector and contain information about the material properties of the sample. The sample and resonator can be laterally rastered with respect to each other to map out images that reveal information about the local properties of the sample.

Role of tip-to-sample interaction. Understanding of the physics of the tip-to-sample interaction is the key for extracting quantitative information about the materials properties of the sample. Research on methods to extract such information is an exciting new area that offers great potential for important discoveries. Microwave circuit theory has well-established ways of characterizing resonators. A more complicated and unresolved problem is the tip-to-sample interaction. One approach is based on the

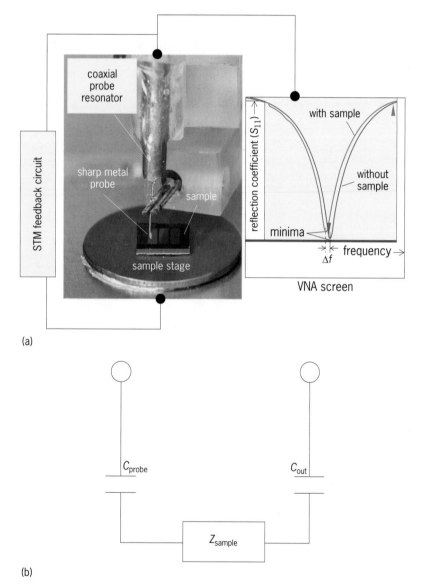

(a)

(b)

Fig. 2. Typical custom-made NSMM, built at the National Institute of Standards and Technology (NIST). (*a*) Photograph of NSMM. The probe is made of platinum-iridium. The "sample stage" is where the material or device under test sits. The screen of the vector network analyzer (VNA) shows a single resonance of the resonator in a plot of S_{11} versus frequency with (red) and without (blue) the sample. The frequency shift is Δf, and change of the minimum amplitudes is related to change of quality factor. (*b*) Circuit diagram for the lumped-element model of the tip-to-sample interaction. The ultimate goal is to construct theoretical models to get the sample impedance, Z_{sample}. In the circuit diagram, C_{probe} is tip-to-sample capacitance, and C_{out} is sample-to-outer-conductor capacitance. The state-of-the-art NSMM is able to measure 1 attofarad (10^{-18} F) capacitance.

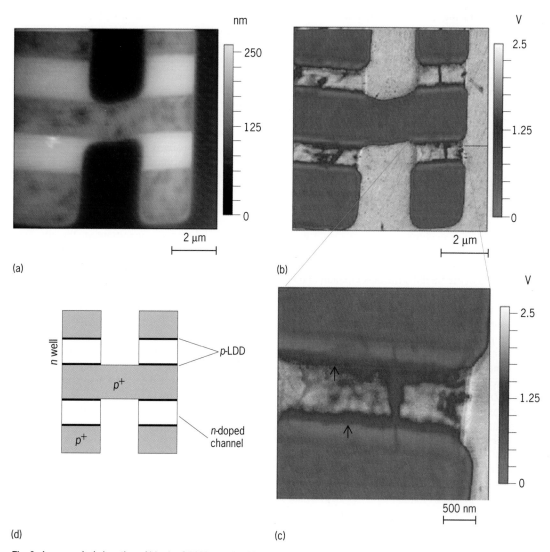

Fig. 3. Images of *p/n* junction of bipolar SRAM sample. (*a*) AFM image, showing the surface topography. (*b*) The *dC/dV*-phase of the same region, acquired simultaneously with topography. Reported units of *dC/dV*-phase are in volts (rather than farads/volt) since the measurement is uncalibrated. (*c*) Zoom (or magnification) of one of the *p/n* junction areas, showing the nanometer-scale spatial resolution in materials contrast. Arrows point to *p*-LDD areas for clarification. (LDD = low dopant density.) (*d*) Diagram showing the layout of the doping concentrations. (*Reprinted with permission courtesy of H. P. Huber et al., Calibrated nanoscale dopant profiling using a scanning microwave microscope, J. Appl. Phys., 111:014301, 2012, copyright © 2012, American Institute of Physics*)

interpretation of this interaction through a lumped-element circuit model, often consisting of local capacitors and resistors (Fig 2*b*). Based on the measured shifts in the resonance frequency and changes in quality factor, one can calculate the local capacitance and resistance of a sample from microwave circuit models. This information is then used to predict the material properties of the sample, such as local complex permittivity or complex permeability. The calculated capacitance and resistance must reflect the physics of the processes in the investigated materials, which may be subsequently used to predict material properties such as charge concentration, mobility, electric polarization, and magnetic moment. An alternative approach is to construct novel, well-understood reference samples, which can be used to calibrate the tip-to-sample interaction experimentally. Both of these approaches are, in part, phenomenological and include some adjustable parame-

ters that have to be obtained independently. Further work on full near-field interaction of electromagnetic waves with the investigated materials needs to be developed for a complete quantitative description of material properties.

Example. As an example, **Fig. 3** shows the results from the NSMM as applied to a *p/n* junction of a hetero-bipolar SRAM sample, imaged by an NSMM at 18 GHz. To get this information one has to measure the derivative of the tip-to-sample capacitance (*dC/dV*) as a function of the frequency and tip-to-sample voltage. The derivative is measured by applying a DC bias voltage to the probe along with an AC voltage modulation while detecting the signal with a lock-in amplifier. The *dC/dV* image clearly has features that do not appear in a simultaneously acquired topographic image. The bipolar *p/n* junction interfaces are visible in Fig. 3*b* and *c*, and the width of interfaces shown is ~100 nm.

Prospects. Moving forward, two important topics will have to be addressed. One is the development of reliable models to understand the relation between the various measured images. Such models will enable local, nanoscale quantitative estimates of material and device properties. The second important topic is developing the local microwave spectroscopy capability of the NSMM. This encompasses several approaches that include both frequency- as well as voltage-dependent spectroscopy. To this end, it has been proposed to measure the derivative of the phase between incident and the reflected signal from the tip-to-sample load (dP/dV) in addition to dC/dV. This technique has the potential to enhance the signal-to-noise ratio significantly, improving sensitivity to water concentrations in soft matter and biological samples. This may enable imaging of living biological cells in saline solution. Another possible extension of the NSMM is the development of the two-probe and multiprobe NSMM, which will enable the measurement of the complex microwave multiport scattering matrix elements and, in turn, potential applications in broadband transport studies of complex systems such as carbon-based nanoelectronics and high-frequency spintronics.

For background information *see* ATOMIC FORCE MICROSCOPY; MICROWAVE; MICROWAVE MEASUREMENTS; OPTICAL MICROSCOPE; SCANNING TUNNELING MICROSCOPE in the McGraw-Hill Encyclopedia of Science & Technology.

Atif Imtiaz; T. Mitchell Wallis; Pavel Kabos

Bibliography. S. M. Anlage, V. V. Talanov, and A. R. Schwartz, Principles of near-field microwave microscopy, in S. V. Kalinin and A. Gruverman (eds.), *Scanning Probe Microscopy: Electrical and Electromechanical Phenomena at the Nanoscale*, Springer, New York, pp. 215–253, 2007; A. Imtiaz and S. M. Anlage, A novel STM-assisted microwave microscope with capacitance and loss imaging capability, *Ultramicroscopy*, 94:209–216, 2003, DOI:10.1016/S0304-3991(02)00291-7; B. T. Rosner and D. W. Van der Weide, High-frequency near-field microscopy, *Rev. Sci. Instrum.*, 73:2505–2525, DOI:10.1063/1.1482150.

Neoadjuvant systemic therapy for breast cancer

The term neoadjuvant is used to describe therapy given prior to a planned surgical resection of a cancer with curative intent. Neoadjuvant systemic therapy (NST) has been explored in many different types of neoplasms, with varying success. Historically, breast cancer has been managed with primary surgery, followed by adjuvant (postoperative) chemotherapy, hormonal therapy, and/or radiation to decrease the chance of locoregional (restricted to a localized region of the body) and distant recurrence. NST for breast cancer was initially used to treat locally advanced breast cancer, which was considered inoperable. More recently, it has been increasingly applied to less advanced, operable breast cancer, and clinical trials have shown that it provides a range of benefits.

Benefits. NST has several clearly proven benefits for operable breast cancer. By shrinking large tumors, NST can increase the likelihood of breast conservation and allows for a reduced extent of resection, which may improve cosmetic outcomes. Moreover, because disease-free survival and overall survival are similar to those achieved with adjuvant therapy, this increase in breast conservation does not have a cost in terms of patient outcomes. By delaying the timing of surgery, administration of NST allows time for the patient and her surgeon to plan the operation, and the patient is also able to undergo genetic counseling and testing and consultation with a plastic surgeon to consider reconstruction options. One potential (but unproven) benefit of NST is the opportunity to monitor the response to therapy. This allows the patient to appreciate the effectiveness of therapy in a way that is not possible in the adjuvant setting when there is no detectable tumor present. Whether changing therapy based on tumor response would be beneficial has not been clearly demonstrated. Results from one large German trial have suggested that this may be the case for some molecular subtypes. NST may also decrease the need for full axillary lymph node dissections, but this remains controversial, especially in women with positive nodes prior to therapy. Finally, the residual cancer burden in the breast and the regional lymph nodes is a powerful indicator of prognosis.

Some benefits of NST relate to clinical research, including the availability of tissue that can be sampled before, during, and after treatment. This, in turn, offers opportunities to identify molecular predictors of response. In addition, new treatments can be tested in the neoadjuvant setting, allowing more rapid assessment of their efficacy, based on tumor response as an end point. However, there is still some debate as to whether this is an accurate surrogate for clinical benefit.

Indications. NST is appropriate in a number of clinical scenarios. It is the preferred approach in patients with locally or regionally advanced breast cancer, such as inoperable stage III or inflammatory cancers. It also can be considered for any patient with a tumor that is greater than 2 cm (0.8 in.) in diameter or with positive nodes, especially if the tumor is hormone receptor–negative (HR–) or overexpresses human epidermal growth factor receptor-2 (HER-2, which is a receptor protein that signals breast cells to grow and divide). In other words, any patient who would clearly need adjuvant chemotherapy based on clinical evaluation or molecular markers can be considered for neoadjuvant chemotherapy. Patients in whom mastectomy is the only option on presentation based on tumor size should be considered for NST.

Marking. In patients treated with NST, it is important to mark the tumor prior to the start of therapy. Using a clip to mark the epicenter of the tumor with imaging guidance will ensure that the appropriate tissue is excised after treatment, even if there should be a complete response (that is, no visible or palpable tumor). Using the clip (or clips) to guide surgery does require preoperative or intraoperative imaging.

Tattoos (dots) on the skin marking the extent of the tumor can also be used to guide surgery, usually without the need for imaging; these dots marking the initial site and extent of the tumor may also help with serial measurements during treatment.

Extent of resection. The appropriate extent of resection after neoadjuvant therapy can be difficult to determine. One should not remove the same volume of breast tissue that would have originally been needed for negative margins (no trace of cancer in the healthy tissue) because this would defeat the purpose of NST. The goal should be to attain negative margins while maintaining good cosmesis (preservation or restoration of physical appearance). Even in the setting of a complete clinical response, based on current evidence and because of the low overall rates (<25–30% for most types) of a pathologic complete response (pCR; that is, no evidence of viable invasive tumor cells), surgical excision should not be omitted. However, as pCR rates increase, especially in certain subsets (for example, HER-2-overexpressing cancers treated with chemotherapy plus HER-2-targeted therapy), interest has developed in trying to determine which patients had a pCR and thus may not require surgical resection. It is also important for the surgeon to inform the pathologist that submitted specimens are from a patient treated prior to surgery, both for appropriate tissue sampling and to understand changes resulting from therapy.

Axillary node sampling. Axillary node sampling is currently routine for most breast cancer patients, with sentinel lymph node (SLN) biopsy having become the standard of care for patients with clinically negative axillary nodes. For patients receiving NST, there is a debate over whether to perform these biopsies before or after neoadjuvant therapy. As an argument against SLN biopsy after neoadjuvant therapy, mapping may fail or may not be accurate, which might result in false-negative results, understaging, and possible undertreatment. One might also lose important prognostic information derived from pretherapy nodal pathology. However, multiple studies have shown that the false-negative rates for SLN biopsy after NST are similar to those for primary surgical resection, and that posttreatment nodal status has at least as much prognostic impact as pretreatment nodes. For clinically node-negative patients, SLN biopsy before or after neoadjuvant therapy is acceptable. If SLN biopsies are positive prior to therapy, this may commit the patient to a complete axillary lymph node dissection (ALND) that may be unnecessary. For patients with clinically or radiographically positive nodes, pretreatment needle biopsies should be performed, and full ALND (after NST has been completed) would then be considered by many to be the standard of care. However, prospective studies of SLN accuracy in such patients have not been conducted; if false-negative rates are acceptable, then up to 40% of patients who initially present with positive nodes may be spared ALND and its morbidities because of the use of NST.

Radiation therapy is commonly used after most partial or some total mastectomies to reduce the risk of locoregional recurrence and to improve survival, and the decisions about who should receive adjuvant regional radiation often depend on pathologic tumor size and nodal status. However, the use of NST alters the final pathological tumor size and the number of positive nodes. Data obtained from the National Surgical Adjuvant Breast and Bowel Project (NSABP) B-18 and B-27 NST trials, which did not allow postmastectomy or regional nodal irradiation, suggest that patients with a pathological complete response and pathologically negative nodes after NST have low risk for regional recurrence; thus, they may not benefit from postmastectomy radiation therapy or regional nodal irradiation after lumpectomy, even if they had clinically positive nodes before chemotherapy.

Hormonal therapy. Neoadjuvant hormonal therapy offers an alternative approach for improving surgical outcomes and has fewer side effects than cytotoxic chemotherapy. It has been shown to improve outcomes in appropriately selected patients with hormone-responsive cancers. Postmenopausal women with hormone receptor–positive disease benefit from neoadjuvant therapy with anastrozole or letrozole, and the use of these therapeutic drugs increases the rates of breast-conserving surgery. Hormonal therapy may even be superior to chemotherapy in these patients.

Conclusions. Neoadjuvant systemic therapy represents an emerging technique that will likely become standard for appropriately selected patients. It relies on multidisciplinary cooperation from the start, which should also become standard in the overall treatment of breast cancer. Neoadjuvant systemic therapy increases the chance of breast-conserving surgery. To achieve better results, both cosmetic and oncologic, the surgeon must make sure that the tumor is appropriately marked, that good preoperative imaging has been completed, and that the patient understands the risks and benefits of breast-conserving surgery. Sentinel lymph node examination after neoadjuvant chemotherapy is controversial; however, pending definitive data, it may be the preferred approach, especially for patients with clinically negative nodes. As the experience with this technique broadens, the use of neoadjuvant hormonal therapy may turn out to be the best option for some patients. Perhaps best of all, neoadjuvant systemic therapy offers opportunities to understand breast cancer biology and improve therapy. Because pretreatment and even mid-therapy tissue is readily available for sampling, NST makes it possible to correlate molecular genetic profiles with response to therapy and patient outcomes.

Neoadjuvant systemic therapy trials, based on response as a primary end point, can be completed in 2–4 years, and only a few hundred patients are necessary (in contrast, 5–10 years and thousands of patients are needed to complete most adjuvant trials). In the future, once a new agent is shown to be effective for NST in a subgroup of tumors, this agent can then be used in smaller adjuvant trials in the same group. This will lead to faster progress than basing large adjuvant trials on results in the metastatic disease setting, with patients and tumors previously treated with systemic therapy. As more subgroups, based on molecular markers and profiles, are explored, more therapies can be individualized on the basis of these profiles, using NST to define the right therapies for the right tumors.

For background information *see* BREAST; BREAST DISORDERS; CANCER (MEDICINE); CHEMOTHERAPY AND OTHER ANTINEOPLASTIC DRUGS; HORMONE; ONCOLOGY; RADIATION THERAPY; SURGERY in the McGraw-Hill Encyclopedia of Science & Technology.

Eric D. Seitelman; Harry D. Bear

Bibliography. H. D. Bear, Neoadjuvant chemotherapy for operable breast cancer: Individualizing locoregional and systemic therapy, *Surg. Oncol. Clin. North Am.*, 19:607–626, 2010, DOI:10.1016/j.soc.2010.04.001; J. S. Chawla, C. X. Ma, and M. J. Ellis, Neoadjuvant endocrine therapy for breast cancer, *Surg. Oncol. Clin. North Am.*, 19:627–638, 2010, DOI:10.1016/j.soc.2010.04.004; B. Fisher et al., Effect of preoperative chemotherapy on the outcome of women with operable breast cancer, *J. Clin. Oncol.*, 16:2672–2685, 1998; L. Gianni et al., Phase III trial evaluating the addition of paclitaxel to doxorubicin followed by cyclophosphamide, methotrexate, and fluorouracil, as adjuvant or primary systemic therapy: European Cooperative Trial in Operable Breast Cancer, *J. Clin. Oncol.*, 27:2474–2481, 2009, DOI:10.1200/JCO.2008.19.2567; K. K. Hunt et al., Sentinel lymph node surgery after neoadjuvant chemotherapy is accurate and reduces the need for axillary dissection in breast cancer patients, *Ann. Surg.*, 250:558–566, 2009, DOI:10.1097/SLA.0b013e3181b8fd5e; M. Kaufmann et al., Recommendations from an International Consensus Conference on the Current Status and Future of Neoadjuvant Systemic Therapy in Primary Breast Cancer, *Ann. Surg. Oncol.*, 19:1508–1516, 2012, DOI:10.1245/s10434-011-2108-2.

New insights on ocean acidification

Since the Industrial Revolution, rising carbon dioxide (CO_2) levels in the atmosphere and increased absorption of CO_2 by the oceans have created an unprecedented ocean acidification (OA) phenomenon that is altering pH levels and threatening a number of marine ecosystems. Although the average oceanic pH can vary on interglacial time scales, the changes are usually of the order of about 0.002 unit per 100 years; however, the current observed rate of change is about 0.1 unit per 100 years, or roughly 50 times faster. Even more disconcerting, regional factors such as coastal upwelling, changes in riverine and glacial discharge rates, and loss of sea ice have created OA "hotspots" where changes are occurring at even faster rates.

While OA is a global problem that will likely have far-reaching implications for many marine organisms, there are areas that will be affected sooner and to a greater degree. Recent observations have shown that one area in particular is the cold and highly productive region of the subarctic Pacific and western Arctic Ocean, where unique biogeochemical processes create an environment that is both sensitive and particularly susceptible to accelerated reductions in pH and carbonate-mineral concentrations. Extraordinary changes caused by natural and anthropogenic perturbations are affecting the availability of carbonate minerals necessary for the formation and maintenance of shells and skeletons of marine calcifying organisms. The OA phenomenon can cause waters to become undersaturated in carbonate minerals and thereby affect extensive and diverse populations of marine calcifiers. Some of these affected organisms are keystone species in the ecosystem and are critical to the fishing industry and subsistence communities. The Pacific-Arctic Region (PAR) is an excellent case study for the potential effects of OA. The region has rapidly changing physical conditions, and, as a result, high susceptibility to OA and economic viability.

The CO_2 problem. Data from ice cores taken from Greenland and Antarctica have shown that CO_2 concentrations in the atmosphere have varied between 200 and 300 parts per million (ppm) over the last 400,000 years when significant anthropogenic forcing was absent. However, over the last 250 years, human activities such as the burning of fossil fuels and changes in land-use practices have resulted in atmospheric CO_2 levels increasing sharply, resulting in an oceanic uptake of over 146 ± 20 petagrams (Pg) of carbon or roughly 1.1 billion tons of CO_2. From 1970 to 2000, CO_2 concentrations in the atmosphere increased at a rate of approximately 1.5 ppm annually. However, since 2000 these increases have accelerated to approximately 2.2 ppm per year. This means that not only has the total content of CO_2 in the atmosphere continually increased, but that the rate at which it is increasing is also accelerating.

Currently, 9 Pg of anthropogenic CO_2 are released into the atmosphere every year. Of this, approximately 7.5 Pg come directly from the burning of fossil fuels and other industrial processes that emit CO_2. The remaining 1.5 Pg are due to changes in land-use practices, such as deforestation and urbanization. These changes are important because terrestrial ecosystems, such as forests and grasslands, provide an import sink for atmospheric CO_2 by removing it from the air during photosynthesis. When human activity removes portions of this biomass, the sink no longer exists, and the result is an accumulation of additional CO_2 in the atmosphere. Of the 9 Pg of anthropogenically produced CO_2 emitted annually, approximately 2.6 Pg (or 29%) are incorporated into terrestrial plant matter. Another 4.2 Pg (or 46%) are retained in the atmosphere, which has led to some planetary warming. The remaining 2.3 Pg (or 26%) are absorbed by the world's oceans.

Why is the ocean more acidic? For many of the same reasons the term "global warming" is not an ideal way to describe global climate change, "ocean acidification" is a similar misnomer. It is highly unlikely that the ocean will ever become acidic on the pH scale of 0–14. This is because seawater has the intrinsic ability to buffer against changes in pH using its high concentrations of bases, which have a negative ionic charge. The concentration of these bases is referred to as total alkalinity. When the ocean absorbs a CO_2 molecule, it undergoes a fairly simply chemical reaction, which produces two positively charged hydrogen ions for every one molecule of CO_2 that is absorbed. The pH, by definition, is the measure of the concentration of hydrogen ions dissolved in a solution. The more hydrogen ions are produced, the lower the pH becomes. Once they are produced, the positively charged hydrogen ions react with the negatively changed bases to buffer the pH of the water. One of these bases is the carbonate ion, which is necessary for shell and skeletal growth in marine calcifying organisms. As the hydrogen ions produced during CO_2 dissolution scour carbonate ions out of seawater, the rate of calcification of shell-building organisms slows down. When too few carbonate ions are available for adequate shell building, calcifying organisms begin to dissolve. Because the real threat to ocean organisms lies in the removal of these carbonate ions, a more correct term for the OA process might be "ocean debasification," but that term does not convey the necessary degree of concern.

The intrusion of anthropogenic CO_2 is not the only way to cause a decrease in carbonate ions in marine waters. Any process that reduces total alkalinity will have the same net effect. Observations in the PAR have shown that increases in tidewater glacial runoff and sea-ice melt, both of which are at least partially accelerated by human activities, are also having an effect on ocean acidification. Both glacial discharge and sea-ice melt have low concentrations of total alkalinity, so when they are discharged or mixed into surface seawater, they cause a dilution of total alkalinity and a reduction in the concentration of bases in solution. When combined with anthropogenic CO_2 intrusion, these dilutions can have a compounding effect on high-latitude regions of the ocean where these processes occur. This makes these areas, and their associated ecosystems, even more susceptible to changes in carbonate-ion concentrations.

What are the effects of a more acidic ocean? Throughout the last 25 million years, the average pH of the ocean has remained fairly constant between 8.0 and 8.2. However, in the last three decades a precipitous drop has begun to occur, and if CO_2 emissions are left unchecked, the average pH could fall below 7.8 by the end of this century. This is well outside the range during any other time in recent geological history. Since the Industrial Revolution, the increase in oceanic CO_2 concentrations has caused the average surface-water pH to decrease by approximately 0.1 unit, with the most dramatic changes occurring in the last half-century due to accelerating emission rates. Because the ocean mixes much more slowly (over 1000 years) than the atmosphere, the accumulation of anthropogenic CO_2 is confined to the upper 10% of the water column in most places. This means that the greatest changes in pH will occur at the top of the water column, where the greatest biological activity and diversity occurs. Indeed,

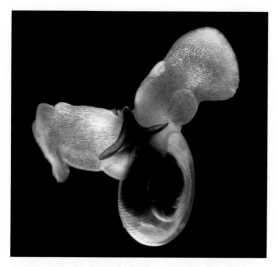

Pteropod species *Limacina helicina*, which is commonly found in subarctic seas. Pteropods are an import prey source for juvenile salmon in the marine environment. (*Photo courtesy of Professor Russ Hopcroft, University of Alaska Fairbanks*)

if CO_2 emission rates continue to rise as projected, the average pH of the surface ocean will decrease by another 0.3–0.4 unit by 2100. Calcifying organisms, such as corals, crabs, clams, oysters, and the tiny free-swimming pteropods (see **illustration**) that form calcium-carbonate shells could be particularly vulnerable, especially during the larval stage.

Many of the processes that cause OA have long been recognized, but the ecological implications of the associated chemical changes have only recently been investigated, with mixed results. Increased CO_2 concentrations and the accompanying changes in ocean chemistry may alter species' compositions and abundance. Such effects in the ocean could be felt by both calcifying and noncalcifying primary producers and microbes, and could ultimately disrupt the cycling of elements, such as iron, which are critical for growth. For these reasons, OA could profoundly affect the most fundamental chemical and biological processes of the sea in the coming decades.

The responses of both calcifying and noncalcifying organisms are by no means uniform, however. Recent studies, which exposed organisms to varying pH levels for short periods of time (often weeks to months), have shown a wide range of responses. Even organisms within the same general classification (that is, crabs) have shown variable responses when exposed to acidified conditions, and there appears to be an inherent ability in some species to tolerate changes in pH, at least over short-term intervals. Some species may also be better able than others to adapt to changing pH levels due to their exposure to environments where pH varies naturally over a wide range. As a result, there will likely be ecological "winners" and "losers" as local competition for resources plays out. Depending on the region, this could lead to changes in ecosystem structure and possible species migration to more suitable habitats, or, in the worst-case scenario, a complete regime shift. In some cases, these ecosystem shifts may reduce the economic viability of a region. However, at this point it is still very uncertain what will be the ecological and societal consequences from any potential losses of keystone species, and how the "winners" will affect the ecosystem or the biogeochemical cycles as a whole. There are places in the ocean where CO_2 levels are naturally high (that is, CO_2 vents). These ecosystems provide a glimpse of what parts of the ocean may look like in the future. In these naturally CO_2-rich areas, photosynthetic species (such as sea grasses) thrive, but the biodiversity in these systems is 30% less than in comparable regions with "normal" CO_2 levels.

Although it is still very difficult to determine what the effects of OA on marine ecosystems will be, because the responses of organisms to experimental changes in pH in laboratory settings have been so variable, we can gain some insight by looking at previous acidification events during the Earth's history. Data from the geologic records suggests that during past OA events, caused by the natural accumulation of CO_2 in the atmosphere and ocean, marine calcification was greatly reduced and mass extinctions occurred. These prior events have allowed economists to begin forecasting the potential financial consequences of OA. Using data from 2007, S. R. Cooley and S. C. Doney found that annual domestic sales and imports of fisheries susceptible to OA topped $70 billion in the United States and added $35 billion to its gross national production. Of the $4 billion in annual domestic sales, Alaska and the New England states contributed the most, at about $1.5 billion and $750 million, respectively. These numbers clearly show that any disruption in the commercial fisheries in these regions as a result of OA could have a cascading effect in the local as well as national economy. In Alaska, where the fishing industry is one of the largest employers and a number of communities rely solely on fisheries for subsistence, the consequences of OA could be dire.

Ocean acidification hotspots. The cold waters of high-latitude oceans, such as those in the PAR surrounding Alaska, are naturally low in carbonate-ion concentration due to the increased solubility of CO_2 at low temperatures, ocean mixing patterns, and unique riverine and glacial inputs. Consequently, seawater pH and carbonate-mineral concentrations are typically lower than in temperate and tropical regions. In most areas, the surface waters of high-latitude oceans are supersaturated in carbonate minerals, meaning that calcification can occur readily, but models project that at current rates of CO_2 emissions, surface waters of the subarctic Pacific and the western Arctic Ocean will become undersaturated in the most soluble carbonate minerals by the end of this century and in some regions as early as 2023. However, recent observations in the eastern Bering Sea have shown that some areas are already experiencing seasonal undersaturations, meaning that some calcifiers, particularly the hypersensitive pteropods, may already be in trouble. Because of the rapid rate of change and ecosystem structure, the PAR will likely be a bellwether of

potential consequences of OA on marine organisms, and a natural laboratory for mechanistic studies, evaluations of potential acclimation and adaptation, and understanding the effects at population and community levels.

Outlook. In terms of the effects of OA, the future is now. There have already been real-world examples of economic disruptions due to OA, the most visible case being the harvest failures in the oyster hatcheries along the coast of the Pacific Northwest. Hatcheries that supply the majority of the oyster spat (larvae) to farms around the county nearly went out of business as they unknowingly pumped acidified water, corrosive to oyster larvae, into their operations. New innovations and interactions with scientists have allowed these hatcheries to monitor the pH of the incoming water, and thereby saved the industry. However, this is only a short-term solution, as the pH of the coastal waters necessary for their operation continues to become even more acidic.

Although several anthropogenic factors are influencing pH change and the reduction of carbonate-mineral concentrations in the ocean, the release of CO_2 is the major driver. As long as CO_2 concentrations increase, the pH of the ocean will continue to drop. This is very disconcerting in a world where over 75% of all energy production is derived from burning fossil fuels. Up until 2005, the United States was the top emitter of CO_2, releasing roughly 1.6 million tons annually. However, China has now surpassed the United States in total emissions, and as of 2007 its discharge rates had passed 1.8 million tons of CO_2 per year. Other industrial nations, such as India, Russia, and Japan, are all well behind the United States and China in their emission rates. However, vast segments of the global population that have previously had little to no effect on global anthropogenic CO_2 production are industrializing at a very rapid rate. In China, the per-capita production of CO_2 is only 25% of what Americans produce, indicating that there is a great potential for an exponential increase in CO_2 emissions as Chinese energy consumption grows.

During the past 20 years, over 440 million Chinese (more than the total population of the United States) have lifted themselves out of poverty to become energy consumers and CO_2 emitters. This accounts for China's ascension to the top global emitter of CO_2, but there are hundreds of millions more people across Asia and Africa who are not far behind in their demand for Western-style energy consumption. Even as the international community struggles to develop mitigation plans, the momentum in the global economy will be difficult to overcome without a radically new, clean (CO_2-free) source of energy. As a result, it appears that OA is a problem that is here to stay for a while.

Although the number-one goal should be broad mitigation of CO_2 emissions, there are steps we can take to protect ourselves from the consequences of OA. Using the oyster hatcheries on the West Coast as a model, we can begin to think about monitoring and management strategies to keep sensitive areas of the marine economy stable. These efforts will not be cheap or easy to carry out. But with few alternatives, they may be our best chance of dealing with the first wave of OA effects. Hopefully, these protective measures can buy us enough time to do something about the underlying problem.

For background information *see* ARCTIC OCEAN; ATMOSPHERIC CHEMISTRY; BIOGEOCHEMISTRY; CARBON DIOXIDE; CARBONATE MINERALS; GLACIOLOGY; GLOBAL CLIMATE CHANGE; MARINE ECOLOGY; MOLLUSCA; PACIFIC OCEAN; PH; SEA ICE; SEAWATER; SEAWATER FERTILITY in the McGraw-Hill Encyclopedia of Science & Technology.   Jeremy T. Mathis

Bibliography. S. R. Cooley and S. C. Doney, Anticipating ocean acidification's economic consequences for commercial fisheries, *Environ. Res. Lett.*, 4:024007, 2009, DOI:10.1088/1748-9326/4/2/024007; V. J. Fabry et al., Ocean acidification at high latitudes: The bellwether, *Oceanography*, 22(4):160–171, 2009, DOI:10.5670/oceanog.2009.105; C. L. Sabine and R. A. Feely, The oceanic sink for carbon dioxide, in D. Reay et al. (eds.), *Greenhouse Gas Sinks*, pp. 31–49, CABI Publishing, Oxfordshire, U.K., 2007; J. C. Orr et al., Anthropogenic ocean acidification over the twenty-first century and its impact on calcifying organisms, *Nature*, 437(7059):681–686, 2005, DOI:10.1038/nature04095.

Nighttime starch degradation, the circadian clock, and plant growth

The steadily rising demand for food and renewable resources has challenged plant breeders and biotechnologists to rapidly increase crop productivity. To realize this goal, a holistic knowledge is required of how plant metabolic pathways are controlled to allow optimal growth. Today, very little is known about the partitioning of photosynthetically assimilated carbon among growth, storage, and respiration. This article describes recent progress in understanding how the model plant *Arabidopsis thaliana* uses its carbon resources to ensure a continuous energy supply for growth during the night.

Diurnal starch turnover in Arabidopsis plants. During the day, plants assimilate CO_2 to produce sugars (photosynthates) in the process of photosynthesis. Plants use these sugars to fuel their metabolism and growth, producing the primary carbon source for almost all nonphotosynthetic organisms. However, not all photosynthates acquired during the day are used for immediate growth. Plants partition a fraction of the assimilated carbon into storage compounds in leaves to support respiration and continued growth during the night when photosynthesis is not possible.

In many plant species, the main carbon storage compound is starch. The synthesis of starch in leaves during the day and its degradation during the night have been studied intensively in the model plant *A. thaliana*. During a normal day, *Arabidopsis* plants store approximately 50% of the carbon

assimilated by photosynthesis as starch granules in the chloroplasts of leaves. During the night, starch is degraded with a near-linear rate such that the starch reserves are almost completely utilized by dawn (**Fig. 1***a*).

This match between the length of time taken to degrade the starch reserves and the length of the night is vitally important for normal plant growth. If the night is artificially extended beyond the normal dawn, the growth rate of the plant drops abruptly. Mutant plants that cannot accumulate starch or that degrade it only very slowly have much lower growth rates than wild-type plants and show a severely reduced overall rate of growth. These reductions in growth rate are accompanied by large changes in gene expression indicating carbon starvation.

Considering the importance of a continuous carbon supply during the night for plant growth, it is not surprising that starch turnover is tightly controlled. *Arabidopsis* plants adjust the rates of starch synthesis and degradation to different environmental conditions (for example, temperature, light levels, and day length). The rate of starch synthesis is inversely related to day length: the shorter the day, the greater the proportion of assimilated carbon that is partitioned into starch. The rate of starch degradation is also adjusted such that a linear and almost complete degradation during the night is achieved for day lengths ranging from 18 h to as short as 4 h. Remarkably, the rate of starch degradation in *Arabidopsis* plants can adjust immediately in response to an unexpected early or late onset of night. If plants grown in 12 h of light/12 h of darkness are subjected to darkness after only 8 h of light, the rate of starch degradation is much slower than on previous nights, but remains constant throughout the 16-h night. These observations imply that plants at dusk integrate information about the amount of starch present in leaves and the anticipated length of the night to set the rate of starch degradation. Recent investigations have revealed that the timing of starch degradation in *Arabidopsis* plants is linked to the circadian clock.

Circadian clock and starch degradation. Almost all organisms possess an endogenous oscillating timer called the circadian clock. This timer keeps track of the estimated position of an organism in the 24-h light–dark cycle. The clock controls physiological processes that function at specific, appropriate times of day and supports the anticipation of dusk and dawn. In plants, the circadian clock affects a wide range of physiological and biochemical processes, including expansion growth, flowering time, stomatal aperture, leaf movement, and responses to drought stress and pathogen attack.

An important hallmark of the circadian clock is its free-running 24-h rhythm. Free running refers to the fact that, once entrained by light signals, the circadian clock maintains a 24-h rhythm in continuous light or continuous darkness, anticipating dusk and dawn according to previously encountered conditions. In fact, the property of being a 24-h timer has revealed the involvement of the circadian clock

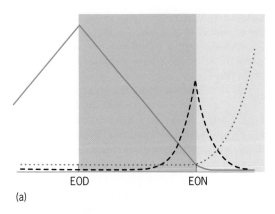

(a)

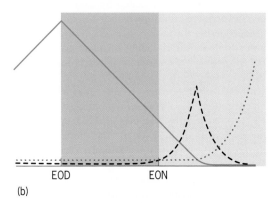

(b)

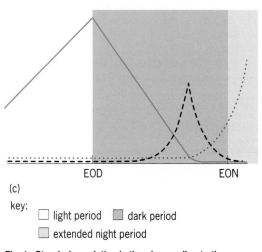

(c)

key:

☐ light period   ▨ dark period

☐ extended night period

Fig. 1. Starch degradation is timed according to the anticipation of dawn by the circadian clock, regardless of the actual length of the night. The graphs show situations in which the period of the clock matches the length of the light–dark (LD) cycle (*a*), is longer than the LD cycle (*b*), or is shorter than the LD cycle (*c*). Starch (solid line) accumulates in a linear fashion during the light period and is degraded in the dark period. The anticipation of dawn by the circadian clock is revealed by a peak of transcript abundance of morning-phased clock-controlled genes (dashed line). Accumulation of transcripts indicative of starvation (dotted line) occurs only when starch supplies are exhausted. This point is reached early in an artificial extension of the night when the period of the clock matches the length of the LD cycle (*a*), only after several hours of extended night when the period of the clock is longer than the LD cycle (*b*), and well before the end of the normal night when the period of the clock is shorter than the LD cycle (*c*). EON = end of night; EOD = end of day. *(Based on data from A. Graf et al., Circadian control of carbohydrate availability for growth in Arabidopsis plants at night, Proc. Natl. Sci. USA, 107:9458–9463, 2010)*

in the control of starch degradation. When plants are grown in light–dark cycles shorter or longer than 24 h, abnormal starch degradation patterns are observed during the night. In 28-h light–dark cycles (14 h of light, 14 h of darkness), starch is degraded extremely fast, so reserves are exhausted before dawn—specifically, at 10 h into the night rather than at the actual dawn after 14 h of night (Fig. 1*b*). Conversely, in 20-h light–dark cycles (10 h of light, 10 h of darkness), starch is degraded too slowly, resulting in the presence of substantial reserves at dawn. If the night is extended beyond dawn, starch is eventually depleted after approximately 14 h of darkness (Fig. 1*c*).

Measurements of clock-related gene transcription can be used to analyze the timing of the *Arabidopsis* clock. So-called morning-phased clock genes show a sharp expression peak at the anticipated dawn. Quantification of transcription level and starch content during the night has revealed that the anticipation of dawn by the circadian clock coincides with the exhaustion of starch reserves in all light–dark cycles (Fig. 1). These results indicate a link between starch degradation and the timing of the circadian clock in *Arabidopsis*, and they offer an explanation for the abnormal starch degradation pattern in light–dark cycles that are longer or shorter than 24 h. Thus, starch degradation is programmed so that reserves would be exhausted 24 h after the previous dawn, regardless of the timing of the actual dawn experienced by the plant throughout its development.

Work on *Arabidopsis* mutant plants, in which the period of the clock is altered, confirmed these findings. The *Arabidopsis cca1/lhy* mutant lacks two transcription factors that control functioning of the clock. This loss does not eliminate clock function, but causes the clock to run fast. Analogous to wild-type plants in light–dark cycles longer than 24 h, *cca1/lhy* mutants fail to correctly anticipate dawn in a 24-h light–dark cycle (Fig. 1*c*). The expression peak of morning-regulated clock genes, indicating the anticipation of dawn, happens 4 h before the actual dawn. At exactly this time point, *cca1/lhy* mutant plants exhaust their starch reserves. Hence, despite the abnormal behavior of the circadian clock in these mutants, the link between the clock and starch degradation remains intact.

Because mobilization of starch reserves is linked to the timing of the circadian clock, the normal starch degradation pattern (that is, linear and near-complete degradation of starch over the course of the night period) can only occur if the length of the light–dark cycle matches the clock period (Fig. 1). It might be expected that abnormal rates of starch degradation and hence suboptimal utilization of carbon reserves affect plant productivity. This is indeed the case. Wild-type plants grown in 28-h light–dark cycles (14 h of light, 14 h of darkness) show symptoms of carbon starvation during the night, and plant growth is reduced. Providing sugar in the growth medium prevents carbon starvation during the night and restores normal plant growth. Carbon starvation in 28-h light–dark cycles can also be prevented by genetic modification of plants. *Arabidopsis* starch-excess mutants have defects in starch-related enzymes and show reduced rates of starch degradation. As a consequence, these mutants do not exhaust their starch reserves during the night in either 24-h or 28-h light–dark cycles. Unlike wild-type plants, starch-excess mutants show no symptoms of carbon starvation and no reduction in growth in 28-h light–dark cycles compared to 24-h light–dark cycles. Taken together, these results reveal that mistiming of starch degradation and the resulting carbon starvation during the night have negative effects on plant growth.

The current knowledge of the relationship between circadian timing, starch degradation, and growth can be summarized in a basic model (**Fig. 2**). According to the model, the circadian clock is entrained by light–dark cycles. A functional circadian clock allows the plant to anticipate the length of night, and starch degradation as a clock output is regulated accordingly. The correct timing of starch degradation ensures a continuous supply of carbon from starch throughout the night, thereby maximizing potential productivity.

Raising questions for the future. How the circadian clock is linked to starch degradation remains to be established. It is unlikely that control of starch degradation happens on a transcriptional level. Although transcripts for many of the enzymes of starch degradation in *Arabidopsis* show strong diurnal and circadian patterns of abundance, most of these enzymes show little or no change in protein abundance during the light–dark cycle. Thus, at least on a diurnal basis, control of starch degradation probably occurs by posttranslational mechanisms. Regulatory mechanisms such as redox activation, allosteric regulation by metabolites, reversible phosphorylation, and protein–protein interactions have been shown to influence the activity of several starch-related enzymes. However, the significance of these mechanisms for control of flux through starch degradation in vivo remains to be discovered. Elucidation of the signaling pathway that links the clock, located in the nucleus, to the posttranslational control of enzymes in the chloroplasts presents an interesting challenge for the future.

In addition, it is still unknown which regulatory mechanisms are used to balance carbon availability and plant growth. When plants are transferred to an unexpected early night, the rate of starch degradation is lower than during the previous night. Thereby, plants ensure a continuous carbon supply from starch throughout the night, despite the lower starch amount at dusk and a longer night. However, because of the lower rate of starch degradation, the carbon availability during the night is reduced, and plant growth needs to be adjusted accordingly. Indeed, measurements of root growth of *Arabidopsis* seedlings grown in 12-h light/12-h dark cycles and shifted to an early night after only 8 h of light show that root growth is adjusted immediately. After the transfer to an early night, root growth slows down and continues at a lower level relative to the previ-

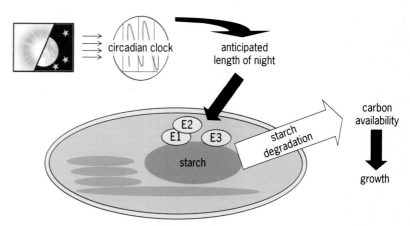

Fig. 2. A proposed relationship between the circadian clock, starch degradation, and growth at night. The circadian clock is entrained by light signals, allowing the anticipation of dusk and dawn, and the length of the night. Information from the circadian clock is transmitted via an unknown signaling pathway to the chloroplast, where it is used to modulate the rate of degradation of the starch granule. Regulation of the starch degradation rate occurs most likely by posttranslational modification of starch-degrading enzymes (schematically represented by E1–E3). The enzyme or enzymes subject to modulation remain to be discovered. Modulation of the rate of starch degradation ensures that substrates for metabolism and growth are available throughout the night.

ous night. How the root perceives signals of lower carbon availability in leaves and how signals are integrated to regulate growth are not known.

Progress made in understanding the timing of starch degradation during the night raises another important question: How do plants measure their starch reserves? To achieve a linear and near-complete utilization of starch during the night, anticipation of the length of the night is not sufficient by itself. Plants also need to sense the amount of starch present in the chloroplasts at dusk. The correct rate of starch degradation must be set by a molecular division of the starch amount according to time. How the plant cell achieves this calculation on the molecular level remains elusive.

In summary, consideration of how plants survive the night raises many new and complex questions about the relationship of photosynthesis, carbon storage and utilization, and plant productivity. The research on *Arabidopsis* starch metabolism thus illustrates the importance of considering these questions in attempting to increase crop productivity.

For background information *see* AGRICULTURAL SCIENCE (PLANT); BIOTECHNOLOGY; CARBON; CIRCADIAN CLOCK (PLANTS); PHOTOPERIODISM; PHOTOSYNTHESIS; PLANT DEVELOPMENT; PLANT GROWTH; PLANT METABOLISM; PLANT PHYSIOLOGY; STARCH in the McGraw-Hill Encyclopedia of Science & Technology. Alexander Graf

Bibliography. A. Graf et al., Circadian control of carbohydrate availability for growth in *Arabidopsis* plants at night, *Proc. Natl. Acad. Sci. USA*, 107:9458-9463, 2010, DOI:10.1073/pnas.0914299107; S. L. Harmer, The circadian system in higher plants, *Annu. Rev. Plant Biol.*, 60:357-377, 2009, DOI:10.1146/annurev.arplant.043008.092054; M. Stitt and S. C. Zeeman, Starch turnover: Pathways, regulation and role in growth, *Curr. Opin. Plant Biol.*, 15:282-292, 2012, DOI:10.1016/j.pbi.2012.03.016.

Optical clocks and relativity

More than 100 years after its introduction, Einstein's theory of relativity continues to strain human imagination. Among physical theories, it is particularly counterintuitive because it challenges assumptions that seem self-evident from our daily experience. An important example of this is the notion that the time elapsed between two events depends on the reference frame of the observer. In relativity the lengthening of a time interval in the frame of one observer compared to another is called "time dilation."

Twin paradox. As a way of explaining these ideas we resort to "gedanken" or "thought" experiments that may be difficult to perform in practice but can be analyzed directly in the context of relativity and tend to exaggerate relativistic effects so that the implications become clear. A famous example is the so-called twin paradox. This story imagines one twin who boards a spaceship and travels at high velocities to a distant star while the other twin stays at home. After reaching the star, the traveling twin turns around and travels back, again at speeds approaching the speed of light ($c = 3.0 \times 10^8$ m/s or 6.7×10^8 mph). On reuniting, the two twins discover an amazing thing—that the twin who stayed home is older than the spacefaring twin. How much older depends on how fast the spaceship traveled and how far away the star is from Earth, but there is no fundamental limit. The spacefaring twin might have taken a day trip while the Earthbound twin lived a lifetime.

The twin paradox is an effect of special relativity, which deals with physical systems in relative motion. The extension of special relativity to include gravitational effects is called general relativity. A different version of the twin paradox arises in this context. This time one twin leaves the other at home near sea level and travels to the top of a mountain to live. When they reunite they discover that the twin who lived on the mountaintop, farther from the center of earth's gravity, is a bit older.

While the idea of the twin paradox is astounding, a quantitative analysis of these relativistic effects tells a different story: For any conditions we will experience in our lives, these effects are tiny. For example, replace the twin's space trip with a round-trip flight from New York to Los Angeles and the difference in age becomes a mere 20 billionths of a second. Likewise, living on the summit of Mount Everest for 30 years would only make someone a millisecond older than their twin at sea level.

Atomic clocks. With the small scale of these relativistic effects, it is clear that detecting them requires extremely accurate clocks. Driven by numerous technological and scientific applications, the accuracy of clocks has steadily improved over time. In general, the most accurate clocks are oscillator clocks, in which the clock-ticks are synchronized to the oscillation of a stable physical system such as the pendulum in a grandfather clock or the microscopic vibration of a quartz crystal. In the search for ever more accurate clocks a clear trend has emerged—the

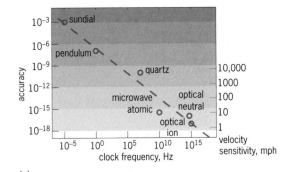

(a)

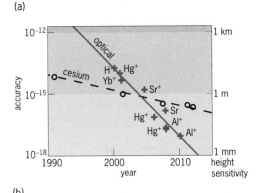

(b)

Fig. 1. Clock accuracy. (*a*) A comparison of different clocks. The clock accuracy is plotted versus its frequency on a log-log scale. The right axis label puts the accuracy in terms of special relativistic time dilation with the minimum resolvable velocity difference for a given accuracy beginning from 10,000 mph (approximately the velocity of a rocket) going down to 1 mph (a slow walk). (*b*) Progress in atomic clock accuracy over the last two decades. The blue crosses represent optical clock measurements and are labeled by the atomic species used in the clock. The right axis label puts the accuracy in terms of gravitational time dilation with the minimum resolvable height difference close to the Earth's surface for a given accuracy.

higher the frequency of oscillation, the more accurate the clock (**Fig. 1a**).

In the 1950s, clocks took a big step toward higher frequency and improved accuracy with the development of atomic clocks operating at as many as 10 billion hertz (1 Hz = 1 cycle per second). An atomic clock looks at the periodic evolution of electrons in atoms as a reference for a stable frequency. There are several important advantages of using atoms as clocks. First, they exhibit electronic resonances with extremely high quality factors, meaning that the oscillation of the electron continues for many cycles before being lost. While a pendulum clock will slow after some number of cycles because of mechanical friction or other effects, modern atomic clocks can undergo as many as a million billion oscillations before skipping a beat. Secondly, every atom of a particular element and isotope is the exact same. Therefore, to the extent that we can isolate the atom from the environment, all clocks based on a particular electronic resonance will operate at exactly the same frequency. This makes them naturally suited for establishing a universal time standard. Finally, the electronic resonances in an atom can be highly isolated from the environment, a fact which is critical for their high accuracy. The measurement of environmental effects on the clock frequency is a central task in improving atomic clock performance, and the steady improvement of cesium clocks over the last half century has been largely due to more careful isolation of the atoms from frequency shifting effects (Fig. 1*b*).

Atomic clocks and time dilation. Atomic clocks have been used in the past to precisely measure time dilation due to relativity. The experiments approach an ideal realization of the twin paradox, in which two nearly identical atomic clocks play the roles of the two twins. Over the years, the traveling "twin" has been placed in everything from an airplane to a particle accelerator and, so far, the results have been consistent with the predictions of relativity. One spectacular and successful experiment in 1976 launched a hydrogen maser clock aboard a Scout rocket while it was compared to an identical clock on the ground. By reaching an elevation of about 10,000 km above Earth's surface, this experiment produced relatively large effects from gravitational time dilation and was able to confirm the theory to greater accuracy than had been achieved before, an achievement with atomic clocks that remains unsurpassed now, more than 30 years later. Since that time, particularly with the development of the Global Positioning System (GPS), the use of atomic clocks in space has become widespread. Accounting for relativistic effects is, in fact, necessary for the GPS system to function.

Optical atomic clocks. It was recognized early during the development of atomic clocks that greater accuracy could be achieved by going to yet higher frequencies. Optical resonances accessible by laser radiation can be more than 100,000 times higher in frequency than their microwave counterparts. Furthermore, some atoms exhibit optical resonances with extremely high quality factors. Everything else being equal, this translates to much higher accuracy, but realizing such high performance presents a new set of technical challenges.

Producing a laser stable enough to probe the narrow optical resonances is a central challenge, which has been met by locking the laser frequency to the resonance of a high-quality-factor, isolated optical cavity. Current state-of-the-art optical cavities can produce laser radiation with a linewidth below 1 Hz at a frequency of 10^{15} Hz.

A second problem arises from the atom's motion, which must be strongly suppressed in order to avoid uncontrolled relativistic effects. For example, a free atom will gain enough momentum on absorbing a single optical photon to cause a significant shift to the clock frequency. The solution here is to tightly confine the atoms in a trap so that the radiation pressure has a negligible effect on the atom's motion. In fact, a similar radiation pressure effect is used to cool the atoms through now well-established laser cooling techniques.

Finally, before stable optical signals can serve as clocks, their oscillations must be counted. But how do we count laser oscillations at a frequency of a million billion hertz? The answer to this question

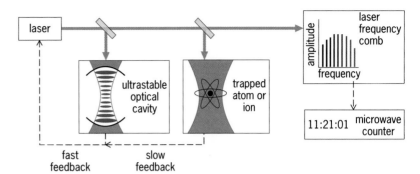

Fig. 2. Schematic diagram of an optical atomic clock.

came with the invention of a specialized laser called a femtosecond frequency comb, which earned the Nobel prize in 2005 for its coinventors John Hall and Theodor Hänsch. This laser produces short pulses (less than 1 ps duration) at a repetition rate stabilized to the optical clock signal. It thereby divides the optical frequency to a countable microwave frequency while maintaining the original optical stability.

With the major technical hurdles overcome, several research groups around the world have succeeded in producing more and more accurate optical clocks (Fig. 1*b*). These clocks all operate on the same basic principle (**Fig. 2**). The clock laser is first stabilized to a high-quality-factor optical cavity and is then used to probe a resonance in one or more trapped atoms. The absorption of photons when the laser is resonant with the atomic frequency provides a signal that can be used to steer the laser frequency. The atom-stabilized laser frequency can then be divided with a femtosecond frequency comb and counted with a microwave counter.

Aluminum ion clock. The current state-of-the-art optical clocks outperform microwave clocks by a factor of about 30 in terms of accuracy. This corresponds to the clock neither gaining nor losing a picosecond (10^{-12} s) over the course of 1 day. The particular clock design that has achieved this accuracy bases its frequency on an optical resonance in a single trapped aluminum ion. Although Al^+ had been proposed as early as 1967 as an excellent choice as a frequency standard, it has the additional technical complication, not present in the other atomic clock species, that it cannot be directly laser cooled. Therefore the clock operates with not one, but two ions. The second ion serves as a tool for laser cooling and for probing the Al^+ state after it interacts with the clock laser. Because of this unique aspect of the aluminum ion clock, it is sometimes called the "quantum logic" clock referring to the transfer of information between the two ion species.

In addition to the high-quality-factor optical resonance that motivated its proposal as a clock atom, recent work on Al^+ has revealed an intrinsically lower sensitivity to ambient electric fields compared to other atoms and ions being studied. Since the effects of stray electric fields in the vicinity of the clock atoms can be a leading limitation to clock accuracy, the low sensitivity of Al^+ is an important advantage,

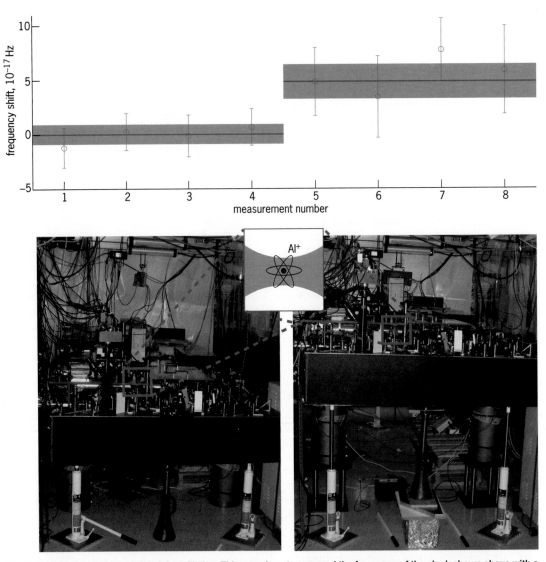

Fig. 3. Measurement of gravitational time dilation. This experiment compared the frequency of the clock shown above with a separate Al$^+$ clock not shown. After the initial set of frequency comparisons (1–4) that effectively synchronized the two clocks, this clock was moved up by 33 cm and a new set of measurements was taken revealing a slight shift due to gravitational time dilation. The horizontal bars represent the mean and standard deviation of all four measurements together for the two heights.

and is largely responsible for its current lead in performance.

Experimental results. While the new generation of optical atomic clocks are not robust enough yet to survive a trip to space, with the increased accuracy and stability, laboratory experiments are capable of posing the question: What is the smallest relativistic effect we can observe?

Experiments measuring this have been performed with two Al$^+$ clocks at the National Institute of Standards and Technology (NIST) in Boulder, Colorado. Like previous time dilation experiments, these measurements approach a real-world realization of the twin paradox. In the first experiment one ion was held nearly motionless while the other was set into harmonic motion, equivalent to the traveling twin taking repeated round-trip journeys. After less than an hour of moving back and forth with an average speed no faster than a person can sprint, the relative youth of the traveling twin became apparent as that

clock had recorded several hundred fewer ticks. If this experiment were extended to the current accuracy limit of the Al$^+$ clock, in the span of one day the time dilation that results from a comfortable stroll could be detected.

In a second experiment, gravitational time dilation was observed on a laboratory scale (**Fig. 3**). For this experiment an initial series of measurements were made to synchronize the two clocks. Then, one clock was elevated by a distance of 33 cm (1.1 ft). A second series of measurements confirmed that the elevated clock ran slightly faster than the clock that stayed closer to the Earth.

Future of optical clocks. These measurements reveal, in a textbook experiment, the effects of Einstein's theory on scales of relative velocity and height that we routinely experience in our lives. More importantly, this shows the capability of a new generation of atomic clocks to make contributions to both fundamental science and technology.

Already, around the world, there are numerous projects underway to extend the observation of gravitational time dilation to greater precision and toward various applications. For example, in Germany, a fiber-optic link capable of maintaining the stability of optical clock signals over 920 km (572 mi) between laboratories in Garching and Braunschweig was recently demonstrated. Other projects aim to build more robust, portable optical clocks that could be transported for the purpose of clock comparisons. Such comparisons between geographically distinct locations will enable new observations of Earth's dynamic gravitational properties. Not only can the static mass distribution of Earth be observed, but its continual redistribution due to geological, hydrological, and atmospheric effects could be measured. Other ambitious projects have been proposed to send an optical clock into space, where comparisons with clocks on the ground could constitute the most stringent test yet to the predictions of relativity.

With the current pace of activity surrounding optical clocks and their use in studying relativistic phenomena, it seems we are at the beginning of a rich new field of research. It is almost certain that optical clocks will continue to improve in accuracy in the coming years allowing them to study relativistic effects on a finer scale. It is also clear that further refinement will carry their signals out of the laboratory and across distances where geophysical effects become more important. What is less certain and what makes this field exciting is the new insights these measurements might bring and, then, what new questions might be asked.

For background information *see* ATOMIC CLOCK; CLOCK PARADOX; FREQUENCY COMB; GRAVITATIONAL REDSHIFT; LASER SPECTROSCOPY; LINEWIDTH; PARTICLE TRAP; Q (ELECTRICITY); RELATIVITY; SATELLITE NAVIGATION SYSTEMS in the McGraw-Hill Encyclopedia of Science & Technology.   David Hume

Bibliography. C. W. Chou et al., Optical clocks and relativity, *Science*, 329(5999):1630–1633, 2010, DOI:10.1126/science.1192720; T. Jones, *Splitting the Second: The Story of Atomic Time*, Institute of Physics Publishing, Bristol, U.K., 2000; D. Kleppner, Time too good to be true, *Phys. Today*, 59(3):10–11, 2006, DOI:10.1063/1.2195297; C. M. Will, *Was Einstein Right? Putting General Relativity to the Test*, 2d ed., Basic Books, New York, 1993.

Origins of cooking

How an animal obtains and consumes its food is a defining characteristic of a species. It also provides information that can be used to broadly categorize animals. Many primates, for example, are omnivorous and consume a variety of food types, including fruits, leaves, and insects. Omnivorous primates have ecological and evolutionary relationships with local plant species that are different from those of local carnivores. Diet also plays an important role in shaping the anatomy and physiology of animals. Some primate species, for instance,

have independently evolved adaptations to digest foliage. Species within the Old World monkey group Colobinae are foregut fermenters with compartmentalized stomachs, whereas hindgut fermenters such as some strepsirrhine folivores (leaf-eating lemurs, including species of Indriidae) have compartmentalized ceca and long and coiled colons. Not all dietary adaptations are found in the gut; changes also occur in the limbs, salivary glands, teeth, and skull. Some primates also use tools to obtain and process food. Although basic tool use (stone hammers and sticks) has been observed in wild capuchin monkeys (New World monkeys of the subfamily Cebinae), it has been extensively documented in great apes (Hominidae). For example, chimpanzees (genus *Pan*) [**Fig. 1**] create termite-collecting sticks by using their teeth to clip off leaves. Although chimpanzees eat fruits, leaves, and insects, they also hunt other primates, and they even use sticks that have been fashioned into spears in this activity. Chimpanzees also process food to a surprising degree, having been observed pounding oil-palm stems, mashing and soaking fruit in water, and chewing raw meat with tough leaves that have no nutritional value (which apparently aids in mastication). Processing food increases the number of freely available calories compared with unprocessed food of the same type. Therefore, food-processing behaviors are important for individual fitness because, over time, increased caloric intake influences reproductive success.

The closest living relatives of chimpanzees are humans (*Homo sapiens*), who use tools and technology to obtain and process food more extensively than any other animal. Perhaps the most emblematic of these tools and technologies is fire. Fire is a quintessential human technology and is at the heart of cooking traditions around the world. Not all of our food is thermally processed by fire (heat), but some food in every culture is processed in some way. In fact,

Fig. 1. **Chimpanzee from Kanyawara community, Kibale National Park, Uganda.** (*Photograph taken by Ronan Donovan and supplied by Richard Wrangham*)

there is good evidence to suggest that our bodies have adapted anatomically and physiologically to eat cooked food. Specifically, modern humans must eat cooked and processed food to obtain enough calories for survival over long periods of time. Cooking also gelatinizes starch, denatures proteins, and mitigates food-borne illness, which are all important benefits that would have increased the reproductive success of the hominins that invented cooking.

When and why did cooking evolve? For animals, eating is about obtaining energy, and this energy is used for all biological functions, from growth and development to movement and other behaviors. The amount of energy that an animal can sequester has a large impact on its survival and reproductive success. Cooking food helps in this regard. Cooking requires fire, though, and there is debate about its oldest controlled use. A widely accepted date for the earliest evidence of controlled fire is 790,000 years ago, at Gesher Benot Ya'aqov in Israel, although various earlier dates have been proposed in Africa's Lower Paleolithic. However, rigorously establishing whether a fire was controlled is difficult, and evidence of fire is not necessarily evidence of cooking. Recent advances in microscopic analysis of sediments from Wonderwerk Cave in South Africa strongly support the hypothesis that controlled fire was a known technology as far back as 1 million years ago. This evidence suggests that *H. erectus* may have been cooking its food, which is a hypothesis supported by phylogenetically analyzing rates of change in feeding time and molar size in primates (see below). If these hypotheses are correct, then modern humans were not the inventors of cooking; instead, it was a skill that we inherited from our ancestors.

Controlled fire use was a critical turning point during the evolution of hominins, but there are other forms of evidence to evaluate when cooking was first practiced by the ancestors of modern humans. For example, fossils of *H. erectus* show signs of reduced chewing-induced strain on the skull, including shortening of the face, which implies a diet of softened food during development. *Homo erectus* also had smaller postcanine teeth and probably had smaller guts (based on the size of the rib cage) compared with earlier hominins, suggesting that *H. erectus* individuals chewed less and consumed more easily digestible food than did their relatives. These data might imply that *H. erectus* had a lower energy budget than its earlier relatives, but other anatomical evidence suggests otherwise, and this incongruity provides more support for the hypothesis that *H. erectus* was cooking. *Homo erectus* had a larger body that was better suited to long-duration aerobic exercise (long-distance running) and had a larger, energy-hungry brain compared with earlier hominins. It is possible that these anatomical changes evolved because of a meat-enriched diet. This hypothesis, however, does not explain the small gut size of *H. erectus*, whose diet would still contain large quantities of plant material. More importantly, increased meat consumption likely occurred in hominins prior to *H. erectus*. For example, stone-tool cut marks for flesh removal and percussion marks for marrow access suggest that hominin (*Australopithecus afarensis*) diets may have shifted to include meat at least 3.39 million years ago.

It is also possible that the smaller guts and teeth of *H. erectus* evolved because of refined food processing. For example, modern hunter-gatherers pound seeds, tubers, and meat, and they allow buried fruits, fish, and meat to partially rot before consumption. These techniques increase the available number of calories for absorption by the body. Importantly, studies have shown that a substantial increase in calorie uptake is provided when these techniques are combined with cooking. Assuming that *H. erectus* had an energy budget similar to modern humans, it is unlikely that *H. erectus* individuals could have survived by consuming raw food alone. People committed to a raw food diet, even with the benefit of modern electronic food-processing equipment, suffer from insufficient caloric intake, resulting in, among other things, cessation of menstruation.

Evolutionary (phylogenetic) analysis. One of the difficulties involved with understanding the evolution of cooking, or any other trait, in hominins is that some hominin relationships are controversial. Evolutionary trees (phylogenies) provide the context for understanding trait evolution (**Fig. 2**). The hominin phylogeny is inferred from anatomical characters, which are notorious for evolving convergently and therefore confusing phylogenetic relationships. Recent advances allow traits to be analyzed over a distribution of many trees. An evolutionary outlier test can be performed that takes into account this phylogenetic distribution (uncertainty). If the value of a trait within a species cannot be explained by its phylogenetic position, branch lengths, and one or more independent variables, then the trait has evolved substantially in the species compared with its close relatives. Importantly, the evolutionary outlier test can be used to study traits (for example, cooking) that evolved only once in single lineages. This approach has been

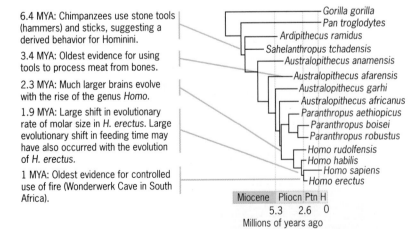

6.4 MYA: Chimpanzees use stone tools (hammers) and sticks, suggesting a derived behavior for Hominini.

3.4 MYA: Oldest evidence for using tools to process meat from bones.

2.3 MYA: Much larger brains evolve with the rise of the genus *Homo*.

1.9 MYA: Large shift in evolutionary rate of molar size in *H. erectus*. Large evolutionary shift in feeding time may have also occurred with the evolution of *H. erectus*.

1 MYA: Oldest evidence for controlled use of fire (Wonderwerk Cave in South Africa).

Gorilla gorilla
Pan troglodytes
Ardipithecus ramidus
Sahelanthropus tchadensis
Australopithecus anamensis
Australopithecus afarensis
Australopithecus garhi
Australopithecus africanus
Paranthropus aethiopicus
Paranthropus boisei
Paranthropus robustus
Homo rudolfensis
Homo habilis
Homo sapiens
Homo erectus

Miocene | Pliocn | Ptn H
5.3 | 2.6 | 0
Millions of years ago

Fig. 2. Evolutionary (phylogenetic) tree of great apes and extinct hominins along the human lineage (*right*). Important milestones related to the evolution of diet and cooking are also noted (*left*). Abbreviations: MYA, million years ago; Pliocn, Pliocene; Ptn, Pleistocene; H, Holocene.

used to study feeding time and molar size in primates. Modern humans have been found to spend an order-of-magnitude less time feeding than predicted by body mass and phylogenetic position (4.7% versus the predicted 48% of daily activity). Because this is a phylogenetic test, the result implies a substantial change in feeding time's evolutionary rate along the human branch of the great ape tree. But when did this change occur? Using the hominin phylogeny, the time that *H. erectus* spent feeding is predicted to be similar to the time that modern humans spent feeding. As noted previously, members of the genus *Homo* have smaller postcanine teeth compared with earlier hominins. A phylogenetic outlier test shows that this change can be explained by the rate of primate craniodental and body size evolution for members of early *Homo* (*H. habilis* and *H. rudolfensis*). However, shifts in the size of postcanine teeth in *H. erectus*, *H. neanderthalensis*, and *H. sapiens* cannot be explained by the rate of primate craniodental and body size evolution. Because postcanine teeth are used for chewing, this result, along with the large evolutionary shift in feeding time, suggests that *H. erectus* was feeding in dramatically different ways compared with earlier hominins. Thus, these results imply that *H. erectus* may have been cooking.

Cooking is about more than calories. An animal can spend less time feeding if the food consumed is richer in freely available calories compared with alternatives. In terms of time commitment, cooking is somewhat circular because part of the time saved by eating cooked food is spent cooking instead. However, the activity of cooking and processing food also has tremendous social value in human communities. It is associated with rituals from birth to death, is a shared part of daily family experience, helps bond communities together, and is a defining cultural attribute. With the controlled use of fire dating to potentially 1 million years ago, the importance of culinary traditions is perhaps more ancient than previously recognized. The invention of cooking, therefore, has implications beyond increasing calorie absorption, but the social implications of cooking in extinct hominins remain unknown.

For background information *see* ANTHROPOLOGY; BEHAVIORAL ECOLOGY; EARLY MODERN HUMANS; FIRE; FOOD; FOSSIL; FOSSIL HUMANS; PHYSICAL ANTHROPOLOGY; PREHISTORIC TECHNOLOGY; SOCIOBIOLOGY in the McGraw-Hill Encyclopedia of Science & Technology.   Chris Organ

Bibliography. A. G. Henry et al., Microfossils in calculus demonstrate consumption of plants and cooked foods in Neanderthal diets (Shanidar III, Iraq; Spy I and II, Belgium), *Proc. Natl. Acad. Sci. USA*, 108:486–491, 2011, DOI:10.1073/pnas.1016868108; S. P. McPherron et al., Evidence for stone-tool-assisted consumption of animal tissues before 3.39 million years ago at Dikika, Ethiopia, *Nature*, 466:857–860, 2010, DOI:10.1038/nature09248; C. Organ et al., Phylogenetic rate shifts in feeding time during the evolution of *Homo*, *Proc. Natl. Acad. Sci. USA*, 108:14555–14559, 2011, DOI:10.1073/pnas.1107806108; R. Wrangham, *Catching Fire: How Cooking Made Us Human*, Basic Books, New York, 2009; R. Wrangham and N. Conklin-Brittain, Cooking as a biological trait, *Comp. Biochem. Physiol. A Mol. Integr. Physiol.*, 136:35–46, 2003, DOI:10.1016/S1095-6433(03)00020-5.

Photoionization, fluorescence, and inner-shell processes

Much of the light (photons) from stars ultimately originates from the photon emission from a particular atom or molecule at an exact wavelength (color). These wavelengths allow us to uniquely identify the atoms and molecules that distant stars are composed of without the considerable effort required to travel there. For example, closer to home, the vivid green and blue colors observed in the aurora borealis (the northern lights) correspond to emissions from oxygen and nitrogen, respectively. Most of the known matter in the universe is in a plasma state (the state of matter similar to gas in which a certain portion of the particles are ionized) and our information about the universe is carried by photons (light), which are dispersed and detected for example by the orbiting *Chandra X-ray Observatory*. When photons travel through stellar atmospheres and nebulae (interstellar clouds of gas and dust), they are likely to interact with matter and therefore with ions. This makes the study of photoionization (the physical process in which an incident photon ejects one or more electrons from an atomic or molecular system) of atoms, molecules, and their positive ions very important for astrophysicists, helping them to interpret stellar data (spectroscopy). To measure the chemical evolution of the universe and then understand its ramifications for the formation and evolution of galaxies and other structures is a major goal of astrophysics today. The answers to these questions ultimately address human and cosmic origins. Our ability to infer chemical abundances relies extensively on spectroscopic observations of a variety of low-density cosmic plasmas including the diffuse interstellar and intergalactic media, gas in the vicinity of stars, gas in supernova remnants, and gas in the nuclei of active galaxies. Cosmic chemical evolution is revealed through emission or absorption lines from cosmically abundant elements.

The quantitative information we have about the cosmos is the result of spectroscopy from the ground and an array of orbital missions: *International Ultraviolet Explorer* (*IUE*); *Hubble Space Telescope* (*HST*); *Extreme Ultraviolet Explorer* (*EUVE*); *Infrared Space Observatory* (*ISO*); Hopkins Ultraviolet Telescope (HUT); *Orbiting and Retrievable Far and Extreme Ultraviolet Spectrograph-Shuttle Pallet Satellite* (*ORFEUS-SPAS*); *Solar and Heliospheric Observatory* (*SOHO*); *Far-Ultraviolet Spectrograph Explorer* (*FUSE*); *Chandra X-ray Observatory*; *XMM-Newton*; *Galaxy Evolution Explorer* (*GALEX*); *Constellation-X*, now the *International X-ray Observatory* (*IXO*); *Suzaku*, a reflight of *ASTRO-E*; and soon, *ASTRO-H*.

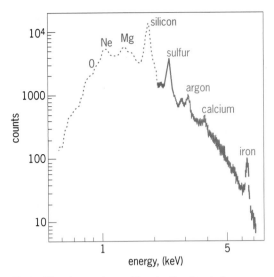

Fig. 1. *Chandra* spectrum of the star Cassiopeia A, illustrating characteristic x-ray peaks.

Satellites launched in 1987 indicated that the x-ray spectra of active galactic nuclei (AGN) were not featureless continua, but contained structure associated with various atomic processes. Iron emission lines from inner shells at wavelengths of 0.1936 and 0.1940 nm (1 nm = 10^{-9} m) were observed. These lines are characteristic of x-ray-illuminated cold material, and act as an indication of the amount of cool material illuminated by the AGN. X-ray satellites, such as *Chandra* and *XMM-Newton*, launched over a decade ago, have provided for the first time the ability to do x-ray spectroscopy at a very high resolution. Elements of prime interest are carbon, nitrogen, oxygen, neon, magnesium, silicon, sulfur, calcium, argon, and iron, primarily at the *K*- and *L*-edge energies, 0.2–12 keV (**Fig. 1**). The *Chandra* and *XMM-Newton* satellites currently provide a wealth of spectral data, which spectral modeling codes struggle to interpret due to a lack of quality atomic data.

Applications of photoionization of atomic elements. Space-based ultraviolet (UV) observations of emission and absorption lines from ions of carbon, nitrogen, oxygen, sulfur, and silicon in photoionized sources play an important role in addressing many of the astrophysical issues listed above. *Hubble Space Telescope* (*HST*) observations of carbon and sulfur ions in clouds have been used to study the origin and chemical evolution of gas at large distances from galaxies. Observations have provided a wealth of photon data on the ionizing spectrum and discriminated between an extragalactic radiation field due to AGN or O-type stars from starburst galaxies. The x-ray spectra of the outflowing gas in Seyfert galaxies have been measured by *Chandra*; absorption edges and dozens of x-ray emission lines are observed in their spectra.

Selected types of atomic processes with photons. The atomic processes occurring for a photon (light) with sufficient energy ($h\nu$, where h is Planck's constant and ν is the frequency) to remove an outer-

or inner-shell electron in a collision with a complex atom or ion containing a specific number of electrons are:

1. Valence (outer) shell single photoionization.

2. Photoexcitation (the mechanism of electron excitation by photon absorption, when the energy of the photon is too low to cause photoionization) and photoabsorption (**Fig. 2**).

3. Photoionization with photoexcitation.

4. Fluorescence (emission of a photon) from the resulting atomic ion (**Fig. 3**). The hole (vacancy in the innermost shell) is caused by the incoming photon having sufficient energy to remove it.

5. Auger processes (atomic transitions that do not emit light), where a secondary electron is ejected due to removal of an inner-shell electron. A photon is ejected by core relaxation of the subsequent ion. The process was named after Pierre Victor Auger.

In the vacuum-ultraviolet wavelength region, the photoionization process is the dominant interaction

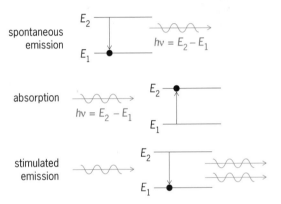

Fig. 2. Energy level diagrams of an atom illustrating the origin of line spectra. The absorption of a photon of energy $h\nu$ causes atomic transition from levels E_1 to E_2. Transition from level E_2 to E_1 causes emission of a photon.

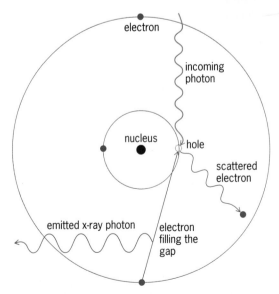

Fig. 3. A pictorial representation, of x-ray fluorescence in atomic systems. A hole is created by removal of an innermost electron by an incoming photon. An outer electron then fills this hole, causing the emission of a second photon, the emitted x-ray photon (illustrated by the purple photon trace), which is the x-ray fluorescence.

between the photon and atoms (or electrons). In the soft x-ray region, that is, photon energies in the range 0.2–2 keV, or wavelengths in the region of approximately 0.5–4.5 nm, the Auger process dominates. Theoretically, sophisticated numerical methods are required to solve the quantum mechanical equations of motion governing the dynamics and interactions of the electron and photons to obtain the atomic collision observables. Continuous or line spectra may be obtained, the latter consisting of emission or absorption lines (**Fig. 4**), for various wavelengths across the electromagnetic spectrum (**Fig. 5**).

Photoionization of low-mass atomic elements. The body of information concerning photoionization processes in atomic ions is mainly theoretical knowledge that requires benchmarking against experimental measurements from modern-day synchrotron facilities, such as the Advanced Light Source (ALS), SOLEIL, ASTRID, and PETRA, which provide electromagnetic radiation of high intensity and coherence. The motivation underlying these studies is to obtain a better understanding of the electron-electron interactions occurring in high-temperature environments, stellar atmospheres, and laboratory plasmas.

Knowledge of the photoionization cross section of the two-electron lithium ion provides an essential benchmark for future photoionization studies on more abundant highly ionized two-electron species such as carbon, nitrogen, oxygen, and neon. Such details are important in determining the mass of missing baryons (heavy subatomic particles made up of quarks, which are elementary particles and a fundamental constituent of matter) in the x-ray spectrum of the warm-hot intergalactic medium (WHIM).

Two-electron promotions in the lithium ion produce doubly excited states (states lying above their lowest energy levels) that can autoionize (undergo transitions that do not emit light) to form a single-electron ion and an outgoing free electron. The strongest process from the ground state of the lithium ion is the population of these doubly excited resonance states. Experimental studies on the lithium ion in the late 1970s resolved the spectrum in the wavelength region spanning 20–5 nm, observing

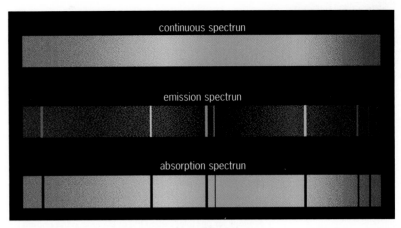

Fig. 4. Continuous spectrum and two types of line spectra.

strong peaks in the absorption spectrum that were attributed to these doubly excited states.

Two decades elapsed before the advent of third-generation synchrotron light sources produced high-resolution measurements on the lithium ion, first at the ALS and then at the Super-ACO facility. These experiments produced extremely high-quality data for several peaks in the spectrum of this system. This vast improvement in the magnitude of photon energy resolution yielded extremely accurate resonance energy positions, widths, and shapes of the resonances, providing stringent tests of theory (**Fig. 6**).

Additional information such as the lifetime of atomic states may be determined from sophisticated numerical methods in quantum physics. Line fluorescence yields ω (emission of light by the system that has absorbed it) may also be determined, assuming statistical population of the states. The fluorescence yield ω, together with branching ratios, serves as a quantitative measure of the decay mode of the atomic state.

Photoionization of atomic elements of intermediate mass. For atomic elements in the periodic table of intermediate charge value Z, relativistic effects become increasingly important. For comparison with high-resolution measurements from synchrotron

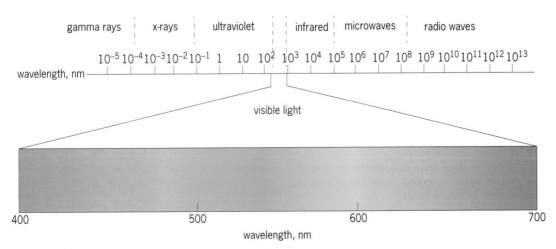

Fig. 5. The electromagnetic spectrum over a selected wavelength range.

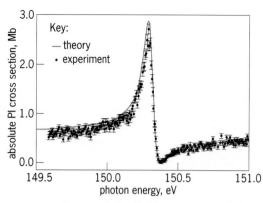

Fig. 6. Comparison of experimental and theoretical cross sections for photoionization (PI) of Li$^+$ ions in the photon energy range 149.5–151 eV. The measurements were obtained with an energy resolution of 38 meV at the Advanced Light Source (ALS). (*Adapted from S. W. J. Scully et al., Doubly excited resonances in the photoionization spectrum of Li$^+$: experiment and theory, J. Phys. B.: At. Mol. Opt. Phys., 39:3957–3968, 2006, DOI:10.1088/0953-4075/39/18/024*)

radiation facilities, state-of-the-art numerical methods that include semi-relativistic effects are necessary. This optimizes experimental beam time, as coarse energy scans may be used in resonance-free regions and fine scans at high resolution in energy ranges with dense resonance structure, as illustrated in recent work on Ar$^+$ ions at the ALS.

The interaction of photons with atomic ions is an important process in determining the ionization balance (the population ratio related to atomic collision processes) of the abundances of elements in photoionized astrophysical nebulae. In nebulae (interstellar clouds of gas and dust, containing hydrogen, helium, and other atoms and molecules, some of which are ionized), the formations of gas, dust, and other materials "clump" together to form larger masses, which attract further matter, and eventually will become massive enough to form stars.

Among atomic elements heavier than iron ($Z = 26$), known as trans-iron elements, neutron (n)-capture elements (for example, selenium, krypton, bromine, xenon, rubidium, barium, and lead) have been detected in a large number of ionized nebulae. Elements are produced by slow or rapid neutron-capture nucleosynthesis (the process of creating new atomic nuclei from preexisting nucleons). With the formation of stars, heavier nuclei were created from hydrogen and helium by stellar nucleosynthesis, a process that continues today. Some elements, those lighter than iron, are delivered to the interstellar medium (the matter that exists in the space between the star systems in a galaxy) in the last stages of evolution of dying low-mass stars in the nonexplosive ejection of the outer envelope gases of planetary nebulae before these stars continue to form white dwarfs (a small star composed of electron-degenerate material that no longer undergoes fusion reactions). Measuring the abundances of these elements helps to reveal their dominant production sites in the universe, and details of stellar structure, mixing, and nucleosynthesis.

The level of enrichment of individual elements is strongly sensitive to the physical conditions in the stellar interior. Uncertainties in the photoionization and recombination (the formation of neutral atoms from the capture of free electrons by the cations in a plasma) cross-section data of n-capture element ions can result in elemental abundance uncertainties of a factor of 2 or more.

Photoionization of atomic elements of heavy mass. For atoms or ions with charge greater than 30, relativistic and electron correlation effects are essential to include, to obtain accurate modeling parameters. Computationally, the number of atomic levels and coupled states involved dramatically increases.

To address the challenge of electrons or photons interacting with heavy atomic systems containing hundreds of levels resulting in thousands of scattering channels, efficient parallel numerical algorithms were developed. This allows the single-photon ionization modeling of trans-iron elements and opens the doorway to comprehensive large-scale calculations along isonuclear sequences (atomic systems comprising of the same nucleus but in successive stages of ionization, for example, Se$^+$, Se^{2+}, Se^{3+},...).

Selenium ions were chosen for study because they have been detected in nearly twice as many planetary nebulae (70 in total) as any other trans-iron element. Experimental photoionization cross-section data results for singly-ionized selenium obtained at the ALS in the near-threshold energy region show excellent agreement with theoretical results obtained from fully relativistic calculations.

Xenon and krypton ions are of importance in human-made plasmas such as light sources for semiconductor lithography (a process used in the electronic industry to selectively remove parts of a thin film), ion thrusters for spacecraft propulsion, and nuclear fusion plasmas. Xenon and krypton ions have also been detected in cosmic objects, planetary nebulae, and the ejected envelopes of low- and intermediate-mass stars. For an understanding of these plasmas, accurate cross sections are required for ionization and recombination processes that govern the charge balance of ions in plasmas. Krypton and xenon ions are also of importance in tokomak (a device that uses a magnetic field to confine a plasma in the shape of torus) plasmas for fusion physics. Photoionization experimental results from the ALS synchrotron radiation facility for singly-charged xenon ions in the near-threshold region show suitable agreement with theoretical results including fully relativistic effects (**Fig 7**).

X-ray spectroscopy of atomic elements. X-rays are short-wavelength electromagnetic radiation produced by the deceleration of high-energy electrons or by electronic transitions in the inner orbitals of atoms. The wavelength range of x-rays is approximately 10^{-15}–10^{-8} m; conventional x-ray spectroscopy is largely confined to the region of about 10^{-11}–2.5×10^{-9} m. X-ray spectroscopy is a form of optical spectroscopy that utilizes emission, absorption, scattering, fluorescence, and diffraction of x-rays.

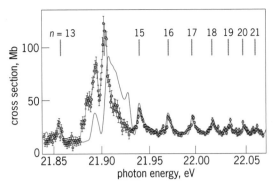

Fig. 7. Advanced Light Source (ALS) Xe$^+$ photoionization experimental cross-section data (green circles) for photon energies in the range 21.84–22.08 eV, at a photon energy resolution of 4 meV. Theoretical results (red line) at 4 meV include fully relativistic effects. The bars mark the energies of the $5p \rightarrow nd$ resonances in the Xe$^+$ spectra. (*Courtesy of B. M. McLaughlin and C. P. Ballance, Photoionization cross section calculations for the halogen-like ions Kr$^+$ and Xe$^+$, J. Phys. B: At. Mol. Opt. Phys., 45:085701, 2012, DOI:10.1088/0953-4075/45/8/085701*)

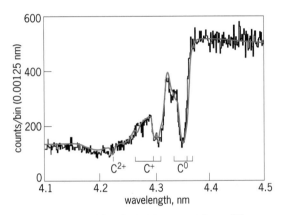

Fig. 8. Modeling of the observed x-ray spectrum of the bright extragalactic x-ray source blazar Makarian 421 near the carbon *K*-edge, obtained from the *Chandra X-ray Observatory*. The series of connected black bars are the observed spectrum. The red curve is the modeling of the observations by M. F. Hasoglu and collaborators. Wavelengths of transitions of carbon and its ions are also indicated. (*Courtesy of M. F. Hasoglu et al., K-shell photoabsorption studies of the carbon isonuclear sequence, Astrophys. J., 724:1296–1304, 2010, DOI:abs/1003.3639*)

Inner-shell (x-ray) photoionization processes play important roles in many physical systems, including a broad range of astrophysical objects as diverse as quasars, the atmospheres of hot stars, nebulae, novae, and supernovae. Inner-shell studies of carbon and its ions indicate that high-quality theoretical data are required to model the observations in the x-ray spectrum of the bright blazar Makarian 421, observed by the *Chandra X-ray Observatory* (**Fig. 8**). Abundances for carbon, nitrogen, and oxygen in their various ionized stages are essential for photoionization models applied to the plasma modeling in a variety of planetary nebulae. Nitrogen abundance in particular plays a fundamental role in studies of planetary nebulae because it is a key tracer of the carbon-nitrogen-oxygen processing.

Inner-shell photoionization of atomic nitrogen is of crucial importance to the energetics of the terres-

trial upper atmosphere and, together with photoionization of atomic oxygen and molecular nitrogen (which is the most abundant species), determines the ion-neutral chemistry and temperature structure of the upper atmosphere, ultimately through the production of nitric oxide (NO). The NO abundance is highly correlated with the soft x-ray irradiance (the power per unit area radiated by a surface), but uncertainties in the photoionization cross sections and solar fluxes remain. Highly accurate experimental results from the ALS for the photoionization cross section of atomic nitrogen in the vicinity of the *K*-edge, where atomic resonances features (peaks) are present in the spectrum, are well reproduced by theory (**Fig. 9**).

PETRA III, in Hamburg, Germany, which began operation in 2009, is the most brilliant storage-ring-based x-ray source in the world. This exceptionally high brilliance offers scientists outstanding experimental opportunities. PETRA III benefits researchers investigating very small samples or those requiring tightly collimated and very short-wavelength x-rays for their experiments. Investigations on the resonant fluorescence from highly charged ions of iron have been studied at energies of 6–7 keV, with signals from 2–6-electron iron ions being observed. Further studies plan to look at photon impact on these highly charged iron ions in the *K*-excitation channel (the case where one of the innermost electrons is excited) with subsequent autoionization.

Conclusions. In the past decade or more, observations by orbiting satellites coupled with major experimental advances from light sources have made it possible to determine accurate results for cross sections of photons interacting with atomic systems in their neutral, singly ionized, or highly ionized states.

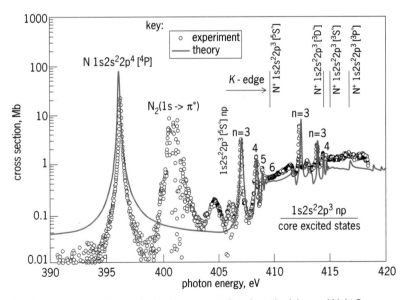

Fig. 9. Atomic nitrogen, photoionization cross sections from the Advanced Light Source (ALS) compared with theory. Theoretical results are convoluted with a 60-meV FWHM (full width at half maximum) Gaussian to simulate experiment. Experimental results include additional molecular components between 400 eV and 406 eV. Spectroscopic notation is to identify the molecular and atomic states of the peaks in the spectra. (*Courtesy of M. M. Sant'Anna et al., K-shell x-ray spectroscopy of atomic nitrogen, Phys. Rev. Lett., 107:033001, 2011, DOI:10.1103/PhysRevLett.107.033001*)

High-resolution cross-section measurements at third-generation light sources have pushed the boundaries of theoretical methods.

Major advances in state-of-the-art theoretical methods coupled with advances in computational architectures have opened the doorway to large-scale computations on elements across the periodic table, allowing the inclusion of fully relativistic effects. Various stages of ionization, necessary for the many applications in astrophysics, can be studied in the absence of experimental values. We therefore look forward to another decade of major advances in experiment and theory and the synergistic relationship between them.

For background information *see* ASTRONOMICAL SPECTROSCOPY; ATOMIC STRUCTURE AND SPECTRA; AUGER ELECTRON SPECTROSCOPY; AURORA; CARBON-NITROGEN-OXYGEN CYCLES; CHANDRA X-RAY OBSERVATORY; FLUORESCENCE; GALAXY, EXTERNAL; HUBBLE SPACE TELESCOPE; INTERSTELLAR MATTER; LIGHT; NEBULA; NUCLEOSYNTHESIS; PHOTOIONIZATION; PHOTON; PLANETARY NEBULA; PLASMA (PHYSICS); SYNCHROTRON RADIATION; ULTRAVIOLET ASTRONOMY; X-RAY ASTRONOMY; X-RAY SPECTROMETRY; X-RAY TELESCOPE; X-RAYS in the McGraw-Hill Encyclopedia of Science & Technology.

Brendan M. McLaughlin; Connor P. Ballance

Bibliography. J. Berkowitz, *Photoabsorption, Photoionization & Photoelectron Spectroscopy*, Academic Press, New York, 1979; A. M. Covington et al., Valence-shell photoionization of chlorine-like Ar$^+$ ions: Experiment and theory, *Phys. Rev. A*, 84:013413 (11 pp.), 2011, DOI:10.1103/PhysRevA.84.013413; G. Drake (ed.), *Springer Handbook of Atomic, Molecular and Optical Physics*, Springer, Berlin, 2006; R. E. Eisberg and R. M. Resnick, *Quantum Physics of Atoms, Molecules, Solids, Nuclei and Particles*, 2d ed., Wiley, New York, 1985; J. Kaastra and F. Paerels (eds.), *High-Resolution X-Ray Spectroscopy: Past, Present and Future*, Springer, Berlin, 2011; B. M. McLaughlin and C. P. Ballance, Photoionization cross section calculations for the halogen-like ions Kr$^+$ and Xe$^+$, *J. Phys. B: At. Mol. Opt. Phys.*, 45:085701, 2012, DOI:10.1088/0953-4075/45/8/085701; T. Ohashi, Future x-ray missions for high resolution spectroscopy, *Space Sci. Rev.*, 157:25–36, 2010, DOI:10.1088/0953-4075/45/8/085701.

Plant MAP kinase phosphatases

Cells are constantly monitoring their environment. If changes are perceived, a series of molecular events are triggered in order to adjust the cellular behavior according to the prevailing conditions. Within a cell, the information gathered by receptors is communicated, by a process called signal transduction, to molecules that directly control cellular functions. The addition or removal of a phosphate group (reversible phosphorylation) is a universal posttranslational protein modification that is used to pass on signals (**Fig. 1***a*). It has often profound effects on the structure and thereby the properties of the target protein, including activity, localization, stability, and molecular interactions. Protein kinases phosphorylate their targets, and protein phosphatases remove phosphate groups. Mitogen-activated protein (MAP) kinase (MAPK) cascades constitute fundamental pillars of signal transduction mechanisms in all eukaryotes, including plants. They are activated in the presence of environmental stresses (for example, ultraviolet radiation, cold, heat, heavy metals, salinity, pathogens, and wounding) or in response to developmental cues (such as hormones and peptide signals). MAPK activity controls a broad range of cellular processes, including transcription, cell division, cell cycle, and metabolism. Importantly, MAPK activity needs to be tightly regulated because unbalanced MAPK signaling can lead to detrimental effects, including cell death.

MAPK cascade principles. MAPK cascades involve sequential phosphorylation of three interlinked kinases: a MAPK kinase kinase (MAPKKK), a MAPK kinase (MAPKK), and a MAPK (Fig. 1*b*). Upon activation, that is, phosphorylation of the threonine and tyrosine amino acid residues within the activation loop, the terminal MAPK phosphorylates effectors that drive the cellular response (Fig. 1*b*). As sessile organisms, plants are unable to escape an ever-changing environment. To cope with such a lifestyle, plants evolved efficient mechanisms to sense and adapt their physiology, growth, and development, employing MAPK signaling pathways in many cases. The genome of the model plant *Arabidopsis thaliana* (thale cress) encodes about 60 MAPKKKs, 10 MAPKKs, and 20 MAPKs. However, MPK3, MPK4, and MPK6 (three members of the MAPK family) emerged as common elements activated by a broad range of stimuli, yet they elicit divergent cellular responses. This raises the important question of how specificity is achieved. MAPKKKs, which outnumber MAPKs by threefold, could mediate specific interactions and responses. Alternatively, transcriptional control or restriction of expression to particular cell types could serve to direct MAPK activity to an adequate target at the right time. In plants, these mechanisms remain to be broadly explored. Similarly, our understanding of the molecular mechanisms upstream of MAPKKK activation remains scarce, and only few downstream MAPK targets have been identified so far in plants.

Negative regulation of MAPKs by MAP kinase phosphatases. Besides specificity, another key aspect determining the output of a signaling pathway is the intensity and duration of MAPK activity. As phosphorylation is a reversible modification, phosphatases constitute crucial negative regulators of MAPKs. The *Arabidopsis* protein tyrosine phosphatase PTP1 and certain PP2C-type serine/threonine phosphatases have been shown to regulate MAPKs by dephosphorylation of the respective single phosphorylation site (either tyrosine or threonine within the activation loop). However, the present overview focuses on the dual-specificity-type MAPK phosphatases (MKPs) that are able to remove both phosphates of active MAPKs. There are five MKPs in *Arabidopsis*:

DUAL-SPECIFICITY PROTEIN TYROSINE PHOS-PHATASE 1 (DsPTP1), MAP KINASE PHOSPHATASE 2 (MKP2), INDOLE-3-BUTYRIC ACID RESPONSE 5 (IBR5), PROPYZAMIDE HYPERSENSITIVE 1 (PHS1), and MAP KINASE PHOSPHATASE 1 (MKP1), which are characterized by their catalytic domain, but differ greatly in their overall size and domain composition (**Fig. 2**). Analyses of plants with mutation in or mis-regulated expression of four of these MKPs (there are no data on DsPTP1 yet) indicate altered responses to several environmental stresses and developmental cues. This highlights the importance of MKPs in a broad range of MAPK signaling networks.

Abiotic stress responses controlled by plant MKPs. Knockout of the *MKP1* gene renders plants hypersensitive to ultraviolet-B (UV-B) radiation, which is a phenotype attributable to sustained MPK3 and MPK6 activation. Remarkably, *mkp1* mutants display increased tolerance to elevated salinity, but the corresponding MAPKs remain unknown in this case. These divergent phenotypes illustrate the complexity of MKP1-mediated signaling, which can promote or inhibit plant survival, depending on the nature of the abiotic stress stimulus. How are these opposing responses kept under control? How is one of them favored or repressed in a particular context? These are questions of extreme significance if we seek to gain in-depth understanding of the MAPK signaling mechanisms. Another response regulated by MKP1 is wounding: rice *mkp1* mutants show elevated expression of wound-responsive genes as a consequence of enhanced rice MPK3 and MPK6 activity. MKP2 is another dual-specificity phosphatase modulating MPK3 and MPK6 activities in vivo. Reminiscent of MKP1 function in UV-B signaling, ozone treatment renders *MKP2*-knockdown plants hypersensitive as a result of prolonged MAPK activation. It is notable that MKP1 localizes to the cytoplasm, whereas MKP2 interacts with MPK3 and MPK6 in the nucleus and cytoplasm. The differential localization of these two MKPs (MKP1 and MKP2) provides a means of specificity toward the regulation of MPK3 and MPK6. Moreover, their structures differ considerably: whereas MKP2 consists mainly of the phosphatase catalytic domain, MKP1 is a rather large protein with several additional domains (Fig. 2). Although further studies are needed to define their roles, these domains are likely responsible for the functional specialization of MKP1.

Biotic stress responses controlled by plant MKPs. The phosphatases MKP1 and MKP2, in concert with the kinases MPK3, MPK4, and MPK6, are important regulators of responses to various plant pathogens. MKP2 is a positive regulator of resistance against necrotrophic fungi (which kill the host for nutrition). Conversely, MKP2 is a negative regulator when challenged with biotrophic bacteria (which feed and reproduce on living host cells). Moreover, MKP2 seems to have no major role in the response to hemitrophic bacteria (which display a biotrophic phase followed by a necrotrophic one). In contrast, MKP1 negatively regulates tolerance to a hemitrophic bacteria. Having explained the similarities and differences be-

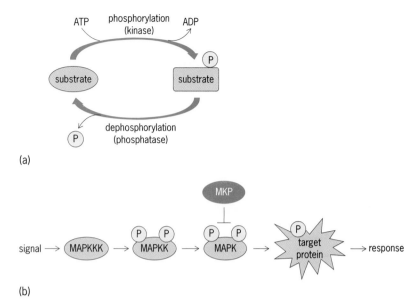

(a)

(b)

Fig. 1. Signal transduction is dependent on protein phosphorylation. (*a*) Kinases are enzymes that add phosphate groups from ATP (adenosine triphosphate) to a given substrate (phosphorylation). The process is reversed by phosphatases (dephosphorylation). (*b*) MAPK signaling involves sequential phosphorylation of three interlinked kinases. The pathway is negatively regulated by MAPK phosphatases (MKPs).

tween MKP1 and MKP2 in the previous section, the same question arises in the context of pathogen attack: what determines which phosphatase is required upon MPK3 and MPK6 activation? Analysis of biotic stress responses will open the possibility of exploring the mechanisms by which MKP1 and MKP2 differentially regulate MPK3 and MPK6, despite their similarities.

Developmental processes controlled by plant MKPs. MPK3 and MPK6 are involved, for example, in stomatal pore, pollen, embryo, and ovule development; cytokinesis (division of the cytoplasm following nuclear division); and response to plant hormones. Whereas our knowledge of phosphatase-mediated deactivation of stress-responsive MPK3 and

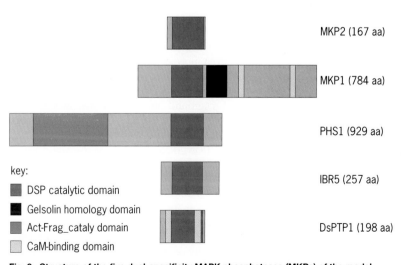

key:

■ DSP catalytic domain

■ Gelsolin homology domain

■ Act-Frag_cataly domain

□ CaM-binding domain

Fig. 2. Structure of the five dual-specificity MAPK phosphatases (MKPs) of the model plant *Arabidopsis thaliana*. Abbreviations: aa, amino acids; DSP, dual-specificity phosphatase; Act-Frag_cataly, actin-fragmin kinase, catalytic; CaM, calmodulin. (*Modified according to S. Bartels et al., Emerging functions for plant MAP kinase phosphatases, Trends Plant Sci., 15:322–329, 2010*)

MPK6 is progressing, our understanding of MKP-MPK3/MPK6 modules relevant to plant development is less developed. Notwithstanding this, IBR5-MPK12 and PHS1-MPK18 have been identified as functional modules in signaling of the plant hormone auxin and cytoskeleton organization, respectively. Auxin directs many aspects of plant growth and development. It was found that *ibr5* mutants are unable to respond effectively to auxin, displaying, for example, fewer lateral roots and abnormal vasculature. It also was found that IBR5 interacts specifically with and dephosphorylates MPK12. Consistent with the function of IBR5 as a positive regulator of auxin responses and a negative regulator of MPK12, *MPK12*-knockdown plants are hypersensitive to auxin. How IBR5-regulated MPK12 controls auxin responses at the molecular level remains to be established. A similar scenario occurs in the case of PHS1-MPK18. PHS1 seems to specifically regulate MPK18, and genetic evidence supports a role for the PHS1-MPK18 complex in microtubule-related functions (microtubules are slender cylindrical structures found in the cytoskeleton). However, the intermediates between MPK18 and microtubule-associated proteins are not yet identified.

Regulation of MKPs. The view of signaling modules as an array of proteins passing on the signal sequentially is a simplified one. The signal transduction machinery should be better described as an intricate network of proteins that modify each other, even themselves, and it includes cross talk among pathways activated by different stimuli. Accordingly, it is not surprising that the MKPs are themselves subject to control. In response to environmental stress, certain MKPs are transcriptionally regulated. Moreover, MKPs can be phosphorylated by their target MAPKs, possibly so as to regulate their activity or stability (or both). Such feedback loops are common in biological systems and allow transient signal outputs, signal amplification, or stabilization. Despite its significance for understanding how plant cells respond to their environment, virtually nothing is known about MKP regulation in plants. Moreover, although MKPs have been implicated in diverse cellular responses (as described above), MAPKs are known to regulate an even more comprehensive set of responses. Clearly, many additional MKP functions are waiting to be discovered.

For background information *see* AUXIN; CELL (BIOLOGY); GENE; PHOSPHATASE; PLANT CELL; PLANT PHYSIOLOGY; PROTEIN; PROTEIN KINASE; SIGNAL TRANSDUCTION; TRANSCRIPTION in the McGraw-Hill Encyclopedia of Science & Technology.

Marina A. González Besteiro; Roman Ulm

Bibliography. B. Alberts et al., Mechanisms of cell communication, pp. 879–964, in *Molecular Biology of the Cell*, 5th ed., Garland Science, New York, 2007; E. Andreasson and B. Ellis, Convergence and specificity in the *Arabidopsis* MAPK nexus, *Trends Plant Sci.*, 15:106–113, 2010, DOI:10.1016/j.tplants.2009.12.001; S. Bartels et al., Emerging functions for plant MAP kinase phosphatases, *Trends Plant Sci.*, 15:322–329, 2010, DOI:0.1016/j.tplants.2010.04.003; M. C. Suarez Rodriguez, M. Petersen, and J. Mundy, Mitogen-activated protein kinase signaling in plants, *Annu. Rev. Plant Biol.*, 61:621–649, 2010, DOI:10.1146/annurev-arplant-042809-112252; A. Schweighofer, H. Hirt, and I. Meskiene, Plant PP2C phosphatases: Emerging functions in stress signaling, *Trends Plant Sci.*, 9:236–243, 2004, DOI:10.1016/j.tplants.2004.03.007.

Plant-produced volatile organic compounds

Perhaps the most striking feature of plants is that they perform photosynthesis: They produce food (sugars) from sunlight, carbon dioxide (CO_2), and water, thereby fueling life on Earth. In addition to this primary metabolism, plants also produce secondary metabolites, which are mostly involved in the interactions of plants with their environment and in adjusting plant growth and behavior through plant hormones. Secondary metabolites include, for example, pigments that color their flowers and components that give taste and smell to plants. Components that provide smell are typically (partly) volatile and are termed volatile organic compounds (VOCs). Plants produce a broad range of VOCs, with the largest groups being terpenoids (compounds with an isoprenoid structure similar to that of the terpene hydrocarbons) and green leaf volatiles (GLVs). GLVs are best known as the smell that is produced by freshly mown grass; this group of volatiles mainly consists of C_6-aldehydes, C_6-alcohols, and their acetates. Terpenoids contribute to many different scents. For example, the smell of pine trees comes from pinene, the smell of ginger is zingiberene, whereas limonene contributes to the taste and smell of many citrus fruits. Terpenoids are synthesized through polymerization of C_5 isoprene units. Monoterpenes (composed from two isoprene subunits), diterpenes (composed from four isoprene subunits), and tetraterpenes (composed from eight isoprene subunits) are produced in the plastids of plants. Sesquiterpenes (composed from three isoprene subunits) and triterpenes (composed from six isoprene subunits) are formed in the plant cell's cytosol and endoplasmic reticulum.

As might be expected from the wide biochemical variety of volatile compounds, VOCs have been found to play many roles in nature. VOCs serve as signals for other plants, plant parts, and organisms in a wide variety of ecological functions; some of these functions will be discussed below.

VOCs as signals in plant–insect interactions. VOCs play important roles in plant–insect interactions. Many pollinators find their host plant based on the composition of the volatile bouquet produced by its flowers. Specialist pollinators are often able to detect their preferred plant species from great distances, even within a field of flowering plants. Another well-studied system in which volatiles are essential is the plant–herbivore–predator system. Tomato plants, for example, increase production of a cocktail of different volatile components when they are under spider mite attack. The emission of these volatiles has been coined a plant's "cry for help" because it attracts predatory mites that feed on the spider mites. Similarly, parasitoids of herbivorous insects make use of herbivore-induced volatiles to find their hosts. For example, the parasitic wasp *Cotesia glomerata* is attracted to volatiles emitted by wild cabbage (*Brassica oleracea*) plants that are infested with small cabbage white caterpillars (*Pieris rapae*). These predators of herbivores thus serve as bodyguards to the attacked plant, and plants emitting herbivore-induced volatiles indeed suffer less damage from herbivory.

VOC cocktails produced by plants are usually very specific. Blends vary not only among plant species but also among different cultivars of one crop species. Different cultivars of barley show differences in attractiveness to the aphid *Rhopalosiphum padi* and the ladybird *Coccinella septempunctata* that feeds on these insects. Furthermore, herbivores of different species induce different volatile blends. These blends thus contain a lot of detailed information. Insects use this information for their own benefit; butterflies not only find their preferred food source in a field of flowers, but they also select plants for oviposition based on scents that are induced by previously laid eggs. Although the composition of the volatile blend is, in most cases, the factor determining insect behavior, sometimes a single component serves as a signal and even its stereoisomeric form is relevant. For example, whiteflies respond to zingiberene (a sesquiterpene) produced by tomato, but not to the one produced by ginger, even though the volatiles differ only in the orientation of one side chain.

VOCs in plant–plant interactions. It has been shown that neighboring plants use herbivore-induced plant emissions for their own benefit. Some of these interactions are termed priming: plants that receive volatiles that were emitted by neighboring plants demonstrate more accurate defenses upon subsequent pathogen or herbivore infestation. There is still discussion about how the emission of VOCs is beneficial for the emitting plant, and there are currently two main theories. The first theory states that the emitter produces volatiles and profits from the attraction of predators of the herbivore. The second theory states that volatile signaling evolved as within-plant signaling, as, especially in trees and shrubs, there is often a discrepancy between vascular distance and absolute distance. Because herbivores are likely to travel between adjacent leaves and branches, volatiles can travel quickly to nearby branches and leaves where they can serve as a signal to increase defenses. In both of these cases, neighboring plants or other organisms may just "eavesdrop" on this signal and use it for their own benefit.

VOCs can also play a more direct role in plant-plant interactions. A striking example is that of purple sage (*Salvia leucophylla*), which is a plant species emitting high amounts of monoterpenes that inhibit root growth and seed germination of neighboring plants. Such allelopathic behavior is a powerful way to prevent competition from surrounding vegetation for resources (note that allelopathy refers to the harmful effect of one plant or microorganism on another owing to the release of secondary metabolic products into the environment). Even within species, plants can influence the growth of conspecifics (individuals or populations of a single species) by emitting VOCs. In barley, the ratio between root and shoot biomass is affected when plants are exposed to volatiles of different cultivars. Exposure to constitutive volatile emissions can also be used to the advantage of a receiving plant. In tobacco (*Nicotiana tobaccum*), shading by neighbors causes the production of the volatile plant hormone ethylene. Wild-type plants outcompete ethylene-insensitive neighbors, even though these ethylene-insensitive plants display a normal shade-avoidance response. In another example, the parasitic plant dodder (*Cuscuta pentagona*) locates its host (tomato) through volatiles produced by the host plant. The parasite then grows toward the host, wraps itself around the host, grows structures into the host's vasculature, and "steals" nutrients from the host.

VOCs and the environment. VOCs play an important role in atmospheric processes, coupling the biosphere and atmosphere. When released by plants into the atmosphere, volatiles are oxidized and gradually degraded to smaller subunits because they react quickly with air pollutants such as ozone and nitrogen oxides (NO_x). Through this reactivity, VOCs affect the oxidation capacity of the atmosphere, thereby controlling the concentration and lifetime of ozone and NO_x. In addition to being important greenhouse gases, both ozone and NO_x also determine air quality. Furthermore, VOC oxidation products can form new particles or condense on existing particles. They might even be the most important factor with regard to formation and growth of secondary organic aerosols. These aerosols may cool the climate by nucleating clouds and by scattering and blocking solar radiation.

The biosphere–atmosphere interaction works from biosphere to atmosphere and vice versa, and VOC emissions are influenced by an array of environmental conditions. First, light is an essential requirement for volatile production. In darkness, volatile emission is reduced almost to zero in many cases, although there are certain flowers that produce scents particularly at nighttime when pollinated by nocturnal pollinators. Second, temperature and CO_2 levels, which are both expected to rise over the coming decades as a consequence of climate change, might greatly affect VOC emissions as well as their longevity in the atmosphere. A rise in temperature exponentially increases the emission rate of most VOCs. Increased temperatures not only increase volatile synthesis by enhancing enzymatic activities, but they also raise VOC vapor pressure and decrease diffusion resistance. Elevated CO_2 levels lead to increased production of most VOCs, at least in the short term. The long-term effect of CO_2 on VOCs is mostly unexplored, although it can be argued that an increase of ambient CO_2 will enhance the production of volatiles because of higher carbon availability. Upon exposure to ultraviolet radiation, volatile production and specifically isoprene production are increased. Isoprene is hypothesized to protect plants from heat, ozone, and other abiotic stresses by quenching ozone and other radicals. However, it has also been shown that, outside the plant, high levels of NO_x, isoprene, and other terpenoids might lead to production of ozone.

Future perspectives. In the last 2 decades, various functions of VOCs in signaling between plants, insects, and pathogens were discovered. Furthermore, the biosynthetic pathways of various classes of volatiles were described. There now are at least two main challenges in VOC research. First, it is necessary to learn more about the perception of VOCs by plants. Although plants are affected by VOCs, it is not known how they perceive volatile signals. The second challenge lies in combining the knowledge of atmospheric interactions with VOCs with the signaling function of VOCs. Although it is known how volatiles react with components such as ozone and NO_x in the atmosphere, it is less clear how this changes the relative compositions of blends of VOCs. By investigating the effects of altered VOC blends on VOC-mediated plant-pollinator, plant-herbivore, and plant-plant interactions, the effects of air pollution on plant ecosystems will be further unraveled.

For background information *see* ALLELOPATHY; CHEMICAL ECOLOGY; ECOLOGY; FLOWER; PLANT-ANIMAL INTERACTIONS; PLANT COMMUNICATION; PLANT GROWTH; PLANT HORMONES; PLANT METABOLISM; PLANT PHYSIOLOGY; POLLINATION in the McGraw-Hill Encyclopedia of Science & Technology.   Wouter Kegge; Paulien Gankema; Ronald Pierik

Bibliography. I. T. Baldwin et al., Volatile signaling in plant-plant interactions: "Talking trees" in the genomics era, *Science*, 311:812-815, 2006, DOI:10.1126/science.1118446; M. Dicke and I. T. Baldwin, The evolutionary context for herbivore-induced plant volatiles: Beyond the "cry for help," *Trends Plant Sci.*, 15:167-175, 2010, DOI:10.1016/j.tplants.2009.12.002; M. Heil and J. C. Silva Bueno, Within-plant signalling by volatiles leads to induction and priming of an indirect plant defense in nature, *Proc. Natl. Acad. Sci. USA*, 104:5467-5472, 2007, DOI:10.1073/pnas.0610266104; W. Kegge and R. Pierik, Biogenic volatile organic compounds and plant competition, *Trends Plant Sci.*, 15:126-132, 2010, DOI:10.1016/j.tplants.2009.11.007; J. Peñuelas and M. Staudt, BVOCs and global change, *Trends Plant Sci.*, 15:133-144, 2010, DOI:10.1016/j.tplants.2009.12.005.

Polarization sensing cameras

Color is intuitive to people, because we can perceive it visually, but polarization is a less obvious property of light. Light is a wave, and therefore it has a direction of travel, wavelength, and polarization. The direction is where the light is headed, the wavelength (color) is the distance between crests of the wave, and the polarization is the orientation of the oscillation (**Fig. 1a**). If the wave oscillates along a

single axis, it is said to be linearly polarized. If the oscillations rotate as the wave travels, it is elliptically or circularly polarized. Although humans are unable to sense polarization, many animals can sense polarized light. This ability is traditionally associated with behavioral tasks such as determining orientation or navigation. However, recently it has also been shown that animals can use this polarization vision for contrast enhancement, camouflage breaking, object recognition, and signaling.

Camera types. Cameras can be classified based on which of these properties they detect. A black-and-white camera measures only the intensity of an optical field. In a digital camera the intensity is measured by a pixelated detector, typically a charge-coupled device (CCD) or complementary metal-oxide-semiconductor (CMOS) image sensor. These detectors can measure only the intensity of the field over the image plane. To sensitize a camera to either the wavelength or the polarization of light, multiple measurements must be collected using different spectral or polarization filters. For instance, a typical digital camera takes a color image using a pixelated array of color filters. The color filter array is a repeating pattern of red, blue, and green filters that allow a set of neighboring pixels to estimate the color spectrum of the incoming light (Fig. 1b). An analogous polarization filter, discussed in detail later, can be used to estimate the incident polarization (Fig. 1c). A color camera can tell us about the material of an object, but a polarization camera can tell us about an object's shape, orientation, and roughness. This information may have important applications in remote sensing, manufacturing, machine vision, and medical diagnostics.

Stokes vector representation. A color camera estimates the incident wavelength with an RGB or CMYK color model. A similar model is necessary for a polarization camera to represent the measured polarization state. Stokes vectors are the most common method of representing the polarization of incoherent light. A Stokes vector representation allows for the partial polarization of light but does not account for its incident phase, because it is based on intensity, not electric field. Jones calculus is commonly used for coherent light (such as from a laser), and describes the amplitude and phase of the field. An imaging polarimeter estimates the incident polarization state by recording multiple intensity measurements through varying polarization elements such as $0°$, $45°$, $90°$, $135°$, right-handed circular, and left-handed circular polarizers. Using these intensity values, the Stokes vector can be calculated for each point in the scene.

A Stokes vector consists of four elements, S_0, S_1, S_2, and S_3, defined in Eq. (1). S_0 denotes the total

$$\vec{S} = \begin{vmatrix} S_0 \\ S_1 \\ S_2 \\ S_3 \end{vmatrix} = \begin{vmatrix} I_{0°} + I_{90°} \\ I_{0°} - I_{90°} \\ I_{45°} - I_{135°} \\ I_{RH} - I_{LH} \end{vmatrix} \quad (1)$$

intensity of incident light, and therefore is greater than zero. It can be calculated from the sum of any

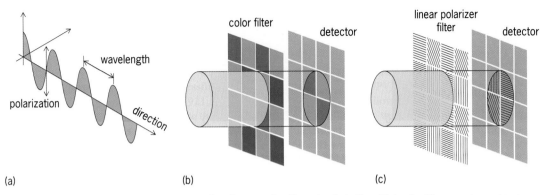

Fig. 1. Wave properties of light. (*a*) Light has a direction, wavelength, and polarization. (*b*) A color filter array is used on a camera to estimate the incident color. (*c*) An analogous polarization filter can be used to measure the incident polarization instead.

two orthogonal polarization states such as linear polarization at $0°$ and $90°$. S_1 and S_2 represent the affinity toward linear polarization. The angle of linear polarization can then be calculated via Eq. (2). S_3 denotes

$$\theta_{\text{linear}} = \frac{1}{2}\tan^{-1}\frac{S_2}{S_1} \qquad (2)$$

the fraction of the intensity that is circularly polarized.

While S_0 must be greater than zero, $S_{1,2,3}$ can range from 0 to $\pm S_0$. The fraction of incident light that is polarized is called the degree of polarization (DOP) and is calculated by Eq. (3). The fraction of light that

$$\text{DOP} = \sqrt{S_1^2 + S_2^2 + S_3^2}\Big/S_0 \qquad (3)$$

is linearly or circularly polarized is called the degree of linear polarization (DOLP) or degree of circular polarization (DOCP).

Polarimeter configurations. As mentioned, multiple measurements through varying polarizer combinations are necessary to estimate the Stokes vector. The method a device uses to capture these different intensities is the primary way by which imaging polarimeters are classified, and the trade-offs between methods need to be examined for the individual application. The simplest design is a division-of-time polarimeter (DoTP), which sequentially captures multiple images through different combinations of polarizers and retarders. This is commonly imple-

mented by rotating a retarder in front of a fixed polarizer (**Fig. 2*a***). The polarization elements may be before or after the imaging optics. The primary disadvantage of this technique is its acquisition time, as at least four images need to be recorded serially and any movement in the camera or scene will result in false polarization signatures. Depending on the design, the polarizer can also be rotated and, if only linear polarization measurements are required, the retarder can be removed from the system.

A division-of-amplitude polarimeter (DoAmP) splits the incident light into multiple optical paths. Each optical path includes a different polarization element and a detector (Fig. 2*b*). The example shown is for a linear imaging polarimeter. Full Stokes images can be achieved by the addition of a retarder in one of the beam paths. The design shown utilizes three 50/50 beam splitters. Polarization beam splitters can also be implemented for more efficient use of the light. The advantage of this system is that the resulting images have full spatial resolution in a single-shot design. However, alignment, optical path length, and aberrations must be matched between the paths.

A third type is a division-of-focal-plane polarimeter (DoFP), which uses micropolarization elements placed immediately prior to the detector. The micropolarizers must be equal in size to a single pixel on the detector. The most common configuration uses four polarizer orientations ($0°$, $45°$, $90°$, and $135°$) and does not measure S_3. Ideally the polarizers

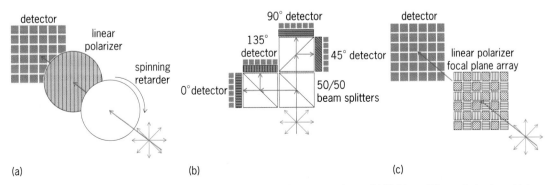

Fig. 2. Different imaging polarimeter configurations. Imaging optics are not shown. (*a*) Division-of-time polarimeter, which records a number of exposures with varying retarder orientations. (*b*) Division-of-amplitude polarimeter, which divides the wave front between multiple paths and detectors. (*c*) Division-of-focal-plane polarimeter, which has a repeating pattern of micropolarizers immediately before the detector. The examples shown in (*b*) and (*c*) measure only linear polarization.

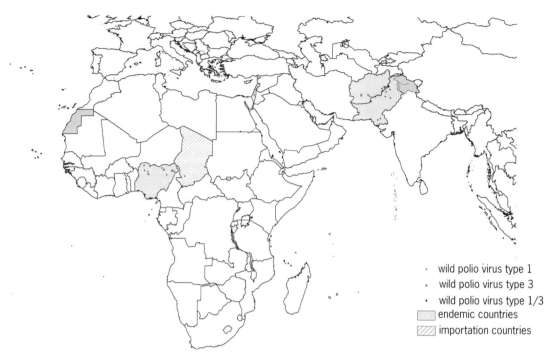

Polio cases in the world in 2012. The data were obtained in March 2012 and exclude viruses detected from environmental surveillance and vaccine-derived polioviruses. (*Courtesy of the Centers for Disease Control and Prevention*)

(see **illustration**). To further emphasize this need, the United Nations, under the leadership of the Secretary-General Ban Ki-moon, has urged the heads of state of these countries to carry out polio vaccinations. In these areas, sanitation is poor and the polioviruses are endemic. They are transmitted by the fecal–oral route. Newborns in these nations are partially protected for 2–3 months until antibodies received from their mothers begin to wane. During this time, the infants are usually exposed to the polioviruses. In addition, the following countries are experiencing outbreaks of polio following importation: China, Congo, Gabon, Guinea, Kenya, Liberia, Mali, Nepal, and Niger. Thus, whether polio can be eradicated completely in developing parts of the world is still very questionable.

For background information *see* BIOLOGICALS; ENTEROVIRUS; EPIDEMIC; INFECTIOUS DISEASE; PICORNAVIRIDAE; POLIOMYELITIS; PUBLIC HEALTH; VACCINATION; VIRUS in the McGraw-Hill Encyclopedia of Science & Technology.   John P. Harley

Bibliography. C. A. De Quadros, Polio, pp. 762–772, in J. Lederberg (editor-in-chief), *Encyclopedia of Microbiology*, vol. 3, 2d ed., Academic Press, San Diego, 2000; A. T. Dufresne and M. Gromeier, Understanding polio: New insights from a cold virus, *Microbe*, 1:13–18, 2006; J. L. Melnick, Current status of poliovirus infection, *Clin. Microbiol. Rev.*, 9(3):293–300, 1996.

Print compliance verification

Print compliance verification is a process for assessing how well graphic arts workflows and printed products conform to nationally or internationally rec-

ognized standards. When verification is done by a competent, impartial third party in the form of an audit and the results pass, the printing organization earns the distinction of being certified.

Printing standards and print compliance programs. There are three types of printing standards: industry, national, and international. G7 is an example of an industry standard and print qualification program that was developed by the IDEAlliance to address the "press calibration" standardization of the North American printing industry. (An industry standard is more a guideline, compared to a document that has been voted on and approved by a standardization body.) A print qualification program is less rigorous than a print certification program. In this instance, one can start by downloading a G7 how-to manual from the IDEAlliance website (http://www.idealliance.org). A G7 expert or professional is required to submit a printed sheet on behalf of a printing company for assessment, and if the results pass, the printer receives the distinction as a G7 Master Printer. There are more than 800 Master Printers in the United States and internationally.

The Committee for Graphic Arts Technology Standards (CGATS) is a graphic arts standardization body accredited by the American National Standards Institute (ANSI) in the United States. CGATS TR006 is a U.S.-sanctioned standard characterization data set for the General Requirements and Applications for Commercial Offset (GRACoL) proofing and printing; and CGATS TR016 (2009) is a U.S.-sanctioned standard for printing tolerance and conformity assessment. The Rochester Institute of Technology (RIT) is developing a printing certification program, the Printing Standards Audit (PSA), to award to printing companies that can demonstrate conformance to characterization data sets, such as CGATS/GRACoL TR006, according to CGATS TR016.

The International Standards Organization (ISO) Technical Committee, ISO TC130, is responsible for developing international printing standards; for example, ISO 12647-2, ISO 15339-1, and so on.

There are a number of European certification bodies, such as Fogra (Germany) and Ugra (Switzerland), that have been granting certifications, known as Process Standards Offset (PSO), to printing companies that demonstrated conformance to printing aims according to ISO 12647-2 and related standards. RIT offers PSA certification according to ISO 15339-1 to printers worldwide.

It is important to recognize that ISO does not sanction what standards should be included in the certification requirements. Thus, certification schemes are open and specific to the market. As such, Brazil offers NBR 15936-1, Italy offers cmyQ and CeriPrint, Japan offers JapanColor, the Netherlands offers SCGM, Sweden offers CGP, and the United Kingdom offers BPiF certification schemes to printers in their respective countries.

Importance of standard or print compliance verification. Printing standards (industry, national, or international) are inputs of a print compliance verification program. Assessment by a competent and impartial third-party auditor is the key activity of print compliance verification. Being qualified or certified is the successful outcome of a print compliance verification program.

One may wonder why standard or print compliance verification is important. Standards and printing certification contribute to the printing industry by ensuring quality and process control, and in building trust.

Aligning quality expectations. Printing standards replace subjective print-quality judgments with measurable and objective criteria. When printers and print buyers agree to use printing standards to define their quality expectations, the waste associated with chasing subjective quality requirements is reduced.

Enabling process control. Standardizing quality criteria enables the use of common process aims. By printing to a single set of process aims (instead of trying to match new aims for different jobs), printers can standardize and optimize their processes. The result is enhanced process control, reduced color-matching downtime, and more competitive production costs.

Building trust. In today's printing industry, international supply chains are becoming the norm. Printers and print buyers who used to build trust through frequent face-to-face meetings now find themselves on opposite sides of the world. When frequent meetings are no longer possible, a new basis for trust is required. Standards provide the basis for objective conformance assessment. Printing certification has become the new basis for building trust in global supply chains.

Achieving print certification status. A printing company may be motivated by the benefits mentioned above. It is clearly a strategic issue and there are costs and efforts involved. Senior management must make a conscientious decision. After making the decision to become certified, a printer will go through the following five steps.

Select appropriate standards to be certified in your workflow. ISO 12647-2 (2004) is the established printing standard developed in the film-based workflow era. ISO/DIS 15339-1 (2011) is an emerging printing standard developed entirely based on printing from digital data. Many think the two standards are in conflict with each other. In fact, the two standards complement each other.

If we define a color-managed workflow from data reception, to prepress and color management, to printing, ISO 12647-2 only focuses on printing process control. ISO 12647-2 conformity assessment begins with a limited number of color patches (solid and tints) and on platemaking, printing, sampling, and measurement, and it ends with process-control-related parameters [color of solids, tone value increase (TVI), and midtone spread]. By calibrating the press and demonstrating conformance to process-control aims, ISO 12647-2 ensures repeatable color. Thus, prepress and color management are outside the scope of ISO 12647-2.

The methodology of ISO/DIS 15339-1 is similar to ISO 12647-5, -6, -7, and -8 and is based on the use of characterization data to define printing. ISO/DIS 15339-1 recognizes characterization data sets and their associated International Color Consortium (ICC) profiles as color exchange spaces. ISO/DIS 15339-1 also recognizes characterization data set-derived printing aims, but is silent on process-control approaches and related tolerances. ISO/DIS 15339-1 conformity assessment begins with many color patches per ISO 12642 (IT8.7/4). By calibrating the press to substrate-corrected printing aims, and by using color management in prepress, ISO/DIS 15339-1 ensures product color conformance.

ISO 12647-2 and ISO/DIS 15339-1 have different scopes. It is natural and reasonable to have a process-control standard nested inside a product conformance workflow standard (see **illustration**).

As such, the ISO 12647-2 process-control standard addresses the printer's needs, and the ISO/DIS 15339-1 product conformance standard addresses the print buyer's needs. Printing aims from ISO 12647-2 may be slightly different than printing aims derived from the ISO/DIS 15339-1 data set, but they are compatible as far as tolerances are concerned.

Prepare for the audit. The first step is to form a project team responsible for planning and executing the steps leading to certification. Typically this team consists of a team leader plus representatives from prepress, platemaking, quality control, and printing. Shortly after being formed, the team must decide on the standards and workflows that it intends to certify.

The project team should contact the certification body and request information on the scope of the audit and details concerning the anticipated audit, such as hardware and software. This process will result in an audit proposal. The certification

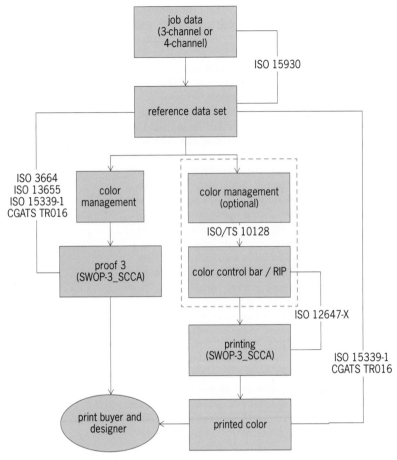

Color-managed workflow and printing standards.

body will send the project team a pre-audit kit that contains comprehensive information for preparing for the audit.

Printers are encouraged to use supplied materials to dry run the audit process before the audit team arrives. With proper preparation, the audit itself is just another print-job run in conformance to the standard selected. The printer may get expert advice to identify the gaps that exist in its workflow and to obtain help in closing them.

Host an on-site audit of the selected workflow. The audit process typically begins with an in-briefing for site management and the audit participants. The in-briefing establishes the schedule of the audit activities, confirms the standard and scope selected, and initiates the audit by providing a set of test files for the printer to process.

Each of the areas identified in the scope of the audit will be assessed according to the schedule. The printer is expected to independently demonstrate its ability to operate a workflow conforming to the standard(s) selected.

At the end of the audit, the audit team will have collected a set of observations and a series of test forms (that is, processed files and physical prints) that will be assessed according to the standards.

After all audit activities have been completed, the audit team will meet with site management and the audit participants for an out-briefing. The result of

the audit will not be known until the test forms have been measured and assessed. Nevertheless, the audit team will confirm whether it was able to complete the full range of audit activities. It will also provide feedback on the work practices it observed during the audit, and establish a timeframe for completion of the full assessment.

Review the results of the audit. When the printed samples are received and measured, the certification body will complete the audit report and send it to the printer in written and electronic form.

Certification. If the audit report demonstrates that the printer met or exceeded the level of performance required to achieve certification, a certificate will accompany the audit report. Certified companies are also recognized on the certification body's website and authorized to use the certification logo.

If the workflow failed to demonstrate conformance, the certification body will request a call to review the audit report with the printer. Analysis of the nonconformance will be discussed during the call.

Trends in printing certification. A number of surveys have been done regarding printing certification activities worldwide. For example, findings from the *Printing Standards: A 2010 Survey Report* (RIT, 2010) indicated that selected standards have regional preference. A case in point is that U.S. printers prefer G7 press calibration methodology and the GRACoL characterization data set, and European printers prefer ISO 12647-2 process-control standard and the Fogra39 data set. Certification programs also differ in scope. For example, IDEAlliance only addresses press calibration conformance, while Fogra and Ugra address both press calibration and production variation conformance.

To harmonize the printing certification requirements and assessment activities, ISO TC130 at its 2010 plenary meeting in Sao Paulo, Brazil, established Working Group 13 to develop the standard, ISO 16761-1, graphic technology—conformity assessment and management system requirements for tone and color quality. This is a three-part ISO standard with part one addressing basic principles, part two addressing printing tolerance and conformity assessment, and part three addressing sector-specific management system requirements. Currently, ISO 16761 is at the "new work item" stage. The projected timeframe for committee draft and beyond is 2013.

When developing and drafting any ISO standard, ISO/IEC (IEC is the International Electrotechnical Commission) directives state that there must be a separation between technical requirements (the responsibility of WG3 process-control standards) and conformity assessment requirement (the responsibility of WG13 printing certification requirements). In terms of conformity requirements, the directives articulate that the standard must adhere to the neutrality principle. In terms of management system requirements, the directives further articulate that a justification study must be submitted and approved before the work of ISO 16761-3 could begin.

For background information *see* PRINTING; PRO-
CESS CONTROL; SUPPLY CHAIN MANAGEMENT in the
McGraw-Hill Encyclopedia of Science & Technology.
Robert Chung

Bibliography. ANSI CGATS/GRACoL TR 006,
*Graphic technology—Color Characterization Data
for GRACoL® Proofing and Printing on U.S. Grade
1 Coated Paper*, ANSI, 2007; ANSI CGATS TR 016,
*Graphic Technology—Printing Tolerance and Con-
formity Assessment*, ANSI, 2012; R. Chung, Assessing print conformance based on ISO 12647-2, *Test
Targets 10*, pp. 2-11, December 2011, http://cias.
rit.edu/~gravure/tt/WB10/index.html; R. Chung,
International printing standards, a value-added
proposition, *Test Targets 9.0*, RIT, pp. 2-7,
November 2009, http://cias.rit.edu/~gravure/
bob/pdf/TT9_international_printing_standards.pdf;
R. Chung and S. Jensen, *Printing Standards: A
2010 Survey Report*, No. PICRM2011-01, RIT, 2010,
http://print.rit.edu/pubs/picrm201101.pdf.

Prism light guides

In the field of illuminating engineering, there are advantages to piping light. These advantages include
the economies of scale arising from applications
associated with using larger, more efficient light
sources and separation of the light sources and their
associated electrical connections from the illuminated space. This separation is advantageous for a
wide range of applications, including tunnel lighting, high ceiling areas, and architectural highlights
on buildings. In addition, there is considerable interest in guiding sunlight from the outdoors to the
interior spaces of buildings to reduce the electrical
energy required for lighting, while also improving
the quality of illumination in the space. Piping light
efficiently has proven to be a technical challenge,
and the prism light guide, a light guidance structure
based on total internal reflection (TIR) in prismatic
microstructures, offers a practical solution for numerous illuminating engineering applications.

Introduction to hollow light guides. In many circumstances, it is desirable to transport light from a bright
source and distribute it to where it is needed. For
such a light guidance and distribution system to be
practical, it must have minimal loss per unit length,
incorporate a mechanism that allows for controlled
extraction of the light, and be relatively lightweight
and inexpensive. These requirements are technically challenging. Solid guides can be effective in
small-scale applications, but they are often unsuitable for large-scale lighting applications because of
cost, mass, and limited efficiency. Consequently, hollow light guides are the preferred solution, provided
the inner surface of the guide has sufficiently high reflectance so that the light loss during propagation is
minimized. Furthermore, if a light guide is intended
to distribute the light within the space that is to be
illuminated, it is desirable for the guide to also incorporate controlled light emission. The prism light
guide provides both of these optical characteristics

since its unique microstructured inner liner reflects
the propagated light efficiently by TIR, while also
allowing controlled extraction of the light into the
regions where it is needed. There have been a wide
variety of installations of prism light guides. A few of
these are shown in **Fig. 1**.

Technical challenges associated with piping light.
Solid dielectric light guides are impractical for a number of reasons related to weight and cost, but most
importantly because dielectric materials in general
absorb a significant portion of the light as it travels
along the guide. The amount of light absorbed by a
material is typically characterized by the absorption
coefficient (the fractional energy loss per unit length,
given in units of m^{-1}) which, by means of the law
of exponential decay, enables the determination of
the transmitted fraction of light (also known as the
transport efficiency) over the distance the light travels in the material. Most reasonably priced dielectric
materials have an absorption coefficient that exceeds
$0.1\ m^{-1}$ (or, equivalently, an attenuation distance less
than 10 m, where the attenuation distance is calculated as 1 divided by the absorption coefficient). Ideally for such a guide, it is desirable for the absorption
coefficient to be less than $0.01\ m^{-1}$, which will ensure that the attenuation does not exceed 10% over
a guide distance of 10 m. Furthermore, in the case of
typical solid dielectrics, the absorption is wavelength
dependent, which means that the light that exits the
guide will have a different spectral distribution than
the light that enters.

The requirement for a low-loss medium makes
transmission through air desirable, and there are
a number of optical configurations using lenses
and mirrors that enable free-space transmission.
However, appropriately low-cost lenses and mirrors
themselves generate unacceptably high light loss,
which often makes these free-space transmission
options impractical.

This leaves hollow, air-filled reflective structures
as the most practical means of piping light for large-
scale applications. The interior surfaces of the guide
must be sufficiently reflective. However, since the
light typically reflects 20–30 times as it travels down
a hollow light guide, even a relatively small amount
of absorption will cause a significant loss of light.
Figure 2 shows how the transport efficiency
changes for different reflectance values.

A metallic reflector, even highly polished silver or
aluminum, is not sufficiently reflective. If the light
reflects 10 times along a guide lined using a 95% reflective material, such as silver, the efficiency of transmission will be $0.95^{10} \cong 0.60$. In other words, about
40% of the light will be absorbed after only 10 reflections. In contrast to metallic reflectors, dielectric
multilayer reflective films can be used effectively in
hollow light-guide applications, since the reflectance
of such a film can exceed 99%. However, these dielectric multilayer films are not useful on their own
for light distribution systems, as will be described
next.

Advantages of the prism light guide. The prism light
guide uses the well-known phenomenon of TIR, an

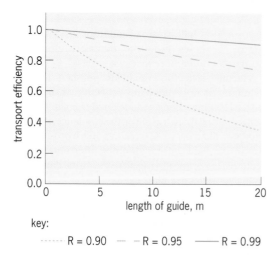

Fig. 1. Various prism light-guide installations. (*Images courtesy of TIR Systems Ltd.*)

entirely different reflection mechanism, to overcome these efficiency challenges; and it is also compatible with the need for light distribution. This is the same mechanism by which light propagates in an optical

key:

····· R = 0.90   — — R = 0.95   —— R = 0.99

Fig. 2. Transport efficiency of a hollow light guide of typical length and cross section with different surface reflectance R values, as indicated.

fiber. But unlike optical fibers, a prism light guide can be hollow and this gives it a number of important practical advantages, as have been described earlier. Light from a high-intensity source is partially collimated so that most of the emitted rays lie within the acceptance angle of the guide, and this light is directed into the guide so that it propagates along its length. A specially tailored light escape mechanism is incorporated into the guide that scatters light beyond the guide's acceptance angle and causes the light to be emitted, as desired, for the particular application, typically fairly evenly along its length.

The prism light guide is a hollow pipe made of optically transparent materials in which the inner surface is lined with a film whose outside surface is textured with very small isosceles right-angle prisms that run parallel to the guide axis, as depicted schematically in **Fig. 3**. The prisms cause light rays within a certain angular acceptance cone that are traveling within the guide to undergo TIR and continue along inside the guide. Figure 3 depicts the maximum off-axis light ray that will undergo TIR as having an angle of α, as measured from the axial direction of the guide. Light rays within the angular acceptance cone defined by α will reflect by TIR and be efficiently transported

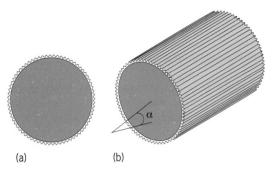

Fig. 3. Basic structure of a prism light guide shown in (*a*) cross section and (*b*) isometric view.

along the guide; and those rays exceeding this angular range will not. Instead, the rays exceeding this angular range will transmit, with little loss, through the side of the guide and enter the region outside the guide. The amount of this light leakage can be easily and carefully controlled, so this angular-dependence property enables the prism light guide to be used as the light distribution system as well as the transport system. This is a unique characteristic of the prismatic film. In contrast, a highly reflective dielectric multilayer film can be used to achieve efficient light transport, but not for the distribution system, as it is highly reflective to light at all angles. Typically, the guide housing is a transparent protective outer structure to keep the guide clean and prevent abrasion of the optical surfaces. Optically precise microstructured films are now readily produced using common polymeric microreplication techniques, so this special prism light-guide geometry enables efficient transport of light at a reasonable cost.

Important physical concepts: TIR and enhanced effective refractive index. Total internal reflection occurs when light reaches an interface with a material of lower refractive index and strikes it at a sufficiently large angle of incidence. TIR is a rich topic in physics because of its many subtleties that lead to deeper physical insights, but for the purpose of this discussion, the simplified explanation of TIR provided here is sufficient.

When a light ray traveling in a medium of higher index of refraction strikes an interface with a second medium of lower index, the ray refracts, or bends, away from the surface normal at an angle defined by a relationship known as Snell's law. TIR occurs whenever Snell's law generates a complex solution for the refracted angle. This occurs for all incident angles greater than a defined critical angle, which depends on the ratio of the refractive index values. (This overly simplified explanation is unsatisfying since it does not explain why TIR occurs. A more complete explanation describes the full electromagnetic treatment of a plane wave incident on an interface and the resultant Fresnel relations provide a complete description of the relationship between the angle of incidence and the intensity of the reflected light.)

A key point is that the critical angle, θ_c, at which TIR occurs depends on the ratio of the refractive index values of the interfacial materials [Eq. (1)],

$$\theta_c = \arcsin(n_2/n_1) \qquad (1)$$

where n_1 is the refractive index of the material in which the light is traveling, and n_2 is the refractive index that is encountered at the interface. This highlights a limitation in a number of applications, as it is difficult to ensure a sufficiently large refractive index ratio to cause TIR to occur for a useful range of angles. In the case of the prism light guide, the lowest refractive index value in the ratio is that of the air inside the guide (1.0003) and the highest value is restricted to that of the polymers that are used for molding precision optical microstructures. High-refractive-index materials would be advantageous since they would reduce the critical angle and therefore increase the angular acceptance range for the light guide. Although there is work underway in the development of high-index polymers, at the moment a practical upper limit is 1.59, which is the refractive index of polycarbonate resin.

Fortunately, the prism light guide embodies an optical technique that effectively enhances the index ratio under certain conditions. **Figure 4** is an isometric drawing of an array of longitudinally oriented prisms, having translational symmetry as shown. Ray A transmits and ray B undergoes TIR, as shown by the projection of these paths onto a cross-sectional plane in Fig. 3*b*.

The propagation of the light ray along the direction of translational symmetry of the system requires a three-dimensional (3D) analysis of the system, and the net result is that the 3D analogue to Snell's law can be described using an effective refractive index ratio that is higher than the refractive index ratio of the materials. This in turn enables TIR to occur over a wider range of azimuthal angles than would be the case without the propagation of the light rays in the direction of translational symmetry. For the application of the prism light guide, this means that light rays traveling in the axial direction along the guide are reflected by TIR, whereas those incident on the film at closer to perpendicular incidence are transmitted through the guide.

Design of a prism light guide. The microstructures lining the guide are typically 200-μm-high right-angle prisms running lengthwise along the pipe. Light rays propagating along the guide will reflect by TIR when

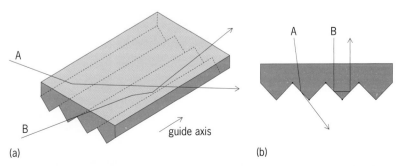

Fig. 4. Light rays in a prism light guide having translational symmetry along the direction of the guide axis: (*a*) isometric view and (*b*) cross-sectional view.

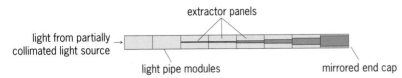

Fig. 5. Extraction from a prism light guide.

they encounter the interface, provided the angle α that the light ray makes with the axial guide direction is within the angular acceptance range given by the maximum angle α_c [Eq. (2)], where n is

$$\alpha_c = \arcsin\sqrt{(3 - 2\sqrt{2})(n^2 - 1)} \qquad (2)$$

the refractive index of the prismatic material. For a typical prism light guide made using polycarbonate resin microstructures that have refractive index value $n = 1.59$, α_c is approximately $30°$. The light emitted from many standard high-intensity light sources can be readily collimated to within this angular range, which makes the prism light guide well suited to large-scale illumination applications. This comparatively large acceptance angle achieved using moderate index polymeric material is a consequence of the index enhancement effect mentioned earlier.

There are some losses associated with the diffraction of light when it encounters the prismatic microstructures, but they are extremely small. The main loss is caused by bulk absorption and scatter in the polymeric microstructure material and optical imperfections of the prismatic surfaces. These losses typically amount to less than 2% per wall reflection and most of the loss is in the form of escaping light rather than absorption. This escaping light is useful for the illumination purpose for which the guide is intended, so the majority of this light is not truly lost. Nevertheless, this intrinsic loss rate limits the number of reflections possible in a light guide that emits light uniformly along its length. In typical designs, the light rays reflect an average number of times ranging from about 3 to 30.

Light extraction. For illuminating engineering applications, it is necessary to incorporate a mechanism that causes the light to controllably escape from the guide. This mechanism is typically known as the light extractor, and it often consists of a light-scattering material that deflects light rays to directions outside the angular acceptance range of the guide. For most applications, it is desirable to have a light output intensity that is relatively constant along the length of the guide. This can be achieved by increasing the density of the extraction material with the distance from the source, to compensate for the decreasing light intensity caused by the induced escape. One such extractor profile is shown in **Fig. 5**.

Uniform light escape is achieved when the product of the extraction rate and the luminous flux in the guide remains constant. This is a complex nonlinear design problem, but it is usually straightforward to find a good solution by design iterations using computer Monte Carlo ray tracing or construction of optical mock-ups.

Current prism light-guide applications. In addition to the piping of light from electric sources, the recent focus of prism light guide has been to bring sunlight deep into buildings. The illuminating engineering industry today is challenged to deliver high-quality natural illumination, while using less electrical energy. To achieve this, it is generally agreed that incorporating daylight within buildings is useful. The prism light guide is an efficient method for piping sunlight, one that has the potential for low net life-cycle costs to enable widespread implementation. A number of demonstrations of core sunlighting systems using prism light guide technology have recently been installed and are under evaluation. Examples of two of these installations are shown in **Fig. 6**.

Conclusion. There are considerable technical challenges associated with piping light, and the prism light guide, a hollow light guide employing linear prismatic microstructures, is one approach that

(a)

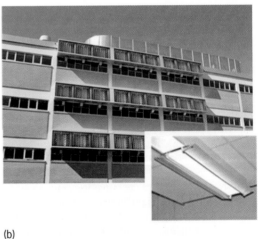

(b)

Fig. 6. Demonstration core sunlighting systems have been installed in (*a*) Burnaby and (*b*) Vancouver, British Columbia, Canada. A concentrated beam of sunlight is directed into prism light guides within the building. The guides are shown in the inset photographs. (*Images courtesy of SunCentral Inc.*)

cost-effectively achieves the benefits of large-scale illumination. Using TIR, the structure guides light within a well-defined range of angles with a very low loss rate, while allowing light rays outside of this angular range to escape, resulting in uniform distribution of light from a remote source. This approach has been widely used with high-intensity electric sources and is currently being adapted to core sunlighting systems that bring sunlight deep into buildings, both to reduce the requirement for electric lighting, and to improve the overall quality of illumination within the space.

For background information *see* ABSORPTION OF ELECTROMAGNETIC RADIATION; DIELECTRIC MATERIALS; ILLUMINATION; LIGHT; LUMINOUS FLUX; OPTICAL PRISM; PHOTOMETRY; REFLECTION OF ELECTROMAGNETIC RADIATION; REFRACTION OF WAVES in the McGraw-Hill Encyclopedia of Science & Technology.

Lorne A. Whitehead

Bibliography. CIE Technical Committee 3-30, *Technical Report on Hollow Light Guides*, March 2005; L. A. Whitehead, New simplified design procedures for prism light guide luminaires, *J. Illum. Eng. Soc.*, 27(2):21–27, 1998; L. Whitehead and M. Mossman, Reflections on total internal reflection, *Opt. Photonics News*, 20(2):28–34, 2009, DOI:10.1364/OPN.20.2.000028; L. Whitehead, M. Mossman, and A. Kotlicki, Visual applications of total internal reflection in prismatic microstructures, *Phys. Can.*, 57:329–335, 2001; L. Whitehead, R. Nodwell, and F. Curzon, New efficient light guide for interior illumination, *Appl. Optics*, 21(15):2755–2757, 1982, DOI:10.1364/AO.21.002755.

Proton radius from muonic hydrogen

The recent measurement of the Lamb shift (2S-2P energy difference) in muonic hydrogen (μp) resulted in a new determination of the root-mean-square (rms) charge radius of the proton, $r_p = 0.8418 \pm 0.0007$ femtometer (1 fm $= 10^{-15}$ m), which is 10 times more accurate than any previous determination from elastic electron-proton (*e-p*) scattering or precision spectroscopy of energy levels in the hydrogen atom (H), but differs by seven standard deviations from these measurements (**Fig. 1**). This discrepancy is now known as the "proton radius puzzle." This article reviews the traditional methods for the determination of r_p, describes the muonic hydrogen experiment, and reviews the attempts to resolve the puzzle, including suggestions for physics beyond the standard model.

Proton charge radius. The proton is a compound object, made from quarks and gluons, held together by the strong nuclear force. For low-energy probes such as elastic electron-proton scattering and atomic systems, however, the proton resembles a "fuzzy ball" of positive charge, with a radius of about 0.8 fm. The radius of the proton r_p is taken as the root-mean-square radius, $r_p = \sqrt{<r_p^2>}$, where $<r_p^2> = \int r^2 \rho(r) \, d^3r$, and $\rho(r)$ is the radial charge distribution of the proton.

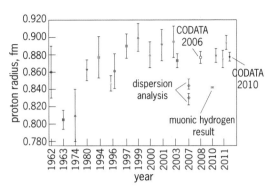

Fig. 1. Measurements of proton charge radius over time from elastic electron-proton scattering (blue, green) and hydrogen atom spectroscopy (red). The muonic hydrogen result disagrees with most previous determinations except for the dispersion analysis of scattering data. The points labeled "CODATA 2006" and "CODATA 2010" are the results of adjustments of the fundamental constants carried out under the auspices of the CODATA Task Group on Fundamental Constants.

Elastic electron-proton scattering. Historically, elastic electron scattering on protons has been used to determine the proton's charge radius. Cross sections are measured for various values of the negative four-momentum transfer squared, Q^2. The Rosenbluth technique allows the separation of the cross sections into electric and magnetic form factors, $G_E(Q^2)$ and $G_M(Q^2)$, which are the Fourier transforms of the charge and magnetization densities, respectively. The rms charge radius is deduced from the slope of $G_E(Q^2)$ at $Q^2 = 0$, which requires a model-dependent extrapolation of measured cross sections to $Q^2 = 0$.

A global Rosenbluth analysis in 2005 of the world data obtained from 1962 until 1980 in Stanford, Orsay, Saskatoon, and Mainz resulted in a value of $r_p = 0.895 \pm 0.018$ fm. This corresponds to an accuracy of 2%. Recently, a new form factor measurement by J. C. Bernauer and colleagues in Mainz has doubled the precision, giving $r_p = 0.879 \pm 0.008$ fm.

However, it should be noted that an alternative way to analyze scattering data using dispersion relations has for a long time provided r_p values around 0.84 fm, in agreement with the muonic value.

Hydrogen spectroscopy. Precision spectroscopy of energy levels in the hydrogen atom is also sensitive to r_p. In the hydrogen atom, r_p is encoded in the Lamb shift, that is, the displacement of energy levels compared to the energy level obtained from the relativistic Dirac equation. Quantum electrodynamics (QED) effects such as self-energy, vacuum polarization, and relativistic and recoil corrections constitute the main contributions to the Lamb shift and have been accurately calculated over the last several decades. The finite size of the proton leads to an additional contribution to the Lamb shift of S states in the hydrogen atom. Atomic S states have the largest wave-function overlap with the nucleus. The electron literally spends some of its time inside the nucleus, where it experiences a reduced Coulomb attraction. The S states are therefore less bound. In the hydrogen atom, this reduction is proportional to the square of the proton charge radius, $<r_p^2>$, and to the square of the electron wave function at the origin, $|\Psi(0)|^2$.

The best measurement of the "classical" Lamb shift, the $2S_{1/2} - 2P_{1/2}$ splitting in the hydrogen atom, performed by S. R. Lundeen and F. M. Pipkin in the 1980s, can be used to extract $r_p = 0.879 \pm 0.027$ fm. This 3% relative accuracy is limited by the natural linewidth of 100 MHz and required a determination of the line center to 9 kHz, that is, 10^{-4} of the linewidth.

More recently, the advent of precision laser spectroscopy allowed accurate measurements of optical two-photon transitions in the hydrogen atom, like $2S$-nl ($nl = 6D, 8S, 8D, 12D$). These states all have much longer lifetimes, leading to significantly narrower linewidths (for example, 572 kHz for the $2S$-$8D$ transition) and hence higher precision. Such a measurement, however, determines not the pure $2S$ Lamb shift but a combination of the Lamb shift and the gross structure of the hydrogen atom (because the transition is between levels with different principal quantum numbers n).

In a simplified picture, the energy levels of S states in the hydrogen atom are given by Eq. (1). The first

$$E_{nS} \approx -\frac{R_\infty}{n^2} + \frac{L_{1S}}{n^3} \qquad (1)$$

term corresponds to the well-known Bohr formula and contains the Rydberg constant R_∞, which governs the gross structure of the hydrogen atom with principal quantum number n. The second term is the Lamb shift of S states containing the finite size effect proportional to $<r_p^2>$, which is maximal for the $1S$ ground state, $L_{1S} \approx 8171.641(4) + 1.5648 <r_p^2>$ MHz, when r_p is given in femtometers. The approximate $1/n^3$ scaling of the Lamb shift term reflects the wave function overlap as explained above.

A measurement of two optical transitions between energy levels with different n can be used to simultaneously extract the two unknowns R_∞ and L_{1S}, and hence r_p (**Fig. 2a**). Here one exploits the different scaling in n of the gross structure ($1/n^2$) and Lamb shift ($1/n^3$). The $1S$-$2S$ transition has been measured with an accuracy of 10 Hz ($4.2 \cdot 10^{-15}$ relative precision), but the determination of r_p is limited by the lack of a sufficiently accurate value of the Rydberg constant. The Rydberg constant R_∞ is mainly determined by several transition frequencies between the $2S$ state and higher excited states (nl) in the hydrogen and deuterium (D) atoms, for example, $2S$-$8S$, $2S$-$8D$, $2S$-$12D$, individually measured with an accuracy on the level of 10 kHz, that is, around 10^{-11} relative precision. The CODATA 2010 global analysis of all measurements in the hydrogen and deuterium atoms gives R_∞ with a relative precision of $7.4 \cdot 10^{-12}$. This results in $r_p = 0.876 \pm 0.008$ fm from spectroscopy in the hydrogen and deuterium atoms, in excellent agreement with the most recent value from elastic e-p scattering.

Muonic hydrogen spectroscopy. In muonic hydrogen, the electron is replaced with a negative muon μ^-. The muon is, like the electron, a point particle,

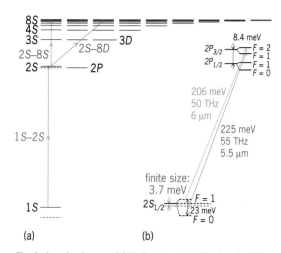

Fig. 2. Level schemes. (a) Hydrogen atom. The Lamb shifts contain information about the proton radius and are indicated as broken lines. Some optical two-photon transitions are shown. **(b)** Muonic hydrogen (μp), with the two measured transitions. The large finite size effect is indicated.

but its mass is about 200 times larger, leading to a 200 times smaller Bohr radius in muonic hydrogen, compared to the hydrogen atom. This increases the wave function overlap, and hence the sensitivity to the finite size of the proton, by $200^3 \approx 10^7$. In muonic hydrogen, the Lamb shift relates to r_p as in Eq. (2)

$$\Delta E_{2S-2P_{1/2}} = 206.0573(45) - 5.2262 r_p^2$$
$$+ 0.0347 r_p^3 \text{ meV} \qquad (2)$$

when r_p is given in femtometers. The last term, proportional to r_p^3, parameterizes higher moments of the charge distribution $\rho(r)$. The main uncertainty is related to the proton polarizability involving two-photon exchange processes with virtual excitations of the intermediate proton. The polarizability contribution amounts to $\Delta E_{\text{TPE}} = 0.015(4)$ meV.

One sees that the finite size effect in muonic hydrogen is as large as 2% of the Lamb shift. This is why a measurement of the Lamb shift in muonic hydrogen has for a long time been considered the most sensitive method to determine r_p. The author and his colleagues have recently determined the level spacing of two hyperfine sublevels in muonic hydrogen to be given by Eq. (3), from which the value

$$\nu \left(2S_{1/2}^{F=1} \rightarrow 2P_{3/2}^{F=2} \right) = 49,881.88(76) \text{ GHz} \qquad (3)$$

$r_p = 0.8418 \pm 0.0007$ fm is extracted.

This experiment (**Fig. 3**) was made possible only by recent advances in accelerator and laser technology. A novel low-energy muon beam line was built at the Paul Scherrer Institute (PSI) in Switzerland, which provides the world's most intense muon source. Low-energy muons with only a few kiloelectronvolts kinetic energy are required to obtain a sufficiently high muon stop probability in a small volume of H_2 gas at a pressure of only 1 mbar (100 Pa). This very low target gas pressure is required to ensure a sufficiently long lifetime of the

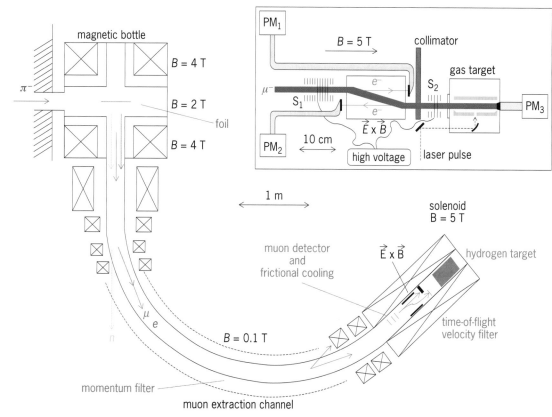

Fig. 3. Muon beam line of the muonic hydrogen experiment. Pions injected into the magnetic bottle decay into muons, which are decelerated in a thin foil until they escape axially. The muon extraction channel selects the muon momentum and guides the muons into the solenoid where the experiment takes place. The inset shows the interior of the solenoid: Muons passing thin carbon foils S_1 and S_2 eject electrons, which are detected in plastic scintillators read out by photomultipliers (PM₁, PM₂, and PM₃). A laser pulse illuminates the muon stop volume in the H_2 target gas.

metastable $\mu p(2S)$ state, $\tau_{2S} \approx 1$ μs. At higher gas pressures the $\mu p(2S)$ atoms are quickly depopulated by collisions with H_2 molecules.

Upon arrival, each muon is detected and a pulsed laser is triggered. It provides laser pulses of 0.2 mJ pulse energy, tunable around $\lambda \approx 6.0$ μm, less than 1 μs after the trigger signal. A few hundred muons per second are stopped in the H_2 gas, but 99% of these end up in the $1S$ ground state. Only 1% forms long-lived $\mu p(2S)$, which can be excited to the $2P$ level when the laser light is on resonance. A multipass mirror cavity ensures efficient illumination of the muon stop volume in the target gas.

Successful $2S \rightarrow 2P$ laser excitation on resonance is signaled by immediate $2P \rightarrow 1S$ deexcitation, accompanied by the emission of a K_α x-ray at 1.9 keV. The number of K_α x-rays as a function of the laser frequency displays the resonance shown in **Fig.** 4. On the peak of the resonance, 6 events per hour were recorded.

Proton radius puzzle. The muonic value is an order of magnitude more precise than any of the previous determinations, but 4% smaller, corresponding to a 7-standard-deviation discrepancy to the 2010 CODATA result for r_p.

All 3 measurements (scattering, and spectroscopy of the hydrogen atom and muonic hydrogen) could be reconciled if the proton's charge distribution $\rho(r)$ had a small core and a large halo, but the resulting

large values of higher moments of $\rho(r)$ are excluded by elastic electron-proton scattering.

It was noted that the observed 2S-2P transition in muonic hydrogen would be shifted by the observed amount by the presence of an additional electron, for example in a molecular ion p-μ-e, but three-body calculations showed that no such stable molecular ions exist. Also, the shape and width of the observed resonance in muonic hydrogen disfavor the p-μ-e ion

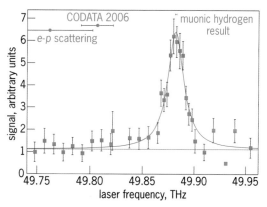

Fig. 4. The measured resonance curve in muonic hydrogen, compared with predicted positions using previously measured proton charge radii (green points with error bars). The discrepancy between the observed and expected position using the CODATA 2006 value is 75 GHz.

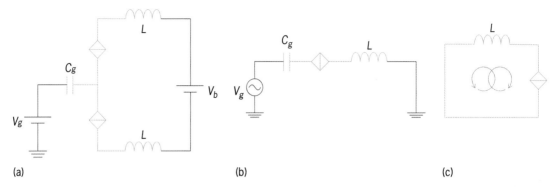

Fig. 3. Applications of QPS. (*a*) dc QPS transistor electrometer. (*b*) QPS box electrometer. (*c*) QPS qubit. The qubit states consist of superpositions of clockwise and anticlockwise circulating currents. Superconducting parts of the circuits are shown in blue. QPS junctions are represented by the diamond elements; *L* represents an inductor, realized via the kinetic inductance of a strip of superconductor; C_g represents a gate capacitor through which the gate voltage V_g controls the charge on the superconducting island.

temperature of 10–20 mK, which is sufficient to observe QPS.

Possible applications. One of the first practical applications of QPS might be a quantum current standard. This would be dual to the Josephson voltage standard (JVS), which is used to maintain the International System of Units (SI) volt. The JVS utilizes the ac Josephson effect, which gives rise to quantized voltage steps known as Shapiro steps when a Josephson junction is driven by a microwave-frequency ac current (Fig. 2*b*). A QPS-based quantum current standard would utilize the dual ac QPS effect, which is predicted to cause quantized current steps known as dual Shapiro steps when a QPSJ is driven by a microwave-frequency ac voltage. These steps are quantized in integer multiples of $2ef$, where f is the ac frequency, because an integer number of Cooper pairs are transported through the QPSJ per cycle of the microwave drive. *See* QUANTUM VOLTAGE STANDARDS.

A quantum current standard of sufficient accuracy would enable redefinition of the ampere, which is presently defined in terms of the force between two parallel wires, but is in practice realized by a combination of the JVS and the quantum Hall resistance standard. In addition, a quantum current standard would enable an intercomparison between the three quantum electrical standards, known as the metrological triangle, which would yield an improved understanding of the fundamental constants h and e and their universality across different materials. A sufficient accuracy to achieve these aims is 1 part in 10^7 or better at a current greater than 100 pA. The latter condition is due to practical considerations.

A QPS quantum current standard would have two advantages over other technologies such as electron pumping through gated semiconductor devices: first, being based on a truly quantum coherent effect, and second, being realized in a superconducting system that could, in the future, be incorporated onto the same chip as the JVS, allowing simplification of the quantum electrical standards.

Another application of QPS is for sensitive electrometry, that is, charge measurement. Two types of QPS electrometer can be envisaged, dual to the

dc and radio-frequency (rf) superconducting quantum interference devices (SQUIDs), which detect magnetic flux. The dc QPS transistor electrometer comprises two QPS junctions in series and is sensitive to the charge on the superconducting island between them (**Fig. 3***a*). A prototype has been demonstrated with Nb_xSi_{1-x} nanowires and exhibits periodic current oscillations as the charge on the island is controlled via a gate electrode. The rf QPS box electrometer, dual to the rf SQUID, has yet to be demonstrated (Fig. 3*b*).

A third application of QPS is the QPS qubit, which consists of a superconducting loop interrupted by a superconducting nanowire (Fig. 3*c*). A QPS event causes a fluxon to tunnel into or out of the loop, changing the direction of the circulating current. Transitions between the qubit states can be excited by a microwave field. This has recently been demonstrated with an InO_x qubit coupled to a coplanar transmission line resonator, which supplies the microwave field. Other superconducting qubits contain Josephson junctions, which harbor spurious two-level charge fluctuators within the insulating layer, causing decoherence (loss of information) of the qubit state. In principle, the QPS qubit avoids this.

For background information *see* ELECTRICAL UNITS AND STANDARDS; ELECTROMETER; JOSEPHSON EFFECT; LOW-TEMPERATURE PHYSICS; MICROLITHOGRAPHY; QUANTUM COMPUTATION; QUANTUM MECHANICS; SQUID; SUPERCONDUCTING DEVICES; SUPERCONDUCTIVITY; UNCERTAINTY PRINCIPLE in the McGraw-Hill Encyclopedia of Science & Technology.
Carol H. Webster

Bibliography. A. Bezryadin, Tunnelling across a nanowire, *Nature*, 484:324–325, 2012, DOI:10.1038/484324b; J. Gallop, The quantum electrical triangle, *Phil. Trans. Roy. Soc. A*, 363:2221–2247, 2005, DOI:10.1098/rsta.2005.1638; D. Haviland, Quantum phase slips, *Nat. Phys.*, 6:565–566, 2010, DOI:10.1038/nphys1747; G. Schön, Superconducting nanowires, *Nature*, 404:948–949, 2000, DOI:10.1038/35010260; M. Tinkham, *Introduction to Superconductivity*, 2d ed., McGraw-Hill, New York, 1996, Dover, Mineola, NY, reprint, 2004.

Quantum voltage standards

A quantum standard of voltage is made possible by the Josephson effect, a remarkable property of a thin junction between two superconducting materials. The effect was predicted theoretically in 1962 by Brian Josephson, after whom it is named, and it has revolutionized the way in which standards of voltage are maintained and disseminated across the world. It relates voltage to frequency with a constant of proportionality that depends only on the Planck constant, h, and the elementary charge, e. Thus, voltage standards can be realized at any time or place with an access to just a frequency reference. It is estimated that in excess of 50 primary standards of voltage are in existence at laboratories around the world, and they work together to provide a consistent and stable basis for voltage measurement in science, engineering, and manufacturing.

Josephson junction. A Josephson junction (**Fig. 1**) is formed when two superconducting materials are separated by a thin insulating barrier, typically less than 100 nm. Superconductivity is characterized by the dissipationless flow of Cooper pairs of electrons. These Cooper pairs can tunnel through the insulating layer from one superconductor to the other, and so the junction is referred to as a weak link. The tunneling gives rise to a direct current (dc), which flows without an associated voltage drop until the critical current of the junction is reached. This is known as the dc Josephson effect. When the critical current is exceeded, microwave voltage oscillations develop across the junction, described by the alternating-current (ac) Josephson effect. It is this second effect that gives rise to a quantum standard of voltage, because it relates the average value of the voltage across the junction to the frequency of the microwave oscillations. Originally it was envisaged that a voltage standard could be made by simply observing the frequency of the oscillations for a given applied voltage. However, it is difficult to achieve a high level of accuracy with this approach, so practical standards rely instead on the application of an external microwave stimulus and the

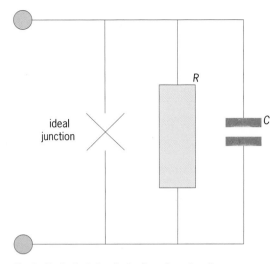

Fig. 2. Equivalent circuit of a Josephson junction.

observation of the mean voltage developed across the junction.

Junction as a voltage standard. When operated as a voltage standard, the mean voltage across the junction is the parameter measured, and this represents the average of the voltage oscillations occurring at gigahertz frequencies. The dynamics of the junction depend on its physical parameters, which can be approximated by an equivalent circuit consisting of an ideal junction in parallel with a resistance and a capacitance (**Fig. 2**). The operation of the junction can be explained by considering its impulse response. The ideal junction is described by nonlinear equations derived from a combination of the dc and ac Josephson equations, with the capacitance and resistance included using normal circuit theory. A numerical simulation of the junction response to a short current pulse, for a range of pulse amplitudes, is shown in **Fig. 3**. The fundamental parameter is the area under the curve of the voltage waveform, which is a constant, even though the shape of the waveform changes. By considering the initial and final states of the junction, it can be shown that this area is quantized in integer multiples n of $h/2e$, where h is the Planck constant and e is the elementary charge (the magnitude of the charge on the electron). This quantity is also known as the flux quantum, Φ_0, and is approximately 2.07×10^{-15} V·s. It is this quantization that leads to a quantum standard of voltage, a constant pulse repetition frequency f giving rise to a constant average voltage $\bar{V} = nhf/2e$. A junction can be biased with a pulse stream or with single-frequency sine waves. The integer n can conveniently be selected by adding a dc bias in parallel with the microwave bias, leading to a current–voltage or $I - V$ characteristic plot as shown in **Fig. 4**. The voltage plateaus indicate the range of bias parameters over which n has a constant value.

Junction parameters. The junction is characterized by its critical current, I_C, its capacitance, C, and the parameter R, which represents the flow of normal electrons or quasiparticles across the junction. All three depend on the details of the fabrication

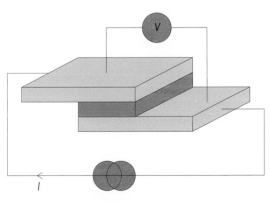

Fig. 1. Schematic diagram of a Josephson junction showing superconducting elements (light blue) separated by an insulating layer (dark blue). The junction properties are measured using a current source and a voltmeter in a 4-wire connection.

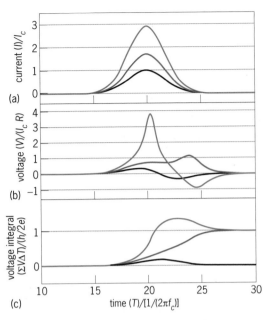

Fig. 3. Numerical simulation of the response of a Josephson junction to a current pulse. (*a*) Current pulse stimulus, normalized to the junction critical current. (*b*) Instantaneous voltage across the junction, normalized to the characteristic voltage. (*c*) Integral of the voltage with time, normalized to $h/2e$. The time axis is normalized to the characteristic frequency. The lowest pulse amplitude (black curves) results in a zero value for the voltage integral. The two higher amplitudes (blue and red curves), though different by 50%, result in the same voltage integral, demonstrating the voltage quantization.

process and can be optimized for different applications. The critical current scales with the junction area, so it is normal practice to quote a critical current density, J_C, for a given junction design. Junctions used for quantum voltage standards have been fabricated over a wide range of J_C (10^1–10^5 A·cm^{-2}). The size of the junctions ranges from a few micrometers square up to 20 μm × 50 μm, and the critical current typically covers a smaller range from 100 μA to 10 mA. Key parameters are the product $I_C R$, known as the characteristic voltage V_C, and the corresponding characteristic frequency, $f_C = V_C(2e/h)$, as f_C determines the useful upper operating frequency of the junction. Values of V_C range from 10 μV to 100 μV, setting a frequency range of approximately

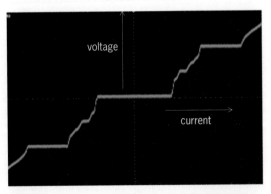

Fig. 4. Example of an $I - V$ curve for a Josephson junction biased with a combination of dc and microwave ac. The central $n = 0$ constant voltage plateau or step is shown together with the $n = \pm 1$ steps.

10–100 GHz for practical voltage standards. Finally, the dimensionless ratio $\beta = 2\pi f_C RC$ determines the damping of the junction, with $\beta \gg 1$ corresponding to underdamped junctions exhibiting an $I - V$ curve with hysteresis, and $\beta \ll 1$ describing overdamped junctions with a single-valued $I - V$ curve. Early Josephson junctions for voltage standards were highly hysteretic, to the extent that constant-voltage plateaus in the $I - V$ curve crossed the zero current axis. More recently, research has concentrated on damped junctions, because their single-valued $I - V$ curve means that they can be uniquely biased to an operating point, which is important for applications such as waveform synthesis.

Josephson junction arrays. Even at a microwave bias frequency of 70 GHz, the first voltage plateau, corresponding to $n = 1$ for a single Josephson junction, is still only approximately 140 μV. A single junction can be operated on at least plateau $n = 10$, but the resulting 1.4 mV is still 1000 times smaller than the voltages that are routinely used by the measurement and test industry. This problem has been solved by connecting a large number of junctions in series with a common microwave and dc bias. Arrays of more than 10,000 junctions have been fabricated, giving an output voltage in the region of 10 V. The fabrication of these arrays is a significant engineering challenge, because the junctions all have to have the same parameters to within a few percent and have to receive a similar level of microwave bias. The junctions are arranged in parallel microwave striplines, and individual lines are joined in series to obtain the total output voltage.

Junction technology. The simplest junction design for a quantum voltage standard consists of a thin-film trilayer to form a superconductor–insulator–superconductor (SIS) junction. Niobium is deposited on a silicon substrate, followed by aluminum which is subsequently oxidized to Al$_2$O$_3$, and then another layer of niobium. The layers are typically 150, 15, and 75 nm, respectively. The individual junctions are patterned using reactive-ion etching and sputter etching and then interconnected using a further niobium wiring layer. Finally, the junctions are covered with a 2-μm layer of silicon dioxide and a niobium layer added to form the ground plane for the microwave stripline. Junctions of this type have the lowest critical current density of about 10 A·cm^{-2} and are hysteretic. Arrays of nonhysteretic junctions can be fabricated using a variety of layer compositions. An aluminum layer can be introduced to form a superconductor–insulator–normal metal–insulator–superconductor (SINIS) junction. Such junctions have a J_C of approximately 100 A·cm^{-2} and a relatively high V_C of 120 μV, enabling effective operation over the frequency range 70–100 GHz. Alternatively, a trilayer of superconductor–normal metal–superconductor (SNS) can be used with, for example, palladium–gold as the normal metal. This yields a high J_C in the region of 10^5 A·cm^{-2} but a lower V_C of 20 μV, so these junctions are suitable for operation at around 20 GHz. Other layers include molybdenum silicide and titanium nitride.

Voltage measurements. A practical quantum voltage standard consists of an array of Josephson junctions operating at a liquid helium temperature of 4.2 K; a microwave source, normally a Gunn diode; and an adjustable dc bias for selecting the required voltage plateau. The frequency of the radiation from the Gunn diode is measured with a counter and is stabilized against a known frequency reference, normally at 10 MHz. The Planck constant and the elementary charge are currently measured quantities in the SI system. Their values are routinely updated and have an associated estimated uncertainty. To ensure consistency of voltage measurements in terms of the Josephson effect at all locations, an agreed value of the voltage-to-frequency conversion coefficient is therefore used and was set in 1990 as $K_{J-90} = 483,597.9$ GHz·V^{-1} with no uncertainty. Industrial voltage standards, normally based on Zener diodes, can be compared with an array of junctions set to the same voltage with an uncertainty of better than 100 nV. The use of quantum voltage standards to measure steady voltages is now routine, and systems for generation and measurement of voltage waveforms are currently under development. For this application, arrays of nonhysteretic junctions, subdivided by junction number in a binary sequence, can be configured as a digital-to-analog converter with perfect linearity and quantum accuracy of the output voltage. Such arrays can be biased with different bit patterns to generate a stepwise-approximated waveform. Alternatively, a single-segment array can be biased with short pulses of varying repetition rate to achieve a continuously variable voltage. This latter technique has the potential for higher accuracy but is currently limited in output voltage to a few hundred millivolts.

For background information *see* ELECTRICAL UNITS AND STANDARDS; FUNDAMENTAL CONSTANTS; JOSEPHSON EFFECT; SUPERCONDUCTING DEVICES; SUPERCONDUCTIVITY in the McGraw-Hill Encyclopedia of Science & Technology.   Jonathan Williams

Bibliography. S. P. Benz and C. A. Hamilton, Application of the Josephson effect to voltage metrology, *Proc. IEEE*, 92(10):1617–1629, 2004, DOI:10.1109/JPROC.2004.833671; C. Kittel, *Introduction to Solid State Physics*, 8th ed., Wiley, Hoboken, NJ, 2005; J. Kohlmann et al., Improved 1 V and 10 V Josephson voltage standard arrays, *IEEE Trans. Appl. Supercond.*, 7(2):3411–3414, 1997.

Quintuple bonding

Bond orders, formally the number of electron pairs that contribute positively to a chemical bond between atoms, matter fundamentally. This becomes obvious by considering the simple hydrocarbons ethane, ethylene, and acetylene, which have a bond order of one, two, and three, respectively. All three consist of two carbon atoms linked via a single, double, or triple bond. Ethane, the molecule having the single bond, is a low-value compound, especially compared to ethylene or acetylene. Because of its

low reactivity (it does not react easily with most compounds), ethane is mainly burned to generate heat. Ethylene, having the carbon–carbon double bond, is much more reactive. It is the basis of the most important synthetic polymer, polyethylene (PE). We are surrounded by PE use in everyday life, including plastic bags, car parts, packing material, water pipes, and artificial hips, to name some applications. Acetylene, having a triple bond, is also of great importance in the chemical industry. It is already too reactive and has to be treated with much care to avoid explosions.

The elements of which these simple hydrocarbons are made belong to the main-group elements. It is broadly accepted in the chemical community that the maximum bond order between such elements is three. And a triple bond does not necessarily mean high reactivity. The dinitrogen molecule, a major component of the air, is as unreactive as a rock and has a triple bond too.

Higher bond orders than three are accessible for the transition metals. Here the availability of s and d orbitals allow bond orders up to six. Theoretically, six electron pairs are available and can be used to contribute positively to link two of such metal atoms (**Fig. 1**).

A breakthrough in terms of bond orders higher than three was published about 50 years ago, when $Re_2Cl_8^{2-}$ was synthesized, characterized, and analyzed by a few chemists and found to have one sigma (σ), two pi (π), and, as the novel feature, one delta (δ) bond. This key compound inspired many scientists to work on quadruply bonded coordination compounds. The still ongoing activities are documented in many (text) books and scientific articles. The next breakthrough in terms of unusually high bond orders was made a few years ago, in 2005, when the research group of P. P. Power reported the first stable molecule having a quintuple bond (**Fig. 2a**).

When new classes of compounds are made, we have to integrate them into our existing chemical concepts. Two issues had to be addressed in this case: the Cr–Cr bond length (Fig. 2a) is not appreciably short, and chromium is not the best choice in terms of "real" high bond orders. In Fig. 2a, the bond length of the quintuple bond is 1.8351(4) Å. It

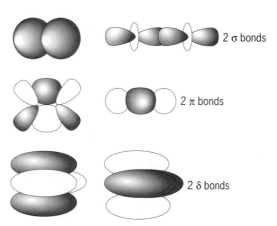

Fig. 1. The six bonds that can be used to link transition metals.

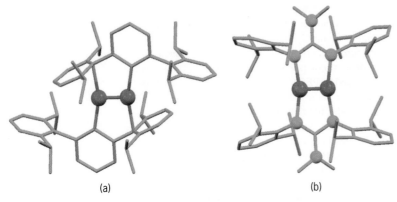

(a)   (b)

Fig. 2. Dichromium complexes having a quintuple bond (atom color code: red = Cr, green = N, orange = C). (*a*) The first example (Cr–Cr distance about 1.84 Å). (*b*) The shortest metal–metal bond observed so far (Cr–Cr distance about 1.73 Å).

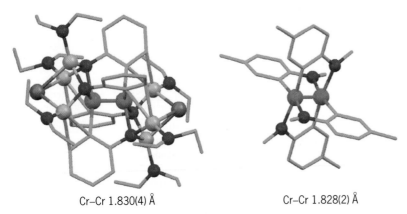

Cr–Cr 1.830(4) Å   Cr–Cr 1.828(2) Å

Fig. 3. Quadruply bonded Cr₂ complexes having very short metal–metal distances (atom color code: red = Cr, orange = C, blue = O, purple = Li, brown = Br).

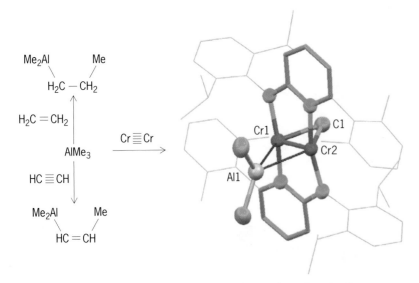

Fig. 4. Carboalumination of a Cr–Cr quintuple bond (atom color code: red = Cr, orange = C, pink = Al).

true for the simple hydrocarbons discussed earlier. The C–C distance in ethane is longer than that of ethylene, which in turn is longer than that of acetylene. Maybe only the main-group elements follow the trend? Many chemists became motivated to address this problem, and in doing so set a new world record for the shortest metal–metal bond. In summary, the metal–metal distance in quintuply bonded dichromium complexes is strongly influenced by the ligand holding the two chromium atoms together. By tuning the steric properties of the ligand, the metal–metal distance can be influenced. A Cr–Cr distance of 1.73 Å is observed for the complex in Fig. 2*b*, the shortest metal–metal bond observed in stable molecules so far.

Quintuple bonding was first defined by P. P. Power and colleagues as follows. "The description 'quintuple bond' is intended to indicate that five electron pairs play a role in holding the metal atoms together. It does not imply that the bond order is five or that the bonding is very strong, because the ground state of the molecule necessarily involves mixing of other higher-energy configurations with less bonding character. This gives lower, usually noninteger, bond orders." For a better understanding one may read the term "bond order" used in the middle of the definition as "effective bond order." This means that quintuple bonding involves a formal bond order of five as defined earlier, but not an effective bond order of five. Effective bond order calculations for dichromium compounds with high bond orders gave noninteger numbers between three and four only. Chromium is not ideally suited to give very high bond orders because of the mismatch in size of the *s* and *d* orbitals involved in multiple-bond formation. The higher homologs of the group, namely, molybdenum (Mo) and tungsten (W), are much better suited. Here, effective bond orders of 5.2 are observed for the transient species Mo₂ and W₂. The synthesis of a stable dimolybdenum complex having a quintuple bond seems to be a good choice for obtaining maximum effective bond orders. For molybdenum the orbital mismatch is less relevant.

Having stable compounds with formal quintuple bonds in hand allows chemists to explore their reaction behavior, to explore their reactivity. Quintuply bonded bimetallic complexes should have a potential for small-molecule activation. Particularly of interest is the diatomic platform that can provide from two to eight (in principle, even 10) electrons. These bimetallic complexes feature not just low-valent metal centers, they are also coordinatively highly unsaturated. All these features are characteristic of high reactivity. Conceptually, one of the most surprising reactions observed yet is the carboalumination of a Cr–Cr quintuple bond.

A chromium aminopyridinate dimer having a quintuple bond was reacted with trimethyl aluminum (**Fig. 4**), causing the quintuple bond to insert into the metal–carbon bond. In the course of this reaction, the bond order was reduced to a quadruple bond. Carbon–carbon double and triple bonds are well known to undergo such carboalumination reactions,

is thus longer than the shortest metal–metal bonds observed for quadruply bonded dichromium complexes (**Fig. 3**).

Chemists were surprised. Maybe the concept that an increase in bond order goes along with a decrease of the bond length was wrong? Anyway, it is

which are very important tools in organic synthesis. Highly functionalized pharmaceuticals are made by applying this methodology. The quite exotic quintuple bond shows reactivity features that are similar to those observed for rather simple carbon–carbon double or triple bonds. Maybe the sketchy picture we draw—five lines between two metal atoms—is not as far from reality as it seemed in the first place.

For background information *see* CHEMICAL BONDING; COORDINATION CHEMISTRY; MOLECULAR ORBITAL THEORY; TRANSITION ELEMENTS in the McGraw-Hill Encyclopedia of Science & Technology.

Rhett Kempe

Bibliography. F. A. Cotton et al., Mononuclear and polynuclear chemistry of rhenium(III): Its pronounced homophilicity, *Science*, 145:1305–1307, 1964, DOI:10.1126/science.145.3638.1305; T. Nguyen, A. D. Sutton, M. Brynda, J. C. Fettinger, G. J. Long, and P. P. Power, Synthesis of a stable compound with fivefold bonding between two chromium(I) centers, *Science*, 310:844–847, 2005, DOI:10.1126/science.1116789; A. Noor et al., Carboalumination of a chromium–chromium quintuple bond, *Nature Chem.*, 1:322–325, 2009, DOI:10.1038/nchem.255; A. Noor et al., Metal-metal distances at the limit: Cr–Cr 1.73 Å—The importance of the ligand and its fine tuning, *Z. Anorg. Allg. Chem.*, 635:1149–1152, 2009, DOI:10.1002/zaac.200900175; A. Noor and R. Kempe, The shortest metal-metal bond, *Chem. Rec.*, 10:413–416, 2010, DOI:10.1002/tcr.201000028; B. O. Roos, A. C. Borin, and L. Gagliardi, Reaching the maximum multiplicity of the covalent bond, *Angew. Chem. Int. Ed.*, 46:1469–1472, 2007; S. Shaik et al., Quadruple bonding in C2 and analogous eight-valence electron species, *Nat. Chem.*, 4:195–200, 2012, DOI:10.1038/nchem.1263; Y.-C. Tsai et al., Journey from Mo–Mo quadruple bonds to quintuple bonds, *J. Am. Chem. Soc.*, 131:12534–12535, 2009, DOI:10.1038/nchem.1263; F. R. Wagner, A. Noor, and R. Kempe, Ultrashort metal-metal distances and extreme bond orders, *Nat. Chem.*, 1:529–536, 2009, DOI:10.1038/nchem.359.

Radioisotopes for medical diagnosis and treatment

Medical radioactive isotopes will continue to play a major role in the advancement of twenty-first-century medicine. They are currently showing outstanding results in both diagnostic and therapeutic medical applications, which will continue to expand for all major diseases (cancer, heart disease, arthritis, Alzheimer disease, and so forth) for the rest of the century. Nuclear medicine uses radiation to provide diagnostic information about the functioning of a person's specific organs, or to treat them. Radiotherapy can be used to treat some medical conditions, especially cancer, using radiation to weaken or destroy particular targeted cells.

Medical radioisotope production. Radioisotopes are artificially produced as unstable atoms of the same chemical element, which have a different number of neutrons in the nucleus, but the same number of protons, the same chemical properties determined by the electrons, and different weights. Radioisotopes can be manufactured in several ways. The most common is by neutron activation in nuclear reactors involving the capture of a neutron by the nucleus of an atom, resulting in an excess of neutrons (neutron-rich isotopes). Some radioisotopes are manufactured by accelerators, such as cyclotrons, in which protons are introduced into the nucleus, resulting in a deficiency of neutrons (proton-rich isotopes). The stable isotope is transmuted into an unstable isotope of the same or a different element. Thus, nuclear reactors and the cyclotrons complement each other in the production process for a full range of radioisotopes. However, a few radioisotopes are exceptions to this rule and can be produced by both facilities, such as iodine-125 (^{125}I) and palladium-103 (^{103}Pd).

Applications of radioisotopes in diagnosis. Nuclear medicine utilizes radioisotopes for diagnostic and therapeutic applications (**Fig. 1**). The most widely used radioisotope for diagnostic applications in nuclear medicine is technetium-99m (^{99m}Tc). Approximately 80% of the 30 million radiodiagnostic procedures are carried out worldwide every year with this single isotope (**Fig. 2**). This percentage share is expected to remain as such in the near future because of its availability from the convenient and cost-effective molybdenum-99 (^{99}Mo)/^{99m}Tc generator. Until recently, there were only five research reactors (**Fig. 3**) irradiating targets to produce 90%–95% of the global ^{99}Mo supply: three in Europe [Belgium Reactor 2 (BR2) in Belgium, High Flux Reactor (HFR) in the Netherlands, and OSIRIS in France], one in Canada [National Research Universal (NRU)], and one in South Africa (SAFARI-1). However, all these reactors are more than 45 years old and subject to extended shutdown periods for maintenance to ensure safe and reliable operation. Between 2009 and 2010, the NRU and HFR reactors

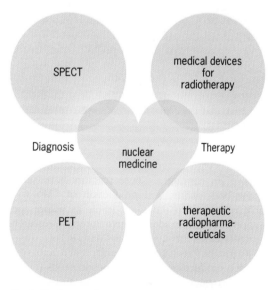

Fig. 1. Nuclear medicine diagnosis and therapy.

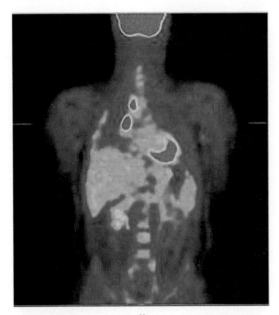

Fig. 2. The decay product of ⁹⁹Mo—the metastable isotope ⁹⁹ᵐTc—is used for scans.

Fig. 3. Top view of the BR2 reactor.

were subject to extended shutdowns, causing global ^{99}Mo/^{99m}Tc shortages. These reactors have since come back online, and production of ^{99}Mo was able to return to the same levels as before the supply shortage. In addition, two European research reactors joined the supply chain (Maria in Poland and LVR-15 in the Czech Republic), and other reactors expanded production beyond domestic needs [Open Pool Australian Lightwater (OPAL) in Australia and RA-3 in Argentina]. Several alternative production

routes are currently being investigated, especially in the United States and Canada, without the use of highly enriched uranium.

In vivo diagnostic techniques. These techniques are based on an approach called the tracer principle. A radionuclide—in a carefully chosen chemical form and duration of being radioactive called a radiopharmaceutical—is administered to the patient to trace a specific physiological phenomenon by means of a special detector, often a gamma camera, placed outside the body. The radiopharmaceutical can be designed to seek out only desired tissues or organs, such as the lungs. Some 100–300 available radiopharmaceuticals, mostly organic in nature and labeled with artificial radionuclides, such as ^{99m}Tc, indium-111 (^{111}In), and gallium-67 (^{67}Ga), are used to study organs and tissues without disturbing them. Nuclear diagnostic methods expose the patient to a small radiation dose. This is minimized further with the use of shorter-lived radioisotopes, which decay to a stable form within hours. The strength of the technique lies in the fact that use can be made of substances, which move to organ systems in very selective ways. The link to radioactive substances enables imaging of the distribution of the tracer in the human body with a gamma camera or a positron emission tomography (PET) scanner. A major advantage is the high sensitivity that enables measuring molecular processes at picomolar level.

Three different modalities are available for this process: planar scintigraphy, single-photon emission computed tomography (SPECT), and PET.

Planar scintigraphy is the simplest available technique yielding a two-dimensional projection image of the activity distribution of the tracer (^{99m}Tc, for example) in the human body. The technique is based on the gamma radiation that is emitted by decay of the radioisotope.

SPECT provides three-dimensional images of the nuclear activity distribution with a resolution of 0.5–1 cm, which is not very high compared to other techniques (**Fig. 4**). Imaging involves an intravenous injection of two substances, a tracer (radioisotope) and a blood flow agent. The radioisotopes usually used in SPECT scans are ^{99m}Tc, iodine-123 (^{123}I), and thallium-201 (^{201}Tl).

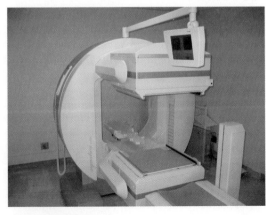

Fig. 4. SPECT scanner.

PET is an imaging technique with radioactive isotopes [fluorine-18 (^{18}F), carbon-11 (^{11}C), nitrogen-13 (^{13}N), oxygen-15 (^{15}O), and copper-64 (^{64}Cu)] that emit positrons during their decay. Compared to SPECT, PET offers better sensitivity, higher resolution, and the possibility of accurately quantifying the examined processes.

In vitro techniques. These techniques allow clinical diagnosis without the patient being exposed to radiation. In fact, the patient need not even to be present. A blood sample taken from the patient is sent to the laboratory and examined through nuclear techniques such as radioimmunoassay (RIA) or immunoradiometric assay (IRMA). These tests measure precisely previous and current exposure to infection by assessing antibodies. Another application of RIA and IRMA techniques includes the detection of tumor markers, which are specific substances secreted by many, but not all, tumors that can indicate the presence of malignancy.

From a research perspective, the utilization of in vivo and in vitro markers will play a vital role in future nuclear medicine. In vitro tests—which include the genetic analysis of blood samples—will be much more important in identifying high-risk patient populations, such as individuals at risk for developing cardiovascular disease or cancer. In vitro tests will be used to predict a patient's response to a certain type of therapy, followed up with molecular imaging. The latter is a research area that is currently supported by the International Atomic Energy Agency (IAEA), which is working on five specific project areas including nuclear medicine imaging in the management of incommunicable diseases; the application of PET in molecular imaging; in vitro nuclear medicine; molecular biology; and genomic studies applied to communicable diseases, cancer, and genetic disorders.

Applications of radioisotopes in radiotherapy. Since internal targeted radionuclide therapy has been considered as a good alternative to gamma external-beam radiotherapy [with radioisotopes such as cobalt-60 (^{60}Co)], a fast-growing demand for the production of beta-emitting radionuclides has been observed at research reactor sites such as the BR2 high-flux reactor operated by the Belgian Nuclear Research Center.

New radionuclide production routes and new radiopharmaceuticals have been developed to deliver selective radiation doses to target tissues in order to minimize damage in healthy tissues. In addition to the existing use of iodine-131 (^{131}I) in the treatment of hyperthyroidism and metastatic thyroid cancer, the beta emitters have found applications in the treatment of primary cancer by localized irradiation [with iridium-192 (^{192}I)] or by the selective administration of radiopharmaceuticals [lutetium-177 (^{177}Lu), rhenium-188 (^{188}Re), yttrium-90 (^{90}Y), phosphorous-32 (^{32}P), holmium-166 (^{166}Ho), and so forth].

Beta emitters are also utilized for pain palliation [rhenium-186 (^{186}Re), ^{188}Re, samarium-153 (^{153}Sm), strontium-89 (^{89}Sr), ^{90}Y, ^{177}Lu, erbium-169 (^{169}Er), and so forth], providing significant improvement in the quality of life of cancer patients suffering from pain associated with bone metastases, as well as for the treatment of joint pain (rheumatoid arthritis). Some radionuclides, encapsulated in a titanium-welded capsule, decaying by electron capture with the emission of characteristic x-rays, find applications in brachytherapy (^{125}I, ^{103}Pd, and so forth) for the treatment of prostate cancer by local seeds implantation.

The next step in the development of new radiopharmaceuticals could be targeted alpha therapy, since localized alpha particle energy deposition minimizes damages to healthy tissues. The ongoing studies are showing great promise in the treatment of leukemia [bismuth-213 (^{213}Bi)] and in bone pain palliation [radium-223 (^{223}Ra)]. These radionuclides can be made available from a generator [actinium-225 (^{225}Ac)/^{213}Bi, actinium-227 (^{227}Ac)/^{223}Ra]. This very attractive way to produce radionuclides is offering a new dimension to the availability of therapeutic radiopharmaceuticals since repeated elution (extraction by washing with a solvent) provides many patient doses in a cost-effective way, as has been already demonstrated by the ^{99}Mo/^{99m}Tc, tungsten-188/rhenium-188 (^{188}W/^{188}Re), and strontium-90/yttrium-90 (^{90}Sr/^{90}Y) generators.

For background information *see* ISOTOPIC IRRADIATION; MEDICAL IMAGING; NUCLEAR MEDICINE; RADIATION THERAPY; RADIOACTIVE TRACER; RADIOISOTOPE in the McGraw-Hill Encyclopedia of Science & Technology.　　　　Yvan Bruynseraede; Bernard Ponsard

Bibliography. B. Ponsard, Production of radioisotopes in the BR2 high-flux reactor for applications in nuclear medicine and industry, *J. Labelled Comp. Rad.*, 50:333–337, 2007.

Rare-earth mining

In industry, the rare-earth elements (REEs) comprise the 14 naturally occurring lanthanide elements, plus yttrium. These elements are generally unfamiliar but are essential for a large number of applications that affect the daily lives of most everyone (see **table**). Up until the early 1980s, the United States was the dominant global producer of REE ores and derived products but has since been displaced by China, even as global REE production has more than doubled. Recent restrictions on REE exports from China have raised concern over the reliability of China as a supplier of REE raw materials and stimulated exploration and new mine development activity for REEs outside of China.

The REEs were first commercialized during the 1880s, when rare-earth oxide mixtures were used to make incandescent mantles for gas lights. The REEs were extracted from monazite mined from coastal heavy-mineral placer deposits in Brazil and India. Similar mixtures found applications in steel making. The discovery of the Mountain Pass REE deposit in California in 1949, one of the largest REE deposits in the world, provided a potential source of supply much larger than was required by industry at the time

Rare-earth elements (REEs) with their principal commercial uses and share of total REE consumption in 2008 as oxide			
Atomic symbol	Element	Percent share in use*	Principal uses in 2008
La	Lanthanum	32.7	Fluid catalytic cracking (46%), battery alloys (16%), glass polishing (13%), metallurgy (8%), glass additives (7%), ceramics (3%), phosphors (2%), automobile catalytic converters (1%), fuel cells
Ce	Cerium	29.9	Glass polishing (25%), glass additives (19%), automobile catalytic converters (16%), metallurgy (14%), battery alloys (10%), fluid catalytic cracking (5%), phosphors (2%), ceramics (2%)
Pr	Praseodymium	6.8	Magnets (70%), metallurgy (7%), glass polishing (7%), ceramics (5%), battery alloys (5%), automobile catalytic converters (2%), glass additives (1%)
Nd	Neodymium	18.5	Magnets (76%), metallurgy (8%), battery alloys (5%), ceramics (4%), glass additives (2%), automobile catalytic converters (1%), superconductors
Sm	Samarium	0.4	Battery alloys (73%), magnets, catalysts, glass additives, ceramics, fuel cells
Eu	Europium	0.3	Phosphors (100%)
Gd	Gadolinium	1.0	Magnets (69%), phosphors (21%), metallurgy, nuclear reactor shielding
Tb	Terbium	0.4	Phosphors (89%), magnets (11%), fuel cells, magnetic sensors
Dy	Dysprosium	1.0	Magnets (100%)
Ho	Holmium	†	Glass additives, magnetic flux concentrators, nuclear fuel control rods
Er	Erbium	†	Glass additives, phosphors, metallurgy
Tm	Terbium	†	Fuel cells, phosphors, magnetic sensors
Yb	Ytterbium	†	Glass additive, gamma ray source, metallurgy, stress gauges
Lu	Lutetium	†	Phosphors, magnetic bubble memory devices, positron emission tomography detectors, medical research
Y	Yttrium	9.0	Phosphors (54%), ceramics (32%), glass additives (2%)

*As rare-earth oxide (REO), percent share in total REO use.
†These elements combined comprise 0.4% of REO consumption.
Source: Data from T. G. Goonan (2011).

(**Fig. 1**). Molycorp, owner of the property, developed a small mine and undertook significant research to find commercial applications for REEs. Among the new uses that emerged at the time were REE catalysts for refining crude oil, cerium oxide for polishing glass, and REE phosphors for color televisions. To meet this emerging demand, Molycorp greatly expanded production during the early 1960s, establishing itself as the dominant producer. Meanwhile, the world's largest REE deposit had been found in China at Bayan Obo in Inner Mongolia, which became the basis for China's emergence as the current dominant producer of REEs, reaching over 90% of world production by 2000 (**Fig. 2**).

New applications for REEs continued to emerge, including REE magnets, ceramic capacitors, optical glass, and nickel-hydride batteries. REEs are generally expected to play an important role in many alternative energy technologies, and demand for REEs is growing rapidly. By some estimates, the overall rate of demand growth is about 8% per year, which is an unusually high rate for a mineral commodity. Meanwhile, China, the dominant producer, has imposed successively smaller quotas on exports of REE raw materials and some intermediate products, placing severe restrictions on supply. Prices of REE materials outside China have risen sharply, causing many consumers to seek alternatives. There is a general

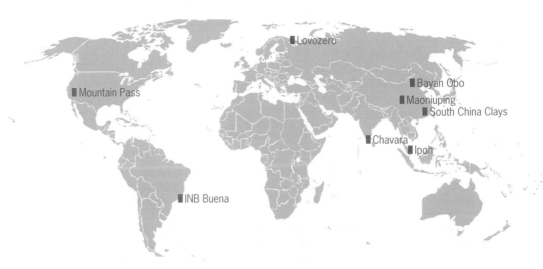

Fig. 1. Locations of rare-earth element (REE) mines as of 2011.

consensus among policy makers in the advanced industrial economies of Asia, Europe, and North America that alternative sources of REE supply need to be developed outside China.

REE mineral deposits. The REEs have very similar chemical properties and thus occur together in nature. Within the Earth's crust, the abundance of each REE generally decreases as its atomic number increases. The REEs are conventionally subdivided into the light REEs (lanthanum through europium) and the heavy REEs (gadolinium through lutetium plus yttrium). Where geologic processes have concentrated REEs into deposits that are economic to mine, light REE-rich deposits are the most common and deposits that are relatively enriched in heavy REEs are rare. Overall, the crustal abundance of the REEs, estimates of which range from 150 to 220 ppm, exceeds that of many other metals that are mined on a large scale, such as copper (55 ppm) and zinc (70 ppm). Despite this favorable crustal abundance, REE deposits are quite uncommon and usually modest in size.

The principal source of REEs are deposits hosted by carbonatites, unusual intrusive rocks composed of at least 50% carbonate minerals. Carbonatites are found emplaced within continental rift zones where a thinned and structurally fractured crust allows melted mantle to reach the surface. At Mountain Pass, California, the REE carbonate mineral bastnaesite (bastnäsite) is an essential rock-forming mineral and the carbonatite contains 8% total REEs. Most known carbonatites contain significantly less REEs, generally at concentrations too low for economic extraction. At Mount Weld, Western Australia, deep weathering has concentrated REEs in parts of the weathered zone to a level similar to that at Mountain Pass. The Bayan Obo deposit in China, the nature and origin of which is still debated, appears to be an overprinting of carbonatite-REE mineralization on parts of an iron deposit.

A broader class of alkalic igneous rocks, sometimes associated with carbonatites, also hosts REE mineralization that has sometimes been mined. These deposits are generally lower in REE grade and are often mineralogically complex, which presents problems for economic extraction. Many of these deposits are quite large and are under active exploration and metallurgical testing. REE-rich monazite is found in many coastal heavy-mineral deposits that are now mined for titanium and zircon, but the monazite is usually enriched in uranium and thorium and few producers are willing to take on the costs of processing the radioactive materials. Deep weathering of some granites in south China has yielded lateritic soil with about 0.3% REEs, often relatively enriched in heavy REEs. Despite the low grade, the clayey material is cheaply extracted and need only be leached to remove the REEs. Exploration for similar deposits outside China is currently underway.

Several key factors determine the economic viability of a REE deposit. Distance from existing infrastructure, especially the electrical grid, significantly affects capital and operating costs. A very remote mine will have to generate the significant power re-

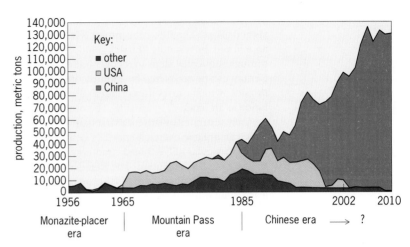

Fig. 2. Growth and distribution of rare-earth element (REE) production since 1900. (*Courtesy of USGS*)

quired for processing REE-bearing material, onsite, with diesel-fired plants. A deposit located at or near the surface can be mined by open-pit methods at considerably less expense than the underground methods required for deeply buried deposits. Grade, the proportion of the material to be mined that contains rare earths, determines how much material must be processed to obtain a given amount of REE product, a key factor in economic efficiency. Size and type of REE minerals is also critical. The smaller the mineral size, the more finely the rock must be crushed and ground to liberate the REE-bearing mineral, at significantly higher energy cost. Some REE minerals are readily leached with relatively inexpensive, commercially available chemicals, while others have no commercially tested treatment processes available.

Mining and mineral processing. REE-bearing rock must first be removed from the ground and the REE content concentrated before chemical methods of REE extraction and separation may be applied. Most rare-earth ores are mined by conventional open-pit methods in which rock is broken by blasting, loaded onto trucks with large shovels, and hauled to a concentration plant. Concentrating is by physical separation of the REE-bearing minerals from all other minerals in the rock. The ore is crushed and ground in multiple stages until most of the rare-earth minerals interlocked with the other minerals are broken free. Next, in a method known as froth flotation, the rare-earth minerals are coated with a chemical that repels water and allows them to float to the surface attached to air bubbles in agitated tanks, where they are skimmed off as a concentrate. The remaining minerals are disposed of as waste and the REE concentrate is treated onsite or sent to another location for extraction and separation. Research is underway to improve the flotation process, perhaps with better reagents, to allow the economic treatment of lower-grade REE ores.

Separation and commercialization. Although there is a commercially available mixed REE oxide product known as mischmetal, most REEs are recovered as individual rare-earth oxides (REOs) and marketed as such or used to make other REE chemicals or metals and alloys. The REE concentrate is leached with an acid and the resulting REE-rich solution processed through sequential steps to recover individual REEs. Cerium can be recovered by addition of sodium hydroxide, which causes the cerium to drop out of solution as an oxide or hydroxide. The other REEs are typically separated by solvent extraction, a process in which an organic chemical specially designed to extract a particular REE is forced countercurrent to the REE-bearing leach solution. The REE desired passes into the organic phase, which is separated from the leach solution, and the REO is recovered by stripping with acid. Multiple solvent extraction steps are required because of low separation efficiencies.

No individual REE plant recovers every REE present in the ore. Many of the REEs present, particularly the heavy REEs, are too low in concentration for economic recovery. Some of the heavy REEs, such as lutetium, have very small markets and are unlikely to be recovered even from one of the rare deposits that are relatively rich in heavy REEs. Research is underway to improve the separation of REEs, particularly through the development of more efficient extractants. Many of the advanced methods developed for recovering actinide elements from radioactive wastes, such as supercritical fluid extraction, should be applicable to the lanthanide elements. However, work in this area has only just started.

The reduction of REOs to rare-earth metals is very difficult because of the high stability of REO compounds. The present method, used at only a few plants around the world, is a highly complex refining method using chlorides as a reductant. Molten-salt electrolysis and other electrorefining methods are under investigation as more economic alternatives.

Environmental challenges. Mining and concentration of REE ores presents conventional problems of waste rock and concentrator waste disposal, which are generally handled through careful engineering of onsite waste-disposal facilities. When mining ceases, all equipment and processing facilities are removed, and waste-disposal facilities are recontoured, capped with topsoil, and revegetated. Water flow onsite is directed to prevent interaction with waste materials. Open pits are typically left unfilled. The rock that was blasted and removed expands by about 40% by volume and thus will not all fit back in. If backfilled with waste rock, any inflowing ground and surface waters will encounter a highly porous and permeable fill, which greatly increases water–rock interaction and, depending on the rock chemistry, has the potential for significant water contamination. In nonarid climates, a pit lake may form and may develop undesirable water chemistry through water–rock interactions, requiring ongoing water treatment. These best practices are widely used outside China, which has only recently begun to address environmental problems in its REE industry.

A modern REE separation plant, such as that constructed at Mountain Pass, California, in 2012, recycles and regenerates chemicals and water used and even produces power from the waste heat generated from the exothermic chemical reactions involved. Any waste generated is rather small and is disposed of onsite with proper sealing to prevent ground- or surface-water contamination. No smelting is involved, so air-pollution issues are limited to dust generation during mining and from any onsite power-generation facilities. The final REE products are shipped in sealed containers of various kinds.

Outlook for mine development. As of 2012, two REE mines were under development outside China, and about a dozen other REE deposits were at various stages of advanced exploration and economic feasibility studies. The two new mines are Mount Weld in Australia and Mountain Pass in California, the latter being a redevelopment of an older, inactive mine (**Fig. 3**). Mount Weld ships its REE concentrates to a plant in Malaysia for extraction and separation. Mountain Pass extracts its REEs onsite and produces a wide range of REE chemical products. As for the other projects, whether and when they will become producing mines depends on the economic viability of the deposit itself, the time and expense required

Fig. 3. Locations of rare-earth element (REE) mine development and advanced exploration projects as of 2011.

for environmental permitting, how much demand increases and Chinese exports decrease in the future, as well as how quickly competitors get their new mines into production.

A number of these other projects have recently released positive economic feasibility studies. This is a necessary, but not sufficient, condition for investment in a new mine. The various sources of financing of new mines can compare projects as well as assess the need for new capacity and will invest where returns are commensurate with risks. A review of the various possibilities by this author found that supplies of REEs will likely remain tight through 2015, if not longer, simply because of the time required to permit and develop new mines. In the long run, there do appear to be substantial REE resources of sufficient quality to compensate for the lost Chinese exports and to meet increasing demand. The challenge for future producers is to keep costs competitive with China's and perhaps invest downstream to obtain captive markets for REE mine products.

For background information *see* CARBONATITE; ELECTROLYSIS; ELECTROMETALLURGY; HEAVY MINERALS; LANTHANIDE CONTRACTION; LATERITE; MONAZITE; OPEN-PIT MINING; RARE-EARTH ELEMENTS; RARE-EARTH MINERALS; YTTRIUM in the McGraw-Hill Encyclopedia of Science & Technology.

Keith R. Long

Bibliography. T. G. Goonan, *Rare Earth Elements— End Use and Recyclability*, U.S. Geological Survey Scientific Investigations Rep. 2011-5094, 2011, http://pubs.usgs.gov/sir/2011/5094; C. K. Gupta and N. Krishnamurthy, *Extractive Metallurgy of Rare Earths*, CRC Press, 2005; G. B. Haxel et al., *Rare Earth Elements—Critical Resources for High Technology*, U.S. Geological Survey Fact Sheet 087-02, 2002, http://pubs.usgs.gov/fs/2002/fs087-02; D. M. Hoatson et al., *The Major Rare-Earth-Element Deposits of Australia*, Geosciences Australia, 2011, https://www.ga.gov.au/products/servlet/controller?event=GEOCAT ` DETAILS&catno=71820; K. R. Long et al., *The Principal Rare Earth Deposits of the United States*, U.S. Geological Survey Scientific Investigations Rep. 2010-5220, 2010, http://pubs.usgs.gov/sir/2010/5220; K. R. Long, *The Future of Rare Earth Elements—Will These High-Tech Industry Elements Continue in Short Supply?*, U.S. Geological Survey Open-File Rep. 2011-1189, 2011, http://pubs.usgs.gov/of/2011/1189.

Reduction of diesel engine particulate emissions

The diesel engine is one of the most efficient power-generating technologies, and therefore it is widely used in applications where fuel economy is the main priority. Diesel engines are used almost exclusively in heavy-duty vehicle and off-road applications and are becoming increasingly popular in passenger car applications.

The combustion of fuel in a diesel engine takes place via compression-ignition. Using high compres-

sion rates, the air trapped in the engine cylinders is heated during the compression stroke and reaches the autoignition point of the diesel fuel injected into the combustion chamber. The combustion stoichiometry is overall "lean," that is, there is always excess air. However, unlike the gasoline engine combustion process, diesel engine combustion is heterogeneous, since the injected fuel does not have enough time to mix and form a homogeneous mixture with the combustion air. The fuel oxidation reactions thus take place at locally variable air–fuel ratios, including "rich" regions with excess fuel. Particulate emissions are solid particles that are produced as a byproduct of the complex reaction process occurring at fuel-rich areas of the combustion chamber. Later in the power stroke, the available oxygen and high temperatures favor oxidation of most of the solid particles formed during the combustion process. However, a substantial percentage of them survive until the oxidation reaction stops and they are expelled from the engine exhaust to the environment.

Diesel particles are mostly agglomerates formed by coagulation of primary spherical particles (**Fig. 1**). These primary particles have diameters in the range of 10–20 nm. The size of the agglomerated particles can be substantially larger and is measured by techniques used to measure aerosols. The most common method is to measure the equivalent "mobility" diameter of the particles. In the exhaust gas of a typical modern diesel engine, the size of the particles follow a log-normal statistical distribution, with average diameters in the range of 50–80 nm. Unfortunately, this particle size is the most harmful for human beings, because of their high penetration rates in the respiratory system.

Because of the well-known adverse health effects of diesel particulates, international legislation has enforced increasingly severe limits for particulate matter (PM) exhaust levels for vehicular as well as off-road and marine engines. In Europe, the PM emission standards have been to a large extent directed by the best available technology approach. This has led to a more than 10-fold reduction of the PM limits in the last 10 years, as a result of the introduction of diesel particulate filters (DPF). The discussion about the health hazards associated with smaller particles has

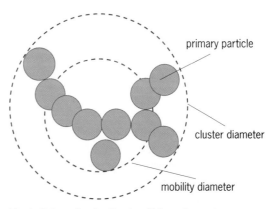

Fig. 1. Schematic of a diesel particle agglomerate.

led to the introduction of additional number-based emission standards in Europe. Although the United States has not followed the number-based emission limit approach, the mass-based standards there are already strict enough to require the use of filtration devices in most diesel engine applications for all sectors, including passenger cars, trucks, and off-road engines.

The preferred emission reduction concept for engine manufacturers is by cleaner combustion in-cylinder, due to lower cost and complexity. Modern fuel delivery technologies, based on high pressure and electronically controlled injection in combination with advanced turbocharging and exhaust gas recirculation, offer great potential for emission reduction. However, manufacturers are typically faced with trade-offs between low emissions and fuel economy or engine noise. For example, when combustion is optimized for fuel efficiency, PM levels are low but the emissions of nitrogen oxides are too high and the noise levels are increasing. In general, heavy-duty applications with the highest priority on fuel economy are designed for low engine output of PM, whereas passenger car applications are designed for lower NOx and noise. In some cases, the engine combustion measures are sufficient to eliminate the need for exhaust aftertreatment of some of the legislated emissions (either PM or NOx). However, it is likely that the increasingly stringent legislative requirements will eventually force the use of diesel particulate filtration across the vast majority of diesel engines. In Europe, the diesel particulate filter is already an established technology for all diesel passenger cars.

Particulate filtration technologies. Diesel particulate filters are devices that physically capture solid particles in the engine exhaust system. Because of their small size (nanometer scale), these particles are very difficult to capture by inertial filtration mechanisms that have reasonable filter structures with collector sizes in the order of 10–100 micrometers (μm). However, these particles can be efficiently captured by diffusion mechanisms and to a lesser extent by interception. As with all filtration processes, the maximization of filtration efficiency and the minimization of pressure drop during operation are major

challenges. Both targets can be met by achieving low filtration velocities, which calls for filters with high surface areas. In this respect, the monolithic honeycomb wall-flow filter has been established as a very efficient filter. It has an extruded ceramic structure with a large number of small parallel channels, similar to the common catalytic converters used for exhaust gas treatment. In contrast to the standard flow-through catalysts, the channels of the wall-flow particulate are alternately plugged, thus forming inlet and outlet channels. The particulate-laden gas entering the inlet channels is forced to flow through the porous walls that act as the filtering medium for the solid particles (**Fig. 2**).

Given the harsh conditions in the engine exhaust system and the durability standards of the automotive industry, only a few materials are able to meet the requirements of diesel particulate filters. Cordierite (magnesium aluminosilicate) is well-known from its use in exhaust catalytic converters as a low-cost material with excellent thermal and corrosion stability. The porosity and the mean pore size of extruded cordierites can be engineered to meet the filtration requirements and are currently widely used as diesel particulate filters, mainly in the heavy-duty sector. In the passenger car sector, materials based on silicon carbide (SiC) play a dominant role because of their higher thermal stability. Other ceramic materials used are aluminum titanate, mullite (aluminum silicate), and silicon nitride (SiN). In all cases, the channel structure is similar, with cell densities in the order of 100–300 cells/in.^{2} and wall thickness in the order of 10–15 mils (0.25–0.38 mm). To achieve high filtration efficiencies and reasonably low pressure drops, the wall porosities are in the range of 45–60%, whereas the optimal pore size range for maximum retention of the diesel particles is in the order of 10–20 μm. It is not only the porosity and the mean pore size of the wall microstructure that affects filter performance but also the pore size distribution. Narrower pore size distributions that are close to the optimum values are more favorable.

Starting from the clean state, the particles passing through the walls are initially collected in the micropores inside the filter wall. Because of their low density, these particles quickly change the apparent morphology of the initial wall microstructure, thus acting as a filtering medium themselves. This results in a rapid increase of the filtration efficiency, which is unfortunately accompanied with a steep increase of backpressure. After a critical soot mass accumulation in the filter wall microporosity, a soot layer starts to form on the surface of the DPF inlet channels (**Fig. 3**). This soot layer acts as a perfect filter for incoming soot and the filtration efficiency reaches 100% with all incoming soot being deposited on the cake. Although the deposited soot inevitably increases the flow resistance and the pressure drop, the permeability of the soot cake per unit soot mass is in general higher compared to the permeability of the loaded wall. It is therefore preferable to design wall microstructures that promote the cake formation as quickly as possible.

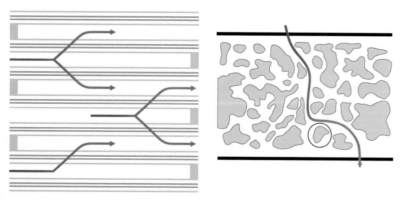

Fig. 2. Wall-flow filtration principle in diesel particulate filters.

DPF regeneration. The filtration efficiency of a completely clean filter is in the order of 60–80%. As mentioned earlier, the filtration efficiency increases immediately after soot is accumulated in the wall pores and reaches 100% after the accumulated soot mass exceeds about 0.2–0.5 grams per liter of filter. These values are typical for commercial filters but may vary, depending on the filter microstructure. As a basis for calculation, the typical soot emission rate of a modern mid-size passenger car with a 2-liter diesel engine would be about 20 mg/km and the accumulation of soot after 1000 km would be approximately 20 g. Considering that the DPF volume is usually selected to be in the same order of magnitude of the engine displacement, the specific soot loading of a 2-l DPF after 1000 km would be 10 g/liter. The backpressure induced at such a soot loading level could already increase the engine pumping losses and thus reduce its power output and fuel efficiency. Even worse, a spontaneous oxidation of the accumulation soot of this quantity could result in an extremely high exotherm, damaging the filter itself. It is therefore necessary to control the amount of soot loading in the filter by periodically removing the accumulated soot by controlled oxidation. This process, known as filter regeneration, involves oxidation of the carbonaceous soot with the oxygen present in the exhaust gas and is kinetically favored above 550–600°C.

It can be easily shown by energy balance considerations that the peak temperature expected during soot oxidation increases with the initial soot loading. The second critical parameter affecting the thermal loading of the filter is its thermal mass, which depends on the filter geometry and material. In this respect, higher density materials such as SiC are thermally more stable and can withstand regeneration with initial soot loadings higher than 10 g/liter. The filters have to be designed to sustain "worst-case" regenerations, which occur when the oxidation is initiated at high temperature and the exhaust flow rate is too low to cool the filter as the reaction exotherm increases in temperature. In practice, this can happen when the engine drops to idle during regeneration and models have been developed that closely describe the experimental data.

To eliminate the chances of DPF overheating and failure, the system has to be designed to regenerate at safe soot loadings. Commercial passenger car DPFs are allowed to operate until the soot loading reaches 6–8 g/l, for the case of SiC materials. The safe soot limit in the case of cordierite filters is roughly 30–50% lower because of its lower heat capacity. The complex interactions between the factors affecting the thermal stresses in DPFs are nowadays studied using mathematical models to describe the mass, momentum, transport, and reaction phenomena in the DPF channels.

The diesel exhaust gas temperature is rarely or never sufficiently high enough to oxidize the accumulated soot. The introduction of electronic fuel injection management, combined with the flexibility offered by the common-rail technology was the

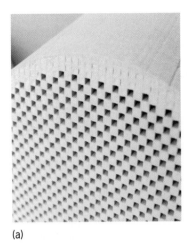 

(a)   (b)

Fig. 3. Close-up view of (*a*) the inlet face of a cordierite DPF with alternately plugged channels. Image of (*b*) cross section of a DPF channel with accumulated soot layer.

technology enabler for DPFs in 2000. Using multiple injections in one engine cycle, part of the fuel may be used to produce heat instead of power, thus increasing the exhaust gas temperature to the desired level. Part of this heat is generated in the combustion chamber but most of it is usually released in an oxidation catalyst located upstream of the DPF that oxidizes the large amounts of hydrocarbons (HC) deliberately emitted during such a forced regeneration mode. However, forcing DPF regeneration too frequently via fuel post-injection is undesirable not only because of fuel economy considerations but also because part of the post-injected fuel adsorbs in the lubricating oil and reduces the oil change interval.

To facilitate DPF regeneration and minimize post injection requirements, it is possible to use catalytic promoters. Metal oxides can substantially promote soot oxidation, provided that they are in contact with the soot. Very intimate oxide-soot contact can be achieved by introducing oxides in the fuel that are emitted in contact with soot particles. This fuel-borne catalyst technology was first introduced by PSA Peugeot-Citroen in the first successful commercial DPF application in 2000. Cerium- and iron-based fuel dopants can decrease the temperature required for soot oxidation by 100–150°C.

One of the drawbacks of the fuel-borne catalyst (FBC) technology is that the active oxides eventually stay in the DPF channels as inactive residual, blocking the active filtration area. This necessitates the periodic cleaning of the DPF by removal at the workshop. For first-generation commercial DPFs, such maintenance was necessary after about 80,000 km (50,000 mi). Recent advances in channel geometry designs and reduction in fuel-borne catalyst concentrations have increased the maintenance interval be equivalent to the useful vehicle life.

Most manufacturers have followed the maintenance-free option of the catalyzed DPF technology. The idea is to promote soot regeneration by coating the DPF with active catalytic materials. Platinum-based catalysts are well known to enhance soot oxidation when Pt is in intimate contact with soot. However, in actual DPF applications the contact between soot and coated wall is almost negligible rendering this direct soot catalysis almost negligible. Interestingly, Pt-based coatings indeed promote soot oxidation at low temperatures via an indirect mechanism involving NO oxidation to NO_2 on Pt. NO_2 is an excellent soot oxidizer at temperatures as low as $300°C$. However, in order to exploit this regeneration mechanism two requirements have to be met: the temperature should be higher than $300°C$ and there needs to be enough available NO_2. As a rule, the critical NOx/soot ratio should be higher than 16 on a mass basis to sustain NOx-based regeneration. This condition can be met for heavy-duty applications in which the engine combustion is calibrated for high NOx and low PM. Many heavy-duty applications, such as long-haul trucks, operate at high loads and exhaust gas temperatures that favor the NOx-based regeneration mode.

The contribution of the NO_2 produced in the catalyzed wall as an oxidant of soot has been questioned due to the fact that most of the soot lies on the channel surface, which is upstream of the catalyzed wall. Counterintuitively, soot is indeed oxidized by the NO_2 produced in the wall and diffuses opposite to the bulk flow due to the developed concentration gradient between the catalyzed wall and the soot layer.

The formation of NO_2 can be achieved with a Pt-based oxidation catalyst (DOC; diesel oxidation catalyst) that can be placed upstream of the DPF. This regeneration concept of combining a DOC with a DPF has been known since the late 1980s and has been trademarked as Continuously Regenerating Trap or CRT®. Applying Pt-coatings in the DPF channels additionally serves as an oxidation catalyst for undesirable carbon monoxide (CO) produced during regeneration, which is converted to carbon dioxide (CO_2). Because of the high (exothermic) heat of this reaction, a catalyzed DPF will accelerate the regeneration process.

DPF durability issues. As mentioned earlier, the accumulation of noncombustible material in the DPF channels can result in gradual blocking of the active filtration area and therefore increase the backpressure. Noncombustible materials are not only associated with the use of fuel-borne catalysts but also with standard fuel combustion. Fuels and, more importantly, lubrication oils contain metal additives (such as P, Ca, and Zn) that accumulate as ash in the DPF channels. Ash accumulation can be a significant issue, especially for heavy-duty engine applications with long useful lives and long service intervals. In order to extend the DPF useful life, special low-ash lubricants are necessary.

Structural failure of DPFs can result from overheating during uncontrolled regeneration modes. In the case of a cordierite filter, overheating can result to local melting when the temperature exceeds $1400°C$. In the case of SiC filters, the typical failure mode is crack formation resulting from temperature gradients in the axial or radial direction. To avoid cracking,

Fig. 4. Segmented SiC DPF.

one has to minimize the thermal stresses resulting from temperature gradients, which is achieved by segmenting the filter in smaller square-shaped pieces that are glued together by a thin cement layer that acts as stress-reliever (**Fig. 4**).

To avoid DPF overheating, it is critical to precisely control the regeneration frequency and ensure that the soot amount in the DPF does not exceed the safe limit. This is one of the major challenges for the control engineers, since the amount of soot in the DPF is not directly measurable onboard. Most manufacturers measure the pressure drop across the DPF as an indicator of accumulated soot. However, pressure drop cannot be directly correlated with soot mass, at least with the sensitivity required for safe regeneration management. Engine soot emission models are sometimes used in combination with pressure drop models to increase the fidelity of the soot loading estimation.

Outlook. The introduction of DPF technology in the last decade has been a great technological challenge and at the same time a remarkable success for the emission-controls industry. The extremely efficient abatement of diesel particulate emissions is rapidly changing the perception of the once "dirty" diesel engine technology. The success has been a result of collaborative efforts in a challenging multidisciplinary field involving engine developers, material and catalyst scientists, and control engineers. Although DPF technology is reaching a level of maturity, there are still challenges to be faced and improvements to be expected.

Current research and development efforts are directed at further improving filtration efficiency with lower pressure drop. This can be achieved by using novel materials and careful design of the filter's wall microstructure. Technologies involving the application of additional membrane layers on the surface of the inlet channel (**Fig. 5**) or even creating a nonuniform wall microstructure are reported to increase the filtration efficiency and accelerate the onset of the surface filtration mechanism. By minimizing the depth filtration mode, the pressure drop associated with blocking the internal wall pores is minimized. In addition, these technologies are reported to allow a better and more linear correlation between pressure

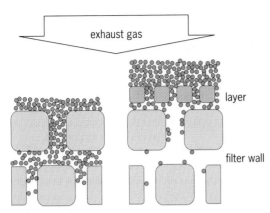

Fig. 5. Principle of applying an additional filtration layer (right) on the channel surface.

drop and soot loading, thus facilitating regeneration management.

The need to reduce the volume, weight, and costs of diesel aftertreatment has motivated efforts to combine multiple aftertreatment functions in single devices. By applying catalytic coatings on DPFs, it is possible to partially or fully substitute separate devices necessary for gaseous emission control (such as CO, HC, and NOx). The integration of catalytic functionalities, especially for NOx reduction on the DPF is currently a topic of major focus and expected to further improve the attractiveness and cost competitiveness of clean diesel engines.

For background information *see* AIR POLLUTION; COMBUSTION; COMBUSTION CHAMBER; DIESEL ENGINE; DIESEL FUEL; FILTRATION; NITROGEN OXIDES; PARTICULATES; TURBOCHARGER in the McGraw-Hill Encyclopedia of Science & Technology.

Grigorios Koltsakis

Bibliography. B. J. Cooper and J. E. Thoss, Role of NO in diesel particulate emission control, SAE Tech. Pap. 890404, 1989, DOI:10.4271/890404; O. A. Haralampous et al., Reaction and diffusion phenomena in catalyzed diesel particulate filters, SAE Tech. Pap. 2004-01-0696, 2004, DOI:10.4271/2004-01-0696; D. B. Kittelson, Engines and nanoparticles: a review, *J. Aerosol Sci.*, 29(5/6):575–588, 1998, DOI:10.1016/S0021-8502(97)10037-4; M. Maricq, Chemical characterization of particulate emissions from diesel engines: A review, *J. Aerosol Sci.*, 38(11):1079–1118, 2007, DOI:10.1016/j.jaerosci.2007.08.001; O. Salvat, P. Marez, and G. Belot, Passenger car serial application of a particulate filter system on a common rail direct injection diesel engine, SAE Tech. Pap. no. 2000-01-0473, 2000, DOI:10.4271/2000-01-0473.

Repressed memories

In 1990, George Franklin stood trial in California for the murder of his daughter's friend, Susan Nason. However, this was unlike most murder trials. The key evidence against George Franklin was the eyewitness testimony of his daughter, Eileen. Eileen had been 8 years old at the time that her friend was murdered, and as an adult she recovered what she thought was a repressed memory of seeing her father strike and kill her friend. Eileen's recovered memory was convincing—George Franklin was convicted of the murder and completed more than 6 years of a life sentence before his conviction was overturned. This case was one of the earliest to raise concerns about psychological theory and clinical practice relating to repressed memories, beginning what became known as the memory wars.

In the nineteenth century, Sigmund Freud described repression as a protective mechanism that pushed disturbing and upsetting memories out of awareness. These memories would allegedly remain intact and hidden away until uncovered later through special therapeutic techniques designed to retrieve them. Years later, Freud revised his theory and suggested that the details his patients were recalling were the result of fantasies rather than real experience. Although many clinicians celebrated Freud's original thesis and practiced recovered memory therapy (RMT; using techniques such as guided imagery, hypnosis, and dream interpretation) in the pursuit of hidden memories of trauma, there is no convincing scientific evidence that repression exists.

Repression phenomenon of the 1980s and 1990s. The notion of repression had been around since the late nineteenth century, but it became clear during the 1980s and 1990s that many clinicians were practicing RMT, and many patients were allegedly discovering memories of distressing, traumatic events. The concept of repression became so popular in the late 1980s that many self-help books offered advice to women who thought they had repressed memories of child sexual abuse (CSA). Some of these books provided readers with a list of symptoms to detect the presence of a repressed memory, and these readers were encouraged to conclude they had a buried memory if their respective lives showed these symptoms. The use of these self-help books and RMT practices soon had serious societal repercussions. Many hundreds of legal cases based on supposedly recovered memories tore families apart and resulted in some people going to prison.

In some cases, however, witnesses came forward and the allegations of CSA were withdrawn in the belief that the memories were indeed false. For example, in 2004, Elizabeth Gale was awarded $7.5 million in Illinois for the malpractice of mental-health professionals who led her to recover memories of being the victim of satanic ritual abuse—events that never happened. Gale's case and other similar cases led many psychological scientists to question whether there was in fact any good evidence for repression and whether the RMT techniques might be building false memories rather than finding real ones.

Search for scientific evidence of repression. Memory researchers and clinicians have searched for evidence of repression using a number of different approaches. A common approach is to ask people who today claim that they experienced a trauma in the past whether there was ever a time when they could not remember the traumatic event. When people answer "yes" to this question, it is often taken as evidence for repression (in one study, 59% of people said "yes").

However, there are many reasons why people might say "yes" to this question. First, people can interpret the question in a number of ways. In one study, many of the people who said "yes" indicated that they meant they avoided thinking about it. Second, people are often mistaken in knowing whether they have remembered something before. This is especially true when people remember an event differently than how they did in the past—they might not realize an event was traumatic until later in their lives, when they reevaluate the meaning of the event; this creates a feeling that they are recovering a new memory, leading people to conclude that they have not remembered it before. Third, if people have tried repeatedly to remember an event from the past (a practice encouraged in RMT), it can often make them feel that there was a time when they could not remember it well. This is not unique to traumatic events. When people work hard at remembering nontraumatic events such as a high-school graduation, they often report having had a poor memory for the event in the past. These findings stand in conflict with a repression mechanism that is supposed to banish traumatic events rather than innocuous or even positive events such as a high-school graduation. Thus, there are a number of alternative explanations for why people say "yes" to the could-not-remember question—that is, repression is not necessary to explain why some people report that there was a period in their lives when they could not remember a traumatic event.

Another approach in the pursuit of evidence of repression has been to use prospective studies. In these studies, researchers track people who have documented past histories of childhood trauma and determine whether these people remember the trauma at the present time. One study found that 38% of women did not disclose an incident of documented CSA when queried about it as an adult some 17 years after the incident. However, this is not necessarily good evidence for repression, because there are a number of alternative possibilities or explanations for why people might not disclose an incident of CSA. First, women who were younger at the time of the incident were less likely to report it—a finding that suggests some of the failures to report could be the result of childhood amnesia (a phenomenon that makes it nearly impossible to remember events from infancy later in life). Second, it has been suggested that people might fail to report the CSA because they do not feel comfortable talking about it. Third, even normal forgetting can account for why some people did not mention the abusive incident.

An alternative approach has been to document clinical case histories. Although this approach avoids some of the interpretive issues with the retrospective reports, it is often impossible to gain access to the data to examine whether the presentation is ac-curate. Taken together, although some studies have offered evidence of what looks like repression (for example, people failing to report trauma, or claiming to have forgotten it at some point in their lives), there are a number of alternative possibilities or explanations for these data that do not require a repression mechanism. However, these studies do not explain why some people came out of therapy with rich detailed memories of horrific acts, and then retracted those memories later on, believing them to be false.

Can RMT techniques lead to false memories? Although therapists used RMT with the best intentions to retrieve buried memories of CSA, some psychological scientists became concerned that RMT techniques (for example, imagination and guided imagery) might be planting false memories of events that never occurred. At the time that George Franklin stood trial, there was some evidence that RMT-like techniques could lead to minor memory errors, but there was little evidence that these techniques could lead to entirely false memories of childhood events.

By the mid-1990s, the results of new research began to accumulate. In the first of a series of studies looking at the dangers of RMT techniques, subjects read descriptions of four childhood events—three true events and one false event (getting lost in the mall). The false event was personalized for each subject and described a trip with his or her family to a local mall. After repeated attempts to recall the events, 25% of subjects "remembered" the false event. Many other studies followed, and some showed even higher rates of developing complete or partial false memories (as high as 80%). Over the next decade, researchers demonstrated that people could come to "remember" all kinds of made-up events, ranging from spilling punch at a wedding to being saved by a lifeguard. Although some researchers suggested that people would not come to remember more implausible events, real-life cases suggest otherwise: for example, people remember being abducted by aliens, witnessing demonic possession, and participating in satanic cults—events classified as extremely unlikely or even implausible.

False memories feel real to the subject: they are full of sensory details and bring about psychophysiological stress responses that match those of true memories. Moreover, these false memories can have similar consequences to true memories—many people who recovered false traumatic childhood memories in therapy pursued criminal charges against family members before exonerating evidence came to light exposing the memories as false.

Conclusions. Although historically accepted as a real phenomenon, decades of research have indicated little evidence of repression. Instead, there are a number of alternative possibilities or explanations for the evidence offered and a wealth of research suggesting that RMT techniques may sometimes plant false memories.

For background information *see* AMNESIA; BRAIN; INFORMATION PROCESSING (PSYCHOLOGY); MEMORY; MOTIVATION; NEUROBIOLOGY; PERCEPTION;

PSYCHOLOGY; PSYCHOTHERAPY in the McGraw-Hill Encyclopedia of Science & Technology.

Eryn J. Newman; J. Zoe Klemfuss; Elizabeth F. Loftus
Bibliography. E. F. Loftus, Make-believe memories, *Am. Psychol.*, 58:864–873, 2005; E. F. Loftus and D. Davis, Recovered memories, *Annu. Rev. Clin. Psychol.*, 2:469–498, 2006, DOI:10.1146/annurev.clinpsy.2.022305.095315; P. R. McHugh, *Try to Remember: Psychiatry's Clash over Meaning, Memory, and Mind*, Dana Press, New York, 2008; R. J. McNally, Searching for repressed memory, in R. F. Belli (ed.), *True and False Recovered Memories: Toward a Reconciliation of the Debate*, Springer, New York, 2012.

RNA mimics of green fluorescent protein

A major class of molecules in cells is ribonucleic acid (RNA). Cellular RNAs comprise ribosomal RNA (rRNA), which is a major component of ribosomes; transfer RNA (tRNA), which is used for protein synthesis; and messenger RNA (mRNA), which is an intermediate between DNA and proteins. In addition to these types of RNA, there is also a wide variety of poorly understood noncoding RNAs, including microRNAs, PIWI-interacting RNAs, and termini-associated RNAs. Understanding the functions of these RNAs is a major challenge in molecular biology research.

The cellular function of many of these RNAs is related to their localization in specific cell compartments. For example, in migrating cells, various mRNAs are localized in the elongating portions of the cell. In neurons, subsets of cellular mRNAs are localized in long processes such as axons or dendrites. In both of these cases, the mRNAs are translated in these distal portions of the cell to produce proteins directly in these sites. In the case of noncoding RNAs, different noncoding RNAs can each function in different parts of the cell and can affect different types of cellular processes. An important way to understand the functions of these different RNAs is to study their localization in cells and how these localizations are affected by experimental treatments, during cellular processes, or in disease states. The importance of RNA localization is supported by the finding that mutations in proteins that control RNA trafficking lead to human disease.

Use of green fluorescent protein. In contrast to RNA, imaging of protein localization in cells is fairly straightforward. Green fluorescent protein (GFP, which is a protein produced by the bioluminescent jellyfish *Aequorea victoria* and used to trace the synthesis, location, and movement of proteins) and other genetically encodable fluorescent protein technologies have revolutionized biomedical research and biotechnology. As a result of GFP, studies that address the trafficking and processing of proteins in relation to specific intracellular organelles and sites within the cell have become commonplace. Because GFP and GFP-tagged proteins are genetically encodable, cells can be made to express these proteins,

thereby overcoming the problem of delivering the tagged protein into cells. Although GFP has considerably advanced studies of protein biology, comparably simple and robust tools for genetically encoding fluorescently tagged RNAs for imaging of RNAs in living cells are lacking.

Use of RNA mimics of GFP. A new strategy for imaging RNAs in living cells is to use a new class of RNA tags referred to as RNA mimics of GFP. These RNA mimics of GFP are RNA sequences that bind small molecules and are termed aptamers. Aptamers can be generated using a technique called systematic evolution of ligands by exponential enrichment (SELEX). Cells can be engineered to express these RNA aptamers or to express an RNA of interest that contains one or more of these aptamers. Analogous to GFP, this approach allows for the genetic encoding of RNA that can be imaged in living cells by fluorescence microscopy.

This strategy for genetically encoding fluorescent RNA utilizes aptamers that bind fluorophores (fluorescent molecules) that have unique chemical properties. In principle, RNAs can be imaged in cells if they are engineered to contain aptamers that bind fluorescent dyes. However, these kinds of aptamers have not proved to be useful in live cell experiments because both bound and unbound dyes are fluorescent and have nearly identical emission spectrum properties. Thus, the signal from the unbound fluorescent dye overwhelms the signal from the fluorescent dye that is bound to the RNA.

In order to avoid this issue of background fluorescence, RNA mimics of GFP rely on a fluorophore that is in a nonfluorescent form when it is not bound to RNA and switches to a fluorescent form only when it is bound to a specific RNA aptamer. A critical feature of these fluorophores is that they are not nonspecifically activated by cellular components, such as deoxyribonucleic acid (DNA), mRNA, or lipids. As a result, after the fluorophore is added to cultured cells that express the RNA aptamer or RNAs that contain the RNA aptamer, any fluorescence seen in cells can be ascribed to binding of the fluorophore to the aptamer.

The dye that is used for RNA imaging is based on the dye that is autocatalytically formed in GFP. After GFP is synthesized, it undergoes a posttranslational, autocatalytic, intramolecular cyclization of an internal Ser-Tyr-Gly tripeptide. The cyclized product is then oxidized to the final 4-hydroxybenzylideneimidazolinone (HBI) fluorophore. After the fluorophore is synthesized, its fluorescence is dependent on whether GFP is in a folded state. For example, denaturing agents such as urea cause GFP to lose its fluorescence, but the fluorescence returns if the protein is renatured. The explanation for this conditional fluorescence lies in the chemical structure of the fluorophore.

After photoexcitation, the fluorophore can dissipate its energy through either a radiative (that is, fluorescent) pathway or a nonradiative pathway, which usually involves vibrational or other intramolecular movements. When the protein is unfolded, the fluorophore dissipates its energy primarily through the nonradiative pathway, which involves various bond rotations, including *cis–trans* isomerizations. However, when GFP is folded, these motions are inhibited; thus, radiative decay (fluorescence) is the sole pathway available to dissipate the energy of the excited-state fluorophore. Therefore, the GFP fluorophore is conditionally fluorescent, with the fluorescence arising when the GFP protein immobilizes the fluorophore.

Spinach tag. RNA mimics of GFP take advantage of the conditional fluorescence of the GFP-like fluorophores. Like denatured GFP, the chemically synthesized GFP fluorophore HBI is not fluorescent in solution. Additionally, HBI and other molecules that are structurally related to HBI do not exhibit fluorescence when they are applied to cells, indicating that they are not nonspecifically activated by cellular components. Based on these properties, RNA aptamers were generated (using the SELEX technique) that specifically bind HBI and HBI derivatives. One of these aptamers binds 3,5-difluoro-4-hydroxybenzylidene-imidazolinone (DFHBI), an HBI derivative, and produces fluorescence that is similar in magnitude to the fluorescence seen with GFP and enhanced GFP. This RNA sequence was named Spinach, after the vegetable, in an analogy to the names of various fruits that have been used to name fluorescent proteins. In addition to Spinach, a series of RNA aptamers and the fluorophores that bind them were developed, resulting in a palette of RNA–fluorophore complexes that exhibit fluorescence emissions that span the visible spectrum (**Fig. 1**). The fluorescence is highly specific because other RNAs (for example, tRNA and cellular RNA) or scrambled aptamer sequences elicit no fluorescence upon incubation with DFHBI.

Spinach can be used to tag RNAs for live cell imaging in a manner analogous to GFP-fusion proteins. For example, 5S is a small noncoding RNA transcribed by RNA polymerase III that associates with the large ribosomal subunit. Cells that have been engineered to express a 5S–Spinach fusion RNA using the endogenous human 5S promoter allow

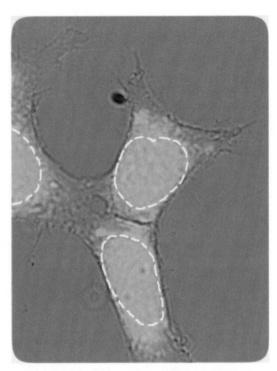

Fig. 2. Imaging 5S RNA in cells using Spinach. To image the noncoding 5S RNA in cells, the cells were engineered to express an RNA fusion comprising 5S and Spinach. After the cells are treated with sucrose (an agent that elicits osmotic stress), 5S appears in clusters in the cytosol and nucleus. The nucleus is indicated with a dotted line.

imaging of the expression of 5S RNA in living cells. 5S–Spinach RNA fluorescence is detected throughout cells, with prominent fluorescence signals appearing in the nucleus and cytosol. Treatment of cells with osmotic stressors, such as sucrose, results in discrete foci of fluorescence signal, indicating clusters of 5S–Spinach RNA (**Fig. 2**). This fluorescence distribution is similar to the patterns seen for endogenous 5S RNA in the same cell type. These data indicate that Spinach–RNA fusions can be imaged in cells and exhibit localizations consistent with endogenous RNA.

The use of Spinach and related RNA tags can provide a simple approach for imaging RNA in cells. Thus, tagging RNAs in cells with Spinach will be useful for gaining insight into the functions of various RNAs in cells.

For background information *see* CELL (BIOLOGY); FLUORESCENCE; FLUORESCENCE MICROSCOPE; GREEN FLUORESCENT PROTEIN; MOLECULAR BIOLOGY; PROTEIN; RIBONUCLEIC ACID (RNA) in the McGraw-Hill Encyclopedia of Science & Technology.

Samie R. Jaffrey

Bibliography. T. A. Cooper, L. Wan, and G. Dreyfuss, RNA and disease, *Cell*, 136:777–793, 2009, DOI:10.1016/j.cell.2009.02.011; J. S. Paige, K. Y. Wu, and S. R. Jaffrey, RNA mimics of green fluorescent protein, *Science*, 333:642–646, 2011, DOI:10.1126/science.1207339; S. Tyagi, Imaging intracellular RNA distribution and dynamics in living cells, *Nat. Methods*, 6:331–338, 2009, DOI:10.1038/nmeth.1321.

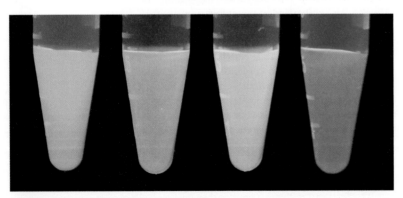

Fig. 1. A palette of RNA–fluorophore complexes acquired under ultraviolet light irradiation. Several different RNA aptamers have been developed that bind and switch on the fluorescence of small molecule fluorophores. Shown are mixtures of fluorophores and the RNA aptamers that specifically bind them. The second image from the left is Spinach, a green fluorescent RNA–fluorophore complex.

Sahara solar breeder project

Energy is the driving force of life, usually accompanying conversion from one form to another, including heat, light, mechanical, electric, and chemical energy as well as changes in mass. Without energy flow, no living things can survive. Throughout human history, energy use (and the rate of consumption) and the advancement of civilization have increased simultaneously.

On Earth, there are three kinds of energy resources available: (1) geothermal and nuclear (radioactive) energy, (2) fossil fuels (crude oil, natural gas, and coal), and (3) renewable energy from incident sunlight, water (hydropower), wind, and biomass.

Both nuclear and fossil fuels will run out in a few hundred years, at the latest. Nuclear power stations are questionable as a main energy source since their vulnerability and serious hazards were realized following the Fukushima Daiichi nuclear disaster on March 11, 2011. Carbon dioxide (CO_2), the main product of fossil-fuel combustion, is a major contributor to climate change. Therefore, we are reaching the point where we must think about alternative energy for the future. The Middle East and North African (MENA) countries, which are presently enjoying economic prosperity from oil and gas production, recognize the situation and have started thinking about future energy strategies. The Sahara Solar Breeder (SSB) plan proposed by the Science Council of Japan is one option that is attracting much MENA attention, partly because of the possibility of using desert sands for photovoltaic power generation to supply the energy needs of the world.

Material aspects of global energy and the environmental problem. In terms of chemistry and thermodynamics, the Earth is in such oxidative conditions (20% oxygen concentration in air) that fossil-fuel combustion inevitably results in accumulation of CO_2 and water (H_2O). Human life is also sustained by the oxidation of food into CO_2, H_2O, and digested mass. In addition, almost all the metallic elements exist in the form of oxides (MO_x) in nature. Since these oxides are energetically stable under terrestrial conditions, we cannot extract effective energy from them, thus making them zero-exergy (that is, no energy available for conversion to useful work) materials. We can argue that the current energy and environmental crisis results from the accumulation of oxidized wastes at a rate much faster than photosynthesis can convert them back to carbohydrates to preserve the energy and material balance of the Earth. Conversion of CO_2, H_2O, and MO_x by new science and technology could be the key to solving the global energy and environment problem. **Figure 1** represents the schemes for approaching this end for three typical zero-exergy materials (that is, materials of little energy value) in gas (CO_2), liquid (biomass), and solid (Si) phases.

Fixation of CO_2 into useful chemicals by reaction with a high-energy compound. Carbon dioxide can be fixed, using a zinc-oxide-based catalyst, into a solid polycarbonate by alternating copolymerization with an epoxide,

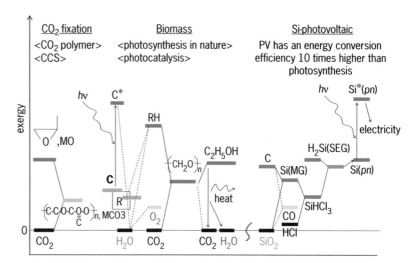

Fig. 1. Exergy profile of zero-exergy oxides and schemes for their conversion. CCS = carbon capture and storage, MO = metal-oxide, MCO_3 = metal carbonate, C = chlorophyll, C* = photoexited chlorophyll, R = NADP (nicotinamide adenine dinucleotide phosphate), RH = NADPH (nicotinamide adenine dinucleotide phosphate hydrogenated), Si(SEG) = semiconductor-grade silicon, Si(MG) = metallurgical-grade silicon. The vertical axis is qualitative and arbitrary.

which has high chemical energy due to a strained three-membered ring. The copolymer can be injection molded into various plastic forms and exhibits unique biodegradability, potentially providing a CO_2 storage mechanism that is much better than conventional carbon capture and storage.

Photosynthesis as the sole mechanism of H_2O and CO_2 renewal in nature. Sunlight can induce a redox reaction through the electronic excited state of a catalyst to oxidize H_2O and subsequently reduce CO_2 to yield O_2 and carbohydrates. Although photosynthesis is crucially important, the efficiency of solar to chemical energy is low (about 1%). Artificial photosynthesis has about the same conversion efficiency level for visible light.

Conversion of silicon dioxide (SiO_2) to semiconductor silicon (Si) for solar cells. Silicon dioxide is the most abundant natural resource on Earth. If we could develop an efficient process for reducing SiO_2 to solar-cell grade silicon (purity level = 99.9999%) and use this to make polycrystalline (polysilicon) solar cells, solar energy could be converted into electricity in about 15% efficiency. This efficiency appears to be lower than conventional power-generation systems (about 40% efficient), but it is substantially higher in view of the fact that sunlight is fuel free.

Sahara solar breeder (SSB) plan for global energy system innovation. The current global energy consumption is about 20,000 TWh per year. By assuming 10–15% energy conversion efficiency, 50% panel coverage of photovoltaics (PV) over a given site, and 10–20% time-averaged operation of a PV power station, this energy consumption could be totally accounted for by solar panels covering an area of 6.4×10^2 (800 × 800 km), which corresponds to only 4% of the desert area in the world. Primary PV contribution to global energy in 20–30 years would require 100 GW of annual solar-cell installation/replacement, judging from the 20-year lifetime of solar panels and increasing energy needs.

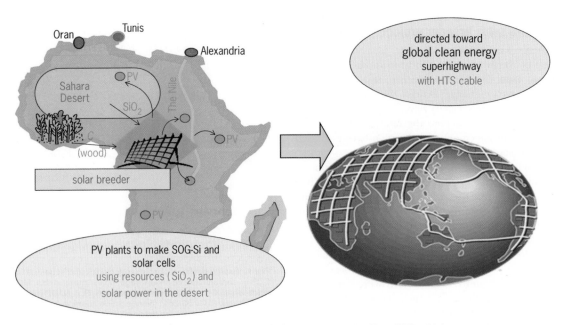

Fig. 2. **Sahara solar breeder plan and its key technologies. SoG-Si = solar-grade silicon. HTS = high-temperature superconductor.**

Although various types of solar-cell materials have been investigated, only Si can supply 100 GW/year of solar cells. Our experience in advanced material and device research inspires us to concentrate first on polycrystalline Si as a high-performance solar-cell material, instead of thin films for which energy projections seem to be overly optimistic. For 100 GW of polycrystalline Si solar cells, it is necessary to produce 1 million tons of Si, preferably at a cost less than 25% of the current price. The main component of desert sands is silica (SiO_2), which is the most abundant natural resource on Earth. At the G8+5 Academies' 2009 meeting in Rome, the Science Council of Japan proposed the "Sahara solar breeder (SSB) plan directed toward a global clean energy superhighway." The Sahara solar breeder model is shown in **Fig. 2**.

In addition to the well-known advantages of the desert—vast land and ample sunshine—the SSB challenges make use of the third valuable commodity of the desert, sand as a source of Si. The serious problems of PV-power fluctuation (such as site, time, and season as well as of needs of electric storage) can be overcome by averaging the power supply on a global scale via direct-current (dc) transmission using superconducting cables. The SSB initiative was inaugurated in 2010 as the Sahara Solar Energy Research Center (SSERC) project by the support of JST (Japan Science and Technology Agency) and JICA (Japan International Cooperation Agency). This SATREPS (Science and Technology Research Partnership for Sustainable Development) project is in cooperation with Algeria. The SSB project could be nicknamed the "Super Apollo project" in view of its combination of solar (Apollo) and superconductivity. Although some may regard SSB as a quixotic fiction, we already have the science and technology to test the possibil-

ity. This dream could come true using building blocks of smart-component technologies through international cooperation in human resources and investment in the future.

Key technologies of SSB. For a PV-dominating society and to facilitate the realization of the SSB dream, there are two key technologies that need to be developed: (1) solar-grade polycrystalline Si production from the desert sands for a low-cost Si feedstock and (2) a supergrid system using liquid-nitrogen-cooled high critical temperature (T_c) superconducting (HTSC) cables to transmit dc electricity generated by PV and conventional grids. Other technologies to support SSB are already available, such as the production of solar cells from polycrystalline Si with a reasonably high quality (99.9999% purity with boron and phosphorus levels below 1 ppm), cell assembling into modules and panels, and rolling out of km-long HTS cables.

To achieve SSB, the first task is the large-scale and low-cost production of Si from desert sands. Although the technology to make solar-grade Si (99.9999% purity) from silica sands and to make solar cells with almost the equivalent conversion efficiency to cells made from semiconductor-grade Si (>99.99999999% purity) was preliminary verified in the 1980s by the New Energy Development Organization (NEDO) as a joint project of Nippon Sheet Glass Co., Ltd. and Kawasaki Steel Co., it was apparently premature and forgotten. We recently reviewed the Si technology to explore new schemes for the Si process, as summarized in **Fig. 3**.

New Si processes have been designed on the basis of thermodynamic calculations and two approaches have been selected for testing. The "More than Siemens" process modifies the current Siemens process by improving the chemical yield of the

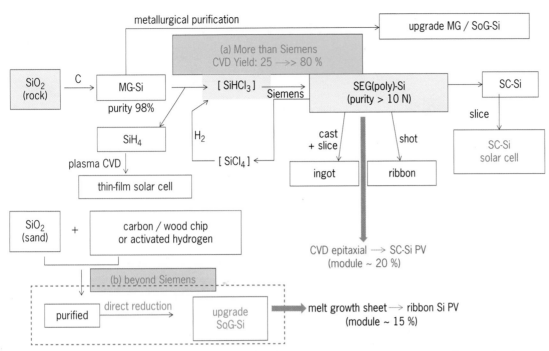

Fig. 3. Two approaches to high-purity Si technologies: (*a*) **More than Siemens** and (*b*) **Beyond Siemens**. Process flow from SiO$_2$ to semiconductor-grade silicon (SEG-Si) and solar-grade silicon (SoG-Si). CVD = chemical vapor deposition. SC-Si = single crystalline silicon. MG-Si = metallurgical-grade silicon (about 98% purity).

Siemens chemical vapor deposition (CVD) step. The main reaction of this step yields Si in only 25% yield, via reaction (1).

$$4SiHCl_3 \rightarrow Si + 3SiCl_4 + 2H_2 \qquad (1)$$

Higher Si yields were expected by introducing H radicals, as follows in reactions (2) and (3).

$$SiHCl_3 + 2H \rightarrow Si + 3HCl \qquad (2)$$

$$SiCl_4 + 4H \rightarrow Si + 4HCl \qquad (3)$$

The process has been verified experimentally using a compact pulsed thermal plasma reactor. The "Beyond Siemens" process is aimed at converting SiO$_2$ directly into Si by carbothermic reduction [reaction (4)], where the energy consuming steps

$$SiO_2 + 2C \rightarrow Si + 2CO \qquad (4)$$

of hydrochlorination and CVD can be skipped. The carbothermic reaction involves the following sub-reactions (5)–(7). Reaction (5) takes place twice for

$$SiO_2 + C \rightarrow SiO + CO \qquad (5)$$

$$SiO + 2C \rightarrow SiC + CO \qquad (6)$$

$$SiO + SiC \rightarrow 2Si + CO \qquad (7)$$

each of reaction (6) and (7). Key points include control of the balance of the subreactions with temperature and feeding purified SiO$_2$ and C under the

conditions of minimum energy consumption and minimized evaporative loss of the intermediate SiO species.

SSB is focusing on the "Beyond Siemens" process using silica sand instead of conventional silica rock because of the following three advantages. (1) Sand is more abundant than rock and available almost everywhere in the world. (2) Advance purification, especially removal of boron and phosphorus, is very difficult after reduction into Si and is possible by a wet chemical (ion exchange) method. (3) Since this process is at the experimental and not the production level, we need to start working together with our Algerian partners from the basic research stage, thus contributing to the education and training of young scientists and engineers.

Energy payback time (EPT) is defined for energy-generating devices as the time to recover the energy used for its construction. Currently for a Si solar cell, EPT is presumed to be about 2 years. If all the PV energy is used for reproduction ("breeding") of PV, the scale of PV power generation can be doubled every 2 years, such that 2 MW of PV will reach 100 GW in 30 years. This breeding pace can be accelerated with the aid of local natural-gas energy at night when solar energy is not available.

The second key technology is the superconducting grid (supergrid) being developed at Chubu University, which has the only superconducting transmission research center in the world. Bi$_2$Sr$_2$Ca$_2$Cu$_3$O$_{10-x}$ powders coated on silver (Ag) tape (4 mm wide × 0.2 mm thick) exhibit superconductivity when cooled with liquid nitrogen (LN) [superconductor T_c = 110 K; liquid nitrogen boiling

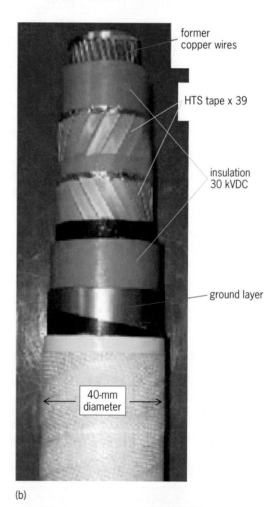

(a)

(b)

Fig. 4. SSB initiative model for (*a*) making solar silicon from Sahara Desert sands and the resulting power generation and transmission technology. (*b*) Prototype of a Bi-Sr-Ca-Cu-O superconducting ($T_c \sim 110$ K) cable.

point 77 K], with a critical current in the superconducting state of 200 A. The coated tapes have been used to make a superconducting cable prototype, consisting of a metal pipe 40 mm in diameter in which 39 tapes were wound around a centrally cooled core (**Fig. 4**). This cable is contained in a 150-mm-diameter pipe, and can transmit 4 kA dc current, which is the equivalent to the current carried by a copper cable 70 mm in diameter. (The copper cable must be equipped with a large heat radiation device.) The advantage of the superconductor cable is that it has zero electrical resistance. A 150-mm-diameter × 200-m-long transmission pipeline, which contains a 40-mm-diameter cable assembly that is cooled by LN and heat-insulated by a vacuum spacer is being tested in Chubu University. The SSERC project cannot afford to install the pipeline in the Sahara, therefore we are planning to collect basic climate data to design the conditions for implementing superconducting pipelines in desert areas and to accept graduate students from Algeria for education and training in supergrid technology.

Other basic research and education programs of the SSERC project include the following targets. (1) A PV system and remote-monitoring installation at Saïda, Algeria (a gateway city to the Sahara) to collect quantitative data on PV performance and problems in the desert area. (2) Installation and promotion of the WebELS e-education and communication system originally developed by the National Institute of Informatics, Japan. (3) A technological assessment of the requirement for PV-panel protection against sand storms.

Outlook. The SSB was proposed based on the quality, quantity, and cost of energy (PV and HTSC) materials as well as the technical feasibility of processes and application systems investigated by IEA (International Energy Association) task force activity reports on "Energy from the desert."

The SSERC project as an initiative of SSB for sustainable development in the future has started in Japan and Algeria, but its budget of $5 million for 5 years, is five orders of magnitude smaller than that of the DESERTEC project. DESERTEC is a huge renewable energy consortium coordinated by a German and international business group. Their goal is to build a massive network of concentrated solar thermal (CSP) power stations and wind farms stretching across the MENA region to provide 15% of the EU's electrical power via a Euro-Mediterranean electricity network using high-voltage direct-current transmission cables (http://www.desertec.org/). Despite a big difference in budget size (€400 billion versus €5 million), we feel SSB surpasses DESERTEC in such aspects as creative technology rather than transfer of known technology of concentrator solar (thermal) power generation and high-voltage dc transmission, scale flexibility and easier access for local people, less maintenance, and greater chance for local employment and promotion of science and technology. The recent attention and developments related to the SSB project worldwide have

Fig. 5. Sahara solar breeder project—quixotic quest or dream come true for the future of the Earth?

created a new context and momentum for developing a new phase of cooperation in establishing a more strategic partnership for research and innovation.

Through an Asia-Arab sustainable energy forum held first in Nagoya, Japan and then in Oran, Algeria in 2011 and 2012, respectively, growing interest and attention is prevailing in Arab (Tunisia, especially) and Asian (Turkmenistan, Russia, Taiwan, and others) countries. The SSB Foundation International (SSBFI) was established in Tunis in May 2011 to promote research and development and globalization of SSB. Japan is so far away from the Sahara Desert that it is not expecting direct benefit or profit from SSB, but hopes to contribute to the health and happiness of those in the Sahara region (**Fig. 5**).

For background information *see* CHEMICAL VAPOR DEPOSITION; DESERT; ENERGY; ENERGY SOURCES; PHOTOSYNTHESIS; PHOTOVOLTAIC EFFECT; SILICA MINERALS; SILICON; SOLAR CELL; SOLAR ENERGY; SUPERCONDUCTIVITY in the McGraw-Hill Encyclopedia of Science & Technology.

Hideomi Koinuma; Kenta Tsubouchi;
Kenji Itaka; Amine B. Stambouli

Bibliography. A. Fujishima and K. Honda, Electrochemical photolysis of water at a semiconductor electrode, *Nature*, 238(5358):37–38, 1972, DOI:10.1038/238037a0; S. Inoue, H. Koinuma, and T. Tsuruta, Copolymerization of carbon dioxide and epoxide, *J. Polymer Sci. B Polymer Lett.*, 7(4):287–292, 1969, DOI:10.1002/pol.1969.110070408; H. Koinuma, Oxide materials research for global environment and energy with a focus on CO_2 fixation into polymers, *Reactive and Functional Polymers*, 67(11):1129–1136, 2007, DOI:10.1016/j.reactfunctpolym.2007.08.010; H. Koinuma, M. Smiya, I. Nakai, Sahara solar breeder plan: Quixotic research for sustainable development of spaceship Earth, *Kagaku* (Chemistry), 66(12):35, 2011 (in Japanese); H. Koinuma et al., SCJ (Science Council of Japan) proposal: Sahara solar breeder plan directed towards global clean energy super highway, G8+5 Academies Meeting, Rome, March 26–27, 2009; K. Kurokawa et al. (eds.), *Energy from the Desert*, vol. 3, Earthscan, 2009; A. Listorti, J. Durrant, and J. Barber, Solar to fuel, *Nature Materials*, 8(12):929–930, 2009, DOI:10.1038/nmat2578; A. B. Stambouli and H. Koinuma, A primary study on a long-term vision and strategy for the realisation and the development of the Sahara Solar Breeder project in Algeria, *Renew. Sustain. Energy Rev.*, 16(1):591–598, 2012, DOI:10.1016/j.rser.2011.08.025.

Saliva-based immunoassay of waterborne pathogen exposure

Water is our most important resource, and ensuring its safety, security, and sustainability is a global priority. Of almost equal importance is the protection of the global community from microbiological and chemical contaminants in our drinking and recreational water sources. The ability to easily identify exposures to waterborne pathogens has a critical role in understanding the public health implications associated with these pathogens. Methods that are rapid, cost-effective, sensitive, use a noninvasively collected matrix, and can detect multiple pathogens simultaneously would greatly facilitate the ability to collect this type of information. The saliva-based multiplex immunoassay method described below may represent significant improvements to these concerns.

Background. Each year, globally, more than a million people die and billions of others become ill from unsafe water, sanitation, and hygiene. According to the World Health Organization (WHO), 9.1% of the disease burden and 6.3% of all deaths worldwide potentially could be prevented by improving water, sanitation, and hygiene. Alarmingly, most of the deaths and illnesses are borne by children 14 years old and younger, particularly those in developing countries. Moreover, the financial cost associated with unsafe drinking and recreational water is astronomical. The WHO estimates that in the United States alone, approximately $84 billion could be saved annually if these factors are improved.

More than one-third of the diseases that contribute to the global disease burden associated with unsafe water are diarrheal diseases. These include the more serious cases of dysentery, typhoid, and cholera and are collectively referred to as acute gastrointestinal illness (AGI). Symptoms of AGI include diarrhea, nausea, and vomiting. A major complication of such diarrhea-related illness is malnutrition and its consequences, which accounts for about 26% of the global total disease burden. Acute respiratory illness (ARI), skin, ear, and eye symptoms are also common. The most frequent waterborne etiologic agents are *Salmonella typhimurium*, *Vibrio cholerae*, *Escherichia coli* O157:H7, *Campylobacter jejuni*, and noroviruses. Intestinal nematode infections, schistosomiasis, lymphatic filariasis, and malaria also

contribute significantly to this global health and financial burden.

To reduce or eliminate these health threats, the global community must be proactive and invest heavily in improving the overall water infrastructure to ensure that the water is safe and that it is used in a sustainable manner. Furthermore, methods to quickly identify source water contamination and human exposure to these contaminants must be developed. Currently, identification of waterborne disease outbreaks and their etiologic agents typically occurs after the incidents have taken place. This practice, while prospectively useful, does not identify the causative agents early enough to constrain the outbreak in its infancy, thereby assuaging its effects. We must be able to develop models that can accurately measure the risks of exposure and have risk-management plans in place to reduce the possibility of human exposure to these waterborne pathogens.

Traditional applications that assess exposure, such as enzyme-linked immunosorbent assays (ELISA) and real-time quantitative polymerase chain reaction (RT-qPCR), excel at high throughput (more than 1000 samples per day), but they lack the ability to multiplex more than one to five tests at a time. Others, such as microarrays that provide high-density solutions (more than 250-plex), lack the reproducibility needed for high-throughput applications. Moreover, many tests rely on collecting blood as the matrix of choice. Blood, though high in antibodies, requires very invasive measures to collect. Many people have an aversion to needle sticks, and children, in particular, find it an extremely frightening and painful experience. For that reason, many epidemiological studies experience difficulty recruiting participants when blood has to be collected. Additionally, blood collection, ELISA, and microarrays are very expensive and can be time and labor-intensive. In recent years, new technologies have emerged that make it possible to achieve both high-throughput and high-density solutions simultaneously using very small sample volumes. Among those technologies is the Luminex® multi-analyte profiling xMAP® platform.

Luminex xMAP technology. This technology combines advanced optics and digital signal processing with microsphere technology in order for multiplexing capabilities to be achieved. Microspheres ($5.6\text{-}\mu\text{m}$ beads) with varying internal concentrations of fluorescent dyes are arranged in unique bead sets that can each be coated with a specific analyte of interest. This allows for the capture and detection of specific analytes from a sample. Using this technology, up to 500 specific analytes can be measured in a single well of a 96-well microtiter plate. Therefore, theoretically, one could multiplex up to 500 antigens in the same test, producing up to 48,000 data points (500 analytes × 96 wells/plate) in about one hour. A Luminex analyzer interrogates the beads by using a light source that excites the internal bead dyes, to determine both the bead type and any reporter dye that has been captured during the assay. To validate the results, many readings are performed on each bead

Known causative agents of waterborne disease outbreaks. These pathogens are prime candidates for use in a multiplex salivary antibody immunoassay.		
Multiplexed waterborne pathogens		
Bacterial	Protozoan	Viral
Escherichia coli O157:H7 *Campylobacter jejuni* *Helicobacter pylori* *Salmonella typhimurium* *Vibrio cholerae* *Shigella species* *Legionella pneumophila*	*Toxoplasma gondii* *Giardia lamblia* *Cryptosporidium parvum*	Norovirus* • Norwalk • VA387 Hepatitis A virus Rotavirus

*Denotes two serogroups of the Norovirus: GI.I (Norwalk) and the GII.4 (VA387).

set. In contrast to traditional methods such as ELISA and microarrays, in which the antigens or antibodies are fixed to the surface of a plate, thereby limiting the kinetics of binding, this platform uses suspension microsphere technology that gives analytes maximum opportunity to bind to the bead surfaces in a liquid matrix.

Method description. The method described here is a Luminex-based multiplex immunoassay that uses human saliva as a source of salivary antibodies that can be used as biomarkers of exposure to waterborne pathogens. This method is rapid, sensitive enough to identify antibodies to multiple pathogens simultaneously, and is performed using saliva, which is an easily collected, noninvasive, and inexpensive matrix. The use of saliva in this multiplex assay ensures that children, some of the most susceptible and vulnerable members of the population, can be included in these exposure studies. Data collected in these types of studies can be used to populate quantitative microbial risk-assessment models.

The immunological rationale for this method is based on the immune system's ability to recognize, respond to, and remember an immunological insult (for example, a pathogen). To summarize the basic principles, when an individual is initially exposed to a pathogen, the humoral (antibody-mediated) arm of the immune system responds by producing specific antibodies against that pathogen. After clearing the initial infection, the immune system produces cells that retain memory (memory cells) that can respond in the event that there is a subsequent exposure to that organism. If there is a subsequent exposure to the pathogen, memory cells are activated, leading to rapid proliferation of clonally identical cells that produce the same antibody as the parent cell. Thus a much more robust secondary antibody response is produced against the pathogen to neutralize it and reduce or eliminate the infection. Some memory capacities, depending on the pathogen and the individual, can be retained for an extended period (years), while others are short-lived (months). Vaccination is based on this immunological principle. Memory cells are produced and remain in the body for several years. Depending on the vaccine, boosters may be needed to maintain protection over a longer period of time.

Antigens to waterborne pathogens (see **table**) were selected for inclusion in the immunoassay, as shown in **Fig. 1**. Briefly, each antigen is coupled to a unique bead set. After all of the antigens are coupled to their respective bead sets, a coupling confirmation test is performed to ensure that the antigens are efficiently bound to the beads. The coupled bead sets are then mixed together and tested for cross-reactivity using commercially available antibodies to each antigen to ensure that they do not interact with each other. Once cross-reactivity has been ruled out, human saliva collected from participants of an epidemiological study is processed to remove debris, serially diluted, and incubated with the beads in a 96-well microtiter plate. To ensure reproducibility and statistical relevance, each assay is done in duplicate. Inter- and intraplate variability is also assessed. Individuals who have been previously exposed should have specific antibodies in their saliva that will bind to the antigens coupled to the beads, forming a bead–antigen–antibody complex. The complex is incubated with a detection antibody: anti-human immunoglobulin G (IgG) antibody

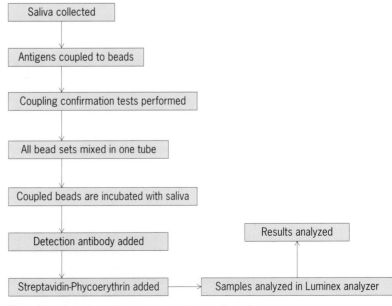

Fig. 1. Workflow chart of the multiplex salivary antibody immunoassay.

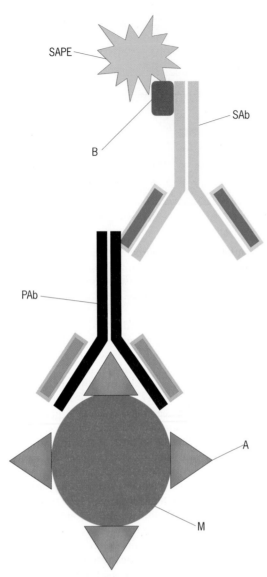

Fig. 2. Illustration of the complete immunoassay complex.
M = microspheres (beads); A = antigen; PAb = primary
antibody in saliva; SAb = secondary (detection) anti-human
IgG antibody; B = biotin; SAPE = streptavidin-
phycoerythrin (reporter).

coupled to biotin. The detection antibody binds to
the salivary antibody, if present, and when exposed
to a reporter dye, streptavidin-phycoerythrin (SAPE),
a quantifiable fluorescent signal is given off (**Fig. 2**).
The protein avidin in SAPE binds very strongly to
biotin.

Data acquisition and analysis. Real-time analysis
and accurate quantification of the assays are performed by the Luminex analyzer. A red laser excites
the internal bead dyes to measure its fluorescence,
while a green laser excites the bead surface to measure the reporter dye fluorescence. The Luminex
xPonent software then identifies the bead set based
on the fluorescence of its two internal dyes. The reporter signal on the bead surface is also calculated.
In total, 100 beads of each set are individually interrogated, and a median value of all events is reported
by the software. Results are reported as median flu-

orescence intensity (MFI) and analyzed for salivary
antibody activity (**Fig. 3**).

Impact of method. The use of salivary antibodies
as biomarkers of exposure to waterborne pathogens
in a multiplexed immunoassay has the potential to
greatly improve our ability to assess human exposure to microbial contamination. The traditional approach is to assess what microbes are in the water, air,
land, or food (occurrence) and then use that information to determine human exposure. This approach
has created significant gaps in both our knowledge
and our mitigation efforts. The new paradigm looks
for biomarkers of exposure in humans (such as proteins, chemicals, and other factors in human bodily
fluids) and then works back to the source. This is
important because it allows us to more clearly and
accurately define what humans are being exposed to
and the potential health effects of such exposures.
This information will enable policy makers to focus
their efforts on reducing that exposure in a more
direct manner.

Environmental researchers now will have the capability to achieve both high-throughput and high-
density testing using very small sample volumes in
a cost-effective, rapid, sensitive, and highly reproducible way. Data collected using this method can
be used to identify specific types of pathogens to
which humans are being exposed. This information
can be used in a variety of ways to study the risks
associated with waterborne pathogens. Examples
include studies to identify/verify possible sources
(such as inadequate sewage treatment, agricultural
activities, and wildlife) of pathogens; indicate the
risks of exposures from wastewater effluents; provide information on what pathogens to focus attention on in terms of detection methods in water; and
to identify pathogens to target for dose–response
models. All of this information will lead to improvements in our ability to estimate risks to public health
and approaches to protect human health. Because
of the noninvasive nature of sample collection, the
use of saliva allows for the potential for greater
participation in epidemiological studies, including
children. Since mitigation actions typically occur
after a waterborne disease outbreak, assessing and
managing risks of exposure could help to reduce the massive global health and financial burden. Moreover, the flexibility and scalability of the
method broadens the scope of what can be studied.
Antigens to foodborne and airborne pathogens can
easily be added to the multiplex immunoassay to determine individual exposure as well as the prevalence of exposure to certain pathogens in specific
populations.

[Disclaimer: The U.S. Environmental Protection
Agency (EPA), through its Office of Research and
Development, funded the research described here.
Although this work was reviewed by the EPA and
approved for publication, it may not necessarily reflect official EPA policy. Mention of trade names
or commercial products does not constitute en-
dorsement or recommendation by the EPA for
use.]

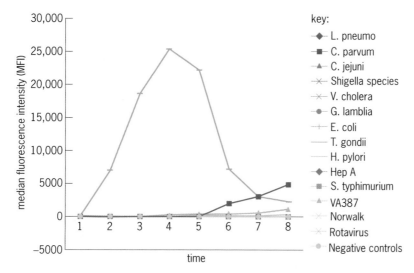

Fig. 3. Multiplexed analysis of individual X's saliva samples during an eight-week study. Human antibody reaction to antigens in the multiplex. Seroconversion to *H. pylori* is clearly demonstrated during the eight-week study, while low-level exposure to *C. parvum* is observed. Salivary antibody responses to the other antigens in the panel appear to be at the same level as the negative controls. (Note: Not actual data. Presented for demonstration purposes only.)

For background information *see* ANTIBODY; ANTIGEN; BIOTIN; IMMUNOASSAY; IMMUNOGLOBULIN; IMMUNOLOGY; INFECTIOUS DISEASE; PATHOGEN; PUBLIC HEALTH; WATER BORNE DISEASE in the McGraw-Hill Encyclopedia of Science & Technology.

Swinburne A. J. Augustine; Kaneatra Simmons

Bibliography. K. Murphy, P. Travers, and M. Walport, *Janeway's Immunobiology*, 7th ed., Garland Science, Taylor & Francis Group, New York, 2008; M. J. Nieuwenhuijsen, *Exposure Assessment in Occupational and Environmental Epidemiology*, Oxford University Press, Oxford, U.K., 2003; W. R. Ott, A. C. Steinemann, and L. A. Wallace, *Exposure Analysis*, CRC Press, Taylor & Francis Group, Florida, 2007.

Self-energizing high-rise towers

In recognition of global warming and regional climate change, we are witnessing the remarkable speed of scientific, technological, and societal developments worldwide in reducing the rate of energy consumption per capita, increasing reliance on generating electricity from renewable natural resources in lieu of fossil fuels, attempting to reduce emissions of carbon dioxide (CO_2) and other greenhouse gases (GHG) globally, and accelerating the movement toward self-energizing high-rise towers.

The American Institute of Architects (AIA) and others have indicated that buildings are responsible for about 40% of GHG emissions, while transportation and industry, including the manufacture and delivery of raw materials and building products, account for about 60% (**Fig. 1**). In recent years, there has been a concerted effort on the part of building owners and managers and design professionals (civil, geotechnical, structural, mechanical, electrical and plumbing engineers) and building contractors to reduce energy use by applying the best

available science and technology in new construction and retrofitting existing buildings. Worldwide, many governments and nongovernmental organizations have established "green" building codes and standards, design criteria, voluntary awards programs, and certifications.

Toward net-zero energy consumption. Developed countries have 21% of the world's population, but they consume three times the energy of underdeveloped countries. The United States, with less than 5% of global population, consumes 24% of the world's energy. The average energy consumed per person in the United States amounts to twice that in Japan, 6 times that in Mexico, 13 times that in China, 31 times that in India, and 370 times that in Ethiopia.

In the past 40 years, oil consumption globally has increased 250%, and coal and natural gas consumption have doubled. Much of the fossil fuels that are not used directly in transportation or industrial applications are used to generate electricity. As a result, CO_2 emissions are increasing (**Fig. 2**).

In 2008, a total of about 600 coal-burning power plants generated 45% of the United States' electricity (**Fig. 3**). Coal is the single largest environmental polluter of ecosystems worldwide. Per year, a

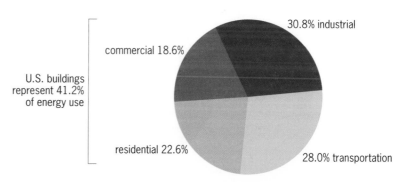

Fig. 1. U.S. 2010 energy consumption of 98,010 trillion BTUs, shown by sector. (*U.S. Department of Energy/NREL*)

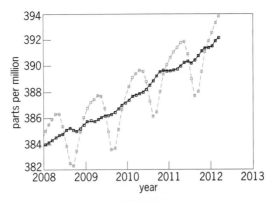

Fig. 2. Global mean atmospheric CO_2 concentration. The Global Monitoring Division of NOAA/Earth System Research Laboratory measures CO_2 and other greenhouse gases at a globally distributed network of air sampling sites. The data for the last year are still preliminary, pending quality-control conformation. (*NOAA, for updates see http://www.esrl.noaa.gov/gmd/ccgg/trends/global.html*)

500-megawatt (MW) coal-burning power plant producing about 3 million tons of CO_2 and requires about 2.2 billion gallons of water from a nearby lake, river, or ocean to create and condense steam to power the turbines that are generating the electricity. The discharged water is overheated and often contains chlorine, which kills aquatic and biotic life.

In 2007, an interdisciplinary faculty group at the Massachusetts Institute of Technology (MIT) issued a report, *The Future of Coal: Options for a Carbon-Constrained World*, which concluded that coal, being the most inexpensive and most plentiful fossil fuel, was here to stay. Therefore, they strongly recommended carbon capture and carbon sequestration as immediate remedies to the CO_2 emissions problem, without addressing water pollution problems from coal-burning power plants.

In 2008, the AIA building sector "green wave" initiative was implemented to reduce GHG emissions by 40–60% below 1990 levels. This represents a concerted effort to involve not only architects, but also engineering consultants, building contractors and owners, materials and equipment suppliers, government leaders, and business entities.

High-rise towers. Architects and structural engineers normally define a high-rise tower as consisting of 10 or more stories with a minimum height of 26 m (85 feet) for residential and 46 m (150 ft) for commercial office towers. In 2010, the Center for

Tall Buildings and Urban Habitat (CTBUH) revised its methodology for determining the heights of high-rise towers. The new CTBUH rules include architectural spires, communications antennae, flagpoles, and highest people-occupied floors, and the criterion for "height to top of roof" has been rescinded (**Fig. 4**).

In capitalist societies, highest and best use often dominates real estate economics. Zoning codes normally permit a specified floor-area ratio (for example, 30:1), allowing the buildable area of land parcels to be multiplied several times. Infrastructure, such as roads, sewers, water, electricity, and fiber-optic cables, geotechnical data and location, accessibility, views, timing, marketability, and financing determine feasibility. In conjunction with the green movement, high-rise towers can prevent unnecessary urban sprawl and reduce commuter transportation time and fuel costs, while conserving agricultural lands, parks, and recreational wilderness areas.

Toward self-energizing high-rise towers. In moving toward self-energizing high-rise towers with net-zero energy consumption and net-zero CO_2 emissions, important first steps include increased natural lighting to reduce the need for artificial lighting, and glazing and shading systems to reduce heat gain, to name a few examples of energy conservation. Solar photovoltaic and wind power currently prevail as sources of renewable energy that can be harvested. Fuel-cell CHP (combined heating and power, also known as cogeneration) systems also promise to significantly reduce energy use and GHG emissions. For example, when construction is completed at the new World Trade Center in New York City, 4.8 MW of electric-generating capacity using fuel-cell CHP technology will be in place.

One of the best examples of an energy-efficient building anywhere in the world is the 71-story Pearl River Tower in Guangzhou, China, designed by Skidmore, Owings & Merrill (SOM) and completed in March 2011 (**Fig. 5**). Its design incorporated many energy-saving and energy-generating strategies, including shaping and orienting the building to maximize daylight and wind patterns, radiant cooling, high-performance glazing, energy-efficient office equipment, high-efficiency lighting, daylight-responsive controls (automated window-blind systems), heat recovery, integrated wind turbines, and

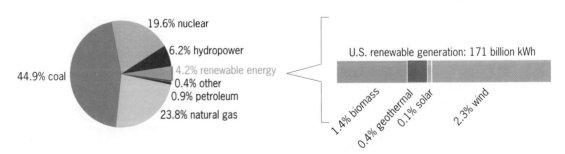

Fig. 3. Net 2012 U.S. electricity generation of 4123 billion kWh, shown by generation method. (*U.S. Department of Energy/NREL*)

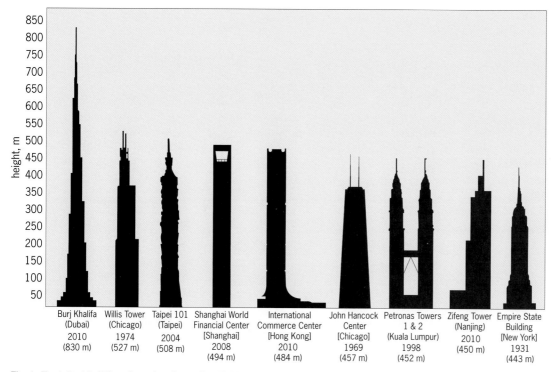

Fig. 4. Ten tallest buildings based on Council on Tall Buildings and Urban Habitat (CTBUH) height-to-tip criteria.

photovoltaics. Perhaps just as important as constructing energy-efficient high-rise towers will be retrofitting the existing ones. Some representative examples of energy savings follow.

Wind. One of the distinct advantages of creating supertall high-rise towers [>300 m (984 ft) in height] is the exponential increase in wind speed with height. For example, in the center of a large city with relatively flat terrain, a wind speed of 50 km/h (31 mi/h) at 30 m (98.4 ft) above ground level can increase to an amazing 160 km/h (100 mi/h) at the top of a high-rise tower 518.3 m (1700 ft) above ground level, as shown in **Fig. 6**, while the power gradient is cubed. The terrain surrounding a high-rise tower has a major effect on increasing wind speeds at the top of a building.

As noted by L. Leung and P. Weismantle of SOM, there are distinct advantages to be gained from

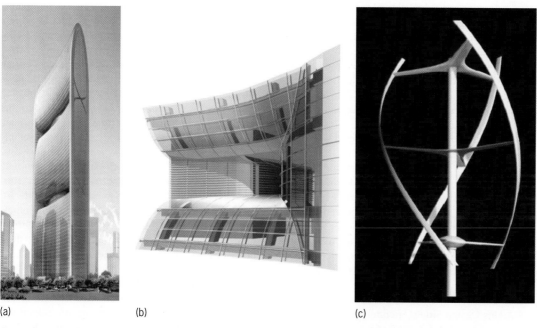

(a)   (b)   (c)

Fig. 5. The Pearl River Tower has (*a*) two openings on each side of the building on the 24th and 48th floors (mechanical floors), (*b*) each of which funnels the wind to (*c*) a vertical-axis wind turbine. (*Courtesy SOM*)

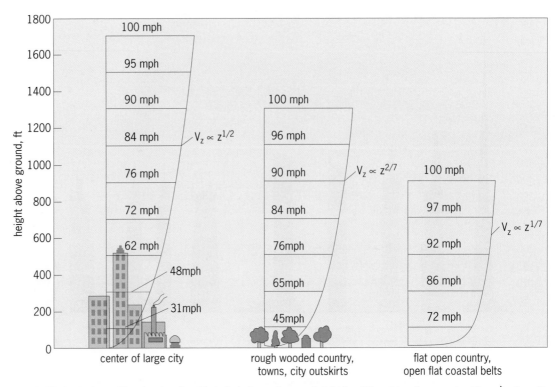

Fig. 6. Wind velocity profiles showing the effect of wind speed versus height for different terrain scenarios. $V_z \propto z^k$, where V_z is the velocity at height z and k is the exponent for the best fitting curve. [*W. A. Dalgliesh and D. W. Boyd, CBD-28: Wind on Buildings, Canadian Building Digests, National Research Council Canada (NRC)*]

supertall buildings other than great increases in wind speed and power with height, including decreased heating and cooling loads on building interiors because of vast temperature differences of 5.5–10°C (10–15°F) or more, depending on height, humidity changes, and inside–outside pressure differentials subject to location and the microclimates encountered.

Taking advantage of the exponentially increased wind velocities and cubed natural renewable energy power with height could provide both economic and environmental benefits (in addition to net-zero carbon emissions) from capturing renewable wind energy from the sky. For example, damping device mechanisms (including pendulums and shock absorbers) used to control appreciable sway in supertall buildings can possibly also provide some useful forms of building-generated energy.

Some engineers concentrating on civil and structural design projects have stated that the erratic behavior of wind is similar to that of water and is therefore controlled by a branch of science known as fluid dynamics. They have noted that the strong uplift and suction forces of wind on roofs in a powerful hurricane can be equivalent to an upside-down waterfall. Some astute architects and engineers are using computational fluid dynamics (CFD) computer simulation modeling techniques in rigorous performance design analyses.

The Bahrain World Trade Center twin towers by WS Atkins Design Studio, architects, planners, and engineers, completed in 2008, were the world's first twin high-rise towers to incorporate integrated large-scale nonfreestanding wind turbines and nacelle housings (for the electrical-generating equipment) attached to sky bridges in socially, culturally, economically, and environmentally well-balanced architecture and advanced engineering replete with masterful site and urban planning (**Fig. 7**). The splayed curvilinear and angularly inclined twin sculptural towers, as designed by Atkins, with cascading terraces and balconies on each of the commercial office floors, are far superior to the more typical mundane, prismatic, rectangular blunt-body buildings, which generate turbulence around their right-angled corners and consequently detract from the wind's exposure range, speed, forces, and electrical generation efficiencies. The CTBUH reported that CFD modeling and wind-tunnel simulation studies showed that a serpentine-channeled wind flow added to the successful outcomes on this project. The 240-m-tall (787-ft) 50-story towers' elliptical and aerodynamically designed sail shapes allow winds to be smoothly accelerated in accordance with established Bernoulli air-movement principles of physics. As they funnel between the towers, this maximizes their thrust to rotate three 29-m (95-ft) in diameter horizontal-axis wind turbines (HAWT), engineered with a wide range of latitude (45° to either side of their ideal due-north orientation). This gives them the unique capability of generating 225 kW per turbine, or 675 kW total, at 50% of the daily regime exposure, for a total electrical-generation contribution of 11–15% of the twin towers' total annual power consumption of 1.1 to 1.3 GW. HAWT are 15–25% more efficient than vertical-axis wind turbines (VAWT) because they capture more energy for the same windswept rotational area.

Retrofits. The sustainable retrofitting of the Empire State Building and the Willis Tower are examples of cost-effective strategies for reducing energy use and GHG emissions and these projects will be discussed in detail.

Empire State Building. On the basis of a sustainability program and retrofit feasibility study of the Empire State Building (ESB), conducted primarily by a real estate marketing and management firm and a buildings interior-climate-control company, with input from the Clinton Climate Initiative and the Rocky Mountain Institute, and in consultation with architectural and engineering firms that had some introspective familiarity with the revered 80-year-old Art Deco structure on Fifth Avenue in New York City, the key finding was that it would be cost-prohibitive to achieve an energy reduction greater than 38%.

Another key finding of the ESB study was that reducing CO_2 emissions beyond a certain point would also be cost-prohibitive. On both of the key findings dealing with energy-consumption reductions and reductions in CO_2 emissions, there is an emphasis on a cost-conscious approach that is inherent in real estate economics and cost/benefit business models. This can create conflicts of interest and result in decisions in which pragmatic considerations override eleemosynary environmental concerns. As a result, eight primary energy-saving retrofit projects were selected (**Fig. 8**).

Two of the primary architectural and engineering retrofit projects for reducing energy consumption and carbon emissions were the complete removal of the existing windows and their replacement with thermally insulated ones, and the removal of the convection-tube radiators below the perforated windowsills and the installation of a "radiative" insulation and vapor barrier on the inner face of the original brick infill panels installed above and below the windows between closely spaced concealed structural columns behind the Art Deco glazed terra cotta panels of the exterior façade. The projected capital cost was $4.5 million for the new windows and $2.7 million for the radiative barriers, with estimated annual energy consumption savings of $410,000 for the window removal and upgrade reinstallation and $190,000 for the radiative barrier installation. The approximately 6500 new custom-manufactured double-hung windows included double insulated glass units and a low-emissivity suspended coated film with gas infill. As a part of the overall energy consumption savings of 38% attributed to the 2008 ESB capitalized sustainability program, the radiative barrier retrofit contributed 3% and the upgraded windows contributed 5%, for a total of 8%, leaving a balance of 30% to come from the remaining six major key retrofit projects. From the standpoint of carbon emissions reduction, both of these retrofits were charted as being almost negligible on the basis of cost per metric ton of CO_2 by individual measure. These two key retrofit projects accounted for 6.7% of the capital budget and resulted

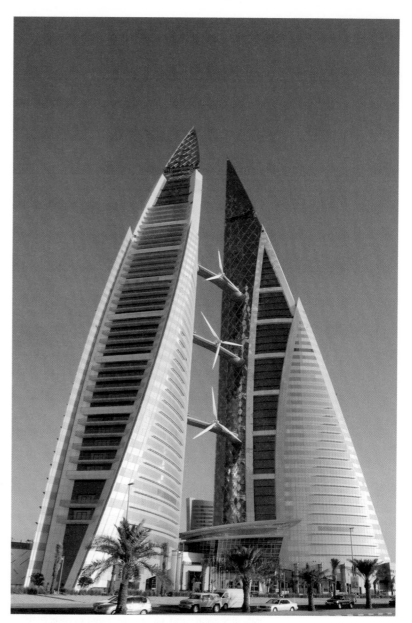

Fig. 7. Bahrain World Trade Center. (*Photo courtesy Atkins*)

in 13.6% of the energy savings budgeted costs, plus they were deemed worthwhile from the standpoint of user thermal comfort and a good sound payback on investment.

The next retrofit item related to the exterior windows was tenant daylighting/plugs. This prospective new tenant retrofit and environmental quality space use improvement item involved ESB occupants being provided with "plug" occupancy load sensors for personal workstations, in addition to the installation of reflective daylighting and artificial illumination baffles to reduce artificial lighting power intensity in foot-candles and density in lumens/ft^2 as well as electrical energy consumption in kilowatt-hours (kWh). The projected capital costs were $24.5 million, with an estimated annual energy savings of $941,000. Again, there was an almost negligible reduction in cost per metric ton of CO_2

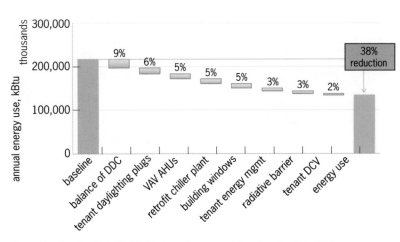

Fig. 8. The Empire State Building's annual energy savings by retrofit project. DDC = direct digital control; VAV AHUs = variable-air-volume air-handling units; DCV = demand-controlled ventilation. (*Courtesy http://www.esbsustainability.com*)

emissions, resulting in an additional 8% in electrical energy consumption savings. This key retrofit project amounted to about 23% of the capital budget. It was the second-largest big-ticket project item, but it resulted in only 21.4% of the energy savings budgeted costs. However, it was deemed very worthwhile from an attractive tenant marketing, tenant user, and educational standpoint within an overall strategy of objective environmental sustainability.

The most expensive retrofit item involved the variable-air-volume air-handling units (VAV AHUs) at a projected capital cost of $47.2 million, or slightly over 44% of the total capital budget, which resulted in less than 16% of the estimated annual energy savings, or a mere 5% of the 38% total energy consumption reduction achieved. This is essentially because of inefficiencies in mechanical air-handling equipment, air distribution systems (ductwork, diffusers, grilles, insulation, and exhaust air fans), electronic sensors, and control systems. This is an area that is ripe for progressive invention in the mechanical, electrical, and electronic engineering realms. This has recently led to movements in innovative directions worldwide, such as water-cooled and floor plenum air-cooled systems. One of the most recent progressive advances in integrated mechanical, electrical, and plumbing (MEP) design coordination with structural and architectural design elements, in order to avoid expensive time and materials clashes and conflicts during construction, has been the implementation of 3D and 4D computerized animated layouts in high-technology visual and quantitative programs on laptop computers in fabricating shops and on project job sites.

With regard to the need for MEP equipment and systems invention, the ESB environmental sustainability program report states, "Value-chain analyses can help determine opportunities for cost reductions for technologies that save significant amounts of energy." The area of thermal comfort for building users has been plagued by a number of issues, resulting in indoor air quality contamination and human health problems. The systems have been prone to dif-

ficulties in balancing and controlling air flows, temperatures, humidification, mold, mildew, and fungicides, equipment breakdowns, water leakage and condensation problems, excessive maintenance requirements including mitigating high-decibel noise, vibrations, and access problems, and have been a major contributor to the Earth's carbon and ecological footprints, especially in the high-energy-usage installations that have too long been promoted by public utility companies. The exemplary movement that the ESB and other prime demonstration building owners and managers are taking in moving initially toward reductions in energy consumption and carbon emissions, and ultimately toward net-zero self-energizing electrical consumption and net-zero carbon emissions, is commendable, but, as the ESB report clearly states, "This project was a great 'test lab,' but what now? If all buildings need to be retrofitted to profitably reduce greenhouse gas emissions by 75% by 2050, we have a lot of work to do in a short amount of time."

The ESB team that espoused taking the whole-systems, dynamic life-cycle approach noted that, "Significant time was spent creating the energy and financial models for this building and then iterating between them. Quicker and simpler (computerized) tools could help accelerate the process." As noted earlier, improvements in this area would also positively affect building performance, human production, and energy production and consumption efficiencies and would reduce CO_2 and GHG emissions. The ESB team learned that more than 60% of the energy savings exist within the tenant spaces, and that the tenant users are responsible for about 20% of the 38% reduction in energy savings, or about 7%. They also learned that energy savings and retrofit service professionals and companies have a greater interest in concentrating on large commercial and large multifamily residential or government buildings. Moreover, they found from their real estate research via available DOE/EIA (U.S. Department of Energy/Energy Information Administration) data that 95% of the total operational U.S. buildings consume about 40% of total energy use. In their report, they include the typical chiller plant upgrade, and mention options for backup generation on-site to include possible cogeneration, natural gas–fired or biofuel-fired generation, fuel cells, natural renewable energy sources such as wind and solar, and even the possibility of purchasing new power capacity from a public utility.

Willis Tower. In June 2009, the owners and managers of the 110-story, 442-m-tall (1450-ft) Willis (formerly Sears) Tower, working with consultants, announced plans for a major retrofit upgrade project, the first since the tower's completion in 1973, to reduce energy consumption and to reduce CO_2 and GHG emissions, and also to spur economic development on-site and in the surrounding neighborhood. As the tallest high-rise tower in the Western Hemisphere, the Willis Tower was to undergo a major economic and environmental sustainability retrofit plan under the direction of AS + GG

Architecture. The owners and their architects decided to surpass the "Leadership in Energy and Environmental Design" (LEED) standards produced by the U.S. Green Building Council. The mutually agreed-upon $350 million program included addressing the following basic items:

1. Upgrade the tower's exterior building envelope and windows, with planned energy service savings of up to 60%.

2. Mechanical systems upgrades (new high-efficiency chillers with improved distribution and new natural gas boilers using fuel-cell technologies).

3. Modernization of the tower's 104 high-speed elevators and 15 escalators, with 40% electrical energy savings.

4. Water conservation measures, including restroom fixture upgrades, condensation recovery systems, and landscaping irrigation efficiencies, with 40% water savings equal to 24 million gallons annually plus electrical energy savings.

5. Lighting upgrades in fixtures, bulbs, and controls as well as daylight harvesting, with 40% electrical energy savings.

6. Renewable energy systems testing, including high-roof wind turbines, solar hot-water panels, and green roofs to improve tenant views and reduce heat-island reflection effects, with added roof insulation and reduced stormwater runoff.

Overall, the project was able to reduce baseline electricity use by 80%. The energy savings reportedly amounted to 68 million kWh annually, resulting in avoiding the purchase and consumption of 150,000 barrels of oil for a cost savings of $11.25 million to $15 million per year, with associated reductions in CO_2 and GHG emissions. These results could have been exceeded if the building owners and managers had accepted the architects' recommendation to apply reflective silver paint to the building exterior in lieu of the existing heat-absorbing black finish. The reason for this recommendation was that metalwork surrounding glazing with a dark bronze or black finish can absorb sunlight heat loads that raise their temperatures up to two times those of the ambient air; for example, for air temperatures of 60–90°F (16–32°C), the metal temperatures could be 120–180°F (49–82°C) at lower elevations.

One of the greatest aspects of the Willis Tower retrofit project was the relationship between the architect and the client, in that the Willis Group Holdings insurance brokerage company executives allowed for experimentation and testing of certain design concepts and innovative ideas on the project site, working toward overall practical goals and objectives for reductions in energy consumption and carbon emissions in conjunction with annual bottom-line projections for operational cost savings. This is in direct contrast to the limited partitioning approach taken on the ESB retrofit project, where an energy-controls service company was the paramount player rather than an architect working with specialized engineering consultants, especially in areas such as the approximately 6500 double-hung

windows that were replaced and the interior side of the exterior wall brick masonry infill panels between fireproofed structural steel columns behind the convection radiators, where reflective insulation and vapor barriers were a part of the ESB retrofit sustainability program. Furthermore, ESB management was in charge of executing the VAV/AHU retrofits and the architect was obscurely listed as playing only a minor role, and this was reflected in the bottom line, which achieved less than half of the energy savings realized in the 43-years-younger Willis Tower.

Another breakthrough that occurred on the Willis Tower energy-savings and carbon reduction project was the decision to test a new product, namely, a revolutionary solar-powered window that includes photovoltaic (PV) monocrystalline cells within the insulated glass unit (PVGU) and has the distinct capability to transform sun-facing windows (east, west, and especially south facades) into electrical-generating power sources (**Fig. 9**). A pilot test project was devised for the south-facing windows on the 56th floor of the Willis Tower, which showed the potential for ultimately delivering up to an additional 2 MW of solar-generated electrical power from the tower's facade if the product is eventually accepted and approved for installation by the Willis team.

The intent is for PVGU to replace the insulated glass units (IGU) because of its low thermal transmission (U) value, low solar heat-gain coefficient (SHGC), and high visual light transmittance (VT). Its claim to high power density is based on its ability to generate up to four times more electricity than thin-film building-integrated photovoltaic (BIPV) solutions; however, some rooftop solar engineers and installers dispute this claim, which is dependent on latitude, solar angles, and PV installation angles as well as on the extent of polluted, smoggy urban air and the amount of tower solar exposure below cloud layers. The PVGU approximate price of $125/ft^2 would typically deliver about a 5-year return on investment (ROI). The first released products are intended for exterior curtain-wall glazing systems and skylights.

The Willis sustainability and modernization plans for the supertall high-rise tower, lobby, and plaza levels, in addition to reducing energy consumption and CO_2 emissions, reportedly resulted in the creation of more than 3600 new jobs in the immediate West Loop area and elsewhere in the Chicago environs. In addition, Sears and Willis had an impressive prior history of energy- and cost-savings consciousness, with both economic and environmental commitments. For example, the tower has reportedly reduced its energy consumption by 34% since 1989, and by even more since 1984. Sears and Willis documented that they have completed energy efficiency improvements that finally have resulted in avoiding dumping as much as 51 million tons of CO_2 emissions into the atmosphere annually, which is the equivalent of removing at least 4400 automobiles from roadways annually.

With more than 1.3 million visitors each year (about 4000 daily), in conjunction with its exemplary

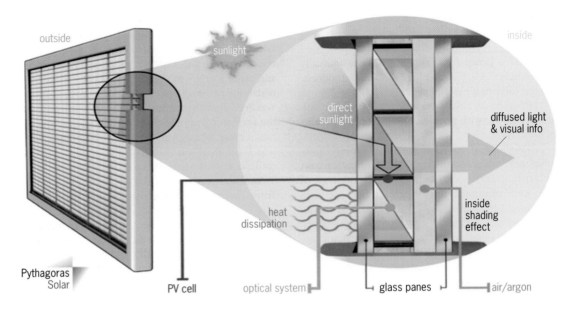

Fig. 9. Photovoltaic glass unit. (*Courtesy Pythagoras Solar*)

2009 energy-savings project, the Willis Tower has established an interactive "Sustainable Technology Learning Center," assisting others in learning how to save energy and money and to reduce CO_2 and GHG emissions by using its urban laboratory experiment as a prime educational tool in achieving success.

Because of its success on the Willis Tower retrofit plan, the AS + GG Architecture firm was able to subsequently produce a Chicago DeCarbonization Plan, addressing the Downtown Loop area, based on the comprehensive 2008 Chicago Climate Action Plan. AS + GG partner Robert Forest stated, "The De-Carbonization Plan looks holistically at development from planning, energy, carbon, finance and construction perspectives." This represents part of an initial movement beyond individual self-energizing high-rise towers and real estate clusters toward self-energizing cities, such as Masdar in the United Arab Emirates, outside its capital of Abu Dhabi.

For background information *see* BUILDINGS; CO-GENERATION SYSTEMS; ENERGY CONVERSION; FUEL CELLS; GLOBAL CLIMATE CHANGE; GREENHOUSE EFFECT; SOLAR POWER; WIND POWER in the McGraw-Hill Encyclopedia of Science & Technology.

Andrew Charles Yanoviak

Bibliography. A. Du Pont, *An American Solution for Reducing Carbon Emissions: Averting Global Warming, Creating Green Energy and Sustainable Employment*, Du Pont Group, Inc., Falls Church, VA, 2009; R. B. Fuller, *Operating Manual for Spaceship Earth*, Amereon, Mattituck, NY, 1998; A. Gore, *An Inconvenient Truth: The Planetary Emergency of Global Warming and What We Can Do about It*, Rodale Press, Emmaus, PA, 2006; M. Guzowski, *Towards Zero-Energy Architecture: New Solar Design*, Laurence King Publishers, London, England, 2010; J. R. McFarland et al., *The Future of Coal: Options for a Carbon-Constrained World, An Interdisciplinary MIT Study*, MIT, 2007; A. Smith, *Toward Zero Carbon: The Chicago Central Area DeCarbonization Plan*, Images Publishing Group Pty., Ltd., Victoria, Australia, 2011; P. K. Takahashi, *Simple Solutions: For Planet Earth*, AuthorHouse, Bloomington, IN, 2007; A. C. Yanoviak, Structural design of high-rise towers, *2010 McGraw-Hill Yearbook of Science & Technology*, McGraw-Hill, New York, 2010.

Sialyl Lewis^X oligosaccharide

Sexual reproduction requires the physical joining of a male gamete (sperm) with a female gamete (egg) to generate a fertilized egg that is usually genetically distinct from its parents. The human egg consists of a single cell (oocyte) that is covered with a specialized coating known as the zona pellucida (ZP). Human fertilization begins when sperm bind to the ZP, penetrate this coating, and come into direct physical contact with the oocyte. A single sperm fuses with the oocyte, generating a fertilized egg (zygote). Another vital adhesion process is the binding of immune and inflammatory cells to the lining of blood vessels (endothelium) in tissues or organs that have become infected or injured. This interaction is required for these cells to respond to bacteria, viruses, parasites, and other organisms that can cause harm, known collectively as pathogens. Cancer cells released into the blood also bind to the endothelium, enabling them to migrate into tissues and organs at distant sites from the original tumor, where they initiate the growth of new tumors (metastases). This process of tumor spread is called metastasis. The sialyl Lewis^X oligosaccharide (SLe^X) and closely related carbohydrate sequences play a vital role in human sperm–egg binding, the movement of immune and inflammatory cells to the sites of infection and tissue injury, and metastasis.

Structure of SLe^X and its relationship to cancer. In 1976, SLe^X was isolated in minute amounts from the human kidney and sequenced. This oligosaccharide

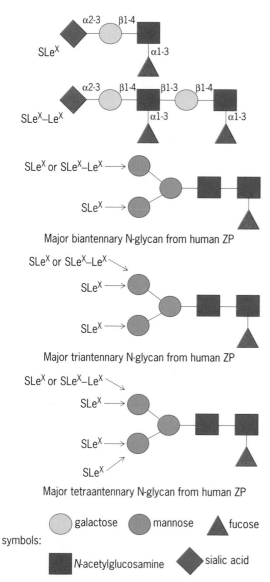

Fig. 1. Carbohydrate sequences and N-glycans associated with the human zona pellucida (ZP).

consists of four different monosaccharides (sugars) that are chemically linked to each other in a very precise arrangement (**Fig. 1**). In 1984, a specific antibody (CSLEX) was developed that could bind to SLe^X. Another antibody (FH6) was simultaneously obtained that could bind to a closely related sequence known as SLe^X-Le^X (Fig. 1). These antibodies were useful for detecting these oligosaccharides on the surface of cells. Both CSLEX and FH6 either weakly bind or do not bind to normal cells and tissues, except for those associated with immune or inflammatory responses. However, CSLEX was avidly bound to approximately 68% of all human tumors, including those isolated from the stomach, colon, lung, esophagus, ovary, female breast, bladder, and pancreas. FH6 was also bound by tumor cells, but not nearly as well as CSLEX. Based on these findings, both SLe^X and SLe^X-Le^X were originally designated as tumor-associated carbohydrate antigens.

SLe^X and its role in immune and inflammatory cell binding. Immune and inflammatory cells constantly move through the circulatory system, acting as sentries that respond to pathogens. Neutrophils are inflammatory cells that usually become activated by either bacterial infection or tissue injury. They bind to the cells lining the endothelium and migrate into affected tissues, where they attack and engulf pathogens. The initial binding of neutrophils to the endothelium is primarily mediated by adhesion molecules known as E-selectin and P-selectin. Selectins are members of a family of carbohydrate-binding proteins (lectins) that are very similar to each other in their protein structure. Another shared characteristic of selectins is that they bind to SLe^X. Immune cells in the circulation known as lymphocytes also circulate through a specialized system of tissues and organs known as the lymphatic system. They bind to the specialized endothelium in these organs by recognizing L-selectin. There is also evidence that these molecules mediate the binding of tumor cells to the endothelium, thereby promoting metastasis.

SLe^X and its role in human sperm–egg binding. For many years, scientists focused their efforts on understanding fertilization in the mouse. These studies confirmed that adhesion primarily depended on the interaction of oligosaccharides presented on ZP glycoproteins with egg-binding proteins that are located on the plasma membrane of sperm. The human model was a substantial challenge for investigation because of the limited number of available eggs and their diminutive size (equivalent in size to a period in the text of this article). Access to human eggs greatly increased in 1978, when the first live human birth as the result of in vitro fertilization (IVF) occurred. The development of techniques for obtaining many eggs during a standard human IVF procedure also greatly increased access to these gametes.

Human zygotes are readily obtained when living human sperm and eggs are mixed together, posing an ethical dilemma for experimentation. Occasionally, human sperm fail to bind to eggs during an IVF procedure. These eggs remain unfertilized and, by the end of the incubation with sperm, they lose their capacity for fertilization. Another method for fertilization is to inject a single sperm directly into the egg by using a procedure known as intracytoplasmic sperm injection (ICSI). If a patient consents, any eggs that fail to fertilize during either procedure can be used in sperm-binding studies. Investigation of human sperm–ZP binding was further maximized by employing a procedure known as the human hemizona assay to assess sperm binding (**Fig. 2**).

In 1982, human sperm–ZP binding was shown to be inhibited by low concentrations of a fucose-containing algal polysaccharide known as fucoidan. Soon after, fucoidan was also found to block the binding of lymphocytes to cells lining the high endothelial venules (small veins). By 1993, it was established that the initial binding of immune and inflammatory cells to the endothelium involved the participation of selectins. Soon afterward, the discovery was made that SLe^X inhibited human sperm–ZP binding in the hemizona assay. These findings provided evidence that the human egg-binding protein was a selectin.

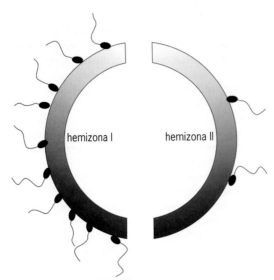

Fig. 2. Analysis of human sperm–zona pellucida (ZP) interaction in the hemizona assay (HZA) reveals deficiencies in sperm-binding capacity. Substandard performance in this assay is almost exclusively associated with the male partner of an infertile couple. Human eggs that have failed fertilization are microbisected by using either a microblade or a laser to separate two matched hemispheres of ZP (hemizona I and II) from the egg cell (oocyte) and then washed. Hemizona I (control) is incubated with sperm from a normal healthy control under optimal conditions for binding. Hemizona II (test) is treated exactly like hemizona I, except that the patient's sperm is added. After incubation for 5 h, each hemizona is thoroughly washed to remove unbound and weakly bound sperm. Bound sperm are treated with a fixative agent to stabilize their binding to the hemizona, and their numbers are counted. In this visual aid, ten sperm are bound to the control half, and two sperm are bound to the test half. The hemizona index (HZI) for the patient is calculated based on the following formula: [number of sperm bound to the test hemizona/number of sperm bound to the control hemizona] × 100. An HZI of 100 indicates that there was no inhibition. An HZI of 0 indicates total inhibition. The HZI in this case is 20, which is subpar binding. The HZA is also used experimentally to test the activity of potential inhibitors and conditions.

However, these adhesion molecules were not detected in human sperm. These results indicated that the human egg-binding protein and selectins could bind to SLe^X, but were different molecules. Based on these results, human sperm–egg binding was proposed to involve a selectin-like interaction, indicating that overlap in the types of carbohydrate sequences that mediate human immune and gamete-binding interactions existed.

This hypothesis was controversial and required definitive proof. The experimental approach that was finally used to confirm this overlap was to determine the structure of the oligosaccharides that were present on the human ZP. If the interaction was selectin-like, then oligosaccharides that could bind to selectins should be present. This goal was very difficult to achieve because of the small amount of human ZP (about 5 micrograms) that was experimentally available. This analysis required the use of highly sensitive mass spectrometers. These devices can measure the mass of ions very precisely. They can also fragment molecules into smaller ions, thereby generating a type of "fingerprint" that enables the

identification of their structures. This approach was used along with other analytical tools to demonstrate that oligosaccharides terminated with the SLe^X sequence were the most abundant carbohydrates on the human ZP. Some of the oligosaccharides were also terminated with SLe^X-Le^X (Fig. 1). The results of more definitive inhibition studies also confirmed this overlap between the immune and gamete-binding interactions.

Why is SLe^X also presented on the human ZP? It is possible that it could be employed as a marker to protect the egg from the mother's immune response. After fertilization, the zygote develops into an embryo. This embryo presents the father's markers for transplant rejection at a very early stage of development. These embryonic cells are foreign to the mother and could cause the embryo to be rejected by her immune system. However, SLe^X also binds to a molecule known as a siglec (sialic acid–binding immunoglobulin-like lectin). Siglecs are specialized lectins that are found on the surface of immune and inflammatory cells. When they bind to a specific carbohydrate sequence, they send a signal into the cell that inhibits its immune response. It is quite possible that the interaction of SLe^X on the ZP with siglecs on immune cells could protect the foreign embryo from the mother's immune response before it emerges from the ZP later during development.

There is also another potential reason why SLe^X is presented on the surface of the human egg. If a woman has an active infection in her Fallopian tubes or uterus (womb), her immune and inflammatory cells will usually become activated. Such activated cells could bind to the ZP via the interaction of SLe^X with their selectins, thus blocking the binding of human sperm. This effect would inhibit fertilization, and prevent the mother from committing her resources to supporting a pregnancy in a hostile uterine environment that could be detrimental to her embryo/fetus.

It is also possible that both mechanisms for the protection of the human egg/embryo and the prevention of a pregnancy under adverse conditions could coexist simultaneously. If true, then the presentation of SLe^X on the ZP could be part of a sensing system that either promotes or ends a pregnancy, depending on the health status of the mother. This system would not only benefit the mother, but also allocate resources to progeny with the best chance of survival.

In summary, SLe^X is an oligosaccharide that has multiple physiological roles in humans. It is likely that continued investigation of this molecule will reveal even more interesting functions for this carbohydrate sequence in the future.

For background information *see* ANIMAL REPRODUCTION; CANCER (MEDICINE); CARBOHYDRATE; FERTILIZATION (ANIMAL); GLYCOPROTEIN; IMMUNOLOGY; INFLAMMATION; LECTINS; OLIGOSACCHARIDE; OVUM; REPRODUCTIVE SYSTEM; SPERM CELL in the McGraw-Hill Encyclopedia of Science & Technology.

Gary F. Clark

Bibliography. D. R. Franken and S. Oehninger, The clinical significance of sperm-zona pellucida binding: 17 years later, *Front. Biosci.*, 11:1227–1233, 2006, DOI:10.2741/1875; J. Magnani, The discovery, biology, and drug development of sialyl-Lea and sialyl-Lex, *Arch. Biochem. Biophys.*, 426:122–131, 2004, DOI:10.1016/j.abb.2004.04.008; P.-C. Pang et al., Human sperm binding is mediated by the sialyl-Lewis^X oligosaccharide on the zona pellucida, *Science*, 333:1761–1764, 2011, DOI:10.1126/science. 1207438; C. A. St. Hill, Interactions between endothelial selectins and cancer cells regulate metastasis, *Front. Biosci.*, 17:3233–3251, 2011, DOI:10.2741/3909; R. Yanagimachi, Mammalian fertilization, pp. 189–317, in E. Knobil and J. D. Neill (eds.), *Physiology of Reproduction*, 2d ed., Raven Press, New York, 1994; A. Zarbock et al., Leukocyte ligands for endothelial selectins: Specialized glycoconjugates that mediate rolling and signaling under flow, *Blood*, 118:6743–6751, 2011, DOI:10.1182/ blood-2011-07-343566.

Space flight, 2011

The year 2011 marked the end of an era, with the final three space shuttle missions capped by *Atlantis*'s last shuttle flight in July. In their history, the four U.S. shuttles, in 30 years of operation, flew a total of 135 missions, carrying 852 people (355 different individuals) into space. Museum locations were finalized for the orbiters, further making it clear that human space flight is in transition. Work continued on replacements for the shuttle to bring American astronauts into space on the Orion and possible Dragon capsules, but most likely not until after 2016. In the meantime the only vehicle capable of bringing humans to the *International Space Station* (*ISS*) will be the Russian Soyuz.

Robotic missions went into orbit around Mercury, explored comets in more detail, started orbiting an asteroid, and continued exploring Mars and the Saturn system. A new probe was launched to Jupiter. The Mars Exploration Rover *Spirit* ended its exceptionally successful mission while the *Mars Science Laboratory* (*Curiosity*) was launched toward Mars.

Human space flight. Space shuttle *Discovery* launched on its last mission (STS-133) on February 24. The shuttle and its crew of six launched from NASA's Kennedy Space Center with the Permanent Multipurpose Module (PMM) onboard. The PMM was on its eighth trip to the station. On the previous seven trips, it traveled as the Multipurpose Logistics Module Leonardo. This time, the astronauts left the module on the station to increase storage and habitable living area. Commander Steve Lindsey, Pilot Eric Boe, and Mission Specialists Alvin Drew, Steve Bowen, Michael Barratt, and Nicole Stott rendezvoused and docked with the station. During *Discovery*'s 7 days at the station, Bowen and Drew conducted two spacewalks to do maintenance work and install new components. On March 9, *Discovery* landed for the final time at Kennedy Space Center after 202 orbits around Earth and a journey of 5,304,140 mi (8,536,186 km). STS-133 was the 39th and final flight for *Discovery*, which spent 365 days in space, orbited Earth 5830 times, and traveled 148,221,675 mi (238,539,663 km).

On March 16, Expedition 26 Commander Scott Kelly and Russian Flight Engineers Alexander Kaleri and Oleg Skripochka safely landed their Soyuz spacecraft on the Kazakhstan steppe after a five-month stay aboard the *ISS*.

On April 12, the 30th anniversary of the launch of Space Shuttle *Columbia* and the 50th anniversary of the first human in space, Russian Yuri Gagarin, NASA announced the permanent museum locations for the orbiters once the shuttle program ended. Shuttle *Enterprise*, the first orbiter built (which was never launched), would move from the Smithsonian's National Air and Space Museum Steven F. Udvar-Hazy Center in Virginia to the Intrepid Sea, Air, and Space Museum in New York. The Udvar-Hazy Center would become the new home for shuttle *Discovery*. The shuttle *Endeavour* would go to the California Science Center in Los Angeles. *Atlantis* would remain at the Kennedy Space Center in Florida, displayed at the Visitor's Complex.

On May 16, space shuttle *Endeavour* launched on its 25th and final mission. *Endeavour* delivered the Alpha Magnetic Spectrometer-2 (AMS-2), the Express Logistics Carrier-3, and spare parts to the *ISS* (**Fig. 1**). Mark Kelly commanded the flight and was joined by Pilot Greg H. Johnson and Mission Specialists Mike Fincke, Drew Feustel, Greg Chamitoff, and the European Space Agency's Roberto Vittori. STS-134 was the last mission for the youngest of NASA's space shuttle fleet. Since 1992, *Endeavor* flew 25 missions, spent 299 days in space, orbited Earth 4671 times, and traveled 122,883,151 mi (197,761,262 km). It returned to Kennedy Space Center for the final time on June 1.

Expedition 27 Commander Dmitry Kondratyev and Flight Engineers Cady Coleman and Paolo Nespoli safely landed their Soyuz spacecraft in Kazakhstan, wrapping up a five-month stay aboard the *ISS*. The trio landed on May 24 (local time) at a site southeast of the town of Dzhezkazgan.

Soyuz TMA-02M/27S was launched on June 7 from the Baikonur Cosmodrome in Kazakhstan. NASA astronaut Mike Fossum, Russian cosmonaut Sergei Volkov, and Japan Aerospace Exploration Agency astronaut Satoshi Furukawa were on board. The trio docked with the *ISS* on June 9 to expand the Expedition 28 crew to six members.

Nearly one million spectators watched shuttle *Atlantis* blast off for the final time from Kennedy Space Center on July 8 on the STS-135 mission. The crew of four veteran astronauts aboard *Atlantis*, Commander Chris Ferguson, Pilot Doug Hurley, and Mission Specialists Sandy Magnus and Rex Walheim, launched on their way to the *ISS* carrying supplies. Then on July 21 at 5:57:54 a.m. EDT, the orbiter *Atlantis* came to a "wheels stop" (quasi-official NASA term) at the Kennedy Space Center, marking the end of the space shuttle era. *Atlantis* brought to an end

Fig. 1. A portion of the *International Space Station* is visible along with the Earth and many stars as photographed by an STS-134 crew member while space shuttle *Endeavour* was docked with the station. (*Photo courtesy of NASA*)

the 135-mission career of the space shuttle program, which spanned 30 years. This was the 33d flight of *Atlantis*, which made its first launch in October, 1985.

On August 24, a crewless Russian *Progress M-12M* space freighter failed to reach the designated orbit due to a rocket engine failure. The robotic spacecraft was supposed to send supplies to the *ISS*. On October 30, the *Progress 45* spacecraft launched successfully to the *ISS* from the Baikonur Cosmodrome in Kazakhstan.

Expedition 29 Commander Fossum and Flight Engineers Furukawa and Volkov landed their Soyuz spacecraft in frigid conditions on the central steppe of Kazakhstan on November 22 (local time). They had spent 167 days in space and 165 days on the *ISS*.

NASA Flight Engineer Don Pettit, Russian Soyuz Commander Oleg Kononenko, and European Space Agency Flight Engineer Andre Kuipers of the Netherlands launched from Kazakhstan aboard their Soyuz TMA-03M craft on December 21 and docked with the *ISS* on December 23.

Robotic solar system exploration. In February, NASA released a solar system portrait taken by the *MESSENGER* spacecraft from approximately the orbit of Mercury. The spacecraft collected a series of images to complete the portrait of our solar system as seen from the inside looking out (**Fig. 2**).

NASA spacecraft *Stardust* made its closest approach to comet Tempel 1 on February 14. The Stardust-NExT mission met its goals, which included observing surface features that changed in areas previously seen during a 2005 Deep Impact mission, imaging new terrain, and viewing the crater generated when the 2005 mission propelled an impactor at the comet. The 12-year-old spacecraft was decommissioned after the encounter in March. During its

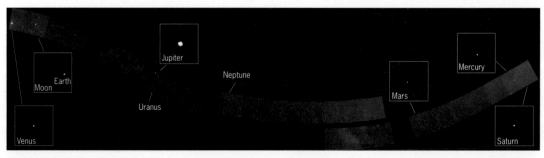

Fig. 2. Solar system "family portrait" of the planets taken by the *MESSENGER* spacecraft from approximately the orbit of Mercury. (*Courtesy of NASA*)

life, the spacecraft collected and returned comet material to Earth and was reused after the end of its prime mission in 2006 to observe Tempel 1.

NASA's *Glory* mission, meant to collect data about Earth's atmosphere for long-term climate studies, launched from Vandenberg Air Force Base in California on March 4 but failed to reach orbit. Telemetry indicated the fairing on top of the Taurus XL rocket did not separate as expected about three minutes after launch. This added weight made it impossible to attain orbital velocity.

On March 17, NASA's *MESSENGER* became the first spacecraft to orbit the planet Mercury. Later in the month, it beamed back the first images ever obtained of the planet from orbit around it.

In March, NASA announced that the *Cassini* spacecraft observed methane rain on Saturn's moon Titan. This was the first time scientists have collected evidence of rain in Titan's equatorial regions. Due to the low Titan temperatures, methane on Titan acts in ways similar to how water interacts with Earth.

Researchers using NASA's *Mars Reconnaissance Orbiter* ground-penetrating radar identified a large, buried deposit of frozen carbon dioxide, or dry ice, at the Mars' south pole. They found that the south pole of Mars contains about 30 times more carbon dioxide than previously estimated to be frozen there.

In late May, NASA gave up trying to communicate with the *Spirit* Mars Exploration Rover spacecraft. It was last able to communicate with Earth on March 22, 2010. *Spirit* landed on Mars on January 3, 2004 for a mission that was designed to last three months but far exceeded expectations. *Spirit* drove 4.8 mi (7.7 km), more than 12 times the goal set for the mission. The rover drove across a plain to reach a distant range of hills that appeared only as bumps on the horizon at landing time, climbed slopes up to $30°$ to be the first robot to summit a hill on another planet, and covered more than half a mile (nearly a kilometer) after *Spirit*'s right-front wheel became immobile in 2006. The rover sent back more than 124,000 images. It ground 15 rock targets and scoured 92 targets with a brush to prepare the targets for inspection with spectrometers and a microscopic imager.

NASA's *Juno* spacecraft lifted off from Cape Canaveral Air Force Station in Florida on August 5 to begin a five-year journey to Jupiter. *Juno* will be the first solar-powered spacecraft to explore Jupiter. It will help reveal Jupiter's origin and evolution. As the most massive planet and almost a failed star, Jupiter can help scientists understand the origin of our solar system and learn more about planetary systems around other stars. *Juno* will arrive at Jupiter in July 2016 and will orbit Jupiter for about one year. Juno will study how much water is in Jupiter's atmosphere; measure composition, temperature, cloud motions, and other atmospheric properties; map Jupiter's magnetic and gravity fields, looking at the planet's deep structure; and study Jupiter's magnetosphere near the planet's poles, especially the

auroras, to help understand Jupiter's massive magnetic field.

In August, the *Dawn* spacecraft entered the first of four planned science orbits during the spacecraft's yearlong visit to asteroid Vesta. The spacecraft started taking detailed observations on August 11, marking the official start of the first science-collecting orbit phase at Vesta, also known as the survey orbit.

Russia's Phobos-Grunt Mars mission launched on November 8 but reliable contact was never made with the spacecraft.

A United Launch Alliance Atlas V rocket carrying NASA's *Mars Science Laboratory* spacecraft was launched on November 26. The car-size, one-ton rover was bound for arrival at Mars on August 5, 2012. A successful landing would start a two-year prime mission of exploration on the surface of Mars.

Other activities. In January, NASA's Kepler mission confirmed the discovery of its first rocky planet, named Kepler-10b. Measuring 1.4 times the size of Earth, it is the smallest planet ever discovered outside our solar system. The discovery of this exoplanet is based on more than eight months of data collected by the spacecraft from May 2009 to early January 2010. Kepler has discovered well over 1000 planets. *See* KEPLER MISSION.

NASA's *Hubble Space Telescope* revealed a majestic disk of stars and dust lanes in an image of the spiral galaxy NGC 2841 released in February. A bright cusp of starlight marks the galaxy's center. Spiraling outward are dust lanes that are silhouetted against the population of whitish middle-aged stars. Much younger blue stars appear in the spiral arms (**Fig. 3**).

A U.S. Air Force Minotaur 1 rocket carrying the Department of Defense Operationally Responsive Space Office's ORS-1 satellite was successfully launched on June 29 from NASA's Launch Range at the Wallops Flight Facility in Virginia. The Wallops launch facilities will likely be busier in the future, with more small satellite launches and launches of Orbital's Cygnus supply craft to the *ISS*. *See* COMMERCIAL SPACE ACTIVITIES TO SUPPORT NASA'S MISSIONS.

NASA's decommissioned *Upper Atmosphere Research Satellite* fell back to Earth on September 24. *UARS* was launched September 12, 1991 aboard space shuttle mission STS-48, and deployed on September 15, 1991. It was the first multi-instrumented satellite to observe numerous chemical constituents in the atmosphere for better understanding of photochemistry. *UARS* data marked the beginning of many long-term records for key chemicals in the atmosphere. *UARS* ceased its scientific life in 2005. On October 23, the German *ROSAT* (*Roentgen Satellite*) fell back to Earth, highlighting the increase in space debris and in satellites falling back to Earth that are large enough for pieces to make it to the Earth's surface.

The first Russian Soyuz rocket launch from the European-run launch complex in French Guiana occurred on October 21. It carried the first two

Fig. 3. Spiral galaxy NGC 2841, as revealed by NASA's *Hubble Space Telescope.* (*Courtesy of NASA*)

European Galileo global navigation satellite system satellites.

Launch summary. Globally there were 84 space launches in 2011, with 18 being commercial. Six launches were failures, of which four were Russian, one Chinese, and one American. American launch vehicles launched no commercial payloads in 2011. Russia had 10 commercial launches. The Sea Launch consortium successfully returned to flight in 2011. Europe had four commercial launches and China two.

Russia once again led in launches in 2011 with 31 (see **table**). Ten of these were commercial and, of the other 21, nine were devoted to the *ISS*. Of these, five were crewless Progress modules launched on Soyuz launch vehicles on *ISS* supply missions,

and four were crewed Soyuz missions. The Russian launch failures were Rockot, Soyuz, Soyuz 2, and Proton M, one each.

Nine Russian launches were for military purposes. Three were civilian missions using the Zenit 3F twice and Zenit 2 once. Launch vehicles used by Russia were Proton-M (9), Soyuz (9), Dnepr M (1), Rockot (1), Soyuz 2 (7), Soyuz U (1), Zenit 3F (2), and Zenit 2M (1).

The three space shuttle launches tied with the Delta II for the most on any U.S. launch vehicle. Other vehicles used included Atlas V 401 (2); Atlas V 501, 541, and 551 (1 each); Delta IV medium + (2); Delta IV Heavy (1); Minotaur IV (2); Minotaur (1); and Taurus XL (1). All launches were successful except for the Taurus XL (*Glory*).

The Chinese use strictly variants of the Long March launch vehicle. Five were Long March 3B; three each for the 2C, 3A, and 4B; two each for the 2D and 2F; one was a 3C. Two were commercial and nine were for geostationary orbit (GEO). There were no Chinese human space flights in 2011. Nine launches carried government civil missions (communications, meteorological, remote sensing, and scientific).

Of the seven European launches five were on Ariane 5 variants, including the second Autonomous Transfer Vehicle (ATV-2) robotic vehicle to carry supplies to the *ISS*. Two Soyuz 2 launches were considered European since they launched from the European French Guiana complex. One carried the two

Space launches in 2011		
Country of launch	Attempts	Successful
Russia	31	27
China	19	18
United States	18	17
Europe	7	7
Japan	3	3
India	3	3
Russian-Ukrainian Zenit-3SL	2	2
Iran	1	1
Total	84	78

previously mentioned Galileo commercial navigation satellites and the other a remote sensing spacecraft (Pleiades HR 1), four French military satellites, and one for Chile.

Japan placed their crewless *ISS* supply spacecraft HTV-2 into orbit on their H-IIB vehicle. Two H-IIA vehicles launched two IGS intelligence satellites into orbit successfully.

India had three successful noncommercial launches of their PSLV launcher.

Iran used the Safir 2 rocket to launch *Rasad*, a remote sensing satellite, on June 15.

Multinational Sea Launch launched the Zenit 3SL from the Pacific Ocean in September and the Zenit 3SLB from Baikonur, Kazakhstan, in October.

For background information *see* ASTEROID; ASTRO-NAUTICS; COMET; EXTRASOLAR PLANETS; GALAXY, EXTERNAL; HUBBLE SPACE TELESCOPE; JUPITER; MARS; MERCURY (PLANET); SATELLITE NAVIGATION SYSTEMS; SATURN; SCIENTIFIC AND APPLICATIONS SATELLITES; SOLAR SYSTEM; SPACE FLIGHT; SPACE PROBE; SPACE SHUTTLE; SPACE STATION; SPACE TECHNOLOGY in the McGraw-Hill Encyclopedia of Science & Technology.
Donald Platt

Bibliography. *Aviation Week & Space Technology*, various 2011 issues; *Commercial Space Transportation: 2011 Year in Review*, Federal Aviation Administration, January 2012; ESA Press Releases, 2011; NASA Public Affairs Office, News Releases, 2011.

Stealth of surface combatants at sea

Surface combatants are required to operate in hostile environments. When underway, combatants are under the constant surveillance of arrays of passive and active sensing devices, designed to detect a range of signals generated or scattered by ships (signatures). The sensing devices are monitored by shipboard and land control centers that identify targets, range them, and direct forces against recognized threats, or activate weapons automatically upon detection. Hence, despite the reduced vulnerability of modern naval vessels to weapon-inflicted damage, deferred detection (reduced detection range) by design is the preferred strategy for increasing combatants' mission effectiveness. Submarines rely very heavily on covertness for their survivability and combat efficacy. The discussion herein is applicable to them as well.

Ship signatures. A ship's observability (susceptibility) is determined by the levels of its signatures as detected by the sensors it encounters. In order to evade detection, surface combatants are required to have intrinsically low susceptibility to an assortment of sensors, commonly known as stealth.

Underwater electric and magnetic (UWEM) signatures. Combatants constructed from conductive ferromagnetic metals develop considerable UWEM signatures that are detectable by highly sensitive underwater multi-axis (vector) magnetometers. The strongest UWEM signatures of ferromagnetic ships are due to their

(a)

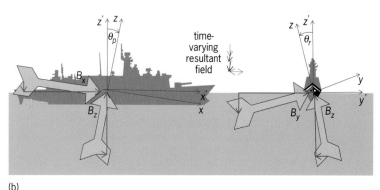

(b)

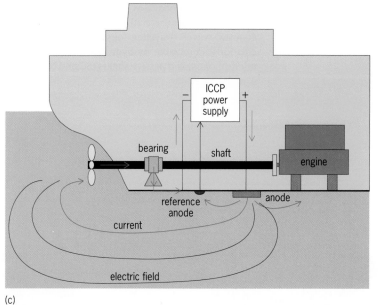

(c)

Fig. 1. UWEM signatures. (*a*) Simplified characteristic distortion of the geomagnetic field by permanent and induced magnetization of ferromagnetic ships and a simplified magnitude distribution of the corresponding vertical distortion (signature). (*b*) The generation of time varying components of the invariant (fixed in time and space) geomagnetic field flux density vector $\vec{B}$ (large arrows) due to pitch (left) and roll (right) ship motions. The components B_x, B_y, and B_z of the geomagnetic field are fixed in the world coordinate system xyz; yet, due to ship pitch and roll motions, θ_p and θ_r respectively, they are viewed as varying in time (red and purple arrows, respectively) in the ship fixed coordinate system $x'y'z'$. For small motions, the observed geomagnetic field variations can be summed to yield the resultant of the time-varying fractions (double headed arrow). This resultant induces eddy currents in the outer surfaces of metallic ships, which in turn generate a time-varying magnetic field around them. (*c*) ICCP currents and the associated electric field (the related magnetic field is not shown). The closed-loop ICCP system–injected currents flow in the seawater from the anode to the propeller and exposed hull surfaces. The return ICCP current passes to the hull through the propeller shaft bearings; thus, it is modulated by the periodically varying bearing-hull electrical resistance as the shaft rotates.

interaction with the Earth's magnetic (geomagnetic) field. The induced magnetization of a ship's ferrous mass, in conjunction with its inadvertent permanent magnetization, creates a local static distortion (anomaly) in the uniform geomagnetic field (**Fig. 1***a*). Additionally, an ultralow-frequency (ULF, below 3 Hz) distortion is created by the combined angular ship motion, which induces periodically varying eddy currents on the outer surfaces of all metallic ships (not necessarily ferromagnetic) [Fig. 1*b*].

The metal dissimilarity of ship hulls (steel) and propellers (bronze), in the presence of seawater, forms a galvanic cell that produces corrosion of unprotected metal surfaces. In large ships, this action is counteracted by an impressed current cathodic protection (ICCP) system that reverses the metal dissimilarity potential difference by injecting currents into the seawater in order to impede the corrosion of exposed hull surfaces. These electric currents induce static and dynamic electric and magnetic fields around the ship's longitudinal axis (Fig. 1*c*).

Additional sources of electric and magnetic signatures are the alternating stray fields generated by onboard electric high-power generation equipment, their distribution cabling, and the electromechanical utility equipment.

Underwater acoustic signature. The sound generated by surface ships propagates in water at the speed of sound (4.3 times faster than in air) in the form of acoustic waves (traveling longitudinal pressure oscillations) detectable by hydrophones (underwater microphones) and passive sonars (SOund Navigation And Ranging).

The major contributor to ship underwater radiated noise (URN) is the propeller. When the propeller rotates faster than what is known as its cavitation onset speed, the static pressure of the seawater flowing around the propeller blades drops below its local boiling point and a large number of small bubbles of vapor form. As the bubbles move to a higher pressure zone they collapse rapidly causing strong audible local shock waves that also excite the rudder, stern plating, and propeller blades, which in turn become re-radiation URN sources in the range of 1–100 kHz.

As propeller blades pass through their self-generated nonuniform pressure field, the blades, stern plating, and propulsion system shafting are excited by the periodic pressure fluctuation and its harmonics as well as by blade-induced turbulence and vortices. These vibrations are transmitted to the hull in the form of structureborne (SB) sound prior to being transmitted to the water by the hull plating.

At low ship speeds, vibrations and airborne (AB) sound of onboard machinery that is transmitted into the water mainly by the plating of the main machinery compartment are the dominant hull contribution to the URN (**Fig. 2**). At high cruising speeds, hull flow noise generated by the boundary-layer pressure fluctuations and stern and bow wave breaking, which is also re-radiated by hull plating, prevails.

Electromagnetic (EM) signatures. EM signatures span the radio-frequency (RF) range, 3 kHz–300 GHz, and the thermal range, 0.1–100 μm (equivalent to 3 THz–3 PHz), of the EM spectrum.

1. *Radar cross section (RCS).* RCS (apparent area) is the measure of the scattering characteristics of RF EM waves (traveling transverse mutually perpendicular electric and magnetic oscillations typically in the 0.75–100-μm band) by electrically conducting targets in the direction of the emitting radar (RAdio Detection And Ranging). RCS is quantified as the cross-sectional area (in m^2) of a perfectly reflecting isotropic (uniform in all directions) sphere that would reflect back to the radar the same total power as echoed by the scatterer. For radar electromagnetic radiation (EMR) with a given frequency (or wavelength) and polarization (the orientation of the electric field plane of oscillation), the magnitude of the radar echo depends on the ship's shape, orientation, and directivity (the ratio of power scattered back to the radar and the power that would be scattered isotropically). For this reason, ship RCS is a function of the radar-ship relative orientation (**Fig. 3**).

2. *Infrared (IR) signature.* IR power (EMR wavelengths in the 0.75–100 μm band) emitted and scattered by ship hulls and exhaust plumes is captured optically by IR detection devices. Modern IR sensors are rectangular staring arrays of solid-state elements that are sensitive to the total (that is, overall wavelengths) IR power projected on their surfaces. Due to atmospheric absorbance (the fraction of radiation absorbed at a given wavelength) characteristics of IR radiation, naval IR sensors are designed to detect EMR mainly in the atmospheric IR windows

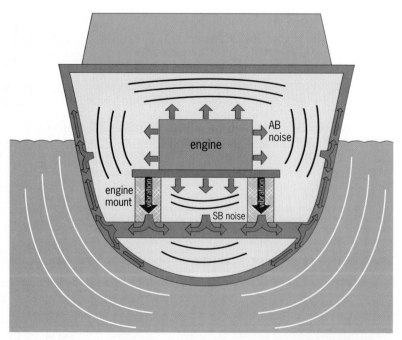

Fig. 2. Machinery-generated URN. Vibration and airborne (AB) sound are converted to structureborne (SB) sound (propagating bending, twisting, compression-extension, and shear waves), which couple into the water, mainly at the plating of the main machinery compartment, to generate underwater acoustic radiation.

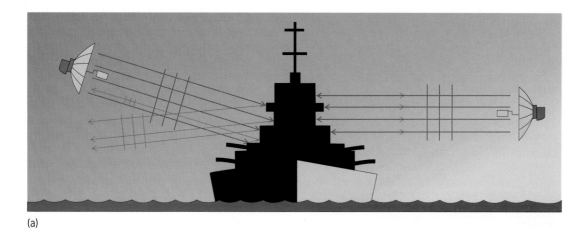

(a)

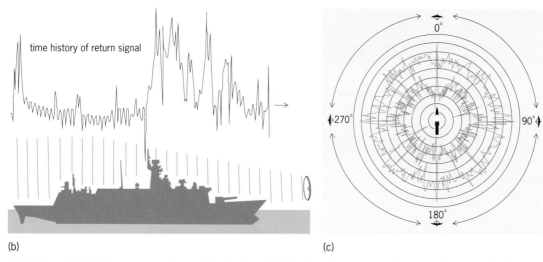

time history of return signal

(b)   (c)

Fig. 3. Monostatic RCS signature of ships. (*a*) The fraction of EMR reflected and captured by a monostatic radar (collocated transmit/receive antennas) depends on the relative radar-ship orientation. (*b*) The radar echo in each radar-ship viewing direction is the measure of the magnitude of the time-dependent compound signal, which is the consequence of the reflection of the radar RF energy off topside surfaces (all ship surfaces above the waterline) whose distances from the radar are different. (*c*) Ship monostatic RCS is presented in all hemispherical directions as azimuth polar plots at elevation cuts. A typical azimuth map of a ship as viewed by a radar at zero elevation is plotted. The signature corresponding to the bare ship (blue-lined curve) deteriorates dramatically (magenta-lined curve) subsequent to considering the topside. Clearly, due to their significant contribution to the RCS signature, antenna and weapon surfaces require shaping and treatment. (*Courtesy of Alion Science and Technology*)

of 3–5 μm (midwave IR, MWIR) and 8–12 μm (longwave IR, LWIR). The IR signature is defined as the difference between the total IR power that originates at the target and the total power that would be captured from its hidden background. Hence, IR signature is a measure of the average contrast of IR emissions from ships and their immediate surroundings in a given viewing direction (radiance scene; **Fig. 4**).

The IR power emitted from ship topside surfaces depends on the fourth power of their temperatures, their emissivity (the ability to emit energy by radiation relative to the perfect emitter at the same temperature, blackbody radiation), and their apparent (projected) area facing the detector. Typically, the contribution of ships' topside surfaces and exhaust ducts to IR emission in the LWIR range is comparable. Conversely, since the constituents of exhaust gasses are mostly water vapor and carbon dioxide gas, the IR radiation from exhaust plumes is in the MWIR range; yet, because of their elevated temper-

ature, the dominant contribution to MWIR comes from exposed exhaust ducts.

3. Visual signature. The EM signature of EMR in the visual range (0.38–0.75 μm wavelength band), known as light, is predominantly due to reflection. Given a light source location, ship orientation, and an observer position in a three-dimensional environment (visual scene), the light reflected off the ship topside is captured by photoelectric staring array sensors. Ship visual signature in a given viewing direction is defined from the visual image as the contrast between the average brightness (the measure of the amount of light incident on the detector) of the ship and its hidden background.

Signature reduction. Stealth of metallic surface combatants is accomplished by means of a variety of passive and active techniques and technologies, known collectively as stealth technologies.

Reduction of UWEM signatures. The most significant passive means of reducing the geomagnetic-field-related signatures is the reduction of plating thickness and

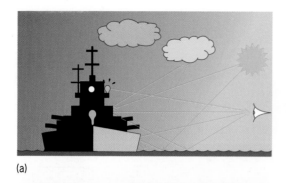

(a)

(b)

Fig. 4. Radiance scene and IR signature. (*a*) Radiation heat exchange between a ship and the environment. Thermal sources on the topside and inside the ship contribute directly and indirectly to the IR signature. (*b*) The dynamically simulated MWIR radiance (the emitted power density that falls within a unit solid angle of the observer sensor aperture, measured in W/m^2/sr) image of a ship against the horizon. (*Courtesy of Alion Science and Technology*)

the use of nonmagnetic alloyed steels in hull and equipment construction.

The most significant active UWEM countermeasure is the ship degaussing (DG) system (**Fig. 5***a*). Advanced DG (ADG) systems are closed-loop, network-based distributed systems incorporating

current-carrying coils that are arranged three-dimensionally along the ship principal directions (longitudinal, athwartship, and vertical) and whose currents are adjusted individually in order to counteract geomagnetic field-related ship signatures (Fig. 5b). Yet, as hull stressing in the presence of the Earth's magnetic field causes ships' permanent magnetization to vary during service, it is reset periodically in a deperming (demagnetization) facility.

ICCP currents are minimized by prudent anode placement, filtering, and closed-loop regulation. An Active Shaft Grounding (ASG) system is employed to act as a regulated short that bypasses the shaft bearings in order to reduce return current modulation.

Stray fields are controlled by placing offending equipment in the upper decks of the vessel or by the application of magnetic shielding in order to reduce EM radiation outside the hull. Cables are run with the energized and return leads in close proximity within magnetically shielded conduits.

Reduction of underwater acoustic signatures. URN reduction is the result of combined strength attenuation and isolation of sources. The propellers are quieted by reducing their load and streamlining the flow around them. Toward this end, highly skewed propellers operated at the lowest possible rotational speed are utilized. This necessitates fine frame line hull designs and the adaptation of bow and stern bulbs in order to reduce hull resistance at cruise speeds (**Fig. 6***a*). Smoothing the wake pattern around the propellers is achieved by stern shaping and keeping propeller tips away from the hull. Propeller noise is further reduced by fitting it with a Prairie air system that supplies cooled air to small holes along the leading edges of the propeller blades

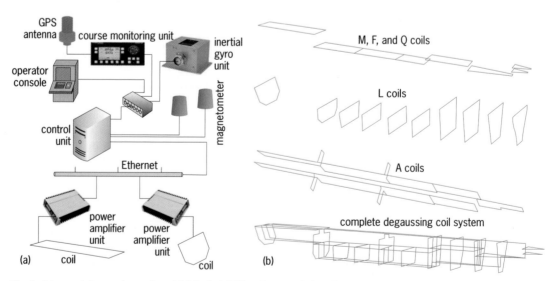

Fig. 5. Advanced degaussing system. (*a*) A typical ADG system consisting of a network-based control unit that uses both ship location data, provided by a course monitoring unit, and feedback readings from a couple of magnetometers in order to determine the required magnitude and direction of the currents in each degaussing coil. (*b*) The coils whose purpose is to oppose the vertical component of the ship magnetic signature encircle the ship hull horizontally near the waterline. These coils are known as main (M) coils. Additionally, forecastle (F) and quarterdeck (Q) coils are placed horizontally just below the forward and after quarters of the weather deck, respectively, in order to remove the fore and aft component of the permanent and induced fields. Enhanced control of fore and aft signature is achieved via the longitudinal (L) coils, which are a series of vertical loops that generate a longitudinal magnetic field when energized. The athwartship (A) coils are installed in a vertical fore-aft plane for the purpose of counteracting the athwartship component of the vessel's field. (*Courtesy of Alion Science and Technology*)

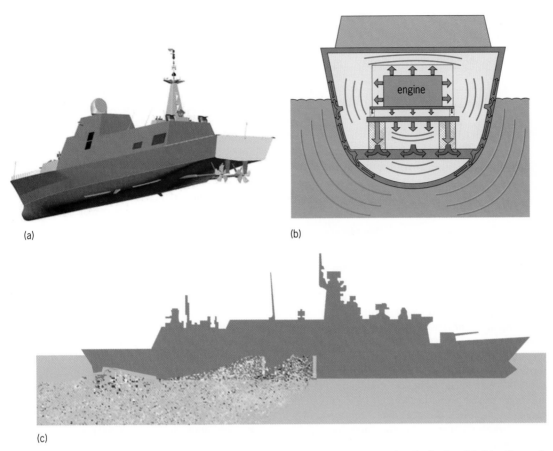

(a)

(b)

(c)

Fig. 6. URN signature reduction. (*a*) Streamlined keel reduces propeller loads and flow-related noise by minimizing flow and wave breaking resistances (*Courtesy of Alion Science and Technology*). (*b*) Reduction of main machinery contribution to URN is accomplished by floating and enclosing propulsion engines and diesel generators. Enclosures provide airborne (AB) noise control, while the intermediate mass and its resilient supports act as a mechanical filter that attenuates the transmission of high-frequency vibration energy into the structure, thereby reducing structureborne (SB) sound. (*c*) The Prairie-Masker system generates air bubbles to quiet propeller cavitation and to isolate the main machinery room hull plating from underwater listeners by means of a bubble curtain.

in order to slow down the collapse of cavitation bubbles.

Machinery noise is reduced by utilizing silenced, inherently quiet equipment (for example, rotating rather that reciprocating), and mounting them on dynamically stiff foundations by means of special high performance resilient isolators. Noisier machinery is placed close to the ship centerline within double hull compartments. Enhanced isolation of up to 40 dB in vibration and 20 dB in AB noise may be achieved when the propulsion engines and generators are mounted in rafted acoustic enclosures to form two-stage isolation systems (Fig. 6*b*). Reduction of AB noise transmission to hull plating is accomplished by covering bulkheads and skin plating of noisy compartments with an acoustic insulation layer; SB noise is dampened (absorbed) by employing constraint-layer panels (cemented polymer layer between metal layers) in bulkheads. Further attenuation of machinery induced URN is attained by employing a Masker system that forces compressed air from underwater belts around the main machinery rooms (Fig. 6*c*). The generated air bubble curtain introduces an acoustic impedance mismatch (isolation) in the path of the acoustic radiation to

the surrounding water; thereby, reflecting the noise back toward its source.

Seawater suction and discharge are performed from carefully designed sea chests in order to avoid radiation of fluid-borne noise induced by pump impulses.

Reduction of radar cross section. Due to its extended range of detection, RCS reduction is the primary means of decreasing ship susceptibility. High-frequency radar beams that incident surfaces at right angles (normal incidence) bounce back directly to their sources, whereas inclined surfaces tend to deflect them away. This principle underlies the means of passive control of RCS signatures of modern warships. Indeed, orienting the surfaces and edges of ship topsides so as to deflect radar energy, while avoiding the creation of corner reflectors (perpendicular flat surfaces that return beams directly to the source), reduces RCS markedly. Of main concern are radar beams that originate from distant ships or sea-skimming anti-ship missiles; accordingly, hull surfaces are inclined so as to create notable local incidence angles with such radar beams. The result is hull forms that feature flared hull sides and bulwarks in conjunction with sloped full-beam superstructure,

Fig. 7. Low RCS topside shaping features. Topside surfaces feature flared hull and sloped full-beam superstructure, mast, and yardarms for the purpose of scattering horizontal radar beams away from their source. (*Courtesy of Alion Science and Technology*)

mast, and yardarms (**Fig. 7**). Nevertheless, in view of the prospect of radar ray multibounce, it is not possible to achieve reduced RCS in all viewing directions, especially in the presence of unconcealed topside equipment (for example, the sizeable spikes in the RCS map in Fig. 3); hence, it is common to create sacrificial sectors within which RCS requirement may not be satisfied. Surfaces that form strong RCS return centers ("hot spots") despite hull shaping are treated with radar-absorbing material (RAM) panels or paint for the purpose of weakening both direct and multibounce returns.

Reduction of IR signature. Ships' skin temperature is determined by solar heating, local heating by the hot exhaust plume gases, and heat losses from internal sources such as propulsion and auxiliary machinery and their internal exhaust piping. Ship surfaces whose temperatures are high with respect to the surroundings are major contributors to the

LWIR signature. Yet, extended warm surfaces can make a significant contribution to the IR signature despite the fact that their temperature may be only a few degrees above the ambient; thus, it is important to apply thermal insulation to all weather boundaries. The temperature of hot spots generated by solar heating is controlled by applying low-solar-absorptivity/low-emissivity paints and coatings, and by means of zonal water-wash or heavy water mist cloud systems, which are the most significant means of hull signature reduction. Prudent separation of cold and warm compartments reduces IR contrasts on external hull surfaces, thereby averting the formation of recognizable patterns in the thermal image.

Heat dissipation from internal sources is further controlled by ventilating machinery rooms and water cooling the thermally insulated IR shielded engine exhaust ducts. Diesel generator exhaust pipes are equipped with water-cooled shrouds in order to obscure their high radiant heat emittance.

The reduction of the direct contribution of hot engine exhaust gasses to the ship MWIR signature is accomplished either by utilizing a shielded IR suppression system (IRSS) to cool the exhaust gases, or by rerouting the water-cooled engine and generator exhaust pipes along the ship for stern exhaustion (**Figs. 8***a* and 8*b*). Due to their high emissive power, it is important that particulates emitted due to incomplete combustion are removed from the exhaust plume.

In order to avoid hot spots, stack design should guarantee that mast surfaces are not impinged upon by the exhaust plume in windy conditions (Fig. 8*c*).

Reduction of visual signature. The objective of visual design of vessels is to minimize the contrast between them and their backgrounds. The largest contrast is observed when highly reflecting topside surfaces are adjacent to shadowed and background areas in the visual image. Thus, reduction of a ship's visual

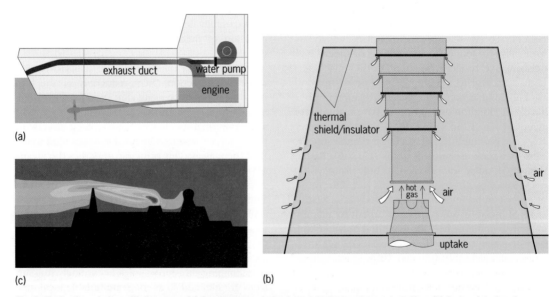

Fig. 8. Reduction of plume IR signature. (*a*) Cooling extended exhaust ducts for stern exhaustion. (*b*) An insulated and shielded diffuser type IRSS induces unassisted suction and mixing of fresh air with the hot exhaust gases in order to reduce the plume temperature. (*c*) Computational-fluid-dynamics- (CFD-) generated thermal image of an exhaust plume from an ill-designed stack impinging on an aft mast in headwind. (*Courtesy of Alion Science and Technology*)

(a)   (b)

Fig. 9. Minimizing visual contrast in a visual image of a ship against the water. (*a*) A uniformly painted ship using light gray paint of reflectivity $\rho = 0.5$ under high sun position in the sky yields top surfaces that are brighter than the water, while the side surface is darker than the sea surface. (*b*) Under the same conditions, a two-tone gray painting scheme of the ship using hazy gray ($\rho = 0.3$) on the top surfaces and very light gray ($\rho = 0.7$) on the side surfaces produces even brightness of both the ship and its surrounding water. As can be observed, the ship's smooth extended surfaces have different size scale than the sea's rippled surface; thus providing visual cues for the ship detection. (*Courtesy of Alion Science and Technology*)

signature for a given visual scene requires the employment of multitone painting schemes of paints whose reflectivities (the fraction of incident radiation that is reflected) produce equalization of the brightnesses of all topside and background surfaces in the visual image (**Fig. 9**). In order to reduce direct reflections and ship-motion-generated glint, glass surfaces are coated with antiglare films. It is worth noting that improper IR signature suppression of exhaust plumes employing water injection might lead to visual exhaust plumes even in the absence of soot.

Summary and outlook. The deployment of advanced and highly sensitive detection devices at sea renders stealth of modern surface combatants essential for their ability to execute their missions. As stealth covers a wide range of acoustic, electric, magnetic, and electromagnetic signatures, its implementation is the consequence of a compromise between many competing objectives. Indeed, compliance with stealth requirements has a profound impact on ship design, construction, and service life as it entails significant complexity, capability (range, performance, and capacity), weight, maintainability, and cost penalties. Due to the rapid pace of advance of detection technology, future surface combatants will face highly sensitive multisensor detectors capable of detecting a wide diversity of signatures; thereby increasing the importance of stealth in the design of surface combatants.

For background information *see* ABSORPTION OF ELECTROMAGNETIC RADIATION; ANTENNA (ELECTROMAGNETISM); CAVITATION; DEMAGNETIZATION; DIFFRACTION; EDDY CURRENT; ELECTRIC FIELD; ELECTROMAGNETIC RADIATION; FERROMAGNETISM; HEAT RADIATION; HYDROPHONE; IMAGE PROCESSING; INFRARED RADIATION; LIGHT; MAGNETISM; MAGNETOMETER; MECHANICAL VIBRATION; POLARIZED LIGHT; RADAR; RADAR-ABSORBING MATERIALS; RADIANCE; RADIATIVE TRANSFER; REFLECTION OF ELECTROMAGNETIC RADIATION; REFLECTION OF SOUND; SONAR; SOUND ABSORPTION; VIBRATION

ISOLATION in the McGraw-Hill Encyclopedia of Science & Technology.   Avigdor Shechter

Bibliography. M. André, T. Gaggero, and E. Rizzuto, Underwater noise emissions: Another challenge for ship design, in C. Guedes Soares and W. Fricke (eds.), *Advances in Marine Structures*, 3d International Conference on Marine Structures, MARSTRUCT 2011, Hamburg, Germany, March 28–30, 2011, Taylor and Francis Group, London, 2011; A. E. Fuhs, *Radar Cross Section Lectures*, American Institute of Aeronautics and Astronautics, New York, 1984; J. J. Holmes, *Reduction of a Ship's Magnetic Field Signatures*, Morgan and Claypool, San Rafael, CA, 2008; B. Morey et al., *Infrared Signature Simulation of Military Targets*, Environmental Research Institute of Michigan, Ann Arbor, 1994; J. Thompson, D. Vaitekunas, and B. Brooking, Lowering warship signatures: Electromagnetic and infrared, in *SMi "Signature Management—The Pursuit of Stealth" Conference*, February 21–22, 2000.

SUMOylation pathway

SUMOs (small ubiquitin-like modifiers) are ubiquitin-like proteins (Ubls) that become conjugated to substrates through a pathway that is biochemically similar to ubiquitination. Ubiquitination is the conjugation of ubiquitin (a 76-amino-acid protein) to cellular proteins and regulates a broad range of eukaryotic cell functions; in particular, it targets proteins for degradation. SUMOylation is involved in many cellular processes, including deoxyribonucleic acid (DNA) metabolism, gene expression, and cell cycle progression. Clinical studies suggest that SUMOylation plays important roles in disease processes (for example, diabetes, viral infection, and carcinogenesis). In this article, a number of topics will be discussed, including the basic mechanism of SUMO modification, the enzymes involved in the process, and how SUMOylation regulates substrate proteins.

SUMO paralogues and isoforms. Yeasts have only one SUMO protein (Smt3p in *Saccharomyces cerevisiae* and Pmt3p in *Schizosaccharomyces pombe*), whereas mammalian cells express four SUMO paralogues (SUMO1–4). Paralogues are genes that diverged through gene duplication, and it is anticipated that the isoforms encoded by these genes differ from each other in function. Among the mammalian SUMO proteins, SUMO1 has approximately 45% identity compared with SUMO2 or SUMO3, while the latter two are approximately 95% identical to each other. Because of this high degree of identity, they are collectively called SUMO2/3. SUMO1–3 proteins are expressed as precursors, with short peptides on their C-termini that must be removed to reveal a di-glycine motif that is required for conjugation (step *a* in the **illustration**). Although SUMO4 has a high degree of sequence identity compared with SUMO2 (approximately 86%), it is only expressed in a subset of tissues. SUMO4 is processed very inefficiently, and it is not clear whether it becomes conjugated to other proteins under physiological conditions; thus, the biological role of SUMO4 remains unclear.

Like ubiquitin, SUMOs can form polymeric chains through the sequential conjugation of SUMOs to each other. In *S. cerevisiae*, the N-terminal domain of Smt3p contains three lysines (K11, K15, and K19) that may serve as internal conjugation sites. In vertebrates, lysine 11 (K11) of SUMO2/3 is primarily used during polymeric SUMO (poly-SUMO) chain extension. SUMO1 lacks a conserved lysine at the equivalent residue and might only become conjugated to the end of SUMO2/3 chains as a "chain terminator."

SUMO conjugation pathway. SUMOylation is mediated through an enzyme cascade similar to ubiquitination (see illustration). Enzymes of this pathway are listed in the **table**. Mature SUMO is activated by the formation of a thiolester bond with SUMO activating enzyme (E1 enzyme) in an adenosine triphosphate (ATP)–dependent manner (step *b*). The SUMO E1 enzyme is a heterodimer composed of SAE1 and SAE2 (also known as Aos1 and Uba2, respectively). SUMO is transferred from SAE1/SAE2 to form a thiolester linkage with the unique SUMO conjugating enzyme (E2 enzyme), known as Ubc9 (step *c*). Ubc9 catalyzes the formation of an isopeptide bond between the SUMO protein and the substrate (step *d*). Isopeptide bonds formed in this reaction are chemically similar to peptide linkages assembled during protein translation, except that they link SUMO's C-terminal carboxyl group and the ε-amino of a substrate lysine side chain (rather than linking to an amino group of a previous amino acid within a polypeptide). There is some preference for lysine residues that reside within a consensus sequence, ψKXE, where ψ is a hydrophobic residue and X is any amino acid. However, many instances have been found in which lysines outside of such canonical sequences are utilized.

Although Ubc9 can be sufficient for in vitro conjugation, in vivo conjugation typically involves SUMO ligases (E3 enzymes). E3 enzymes are likely to play a key role in determination of the overall spectrum of SUMOylated species, especially because conjugation of all SUMOs and all substrates utilize the same E1 and E2 enzymes. E3 enzymes facilitate conjugation either by bringing the Ubc9–SUMO thiolester and substrate together or by stimulating the transfer of SUMO from Ubc9 to the substrate. The number of known SUMO E3 ligases remains much smaller than those of ubiquitin E3 ligases. For some of these enzymes, specificity has been demonstrated for substrate recognition or for particular SUMO isoforms.

SUMO E3 enzymes can be broadly categorized into two groups. The first group is composed of SUMO E3 enzymes possessing RING-fingerlike domains (Siz/PIAS-RING; SP-RING). These enzymes are proposed to function in a manner analogous to ubiquitin RING E3 ligases, mostly by juxtaposition of substrates to the Ubc9–SUMO thiolester. This class of SUMO ligases is found in both yeast and vertebrates. The second class consists of a diverse and growing set of vertebrate proteins that are simply categorized by the absence of any SP-RING domain. The putative SUMO ligase domains of these proteins lack common features, and their enzymatic mechanisms are generally not well studied. Indeed, there is no reason to anticipate that they will utilize shared mechanisms given their structural diversity.

SUMO proteases and SUMO-targeted ubiquitin ligases. The collective activity of SUMO proteases enables SUMOylation to be a highly dynamic modification. Both post-translational processing (step *a*) and the deconjugation of SUMOylated species (steps *e* and *g*) are mediated by a family of SUMO proteases known as ubiquitin-like protein-specific proteases in budding yeast (Ulp1p and Ulp2p) or as sentrin-specific proteases in mammals (SENP1–3 and SENP5–7). By comparing their protein sequences, these SUMO proteases can be divided into two branches. Vertebrate SENP1, SENP2, SENP3, and SENP5 are most closely related to Ulp1p, whereas SENP6 and SENP7 are more closely related to Ulp2p.

Ulps/SENPs show distinct substrate specificities, and their activities are regulated by localization. Ulp1p and SENP2 are primarily localized to nuclear pores; in contrast, a number of patterns of SENP1 localization have been reported. Ulp1, SENP1, and SENP2 catalyze both deconjugation of SUMO and processing, whereas SENP1 and SENP2 act on SUMO1 and SUMO2/3. SENP3 and SENP5 concentrate within the nucleolus. In vitro, they are active in both processing and deconjugation reactions, and they have higher activity towards SUMO2/3 substrates. Ulp2p, SENP6, and SENP7 localize within the nucleoplasm. They are particularly active in trimming of poly-SUMO chains (step *g*), and SENP6 and SENP7 show a strong preference for SUMO2/3-conjugated species. Non-Ulp/SENPs have recently been reported as novel SUMO proteases, including a metalloprotease that may edit SUMO chains in budding yeast (Wss1p), as well as two vertebrate cysteine proteases (DeSI-1 and DeSI-2).

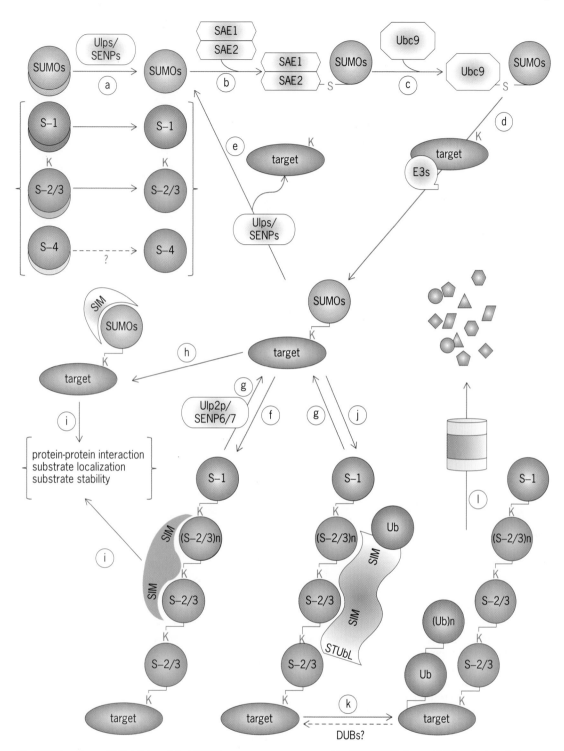

The SUMO pathway. (*a*) Newly translated SUMO precursors are processed by Ulps/SENPs to reveal C-terminal di-glycine motifs. (*b*) Mature SUMOs undergo ATP-dependent activation and form a thioester linkage with the E1 enzyme, SAE1/SAE2. (*c*) The thioester linkage is transferred to the E2/conjugating enzyme, Ubc9. (*d*) Ubc9 conjugates SUMOs onto substrates through an isopeptide linkage between the SUMO C-terminal glycine and the ε-amino group of the receptor lysine. Conjugation is typically facilitated by SUMO ligases (E3 enzymes). (*e*) Ulps/SENPs can remove SUMO from the conjugated species. (*f*) In some cases, SUMOs can form polymeric chains. (*g*) Polymeric chains are primarily disassembled by Ulp2p family proteases. (*h*) Proteins containing SIM motifs interact with SUMO or SUMO conjugates, thereby regulating substrate functions. (*i*). STUbLs are an important set of SIM-containing proteins that recognize poly-SUMO conjugates (*k*) and promote their degradation by proteasomes through ubiquitination (*l*).

Poly-SUMO chains can be recognized by SUMO-targeted ubiquitin ligases (STUbLs) [step *k*]. STUbLs mediate ubiquitination of the poly-SUMOylated species, leading to their degradation by protea-somes (large proteolytic complexes that degrade intracellular proteins; step *l*). As mentioned earlier, poly-SUMO chains are trimmed by Ulp2p-related proteases, which thus antagonize this process.

SUMO isoforms and SUMO pathway enzymes

Activity	Homo sapiens	Saccharomyces cerevisiae	Schizosaccharomyces pombe
SUMO isoforms	SUMO1 SUMO2 SUMO3 SUMO4	Smt3p	Pmt3p
Activating enzyme (E1)	Uba2/SAE2 + Aos1/SAE1	Uba2p + Aos1p	Uba2/Fub2 + Rad31
Conjugating enzyme (E2)	Ubc9	Ubc9	Hus5
SP-RING SUMO ligases (SUMO E3)	PIAS1 PIAS3 PIASXa PIASXb PIASy Mms21	Siz1p Siz2p/Nfi1p Mms21p Zip3p	Pli1 Nse2p
Non-SP-RING SUMO ligases (SUMO E3)	Topors RanBP2 PC2 CBX4 HDAC4 Krox20* Cyclin B1 interacting protein-1 (CCNB1IP1) Stanniocalcin-1 (STC1) TRIM family proteins		
SUMO-targeted ubiquitin ligases (STUbLs)	RNF4	Hex3p/Slx5p + Slx8p Ris1p/Uls1p	Rfp1p + Slx8p Rfp2p + Slx8p Rad60
SUMO proteases	SENP1 SENP2 SENP3 SENP5 SENP6 SENP7 DeSUMOylating isopeptidase 1 (DeSI-1) DeSI-2 USPL1	Ulp1p Ulp2p/Smt4p Wss1p	Ulp1p Ulp2p

*Activity demonstrated for mouse homologue.

Effects of SUMOylation on substrates. A large number of potential SUMOylation substrates have been identified through proteomic studies. SUMOylation promotes a variety of fates for individual substrates, frequently reflecting the gain or loss of protein–protein interactions upon SUMOylation. Many cellular proteins possess short SUMO-interacting motifs (SIMs) that form low-affinity interactions with SUMOs. These SUMO–SIM contacts promote recognition of SUMOylated species, sometimes in an isoform-specific manner (step *i*). Moreover, numerous proteins that recognize poly-SUMOylated species, such as Ulp2-related proteases or STUbLs, have many SIMs repeated in tandem.

Changes in substrate fate after SUMOylation include alterations of localization, stability, or activity. For example, SUMO modification of the PML (promyelocytic leukemia gene product) protein is important for the organization of subnuclear foci called PML nuclear bodies (PML-NBs), as well as for the recruitment of other proteins into PML-NBs. Furthermore, poly-SUMOylation of PML in response to arsenic trioxide treatment leads to its STUbL-mediated degradation. The budding yeast proliferating cell nuclear antigen (PCNA) is another well-studied substrate that can be modified by either ubiquitin or SUMO. Ubiquitination of PCNA is involved in DNA damage responses, whereas SUMOylation of PCNA mediates interactions with DNA helicase Srs2p (an enzyme that unwinds the DNA double helix) and prevents recombination during DNA replication.

Alternatively, SUMOylation can act in antagonism to ubiquitination in some cases. For instance, IκBα, a regulator of the transcription factor NF-kB, can be SUMO modified on its lysine (K21). This modification either competes with ubiquitin conjugation on the same lysine or blocks the access of another ubiquitin conjugation site and therefore prevents IκBα degradation.

SUMO and diseases. SUMOylation has been implicated in human diseases. SUMOylation substrates play prominent roles in neurodegenerative diseases, including SOD1 in amyotrophic lateral sclerosis; huntingtin in Huntington's disease; ataxin-1 in spinocerebellar ataxia type 1; tau, α-synuclein, and DJ-1 in Parkinson's disease; and tau and APP in Alzheimer's disease. Additionally, Ubc9 polymorphisms may be associated with an elevated risk of Alzheimer's disease in Korean patients. Similarly, non-SUMOylatable mutants of lamin A (a type of structural protein) are associated with familial dilated cardiomyopathy and conduction system disease. A single mutation in SUMO4 (M55V) is associated with an increased risk of type-1 and type-2 diabetes. The reasons for this phenotype are not obvious, however, given that (1) SUMO4 messenger RNA (mRNA) is mainly expressed in kidneys and (2) its extent of conjugation may be limited.

Studies in many organisms indicate that SUMOylation is required for genomic stability, transcriptional regulation, and cell cycle control, all of which are important factors in carcinogenesis. Moreover, dysregulated SUMOylation has been observed in human cancers. For example, elevated Ubc9 levels have been found in ovarian, cervical, prostate, and lung cancers, as well as in multiple myelomas, whereas persistent elevations of SENP1 and SENP3 levels have been found in prostate cancers. These observations may indicate that SUMOylation and de-SUMOylation must be carefully balanced; thus, the SUMO pathway may be a potential target for cancer therapy. Consistent with this idea, arsenic compounds have been effectively used to treat acute promyelocytic leukemia (APL) patients in Chinese traditional medicine. The discovery of how PML is controlled by SUMOylation provides a molecular mechanism for understanding the effectiveness of this treatment. Recent reports also suggest that deregulation of poly-SUMO chain formation may promote sensitivity to genotoxic stress caused by radiation or chemotherapy in cancer cells.

Conclusions. The understanding of SUMO modification has greatly expanded since its discovery, and many basic principles through which it operates have been determined. Nevertheless, many aspects of SUMO regulation and function remain to be explored. The answers to these questions will greatly help in the comprehension of cell function as well as providing clinical targets for disease treatments.

For background information *see* BIOCHEMISTRY; CANCER (MEDICINE); CELL BIOLOGY; CHEMOTHERAPY AND OTHER ANTINEOPLASTIC DRUGS; GENE; PROTEASOME; PROTEIN; PROTEIN DEGRADATION; PROTEIN ENGINEERING; UBIQUITINATION in the McGraw-Hill Encyclopedia of Science & Technology.

Ming-Ta Lee; Mary Dasso

Bibliography. J. R. Gareau and C. D. Lima, The SUMO pathway: Emerging mechanisms that shape specificity, conjugation and recognition, *Nat. Rev. Mol. Cell Biol.*, 11:861–871, 2010, DOI:10.1038/nrm3011; R. Geiss-Friedlander and F. Melchior, Concepts in sumoylation: A decade on, *Nat. Rev. Mol. Cell Biol.*, 8:947–956, 2007, DOI:10.1038/nrm2293; E. S. Johnson, Protein modification by SUMO, *Annu. Rev. Biochem.*, 73:355–382, 2004, DOI:10.1146/annurev.biochem.73.011303.074118; D. Mukhopadhyay and M. Dasso, Modification in reverse: The SUMO proteases, *Trends Biochem. Sci.*, 32:286–295, 2007, DOI:10.1016/j.tibs.2007.05.002; K. D. Sarge and O. K. Park-Sarge, SUMO and its role in human diseases, *Int. Rev. Cell Mol. Biol.*, 288:167–183, 2011, DOI:10.1016/B978-0-12-386041-5.00004-2.

Tet proteins

At the beginning of life, one single cell, the fertilized oocyte, has the capacity to generate all cell types of the body through cellular division and execution of differentiation programs. As the deoxyribonucleic acid (DNA) sequence in most of the cells of an organism is identical, cellular identity relies on the interpretation of genomic information. The driving force in this process is a correct spatiotemporal regulation of gene expression that prevents errors causing developmental failure and disease. Partly, this is controlled by cell-specific enzymatic modifications of chromatin, which is the tight association of acidic DNA and basic protein complexes (the histone-containing nucleosomes). Enzymes involved in the shaping of chromatin can, for example, posttranslationally modify amino acids of histones by adding methyl or acetyl groups. Furthermore, small proteins such as ubiquitin or small ubiquitin-like modifier (SUMO) proteins can be enzymatically coupled to histones, regulating the interpretation of the underlying genomic information. Other enzyme classes directly add functional groups to DNA bases. Various combinations of these so-called epigenetic modifications activate or repress gene expression and define a cell-specific genome function, which is heritable from one cell to another on cell division. Epigenetic configurations of cell types dynamically respond to environmental cues and vary among individuals, as has been shown for monozygotic twins (whose DNA sequences are nearly identical).

Among the epigenetic marks present on DNA, methylation (addition of a methyl group) of the cytosine base at the fifth position of the pyrimidine ring (5-methylcytosine, 5mC, also called the "fifth base") has been extensively studied. Being generally associated with stable gene repression, 5mC is essential for cellular integrity by ensuring X-chromosome inactivation in females, repression of transposable elements (genetic elements capable of moving to a new chromosomal location), and parent-of-origin related expression of alleles (alternative forms of a gene), called genomic imprinting. Moreover, differentiated cells typically display higher levels of 5mC, indicating the need for a stable repression of genes that could potentially counteract the fate of a given cell type.

The search for pathways removing DNA methylation. Although DNA methylation is crucial for maintaining a healthy organism, some natural processes require its reversal. For example, developing germ cells erase the methylation marks underlying X-inactivation and imprinting, thereby resetting the genomic information for proper transmission to the offspring. On fertilization of the oocyte, the hypermethylated sperm genome gets rapidly demethylated before the cell division to the two-cell stage. Additionally, nondividing postmitotic neurons activate defined genes previously marked by 5mC, altogether arguing for an active mechanism of demethylation. In plants, removal of 5mC requires a base-excision repair machinery that enzymatically cuts out the modified base and replaces it by an unmodified cytosine. In animals, this mechanism has been highly controversial, although some studies describing a similar process in early developing germ cells exist.

An alternative proposed mechanism involves inhibition of DNA methylation maintenance at each replication and subsequent cell division; however, this mechanism cannot hold true for nonreplicative germ cells, fertilized oocytes, and neurons.

Tet proteins: a new enzyme class modifying methylcytosine bases. Recently, the search for players in active DNA demethylation revealed new candidates: the ten-eleven translocation (Tet) protein family of 2-oxoglutarate (2OG)–dependent and Fe(II)-dependent enzymes are a conserved enzyme family generating the "sixth base," that is, 5-hydroxymethylcytosine (5hmC). This modified base, originally identified as a nucleic acid component in viruses, was subsequently discovered in the genome of mammalian cells. Tet enzymes can modify 5mC to 5hmC by hydroxylation of the methyl group of the base (see **illustration**). This finding raised many questions about the putative function of Tet proteins and 5hmC in epigenetic gene regulation. Could 5hmC function as an intermediate product that is ultimately replaced by unmodified cytosine, neutralizing 5mC's function as an epigenetic mark? If true, what mechanism triggers this downstream event and in which cell types? Alternatively, could Tet proteins and 5hmC be regulators of chromatin, possibly partly independent of each other? Could Tet proteins play an additional, nonenzymatic role in gene expression? Recently, several groups of researchers unraveled some features of the unnoticed members of the enzyme family and its enzymatic product.

Tet proteins and hydroxymethylcytosine in development. The Tet enzyme family consists of three members, Tet1, Tet2, and Tet3, which have a distinct expression pattern throughout development, but tend to be expressed highest in undifferentiated, pluripotent cell types (pluripotent cells are capable of differentiating into most cell types found in an organism, but are not capable of forming a functional organism). Tet3 levels are highest in oocytes persisting after fertilization and decreasing with the first two embryonic cleavages, when Tet1 and Tet2 become activated. Whereas Tet1 and, to a lesser extent, Tet2 become abundant in preimplantation embryos, their protein levels decrease in cells undergoing differentiation in implanted developing embryos (with the exception of some cell types of the brain). In adults, Tet2 is highly expressed in cells of the hematopoietic system (the blood-making organs, principally the bone marrow and lymph nodes), and it is frequently deregulated or mutated in related cancer cells, which often show an abnormal loss of Tet2 function. Concomitant with the presence of Tet family proteins, 5hmC is relatively abundant in oocytes, the early embryo, and some brain cells. In contrast, most mammalian adult tissues show very low levels of 5hmC. Hence, the presence of Tet proteins and 5hmC correlates inversely throughout the genome with cellular differentiation and high DNA methyla-

tion levels. However, the function of Tet proteins and 5hmC in maintenance of pluripotency is still debated. Pluripotent embryonic stem cells (ESCs) derived from preimplantation embryos, which are a popular ex vivo model system for studying gene function, express high levels of Tet1 and lower levels of Tet2. Some of the research groups that conducted Tet1 (and/or Tet2) loss-of-function experiments in ESCs claim that Tet proteins are involved in the maintenance of the pluripotent ESC state, whereas others found that the enzymes are not necessary. The latter was confirmed by a Tet1 knockout approach using ESCs lacking Tet1 and most of their 5hmC. These cells were analyzed for their potential of contributing to the formation of embryos, which is usually the most stringent test for pluripotency of genetically engineered ESCs. Despite the complete lack of the Tet1 gene and its products, the cells not only contributed to embryonic development, but gave rise to liveborn largely normal offspring. However, this finding does not exclude a putative redundant function of other Tet family members, balancing out the missing Tet1 in vivo. This possibility remains to be clarified by additional genetic approaches in the future.

Functional roles in gene regulation. Although the absence of Tet1 might not abolish the (somewhat artificial) ESC-specific state, the protein family seems to play important roles in gene regulation (see illustration). Careful analysis of Tet1 and 5hmC localization in the genome of ESCs by several research groups provided valuable insights into mechanistic principles. DNA methylation preferentially occurs on CpG dinucleotides [where cytosine (C) precedes a guanosine (G) in the DNA sequence] throughout the genome. However, clusters of CpG dinucleotides [the so-called CpG islands (CGIs)] that are found in approximately 60% of all mammalian gene promoters are exceptionally devoid of 5mC marks. Furthermore, these sites are enriched for the epigenetic histone modification, trimethylated lysine 4 on histone 3 (H3K4me3), that is abundant in transcriptionally active genes. Some of the CGIs are doubly (bivalently) marked by trimethylated lysine 27 on histone 3 (H3K27me3) and H3K4me3, which together maintain a repressive gene expression status. CGIs seem to be the preferential binding sites of Tet1, with highest enrichments around transcriptional start sites. Depletion of Tet1 in ESCs resulted in gene repression of a fraction of Tet1-associated genes, which is in line with the expectation of Tet1's function of counterbalancing against silencing 5mC.

These Tet1-binding sites were highly correlated with the presence of the activating H3K4me3 mark alone. Surprisingly, similar numbers of genes were transcriptionally activated on loss of Tet1 at a distinct set of binding sites, unraveling Tet1's additional repressive function in gene regulation. This second group of Tet1-bound targets is bivalently marked by H3K27me3 and H3K4me3, maintaining a silent gene expression status. Transcriptional repression is partly a result of Tet1's physical association with the Sin3A corepressor complex, which silences transcription, but the exact mechanism of positive

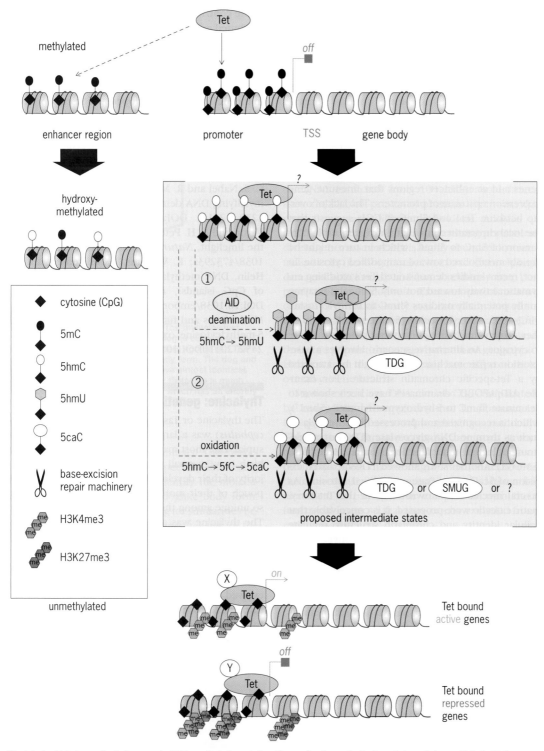

Models for Tet-dependent changes in DNA methylation states. Genomic elements that contain an intermediate to high density of CpG dinucleotides recruit Tet enzymes to oxidize 5-methylcytosine (5mC) to 5-hydroxymethylcytosine (5hmC). This initial event causes the transition from a methylated to a hydroxymethylated or unmethylated state. Enhancer regions with intermediate CpG content remain hydroxymethylated, whereas CpG-rich promoters bind Tet proteins that generate 5hmC as transient demethylation intermediates predominantly at transcription start sites (TSS). 5hmC has been proposed to be subsequently (1) deaminated to 5-hydroxymethyluracil (5hmU), for example, by AID (a deaminase) or (2) oxidized to 5-formylcytosine (5fC) and finally to 5-carboxylcytosine (5caC). 5hmU and 5caC are putative substrates for the base-excision repair machinery, which contains either TDG or SMUG glycosylases and generates unmodified cytosine. Alternatively, an unknown decarboxylating mechanism might convert 5caC to cytosine. Genes bearing unmethylated CGI promoter regions are usually enriched for Tet proteins. These genes are either active, when H3K4me3 histone modifications co-occur at these sites, or silent, when additional H3K27me3 marks are present. X and Y represent unknown cofactors that control gene expression states.

convergent evolution with dogs. In fact, both the isolation and examination of DNA from extinct animals have been done in a number of species, most notably the Neandertals and mammoths. Although early studies were limited to the examination of the DNA code of extinct species, recent advances have taken this further and it is now possible to resurrect extinct gene function in living model systems. The functions of Neandertal and mammoth DNA fragments have been examined in cell lines contained in a culture dish. However, to fully appreciate the function of a gene and what role it may have played in the development of an extinct species, it is essential to examine the gene's function in an entire embryo. Using this technique, it is possible to examine when a gene is turned on in development, and in which tissues.

This technique was first pioneered with thylacine DNA and was used to examine the function of the thylacine proα1(II) collagen gene (*Col2a1*). This gene is known to be essential for the formation of collagen in the mouse and is produced in a cell type known as the chondrocytes (cartilage cells). Among other things, the *Col2A1* gene is essential for the formation of the skeleton of the animal. An enhancer region of the thylacine *Col2A1* gene promoter was coupled to a reporter molecule (LacZ) that can produce a blue pigment when the DNA is switched on during development. This DNA fragment was injected into early mouse embryos, and the embryos were allowed to develop. After two weeks of pregnancy, the embryos were analyzed to see if the DNA had been reactivated in the mouse. Clear blue staining, restricted to the developing mouse chondrocytes, indicates that the function of this extinct DNA thylacine fragment was resurrected (**Fig. 4**).

Because this method confirmed that the thylacine *Col2A1* gene played an essential role in the formation of its skeleton, it provides the first step in assessing the biological function of genes and DNA elements from an extinct species. Using these same techniques, genes can be examined from any extinct specimen where there is sufficient intact DNA. Thus, it is possible to begin to learn about the actual biology of extinct animals.

Recent technical advances in DNA sequencing have greatly reduced the difficulty, time, and cost of determining the sequence of an entire genome, including even those genomes that are highly fragmented. This has allowed a large amount of genomic DNA resources to be available for analysis. Every genome that is sequenced holds exciting information about the development and evolution of that species. The recent analysis of the entire platypus genome revealed a whole suite of novel antimicrobial peptides that could have potential biomedical uses as powerful antibiotics in humans. In the same way, extinct genomes could also hold many potentially beneficial genes for human biomedicines. Extant species (that is, species alive today) represent less than 1% of the genetic diversity that has existed in the animal kingdom. Therefore, advances in ancient DNA technologies open up a vast array of possibilities for exploring animal diversity. Extinction rates are increasing at an

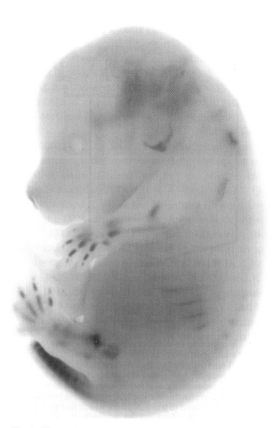

Fig. 4. Mouse fetus containing the thylacine *Col2A1* gene promoter reporter construct, on day 14.5 of gestation. Blue pigment staining indicates the sites of thylacine gene expression. The staining is consistent with expression of this element within the developing cartilage of the mouse.

alarming rate in many species, especially mammals. Many efforts, such as those being undertaken by the Frozen Zoo (San Diego Zoological Society Center for Conservation and Research for Endangered Species), are working to cryoarchive cell and tissue resources from a diverse range of threatened species to protect their genetic information. However, for those species that have already become extinct, access to their genetic biodiversity is no longer completely lost.

For background information *see* BIODIVERSITY; DEOXYRIBONUCLEIC ACID (DNA); EXTINCTION (BIOLOGY); GENE; GENETIC CODE; GENETIC ENGINEERING; GENETIC MAPPING; GENOMICS; MAMMALIA; MARSUPIALIA; POPULATION VIABILITY in the McGraw-Hill Encyclopedia of Science & Technology.

Andrew J. Pask; Brandon R. Menzies
Bibliography. B. R. Menzies et al., Limited genetic diversity preceded extinction of the Tasmanian tiger, *PLoS One*, 7:e35433, 2012; W. Miller et al., The mitochondrial genome sequence of the Tasmanian tiger (*Thylacinus cynocephalus*), *Genome Res.*, 19:213–220, 2009, DOI:10.1101/gr.082628.108; R. Paddle, *The Last Tasmanian Tiger*, Cambridge University Press, Cambridge, U.K., 2000; A. J. Pask, R. R. Behringer, and M. B. Renfree, Resurrection of DNA function in vivo from an extinct genome, *PLoS ONE*, 3:e2240, 2008, DOI:10.1371/journal.pone.0002240.

Time transfer using a satellite navigation system

Several satellite navigation systems are currently in operation or are scheduled for deployment in the near future. These include the United States Global Positioning System (GPS), the Russian Global Navigation Satellite System (GLONASS) satellites, and the European Galileo satellites. In addition to determining the position and velocity of a receiver, signals from these satellites can be used to distribute time and frequency information and to compare clocks at remote stations. A number of different techniques are commonly used for these purposes, and we will describe these techniques in this article. Although the different satellite systems differ in technical details, the methods we will describe can be used with any one of them with only minor modifications.

In the following discussion, we will assume that the receiving stations are at known positions in the coordinate system used by the navigation satellites and are not moving in an Earth-fixed coordinate system. In addition, the uncertainties in the coordinate values should be consistent with the requirements of the timing application. The details will vary with the geometry, but time transfer with an uncertainty of a few nanoseconds generally requires a position uncertainty of less than 1 m. In practice, the vertical position is usually the one with the largest uncertainty because there is usually a significant correlation between the vertical position of the station and the time offset of its clock. Therefore, the vertical position uncertainty often limits the accuracy of the time-difference measurement. This uncertainty can be minimized by the use of data from satellites that are uniformly distributed in elevation with respect to the receiver.

Signals transmitted by the satellites. The satellites transmit two signals that are important for time transfer. The first is a pseudorandom code—a series of binary 1s and 0s that is generated by a deterministic algorithm driven by the atomic clock onboard. Although the sequence is fully deterministic, it satisfies many of the statistical tests for randomness. It is not feasible to compute the next bit in the series given the previous values nor is it feasible to invert the series to determine the parameters of the generating algorithm. The second signal is a data stream that includes an ephemeris message, a group of parameters that characterize the orbit of the satellite and that can be used to determine the position of the satellite at any time. The ephemeris message also contains an estimate of the time and frequency offsets of its clock relative to the system time scale (to be defined in the following) and Coordinated Universal Time (UTC), the international time scale computed by the International Bureau of Weights and Measures (BIPM, in French). In the case of the GPS satellites, the system time scale is called GPS time, and the ephemeris message includes an estimate of the difference between this timescale and UTC as maintained at the U.S. Naval Observatory. These data are transmitted using a carrier frequency of about 1.5 GHz.

A second pseudorandom code is generally transmitted on a second frequency, but access to this code is often restricted to authorized users.

Pseudorange. All of the methods we will discuss start from the pseudorange—a measurement of the apparent time difference between the clock in the satellite and the clock at the receiver. To measure this quantity, the receiver generates a copy of the pseudorandom code transmitted by the satellite and varies the offset time of the code generator to maximize the correlation between the local and received copies of the code. The period of the pseudorandom code used by the GPS satellites is approximately 1 μs, and the cross-correlation hardware can determine the time offset that maximizes the correlation to a few percent of the period, so that the noise in the correlation process is on the order of tens of nanoseconds. The pseudorandom code implicitly defines the satellite time, and the receiver combines this measurement with the data in the navigation message to compute the pseudorange. The time offset of the station clock from the clock in the satellite is computed by applying various corrections to the pseudorange as described in the next sections. The clock-parameter data in the ephemeris message can then be used to refer this computed time difference back to the system time and ultimately to UTC.

Receiver outputs. Timing receivers use the corrected pseudorange computation in several different ways. Some receivers generate an output pulse that is derived from the average of the times extracted from the pseudorange calculations for all of the satellites in view. In addition to the composite output pulse, these receivers often produce a separate data stream containing the contribution of each satellite in view to the computed average time. (This is useful for detecting a bad satellite, whose computed time difference is very different from the computations using the others.) In many cases, the output pulse is offset from the time extracted from the pseudorange data, and this offset is also included in the data stream. Other timing receivers accept an input pulse and measure the time difference between this pulse and the satellite time scale, again using all satellites in view.

System timescale. In positioning or navigation applications, the pseudorange locates the receiver on a spherical surface centered on the satellite. To compute a unique position solution, the receiver uses at least four pseudorange observations to solve for the four unknowns: the three coordinates of its location and the offset of its clock. (There is generally a correlation between the solutions for the position and for the clock, so that timing receivers generally use a fixed position that is treated as a known quantity with no uncertainty.) Since each pseudorange is determined by the difference between the receiver clock and the clock in the satellite being observed, the times of the various satellite clocks must be linked together so that the receiver needs to solve for only one clock offset. This linkage of the satellite clocks is computed on the ground as a weighted sum of the clocks in the system, and the predicted offset of the time and frequency of each clock from this average "system time" is uploaded to the satellite periodically and is then broadcast as part of the ephemeris message. For the GPS satellites, the satellites also broadcast a prediction of the time difference between the system time scale and UTC as maintained at the U.S. Naval Observatory. Other satellite systems will transmit similar information. The existence of a system timescale is required for position and timing applications, but the linkage to UTC is needed only for timing applications, and the details of how the connection to UTC is realized will vary from one satellite system to another.

One-way time transfer. This is the simplest situation. It is used to synchronize the local clock to system time and also is the basis for more sophisticated techniques.

The first step in the analysis is to correct the pseudorange for the time of flight of the signal from the satellite to the receiver. The geometrical delay is about 65 ms (0.065 s) and is estimated using the ephemeris transmitted by the satellite and the known position of the receiver. The computation is usually performed in an Earth-centered-inertial (ECI) frame that corresponds to the Earth-centered-Earth-fixed (ECEF) coordinate system at the instant that the signal is received. This coordinate system simplifies the computations when signals from several satellites are received simultaneously. The calculation must be done by iteration, since the satellite, which is moving with an orbital speed of about 4 km/s, has moved almost 300 m during the time of flight. Depending on the latitude of the receiver, it has also moved by several tens of meters during the time of flight. Therefore, using the position of the receiver at the instant of reception as the origin of the coordinate system, we imagine that the coordinate system is rotated "backward" to compute the satellite position in the ECI frame at the time of transmission. The first-order time of flight is calculated using the geometrical distance between the position of the satellite at the instant of transmission, and the position of the receiver at the instant of reception, keeping in mind that both the receiver and the transmitter are moving during the time of flight. The uncertainty in the time of flight is much smaller than the magnitude of delay itself and is determined by the uncertainty in the position of the receiver and in any errors in the broadcast ephemeris. An uncertainty of 1 m in position translates into an error of about 3 ns in time, so that it is relatively easy to reduce the impact of an uncertainty in the position to tens of nanoseconds.

The second step is to remove the additional time delay due to the path through the ionosphere. The magnitude of this contribution varies, but is typically of order 65 ns. The magnitude of this delay is proportional to the density of charged particles in the ionosphere and varies as the inverse of the square of the carrier frequency, so that it is possible to estimate the delay by measuring its dispersion—the apparent difference in the transit times of signals at two different frequencies. All of the satellite systems transmit a pseudorandom code on two frequencies, but not all receivers can process the second transmission. Simpler, single-frequency receivers can use the estimate of the contribution of the ionosphere to the time of flight using a model broadcast by the satellite in the ephemeris message. This is better than nothing, but is usually not as accurate as the two-frequency technique. Although the magnitude of the correction due to the ionosphere is much smaller than the time of flight, the uncertainty in this correction can be quite large. For example, a single-frequency receiver that does not use any correction for the ionosphere can have a timing error that has a roughly diurnal periodicity with an amplitude that may reach 100 ns. A single-frequency receiver that uses the broadcast estimate of the ionosphere may also have a diurnal timing error, but the amplitude will be smaller, perhaps of order 50 ns. These values depend on the location and on the state of the ionosphere, which is driven by many factors including sunspot activity. A full dual-frequency receiver that measures the dispersion due to the ionosphere will have a much smaller residual uncertainty, as little as a few nanoseconds if the receiver is well calibrated.

The final step is to correct for local effects—the additional delay caused by the refractivity of the troposphere (typically about 6 ns at the zenith, increasing as the reciprocal of the cosine of the zenith angle for satellites at lower elevations due to the increase in the slant path), the delay through the receiving hardware (typically tens of nanoseconds within the receiver itself and about 5 ns per meter of cable between the receiver and the antenna), and small changes in the position of the station due to Earth tides, polar motion, and similar effects. These effects are much smaller than the effects discussed above, but they are more difficult to estimate, so that uncertainties in the magnitudes of these effects make an appreciable contribution to the overall error budget. They are typically of order a few nanoseconds. The effects of multipath reflections, which are discussed in the next section, must also be considered.

Finally, the data in the ephemeris message are used to relate the time transmitted by the clock in the satellite to the system time of the satellite system and then to an international time scale. In the case of the GPS system, the message can be used to relate the time difference to UTC as maintained by the U.S. Naval Observatory (USNO). The values in these messages transmitted by the satellite are predictions calculated on the ground and uploaded into the satellite periodically. The uncertainties in these corrections therefore depend on the stability of the clock in the satellite and on the time that has elapsed since the last upload. These uncertainties generally do not exceed 25 ns and are often much smaller than this value.

The accuracy of the time difference is limited by two sets of effects: the random fluctuations in the cross-correlation of the pseudorandom code, and the systematic errors that result from errors in the position of the satellite or in the magnitudes of the various corrections discussed above. The contribution of the random errors is attenuated by averaging, and the improvement that can be realized by averaging implicitly assumes that the clock in the receiver is sufficiently stable so that the random fluctuations in its time and frequency can be neglected relative to the measurement noise during the averaging time. These considerations can be used to define the range of optimum averaging times—long enough to attenuate the random measurement errors and short enough so that the fluctuations in the time and frequency of the clock do not make an appreciable contribution to the variance of the data. The techniques discussed in the following sections are designed to attenuate the systematic errors.

Multipath reflections. The antennas used with satellite receivers cannot have a strong directional sensitivity, since the satellites are moving with respect to the receiving station and also because a robust position solution depends on observing multiple satellites that are uniformly distributed in elevation and azimuth with respect to the receiver. The antennas are therefore also sensitive to multipath reflections—signals that reach the antenna after reflection from a nearby object. These signals, which combine with the direct signals in the receiver, always travel a longer path than the direct ones and therefore bias both the position and timing solutions.

The amplitude of the multipath signal and its time variation depend in a complicated way on the position of the satellite with respect to the ground station antenna and nearby reflectors. Amplitudes of 5–10 ns and variations on the order of minutes are not unusual, so that the effect of multipath reflections is often the largest unmodeled systematic error for a timing receiver.

One solution to mitigate this effect is to mount the antenna as high as possible above any local reflectors, to use a ground plane, which blocks signals arriving from below the antenna, and choke rings around the antenna, which attenuate signals arriving at very low elevation angles.

The geometrical configuration of the satellite, the ground-station antenna and the reflectors repeats with a nearly sidereal period (about 23 h 56 m), so that the multipath reflection from any satellite also has this periodicity. The BIPM tracking schedule (discussed below) exploits this periodicity by advancing the tracking times used by timing laboratories by 4 min every day. This has the advantage that the multipath effect is the same (in first order) every day, but the disadvantage is that the variation is converted to a systematic offset that is hard to estimate and remove.

Common-view time comparisons. In the common-view method, two (or more) receivers observe the same satellite at the same time. Each receiver computes the one-way time difference as described in the previous section, and these measured time differences are then subtracted to compute the time difference of the clocks at the two stations. The method is illustrated in **Fig. 1**. Receivers 1 and 2 receive a signal transmitted by the satellite at time S. The signals reach the receivers after a time δ, which is the same for both paths. The receivers measure the time dif-

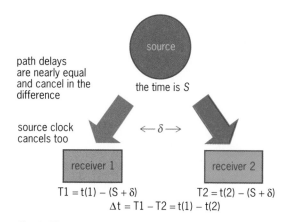

$$T1 = t(1) - (S + \delta) \qquad T2 = t(2) - (S + \delta)$$
$$\Delta t = T1 - T2 = t(1) - t(2)$$

Fig. 1. The common-view method. Receivers 1 and 2 receive a signal transmitted by the satellite at time S. The signals reach the receivers after a time δ, which is the same for both paths. The receivers measure the time difference between the local clock, t, and the received signal, and they then compute the time difference between the local clocks by subtracting these values as shown. Since the path delays are equal, both the path delay and the time of the source cancel in the difference.

ference between the local clock, t, and the received signal, and they then compute the time difference between the local clocks by subtracting these values as shown. Since the path delays are equal, both the path delay and the time of the source cancel in the difference.

If the stations are not too far apart, then many of the one-way corrections discussed in the previous section are almost the same for both stations and are therefore attenuated in the subtraction. Therefore, the common-view method is much less sensitive to errors in the satellite ephemeris, in the accuracy of the satellite clock, and in the refractivity of the ionosphere. The common-view method is less effective in attenuating the effect of the troposphere, since it is usually not so well correlated between stations and it has no effect on the local station-dependent effects discussed in the previous section. The method is also not as useful when the geometrical ranges to the receivers are very different. If the stations are not too far apart, a common-view time comparison can have an uncertainty of less than 25 ns. Timing laboratories with well-calibrated receivers located at well-known positions can realize uncertainties of 1–2 ns using this method.

All-in-view time comparisons. The common-view method depends on the fact that a single physical signal is observable at all of the participating receivers. This requirement limits the maximum distance between the receivers, and there comes a point where the method fails because the satellite is not simultaneously visible at the receivers. For example, common view cannot be used between locations in Australia and most parts of the United States.

The physical common view can be replaced with a logical common view in this case. In this method, each station measures the difference between its clock and the satellite system time—not the physical time of the satellite clock. Since the ephemeris messages transmitted by all of the satellites link the

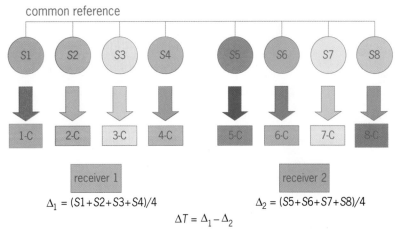

common reference

$\Delta_1 = (S1+S2+S3+S4)/4$

$\Delta_2 = (S5+S6+S7+S8)/4$

$\Delta T = \Delta_1 - \Delta_2$

Fig. 2. The all-in-view melting pot method. Receiver 1 observes satellites S1 through S4 and uses the data from these satellites to compute the difference between the local clock and the system time. The receiver combines the time offset of the local clock with respect to the satellite clock computed from the pseudorange with the offset of the satellite clock from the system time that is broadcast as part of the ephemeris message. Receiver 2 does the same computation using satellites S5 through S8. The time difference is then the difference between these two computations. The result is a logical common-view using the common system time as the reference rather than the physical signal from a single satellite as in Fig. 1.

physical satellite clock to the system time scale, it is possible to compute the common-view time difference between each station and the system time, even though this is a time scale whose time is not realized by any physical clock as discussed above. The method is shown in **Fig. 2**. Receiver 1 observes satellites S1 through S4 and uses the data from these satellites to compute the difference between the local clock and the system time. The receiver combines the time offset of the local clock with respect to the satellite clock computed from the pseudorange with the offset of the satellite clock from the system time that is broadcast as part of the ephemeris message. Receiver 2 does the same computation using satellites S5 through S8. The time difference is then the difference between these two computations. The result is a logical common view using the common system time as the reference rather than the physical signal from a single satellite as in the previous discussion.

The advantage of this method is that all of the satellites in view can be used to compute the time difference between the two stations even when no satellite is simultaneously in view at both sites. There will be multiple pseudorange measurements at each epoch, and we would expect that the measurement noise would be attenuated. Conversely, the accuracy of the relationships between the physical satellite clocks and the system time scale is a requirement of the all-in-view method but is not important for common view; the accuracy of the satellite ephemerides is also more important for the all-in-view method, since the errors in the ephemeris are not attenuated as they are in the common-view differences. Orbital and clock parameters broadcast by the each satellite in the ephemeris message are often not sufficiently accurate for all-in-view measurements, and postprocessed values are often used.

At very short baselines the two types of common view are effectively equivalent, since the same satellites and physical signals are in view at both stations. The all-in-view method is the only choice at very long baselines. The comparison at baselines of intermediate length depends on the magnitudes of the different error contributions discussed above. The full advantage of all-in-view methods can be realized only with post-processed ephemerides, which are available only after some delay, so that they are not suited to real-time clock comparisons. Both methods are limited by local effects, especially those that cannot be attenuated by averaging, such as the effect of the troposphere, uncertainties in the calibration of the receiver delay, and similar problems. Even the advantages of averaging are useful only if the local clock is sufficiently stable to support longer averaging times.

To facilitate common-view and all-in-view analyses, the BIPM has defined a standard format for publishing the data, and all timing laboratories use this format for data interchange. The format specification uses a 13-min average of the time difference between the satellite and receiver clocks, and this average is computed by the use of 52 1-s time-difference measurements. The averaging algorithm and the number of points used to compute the average were based on considerations that were appropriate for the first generation of satellite receivers, and are no longer relevant for newer receivers that can observe multiple satellites at the same time. A shorter averaging time would be appropriate in many situations. For example, many geodetic receivers, which measure the phase of the carrier (to be discussed in the next section), use an averaging time of 30 s.

Using the phase of the carrier. All of the methods described above depend on the pseudorange value, which is the time difference measured using the transmitted code. The transmitted carrier is derived from the same clock as the code, so that it has the same stability. Therefore, the pseudorange could also be estimated using the phase difference between the received carrier and the local receiver clock. Since there is no way of distinguishing one cycle of the carrier from another one, the resulting time-difference measurement would be ambiguous modulo the period of the carrier, which is somewhat less than 1 ns—about 1000 times smaller than the period of the civilian pseudorandom code that is used by most receivers. Since the measurement noise in a time difference is typically some fraction of the period of the signal that is being observed, carrier-phase measurements can have more resolution than code-based estimates, assuming that the integer cycle count can be determined and that cycle jumps can be detected and removed. However, there is usually no way of determining the correct integer cycle count using the carrier phase data alone, and most analyses use the code-based time difference to assist in identifying the correct integer cycle count. Thus, while carrier-phase measurements have greater resolution than those based on the code, the accuracy is often no better. The difficulty of detecting cycle slips in the carrier-phase data depends on the stability of the clock in the receiver, since a more stable clock makes it easier to detect time steps in the data that are due to cycle slips. Comparing measurements using multiple satellites is also used to detect a cycle slip in the data from one of them.

The increase in resolution that can be realized by the use of the phase of the carrier requires a corresponding increase in the accuracy of the various corrections that we discussed above in connection with the pseudorange. In practice, the carrier-phase methods are used with post-processed ephemerides, since the parameters broadcast by the satellites are not accurate enough to take full advantage of the increased resolution that is possible using the phase of the carrier. The post-processed ephemerides are available from the International Geophysical Service (IGS). The IGS has a number of different products with varying delays from real time. Under favorable conditions, the carrier-phase method has a statistical uncertainty of about 50–100 ps using an averaging time on the order of minutes. The accuracy of the data is generally somewhat poorer than this value, since the integer cycle must be determined from the code data, and the long-term accuracy is generally no better than what can be achieved using a code-based analysis.

The time delay in the availability of post-processed ephemerides has been slowly decreasing as more sophisticated computational methods are developed at the analysis centers. There are now "ultra-rapid" ephemerides with delays of hours.

Frequency transfer. Frequency comparisons are generally performed by observing the evolution of the time difference between two clocks over some averaging time. The frequency difference is then expressed as seconds per second, which is a dimensionless parameter. For example, a clock that gained $1\ \mu s$ per day with respect to a reference device would have a frequency of $1\ \mu s/86{,}400\ s = 1.16 \times 10^{-11}$. Using the same ideas, the accuracy of a frequency comparison is determined by the residual noise of the time-difference process divided by the averaging time between observations. For example, if the residual noise in the time-difference measurements is 50 ns, then the frequency transfer noise using a 1-h (3600-s) average will be about 1.39×10^{-11}. For a given uncertainty in the measured time differences, the best frequency transfer will be determined by the maximum averaging time, which is the time over which the parameters of the local clock are constant. This maximum averaging time ranges from seconds for quartz-crystal oscillators to days for hydrogen masers. In practice, the stochastic fluctuations in the systematic corrections we have discussed limit the maximum averaging time to about a month even under ideal conditions (which cannot be realized routinely), and this limits the minimum uncertainty of frequency comparisons to about 3×10^{-16}. This limit is much better than is needed for almost all current applications, but it will not be small enough to compare the next generation of primary frequency standards, which will have stabilities that are smaller than this limit.

Applications. Satellite time signals are used in numerous applications—far too many to enumerate in this article. The signals are widely used to synchronize the time of computer systems and network elements. A number of commercially-available devices can provide time signals in the Network Time Protocol (NTP) format (an Internet standard that defines messages used to transmit time over digital networks) that are synchronized using signals from a satellite constellation.

More demanding applications include providing a frequency reference for the telecommunications network, which requires a frequency accuracy of 1×10^{-11}, and the NASA deep-space network, which requires time synchronization on the order of nanoseconds. Radio astronomy observatories, such as Arecibo in Puerto Rico, have similar requirements.

The data are used as one of the primary techniques for comparing the times of National Metrology Institutes and timing laboratories such as the National Institute for Standards and Technology and the Naval Observatory in the United States and similar laboratories in other countries. These comparisons depend on timing accuracies on the order of 1 ns and frequency comparisons with an uncertainty of less than 5×10^{-15}. These time comparisons form the basis for the computation of TAI and UTC. These comparisons are important for the international interoperability of timing, communications, and navigation systems and for evaluating the next generation of primary frequency standards, which are being developed in many laboratories in different countries. *See* OPTICAL CLOCKS AND RELATIVITY.

The future. There will be a significant increase in the number of satellites that are available for time and frequency comparisons in the next few years. In addition to the U.S. Global Positioning System, the Russian GLONASS system is currently operational, the European Galileo system will be operational in a few years, the Chinese are planning a system, and a number of other countries are planning either regional or global satellite systems that will be useful for time and frequency distribution using the methods that have been described. An important aspect will be interoperability—the ability to combine the data from the different satellite systems in a single time or frequency comparison. The accuracy of the time and frequency data from all of these systems will be limited by the considerations have been discussed, so that improving the accuracy of time and frequency comparisons will depend on how well these limitations are addressed by all of the satellite operators. Improving the accuracy of the broadcast ephemerides and the stability of the clocks on the satellites will facilitate real-time clock comparisons, and more sophisticated receivers that can reject multipath signals and have greater immunity to local environmental perturbations will also be very useful.

The jamming of satellite signals, either inadvertently or intentionally, is also likely to become important in the future, and receivers that have greater immunity to extraneous signals will become more important.

For background information *see* ATOMIC CLOCK; ATOMIC TIME; EPHEMERIS; FREQUENCY MEASUREMENT; RADIO ASTRONOMY; SATELLITE NAVIGATION SYSTEMS; SPACECRAFT GROUND INSTRUMENTATION; TIME in the McGraw-Hill Encyclopedia of Science & Technology.   Judah Levine

Bibliography. C. Hackman and D. B. Sullivan (eds.), *Time and Frequency Measurement*, American Association of Physics Teachers, College Park, MD, 1996; J. Levine, Introduction to time and frequency metrology, *Rev. Sci. Instrum.*, 70:2567–2596, 1999, DOI:10.1063/1.1149844.

Trace elements in wild growing mushrooms

Fungi are eukaryotic organisms classified as a single group, the kingdom of Fungi, which is independent of plants, animals, and bacteria. The macrofungi represent a specific group of fungi that produce easily visible fruiting bodies, such as mushrooms or puffballs. Fungi are heterotrophic, obtaining nutrients from existing organic matter. According to their lifestyle, fungi are recognized as saprotrophs (feeding on dead organic matter), parasites (deriving nutrients from other living organisms with no benefit to them), and mutualists (living in association with vascular plants: ectomycorrhizal symbiosis).

Fungi have important biogeochemical roles in the biosphere and are intimately involved in the cycling of elements and in transformations of both organic and inorganic substrates. The study of the role played by fungi in geochemical processes is termed geomycology. The fundamental importance of fungi involves organic and inorganic transformations and element cycling, rock and mineral transformations, bioweathering, mycogenic mineral formation, fungal–clay interactions, and metal–fungal interactions (**Fig. 1**).

Macrofungi are able to effectively take up various trace elements (including toxic heavy metals), to translocate them through mycelia, and to deposit them in their fruiting bodies. The ability of organisms to take up a particular element is described on the basis of the so-called bioaccumulation factor (BAF), which represents how much of an element is in a tissue relative to how much of that element exists in the environment: $BAF = concentration_{[fruiting\ body]}/concentration_{[soil]}$.

More specifically, BAF may be related to mobility (extractability) of a particular element in soil. When $BAF > 1$, the element is accumulated (bioaccumulation); when $BAF < 1$, the element is excluded (bioexclusion). An extraordinarily high ability to accumulate a trace element in biomass is called hyperaccumulation. To be defined as hyperaccumulating, the amount of a particular element should represent a concentration approximately 100 (or more) times higher than the values expected to be found in nonaccumulating macrofungi growing on the same substrate.

Factors influencing bioaccumulation of trace elements in macrofungi and the biological importance of the accumulation process itself are poorly understood. However, five fundamentals have been recognized: (1) Natural factors (bedrock geochemistry): The characteristics of the geological bedrock and soil influence the concentration of trace elements,

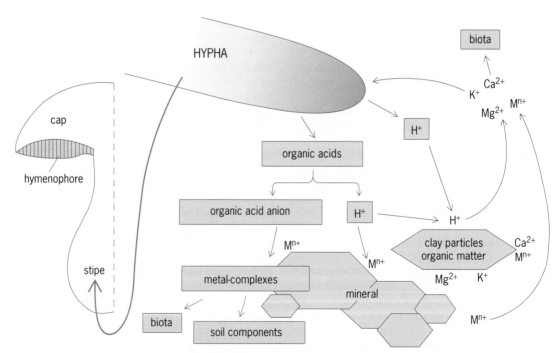

Fig. 1. Proton- and organic acid ligand–mediated dissolution of metals of soil components/minerals and their transfer to fruiting bodies. Proton release results in cation exchange with sorbed metal ions on clay particles, colloids, and so on, and metal displacement from mineral surfaces. Released metals can react with other environmental components or can be taken up by both mycelia and other biota. M^{n+} = heavy metal. [*Modified from G. M. Gadd, Mycotransformation of organic and inorganic substrates, Mycologist, 18(2):60–70, 2004*]

pH and Eh (redox potential) conditions, and consequent mobility of the elements in soil solution. (2) Metalliferous areas and environmental pollution: Naturally or artificially elevated levels of metals in soils usually result in higher concentrations in fruiting bodies. (3) Fungal lifestyle: Concentrations of heavy metals (Hg, Cd, and Pb) and metalloids (Se and As) are usually higher in terrestrial saprotrophs compared to those in ectomycorrhizal fungi. On the other hand, concentrations of alkali metals (Rb and Cs) are generally higher in ectomycorrhizal species. (4) Bioaccumulation of metals in fruiting bodies is species-specific: Some macrofungal species are able to take up elements much more effectively than others; the ability to accumulate a particular element is a characteristic feature of particular macrofungal species. (5) The accumulation process can be highly element-specific: Macrofungi may even discriminate elements with similar properties and chemical behavior (homologs) such as As and Sb. On the other hand, concentrations of the chemical homologs Rb and Cs may be positively correlated in ectomycorrhizal fungi.

Distribution of accumulated elements is not homogeneous within the fruiting body. The highest concentrations are usually found in the hymenophore (consisting of lamellae or tubules), cap skin, and cap flesh, whereas the lowest concentrations are found in the stipe (the short stalk or stem of the fruiting body); a rapid decrease of metal concentrations can be observed in spores. On the other hand, the highest concentrations of nonaccumulated (excluded) elements are often found in the stipe.

Furthermore, concentrations of elements may change during the growth of the fruiting body. A case study of the common puffball (*Lycoperdon perlatum*) focusing on specimens from an auriferous (gold-containing) area has shown that accumulation of Au is a continuous process that, however, slows down significantly during the development (age) of the fruiting body; this is in contrast to accumulation of Ag, As, and Se. Concentrations of elements in fungal mycelia are unknown. However, some studies have reported high levels of metals in ectomycorrhizal roots and have suggested the possible formation of a protective fungal biological barrier that reduces movement of toxic metals in symbiotic tree tissues.

Hyperaccumulators. The fly agaric (*Amanita muscaria*) was the first recognized fungal hyperaccumulator. This temperate-zone species (**Fig. 2***b*), which has a trans-Atlantic distribution, and its relatives, *A. regalis* and *A. velatipes*, hyperaccumulate vanadium. Whereas concentrations of this metal in other macrofungi hardly exceed 1 mg/kg (related to dry matter), *A. muscaria* contains hundreds of milligrams per kilogram. In fruiting-body biomass, vanadium is present as amavadin, an eight-coordinate vanadium complex of unclear biological function.

The pink crown (*Sarcosphaera coronaria*) [Fig. 2*c*], an ectomycorrhizal ascomycete growing on calcareous bedrock in temperate climate regions, hyperaccumulates arsenic. Despite being relatively low naturally in soils, As concentrations are commonly as high as 1000 mg/kg in fruiting bodies of this species, with the highest value ever reported being 7090 mg/kg. The main As compound identified in biomass was methylarsonic acid. Although this compound is less toxic than arsenic trioxide, it is still relatively dangerous. Therefore, the pink crown should not be eaten, although it used to be considered edible in Europe.

Other recently discovered hyperaccumulators include *Amanita* species of the section *Lepidella*. The European *Amanita strobiliformis* (Fig. 2*a*) hyperaccumulates silver. Despite very low natural levels in soils (lower than 1 mg/kg), Ag concentrations in *Amanita* species may be more than 2000 times higher and commonly amount to hundreds of milligrams per kilogram; the highest reported value was 1253 mg/kg. Investigation on the chemical form of Ag has revealed that *A. strobiliformis* employs cysteine-rich, low-molecular-weight proteins (metallothioneins) to sequester intracellular Ag in its fruiting bodies and extraradical mycelium.

The biological importance of elemental hyperaccumulation in macrofungi is unclear; however, in theory, this extraordinary feature can be thought of as being attributable to the "defense hypothesis." It has been shown in vascular plants that elemental hyperaccumulation may have several functions, including plant defense against natural enemies; at least some tests have demonstrated defense by hyperaccumulated As, Cd, Ni, Se, and Zn in plants. In view of this fact, the high Ag concentrations in hyperaccumulating macrofungal species might have some protective effect against pathogenic microfungi, bacteria, insect larvae, or gastropods, but this has yet to be tested.

Accumulation of metals. The ability of macrofungi to accumulate trace elements is generally higher than that of vascular plants. Furthermore, the ability to accumulate trace elements is not primarily determined by the concentrations of such elements in soil; high levels of heavy metals in fruiting bodies can be found commonly in pristine areas (see **table**).

In many European countries, and also in several regions of the United States and Canada, mushroom picking in the wild is a popular hobby. High concentrations of heavy metals (especially Cd, Hg, and As) in mushrooms are thus considered a possible health risk for mushroom consumers. Therefore, species with a high accumulation ability collected in the wild, especially the horse mushroom (*Agaricus arvensis*) and allies, are sometimes not recommended for common consumption. Despite the high accumulation ability of macrofungi, some metals, including Ni, Cr, Tl, Pb, Th, and U, are usually not taken up by these fungi (avoiding hazardous levels of these metals), even if such mushrooms grow on polluted sites. On the other hand, macrofungi from pristine sites may well represent a nutritional supplement because they contain such essential elements as Zn, Cu, Fe, Mn, and Se.

In the aftermath of the Chernobyl nuclear accident in 1986, Cs-accumulating ectomycorrhizal

(a)

(b)

(c)

Fig. 2. Hyperaccumulating macrofungi. (*a*) *Amanita strobiliformis* (European pine cone *Lepidella*), silver hyperaccumulator. (*b*) *Amanita muscaria* (fly agaric), vanadium hyperaccumulator. (*c*) *Sarcosphaera coronaria* (pink crown), arsenic hyperaccumulator. (*Photos courtesy of Jan Borovička*)

fungi in Europe showed a highly elevated radioactivity caused by the ^{137}Cs radioisotope. Interestingly enough, recent investigations of the pathways leading to high ^{137}Cs activity in the meat of the wild boar (*Sus scrofa*) have revealed that ^{137}Cs-rich deer truffles (*Elaphomyces* spp.), widely consumed by wild boars, might function as the main source of radiocesium. However, common consumption of edible macrofungi (up to 10 kg of fresh mushrooms per year) in Central and Western Europe is not considered a health risk.

For their metal-accumulation ability, macrofungi used to be considered bioindicators of environmental pollution or mineral deposits. However, it is difficult to implement a practical application because the fruiting bodies are short-lived; their occurrence at

Macrofungi that accumulate/hyperaccumulate trace elements, and the highest concentrations (mg/kg, dry weight) of such elements in fruiting bodies from pristine areas and polluted sites (including metalliferous areas)

Element	Accumulating families/genera/species	Clean, mg/kg	Polluted, mg/kg
Au	*Lycoperdon perlatum*, *Agaricus* spp., *Boletus edulis*	0.0X–0.X	X
Ag	*Amanita* spp., *Agaricus* spp., *Boletus* spp.	X0–X000	X00
As	*Sarcosphaera coronaria*, *Laccaria* spp., *Russula pumila*, *R. alnetorum*, *Inocybe* spp.	X00–X000	X000
Br	*Amanita phalloides*	X0–X00	—
Cd	*Agaricus* spp., *Macrolepiota* spp.	X–X0	X0–X00
Cl	*Amanita* spp., *Agaricus* spp.	X000–X0,000	—
Cs	*Boletus* spp., *Xerocomus* spp., *Tylopilus felleus*, *Cortinarius* spp., *Laccaria* spp., *Paxillus involutus*, *Elaphomyces* spp.	X0–X00	—
Cu	*Macrolepiota* spp., *Agaricus* spp., Lycoperdaceae	X0	X0–X00
Fe	*Suillus variegatus*, *Hygrophoropsis aurantiaca*, Phallaceae, *Panaeolus* spp.	X00–X000	—
Hg	*Agaricus* spp., *Macrolepiota* spp., *Boletus* spp.	X–X0	X0–X00
Mn	Phallaceae, *Panaeolus* spp.	X00–X000	—
Pb	Lycoperdaceae (*Lycoperdon perlatum*), *Agaricus* spp.	X–X0	X0–X00
Rb	*Boletus* spp., *Xerocomus* spp., *Tylopilus felleus*, *Cortinarius* spp., *Laccaria* spp., *Sarcodon* spp., *Paxillus involutus*, *Elaphomyces* spp.	X00–X000	—
Sb	*Suillus* spp., *Chalciporus piperatus*	X–X0	X00–X000
Se	*Boletus pinophilus*, *Albatrellus pes-caprae*, *Amanita strobiliformis*	X–X0	—
V	*Amanita muscaria*, *A. regalis*, *A. velatipes*	X0–X00	—
Zn	*Russula atropurpurea* and allies, *Hebeloma mesophaeum*	X00–X000	—

Terms: X represents units, X0 represents tens, X00 represents hundreds, X000 represents thousands, X0,000 represents tens of thousands, 0.X represents tenths, and 0.0X represents hundredths.

investigated sites depends on the season and weather conditions; they are usually insufficiently distributed in the surveyed area (low density or even absence of some biotopes); their proper identification requires assistance from an experienced mycologist; there are substantial differences in the ability of macrofungal species to accumulate metals; and, moreover, a wide metal content variation is typical.

Future studies in this field will likely focus on mechanisms of metal uptake, transport, and sequestration in fungal tissues because there is a potential for applications in biotechnology. Furthermore, investigations on the chemical form of metals in fungal fruiting bodies will allow for a better understanding of the possible health risks to consumers. Last but not least, investigations on fungal interactions with geological bedrock and metals, including the role of ectomycorrhizal symbiosis, will help to reveal the role of macrofungi in the environment.

For background information *see* BIODEGRADATION; ECTOMYCORRHIZAL SYMBIOSIS; ENVIRONMENTAL TOXICOLOGY; FUNGAL ECOLOGY; FUNGI; INDUSTRIAL ECOLOGY; MUSHROOM; MYCOLOGY; SOIL; SOIL CHEMISTRY; SOIL MICROBIOLOGY in the McGraw-Hill Encyclopedia of Science & Technology.

Jan Borovička

Bibliography. J. Borovička et al., Hyperaccumulation of silver by *Amanita strobiliformis* and related species of the section *Lepidella*, *Mycol. Res.*, 111(11):1339–1344, 2007, DOI:10.1016/j.mycres. 2007.08.015; G. M. Gadd, Geomycology: Biogeochemical transformations of rocks, minerals, metals and radionuclides by fungi, bioweathering and bioremediation, *Mycol. Res.*, 111(1):3–49, 2007, DOI:10.1016/j.mycres.2006.12.001; G. M. Gadd, Mycotransformation of organic and inorganic substrates, *Mycologist*, 18(2):60–70, 2004, DOI:10.1017/S0269-915X(04)00202-2; P. Kalač, Trace element contents in European species of wild growing edible mushrooms: A review for the period 2000–2009, *Food Chem.*, 122(1):2–15, 2010, DOI:10.1016/j.foodchem.2010.02.045.

Tree seasonality in a warming climate

In the northern boreal areas and temperate regions, climate is characterized by significant annual variation between the summer and winter seasons. Adaptation to this climatic variation has resulted in tree seasonality. This seasonality is visible as annual cycles of growth and inactivity (dormancy). As growth cessation is a prerequisite for significant frost tolerance in northern tree species, the growth-dormancy cycles are correlated closely with seasonality with regard to hardiness and stress tolerance. Proper timing of these seasonal cycles, with respect to prevailing temperature conditions, is critical for survival under these climatic conditions. The seasonal cycles are controlled by an interaction of genetic and climatic factors, but the physiological and molecular mechanisms for the two cycles are quite different. Seasonality in northern tree species is a result of genetic adaptation to climate, and changes in the climatic conditions will affect the control of seasonality and may have consequences for survival and distribution of tree species.

Seasonality of growth activity and dormancy. Two developmental processes characterize the phenological (periodic biological) seasonality of northern tree species: growth cessation followed by bud set in the fall, and bud burst in the spring (see **illustration**). These events mark the beginning and end of significant frost tolerance, and the timings of bud set and bud burst are often critical for winter survival in northern tree species. With respect to bud set, two main growth patterns can be distinguished: free growth and determined growth. In free growth,

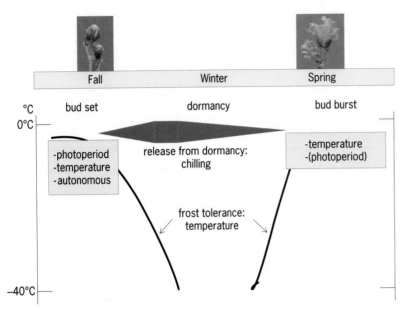

Seasonality in phenology and frost tolerance in northern tree species. Bud set is controlled, depending on the species, by photoperiod, temperature, and/or autonomous properties in the species. Depth of the dormancy increases gradually during the fall, and the dormancy is released gradually by an exposure to chilling temperatures [0–10°C (32–50°F)]. Bud burst is primarily controlled by temperature. Significant increases in frost hardiness take place after bud set, as a response to decreasing temperatures, below 10°C (50°F). Exposure to increasing temperatures leads to a decrease in frost tolerance, which is lost by the time of bud burst. In addition, the frost-tolerance cycle may have a component of autonomous control as well.

production of new leaves (nodes) and stem elongation (internodes) take place simultaneously, and shoot growth continues as long as the external conditions are favorable. Bud set is induced by photoperiod (the duration of an organism's daily exposure to light) or temperature (or a combination of these two factors). Free growth is typical for many hardwood species, including birch (*Betula*), poplar (*Populus*), and willow (*Salix*), but also first-year seedlings of spruce (*Picea*) have a free-growth pattern. In these species, the short photoperiod early in the fall is the primary signal for induction of growth cessation and bud set. In many species of the rose family, such as rowan (*Sorbus*), apple (*Malus*), and chokecherry (*Prunus*), low temperature (and not the photoperiod) induces the growth cessation.

In determined growth, which is typical for conifers such as pine (*Pinus*) and spruce (*Picea*), bud set is controlled by autonomous conditions, and the effect of climatic factors is limited. In addition, some hardwoods have a determined-growth pattern, and typically even trees with a free-growth pattern change over toward a determined-growth pattern with increasing age.

Bud set is followed by development of dormancy (see illustration), with the meristem temporarily losing its ability for growth. The apical meristem in the shoot apex is the tissue where cell division and cell elongation take place. The apical meristem is responsible for elongation growth, whereas the cambium (meristem in the stem) is responsible for thickness growth. Growth activity in the cambium ceases usually shortly after bud set. During bud formation, stem elongation is arrested, and there is also cessation

of formation of new leaf primordia, or floral organs in flower buds. Dormancy is a dynamic phase, and depth of the dormancy increases and decreases gradually during the winter (see illustration). In deciduous trees, bud set and dormancy development are followed by senescence and abscission of the leaves. These processes are often favored by low temperatures and delayed by high temperatures.

The apical meristem is released from dormancy after a sufficient period of exposure to low temperatures; the effective temperature range is 0–10°C (32–50°F). The chilling requirement for bud burst is satisfied early in the winter, but extended chilling enhances the rate of bud burst and reduces the temperature requirement for bud burst. This requirement for chilling has a genetic basis and varies among species and genotypes, but it is significantly affected by temperature conditions during the fall, winter, and spring. In northern tree species, high autumn temperatures enhance dormancy and increase the chilling requirement; in contrast, high spring temperatures can in part compensate for chilling. After reactivation of the meristem through chilling, the development toward bud burst is primarily dependent on temperature. However, particularly in northern tree species, development toward bud burst can proceed at the same temperatures as those giving the chilling effect. Buds of northern birch, for example, are able to burst at temperatures close to 0°C (32°F), but the rate of bud burst is strongly increased with increasing temperature. Long photoperiod may also enhance bud burst, particularly if chilling is insufficient.

Seasonality of frost tolerance. Northern tree species are able to tolerate minor frost events even when the trees are in active growth, but bud set is a prerequisite for significant cold acclimation (see illustration). Short photoperiod is a significant signal for cold acclimation in species with free-growth patterns, but it can initiate cold acclimation also in *Picea* and *Pinus*. After bud set, frost tolerance increases rapidly, as a response to decreasing temperatures, and remains high through most of the winter (see illustration). Usually temperatures below approximately 10°C (50°F) induce cold acclimation. Several boreal tree species are able to tolerate liquid nitrogen [–196°C (–321°F)] during midwinter, and the midwinter frost tolerance is usually not a limiting factor for the native trees. Increasing temperatures during the spring induce deacclimation; and shortly before or by bud burst, the frost tolerance is decreased to the summer level (see illustration). Mild spells during winter may reduce frost tolerance, and the effects of such warm periods increase with approaching spring. Of course, severe frost after such warm spells may lead to winter injuries.

Effects of global warming. How are the aforementioned seasonal changes affected by warming climates? There is no general answer to this question because the effects are dependent not only on the degree of global warming, but also on the timing of the changes. Temperature increase in the summer season will not necessarily have any significant effects on the activity–dormancy cycle or on the timing of cold acclimation. In addition, the effects of a given change will be different in the various parts of the distribution area of a tree species. So far, there is a lack of adequate regional and seasonal estimates of expected changes. Furthermore, trees must cope with several extreme climatic situations during their long life span; they are robust and plastic in their responses. In addition, most of the northern tree species have an extended area of distribution, covering various types of climates, and this provides a genetic basis for adaptation.

The timing of processes that are primarily controlled by photoperiod will not be significantly affected by changes in the temperature climate, although the rate of bud development, for example, can be enhanced by increased temperature. Increased temperature during the fall season, though, will enhance the development of dormancy and increase the chilling requirement in many trees. Such temperature increase will also delay leaf senescence, at least for some species. Moreover, cold acclimation, induced by low temperature, can be delayed by increased fall temperature. However, if the occurrence of freezing temperatures is delayed as well, there might not necessarily be any significant consequences for the native tree species.

Increased winter temperatures may have consequences for the release from dormancy and for cold acclimation. In the southern marginal areas of the distribution, increased winter temperatures may result in an insufficient chilling for dormancy release and, consequently, disturbed bud burst in the spring. In the current climate, such disturbances are known to affect orchards in Mediterranean regions. However, in the boreal areas, these effects are not likely to occur, even with the largest estimates of warming.

An increased frequency of warm spells during the winter season may temporarily reduce the frost tolerance, and sudden freezing temperatures after such warm spells may be harmful. During midwinter, though, the frost tolerance of the native trees is rather stable; however, frost-tolerance sensitivity increases during the spring during deacclimation and with decreasing dormancy.

Increased temperature during late winter and spring will result in earlier bud burst in northern tree species, except in those trees with strong photoperiodic control [for example, beech (*Fagus*)]. Annual variations in the timing of bud burst clearly demonstrate the effect of temperature on bud burst, and some phenological observations suggest a trend toward earlier bud burst. The magnitude of these effects will vary among species; moreover, with the same amount of warming, the effect will be larger in the marginal areas compared to the southern and central areas of species distribution.

Conclusions. Seasonality in trees is controlled by an interaction between genetic background and climatic conditions. As long-living organisms, trees must adapt to changing climatic conditions and must develop robust and plastic mechanisms to control

the seasonality. Changing climate will have varying effects on many species, and the effects will also be different in various parts of the distribution range of a species. Probably the most common effect of enhanced warming will be enhanced bud burst in the spring, and delayed cessation of the growth season in the fall.

For background information *see* APICAL MERISTEM; CLIMATE MODIFICATION; COLD HARDINESS (PLANT); DORMANCY; FOREST AND FORESTRY; GLOBAL CLIMATE CHANGE; PHOTOPERIODISM; PHYSIOLOGICAL ECOLOGY (PLANT); PLANT METABOLISM; SEASONS; TREE; TREE GROWTH; TREE PHYSIOLOGY in the McGraw-Hill Encyclopedia of Science & Technology.

Olavi Junttila

Bibliography. J. E. K. Cooke, M. E. Eriksson, and O. Junttila, The dynamic nature of bud dormancy in trees: Environmental control and molecular mechanisms, *Plant Cell Environ.*, in press, 2012, DOI: 10.1111/j.1365-3040.2012.02552.x; H. Hänninen and K. Tanino, Tree seasonality in warming climate, *Trends Plant Sci.*, 16:412–416, 2011, DOI:10.1016/j.tplants.2011.05.001; O. M. Heide, High autumn temperature delays spring bud break in boreal trees, counterbalancing the effect of climatic warming, *Tree Physiol.*, 23:931–936, 2003, DOI:10.1093/treephys/23.13.931; O. Junttila and H. Hänninen, The minimum temperature for budburst in *Betula* depends on the state of dormancy, *Tree Physiol.*, 32:337–345, 2012, DOI:10.1093/treephys/tps010; M. B. Murray, M. G. M. Cannell, and R. I. Smith, Date of budburst in fifteen tree species in Britain following climatic warming, *J. Appl. Ecol.*, 26:693–700, 1989.

Turbulent combustion

The predominant fraction of matter in the universe is in a fluid state, and the predominant fraction of that is turbulent. Turbulence mixes fluids efficiently, bringing reactants together that can ignite and burn. On a cosmic scale, such reactions can be chemical or nuclear. Chemical combustion in turbulent flow, also known as turbulent combustion, combines the disciplines of chemistry and flow physics and must also deal with the effects on each other: the unsteady and chaotic dynamics of turbulent flow, the complexity of chemical reactions, the effects of the heat released by combustion on the turbulence, and, in turn, the effects of the turbulent environment on combustion itself. Any analysis of turbulent-combustion phenomena must deal with the flow and chemical reactions as coupled processes.

Turbulence and turbulence scaling. An important parameter in turbulence is the Reynolds number, $Re = \rho L U / \mu = L U / \nu$, where ρ is a characteristic fluid density, L is the spatial extent of the turbulent region (usually referred to as the outer or integral scale of the turbulence), U is the characteristic flow velocity (proportional to the turbulent velocity fluctuations), μ is the dynamic or shear viscosity, and $\nu = \mu / \rho$ is the so-called kinematic viscosity (units of area per unit time). For air at normal conditions, for example, $\nu_{\text{air}} \cong 0.15$ cm^2/s. Unsteady flow and turbulence is the normal dynamical state for high values of Re. While with care, steady laminar flow can be extended to high Re, such a flow is unstable and will transition to turbulence spontaneously, induced by minute disturbances. Conversely, once turbulent, the flow cannot revert to a laminar state for $Re \gtrsim 10^4$. By way of example, the Reynolds number for the flow from a 1-m-diameter duct ($L = 1$ m) discharging air at a speed $U = 10$ m/s is $Re \cong 7 \times 10^5$, well above the threshold for sustaining fully developed turbulence. The flow Reynolds number also parameterizes the range of spatial scales spanned by turbulence. The ratio of the largest scale, L, to the smallest flow scale called the Kolmogorov scale, λ_{K}, is given by $L/\lambda_{\text{K}} \cong c_{\text{K}} Re^{3/4}$, with c_{K} a proportionality constant of order unity. The smallest chemical species-diffusion scale, λ_D, is close to λ_{K} for gases, but much smaller for liquids. For the discharging-duct example above, we have $L/\lambda_{\text{K}} \cong 2.3 \times 10^4$.

Chemical combustion. Combustion focuses on processes involved in the reaction of an oxidizer, such as air, with a fuel. Typical fuels include hydrogen, methane, butane, propane, as well as higher hydrocarbons, such as gasoline and diesel fuel. Chemical reactions in combustion are exothermic (sensible energy is released). A common goal of combustion is to produce useful work from this heat release either directly, or through heat transfer to an intermediate fluid. In this discussion, we concentrate on gas-phase combustion, in which the fuel is a gas at normal conditions or has evaporated from a liquid before combustion. We will not discuss other important technical areas such as spray or pulverized-coal combustion.

Chemical kinetics. It is typical for combustion processes to involve tens to hundreds of chemical species and hundreds to thousands of elementary reactions. For complex fuels, the number of elementary reactions in detailed chemical-kinetic models is, roughly, 10 to 100 times the number of species. Determination of large chemical-kinetic mechanisms must be performed by hand, relies strongly on the experience of kineticists, and remains more art than science. In these mechanisms, some species are stable, or long-lived, while others are short-lived and termed radicals. The dependence of elementary reaction rates on temperature is strong and usually modeled by Arrhenius forms, that is, exponential functions of temperature that approximate the dependence on the energy barrier of each reaction as a function of temperature. The strong nonlinearity of these reaction-rate terms poses one of the challenges to theoretical progress in the field.

Combustion classification. Combustion is usually classified into modes. This classification has enabled theoretical progress through a combination of analytical, computational, and experimental techniques. The first classification criterion considers whether the fuel and oxidizer are either perfectly mixed or perfectly unmixed before chemical reactions occur. In the former case, we speak of premixed combustion while in the latter we speak of nonpremixed combustion. As discussed below, these regimes are further partitioned according to the dominant physical process. The second classification criterion considers whether the processes can nominally be approximated as zero-, one-, or two-dimensional, or even three-dimensional. In this context, what matters is the number of dominant dimensions required for an adequate description of reaction zones. Weak variations along some dimensions may be neglected in a first pass at understanding the structure of the reactive zone. In turbulent combustion, the reactive zones may be highly convoluted, sheared, and distorted at any time throughout space. However, appropriate local low-dimension coordinates may yield an adequate parameterization.

Following the first classification criterion, a premixed mode may arise in a zero- or one-dimensional situation. In the first case, a cold mixture may ignite by the buildup of a radical pool followed by thermal runaway (autoignition), or by the action of an external heat source (ignition). In the second case, called a premixed flame, a thin reactive zone can propagate into a region of unreacted species as a wave with large changes across it, for example, in temperature, density, composition, and so forth. Part of the heat released is transported (conducted) ahead of the flame and preheats the reactant mixture to a state that enables self-sustained autonomous combustion. This wave generally travels at speeds in the frame of unreacted species of order 1 m/s and is referred to as an advancing flame, or deflagration, to differentiate it from its combustion-driven shock-wave counterpart, called a detonation, which is not discussed in this article.

The nonpremixed mode is usually differentiated depending on the relative strength of the chemically active zone (also termed a diffusion flame) with respect to flow straining or shearing. Such flames can be established where the interdiffusing reactant fluxes attain stoichiometric proportions. Again, heat released diffuses away from the flame toward as yet chemically inert regions. If reactive zones are thinner than other relevant flow scales, they are usually termed flamelets; both premixed and nonpremixed regimes. This combustion structure is a quasi-one-dimensional structure with much larger variations of thermochemical properties along a line locally perpendicular to the flamelet than along it. At high turbulence intensities leading to high mixing or strain rates, local quenching or extinction can occur, producing dynamical flame holes that can grow or shrink depending on instantaneous local flow conditions. This situation is usually termed the extinction-reignition regime. The boundary that separates an actively burning flame region from a quenched region resembles a two-dimensional combustion structure termed an edge flame that is usually composed of two curved premixed branches, one fuel-lean and one fuel-rich, trailed by a diffusion flame. Edge flames can arise in premixed and nonpremixed flows and have propagation properties similar to those of premixed flames with response characteristics similar to diffusion flames. **Figure 1** shows a diffusion flame in a high-strain flow where flame holes have formed.

In practical combustion environments and devices, a clear distinction between premixed and nonpremixed combustion is not always possible, and regions of the combustor may be either a combination of the two basic regimes, or in a mixture of intermediate states. For this reason, it is useful to introduce the Damköhler number (Da). It measures the ratio of the relevant flow/mixing characteristic time to the characteristic chemical reaction time. For a given chemistry timescale, fast flow (short timescale) has a low Da, corresponding to combustion whose rate is limited by chemical kinetics, while long flow timescales have a high Da, corresponding to mixing-limited combustion. Real chemistry is complex, however, and each elementary reaction has its own Da, with different timescales associated with radical production and termination reactions, and thousands of reactions for practical-fuel kinetic networks. Da scaling aids theoretical understanding of key physical phenomena, such as heat release that is parameterized in terms of a single chemical-conversion time. **Figure 2** provides an example and shows the variation of flame temperature with Da in a reactive mixture, called an S-shape curve. Temperature is high at high Da, approaching a maximum determined by thermodynamics. From this state, decreasing Da yields a sudden drop in temperature to an extinguished state with no combustion. Conversely, starting with a cold mixture at low Da, increasing Da yields a sudden increase in temperature at the so-called ignition transition. The value of Da for this transition is different from that of the extinction transition. The intermediate branch in the hysteresis cycle (the dashed segment in Fig. 2) is unstable and cannot be observed or sustained experimentally.

Turbulent combustion scaling. **Figure 3** depicts turbulent-combustion regimes for premixed and nonpremixed flames. Overlaid on the diagram, by way of example, are two operational envelopes typical of diesel-engine and gas-turbine conditions. At high Da, a flamelet regime is expected with flames in continuous (but possibly highly deformed) surfaces. Lowering Da leads to broken flamelet (premixed) or flame-hole (nonpremixed) regimes. Further decreases in Da lead to distributed reactions (premixed) or flame-extinction (nonpremixed) regimes. The figure includes additional lines representing combustion processes at the integral and Kolmogorov scales of the turbulence. These are useful to differentiate between mildly and strongly corrugated flame conditions.

Numerical simulation of turbulent combustion. Turbulence and combustion are multiscale phenomena, governed by coupled dynamics spanning a large combined parameter space. Only under the most drastic assumptions and simplifications is it possible to solve the underlying equations governing these flows without modeling assumptions, by means of a first-principles-based approach termed direct

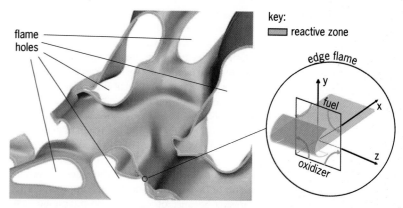

Fig. 1. Flame holes and edge flames from a three-dimensional simulation of a lifted jet-diffusion flame. The inset on the right depicts the local structure of an edge flame at the boundary separating burning from quenched regions. (*Created from data produced for C. Pantano, Direct simulation of non-premixed flame extinction in a methane-air jet with reduced chemistry, J. Fluid Mech., 514:231–270, 2004, DOI:10.1017/S0022112004000266*)

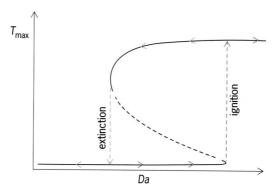

Fig. 2. Flame ignition-extinction S-curve.

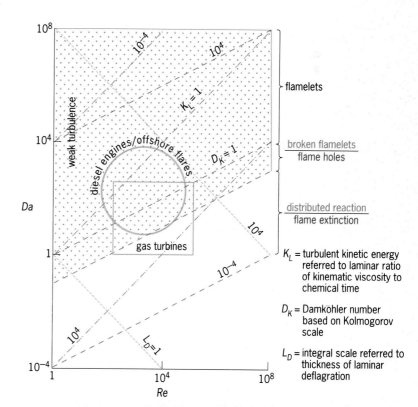

K_L = turbulent kinetic energy referred to laminar ratio of kinematic viscosity to chemical time

D_K = Damköhler number based on Kolmogorov scale

L_D = integral scale referred to thickness of laminar deflagration

Fig. 3. Turbulent-combustion regimes. (*Adapted from F. A. Williams, Turbulent combustion, in J. D. Buckmaster, ed., The Mathematics of Combustion, SIAM, Philadelphia, PA, pp. 97–131, 1985*)

numerical simulation (DNS). If we were to perform a numerical simulation of turbulent flow, a doubling of the linear spatial resolution would incur an increase in computational effort of a factor of $2^4 = 16$; three factors of 2 for space and one for time. In terms of Re, the computational effort increases as Re^3. Fully resolving the range of spatial scales in three-dimensional, unsteady flow for our air duct example will remain beyond the reach of even high-performance computing (HPC) resources for some time to come, even for idealized, uniform-density flows. In engineering devices, the value of Re and possibly of Da are large, and the DNS approach is not promising because of the high computational cost. The preferred alternative is multiscale modeling called large-eddy simulation (LES). Here, the large scales of the flow are resolved, along with geometrical features of the flow device, while smaller unresolved scales are modeled. Subgrid-scale (SGS) models that exploit multiscale physics and similarity must capture the dynamics at flow and combustion scales unresolvable by LES. This approach yields unsteady and three-dimensional predictions of the most energetic structures of the flow. The development of SGS models for turbulence and combustion is an active research area. While the most-promising approach today, a challenge in the design of such models for combustion is that some rate-limiting processes that control SGS contributions may prove not to be describable in terms of resolved scales.

For background information *see* CHEMICAL KINETICS; COMBUSTION; FLAME; GAS DYNAMICS; REYNOLDS NUMBER; TURBULENT FLOW in the McGraw-Hill Encyclopedia of Science & Technology.

Paul E. Dimotakis; Carlos Pantano

Bibliography. J. D. Buckmaster, Edge-flames, *Prog. Energ. Combust. Sci.*, 28:435-475, 2002, DOI:10.1016/S0360-1285(02)00008-4; J. H. Chen, Petascale direct numerical simulation of turbulent combustion—fundamental insights towards predictive models, *Proc. Combust. Inst.*, 33:99-123, 2011, DOI:10.1016/j.proci.2010.09.012; P. E. Dimotakis, Turbulent mixing, *Ann. Rev. Fluid Mech.*, 37:329-356, 2005, DOI:10.1146/annurev.fluid.36.050802.122015; P. Moin and K. Mahesh, Direct numerical simulation: A tool in turbulence research, *Ann. Rev. Fluid Mech.*, 30:539-578, 1998, DOI:10.1146/annurev.fluid.30.1.539; C. Pantano, Direct simulation of non-premixed flame extinction in a methane-air jet with reduced chemistry, *J. Fluid Mech.*, 514:231-270, 2004, DOI:10.1017/S0022112004000266; J. M. Simmie, Detailed chemical kinetic models for the combustion of hydrocarbon fuels, *Prog. Energ. Combust. Sci.*, 29:599-634, 2003, DOI:10.1016/S0360-1285(03)00060-1; L. Vervisch and T. Poinsot, Direct numerical simulation of non-premixed turbulent flames, *Ann. Rev. Fluid Mech.*, 30:655-691, 1998, DOI:10.1146/annurev.fluid.30.1.655; F. A. Williams, Progress in knowledge of flamelet structure and extinction, *Prog. Energ. Combust. Sci.*, 26:657-682, 2000, DOI:10.1016/S0360-1285(00)00012-5; F. A. Williams, Turbulent combustion, in J. D. Buckmaster (ed.), *The Mathematics of Combustion*, SIAM, Philadelphia, PA, pp. 97-131, 1985.

2012 transit of Venus

One of the rarest of predictable astronomical events, a transit of Venus across the face of the Sun, was observed by tens of millions of people around the world on June 5 and 6, 2012. Only six previous times had transits of Venus been observed from Earth: in 1639, when Jeremiah Horrocks and one correspondent saw it; in 1761 and 1769, when Captain James Cook and hundreds of others went to the ends of the Earth, so to speak, to view the event in order to measure the size and scale of the solar system; in 1874 and 1882, when newfangled photography could be used for the first time; and in 2004, the first such transit of our millennium (see **table**). Transits of Venus from Earth occur in our era in pairs separated by 8 years, with then alternate gaps of 105.5 or 121.5 years before the beginning of the next pair.

The 2004 transit was a novelty, because nobody alive on Earth had seen a transit of Venus. Stories abounded about the explorers of the past. By the time of that transit, Glenn Schneider (Steward Observatory of the University of Arizona) and the author, Jay Pasachoff, had solved the problem of the black-drop effect, an apparent linkage between the sky and Venus's black silhouette even when it was completely in front of the solar disk. They had shown, from spacecraft observations of a transit of Mercury, that the black-drop effect did not come from Venus's atmosphere, after all, in spite of what many sources attested; rather, it had to do not only with the optical limitations of telescopes but also with the Sun's limb darkening, its growing less bright from the center of its disk to its edge, something that is extreme at the very edge.

At the 2004 transit of Venus, however, as photographed from a NASA imaging spacecraft, it became clear that Venus's atmosphere could be seen very well; it had been visually detected as Venus crossed the solar limb by nineteenth-century observers, though not all eighteenth-century reports were actually of the atmosphere. It appeared about halfway through Venus's entry onto the solar disk as an arc on the edge of Venus that was outside the Sun,

Transits of Venus, 1631–2255
December 7, 1631
December 4, 1639
June 6, 1761
June 3–4, 1769
December 9, 1874
December 6, 1882
June 8, 2004
June 5–6, 2012
December 10–11, 2117
December 8, 2125
June 11, 2247
June 9, 2255

as it bent (refracted) sunlight toward us on Earth. Similarly, it became visible as Venus started to leave the Sun's disk, and remained visible for over 10 min, more than half the time required for that egress. As a result, scientists around the world and a variety of spacecraft prepared to concentrate on studies of Venus's atmosphere at the 2012 transit.

Ground-based observations. The author's group of scientists and students from Williams College and elsewhere observed the 2012 transit of Venus from the top of the Haleakala volcano on Maui, a Hawaiian island, which is considered the best solar observing site in the world. Haleakala has a science city just outside the Haleakala National Park; several telescopes are located there, and it is the site of the next big solar telescope, the National Science Foundation's Advanced Technology Solar Telescope, hoped for in about 2018. In the United States, only from Hawaii and Alaska could all the more than 6 h of transit be seen. In the continental United States, the beginning of the transit could be seen in the afternoon, and then the Sun set with Venus still silhouetted on it. At Haleakala, during the transit, the sky was exceptionally clear, and the Williams College Expedition obtained hundreds of thousands of images to assemble in their computers in the months to come (**Fig. 1**).

At Haleakala, the Williams College Expedition had one of nine identical coronagraphs—devices that could block out the everyday, bright solar surface to help faint things outside the solar limb become visible. These coronagraphs, designed and built by French astronomers Thomas Widemann and Paolo Tanga, enabled our viewing of the light bent around Venus's edge, giving us views of what twilight on Venus was like. The other coronagraphs were in Kazakhstan, in Japan, in India, at the Lowell Observatory in Arizona, in Svalbard in the Arctic, and in Australia and the Marquesas Islands at southern latitudes. This Venus twilight experiment was aimed toward using the differences in the atmosphere's appearance in different colors to detect the presence of Venus's atmosphere and to compare the results from Earth's surface with those simultaneously obtained by the European Space Agency's *Venus Express* spacecraft, which is in orbit around Venus. The hope is to learn more about Venus's very rapid winds.

At the Sacramento Peak Observatory of the National Solar Observatory's site in Sunspot, New Mexico, the giant Richard B. Dunn Solar Telescope towered into the sky, where mirrors mounted high in a tower reflected the images of the Sun and Venus into other mirrors even deeper underground than the first mirrors are high. The telescope acts like a 55,000-mm telephoto lens, compared with a 50-mm "normal" lens on an ordinary camera, providing apparently close-up observations at large scale (**Fig. 2**). The Interferometric Bidimensional Spectrometer (IBIS) recorded the transit through a filter that passed the wavelengths absorbed by the carbon dioxide gas that makes up most of Venus's atmosphere.

At the Big Bear Solar Observatory in California, their New Solar Telescope, which uses thermal

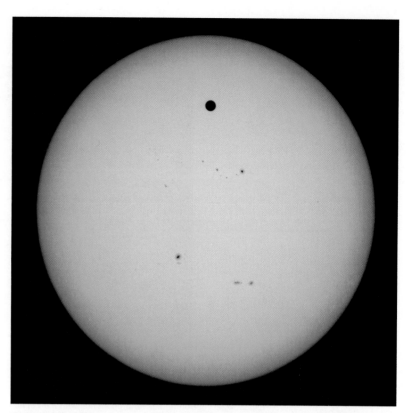

Fig. 1. Venus in transit across the face of the Sun on June 5, 2012, imaged from the Haleakala Observatory on Maui, Hawaii. (*Courtesy of Ron Dantowitz/Clay Center Observatory/Dexter-Southfield Schools and Jay M. Pasachoff/Williams College, as part of the Williams College Expedition, sponsored by the Committee for Research and Exploration of the National Geographic Society*)

control to limit blurring of the images by motions in the air that are heated by the solar beam and adaptive optics, recorded a remarkable series of images, though only for a time span that lasted through first and second contacts and ended partway through Venus's path across the solar disk, when the telescope could not be pointed any lower.

On Mt. Wilson in California, observers brought several antique telescopes from hundreds of years before to see if the instruments were capable of observing Venus's atmosphere. That Cytherean atmosphere was not visible, probably because of turbulent air or the telescopes' capabilities.

NASA's Sun-Earth Day for 2012, organized by scientists at NASA's Goddard Space Flight Center in Greenbelt, Maryland, was devoted to the transit of Venus. They helped organize and keep track of observing parties across the country and around the world. Their final records show tens of millions of "hits" on their websites, which included a live broadcast from Mauna Kea Observatory in Hawaii, and millions of people who individually saw the transit themselves.

Observations from space. NASA and the Japan Aerospace Exploration Agency had several spacecraft that observed the transit, although the transit was not visible from the European Space Agency's *Solar and Heliospheric Observatory*, which is located 1.5 million kilometers (930,000 mi) toward the Sun from us. NASA's most spectacular solar observatory now is the *Solar Dynamics Observatory*

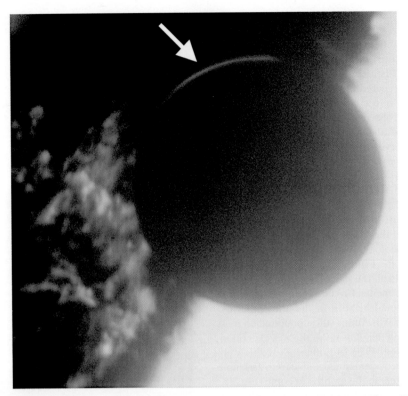

Fig. 2. Venus in front of a solar prominence with a bit of its arc of atmosphere showing, in an image taken with the 55,000-mm-focal-length Richard B. Dunn Solar Telescope of the National Solar Observatory's Sacramento Peak Observatory in Sunspot, New Mexico. (*Courtesy of Kevin Reardon and Glenn Schneider, working with Jay M. Pasachoff and Bryce A. Babcock, as part of the Williams College Expedition, sponsored by the Committee for Research and Exploration of the National Geographic Society*)

(*SDO*), which is in geosynchronous orbit in range of a ground station in New Mexico so it can send back massive amounts of data. One instrument runs through eight filters imaging the entire Sun every 12 s, day in and day out, with only minor exceptions during a year. Another instrument makes 24 images every 25 s.

The SDO's Atmospheric Imaging Assembly (AIA) sent back images in a variety of ultraviolet and visible-light filters (**Fig. 3**). It has pixels about half an arcsec-

Fig. 3. Venus in transit, imaged from the Atmospheric Imaging Assembly on NASA's *Solar Dynamics Observatory*, through a filter that passed only light given off by gases of millions of kelvins in the solar corona. (*Courtesy of LMSAL/NASA/SAO*)

ond across, about 430 km (270 mi) as projected on the solar surface and corresponding to about 120 pixels across Venus's diameter. The SDO's Helioseismic Magnetic Imager (HMI) has pixels slightly smaller than AIA's; it operates in visible light and also sent down images throughout the transit (**Fig. 4**). The Williams College Expedition had worked with the HMI principal investigator to optimize the observations at the times of ingress and egress to show Venus's atmosphere outside the solar disk refracting sunlight toward us.

The Japanese Hinode mission carries a Solar Optical Telescope (SOT), with NASA participation. Its pixels are only 0.05 arcsecond across, more than 10 times as fine as those on the *SDO;* at third contact the full sampling was used, while at second contact the pixels were binned 2 × 2, making the resultant superpixel 0.1 arcsecond across. The SOT's images of Venus's atmosphere outside the solar limb are, therefore, the best ever obtained (**Fig. 5**).

Hinode also carries an X-ray Telescope (XRT) built by the Smithsonian Astrophysical Observatory. Although its downloaded animation is not continuous, with a series of small regions of the Sun shown, each containing Venus's silhouette, it provided a series of images showing Venus silhouetted against various parts of the solar corona on and off the solar disk (**Fig. 6**).

Two NASA spacecraft monitor the total solar irradiance (TSI), a term that used to be called the solar constant until it turned out to be varying. The TSI is the total energy received from the Sun per unit area per second if the Earth were 1 astronomical unit from the Sun. Because Venus's angular diameter is about 1/30 the angular diameter of the Sun, it covers an area about 0.1% that of the Sun's disk. Both the Active Cavity Radiometer Irradiance Measurement instrument, in its third version, aboard the *ACRIMsat*, whose principal investigator is based in Coronado, California, and the Total Irradiance Measurement (TIM) instrument aboard the *Solar Radiation and Climate Experiment* (*SORCE*) spacecraft, controlled from the University of Colorado, recorded that 0.1% dip in the TSI. Confirming earlier measurements of the 2004 transit dip by ourselves with *ACRIMsat* and other scientists with *SORCE*/TIM, they were also able to monitor the gradual rather than abrupt dip of that 0.1%, as well as the effect resulting from the fact that the Sun's surface is less bright the closer to its edge you look, the limb darkening. Dips even greater than that have been reported when huge sunspots appear on the solar disk; several sunspots were visible on the disk during the 2012 transit, a difference from the spotless look of the disk at the 2004 transit.

Study of the 2012 transit of Venus was particularly interesting because NASA's *Kepler* spacecraft and the French *CoRoT* spacecraft as well as telescopes on Earth have discovered thousands of exoplanet candidates through the dips in starlight that occur when the exoplanet candidates go into transit. However, perhaps only 90% of these candidates are really exoplanets. Some complications of the observations, largely the possible presence of a background star in the field of view, can mimic the transit of an exoplanet. So *Kepler* finds thousands of exoplanet candidates, and most of them are really exoplanets, but some small percentage are not, and—at the level of faintness that the *Kepler* spacecraft observes—it may never be able to tell which of the candidates are really exoplanets and which are not. Certainly it behooves us to study the transit of Venus, which occurs with explicitly known circumstances and which can be observed with spatial detail, in order to understand the transits of exoplanets, which are perceived with no spatial detail as mere dips in the light received. Further, the 2012 transit of Venus was observed through a variety of filters, both broad-band and narrow-band at, for example, the wavelength of carbon dioxide, which is known to be present at high concentrations in Venus's atmosphere. Much effort is going into discovering atmospheres around exoplanets, in a search for habitable planets around other stars, however far away, and testing the methods on the transit of Venus within our own solar system could prove invaluable for convincing ourselves that we really have discovered a habitable planet around another star. *See* KEPLER MISSION.

Future transits. Surely, when the next pair of transits of Venus are visible from Earth in 2117 and 2125, instruments will have advanced so in sophistication and power that our observations of 2004 and 2012 will seem puny, but we had to try to do the best we could to provide baseline information to the

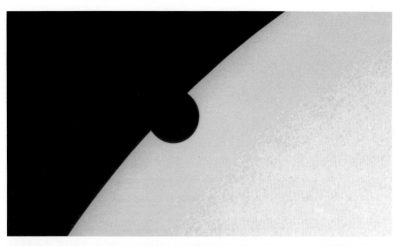

Fig. 4. Venus in transit, imaged from the Helioseismic Magnetic Imager on NASA's *Solar Dynamics Observatory* in visible light, with some of Venus's atmosphere showing in the part of the image above the solar limb. (*Courtesy of Stanford Solar Observatory/NASA, Philip Scherrer and Jesper Schou*)

scientists of the future. We will also be able to observe several transits of Mercury in that interval, the next one on May 9, 2016, but Mercury has sufficient atmosphere for us to detect its aureole at ingress and egress.

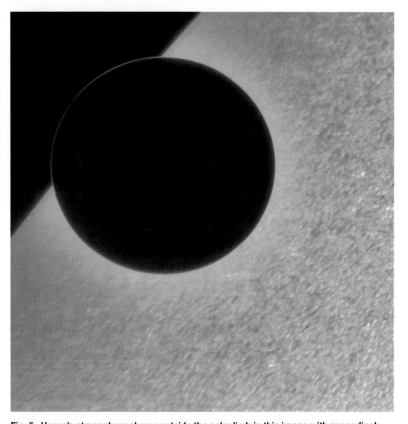

Fig. 5. Venus's atmosphere shows outside the solar limb in this image with exceedingly high spatial resolution made with the Solar Optical Telescope (SOT) on the Japanese Space Exploration Agency's *Hinode* (*Sunrise*) spacecraft. The yellowish atmosphere of Venus may result from differential scattering in Venus's atmosphere as it refracts sunlight toward the spacecraft, which is in Earth orbit. The bluish limb of Venus on the solar disk and the yellowish limb of the Sun result from the fact that three separate exposures at slightly different times were taken through different filters and later assembled into this full-color image. (*Courtesy of Japan Aerospace Exploration Agency/National Astronomical Observatory of Japan/NASA/Lockheed Martin Solar and Astrophysics Laboratory; in collaboration with Jay Pasachoff, Williams College, Alphonse Sterling, NASA's Marshall Space Flight Center, and Kevin Reardon, National Solar Observatory*)

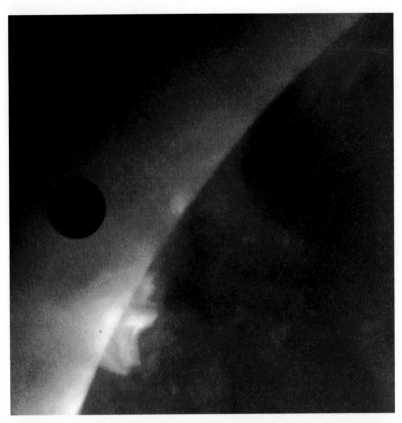

Fig. 6. Venus silhouetted against the solar corona seen in x-rays from the Smithsonian Astrophysical Observatory's XRT on the JAXA *Hinode* (*Sunrise*) spacecraft. (*Courtesy of Leon Golub, Patrick McCauley, and Kathy Reeves, Harvard-Smithsonian Center for Astrophysics*)

NASA's *Cassini* spacecraft is now orbiting in the Saturn system, in place to observe the nearly 10-hour transit of Venus that will be visible from there on December 21, 2012. An international group from the United States and France, including astronomers from Williams College, the University of Arizona, and Cornell University, has an agreement to observe this transit with the VIMS spectrograph on *Cassini*, with a choice of filters meant to try to detect an effect from Venus's atmosphere. Several other opportunities to observe transits from the outer planets will also arise before the next transit of Venus from Earth rolls around in 2117. On September 20, 2012, the Pasachoff-Schneider group with colleagues from France obtained 22 h of observations of Jupiter with the *Hubble Space Telescope* in hope of detecting a tiny drop in its intensity and a slight spectral blue-red spectral difference from Venus's atmosphere during a transit that was visible only from that outer planet; data reduction is under way to see if the tiny effect was detectable, as an analog of exoplanet-transit observations. Perhaps by 2117, people will have been on other planets making their own observations of planets interior to their location in our solar system.

For background information *see* CORONAGRAPH; EXTRASOLAR PLANETS; MERCURY; SOLAR CONSTANT; SPACE PROBE; SUN; TELESCOPE; TRANSIT (ASTRONOMY); VENUS in the McGraw-Hill Encyclopedia of Science & Technology.   Jay M. Pasachoff

Bibliography. N. Lomb, *Transit of Venus: 1631 to the Present*, New South Publishing, Sydney, and The Experiment Publishing, New York, 2011; J. M. Pasachoff, G. Schneider, and L. Golub, The black-drop effect explained, in D. W. Kurtz (ed.), *Transits of Venus: New Views of the Solar System and Galaxy*, Cambridge University Press, Cambridge, U.K., pp. 242–253, 2005; J. M. Pasachoff, G. Schneider, and T. Widemann, High-resolution satellite imaging of the 2004 transit of Venus and asymmetries in the Cytherean atmosphere, *Astron. J.*, 141:112–120, 2011, DOI:10.1088/0004-6256/141/4/112; J. M. Pasachoff, Transit of Venus: Last chance from Earth until 2117, *Phys. World*, 25(5):36–41, 2012; J. M. Pasachoff and W. Sheehan, Lomonosov, the discovery of Venus's atmosphere, and eighteenth-century transits of Venus, *J. Astron. History Heritage*, 15(1):3–14, 2012; P. Tanga et al., Sunlight refraction in the mesosphere of Venus during the transit on June 8th, 2004, *Icarus*, 218:207–219, 2012; J. Westfall and W. Sheehan, *Celestial Shadows: Eclipses, Transits, and Occultations*. Springer, New York, 2013.

Ultracold molecules

One of the long-standing goals in chemistry is to control and manipulate the outcome of chemical reactions to yield desired products. While some progress has been achieved in controlling product branching in photofragmentation of diatomic and triatomic molecules, the absolute control of bimolecular chemical reaction dynamics, even in simple atom-diatom exchange reactions (such as $A + BC \rightarrow AB + C$ or $AC + B$, where BC is the reactant molecule and AB and AC are possible product molecules), has yet to be realized. Precise control of molecular encounters and chemical reactions requires initial preparation of molecules in well-defined internal quantum states. However, at ordinary temperatures, molecules exist in a thermal population of internal quantum states, corresponding to different vibrational, rotational, and hyperfine levels. In typical diatomic molecules such as N_2 or O_2, the energy spacing between vibrational levels is on the order of a thousand kelvins (in this article, energy will be given as E/k_B expressed in kelvins, where E is the kinetic energy and k_B is the Boltzmann constant), the spacing between rotational energy levels is on the order of a few kelvins, and the spacing between hyperfine levels is a small fraction of a kelvin. The energy level separation becomes smaller for heavier molecules. To prepare molecules in specific internal quantum states, their translational temperature T (kinetic energy) must be reduced to significantly below 1 K so that the thermal energy k_BT is smaller than the tiniest energy separation between internal quantum states. Thus, to enable control of reaction dynamics, molecules must be cooled to temperatures lower than a millikelvin or microkelvin, depending on their mass and energy-level structure.

Making of cold and ultracold molecules. During the past 10–15 years, tremendous progress has been achieved in creating translationally cold molecules in well-defined quantum states. The efforts began with the successful creation of Bose-Einstein condensates (BECs) of alkali metal atoms. Atomic BECs are formed when atoms with integer nuclear spins (bosons) are cooled to temperatures close to a billionth of a degree above absolute zero, at which they occupy the lowest quantum state (energy level). Since 2000, dramatic progress has been achieved in creating cold (loosely defined as translational temperatures between 1 K and 1 mK) and ultracold (translational temperatures lower than 1 mK) molecules. The techniques employed in creating cold and ultracold molecules involve cooling preexisting molecules as well as creating new molecules either by photoassociating or magnetoassociating a pair of ultracold atoms, for example, from a BEC. Cooling preexisting molecules is achieved either by elastic collisions (collisions that exchange kinetic energy but do not lead to a change in internal quantum states of molecules) with cold helium buffer gas atoms, or by subjecting them to a time-varying electric field, a technique referred to as Stark deceleration. Both techniques are capable of producing molecules in the lowest vibrational, rotational, and hyperfine states but with kinetic energies in the cold regime. Methods based on photoassociation and magnetoassociation (also referred to as the Feshbach resonance method) can yield ultracold molecules with translational energies below the microkelvin regime. In photoassociation, light of appropriate wavelength is used to bind a pair of atoms to their constituent molecule while in magnetoassociation a magnetic field is used to fuse atoms into molecules. A drawback of these techniques is that the molecules are produced primarily in highly vibrationally excited levels, but more recent advances have allowed efficient transfer of these molecules to their absolute ground state. Such molecules are stable against collisional relaxation and amenable to long observation times. **Figure 1** shows a schematic illustration of the temperature regimes in which cold and ultracold molecules are formed and includes temperatures in other environments that prevail in nature.

Ultracold chemistry. One of the fascinating questions that has arisen in the last decade after successful creation of a variety of ultracold molecules is whether they can be used to study chemistry in the ultimate quantum limit where a full control of internal and translational degrees of freedom of the molecule is possible. Intriguing questions (for example, "Do chemical reactions occur at temperatures vanishingly above absolute zero?" and "How does quantum tunneling influence chemical reactivity at ultracold temperatures?") began to receive considerable experimental and theoretical attention. Over the last several years, experimentalists and theorists have addressed different aspects of these questions with surprising findings that led to the development of a new area of research called ultracold chemistry. Because of the small kinetic energies involved, only exoergic chemical reactions are possible in ultracold collisions. Many exoergic reactions involve an energy barrier (activation energy) as the

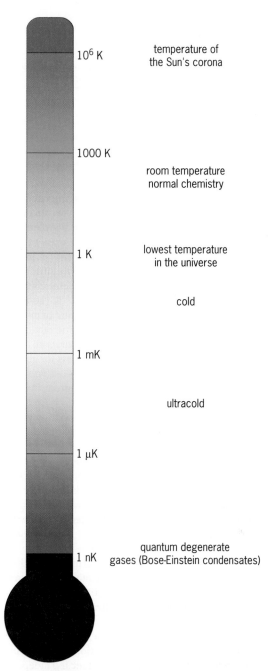

Fig. 1. Schematic illustration of the temperature regimes (in a logarithmic scale) corresponding to the formation of cold and ultracold molecules compared to temperatures in normal and other extreme environments that exist nature.

reactants approach each other. They proceed almost exclusively through quantum tunneling at low energies. The $F + H_2 \rightarrow HF + H$ reaction has long served as the benchmark for tunneling-dominated reactions. Theoretical calculations indicated that the reaction occurs efficiently at ultracold temperatures despite an energy barrier of about 400 K. What differentiates "ultracold" chemistry from "warm" or "hot" chemistry that occurs at ordinary and elevated temperatures is not the "ultracold" aspect of it but the large de Broglie waves associated with collisions at low temperatures. In typical "warm" chemical reactions, as coined by Dudley Herschbach, the range of interaction is about 1 nm, roughly a factor

Thermal de Broglie wavelength of molecular oxygen as a function of the temperature	
Temperature, K	Thermal de Broglie wavelength, nm
100	0.031
1.0	0.31
10^{-2}	3.1
10^{-4}	31
10^{-6}	310
10^{-8}	3100

of 5–10 larger than the equilibrium internuclear separation in typical diatomic molecules. In ultracold collisions, the thermal de Broglie wavelength, $\lambda_{DB} = h/(2\pi m k_B T)^{1/2}$, where m is the mass of the molecule and h is the Planck's constant, is much larger than the interparticle separation. As illustrated for molecular oxygen in the **table**, λ_{DB} increases from 0.031 nm at 100 K to 3100 nm at 10^{-8} K (10 nK). This leads to a completely different picture of molecular encounters in which the range of interaction is much larger than the interparticle separation. As a result, cross sections and rate coefficients for chemical reactions at cold and ultracold temperatures display dramatic quantum effects and quantum threshold phenomena.

Features of ultracold reactions. In comparing ultracold collisions with thermal-energy collisions, it is important to recognize that the distinction is primarily made in how the reactants are prepared. Ultracold collisions are characterized by reactants in specific internal quantum states with a tiny kinetic energy and a large λ_{DB}. Only the lowest allowed orbital angular momentum partial wave (denoted by quantum number l) contributes to the overall cross section ($l = 0$ for s-wave collisions and $l = 1$ for p-wave collisions). In warm collisions, a thermal population of internal quantum states and a range of l values contribute to the overall reaction cross section. For s-wave scattering in the ultracold regime, the inelastic and reactive cross sections vary inversely as the incident velocity. Consequently, the reaction rate coefficients, given by cross section multiplied by the incident velocity becomes constant and independent of the temperature as $T \to 0$ K. This is the limiting case for collisions involving identical bosons. However, for collisions involving identical fermionic species (half-integer spins), the lowest allowed value of l is equal to 1 and the associated centrifugal barrier will suppress reactivity as $T \to 0$ K. In this case, reaction occurs by tunneling through the centrifugal barrier.

The importance of quantum statistics (bosonic and fermionic characteristics) as well as the effect of centrifugal energy barriers in ultracold collisions has been elegantly demonstrated in a landmark experiment by S. Ospelkaus and her colleagues. They prepared ultracold $^{40}K^{87}Rb$ molecules in their absolute ground (vibrational, rotational, and hyperfine) state at a translational temperature of a few hundred nanokelvins. The $^{40}K^{87}Rb$ molecules are composite fermions and, in collisions of two identical fermions

(prepared in the same quantum state), the lowest allowed angular momentum partial wave is a p-wave. A chemical reaction leading to $K_2 + Rb_2$ products occurs via tunneling through the p-wave barrier at collision energies of a few hundred nanokelvins. When two KRb molecules in different initial hyperfine levels are brought together, an s-wave collision is allowed and no angular momentum barrier is present for the reaction. **Figure 2** schematically illustrates the presence and absence of a centrifugal barrier in p-wave and s-wave collisions. The absence of a centrifugal barrier in s-wave collisions led to an order-of-magnitude increase in reactivity, measured by the depletion rate of the KRb molecules. The same occurs in collisions of K and KRb whereas no reaction occurs in Rb + KRb, collisions for which no exothermic reaction channel exists at the temperatures of the experiments. The experiment is a direct demonstration that a chemical reaction may occur efficiently at ultracold temperatures and that its rate

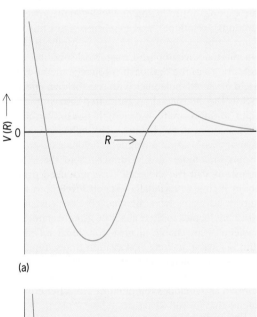

(a)

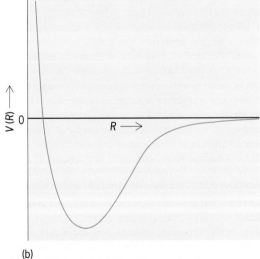

(b)

Fig. 2. Schematic illustrations of (a) a p-wave barrier in ultracold collisions, and (b) a barrierless reaction involving s-wave collisions. The intermolecular distance R and the interaction potential $V(R)$ are in arbitrary units.

can be controlled by merely flipping the spin of one of the reactant molecules.

Applications and future prospects. Ultracold molecules allow sensitive studies of molecular interactions and chemical reactivity with unprecedented resolution. Since the thermal energy k_BT is much smaller than the splitting between energy levels, including hyperfine levels, the reaction outcome can be controlled by applying an external electric or magnetic field, or a combination thereof, which can modify the splitting between hyperfine levels. Also, as demonstrated in the KRb experiment, merely flipping the spin of one of the reactant molecules can control the outcome of a chemical reaction. An array of ultracold polar molecules in a one-dimensional or two-dimensional lattice has been proposed as a model for quibits in quantum computing devices. In this case, the molecules interact through dipole-dipole interactions that in turn can be modified by applying an external electric field.

Currently, the variety of molecules that can be cooled and trapped to quantum degeneracy is restricted to a few alkali metal atom systems and their dimers. BECs and Fermi degenerate gases of several other nonalkali metal atoms, such as strontium (Sr), ytterbium (Yb), chromium (Cr), and dysprosium (Dy), have also been attained in recent years. Thus, it is likely that many more varieties of ultracold molecules will become available in the coming years and ultracold reactions of nonalkali metal atom systems will be experimentally realizable. Further progress in buffer gas cooling and Stark decelerator methods may also allow the cooling and trapping of a wider variety of molecules, including polyatomic molecules. Already molecules such as NH, OH, NH_3, and ND_3 have been cooled to millikelvin temperatures by these techniques, and it is expected that further reduction in translational temperatures and higher densities of these trapped molecules will be achievable in the near future. An experimental challenge that remains to be tackled is to detect and analyze products of ultracold chemical reactions. The prospects of ultracold molecules and ultracold chemistry appear to be bright.

For background information *see* ATOMIC FERMI GAS; BOSE-EINSTEIN CONDENSATION; DE BROGLIE WAVELENGTH; KINETIC THEORY OF MATTER; MOLECULAR STRUCTURE AND SPECTRA; QUANTUM COMPUTATION; QUANTUM MECHANICS; QUANTUM STATISTICS; RESONANCE (QUANTUM MECHANICS); SCATTERING EXPERIMENTS (ATOMS AND MOLECULES) in the McGraw-Hill Encyclopedia of Science & Technology.

This work is supported in part by NSF grant No. PHY-0855470.   Balakrishnan Naduvalath

Bibliography. N. Balakrishnan and A. Dalgarno, Chemistry at ultracold temperatures, *Chem. Phys. Lett.*, 341:652–656, 2001, DOI:10.1016/S0009-2614(01)0515-2; L. D. Carr et al., Cold and ultracold molecules: Science, technology and applications, *New J. Phys.*, 11:055049 (87 pp.), 2009, DOI:10.1088/1367-2630/11/5/055049; D. Herschbach, Molecular collisions, from warm to ultracold, *Faraday Discussions*, 142:9–23, 2009, DOI:10.1039/b910118g; R. Krems, W. C. Stwalley, and B. Friedrich (eds.), *Cold Molecules: Theory, Experiment, Applications*, CRC Press, Boca Raton, FL, 2009; S. Ospelkaus et al., Quantum-state controlled chemical reactions of ultracold potassium-rubidium molecules, *Science*, 327:853–857, 2010, DOI:10.1126/science.1184121.

Ultralow-density materials: from aerogels to microlattices

Over the past several decades, the need to increase the fuel efficiency of automotive and aerospace vehicles has driven the development of new lightweight materials. Among other innovations, this need has led to the emergence of a variety of new low-density porous materials. Whereas solid materials are limited to densities greater than approximately 1 g/cm^3, materials that include significant porosity, termed porous or cellular materials, can exhibit one-tenth that density, or even lower. Examples include manufactured cellular materials (for example, polymer foams and honeycombs) and naturally occurring cellular materials (for example, wood). Cellular materials are commonly used in applications that take advantage of their low density and unique properties. In addition to vehicle applications, cellular materials are commonly used for packaging materials, padding and other sporting goods, building materials, and acoustic baffling. Most cellular materials fall in the density range of 0.01–0.3 g/cm^3, and these materials are generally considered to be technologically mature. Materials with densities below 0.01 g/cm^3 (10 mg/cm^3), termed ultralow-density materials, are rare primarily because of the difficulties associated with forming and maintaining a stable material. Ultralow-density materials hold considerable promise to further improve the acoustic, thermal, and fuel efficiency characteristics of future vehicle systems, and as such are the focus of current research.

Influence of cellular architecture on properties. The properties of a cellular material are defined by the properties of the solid constituent phase, the volume fraction of the solid phase within the cellular material (relative density), and the cellular architecture. The cellular architecture refers to the spatial configuration of the solid constituent phase within the cellular material. Reduction of the density of a cellular material can be achieved by using a lower-density solid constituent or increasing the amount of porosity. However, because of the lower density limitation on the solid constituent (about 1 g/cm^3), increasing the porosity of a material is required to achieve the ultralow-density regime. As the porosity in a cellular material is increased (that is, the overall density is decreased), the influence of the cellular architecture on the mechanical properties also increases.

Until recently, all existing ultralow-density materials were constructed with a random (stochastic) cellular architecture, with mechanical performance

polymer foam

metallic microlattice

Fig. 1. Materials that include significant porosity are termed porous, or cellular, materials. Polymer foam is an example of a cellular material that has a random, or stochastic, cellular architecture, whereas the metallic microlattice is a cellular material that exhibits an ordered cellular architecture. The cellular architecture becomes an increasingly important parameter that influences the material properties of a cellular material as the porosity increases.

controlled by bending-dominated behavior of the solid constituent phase. Bending-dominated behavior refers to bending of the ligaments or cell walls when a mechanical load is applied and transferred through the solid phase of the cellular material. Low-density cellular materials that exhibit bending-dominated behavior have specific properties that are far below those of the bulk constituent. Examples of stochastic, ultralow-density materials include silica aerogels with densities down to 1 mg/cm^3, carbon nanotube forests with a density of 4 mg/cm^3, metallic foams with densities as low as 10 mg/cm^3 and polymer foams reaching 6 mg/cm^3.

In contrast to stochastic architectures, ordered cellular architectures can be designed to suppress the bending-deformation mode of the solid ligaments, thereby encouraging stretch-dominated behavior and increasing the overall mechanical properties of the cellular network. Prior research on the mechanics of cellular materials has shown that the mechanical properties, such as the elastic modulus and strength of a cellular material, are proportional to $\bar{\rho}^n$, where $\bar{\rho}$ is the relative density, or solid volume fraction, and n is a scaling exponent related to the cellular architecture and the loading condition. Generally, an ordered cellular architecture is designed to suppress bending-dominated behavior, which will decrease the scaling exponent. The minimum scaling exponent, $n = 1$, is achieved if the ordered cellular architecture exhibits complete stretch, or axial load transfer through the internal ligaments of the cellular material. This is akin to larger-scale truss architectures, such as seen in the Eiffel tower, that are designed to carry purely tensile or compressive loads in the structural members.

As the relative density $\bar{\rho}$ of a cellular material is decreased to the ultralow-density regime, the influence of the scaling exponent becomes more prominent on the resulting mechanical properties, and hence the stability of the cellular material.

Processing methods and their effect on cellular architecture. Aside from naturally occurring cellular materials, the porosity in a cellular material must generally be introduced into the constituent phase, and as such, the approach used to introduce porosity will ultimately define the cellular architecture. The vast majority of cellular materials are formed using bulk processing techniques that are essentially random in nature (for example, mixing of foaming agents into a resin). These techniques result in a material with a stochastic or random cellular architecture, as exemplified by traditional foams such as those shown in **Fig. 1**.

To create an ultralow-density material, the solid phase of the cellular material must be able to withstand all processing and ambient loading conditions, including gravity. Using the scaling argument described above, decreasing the density of a cellular material is possible by designing an ordered cellular architecture with greater mechanical stability. However, because of the limitations in available manufacturing techniques, only few ordered cellular materials exist. Honeycomb is an example, but it is highly anisotropic and reaches its fabrication limit at around 10 mg/cm^3. Cellular materials with ordered truss or lattice-based architectures have also been developed but generally have been limited to cell sizes greater than 1 cm and densities much greater than 10 mg/cm^3 because of the fabrication approaches used.

Recently, new lattice materials in the ultralow-density regime have been developed and demonstrated; these are termed ultralight metallic microlattices. These materials are fabricated by first making a polymer microlattice template, coating the template with a thin layer of nickel using electroless plating, and subsequently removing the polymer template. The density of the resulting metallic microlattice is defined by the nickel coating thickness and the cellular architecture of the polymer template.

The templating method described above requires a continuous and robust film that can withstand the mechanical forces associated with removal of the polymer template. Coatings as thin as 100 nm are robust and can generally withstand the required etching, handling, and drying processes. Using this method, a metallic microlattice with a density of 0.9 mg/cm^3 has been demonstrated. The density is calculated using the weight of the microlattice as measured in air divided by its volume. This method does not include the weight of the air in the pores, adhering to standard measurement practice for cellular materials. The contribution of the density of air at ambient conditions, 1.2 mg/cm^3, would need to be included to calculate the density of the solid-air composite.

Ultimate density limit for microlattice materials. Although the lowest-density metallic microlattice material exhibits the lowest density for any reported cellular material to date, even lower-density materials could be fabricated by following the same methodology. The density of the synthetic microlattices with an architecture similar to that shown in Fig. 1 can be approximated by the equation below,

$$\bar{\rho} = \frac{8\pi}{\cos^2\theta\ \sin\theta}\left(\frac{D}{L}\right)\left(\frac{t}{L}\right)$$

where $\bar{\rho}$ is the relative density, D is the lattice member diameter, L is distance between nodes, θ is the angle of the lattice member, and t is the thickness of the nickel film. Assuming 100 nm is the thinnest possible nickel film that is stable during processing, the equation can be used to estimate the lowest possible achievable density for a given unit cell size, L, if D/L is held constant. **Figure 2** is a plot of the theoretical density for nickel microlattices based on the equation, along with comparative experimental results.

After processing, the mechanical stability of the microlattice will be based on the buckling strength of the hollow tubes, which is proportional to the elastic modulus of the solid constituent phase (E_s). The density could be pushed lower with a constituent thin film material that exhibits a higher specific stiffness, as well as a cellular architecture that exhibits stretch-dominated behavior (Fig. 2). Diamond may be among the best candidates with its extremely high stiffness and strength and well established vapor deposition routes enabling films with <50-nm thickness. For a diamond microlattice with a 50-nm wall thickness, a fivefold reduction in density should be possible (Fig. 2).

As expected from the equation, the relative density decreases with increasing cell size, or lattice member length L, suggesting that it is possible to attain further reductions in density with even larger cell sizes (**Fig. 3**). With increasing unit cell size, the lattice eventually turns into an entity most observers would describe as a structure. The Eiffel tower is a

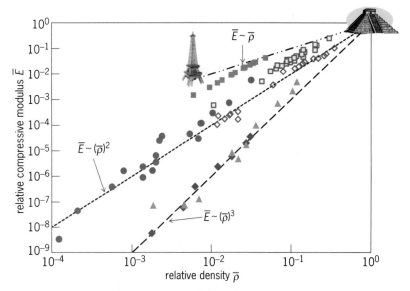

Fig. 3. The relative compression modulus ($\bar{E}$) is the elastic compression modulus of a cellular material normalized by the elastic compression modulus of the solid constituent. Likewise, the relative density ($\bar{\rho}$) is the density of cellular material normalized by the density of the solid. Ancient pyramids have essentially a solid construction, and thus would exhibit a relative compression modulus and relative density equal to 1. The Eiffel tower has a relative density of approximately 6×10^{-3} and a cellular architecture that is designed to exhibit a stretch-dominated behavior in the structural members, which leads to a relative compression modulus that scales proportionally with the relative density. For all known cellular materials in the ultralow-density regime, the relative modulus is proportional to ($\bar{\rho}^n$), where the scaling parameter $n > 1$. These materials include nickel microlattices (●), closed-cell polymer foams (□), open-cell polymer foams (◇), carbon nanotube forests (◆), aluminum honeycomb (■), and silica aerogels (▲) (*Adapted from T. Schaedler et al., Ultralight metallic microlattices, Science, 334:962–965, 2011, DOI:10.1126/science.1211649*)

classic example of a structure that possesses a density comparable to the microlattices, but would never be described as a material.

One methodology to describe the delineation between materials and structures is to require materials to exhibit a minimum shear strength (to exclude fluids and gels), common topological features found throughout an extended volume (as found in crystals, foams and composites), and critical dimensions that are "small" with respect to smallest length scale associated with the constraints put on the lattice, for example, applied loads and boundary conditions. Under these conditions, continuum approximations, such as homogenization techniques, can be used to describe the response of the material in an application. At larger sizes, the lattice is better classified as a structure, where stress and displacement fields vary significantly between unit cells and discrete treatment is necessary. Accordingly, microlattices with a cell size <0.1 m can generally be defined as materials.

At the lower limit, a density of 0.01 mg/cm³ could be achieved with a 50-nm-thick diamond film and a microlattice cell size that still classifies the diamond micro-lattice as a material. The curve representing the theoretical density of diamond microlattices is faded out at this density, which might constitute the general density limit for any material.

For background information *see* DENSITY; DIAMOND THIN FILMS; FOAM; GEL; NICKEL; POLYMER in the McGraw-Hill Encyclopedia of Science & Technology.

A. J. Jacobsen; T. A. Schaedler; W. B. Carter

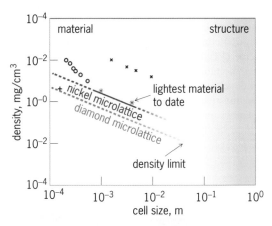

Fig. 2. Limits of ultralow-density microlattices. Calculated density of nickel and diamond microlattices with constant aspect ratio $L/D = 8$ and wall thickness of 100 nm and 50 nm, respectively, as a function of cell size compared to measured densities of nickel microlattices (*), aluminum honeycombs, polyurethane foams (o) and carbon nanotube aerogels (+). Lattices with cell sizes >0.1 m are classified as structures instead of materials.

Bibliography. M. A. Aegerter, N. Leventis, and M. M. Matthias (eds.), *Aerogels Handbook*, Springer, 2011; L. J. Gibson and M. F. Ashby, *Cellular Solids: Structure and Properties*, 2d ed., Cambridge University Press, Cambridge, U.K., 1997; R. S. Lakes, Materials with structural hierarchy, *Nature*, 361:511–515, 1993, DOI:10.1038/361511a0; M. Mecklenburg et al., Aerographite: Ultra lightweight, flexible nanowall, carbon microtube material with outstanding mechanical performance, *Adv. Mater.*, 24:3486–3490, 2012, DOI:10.1002/adma. 201200491; T. A. Schaedler et al., Ultralight metallic microlattices, *Science*, 334:962–965, 2011, DOI:10. 1126/science.1211649.

Vascular development

Blood vessels assume a critical function in organisms too large in size to sufficiently supply their tissues with oxygen and nutrients and to remove toxic waste products by diffusion alone. Therefore, during the embryonic development of a vertebrate, the vascular system (that is, the entirety of all blood vessels) develops first and is a prerequisite for all other organ systems. Genetically engineered mice that are unable to form a functional vascular system die at embryonic (developmental) day E10 without formation of functional organs.

However, vessels do not grow only during ontogenesis (the developmental history of an organism from its origin to maturity); they also grow in adult life. One example is the monthly regeneration of the arteries and veins of the growing uterine lining during the female menstrual cycle. Moreover, a number of acute and chronic stress situations stimulate the organism to, in regenerative attempts, either grow out existing blood vessels or form new ones to ensure adequate oxygenation and nutrient supply in situations where existent blood vessels have become insufficient to do so.

Cancer and vascular occlusive disease are presently major causes of death and disability worldwide. Their pathogenesis critically involves deregulated or inadequate vascular growth and development. Therefore, therapeutic induction, modulation, and termination of vascular growth are major topics in modern biomedical science and therapeutic research. To develop strategies influencing blood vessel development, a thorough understanding of the mechanisms and forms of vascular growth is critical.

In general, three forms of vascular development exist: vasculogenesis, angiogenesis, and arteriogenesis. Each form is characterized by a typical set of stimuli that induce its onset and progression. It remains unclear, however, if all three forms of vascular growth occur during all phases of the vertebrate life cycle or if they are restricted to specific stages.

Vasculogenesis. Vasculogenesis refers to vessel growth during early embryonic development. At approximately embryonic day E18, the earliest blood vessels in a developing human form from clusters of progenitor cells. These so-called hemangioblasts give rise to both the primary vessel wall and the blood cells within the vessel lumen. Hemangioblasts originate from the mesoderm (the middle primary germ cell layer of the earliest embryo) and cluster in the form of blood islets within both the yolk sac and the embryo proper (**Fig. 1**). Extra- and intraembryonic progenitors apparently differ slightly in their capability to differentiate. For example, in blood islets of the yolk sac, the progenitors give rise to both vessel wall cells and blood cells in a close spatial and temporal relation. In contrast, intraembryonic blood islets first form early vessel wall cells of hemogenic potential, and a number of these will later transdifferentiate (transform) to blood cells. Early vessel wall cells that form during vasculogenesis are called primary endothelial cells (ECs). In general, ECs line the inner circumference of a larger blood vessel or form the smallest blood vessels (capillaries). During development, ECs are formed first. The earliest vessels form honeycomb-like primary networks, which later mature to the arteriovenous tree; during this process, mural cells (contractile cells that surround vascular ECs) are recruited to the external vessel wall.

Vasculogenesis is initiated and regulated mostly by genetic predetermination. Modifying environmental factors gain influence at later stages of vessel development. Molecular signals that probably govern early vasculogenesis include the Hedgehog protein family and growth factors such as fibroblast growth factors (FGFs) and the transforming growth factor-β family (TGF-β-like factors). Once blood islets have differentiated into early vascular cells and early blood cells, angiopoietins and the vascular endothelial growth factor (VEGF) family, which are better known for their dominant role in another mechanism of vessel development, namely angiogenesis (see below), become involved in the regulation of vasculogenesis.

Embryonic vasculogenesis forms a rather stereotypical vascular plexus according to a predetermined

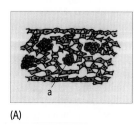 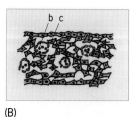 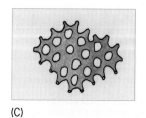 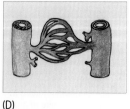

(A)   (B)   (C)   (D)

Fig. 1. Vasculogenesis. (*A*, *B*) Formation of blood islets (a) is followed by their differentiation into early blood cells (b) and primary endothelial cells (c). (*C*, *D*) The resulting primary vascular plexus subsequently matures toward an arteriovenous tree.

genetic program. With the onset of a circulation, that is, a beating heart, physical influences such as blood flow and tissue oxygenation gain in influence and aid vascular plexus maturation and arteriovenous determination. Vasculogenesis is then largely superseded by angiogenesis.

In tissues derived from the outer primary germ cell layer (the ectoderm), including the nervous system, vasculogenesis plays a subordinate role, and primary vascular development occurs mostly by angiogenesis.

The detection of circulating cells, which share significant molecular characteristics with hemangioblast progenitors, in the adult circulation of different vertebrate species (including humans) has stimulated efforts to utilize these cells for vascular regenerative therapies or, on the other hand, to block adult vasculogenesis in growing malignancies to starve a tumor. However, there is conflicting in vitro and in vivo evidence on whether these circulating progenitor-like cells are indeed able to structurally integrate into a regenerating vessel, or if they rather act as suppliers of mediators, and, in general, if vasculogenesis per se does occur in adult organisms at all. Current research efforts focus, for example, on a more thorough characterization of those circulating progenitor-like cells, the regulation of their release from the bone marrow, the tracing of their potential incorporation into the vessel wall, and the functional relevance of these processes.

Angiogenesis. Angiogenesis is the growth of capillaries, which are the smallest vessels in the organism where the exchange of oxygen, carbon

dioxide, nutrients, and metabolic waste takes place. Especially in development, angiogenesis can be followed by vessel maturation, stabilization, specialization, and network maturation, constituting a series of processes that will lead to the development of the mature tree-like arteriovenous system. Angiogenesis proceeds by either sprouting from or division of a preexistent vessel; therefore, it is possible to distinguish between sprouting angiogenesis and intussusceptive (splitting) angiogenesis (**Fig. 2**). Sprouting angiogenesis seems to be the more common process and therefore has been investigated more intensely; however, in certain tissues and organs (for example, the lung), intussusception assumes the leading role.

The major stimulus for sprouting angiogenesis is a lack of tissue oxygenation. This leads to the upregulation of transcription factors such as hypoxia-inducible factor-1 (HIF-1), which is a major signal in angiogenesis initiation. HIF-1 controls the expression of the VEGFs, which comprise the driving force of angiogenesis and which, in a complex interplay with the angiopoietin/Tie-2 (Ang/Tie-2) system, regulate all angiogenic processes that have been observed so far.

Sprouting angiogenesis starts with a controlled, local disintegration of the basement membrane that surrounds the mother vessel. Plasma proteins penetrate into the extravascular space and form a scaffold for the growing vessel. Some ECs become angiogenic, that is, migratory or mitotically active (or both). A pioneering tip cell is designated, which migrates in the direction of gradients of proangiogenic factors such as VEGF, and is followed by mitotically

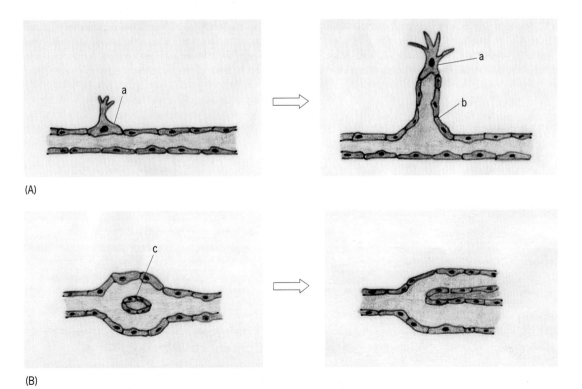

Fig. 2. Angiogenesis. (*A*) In sprouting angiogenesis, a pioneering tip cell (a) starts to migrate toward a gradient of proangiogenic factors. Mitotically active stalk cells (b) follow. (*B*) In intussusceptive (splitting) angiogenesis, tissue pillars (c) are inserted into the lumen of preexistent capillaries.

active stalk cells (Fig. 2*A*). VEGF stimulates most stages of this process, fostering, for example, the initial vascular leakage and EC activation. Ang/Tie-2 counterbalances the effects of VEGF to avoid over-activation. Interestingly, there is an overlap between the factors regulating angiogenesis and neurogenesis: Sprouting capillaries and sprouting neural axons share guidance cues. EC cords form, which are lumenized either during or shortly after sprouting in a pinocytosis-like process or by extracellular lumen formation (note that the lumen is the interior space within a tubular structure, such as within a blood vessel).

In intussusception, a slender tissue pillar is inserted into a vessel; then, it expands and thereby splits the vessel in two (Fig. 2*B*). Thus, it divides the existing vessel lumen. The molecular regulation of intussusceptive angiogenesis also relies on VEGF, the angiopoietin system, and other signaling systems.

Angiogenesis regularly occurs in adult life, and deregulated angiogenesis is critically involved in the pathogenesis of a number of life-threatening and disabling diseases. Malignant tumors become, for example, vascularized by angiogenesis, and atherosclerotic vessel occlusions change tissue oxygenation, which may stimulate angiogenesis, which, however, is usually not sufficient to compensate for the loss of a conductance vessel. Thus, intense research efforts are devoted to therapeutically interfere with angiogenesis.

To inhibit angiogenesis, its master regulator VEGF was an obvious target. In 2004, the U.S. Food and Drug Administration (FDA) approved bevacizumab (the first clinically available humanized monoclonal antibody that inhibits VEGF-A) as a drug to slow angiogenesis for adjuvant use in various cancers. Since then, a number of derivatives and VEGF receptor antagonists have been developed and evaluated. Currently, large-scale clinical trials are in progress to evaluate their use in nonmalignant diseases and in curative settings.

Stimulation of angiogenesis, although possible, poses a number of potential problems due to (1) the wide and complex spectrum of action of the probable target molecules and (2) the fact that the loss of upstream vessels, as occurs in clinically relevant atherosclerotic disease, cannot be compensated for by capillary growth alone.

Arteriogenesis. Arteriogenesis refers to the positive outgrowth of a preexistent arteriolar vessel that bypasses an occluded artery (**Fig. 3**). While angiogenesis is induced by a lack of tissue oxygen, arteriogenesis starts when blood flow through a given vessel suddenly and persistently increases. When an artery becomes occluded, and if intact interconnecting collateral vessels (that is, biological bypasses) are present, a pressure gradient develops between the regions above and below the occlusion. Blood flow increases, and ECs lining the collateral vessels perceive this biophysical stimulus as a pulsatile increase in fluid shear stress (FSS). Thus, they are being deformed and thereby activated metabolically. In a short-term response, ECs release vasodilating factors such as nitric oxide (NO) to relax the underlying smooth muscle cell (SMC) layer. If FSS stays elevated, ECs prompt a permanent enlargement of the vessel through activation of growth processes: Activated ECs attract circulating immunocompetent monocytic cells, which invade the vessel wall and supply factors (for example, matrix metalloproteinases) that degrade the vessel wall in a local and controlled fashion. This allows SMCs to become mitotic, divide and proliferate, and enlarge the vessel. An outgrowth of collateral diameter up to 50-fold has been observed in vivo, in an effort to form a sufficient arterial bypass that compensates for the stenotic or occluded conductance artery. In human patients, the ability to collateralize

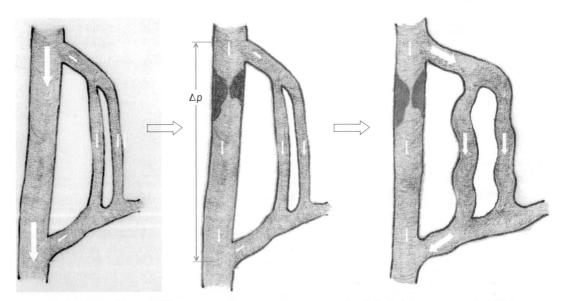

Fig. 3. Arteriogenesis. When a conductance artery is occluded by an atherosclerotic lesion, a pressure gradient (Δp) develops over preexistent interconnecting arterioles between the regions above and below the occlusion. Collateral flow increases and prompts the positive outward remodeling of the biological bypasses, enabling them to partially compensate for the occluded vessel.

stenotic vessels is, at least in part, genetically determined and has significant consequences regarding the severity of the symptoms of atherosclerotic disease. Attempts at therapeutic stimulation of arteriogenesis were first directed toward the stimulation of vessel wall remodeling by, for example, growth factors that attract and stimulate monocytic cells, including granulocyte colony-stimulating factor (G-CSF), granulocyte-monocyte colony-stimulating factor (GM-CSF), monocyte chemotactic protein-1 (MCP-1), and TGF-β. Although they are effective in different vascular regions such as the heart, the brain, and the extremities in experimental settings, most of these factors have severe side effects or are involved in the pathogenesis of atherosclerosis, rendering them less suitable as medical pro-arteriogenic therapies. Alternatively, current research efforts have focused on a targeted elevation of pulsatile FSS by, for example, interventional therapeutic revascularization of an occluded vessel of the lower extremity by percutaneous transluminal angioplasty (PTA) and stenting to restore arterial inflow to the leg, followed by an exercise regimen to keep FSS elevated and aid collateralization in the subordinate vascular regions. Positive results have also been achieved using external counterpulsation therapy in patients unable to exercise actively.

A thorough comprehension of vessel growth and development is the critical prerequisite to understand the pathogenesis of the major causes of death and disability worldwide, and to develop therapies that target them efficiently.

For background information *see* ANGIOGENESIS; BLOOD VESSELS; CANCER (MEDICINE); CARDIOVASCULAR SYSTEM; CELL DIFFERENTIATION; CIRCULATION; DEVELOPMENTAL BIOLOGY; GERM LAYERS; GROWTH FACTOR; VASCULAR DISORDERS in the McGraw-Hill Encyclopedia of Science & Technology.

Anja Bondke Persson

Bibliography. A. Bondke Persson and I. R. Buschmann, Vascular growth in health and disease, *Front. Mol. Neurosci.*, 4:14, 2011, DOI:10.3389/fnmol.2011.00014; P. Carmeliet, Mechanisms of angiogenesis and arteriogenesis, *Nat. Med.*, 6:389–395, 2000, DOI:10.1038/74651; A. S. Chung, J. Lee, and N. Ferrara, Targeting the tumour vasculature: Insights from physiological angiogenesis, *Nat. Rev. Cancer*, 10:505–514, 2010, DOI:10.1038/nrc2868; W. Risau and I. Flamme, Vasculogenesis, *Annu. Rev. Cell Dev. Biol.*, 11:73–91, 1995, DOI:10.1146/annurev.cb.11.110195.000445; W. Schaper, Development of the collateral circulation: History of an idea, *Exp. Clin. Cardiol.*, 7:60–63, 2002; G. L. Semenza, Vasculogenesis, angiogenesis, and arteriogenesis: Mechanisms of blood vessel formation and remodelling, *J. Cell. Biochem.*, 102:840–847, 2007, DOI:10.1002/jcb.21523.

Xyloplax

The echinoderms include familiar marine organisms such as sea stars and sea urchins, as well as other extant forms such as sea lilies (crinoids), sea cucum-

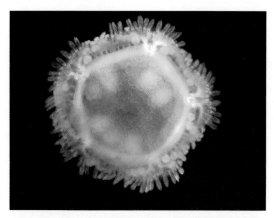

Fig. 1. Oral view of a brooding female of *Xyloplax janetae* [approximately 0.5 mm (0.02 in.) in diameter]. This specimen was collected in July 2010 from wooden blocks that were deployed in August 2007. The experimental substrate was located about 2 m (6.6 ft) from tubeworms (*Ridgeia piscesae*) and diffuse hydrothermal flows of the Main Endeavour vent segment on the Juan de Fuca Ridge [located off the coast of Washington; depth of 2202 m (7225 ft)]. [*Photo courtesy of Benjamin Grupe (Scripps Institution of Oceanography); cruise funded by NSF-0623554 awarded to Raymond Lee (Washington State University)*]

bers, and brittle stars. Lineages of these forms represent the five living classes of echinoderms. The notion of only five classes was recently challenged by the discovery of *Xyloplax* (**Fig. 1**). *Xyloplax* is a genus of small, disk-shaped echinoderms that were first discovered in 1984 based on specimens collected from sunken wood recovered from depths of 1100 m (3600 ft) off New Zealand. Other species of *Xyloplax* have been found on wood from the deep seas of the Northeast Pacific off the United States and in the Tongue of the Ocean (a deep ocean trench) off the Bahamas. The scientists that described the first specimens of *Xyloplax* indicated that it was clearly an echinoderm. However, because *Xyloplax* was so distinctive compared to other echinoderms, it represented a previously undiscovered lineage—that is, a sixth class of extant echinoderms. Lineages of echinoderms are important because echinoderms share a common ancestor with other deuterostomes, including chordates. Because of this link, echinoderms provide clues to understanding the tree of life and the deep history of our species.

Echinoderm design. Two features are widely used to identify echinoderms: the water-vascular system and the underlying microstructure of the skeleton (called the stereom). The water-vascular system of extant echinoderms is made from a circular tube derived from the coelom surrounding the mouth with five radial canals that bear rows of tube feet. Skeletal elements composed of the stereom and epidermal tissues surround the internal organs and water-vascular system of echinoderms. The stereom is composed of distinctive crystals of calcium carbonate.

As most echinoderms grow by adding to radial canals, tube feet, and the skeleton, this provides the familiar five-sided shape to echinoderms called pentaradial symmetry. Some lineages of echinoderms have superimposed other forms of symmetry on top

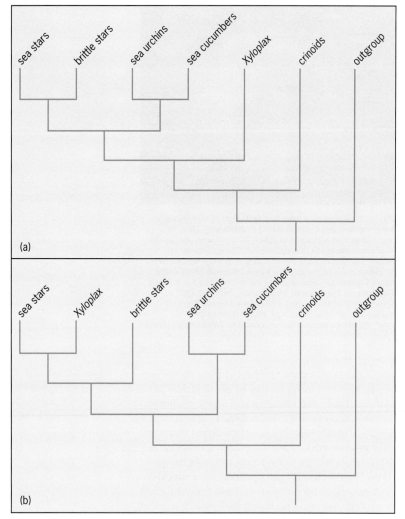

Fig. 2. Panel *a* represents a phylogenetic tree consistent with the hypothesis that *Xyloplax* is a distinct taxonomic class of echinoderm. Panel *b* represents a phylogenetic tree consistent with a competing hypothesis that states that *Xyloplax* is a sea star and thus cannot represent a taxonomic class on its own. Recent genetic and developmental research clearly supports the second hypothesis (panel *b*).

Phylogeny of echinoderms. Although it is easy to identify echinoderms, their ancient history and complex body plans create difficulties in understanding their phylogeny. Moreover, the proposal of a putative sixth extant class (Concentricycloidea) based on the discovery of *Xyloplax* in the 1980s further complicated the problem. In the 1990s, though, the concept of the sixth class began to erode. At that time, scientists uncovered typical echinoderm pentaradial symmetry in *Xyloplax*, as well as shared features of skeletal anatomy among *Xyloplax* and sea stars. In addition, genetic links were found between *Xyloplax* and sea stars. Under the logic of phylogenetics, *Xyloplax* cannot, on its own, represent a distinct class if it is embedded in a known class, that is, the Asteroidea (sea stars) [**Fig. 2**]. Nevertheless, the class distinction of *Xyloplax* has persisted in many references.

Genetic and developmental studies of Xyloplax. Proper preservation of specimens of a new species *Xyloplax janetae* (Fig. 1), discovered in 2004 in Monterey Canyon off California, allowed extended genetic and developmental studies. Using genetic sequences and developmental data for *Xyloplax*, scientists have demonstrated unequivocally that *Xyloplax* is a sea star. Even though *Xyloplax* does not represent an echinoderm class, the animal remains interesting because of the evolution of its development. A typical echinoderm develops via a small bilaterally symmetrical larva that metamorphoses into a pentaradially symmetrical juvenile that grows into an adult. At metamorphosis, the newly formed juvenile reshapes, absorbs, or casts off portions of the larval body. Fascinating modifications in this basic life cycle have evolved among echinoderms. These include a large diversity of larval forms, various dispersal strategies, asexual reproduction, clonal development (larval budding), and direct development.

It has been hypothesized that *Xyloplax* evolved via progenesis, that is, the truncation of somatic growth at a juvenile body plan but with gonadal growth to maturity. In support of this hypothesis, the best estimate with the data at hand is that *Xyloplax* is closely related to a family of sea stars called the Pterasteridae. Members of the Pterasteridae brood their young in deep and polar seas. The young are held in specialized adult brood chambers until they are released as free-living individuals. Some species, such as *Pteraster tesselatus*, have become direct developers with no larval anatomy. In contrast, most other echinoderms have complex larval forms with bilateral symmetry and require metamorphosis to take on the pentaradial shape in the juvenile stage. The extreme direct developers among the Pterasteridae form the pentaradial body plan immediately from the embryo stage. Other pterasterids, such as *Hymenaster pellucidus*, have embryonic bilateral symmetry and structures (brachiolar arms) that are used by the larvae to attach to surfaces. *Xyloplax* fits along this continuum from larval to direct development within the Pterasteridae. The early embryos of *Xyloplax* have bilateral symmetry, but no attachment structures, and become radial later in development.

of the five-sided symmetry provided by the water-vascular system. Examples include bilateral symmetry in sea cucumbers and sand dollars, and extra arms or stubby arms in sea stars, which provide sun-like or ball-like appearances, respectively.

The five extant classes of echinoderms share ancestry with 16 extinct classes of echinoderms that date back to the Cambrian Period. These extinct classes have an even wider range of body symmetries than extant forms, including pentaradial, circular, asymmetrical, helical, and triangular types. Even if their water-vascular systems are not well preserved, fossils can be clearly identified as echinoderms by the presence of the stereom.

Because it has a water-vascular system and stereom, *Xyloplax* was easily identified as an echinoderm; however, its overall design raised interest. *Xyloplax* was described as having a circular rather than five-sided water-vascular system. In other words, instead of having the tube feet borne along the spokes of a wheel, as is typical of sea stars, the tube feet of *Xyloplax* appear on the wheel itself.

More data are required to fill in the evolutionary developmental history of *Xyloplax*.

Life-cycle adaptations. *Xyloplax* is a rare example of an animal lineage taking on the juvenile shape of its ancestors as an adult body plan. Progenesis is a pattern, not a mechanism, but it does make us think about adaptations of the life cycle itself. One hypothesis is that the small, disk-like shape of *Xyloplax* allows it to attach to sunken timbers and prey on smaller animals and microorganisms. Brooding in *Xyloplax* is also a reasonable adaptation to a substrate of sunken timbers as it allows young to be released in their adult habitat, which may be rare on the seafloor. This lifestyle leads to a perplexing question: How does the next generation of *Xyloplax* get from a timber that is almost disintegrated to a fresh one? Means of dispersal for adult *Xyloplax* may include drifting on sunken wood or swimming like jellyfish. An alternative hypothesis is that *Xyloplax* may have larval stages that have not yet been observed. Some pterasterids use two reproductive strategies, that is, (1) brooding of some young and (2) releasing of others into the plankton, providing a hedge against local habitat loss. Consistent with the hypothesis that *Xyloplax* may have a pelagic larval stage is the fact that various species in the genus have been found in very distant parts of the ocean (off Andros Island of the Bahamas, off the North Island of New Zealand, and off California and the Pacific Northwest of the United States). Overall, much more information needs to be gathered about the life history and behavior of *Xyloplax*. One step would be to sample regions such as the Amazon Canyon, in which there should be timber, such that *Xyloplax* can be recovered in numbers reliable enough to make key observations and design experiments that might help us to better understand *Xyloplax* and the evolution of its life cycle.

For background information *see* ANIMAL EVOLUTION; ASTEROIDEA; CONCENTRICYCLOIDEA; DEEP-SEA FAUNA; DEVELOPMENTAL BIOLOGY; ECHINODERMATA; MORPHOGENESIS; PHYLOGENY in the McGraw-Hill Encyclopedia of Science & Technology.

Daniel A. Janies

Bibliography. D. Janies and L. R. McEdward, Highly derived coelomic and water-vascular morphogenesis in a starfish with pelagic direct development, *Biol. Bull.*, 185:56–76, 1993, DOI:10.2307/1542130; D. Janies and R. Mooi, *Xyloplax* is an asteroid, pp. 311–316, in M. D. Candia Carnevali and F. Bonasoro (eds.), *Echinoderm Research 1998*, Balkema, Rotterdam, 1999; D. Janies, J. Voight, and M. Daly, Echinoderm phylogeny including *Xyloplax*, a progenetic asteroid, *Syst. Biol.*, 604:420–438, 2011, DOI:10.1093/sysbio/syr044; D. McClary and P. Mladenov, Reproductive pattern in the brooding and broadcasting sea star *Pteraster militaris*, *Mar. Biol.*, 103:531–540, 1989, DOI:10.1007/BF00399585; F. E. W. Rowe, A. N. Baker, and H. E. S. Clark, The morphology, development and taxonomic status of *Xyloplax* Baker, Rowe and Clark (1986) (Echinodermata: Concentricycloidea), with the description of a new species, *Proc. R. Soc. B Biol. Sci.*, 233:431–459, 1988, DOI:10.1098/rspb.1988.0032; J. Voight, First report of the enigmatic echinoderm *Xyloplax* from the North Pacific, *Biol. Bull.*, 208:77–80, 2005, DOI:10.2307/3593115.

Nobel Prizes for 2012

The Nobel prizes for 2012 included the following awards for scientific disciplines.

Chemistry. The chemistry prize was awarded to Robert J. Lefkowitz of the Howard Hughes Medical Institute and Duke University Medical Center, Durham, North Carolina, and Brian K. Kobilka of the Stanford University School of Medicine, Stanford, California, for studies of G-protein–coupled receptors.

G-protein–coupled receptors are found in cell membranes and extend from the interior to the exterior of cells. Outside the cell the receptors bind with signaling molecules such as neurotransmitters, hormones, and messengers for taste, smell, and sight. This binding reaction activates G-protein molecules bound to the receptors on the inside of the cell, which trigger a sequence of reactions, or processes, that result in a physiological response. For example, when the hormone epinephrine (adrenaline) binds with G-protein–coupled receptors in the body, activated G proteins inside cells initiate intercellular reactions that raise blood pressure and heart rate.

Over the years, Nobel prizes have been awarded to researchers for reasons ranging from creativity to serendipity. But this year's prize in chemistry was largely due to persistence and a fortuitous hiring decision. Lefkowitz had started searching for the β-adrenergic receptor (the receptor for adrenaline) in the 1960s, and by 1970 had found it by using radioactive-iodine-labeled hormones. Not long afterward, G protein and its function within the cell were discovered by Martin Rodbell and Alfred G. Gillman, for which they were eventually awarded the Nobel Prize in Physiology or Medicine in 1994. In 1980, Lefkowitz and coworkers proposed a mechanism for the activity of the membrane-spanning β-adrenergic receptor, with a signaling molecule bound to it outside and G protein bound to it inside the cell. But the question of how to prove this model remained.

The answer arrived in 1984 when Lefkowitz hired Brian Kobilka, just out of medical school, to clone and sequence the gene for the β-adrenergic receptor. Kobilka and coworkers not only succeeded in accomplishing this task, but their subsequent analysis also revealed that the receptor consisted of seven long-chain fatty helices that probably crossed the cell membrane seven times. So too did the receptor for rhodopsin, the light-sensing receptor in the retina, which prompted Lefkowitz's realization that all G-protein–coupled receptors have a similar structure. In 2011, using x-ray crystallography, Kobilka went on to capture the first image of the β-adrenergic receptor at the exact moment it was being activated by a hormone on the outside and coupled with G protein on the inside.

Most physiological processes depend on the G-protein–coupled receptors, as do the actions of many medications. Today, around 1000 such receptors with specific physiological functions are known in humans. Kobilka's recent imaging of a receptor in action after a nearly 20-year epic struggle will give future researchers a better understanding of structure and function. In addition to understanding the physiological processes, Lefkowitz' and Kobilka's discoveries of G-protein–coupled receptors are important for understanding defects in signal transduction, such as is the case for color blindness, and for the development of new drugs that target receptors responsible for disease.

For background information *see* CELL (BIOLOGY); CELL MEMBRANES; EPINEPHRINE; HORMONE; MOLECULAR BIOLOGY; NORADRENERGIC SYSTEM; PROTEIN; SIGNAL TRANSDUCTION in the Encyclopedia of Science & Technology.

Physics. The physics prize was awarded to Serge Haroche of the Collège de France and École Normale Supéreure in Paris and David J. Wineland of the National Institute of Standards and Technology (NIST) and University of Colorado in Boulder for ground-breaking experimental methods that enable measuring and manipulation of individual quantum systems.

Quantum mechanics describes a realm where events occur contrary to our expectations and experiences with phenomena in the macroscopic world. However, this strange behavior is manifested by particles in extremely fragile states, and they lose their quantum properties as soon as they come in contact with the outside world. Thus, for many years quantum phenomena could not be observed directly; they could be described only in "thought experiments" that were performed solely in physicists' imaginations.

Haroche and Wineland are leaders in work that has made it possible to directly observe and manipulate quantum particles while preserving their quantum-mechanical properties. They have pursued complementary approaches.

In the experiments conducted by Wineland and his co-workers, charged atoms, or ions, are kept within a trap by surrounding them with electric fields, and are isolated by performing the experiments in a vacuum at extremely low temperatures. The particles are observed and controlled with light, or photons, usually from pulsed lasers. The lasers have been used to suppress the ion's thermal motion and put the ion in its lowest energy state (ground state), enabling the study of quantum phenomena. Precisely tuned laser pulses have also been used to put the ion in a superposition state, in which the ion exists simultaneously in two different states.

In the experiments conducted by Haroche and his co-workers, the roles of photons and atoms are reversed: Photons are trapped and atoms are the probes. The microwave photons bounce back and forth inside a small cavity between two mirrors, about 3 cm (1 in.) apart. The mirrors are made of superconducting material and are highly reflective, so

much so that a photon can bounce back and forth for about 0.1 s and travel 40,000 km (25,000 mi, the circumference of the Earth) before it is absorbed. Then, specially prepared Rydberg atoms are sent through the cavity. These atoms are in highly excited states with principal quantum numbers $n \sim 50$, and have radii of about 125 nm, about 1000 times as large as ordinary atoms. The interaction of the photons with the Rydberg atoms modifies the quantum states of the atoms in ways that make it possible to investigate the state of the photons in the cavity, for example, to reveal the presence or absence of a single photon, to count the number of photons in the cavity, or to follow the evolution of an individual quantum state.

Both Haroche and Wineland have carried out experiments that are analogous to a famous thought experiment with a cat, proposed by Erwin Schrödinger in 1935. Inside a sealed box are the cat, a radioactive atom, and a vial of lethal cyanide, which is released only upon the atom's decay. Since the radioactive decay is governed by the laws of quantum mechanics, according to which the atom is in a superposition state of having both decayed and not yet decayed, the cat must also be in a superposition state of being both dead and alive. An observer who peeks inside the box will either kill the cat or enable it to survive because the attempt to observe the cat will "collapse" the "cat state" into one of two states in which it is either dead or alive. This "experiment" with its seemingly paradoxical result explores the boundary between the realms of quantum and classical phenomena. Both Haroche and Wineland have devised experiments that show in great detail how the act of measurement causes a quantum state to collapse and lose its superposition character.

The concept of a quantum computer based on trapped ions was suggested in 1995. While the smallest unit of information in a classical computer is a bit whose value is either 0 or 1, the basic unit of information in a quantum computer—a quantum bit or qubit—can be 0 and 1 at the same time. Thus, each additional qubit doubles the number of possible states, and a quantum computer with 300 qubits could hold $2^{300} \approx 10^{90}$ values simultaneously, more than the number of atoms in the observable universe. Wineland's group was the first to carry out a two-qubit operation. Quantum computer technology has been demonstrated with up to 14 qubits and a series of gates and protocols. While quantum computers are still far from reality, it has been suggested that they may change our lives in this century in the same radical way as classical computers in the last century.

Optical clocks based on trapped ions operate at frequencies of the order of 100,000 times higher than the cesium-based microwave atomic clocks that provide the current time standard, and this higher frequency enables them to achieve higher accuracies. An optical ion clock requires an ion with both a narrow (forbidden) transition to achieve the highest possible accuracy and a broad allowed transition for efficient cooling and detection. Wineland's group developed a clock with two ions of different species in a trap, one to provide the spectroscopy transition and the other to provide the cooling transition, and they developed a technique called quantum logic to transfer information between the two ions. This technique is based on a quantum effect called entanglement, whereby two quantum particles that have no direct contact can still affect each other's properties. A precision of about one part in 10^{17} has been achieved, about 30 times better than the best current microwave clock. This precision enables the measurement of relativistic effects such as time dilation at speeds of a few kilometers per hour or gravitational time dilation between two points with a height difference of only 33 cm (1.1 ft). Potential applications include measurements of Earth's dynamic gravitational properties and the most stringent tests so far of the theory of relativity. *See* OPTICAL CLOCKS AND RELATIVITY.

For background information, *see* ATOMIC CLOCK; GRAVITATIONAL REDSHIFT; LASER COOLING; PARTICLE TRAP; QUANTUM COMPUTATION; QUANTUM MECHANICS; QUANTUM TELEPORTATION; QUANTUM THEORY OF MEASUREMENT; RELATIVITY; SUPERPOSITION PRINCIPLE in the McGraw-Hill Encyclopedia of Science & Technology.

Physiology or medicine. The prize in physiology or medicine was awarded jointly to Sir John B. Gurdon of the Gurdon Institute in Cambridge, United Kingdom, and Shinya Yamanaka of Kyoto University in Japan and the Gladstone Institute in San Francisco, California, United States, for the discovery that mature cells can be reprogrammed to become pluripotent.

Pluripotent cells have the ability to become many different cell types. Among the best studied are embryonic stem cells, which also have the ability to self-replicate and to give rise to mature cells. During the process of embryonic development, a single fertilized egg produces the many cell types found in the adult organism. Over time, the cells of the embryo divide, grow, change in shape and position, and finally take on one of many differentiated fates, such as bone tissue or heart muscle. Each differentiated cell is specialized to carry out particular tasks in the adult body.

Almost all cell types share the same genes. Cell differentiation arises from the varied expression of those genes. For example, a myoblast (precursor cell to a muscle fiber) becomes a muscle cell because genes responsible for building muscle-specific structures, such as the contractile apparatus, are selectively activated. Conversely, muscle-specific genes remain inactive in other cell types, such as nerve cells.

As a general rule, there is broad differentiation potential in somatic (body) cells of the early embryo, and this potential becomes progressively constrained in daughter cells as development proceeds. In later developmental periods and into the adult organism, some undifferentiated somatic cells remain for growth and replacement of several tissues. Most of these adult stem cells, although undifferentiated, have restricted potential, meaning that their range of possible fates is often limited. However, the work of Gurdon and Yamanaka has changed the view that adult cells are irreversibly committed to their fate.

The research of both biologists aims to elucidate the molecular and genetic mechanisms that underlie embryonic development and tissue regeneration. Gurdon studies the reprogramming of gene expression by nuclear transfer. Yamanaka focuses on the genetic control of embryonic stem cells and induced pluripotent stem cells.

The announcement of the 2012 Nobel award coincided with the 50th anniversary of Gurdon's groundbreaking discovery that nuclei from adult frog cells can be transplanted into enucleated egg cells to produce normally developing frogs. This work, which began in 1958, demonstrated that changes in gene activity induced by nuclear transplant surgery are indistinguishable from those that occur in normal early development. By 1962, Gurdon had established the principle of gene conservation in differentiated cells, showing that cell differentiation does not necessarily involve any permanent change or irreversible inactivation of genes. His experiments also showed that nuclei of a differentiated cell can be fully reprogrammed (by egg cytoplasm) to allow expression of all genes needed for the formation of a fertile adult. The same technique of nuclear transplantation was used in embryonic cloning 35 years later.

The year 1962 was remarkable for other reasons as well. Not only was it the same year the Nobel Prize in Physiology or Medicine was awarded for the discovery of the molecular structure of DNA, but it was also the year that Shinya Yamanaka was born.

Building on the earlier work of Gurdon and others, Yamanaka, about a decade ago, started researching the possibility of converting adult somatic cells into pluripotent stem cells. Yamanaka focused on a group of genes that are uniquely or highly expressed in embryonic stem cells. In 2006, his team inserted various genes into adult mouse cells and found that four of these genes were sufficient to induce nuclear reprogramming. This brought the adult cells "back in time" to a primordial, pluripotent state resembling embryonic stem cells. To distinguish these reprogrammed cells from their embryonic counterparts, Yamanaka called them induced pluripotent stem cells, or iPS cells. Within a few months, his team and ones in other laboratories showed that the same principles could be applied to reprogram human somatic cells.

The research of these two Nobel Laureates has increased our understanding of how cells work, and provides clues for treatment of disease using easily obtainable cells of adult individuals. For instance, through nuclear replacement therapy, it may be possible to eventually replace lost brain or heart cells from a patient's own skin cells. Gurdon and Yamanaka's work promises to revolutionize the field of regenerative medicine by avoiding the need for lifetime immunosuppression when patients receive replacement tissues or organs from allogenic (genetically different) donors.

Technology using iPS cells also represents an entirely new approach to fundamental studies in developmental biology. Until now, scientists have used model organisms such as yeast, flies, amphibians, and mice to study development and disease. Now, laboratories can make iPS cells from human patients with a specific disease, such as cystic fibrosis. The cells contain a complete set of the genes that resulted in that disease—representing the potential of a nearly flawless disease model for studying new drugs and treatments.

For background information *see* CELL (BIOLOGY); CELL BIOLOGY; CELL DIFFERENTIATION; CELL FATE DETERMINATION; CELL LINEAGE; CLONING; DEVELOPMENTAL BIOLOGY; DEVELOPMENTAL GENETICS; FATE MAPS (EMBRYOLOGY); GENE; STEM CELLS in the McGraw-Hill Encyclopedia of Science & Technology.

Contributors

Contributors

The affiliation of each Yearbook contributor is given, followed by the title of his or her article. An article title with the notation "coauthored" indicates that two or more authors jointly prepared an article or section.

A

Aguilera, Dr. Ana. *Professor of Computer Sciences, Universidad de Carabobo, Valencia, Venezuela.* FUZZY DATABASES—coauthored.

Alba, Dr. David M. *Department of Paleoprimatology and Human Paleontology, Institut Català de Paleontologia Miquel Crusafont, Universitat Autònoma de Barcelona, Cerdanyola del Vallès, Barcelona, Spain.* ANOIAPITHECUS AND PIEROLAPITHECUS.

Augustine, Dr. Swinburne A.J. *Microbial Exposure Research Branch, Microbiological and Chemical Exposure Research Division, National Exposure Research Laboratory, U.S. Environmental Protection Agency, Cincinnati, Ohio.* SALIVA-BASED IMMUNOASSAY OF WATERBORNE PATHOGEN EXPOSURE—coauthored.

B

Baer, Prof. Howard. *Department of Physics and Astronomy, University of Oklahoma, Norman.* NATURAL SUPERSYMMETRY.

Baker, Prof. Matthew. *Alcatel-Lucent, Westlea, Swindon, Wiltshire, United Kingdom.* LONG-TERM EVOLUTION (LTE).

Ballance, Dr. Connor P. *Department of Physics, Auburn University, Alabama.* PHOTOIONIZATION, FLUORESCENCE, AND INNER-SHELL PROCESSES—coauthored.

Battaglia, Joseph. *LGS Innovations, Florham Park, New Jersey.* LIMITATIONS ON INCREASING CELLULAR SYSTEM DATA RATES—coauthored.

Baute, Audrey J. *Department of Biological Sciences, Eastern Kentucky University, Lexington.* GENETIC EXCHANGE AMONG BACTERIA—coauthored.

Bear, Dr. Harry D. *Department of Surgery, Division of Surgical Oncology, Virginia Commonwealth University, Richmond.* NEOADJUVANT SYSTEMIC THERAPY FOR BREAST CANCER—coauthored.

Beaumont, Fabien. *Groupe de Recherche en Sciences Pour l'Ingénieur (GRESPI), Université de Reims Champagne-Ardenne, France.* FLOW PATTERNS IN CHAMPAGNE GLASSES—coauthored.

Benjamin, Dr. Emelia J. *Framingham Heart Study, National Heart, Lung, and Blood Institute, Framingham, Massachusetts.* ATRIAL FIBRILLATION—coauthored.

Bernstein, Dr. I. Leonard. *Department of Internal Medicine, Division of Immunology/Allergy Section, University of Cincinnati College of Medicine, Ohio.* ALLERGENICITY OF CYANOBACTERIA—coauthored.

Bernstein, Dr. Jonathan A. *Department of Internal Medicine, Division of Immunology/Allergy Section, University of Cincinnati College of Medicine, Ohio.* ALLERGENICITY OF CYANOBACTERIA—coauthored.

Bondke Persson, Dr. Anja. *Institut für Vegetative Physiologie, Charité Universitaetsmedizin Berlin, Germany.* VASCULAR DEVELOPMENT.

Borgonie, Dr. Gaetan. *Department of Biology, Nematology Section, Ghent University, Belgium.* DEEPEST NEMATODES.

Boron, Prof. Walter F. *Department of Physiology and Biophysics, Case Western Reserve University School of Medicine, Cleveland, Ohio.* BICARBONATE PHYSIOLOGY AND PATHOPHYSIOLOGY—coauthored.

Borovička, Dr. Jan. *Department of Nuclear Spectroscopy, Nuclear Physics Institute, Academy of Sciences of the Czech Republic, Řež near Prague, Czech Republic.* TRACE ELEMENTS IN WILD GROWING MUSHROOMS.

Braovac, Susan. *Department of Conservation, Museum of Cultural History, University of Oslo, Norway.* CONSERVATION OF ARCHEOLOGICAL WOOD: THE OSEBERG FIND—coauthored.

Breitborde, Dr. Nicholas J. K. *Department of Psychiatry, Early Psychosis Intervention Center (EPICENTER), University of Arizona, Tucson.* EXPRESSED EMOTION.

Bruynseraede, Prof. Yvan. *Department of Physics and Astronomy, Katholieke Universiteit Leuven, Belgium.* RADIOISOTOPES FOR MEDICAL DIAGNOSIS AND TREATMENT—coauthored.

C

Camarero, Dr. Julio. *IMDEA Nanosciencia (Madrid Institute for Advanced Studies in Nanoscience) and Universidad Autónoma de Madrid, Campus Universitario de Cantoblanco, Madrid, Spain.* CANCER TREATMENT USING MAGNETIC NANOPARTICLES—coauthored.

Carter, Dr. William B. *Sensors and Materials Laboratory, HRL Laboratories, LLC, Malibu, California.* ULTRALOW-DENSITY MATERIALS—coauthored.

Cheung, Prof. Kingman. *Department of Physics, National Tsing Hua University, Hsinchu, Taiwan.* HIGGS BOSON DETECTION AT THE LHC.

Chiao, Dr. Chuan-Chin. *Department of Life Science, National Tsing Hua University, Hsinchu, Taiwan.* ANIMAL CAMOUFLAGE—coauthored.

Chu, Prof. Jian. *Professor and James M. Hoover Chair for Geotechnical Engineering, Department of Civil, Construction & Environmental Engineering, Iowa State University, Ames.* APPLICATION OF MICROBIOLOGY IN GEOTECHNICAL ENGINEERING—coauthored.

Chung, Prof. Robert. *School of Print Media, Rochester Institute of Technology, New York.* PRINT COMPLIANCE VERIFICATION.

Churchill, Dr. Celia K. C. *Museum of Zoology and Department of Ecology and Evolutionary Biology, University of Michigan, Ann Arbor.* BUBBLE-RAFTING SNAILS—coauthored.

Clark, Dr. Gary F. *Division of Reproductive and Perinatal Research, Department of Obstetrics, Gynecology, and Women's Health, University of Missouri, Columbia.* SIALYL LEWIS^X OLIGOSACCHARIDE.

Cominsky, Prof. Lynn. *Department of Physics and Astronomy, Sonoma State University, Rohnert Park, California.* GAMMA-RAY BURSTS.

Convertini, Dr. Paolo. *Department of Molecular and Cellular Biochemistry, University of Kentucky, Lexington.* ALTERNATIVE RNA SPLICING—coauthored.

Cruickshank, Dr. Michael J. *Consulting Marine Mining Engineer, Honolulu, Hawaii.* DEEP SEABED MINING.

Crum, Kyle A. *Alcoa Technical Center, Alcoa Center, Pennsylvania.* ALUMINUM SHIPBUILDING.

Cséfalvay, Dr. Edit. *Department of Chemical and Environmental Process Engineering, Budapest University of Technology and Economics, Hungary.* CHEMICALS FROM RENEWABLE FEEDSTOCKS—coauthored.

Cyganski, Prof. David. *Department of Electrical and Computer Engineering, Worcester Polytechnic Institute, Worcester, Massachusetts.* INDOOR NAVIGATION FOR FIRST RESPONDERS—coauthored.

D

Dasso, Dr. Mary. *Program in Cellular Regulation and Metabolism, National Institute of Child Health and Human Development, National Institutes of Health, Bethesda, Maryland.* SUMOYLATION—coauthored.

Deikman, Dr. Jill. *Monsanto Company, Davis, California.* DEVELOPMENT OF DROUGHT-TOLERANT CROPS THROUGH BREEDING AND BIOTECHNOLOGY—coauthored.

del Puerto Morales, Dr. María. *Instituto de Ciencia de Materiales de Madrid, Consejo Superior de Investigaciones Científicas (CSIC), Campus Universitario de Cantoblanco, Madrid, Spain.* CANCER TREATMENT USING MAGNETIC NANOPARTICLES—coauthored.

Dierick, Dr. Manuel. *Department of Physics and Astronomy, Ghent University, Belgium.* MICROTOMOGRAPHY.

Dimotakis, Prof. Paul E. *Graduate Aerospace Laboratories, California Institute of Technology, Pasadena, California.* TURBULENT COMBUSTION—coauthored.

Dowling, Dr. Richard D. *Kittelson & Associates, Inc., Oakland, California.* ACTIVE TRAFFIC MANAGEMENT.

Drake, Dr. David. *Department of Forest and Wildlife Ecology, University of Wisconsin–Madison.* BATS AND WIND ENERGY.

Dror, Dr. Itiel. *Institute of Cognitive Neuroscience, University College London, United Kingdom.* COGNITIVE TECHNOLOGY.

Dubos, Dr. Christian. *Institut Jean-Pierre Bourgin, Institut National de la Recherche Agronomique–AgroParisTech, Versailles, France.* MYB TRANSCRIPTION FACTORS IN PLANTS.

Duckworth, Dr. R. James. *Department of Electrical and Computer Engineering, Worcester Polytechnic Institute, Worcester, Massachusetts.* INDOOR NAVIGATION FOR FIRST RESPONDERS—coauthored.

E

Ellegood, Edward. *Research, Embry-Riddle Aeronautical University, Cocoa Beach, Florida.* COMMERCIAL SPACE ACTIVITIES TO SUPPORT NASA'S MISSIONS.

F

Fallahi, Prof. Mahmoud. *College of Optical Sciences, University of Arizona, Tucson.* HIGH-POWER DIODE LASERS.

Fanelli, Dr. Michael N. *Astrophysics Division, NASA Ames Research Center, Mountain View, California.* KEPLER MISSION.

Farnham, Dr. Peggy J. *Department of Biochemistry and Molecular Biology, Norris Comprehensive Cancer Center, University of Southern California, Los Angeles.* KAP1 PROTEIN.

Frank, Dr. Markus H. *Transplantation Research Center, Children's Hospital Boston, Massachusetts.* CANCER STEM CELLS—coauthored.

Frank, Dr. Natasha Y. *Division of Genetics, Brigham and Women's Hospital, Boston, Massachusetts.* CANCER STEM CELLS—coauthored.

Friščič, Prof. Tomislav. *Department of Chemistry, McGill University, Montreal, Quebec, Canada.* CHEMISTRY THROUGH BALL MILLING.

G

Gallop, Prof. John. *Quantum Detection Group, National Physical Laboratory, Teddington, Middlesex, United Kingdom.* NANO-ELECTRO-MECHANICAL SYSTEMS (NEMS)—coauthored.

Gankema, Paulien. *Plant Ecophysiology Group, Institute of Environmental Biology, Utrecht University, the Netherlands.* PLANT-PRODUCED VOLATILE ORGANIC COMPOUNDS—coauthored.

Gazzaniga, Dr. Michael S. *Sage Center for the Study of the Mind, Department of Psychology, University of California, Santa Barbara.* FREE WILL AND THE BRAIN.

Glucksberg, Prof. Sam. *Department of Psychology, Princeton University, New Jersey.* FALSE-BELIEF REASONING AND BILINGUALISM—coauthored.

González Besteiro, Dr. Marina A. *Department of Botany and Plant Biology, University of Geneva, Switzerland.* PLANT MAP KINASE PHOSPHATASES—coauthored.

Gradl, Paul R. *NASA Marshall Space Flight Center, Huntsville, Alabama.* GAS SHIELDING TECHNOLOGY FOR WELDING—coauthored.

Graf, Dr. Alexander. *Department of Plant Biotechnology, ETH Zürich, Switzerland.* NIGHTTIME STARCH DEGRADATION, THE CIRCADIAN CLOCK, AND PLANT GROWTH.

Grigorenko, Dr. Leonid V. *G. N. Flerov Laboratory of Nuclear Reactions, Joint Institute for Nuclear Research, Dubna, Moscow Region, Russian Federation.* BREAKDOWN OF SHELL CLOSURE IN HELIUM-10—coauthored.

H

Halpern, Dr. Georges M. *Distinguished Professor of Pharmaceutical Sciences, Department of Applied Biology and Chemical Technology (ABCT), Hong Kong Polytechnic University, Hung Hom, Kowloon, Hong Kong, China.* FUNGAL β-GLUCANS.

Hamilton, Prof. Joseph H. *Landon C. Garland Distinguished Professor of Physics, Department of Physics and Astronomy, Vanderbilt University, Nashville, Tennessee.* DEFINITIVE EVIDENCE FOR NEW ELEMENTS 113 AND 115—coauthored.

Hanlon, Dr. Roger T. *Marine Biological Laboratory, Woods Hole, Massachusetts.* ANIMAL CAMOUFLAGE—coauthored.

Hao, Prof. Ling. *Quantum Detection Group, National Physical Laboratory, Teddington, Middlesex, United Kingdom.* NANO-ELECTRO-MECHANICAL SYSTEMS (NEMS)—coauthored.

Harley, Dr. John P. *Department of Biological Sciences, Eastern Kentucky University, Richmond.* POLIO ERADICATION.

Harton, Dr. Jonathan A. *Center for Immunology and Microbial Disease, Albany Medical College, New York.* INFLAMMASOMES.

Hazen, Dr. Robert M. *Geophysical Laboratory, Carnegie Institution of Washington, Washington, D.C.* MINERAL EVOLUTION.

He, Jia. *School of Civil and Environmental Engineering, Nanyang Technological University, Singapore.* APPLICATION OF MICROBIOLOGY IN GEOTECHNICAL ENGINEERING—coauthored.

Hebner, Prof. Robert E. *Director, Center for Electromechanics, University of Texas at Austin.* INTELLIGENT MICROGRIDS—coauthored.

Horváth, Prof. István T. *Department of Biology and Chemistry, City University of Hong Kong, Kowloon, Hong Kong SAR, China.* CHEMICALS FROM RENEWABLE FEEDSTOCKS—coauthored.

Hume, Dr. David. *Kirchhoff Institute for Physics, University of Heidelberg, Germany.* OPTICAL CLOCKS AND RELATIVITY.

I

Imtiaz, Dr. Atif. *Physical Measurement Laboratory, National Institute of Standards and Technology, Boulder, Colorado.* NEAR-FIELD SCANNING MICROWAVE MICROSCOPE (NSMM)—coauthored.

Itaka, Prof. Kenji. *North Japan Research Institute for Sustainable Energy, Hirosaki University, Aomori, Japan.* SAHARA SOLAR BREEDER PROJECT—coauthored.

Ivanov, Dr. Volodymyr. *School of Civil and Environmental Engineering, Nanyang Technological University, Singapore.* APPLICATION OF MICROBIOLOGY IN GEOTECHNICAL ENGINEERING—coauthored.

J

Jacobsen, Dr. Alan J. *Sensors and Materials Laboratory, HRL Laboratories, LLC, Malibu, California.* ULTRALOW-DENSITY MATERIALS—coauthored.

Jaffrey, Dr. Samie R. *Department of Pharmacology, Weill Medical College, Cornell University, New York, New York.* RNA MIMICS OF GREEN FLUORESCENT PROTEIN.

Janies, Dr. Daniel A. *Department of Bioinformatics and Genomics, University of North Carolina, Charlotte.* XYLOPLAX.

Junttila, Dr. Olavi. *Department of Biology, University of Tromsø, Norway.* TREE SEASONALITY IN A WARMING CLIMATE.

K

Kabos, Dr. Pavel. *Physical Measurement Laboratory, National Institute of Standards and Technology, Boulder, Colorado.* NEAR-FIELD SCANNING MICROWAVE MICROSCOPE (NSMM)—coauthored.

Kafka, Richard. *Retired; formerly, Pepco Holdings, Inc., West Bethesda, Maryland.* INDEPENDENT SYSTEM OPERATOR.

Kegge, Wouter. *Plant Ecophysiology Group, Institute of Environmental Biology, Utrecht University, the Netherlands.* PLANT-PRODUCED VOLATILE ORGANIC COMPOUNDS—coauthored.

Kemp, Dr. Martin. *Centre for Process Innovation, NanoKTN, Wilton, Redcar, United Kingdom.* NANOCOMPOSITES IN AERONAUTICS.

Kempe, Prof. Rhett. *Department of Inorganic Chemistry II, Universität Bayreuth, Germany.* QUINTUPLE BONDING.

Kerrigan, Dr. Richard W. *Agaricus Resource Program, Kittanning, Pennsylvania.* BREEDING THE BUTTON MUSHROOM (AGARICUS BISPORUS).

Kim, Dr. Hyun Dae. *NASA Glenn Research Center, Cleveland, Ohio.* DISTRIBUTED PROPULSION—coauthored.

Klemfuss, Dr. J. Zoe. *Department of Psychology and Social Behavior, University of California, Irvine.* REPRESSED MEMORIES—coauthored.

Koinuma, Dr. Hideomi. *Graduate School of Frontier Sciences, University of Tokyo, Japan.* SAHARA SOLAR BREEDER PROJECT—coauthored.

Koltsaki, Dr. Grigorios. *Department of Mechanical Engineering, Aristotle University Thessalonik, Greece.* Reduction of diesel engine particulate emissions.

Krebs, Prof. Frederik C. *Department of Energy Conversion and Storage, Technical University of Denmark, Roskilde, Denmark.* Fabrication of flexible polymer solar cells roll-to-roll—coauthored.

Krishfield, Richard. *Physical Oceanography, Woods Hole Oceanographic Institution, Massachusetts.* Arctic Ocean freshwater balance—coauthored.

Kutzke, Dr. Hartmut. *Department of Conservation, Museum of Cultural History, University of Oslo, Norway.* Conservation of archeological wood: the Oseberg find—coauthored.

L

Lalueza-Fox, Dr. Carles. *Paleogenomics Laboratory, Institute of Evolutionary Biology (CSIC-UPF), Barcelona, Spain.* Denisovans.

Larsen-Olsen, Thue T. *Department of Energy Conversion and Storage, Technical University of Denmark, Roskilde, Denmark.* Fabrication of flexible polymer solar cells roll-to-roll—coauthored.

Lawson, Dr. Mark. *Monsanto Company, St. Louis, Missouri.* Development of drought-tolerant crops through breeding and biotechnology—coauthored.

Lee, Dr. Ming-Ta. *Program in Cellular Regulation and Metabolism, National Institute of Child Health and Human Development, National Institutes of Health, Bethesda, Maryland.* SUMOylation—coauthored.

Levine, Dr. Judah. *Fellow, JILA, and Time and Frequency Division, National Institute of Standards and Technology, Boulder, Colorado.* Time transfer using a satellite navigation system.

Liger-Belair, Prof. Gérard. *Groupe de Spectrométrie Moléculaire et Atmosphérique (GSMA), Université de Reims Champagne-Ardenne, France.* Flow patterns in champagne glasses—coauthored.

Lilley, Prof. David G. *Department of Mechanical and Aerospace Engineering, Oklahoma State University, Stillwater; and Lilley & Associates, Stillwater, Oklahoma.* Fluid mechanics of fires.

Linzey, Prof. Donald W. *Department of Biology, Wytheville Community College, Wytheville, Virginia.* All Taxa Biodiversity Inventory.

Liu, Wenying. *Chemical & Biomolecular Engineering, Georgia Institute of Technology, Atlanta.* Electrospun polymer nanofibers—coauthored.

Loftus, Dr. Elizabeth F. *School of Social Ecology, University of California, Irvine.* Repressed memories—coauthored.

Long, Dr. Keith R. *U. S. Geological Survey, Tucson Arizona.* Rare-earth mining.

Lu, Ping. *Bioengineering, Georgia Institute of Technology, Atlanta. Georgia Institute of Technology, Atlanta.* Electrospun polymer nanofibers—coauthored.

Lum, Dr. Lawrence. *Department of Cell Biology, University of Texas Southwestern Medical Center, Dallas.* Hedgehog signaling proteins—coauthored.

M

Magnani, Dr. Jared W. *Section of Cardiovascular Medicine, Department of Medicine, Boston University, Massachusetts.* Atrial fibrillation—coauthored.

Martin, Dr. Joel W. *Invertebrate Zoology Section (Crustacea), Research and Collections Branch, Natural History Museum of Los Angeles County, California.* Arthropod evolution and phylogeny.

Mata-Toledo, Dr. Ramon A. *Professor of Computer Science, James Madison University, Harrisonburg, Virginia.* Fuzzy databases—coauthored.

Mathis, Dr. Jeremy. *Ocean Acidification Research Center, University of Alaska Fairbanks.* New insights on ocean acidification.

McGuire, Dr. Jenny L. *School of Environmental and Forest Sciences, University of Washington, Seattle.* Extinction and the fossil record.

McLaughlin, Dr. Brendan M. *Centre for Theoretical Atomic, Molecular and Optical Physics (CTAMOP), School of Mathematics and Physics, Queen's University Belfast, Northern Ireland, United Kingdom.* Photoionization, fluorescence, and inner-shell processes—coauthored.

McManus, Dr. David D. *Cardiac Electrophysiology Section, Department of Medicine, University of Massachusetts Medical School, Worcester.* Atrial fibrillation—coauthored.

Menzies, Dr. Brandon R. *Leibniz Institute for Zoo and Wildlife Research, Berlin, Germany.* Thylacine: genetics of an extinct species—coauthored.

Miranda, Dr. Rodolfo. *IMDEA Nanociencia (Madrid Institute for Advanced Studies in Nanoscience) and Universidad Autónoma de Madrid, Campus Universitario de Cantoblanco, Madrid, Spain.* Cancer treatment using magnetic nanoparticles—coauthored.

Myhre, Dr. Graham. *College of Optical Sciences, University of Arizona, Tucson.* Polarization sensing cameras—coauthored.

N

Naduvalath, Prof. Balakrishnan. *Department of Chemistry, University of Nevada, Las Vegas.* Ultracold molecules.

Neu, Dr. Wayne L. *Department of Aerospace and Ocean Engineering, Virginia Polytechnic Institute and State University, Blacksburg,* Flettner rotor ship.

Newman, Eryn J. *School of Psychology, Victoria University of Wellington, New Zealand.* Repressed memories—coauthored.

Nunes, Jr., Arthur C. *NASA Marshall Space Flight Center, Huntsville, Alabama.* Gas shielding technology for welding—coauthored.

O

Ó Foighil, Dr. Diarmaid. *Museum of Zoology and Department of Ecology and Evolutionary Biology, University of Michigan, Ann Arbor.* Bubble-rafting snails—coauthored.

Oganessian, Prof. Yuri Ts. *Scientific Leader, Flerov Laboratory of Nuclear Reactions, Joint Institute for Nuclear Research, Dubna, Moscow Region, Russian Federation.* DEFINITIVE EVIDENCE FOR NEW ELEMENTS 113 AND 115—coauthored.

Organ, Dr. Chris. *Department of Anthropology, and Department of Paleontology, Natural History Museum of Utah, University of Utah, Salt Lake City.* ORIGINS OF COOKING.

Ovsyanikov, Dr. Nikita G. *Wrangel Island State Nature Reserve, Chukotskyi, Russia, and Severtsov's Institute of Problems of Ecology and Evolution, Russian Academy of Sciences, Moscow, Russia.* EFFECTS OF GLOBAL WARMING ON POLAR BEARS.

P

Pantano-Rubino, Prof. Carlos A. *Department of Mechanical Science and Engineering, University of Illinois at Urbana-Champaign.* TURBULENT COMBUSTION—coauthored.

Paquin, Prof. Jean-François. *Department of Chemistry, Université Laval, Québec, Canada.* FLUORINATION OF ORGANIC COMPOUNDS.

Parker, Dr. Mark D. *Department of Physiology and Biophysics, Case Western Reserve University School of Medicine, Cleveland, Ohio.* BICARBONATE PHYSIOLOGY AND PATHOPHYSIOLOGY—coauthored.

Pasachoff, Prof. Jay M. *Director, Hopkins Observatory, and Field Memorial Professor of Astronomy, Williams College, Williamstown, Massachusetts.* 2012 TRANSIT OF VENUS.

Pask, Dr. Andrew J. *Department of Molecular and Cellular Biology, University of Connecticut, Storrs.* THYLACINE: GENETICS OF AN EXTINCT SPECIES—coauthored.

Paterson, Dr. John R. *Division of Earth Sciences, School of Environmental and Rural Science, University of New England, Armidale, New South Wales, Australia.* COMPLEX EYES OF GIANT CAMBRIAN PREDATOR.

Pau, Prof. Stanley. *College of Optical Sciences, University of Arizona, Tucson.* POLARIZATION SENSING CAMERAS—coauthored.

Paul, Dr. Edward S. *Department of Physics, University of Liverpool, United Kingdom.* EVOLUTION OF NUCLEAR STRUCTURE IN ERBIUM-158—coauthored.

Pierce, Dr. Marcia M. *Department of Biological Sciences, Eastern Kentucky University, Richmond.* LISTERIOSIS OUTBREAK; MEASLES OUTBREAK.

Pierik, Dr. Ronald. *Plant Ecophysiology Group, Institute of Environmental Biology, Utrecht University, the Netherlands.* PLANT-PRODUCED VOLATILE ORGANIC COMPOUNDS—coauthored.

Piper, Dr. Robert C. *Department of Molecular Physiology and Biophysics, University of Iowa, Iowa City.* MULTIVESICULAR BODIES.

Platt, Dr. Donald. *Micro Aerospace Solutions, Inc., Melbourne, Florida.* SPACE FLIGHT, 2011.

Pohl, Dr. Randolf. *Laser Spectroscopy Division, Max Planck Institute of Quantum Optics, Garching, Germany.* PROTON RADIUS FROM MUONIC HYDROGEN.

Polidori, Prof. Guillaume. *Groupe de Recherche en Sciences Pour l'Ingénieur (GRESPI), Université de Reims Champagne-Ardenne, France.* FLOW PATTERNS IN CHAMPAGNE GLASSES—coauthored.

Polly, Dr. Paul David. *Department of Geological Sciences, Indiana University, Bloomington.* MORPHOMETRICS IN PALEONTOLOGY.

Ponsard, Dr. Bernard. *Reactor Physicist, BR2 Reactor, SCK•CEN, Mol, Belgium.* RADIOISOTOPES FOR MEDICAL DIAGNOSIS AND TREATMENT—coauthored.

Povinelli, Dr. Louis A. *NASA Glenn Research Center, Cleveland, Ohio.* DISTRIBUTED PROPULSION—coauthored.

Proshutinsky, Dr. Andrey. *Physical Oceanography Department, Woods Hole Oceanographic Institution, Massachusetts.* ARCTIC OCEAN FRESHWATER BALANCE—coauthored.

Puzio-Kuter, Dr. Anna M. *Department of Pediatrics, Cancer Institute of New Jersey, New Brunswick.* CANCER CELL METABOLISM.

R

Rauchwerk, Michael D. *LGS Innovations, Florham Park, New Jersey.* LIMITATIONS ON INCREASING CELLULAR SYSTEM DATA RATES—coauthored.

Rienstra, Dr. Michiel. *Department of Cardiology, University Medical Center Groningen, University of Groningen, the Netherlands.* ATRIAL FIBRILLATION—coauthored.

Riley, Prof. Mark A. *Department of Physics, Florida State University, Tallahassee.* EVOLUTION OF NUCLEAR STRUCTURE IN ERBIUM-158—coauthored.

Rowell, Dr. Roger M. *Professor Emeritus, Department of Biological Systems Engineering, University of Wisconsin, Madison.* ACOUSTICAL PROPERTIES OF NONTRADITIONAL WOODS.

Rubio-Fernández, Dr. Paula. *Department of Psychology, Princeton University, New Jersey.* FALSE-BELIEF REASONING AND BILINGUALISM—coauthored.

Rykaczewski, Dr. Krzysztof P. *Physics Division, Oak Ridge National Laboratory, Oak Ridge, Tennessee.* DECAY HEAT.

S

Schaedler, Dr. Tobias A. *Sensors and Materials Laboratory, HRL Laboratories, LLC, Malibu, California.* ULTRALOW-DENSITY MATERIALS—coauthored.

Seitelman, Dr. Eric D. *Department of Surgery, Division of Surgical Oncology, Virginia Commonwealth University, Richmond.* NEOADJUVANT SYSTEMIC THERAPY FOR BREAST CANCER—coauthored.

Sham, Dr. S H Robin. *Global Long Span and Specialty Bridges Director, AECOM.* FRONTIERS OF LARGE CABLE-STAYED BRIDGE CONSTRUCTION.

Shechter, Dr. Avigdor. *Manager of EMX, Integration Solutions Group, Alion Science and Technology, Alexandria, Virginia.* STEALTH OF SURFACE COMBATANTS AT SEA.

Shladover, Dr. Steven E. *California PATH Program, University of California, Berkeley.* CONNECTED VEHICLES.

Shors, Dr. Teri. *Department of Biology and Microbiology, University of Wisconsin-Oshkosh.* H5N1 VIRUS (BIRD FLU) CONTROVERSY.

Sibille, Dr. Etienne. *Department of Psychiatry, University of Pittsburgh, Pennsylvania.* GENOMICS OF DEPRESSION.

Sidorchuk, Dr. Sergey I. *Scientific Secretary, G. N. Flerov Laboratory of Nuclear Reactions, Joint Institute for Nuclear Research, Dubna, Moscow Region, Russian Federation.* BREAKDOWN OF SHELL CLOSURE IN HELIUM-10—coauthored.

Sime-Ngando, Dr. Télesphore. *Laboratoire Microorganismes: Génome et Environnement, Université Blaise Pascal, Aubière Cedex, France.* FUNGAL ZOOSPORES IN AQUATIC ECOSYSTEMS.

Simmons, Dr. Kaneatra. *Microbial Exposure Research Branch, Microbiological and Chemical Exposure Research Division, National Exposure Research Laboratory, U.S. Environmental Protection Agency, Cincinnati, Ohio.* SALIVA-BASED IMMUNOASSAY OF WATERBORNE PATHOGEN EXPOSURE—coauthored.

Simpson, Prof. John. *STFC Daresbury Laboratory, Daresbury, Warrington, United Kingdom.* EVOLUTION OF NUCLEAR STRUCTURE IN ERBIUM-158—coauthored.

Smith, Dr. Eboni. *Meyer Center for Developmental Pediatrics, Department of Pediatrics, Section of Developmental Pediatrics, Baylor College of Medicine, Houston, Texas.* FACTORS RELATED TO RISK OF AUTISM—coauthored.

Stambouli, Prof. Amine B. *Department of Electronics, University of Science and Technology of Oran, Algeria.* SAHARA SOLAR BREEDER PROJECT—coauthored.

Stamm, Dr. Stefan. *Department of Molecular and Cellular Biochemistry, University of Kentucky, Lexington.* ALTERNATIVE RNA SPLICING—coauthored.

Stigall, Dr. Alycia L. *Department of Geological Sciences, and Ohio Center for Ecology and Evolutionary Studies, Ohio University, Athens.* INVASIVE SPECIES DURING THE LATE DEVONIAN BIODIVERSITY CRISIS.

Stott, Dr. Jeffrey L. *Department of Pathology, Microbiology and Immunology, School of Veterinary Medicine, University of California, Davis.* FOOTHILL ABORTION.

Strathearn, Dr. Lane. *Meyer Center for Developmental Pediatrics, Department of Pediatrics, Section of Developmental Pediatrics, Baylor College of Medicine, Houston, Texas.* FACTORS RELATED TO RISK OF AUTISM—coauthored.

T

Tanis, Prof. John A. *Department of Physics, Western Michigan University, Kalamazoo.* CHARGED-PARTICLE TRANSMISSION THROUGH INSULATING CAPILLARIES.

Ter-Akopian, Prof. Gurgen M. *G. N. Flerov Laboratory of Nuclear Reactions, Joint Institute for Nuclear Research, Dubna, Moscow Region, Russian Federation.* BREAKDOWN OF SHELL CLOSURE IN HELIUM-10—coauthored.

Teran, Dr. Francisco José. *IMDEA Nanociencia (Madrid Institute for Advanced Studies in Nanoscience), Campus Universitario de Cantoblanco, Madrid, Spain.* CANCER TREATMENT USING MAGNETIC NANOPARTICLES—coauthored.

Timmermans, Dr. Mary-Louise. *Department of Geology, Yale University, New Haven, Connecticut.* ARCTIC OCEAN FRESHWATER BALANCE—coauthored.

Tokitoh, Prof. Norihiro. *Institute for Chemical Research, Kyoto University, Japan.* MAIN GROUP MULTIPLE BONDS.

Toole, Dr. John M. *Physical Oceanography, Woods Hole Oceanographic Institution, Massachusetts.* ARCTIC OCEAN FRESHWATER BALANCE—coauthored.

Tryon, Dr. Christian A. *Department of Anthropology, New York University.* ACHEULEAN.

Tsubouchi, Dr. Kenta. *Graduate School of Frontier Sciences, University of Tokyo, Japan.* SAHARA SOLAR BREEDER PROJECT—coauthored.

U

Ulm, Dr. Roman. *Department of Botany and Plant Biology, University of Geneva, Switzerland.* PLANT MAP KINASE PHOSPHATASES—coauthored.

Uriarte, Dr. Fabian M. *Center for Electromechanics, University of Texas at Austin.* INTELLIGENT MICROGRIDS—coauthored.

Utyonkov, Dr. Vladimir K. *Flerov Laboratory of Nuclear Reactions, Joint Institute for Nuclear Research, Dubna, Moscow Region, Russian Federation.* DEFINITIVE EVIDENCE FOR NEW ELEMENTS 113 AND 115—coauthored.

V

Véron, Dr. Nathalie. *Friedrich Miescher Institute for Biomedical Research, Basel, Switzerland.* TET PROTEINS.

Villanueva, Dr. Ángeles. *IMDEA Nanociencia (Madrid Institute for Advanced Studies in Nanoscience) and Universidad Autónoma de Madrid, Campus Universitario de Cantoblanco, Madrid, Spain.* CANCER TREATMENT USING MAGNETIC NANOPARTICLES—coauthored.

Volkmar, Dr. Fred R. *Yale Child Study Center, Yale University School of Medicine, New Haven, Connecticut.* AUTISM AND THE SOCIAL BRAIN.

W

Waikel, Dr. Rebekah L. *Department of Biological Sciences, Eastern Kentucky University, Lexington.* GENETIC EXCHANGE AMONG BACTERIA—coauthored.

Wallis, Dr. Thomas Mitchell. *Physical Measurement Laboratory, National Institute of Standards and Technology, Boulder, Colorado.* NEAR-FIELD SCANNING MICROWAVE MICROSCOPE (NSMM)—coauthored.

Wang, Dr. Xiaofeng. *Department of Physics, Florida State University, Tallahassee.* EVOLUTION OF NUCLEAR STRUCTURE IN ERBIUM-158—coauthored.

Webster, Dr. Carol H. *Quantum Detection Group, National Physical Laboratory, Teddington, Middlesex, United Kingdom.* QUANTUM PHASE SLIP.

Whitehead, Prof. Lorne. *Structured Surface Physics Laboratory, University of British Columbia, Vancouver, Canada.* PRISM LIGHT GUIDES.

Williams, Dr. Jonathan. *Time, Quantum and Electromagnetics Division, National Physical Laboratory, Teddington, Middlesex, United Kingdom.* QUANTUM VOLTAGE STANDARDS.

Wiltshire, Dr. Patricia E.J. *Consultant, Milford House, The Mead, Ashtead, Surrey, United Kingdom.* FORENSIC MYCOLOGY.

Wu, Dr. Xiaofeng. *Department of Cell Biology, University of Texas Southwestern Medical Center, Dallas.* HEDGEHOG SIGNALING PROTEINS—coauthored.

X

Xia, Dr. Younan. *Department of Biomedical Engineering, Georgia Institute of Technology, Atlanta.* ELECTROSPUN POLYMER NANOFIBERS—coauthored.

Y

Yanoviak, Dr. Andrew Charles. *Environmental Systems Planning and Design Consultants, Honolulu, Hawaii.* SELF-ENERGIZING HIGH-RISE TOWERS.

Index

Index

Asterisks indicate page references to article titles.

H

I